Mud Volcanoes of the Black Sea Region and their Environmental Significance

Evgeny Shnyukov
Valentina Yanko-Hombach

Mud Volcanoes of the Black Sea Region and their Environmental Significance

Springer

Evgeny Shnyukov
Department of Marine Geology and the Sedimentary Ore Formation (OMGOR NASU), recently renamed the Center for Problems of Marine Geology, Geoecology and Sedimentary Ore Formation of the NASU
National Academy of Sciences of Ukraine (NASU)
Kiev, Ukraine

Valentina Yanko-Hombach
Department of Physical and Marine Geology
Odessa I.I.Mechnikov National University
Odessa, Ukraine

ISBN 978-3-030-40315-7 ISBN 978-3-030-40316-4 (eBook)
https://doi.org/10.1007/978-3-030-40316-4

This Springer imprint is published by the registered company Springer Nature Switzerland AG.
The registered company address is: Gewerbestrasse 11, 6330 Cham, Switzerland

The book is dedicated to the centenary of the founding of the National Academy of Sciences of Ukraine

Foreword

Humans have known about the existence of mud volcanoes for a very long time. Archaeologists have suggested that early Paleolithic groups migrated to the Black Sea region of Eurasia via mud volcanic provinces. They base this idea on comparisons made between maps showing the locations of mud volcanoes and showing early Paleolithic sites. Paleolithic localities such as Sinyaya Balka, Il'skaya, and Bogatyr are found in proximity to mud volcanoes on the Taman Peninsula of Crimea (Zenin 2012). Furthermore, stone tools have been discovered within some mud volcanoes; for example, a Paleolithic scraper was recovered from within the Akhtarma-Pashalinskaya mud volcano in Azerbaijan (Kovalevskiy 1935). Mud volcanic landscapes could have appealed to ancient people due to the presence of erupted breccia that could be used for making stone tools, as well as the hot water and gas flares produced by mud volcanoes.

The use of the term "mud volcano" dates back to the nineteenth century. It was introduced by Helmersen and Schrenck (1885) as a translation from the German *mudevulkan* previously used by Abikh (1873). Pallas (1795) provided the first scientific description of mud volcanic events when he described the frequently exploding Golubitsky mud volcano in the Sea of Azov and connected its activity to earthquakes.

The intensive study of mud volcanoes in the Black Sea region (largely on the Kerch and Taman Peninsulas) began in the early twentieth century with the work of Vernadsky and Popov (1899–1900), Felitsyn (1902), Borisyak (1907), Chirvinskiy (1908), Zhivilo (1909), Yushkin (1909), Steber (1909–1910), Gubkin (1913), and many others.

After the Second World War, the Kerch-Taman mud volcanoes were investigated as part of a geological survey focused on oil and gas exploration. For the first time, the geological complexity of the anticlinal structures within which mud volcanoes develop was explored (Lychagin 1952), and their composition and geochemistry were summarized (Ronov 1951). In addition, estimated ages of erupted gases as well as the oil and gas potential of mud volcanic hearths were studied (Gattenberg 1954), and the first overview of Tertiary mud breccia (from Maikopian to Pontian on the geological section) from the Kerch Peninsula was published (Alyaev 1947, Maymin 1951). Particular attention was focused on the origins and roots of mud volcanoes of the Taman Peninsula (Shardanov et al. 1962; Shardanov and Znamenskiy 1965); these authors concluded that the roots of the Taman mud volcanoes begin in Lower Cretaceous rocks and that they contain large reservoirs of oil and gas. Later, these conclusions were confirmed by Shnyukov and Netebskaya (2013). The hydrogeology of the Kerch Peninsula's mud volcanoes was also described (Kurishko et al. 1968).

Research on the Kerch-Taman mud volcanoes was renewed in the 1960s due to increased industrial interest in rare elements—mercury, lithium, arsenic, and boron.

The most important breakthrough in the study of the Kerch-Taman mud volcanoes came when ore-bearing deposits in compensated, or recessed, geosynclines formed during the Kimmerian orogenic cycle were discovered. This new finding initiated serious geological exploration carried out by the Department of Mining and Geology of the USSR. A large field of oolitic iron ores was located within the Novoselovsky mud volcanic hearth. Called the Novoselovskoe field, it is estimated to contain 125 mln t of conditional iron ore, and together with non-conditional iron ores, up to 200 mln t were recognized by the USSR Commission of Reserves as being an important source of iron ore. Data obtained through this research was summarized in a fundamental monograph by Shnyukov et al. (1971).

Along with applied investigations into mud volcanoes, its classical geological study has also been ongoing (Nesterovskiy 1990; Shnyukov et al. 2013a; Shnyukov 2016).

At the end of the twentieth century, the "center of gravity" for mud volcanism research moved to the Black Sea. Over the course of dozens of marine expeditions conducted on board different research vessels (e.g., "Mikhail Lomonosov," "Kiev," "Professor Vodyanitskiy," "Vladimir Parshin," and some others), a substantial number of underwater mud volcanoes was discovered (Shnyukov et al. 2005).

Recent studies of Black Sea mud volcanoes have been conducted during the course of multiyear programs pursued by many Russian and Ukrainian research teams from various organizations—such as "Yuzhmorgeologiya," Moscow State University, the National Academy of Sciences (e.g., the Geological Institute, Department of Marine Geology and Sedimentary Ore-Formation, Institute of Biology of the Southern Seas, Geophysical Institute, and Marine Hydrophysical Institute), Odessa I.I. Mechnikov National University, and others. Significant contributions have also been made by Bulgarian, Romanian, German, and other European scientists, who worked together with Russian and Ukrainian specialists.

This volume contains an introduction, 11 chapters, a conclusions section, and an extensive reference list for each chapter. Chapter 1 presents a history of mud volcano studies and details the basic investigative methods used within the Black Sea region. Chapter 2 provides an overview of the study area that includes the Black Sea, the Sea of Azov, the Kerch and Taman peninsulas, and the adjacent northern Caucasus. Chapter 3 reviews the materials and methods used in the study of terrestrial and offshore mud volcanoes of the Black Sea region. Chapter 4 revises modern ideas about mud volcanism. Chapter 5 describes the morphostructures related to mud volcanoes, including compensated, or recessed, geosynclines. The massive Chap. 6 is illustrated by 205 figures and covers the entire panorama of mud volcanoes within the Black Sea region. Chapter 7 provides an overview of other types of degassing in the Black Sea, such as gas seeps, acoustic plumes, or gas torches.

Chapter 8 discusses the origin of mud volcanoes in the Black Sea region. Chapter 9 explores the connection between mineral resources and mud volcanism. Chapter 10 explores the impact of mud volcanoes on the environment and, in particular, the dangers that may arise from mud volcanic activity on land and underwater. Chapter 11 is a case study that examines the relationship between meiobenthos distribution and concentrations of hydrocarbon gases, primarily methane, in seafloor sediments of the northwestern Black Sea.

As can be seen, this book covers a wide range of aspects relating to the science of mud volcanoes in the Black Sea region, including their geology, structure, and dynamics. For example, it describes new discoveries of iron ore deposits in mud volcanic structures of the Kerch Peninsula, forecasts future promising areas of research, outlines ways to make greater use of mud volcano resources, puts forward a new idea about the deep origin of mud volcanoes, proposes to open a multilocation mud volcanic reserve in order to preserve the natural phenomena and reduce the destructive consequences of a catastrophic eruption in densely occupied areas, offers an explanation of the causes for some shipwrecks and a suggestion to modify the sea-lanes of maritime transport when they approach too closely to mud volcanic foci, and shows the important environmental significance of both terrestrial and underwater mud volcanoes in the Black Sea region. The spectrum of issues addressed is extremely broad, and it is clearly apparent that the authors have brought years of experience to the subject. This unusually informative work, however, is based not only on the 50 years of geological work by the authors but also on the results produced by hundreds of their colleagues and scholarly predecessors, who traveled thousands of kilometers on land and thousands of nautical miles on scientific research vessels.

The results of this work are impressive. It is an encyclopedia of mud volcanoes in the Black Sea region and serves not only to document our current knowledge of a particularly significant region but also to lay the groundwork for future research.

It is my great pleasure to introduce the book. I strongly believe that readers will appreciate this fundamental compilation of useful information, much of which appears here in English for the first time.

Academician of the National Academy of Sciences of Ukraine,
Professor, Director of Geological Institute
Kyiv, Ukraine

Peter Gozhik

References

von Abich GV (1873) Geologicheskiy Obzor poluostrovov Kerchi i Tamani (geological survey of the Kerch and Taman peninsulas). Zapiski Kavkazskogo otdeleniya russkogo geograficheskogo obshchestva 8:3–160 (in Russian)

Alyaev SE (1947) Novye dannye o tektonike Kerchenskogo poluostrova (new data about tectonics of the Kerch peninsula). Izvestya AN SSSR Seriya Geol 6:97–99 (in Russian)

Borisyak AA (1907) Doklad po voprosu osmotra gryazevoy sopki bliz Vladislavovkiy (Report on the inspection of the mud hill near Vladislavovka). Izvestiya Geolkoma 26(3):34–36 (in Russian)

Chirvinskiy PN (1908) Zametka o gryazevykh sopkakh Kerchenskogo poluostrova (a note on the mud hills of the Kerch peninsula). Zapiski Kievskogo obshchestva estestvoispytateley 20(2):791–797 (in Russian)

Felitsyn EV (1902) Nekotorye svedeniya o gryazevykh vulkanakh Tamanskogo poluostrova (some information about the mud volcanoes of the Taman peninsula).. Izvesiya obshchestva lyubiteley izucheniya Kubanskoy oblasti (in Russian)

Gattenberg YuP (1954) K voprosu o vozraste gazov gryazevykh vulkanov Kerchenkogo poluostrova (On the question of the age of gases of mud volcanoes of the Kerch Peninsula). Paper presented at the 8th scientific-practical conference of the scientific student society of the Oil Institute. Moscow, 1954 (in Russian)

Gubkin IM (1913) Obzor geologicheskikh obrazovaniy Tamanskogo poluostrova (an overview of the geological formations of the Taman peninsula). Izvesiya Geolkoma 32:8 (in Russian)

Helmersen G, Schrenck L (1885) BAND VIII. Gemischten Inhalts. Beiträge zur Kenntnis des russischen Reiches und der angränzenden Länder Asiens; Folge 2, mit 15 lithographischen Tafeln. Kaiserlichen Akademie der Wissenschaften, St-Petersburg, Riga, Leipzig

Kovalevskiy SA (1935) Gazovyy vulkanizm (vulkany i vulkanoidy) (Gas volcanism (volcanoes and volcanoids)). Azerbaydzhanskoe neftyanoe khozyaystvo 1 (in Russian)

Kurishko VA, Mesyats IA, Tverdovidov AS (1968) Gidrogeologiya gryazevogo vulkanizma Kerchenskogo poluostrova (hydrogeology of mud volcanism of the Kerch peninsula). Geologicheskiy zhurnal 28(1):49–59 (in Russian)

Maymin ZL (1951) Tretichnye otlozheniya Kryma (tertiary deposits of Crimea). Gostoptekhizdat, Moscow-Leningrad (in Russian)

Nesterovskiy VA (1990) Aktivizatsiya gryazevykh vulkanov Kerchensko-Tamanskoy oblasti (activation of mud volcanoes of the Kerch-Taman region). Geologicheskiy zhurnal 1:138–143 (in Russian)

Pallas PS (1795) Kratkoe fizicheskoe i topograficheskoe opisanie Tavricheskoy oblasti (Brief physical and topographical description of the Tauride region) St Petersburg (in Russian)

Ronov AB (1951) K voprosu o gryazevom vulkanizme yugo-vostochnogo Kavkaza (on the issue of mud volcanism in the southeastern Caucasus). Doklady AN SSSR 77(6):268–284 (in Russian)

Shardanov AN, Znamenskiy VA (1965) Gryazevoy vulkanizm i perspektivy neftenosnosti Tamanskogo poluostrova (mud volcanism and oil potential of the Taman peninsula). Geologiya nefti i gaza, pp:18–20 (in Russian)

Shardanov AN, Malyshok VG, Peklo VP (1962) O kornyakh gryazevykh vulkanov Tamani (On the roots of mud volcanoes of Taman'). Trudy Krasnodarskogo filiala VNIINeft 5(1):53–66 (in Russian)

Shnyukov EF (2016) Flyuidogennaya mineralizatsiya gryazevykh vulkanov Azovo-Chernomorskogo regiona (Fluidogenic mineralization of mud volcanoes of the Azov-Black Sea region). Logos, Kiev (in Russian)

Shnyukov EF, Netrebskaya EYa (2013) Korni Chernomorskikh gryazevykh vulkanov (the roots of the Black Sea mud volcanoes). Geologiya i poleznye iskopaemye Mirovogo okeana 14:87–92 (in Russian)

Shnyukov EF, Starostenko VI, Ivannikov AV et al (2005) Gazovyy vulkanizm Chernogo morya (Black Sea gas volcanism). OMGOR NAN Ukrainy, Kiev (in Russian)

Shnyukov EF, Kobolev VP, Pasynkov AA (2013a) Gazovyy vulkanizm Chernogo morya (gas volcanism of the Black Sea). Logos, Kiev (in Russian)

Steber EA (1909/1910) Gryazevoy vulkan Karabetova gora bliz Tamani (Mud volcano Karabetova mountain near Taman). Izvestia Kavkazskogo otdeleniya Russkogo heograficheskogo obshchestva 20 (in Russian)

Vernadskiy VI, Popov SP (1899–1900) Yenikalskie gryazevye vulkany (Enikal mud volcanoes). Byulleten' moskovskogo obshchestva ispytateley prirody (Protokoly zasedaniy. Prilozheniya), pp 37–41 (in Russian)

Yushkin EM (1909) O neftyanykh mestorozhdeniyakh Tamanskogo poluostrova, a takzhe o tamanskoy rude (on the oil fields of the Taman peninsula, as well as on the Taman ore). Groznyy (in Russian)

Zhivilo K (1909) Ekskursiya na Tamanskiy poluostrov (Excursion to the Taman Peninsula). Kubanskiy sbornik. Trudy publikatsiy obl. stat. otd. 14, Yekaterinodar, pp 3–17 (in Russian)

Preface

A lengthy book does not need a lengthy preface, so these introductory words will convey only some essential matters, including the circumstances that led to the present publication, some of the background to the research it contains, and words of gratitude to those who helped in the effort.

In these prefatory paragraphs, the authors want to present some necessary information that includes the background for the creation of this book, the process by which it came to be, and appreciative words for those who made this publication possible.

The Black Sea region encompasses the Black Sea, the Sea of Azov, and their coasts, and the area has been inhabited by humans for at least the last 1.8 million years. It was formed at the end of the Mesozoic as a back-arc basin of the early Cretaceous (revealed as a result of continental rifting at the end of the Albian). This led to the splitting of the crust along the axis of the Albian volcanic arc and the subsequent opening of the Cenomanian-Coniacian deepwater trough, a severe thinning of the continental and/or oceanic crust, and the separation of two basins, which became the Western and the Eastern Black Sea. Starting from the end of the Santonian and before the end of the Paleocene, the Black Sea depression experienced compressive phases. Within the Eastern Black Sea depression during the Eocene, a new phase of stretching began, which led to the formation of the Adzharo-Trialetskiy rift. Since the end of the Eocene and into the present, the bottom of the basin has been in a compressive phase, and as a result, it is broken by numerous cracks (Nikishin et al. 2003).

One of the most interesting features of the region is the extensive development of mud volcanism—a geological phenomenon that is widespread across the planet. The term "mud volcano" is generally applied to a more or less violent eruption or surface extrusion of watery mud or clay that is almost invariably accompanied by methane gas. The most ancient Early Paleozoic mud volcanoes are known from Decaturville, Missouri, in North America (Zimmermann and Amstutz 1972).

According to archaeological evidence, mud volcanoes of the Black Sea region have attracted the attention of humans since antiquity. The history of investigation into mud volcanoes goes back to the tenth century AD when Byzantine Emperor Constantine VII Porphyrogenitus, also called Constantine VII Flavius, described some rocks that spewed oil located near the town of Tamatarki on the Taman Peninsula. The first mud volcano described in the scientific literature was dubbed "Golubitsky" by Pallas (1795); it was located in the Sea of Azov near Temryuk. Subsequently, the study of mud volcanoes was conducted mostly by Russian scientists, e.g., V.V. Belousov and A.D. Arkhangelsky. Research intensified in the second half of the twentieth century as the search for oil, gas, and other mineral resources accelerated.

Currently, 180 mud volcanoes are known within the Black Sea region. Of these, 100 have been found on land (on the Kerch and Taman Peninsulas and in northwestern Caucasus), and 68 subaqueous mud volcanoes are present in the southeastern part of the Sea of Azov; in the northwestern Black Sea, including the Sorokin, Kerch-Taman, and Tuapse troughs; and in the western Black Sea depression. Judging by their geomorphology, two additional seafloor locations are presumably mud volcanoes, but future investigation is required to confirm this identification.

Multidisciplinary studies undertaken since 1990 enabled E.F. Shnyukov, the first author of this publication, to discover large fields of oolitic iron ores—including the Novoselovskoe field on the Kerch Peninsula, with a reserve of 150 million tons—that were formed by the activity of mud volcanoes. Shnyukov was also able to connect these ores to compensated or compressed geosynclines formed during the Cimmerian Orogeny cycle, as he had initially predicted based on scientific evidence.

All evidence discussed here supports the main ideas of Kropotkin and Valyaev (1979, 1980, 1984), Valyaev (2011), Dmitrievsky and Valyaev (2002), Pikovskiy (2002), and Lukin (2013), whose hypotheses proposed the degassing of a "cold" non-magmatic Earth. Letnikov et al. (2010) considered the development of the Earth as a monotonous extinguishing process, in which the depletion of high-temperature fluids into the lithosphere was followed by periodic pulses of intense degassing. The main transmitter of fluids is hydrogen, which forms two fluid systems—hydrogen-carbon and hydrogen-sulfide—that are located at different levels within the liquid core. The former may be responsible for the development of mantle plumes and mud volcanic activity. The dynamics of mud volcanoes are caused by the action of these hydrogen-carbon plumes, which generate fluid streams that break up as they transition first into a liquid state and then second into a gaseous state, eventually leading to the unusual mineralizations associated with mud volcanoes. High-temperature mantle plumes raise the flow rate of the fluids. Their modern analogue is lithospheric hydrothermal plumes, which are especially pronounced in spreading zones.

In general, the phenomenon of mud volcanism presents many dimensions, the study of which contributes significantly to improving the human condition, as it may aid in maintaining environmental integrity as well as providing a means for sustainable development of the region. Outbursts of submarine mud volcanoes located below 600–700-m isobaths commonly contain ice-like aggregates of gas hydrates (largely methane), which indicate the presence of mud volcanoes under the seafloor. The presence of both ore formations and gas hydrates is attributable to the compensated or compressed geosynclines, which can be used as reliable indicators in prospecting for mud volcanoes.

The dynamics of mud volcanoes pose many potentially serious risks to maritime activity and environmental security. Gas outbursts from offshore mud volcanoes affect the hydrochemical regime of the sea, producing currents and acoustics, and substantially affecting ecosystems, including especially seasonal fish migration routes. The entire biota of the Black Sea is to a certain degree determined by fluctuations in hydrogen sulfide and methane exposure levels (Yanko-Hombach et al. 2009, 2017). Mud volcanoes can also cause great damage to the physical environment, as powerful eruptions create ground subsidence in nearby areas, thereby presenting a serious threat to nearby urban agglomerations, for example, Kerch in Crimea and Temryuk in Krasnodar territory. Contamination of the air with mercury and other elements can also be hazardous to people. Mud volcanoes can pose a significant threat to maritime traffic, especially in narrow waterways. For instance, 7 mud volcanoes have been found offshore within the Kerch Strait, and every year, nearly 10,000 ships cleave their way through the strait. There have been cases of ships running aground even though they were within the navigational channel (e.g., S/S Caesar in 1914 and some others). The shoals that grounded them proved to be mud volcanic in nature. At times, mud volcanoes located within the strait have formed small islets composed of ejected materials. Some researchers have proposed that methane outbursts were the cause of ship loss in the Bermuda Triangle. Accidents of this kind are likely to happen in the Black Sea as well. The probability of such accidents has been shown experimentally. Mud volcanoes serve not just as regional signs for petrochemical prospection, but sometimes, they can be used for precise localization of oil traps, as they are good indicators of oil- and gas-bearing provinces. This generality can be exploited in the future for the development of earthquake forecasting criteria.

In general, mud volcanism in the Black Sea region is an extremely interesting phenomenon of multidimensional importance, deserving in-depth study primarily as an indicator of the Earth's oil- and gas-bearing capacity. Today, mud volcano studies are largely focused on their scientific rather than applied aspects. To date, no surveys of mud volcanoes as markers for gas hydrates have been performed, and no calculations of their contribution to the total degassing of the seafloor have been performed on a basin-wide scale. Likewise, no research on the hazards they pose has been conducted. At the same time, mud volcanoes are the seafloor's expression of endogenic processes and a "cheap window" (Golubyatnikov 1923) into the deep geosphere; they may be considered valuable tools for future industrial applications.

This book contains an introduction, 11 chapters, a conclusions section, and an extensive reference list for each chapter. Each chapter underwent a lengthy review process (two reviewers per chapter as a rule) as well as extensive editing of both language and graphics. The complexity of the graphics editing (done by Nikolay Maslakov and Alexander Paryshev) and that of the text editing (done mostly by Allan Gilbert) took longer than anticipated, leading to the unexpected 2-year delay in getting the volume published.

Chapter 1 includes the history of mud volcano studies and the basic investigative methods used within the Black Sea region. It contains a detailed overview of previous investigations and emphasizes the main achievements as well as the present gaps in our knowledge.

Chapter 2 provides an overview of the study area that includes the Black Sea, the Sea of Azov, the Kerch and Taman Peninsulas, and the adjacent northern Caucasus.

Chapter 3 provides an overview of the materials and methods used in the study of terrestrial and offshore mud volcanoes of the Black Sea region. The general strategies include geomorphological, geological, geophysical, gas-geochemical, paleontological, and micropaleontological dimensions.

Chapter 4 includes an overview of modern ideas about mud volcanism, including their morphology, geological structure, and the composition of their mud breccias, water, gases, and terrigenous material, as well as the main characteristics of their dynamics.

Chapter 5 is devoted to a description of the morphostructures related to mud volcanoes, including compensated or compressed geosynclines.

Chapter 6 covers the entire population of mud volcanoes that have developed within the Black Sea region, in particular the Kerch and Taman Peninsulas, the northwestern Caucasus, the Black Sea itself, and the southeastern part of the Sea of Azov.

Chapter 7 provides an overview of other types of degassing in the Black Sea, which are referred to as gas seeps, acoustic plumes, or gas torches.

Chapter 8 covers the origin of mud volcanoes in the Black Sea region.

Chapter 9 explores the connection between mineral resources and mud volcanism. Mud volcanoes can be of practical assistance in the search for fossil fuels, iron ores, rare chemical elements, nonmetallic raw materials, therapeutic mud, and much more.

Chapter 10 discusses the impact of mud volcanoes on the environment and, in particular, the dangers that may arise from mud volcanic activity.

Chapter 11 represents another case study on the relationship between meiobenthos distribution and concentrations of hydrocarbon gases, primarily methane, in the sediments of the northwestern part of the Black Sea, including gases released by mud volcanoes and gas seeps.

In the Conclusions chapter, the authors propose some principal directions for future investigations in mud volcano research in the Black Sea region, together with possible applications to other basins. It is shown that mud volcanism is a complex and multidimensional phenomenon, the study of which requires a multidisciplinary and in-depth approach. It appears that there is a clear interrelationship between mud volcanoes and methane gas hydrates, which allows mud volcanoes to be used in the search for the latter below isobaths 600–700 m.

A reference list is appended to each chapter; the sources include numerous items, many of which were published in regional languages, and as such are not well known in the west.

The book is illustrated with 473 figures. It contains detailed descriptions and images of major Black Sea mud volcanoes, together with their coordinates, types of geomorphological structures, and the contents of their breccia—including unique chemical elements and minerals indicating that the deep roots of mud volcanoes extend downward to the mantle. The coverage incorporates a wide geographic region, encompassing both terrestrial and underwater areas, and a broad approach, ranging from geological subjects to environmental applications.

The authors have studied the mud volcanoes of the Black Sea region for more than 50 years, and the results of their research are presented in a significant number of monographs and articles, largely published in Russian. This book summarizes the authors' findings. Unfortunately, the recent change in the legal status of Crimea has precluded the opportunity to update a number of photographs, so that those appearing here had to be taken mostly from previous publications, e.g., Shnyukov et al. (2006, 2013b). Yet, this monograph represents the results of research conducted not only by the authors but also by a large informal team of geologists from many scientific and applied geological organizations. The contributions of individual experts are acknowledged in the text. Among these experts, it is especially appropriate to note the contribution of P.I. Naumenko, the head of the association "Ukrchermetgeologiya" in the Ministry of Ferrous Metallurgy of Ukraine. As a result of the close contacts between this organization and the NAS in Ukraine, many hundreds of wells were drilled, and new mineral fields were discovered and explored, and this collaboration has led to new and unique discoveries on the influence of mud volcanoes upon the environment.

Studies of recovered materials that provide the evidentiary basis for this book were conducted under the auspices of the National Academy of Ukraine in the Department of Marine Geology and the Sedimentary Ore Formation (OMGOR NASU, recently renamed the Center for Problems of Marine Geology, Geoecology and Sedimentary Ore Formation of the National Academy of Sciences of Ukraine, Institute of Geological Sciences) and Odessa I.I. Mechnikov National University.

All transliterations of cited sources published in languages using the Cyrillic alphabet comply with the requirements of the international standards for bibliographic references according to the US Library of Congress (https://www.loc.gov/catdir/cpso/romanization/russian.pdf). Exceptions are the names of the authors, which are left in their own preferred transliterations, as well as geographical names as presented most commonly in the majority of English papers.

The authors are grateful to all of the colleagues who have supported their work. Particular gratitude must be given to Dr. Nikolay Maslakov and Dr. Alexander Paryshev (Center for Problems of Marine Geology, Geoecology and Sedimentary Ore Formation of the National Academy of Sciences of Ukraine) for their assistance in drawing the illustrations.

They are also grateful to Prof. Allan Gilbert from Fordham University, USA, for editing the English text and for his valuable comments and to Alexander Motnenko, University of Manitoba, for the initial editing of Chaps. 1, 4, and 5.

Dr. Derman Dondurur (Dokuz Eylül University, Institute of Marine Sciences and Technology, Turkey) is thanked for his contribution on mud volcanoes in the Turkish sector of the Black Sea.

The authors are grateful to Peter Gozhik, Academician of the NASU, Professor, Director, Geological Institute of NASU, for agreeing to introduce this book.

Grateful acknowledgment is offered for the thoughtful efforts of external reviewers—top scientists in the field—Vladimir Kobolev, Professor, Doctor of Geological Sciences, Chief Scientist of the Institute of Geophysics of NASU, and Adil Abbas Ogly Aliyev, Professor, Doctor of Geological Science, Head of the Department of Mud Volcanism of the Institute of Geology and Geophysics of the National Academy of Sciences of Azerbaijan, Laureate of the State Prize of Azerbaijan for the Atlas of Mud Volcanoes of the World.

Dr. Revinder Sidhu, Microscopy and Materials Characterization Facility Manager (Manitoba Institute for Materials, University of Manitoba, Canada), is sincerely thanked for her help in imaging of microfauna by SEM (Chap. 11).

Lastly, the authors thank the managing team at Springer, Ms. Malini Arumugam, Project Coordinator (Books), Margaret Deignan, and Petra van Steenbergen for, above all, their patience in awaiting the delivery of the finished manuscript.

The work was executed with the financial support of the EU-FP6 project HERMES (Hotspot Ecosystems Research on the Margins of European Seas) 2006–2010; the State Budget Themes # 0117U000775 "Mud Volcanism: Emergencies, Study, Forecasting and Minimization," #0113U004857 "Lithologic-Mineralogical Composition of the Muddy Breccias of Mud Volcanoes of the Kerch-Black Sea Region," #0117U000776 "Gas Volcanism in the Azov-Black Sea Province and Related Minerals," and some others financed by the National Academy of Sciences of Ukraine; and the State Budget Themes #539 "Study of the Formation Processes and Spatial Distribution of Methane in the Black Sea and Theoretical Considerations of Their Influence on Basin Eco- and Geosystems" and #590 "Development of Predictive Search Criteria for Hydrocarbon Deposits in the Black Sea Based on the Theory of Fluidogenesis" financed by the Ministry of Education and Science of Ukraine.

This study is a contribution to UNESCO-IUGS-IGCP 610 project "From the Caspian to Mediterranean: Environmental Change and Human Response During the Quaternary."

Kiev, Ukraine — Evgeny Shnyukov
Odessa, Ukraine — Valentina Yanko-Hombach

References

Dmitrievskiy AN, Valyaev BM (2002) Uglevodorodnaya degazatsiya cherez dno okeana: lokalizovannye proyavleniya, masshtaby, znachimost' (Hydrocarbon degassing through the ocean floor: localized manifestations, extent, significance). Materialy vserossiyskoy konferentsii "Degazatsiya Zemli i genezis uglevodorodnykh flyuidov i mestorozhdeniy," GEOS, Moscow (in Russian)

Golubyatnikov DV (1923) Iskopaemyy gryazevoy vulkan na promysle "Ilych" (Fossil mud volcano in the "Ilyich" mining). Neftyanoe i slantsevoe khozyaystvo 5(78): 9–74 (in Russian)

Kropotkin PN, Valyaev BM (1979) Glubinnye razlomy i degazatsiya Zemli (Deep faults and degassing of the Earth). Tektonicheskoe razvitie zemnoy kory i razlomy, Nauka, Moscow, pp 257–267 (in Russian)

Kropotkin PN, Valyaev BM (1980) Geodinamika gryazevulkanicheskoy deyatelnosti (v svyazi s netegazonosnostyu) (Geodynamics of mud volcanic activity (due to petroleum potential)). Geologicheskie i geokhimicheskie osnovy poiskov nefti i gaza. Naukova Dumka, Kiev, pp 148–178 (in Russian)

Kropotkin PN, Valyaev BM (1984) Tektonicheskiy kontrol protsessov degazatsii Zemli i genezis uglevodorodov//Tr. XXVII Geol. kongressa. – T. 13. – M.: Nauka, S 13–25

Letnikov FA, Zaechkovskiy NA, Letnikov AF (2010) K voprosu o geokhimicheskoy spetsializatsii glubinnykh vysokouglerodnykh sistem (On the issue of geochemical specialization of deep-carbon systems). Doklady AN Rossii 433(3):374–377 (in Russian)

Lukin A Ye (2013) Mineral'nye sferuly – indikatory flyuidnogo rezhima rudoobrazovaniya i naftidogeneza (Mineral spherules – indicators of the fluid regime of ore formation and naphthidogenesis) Geofizicheskiy zhurnal 35(6):10–53 (in Russian)

Nikishin AM, Korotaev MV, Ershov AV (2003) The Black Sea basin: tectonic history and Neogene-Quaternary rapid subsidence modelling. Sediment Geol 156:149–168

Pallas PS (1795) Kratkoe fizicheskoe i topograficheskoe opisanie Tavricheskoy oblasti (Brief physical and topographical description of the Tauride region). St Petersburg (in Russian)

Pikovskiy Yu I (2002) Flyuidnye plyumy litosfery kak model nefteobrazovaniya i neftegazonakopleniya (Fluid plumes of the lithosphere as a model of oil formation and oil and gas accumulation). Degazatsiya Zemli i genezis neftegazovykh mestorozhdeniy. GEOS, Moscow, pp 254–268 (in Russian)

Shnyukov EF, Sheremet'ev VM, Maslakov NA et al (2006) Gryazevye vulkany Kerchensko-Tamanskogo regiona (Mud volcanoes of the Kerch-Taman region). Glavmedia, Krasnodar (in Russian)

Shnyukov EF, Kobolev VP, Starostenko VI (2013b) Gazovyy vulkanizm Chernogo morya (Gas volcanism of the Black Sea). Logos, Kiev (in Russian)

Valyaev BN (2011) Uglevodnaa degazatsia Zemli, geotektonika i proiskhozhdenie nefti i gaza (Hydrocarbon degassing of the Earth, geotectonics, and the origin of oil and gas). Degazatsia Zemli i genezis uglevodorodnykh fluidov i mestorozdeniy. GEOS, Moscow, pp 10–32 (in Russian)

Yanko-Hombach V, Shnyukov E, Konikov E et al (2009) Response of biota to methane emissions in the Black Sea: Preliminary results from complex geological, geochemical, palaeontological, and biological study. In: Gilbert A, Yanko-Hombach V (eds) Extended Abstracts of the Fifth Plenary Meeting and Field Trip of IGCP 521-INQUA 0501, Izmir-Çanakkale (Turkey), August 22–31, 2009. DEU Publishing House, Izmir, pp 181–184

Yanko V, Kravchuk A, Kulakova I (2017) Meyobyentos myetanovykh vykhodov Chyernogo morya (Meiobenthos of methane outlets of the Black Sea). Phenix, Odessa (in Russian)

Zimmermann RA, Amstutz GC (1972) The Decaturville sulfide breccia—a Cambro-Ordovician mud volcano. Chemie der Erde 17(31).253–273

Contents

About the Authors

Evgeny Shnyukov is an eminent Ukrainian scientist in geology, geochemistry, lithology, sedimentary ore-formation, marine geology, and minerals of the World Ocean. He is Academician of the National Academy of Sciences of Ukraine (NASU), Doctor of Geological-Mineralogical Sciences, Professor, Member of the International Academy of Sciences of Eurasia, Honorary Director of the Center for Problems of Marine Geology, Geoecology and Sedimentary Ore Formation of the NASU (formerly Department of Marine Geology and Sedimentary Ore-Formation of NASU), and Head of the Ukrainian Lithology Committee. He has received many awards, including the State Award of Ukraine in Science and Technology, two State Prize of Ukraine for a series of works devoted to the World Ocean, and gold medal "Leonardo da Vinci" of the International Academy of Sciences of Eurasia, among many others. He has authored, coauthored, and edited over 700 scientific publications, including 50 books, and supervised over 30 PhD students. As the first scientist to recognize the need for marine geological research in Ukraine, he established the School of Marine Geology and Sedimentary Ore-Formation and initiated a new stage in the study of sedimentary iron and manganese ores, geology of the Black and Azov Seas, as well as comprehensive geological–geophysical and metallogenic investigations of the World Ocean's bottom sediments. He embraces a wide range of geological problems and represents a striking example of the fruitful union of science and practice. He supervised and actively participated in building of the research vessel "Geochimik" and headed a series of geological–geophysical expeditions on it and other research vessels. Materials from these expeditions led to the discovery of a submarine massif of crystalline rocks (Lomonosov Ridge), tens of mud volcanoes, and hundreds of gas seeps suggesting the presence of oil and gas beneath the Black Sea. Some of his books, such as *The World of Minerals and Catastrophes in the Black Sea* as well as many others, have gained wide recognition and are already bibliographic rarities.

Valentina Yanko-Hombach is the world leading scientist in the field of marine geology and micropaleontology (foraminifera). She is Doctor of Geological-Mineralogical Sciences, Professor, Head of the Department of Physical and Marine Geology of Odessa I.I.Mechnikov National University, Head of the Scientific and Educational Center of Geoarchaeology, Marine and Environmental Geology, Odessa I.I. Mechnikov National University, Ukraine, and President of the Avalon Institute of Applied Sciences, Winnipeg, Canada. She was a founder and past-President of the International Society of Environmental Micropalaeontology, Microbiology and Meiobenthology, and she has been president/executive director/plenary speaker at numerous international conferences, head of numerous international projects and research cruises, and chairman of special sessions and symposia at IGU, EGU, GSA, INQUA congresses, and other international fora. She has received many awards including Certificate and Medal of the All-Ukrainian Union of Geologists, award of Excellence from Universities of Palermo and Isparta, award of Excellence for Sustainable Development, Manitoba, Canada, and some others. In 1990 she initiated a multidisciplinary program in environmental micropalaeontology, through which she carried out pioneering research in both field and experimentally in the "Application of microorganisms to environmental change." The geographical area of her investigations covers the southern

European Seas (Mediterranean Sea, Sea of Marmara, Black Sea, Sea of Azov, Caspian Sea, and Aral Sea) from the Quaternary to the present. She is author/coauthor of more that 350 scientific papers including 11 monographs. Fluent in several languages and working easily in both East and West, she is devoting her energies to bringing together internationally recognized experts on environmental micropalaeontology and the geological and archaeological history of the Black Sea and adjacent basins to integrate sedimentological, paleontological, and archaeological data, synthesize a vast amount of non-English literature, and bridge perceived and actual barriers between eastern and western researchers.

1 History of the Geological Study of Mud Volcanoes in the Black Sea Region

Summary

This chapter discusses the history of the study of mud volcanoes in the Black Sea region from ancient times to the present. In ancient times, there were random, isolated descriptions of anomalous phenomena associated with mud volcanoes. Serious scientific study of mud volcanic activity began at the end of the eighteenth century. One of the first naturalists to describe a catastrophic eruption (of the Golubitskiy mud volcano in the Sea of Azov) was P. Pallas. By the nineteenth century, mud volcanoes had already been described by many researchers.

These early descriptions were scattered and sporadic, however, and there were no systematic studies. Only in the twentieth century were the mud volcanoes of the Kerch-Taman region systematically studied by V.V. Belousov and L.A. Yarotskiy. In a number of subsequent works, the presence of recessed synclines near mud volcanoes was noted by K.A. Prokopov, G.A. Lychagin, Z.L. Maimin, and others.

In the second half of the twentieth century, iron ores were discovered in recessed synclines near mud volcanoes (e.g., Novoselovskaya, Kamenskaya, Uzunlarskaya, Baksinskaya, and others) by E.F. Shnyukov and P.I. Naumenko.

In the last quarter of the twentieth century, as a result of numerous marine expeditions conducted by Moscow State University, "Yuzhmorgeologiya" (http://www.ymg.ru/en), institutes of the National Academy of Sciences of Ukraine, and Odessa State (now National) I.I. Mechnikov University, E.F. Shnyukov, M.K. Ivanov, and many others managed to discover up to 80 mud volcanoes within the Black Sea and many thousands of gas torches in different areas of the basin. An important role in these discoveries was played by the European participants of these expeditions (e.g., H. Bormann, P. Dimitrov, and N. Panin). Also, gas hydrates were recovered from the erupted materials of these mud volcanoes.

In general, studies of mud volcanoes have led to many geological discoveries. A new type of iron ore deposit has been identified on the Kerch Peninsula, and gas hydrates have been found in the depths of the basin.

1.1 Ancient Times

In the fifth century BC, the Greek historian Herodotus wrote: "the Black Sea is the most remarkable sea in the world." Whether Herodotos was aware of it or not, one of the most remarkable and peculiar features of this inland sea is the presence of huge quantities of methane gas lying beneath the seafloor. Its presence is evident from the outbursts of submarine mud volcanoes that often contain ice-like aggregates of gas hydrates (largely methane) as well as many high-intensity gas seeps and gas bogs, all of which release great

E. Shnyukov, V. Yanko-Hombach, *Mud Volcanoes of the Black Sea Region and their Environmental Significance*, https://doi.org/10.1007/978-3-030-40316-4_1

volumes of methane (e.g., 4.95–5.65 Tg $year^{-1}$) into the water column (Kessler et al. 2006).

Humans have known of mud volcanoes for a very long time. Archaeologists suggest that early Paleolithic groups migrated to the Black Sea region of Eurasia via mud volcanic provinces. They base this conclusion on comparisons made between maps showing the locations of mud volcanoes and those showing early Paleolithic sites. Paleolithic localities such as Sinyaya Balka, Il'skaya, and Bogatyr are found together with mud volcanoes on the Taman Peninsula of Crimea (Zenin 2012). Within some mud volcanoes, stone tools have been discovered; for example, a Paleolithic scraper was recovered from within the Akhtarma-Pashalinskaya mud volcano in Azerbaijan (Kovalevskiy 1935). Mud volcanic landscapes could have appealed to ancient people due to the presence of erupted breccia that could be used for making stone tools, as well as the hot water and gas flares produced by mud volcanoes.

According to the noted Russian geologist Kovalevskiy (1935), a passage from the epic poem, The Odyssey, written by the Greek poet Homer may indicate that people inhabiting the Taman Peninsula were familiar with the activities of mud volcanoes during the eighth century BC when the mythical voyage of the "Argo" took place. He based his conclusion on the following passage (translation by Lombardo 2000):

> One route takes you past beetling crags
> Pounded by blue-eyed Amphitrítê's seas.
> The blessed gods call these the Wandering Rocks.
> Not even birds can wing their way through.
> Even the doves that bring ambrosia to Zeus
> Crash and perish on that slick stone,
> And the Father has to replenish their numbers.
> Ships never get through. Whenever one tries,
> The sea is awash with timbers and bodies
> Blasted by the waves and the fiery winds.
> Only one ship has ever passed through,
> The famous Argo as she sailed from Aeëtes,
> And even she would have been hurled onto those crags.
> The Odyssey (XII, 65, 70)

Kovalevskiy (1935) suggested that the phrase "Blasted by the waves and the fiery winds" could describe the Argonauts bypassing the "fire-breathing mud volcanoes" of Taman on the way to Meotida (known today as the Sea of Azov).

It should be noted that the sailing route taken by the "Argo" to the Caucasus is relatively well-known, but the way back is quite unclear and has drawn much attention since antiquity. Some authors brought the Argonauts back home through Asia and Libya, while others took them through the Danube or around the Mediterranean Sea. According to one version, the Argonauts passed through the upper reaches of Tanais on their return. Tanais was an ancient city of the third century BC to the fifth century AD located at the mouth of the Don River, and it was here according to the reconstructed story where the Argonauts took their ship out of the water. From there, they sailed along the western Atlantic coasts of Europe, passing through the Pillars of Hercules (Gibraltar Strait) into the Mediterranean. This route through Tanais was confirmed by Dionysius Scytobrachion, a mythographer, who wrote his *Argonautika* in the second to first centuries BC. His proposed route from the Caucasus passed through the Kimmerian Bosporus (Kerch Strait) to Tanais. Although this route was criticized by Strabo and other ancient writers, it is no longer possible to ignore the possibility.

It may be that the sailing route of the "Argo" did not follow the main stream of the Kerch Strait that we know today. It could have shoaled or been blocked by the Chushka Spit. If this was the case, the "Argo" would have been forced to take a different route via Taman Bay, an arm of the Kuban River, the Ahtanizovskiy estuary, and the Sea of Azov, in close proximity to the town of Temryuk. This would mean that the Argonauts were sailing along a chain of exploding mud volcanoes, such as Gorelaya Sopka, Karabetova Gora, Cymbals, Boris and Gleb, Ahtanizovskiy, Blevaka, and others (Fig. 1.1).

The Byzantine emperor Constantine VII Porphyrogenitus (tenth century AD) wrote in his book *De administrando imperio* (*On the Management of the Empire*) that in the city of Tamatarha (in today's Taman district) and in Zikhia (Cherkessia), numerous sources that spewed out oil were present. These were likely

Fig. 1.1 A possible route of "Argo": (1) "Argo," (2) Navy ships of Aeëtes, King of Colchis, (3) route of "Argo," (4) mud volcanoes

mud volcanoes, and the oil was most definitely needed for the production of "Greek fire."

1.2 Late Eighteenth to Nineteenth Centuries

The first scientific description of mud volcano activity belongs to Pallas (1795) (Fig. 1.2).

Pallas was born on September 22, 1741, in Berlin into the family of a German professor of surgery. His mother was French. Home taught until the age of 13, Pallas mastered Latin and modern European languages, which facilitated his scientific activity, especially when compiling dictionaries and in developing scientific terminology. Invited to Russia by Catherine the Great in 1767, he described the frequently exploding Golubitskiy mud volcano in the Sea of Azov and connected its activity to earthquakes. He also provided important details on a large eruption of the Dzhau-Tepe mud volcano (Kerch Peninsula) that occurred during the late eighteenth century.

Fig. 1.2 Peter Simon Pallas (1741–1811) (in Russia he was called Peter Semenovich)

According to him, a key reason for the eruption was the slow burning of coal or oil shale underneath the Taman and Kerch peninsulas, followed by the evaporation of seawater, which raised the mud breccias to the surface (Pallas 1883).

In their article on the geology of the Taman Peninsula, Voskoboynikov and Gur'ev (1832) provided information on mud volcano localities, morphologies, and activities. In particular, they described an eruption of Gorelaya Gora in 1794 as well as an eruption of another mud volcano located 4 km from Taman (in the opinion of the present authors, this was Karabetova Gora). They considered that the underlying oil reservoirs were the primary reason for mud volcano activity in the Baku and Taman areas.

Dubois de Montpéreux (= Dyubua de Monpere 1837) mapped a number of mud volcanoes in the Taman and Kerch peninsulas. He also described the fiery eruptions of Gorelaya Gora (Cuku-Oba) in 1794, Ahtanizovskiy (Kusu-Oba) in 1818, Gnilaya Gora in 1815, and Karabetova Gora in 1835. Additionally, he pointed to the appearance of islands in the Sea of Azov subsequent to mud volcano eruptions that occurred between 1799 and 1811.

Abryutskiy (1853) reported on the simultaneous eruptions of Karabetova Gora and Ahtanizovskiy in August of 1853, as well as previous eruptions of the former in 1818 and 1833.

Otto Wilhelm Hermann von Abich (Fig. 1.3) described the geology of the Kerch and Taman peninsulas, supplemented by detailed descriptions of their mud volcanoes (Abich 1873).

Otto Wilhelm Hermann von Abich was a German mineralogist and geologist and full member of the St. Petersburg Academy of Sciences (honorary member from 1866). He was born in Berlin and educated at the local university. His earliest scientific work was related to spinels and other minerals. Later, he conducted special studies of fumaroles, the mineral deposits around volcanic vents, and the structure of volcanoes. In his opinion, mud volcano gas eruptions could not be explained by the influence of high temperatures on organic debris. Instead, the same endogenic forces that formed the Great Caucasian Ridge were responsible for the phenomenon of mud volcanism. Based on the presence of petroleum in metamorphic and volcanic rocks from the Kerch Peninsula, he suggested that hydrocarbons had a volcanic origin. He drew attention to the connection between mud volcanoes and earthquakes and pointed out that mud volcanoes comprise the same sedimentary rocks that underlie the places where they form. Based on their internal structure, these rocks are almost identical to those erupted by real volcanoes; they were created by fiery liquid processes, and as such, the same internal forces must be responsible for the activities of both types of volcanoes. He drew attention to the linear arrangement of mud volcanoes and their close association with the nodes of folding system intersections, and he considered that all mud volcanoes were confined to the zones of faults in the Earth's crust.

Filippov (1857) considered that the mud volcanic manifestations on both sides of the Kerch Strait are intense and function on a large scale, surmising that the Taman Peninsula might have been formed by the activity of mud volcanoes.

Helmersen (1864) (Fig. 1.4) stressed a possible relationship between the geological structures

Fig. 1.3 Otto Wilhelm Hermann von Abich (1806–1886)

Fig. 1.4 Grigoriy Petrovich Helmersen (Georg von Helmersen) (1803–1885)

of the Kerch and Taman peninsulas and mud volcano activities.

He described the Taman and Kerch mud volcanoes (Dzhardzhava, Bulganakskiy, Yenikalskiy) and drew attention to their possible relation to petroleum resources present in the valleys between anticlinal structures. He explained it by reference to fragmentation in the middle parts of anticlinal folds.

Land (1866) described Taman mud volcanoes, particularly the volcano Gnilaya Gora, and raised a question about the balneological benefits of mud breccias.

Ansted (1868) described the appearances of a number of mud volcanoes on the Kerch Peninsula.

In his review of archaeological research on the Taman peninsula, Gerts (1876) reported on an eruption of the Karabetova Gora mud volcano in August of 1856.

Alekseev (1880) described some Kerch and Taman mud volcanoes, mainly paying attention to the composition of erupted gases.

Dubinevich (1885), Kramarevskiy (1866), and Morozevich (1886) provided an analysis of the mechanical and chemical composition of Kerch mud volcanic breccia.

Pototskiy (1888) related mud volcanoes to petroleum sources and mining.

Nikolay A. Golovkinskiy (Fig. 1.5) was a Russian geologist, who attributed mud volcanoes to anticlinal folds (Golovkinskiy 1889).

He described clastic material in the breccia of the Dzhau-Tepe mud volcano, indicating that this material is composed of fine-grained sandstones, clay iron ore, limestone, iron pyrite, calcareous spar, lignite, and white marl. He explained the appearance of Sarmatian limestones in the cores of Maikopian anticlinal structures as a result of subsidence due to removal of material from depth and exposure of limestones due to mud volcano activity.

Nicolai I. Andrusov was a Russian geologist, stratigrapher, and paleontologist. He described the geological structure of the Kerch Peninsula and provided a description of the Tarkhan group of mud volcanoes (e.g., Burashskiy, Dzhau-Tepe, etc.). This work is regarded as among the most fundamental and classic studies of the Kerch Peninsula's geology (e.g., Andrusov 1884, 1893) (Fig. 1.6).

Samoylov (1898) conducted an expedition to the Yenikalskiy mud volcano and provided a detailed description of its gas emissions.

1.3 Early Twentieth Century

Vernadskiy and Popov (1899–1900) visited the Yenikalskiy, Bulganakskiy, and Tarkhanskiy mud volcanoes and provided a visual observation of the first one. The discovery of boron compounds, such as borax, in the mud breccia was the main result of their observations.

Felitsyn (1902) described the Taman mud volcanoes using the reports of previous investigators—Pallas, Abikh, Hertz, and Land.

Fig. 1.5 Nikolay Alekseevich Golovkinskiy (1834–1897)

Fig. 1.6 Nicolai Ivanovich Andrusov (1861–1924)

Borisyak (1907) examined and described the Hartsyz-Shiban mud volcano near Vladislavovka village on the Kerch Peninsula.

Chirvinskiy (1908) observed some changes in the Samoylov mud volcano that occurred subsequent to the observations by Vernadskiy and Popov (1899–1900).

Zhivilo (1909) described the topographic development of the Taman Peninsula in connection with mud volcano activities.

Yushkin (1909) considered the petroleum fields of Neftyanaya Gora and Ocheretin as extinct mud volcanoes.

Steber (1909/1910) described a number of violent eruptions of the Karabetova Gora mud volcano in 1835, 1852, and 1882. He offered some notes on changes in its appearance, which were supplemented by an overview of the chemical and mineralogical composition of its breccias.

Gubkin (1913) was a Russian geologist and president of the 1937 International Geological Congress in Moscow (Fig. 1.7). He was a petroleum geologist particularly interested the region between the Volga and the Urals.

Gubkin examined the geological structure of the Taman Peninsula and the Kuban region. He described the Gnilaya Gora and Miska mud volcanoes, provided a chemical analysis of erupted gases, and underlined the confinement of mud volcanoes and oil spills to brachyanticlines that form orographic terrain hills. Under the supervision of Gubkin (1940), petroleum geologists studied the mud volcanoes of Azerbaijan, Georgia, and the Taman and Kerch peninsulas and established a link between mud volcanoes and oil and gas mineral resources.

Fig. 1.7 Ivan Mikhaylovich Gubkin (1871–1939)

Dvoychenko (1914), Klepinin (1914), Gembitskiy (1914), and Sedelshchikov and Kulchavov (1914) documented the catastrophic eruption of the Dzhau-Tepe mud volcano on March 18, 1914, and in part, its earlier eruption in March of 1909.

Steber (1915) studied the radioactivity of Kerch-Taman mud volcanoes. He found that the erupted mud breccia was not radioactive, whereas the volcanic gases of the mud volcanoes Bulganakskiy, Yenikalskiy (Kerch Peninsula), and Gnilaya Gora (Taman Peninsula) were (up to 0.45 ME).

Ergart (1916) recorded the radioactivity (1.33 ME) of water that erupted from the Gnilaya Gora mud volcano. He also described explorations into combustible gas in the Gnilaya Gora and Miska mud volcanoes, located in the town of Temryuk.

Beketov (1916) presented data on the concentration of iodine, bromine, and boric acid found in eruptions of the Kerch-Taman mud volcanoes. Results showed that erupted water contained, on average, 3 mg of iodine, 2 g of anhydrous borax, and 3.5 g of anhydrous soda per liter of discharged water. However, their flow rate was too small for exploration.

Izgaryshev and Sludskiy (1917) provided information on boron concentrations found in the clastic mud breccia of different Kerch-Taman mud volcanoes. He evaluated the boron reserves of the Gnilaya Gora and Shugo mud volcanoes as 25,200 t and 390 t, respectively.

Khlopin (1919) provided some details on the amounts of boron found in clastic breccia from mud volcanoes on the Kerch Peninsula but mainly sourced this information from literature.

Stopnevich (1920) reported some data on the production of iodine, borax, potash, and bromine from the Kerch-Taman mud volcanoes.

Obruchev (1921) listed oil and gas outbursts from the Kerch Peninsula and briefly described almost all of the mud volcanoes located there.

Pyatnitskiy (1925) described the tectonics of the Taman Peninsula. He concluded that its dome-shaped hills are similar to laccoliths but that they were built up by erupted mud volcanic breccia and gases that were not of magmatic origin. He also provided a description of the Gora Gnilaya mud volcano and estimated its reserves of iodine as 192 t.

Arkhangel'skiy (1925) considered the tectonic origin of the clay breccias of mud volcanoes, suggesting that they developed during the formation process of diapir folds due to friction that squeezed the cores of the folds between their wings.

Obruchev (1926) provided a summary of numerous mud volcanoes and carbonated springs along with a description of the oil and gas potential of the Kerch-Taman region.

Murzaev (1928) stated that the eruption of the Dzhau-Tepe mud volcano in September 1927 corresponded to the famous Crimean earthquake shocks of the same year.

Arkhangel'skiy et al. (1930) provided a complete description of the Kerch Peninsula's geology as well as its oil and gas resources. They considered the manifestations of mud volcanism as a result of certain tectonic conditions related to the development of clay breccia, which represents a purely tectonic formation. The latter was created by the overflow of sediments under the development of diapirs, thrusts, and faults. Knoll breccias are the same as tectonic breccias, but they have been softened by water to a free-flowing state. The authors produced a geological map of the Kerch Peninsula, including all known mud volcanoes at that time.

Turley (1930) provided results from his observations of the Bulganakskaya group of hills. He noted that mud volcanoes are most active in spring and presented some data on a mechanical analysis of mud breccias and the composition of large fragments. He reported that about 5 t of borax and a few kg of iodine were mined in 1925 and that these numbers had increased to 3150 t of borax and 3.4 t of soda in 1926.

Grechishkin (1931) described his study of petroleum fields in the northern and central part of the Kerch Peninsula and passed on details of the Burashskiy mud volcano.

Prokopov (1931) documented the relationship between mud volcanic activity and the development of cup-shaped depressions in the anticlinal folds of the Kerch Peninsula (Fig. 1.8).

Prokopov (1931) concluded that the knoll gases of the northwestern part of the North Caucasus originated in sedimentary rocks no younger than the Lower Cretaceous period. Thus, they are much older than the oil field gases of the younger Eocene epoch. He agreed with the ideas of Golovkinskiy (1889) that the main cause of the formation of such depressions was the activities of mud volcanoes, which lead to the creation of cavities at greater depths. He particularly emphasized the activity of mud volcanoes and the formation of cavities within Sarmatian sedimentary rocks. He stressed that the ideas of Andrusov (1893) and Obruchev (1926) regarding secondary geosynclines embedded within anticlinal folds, which they explained as the result of tangential pressure, are insufficient to explain the uniqueness of such structures.

Sivere (1931) described a strong eruption of the Dzhardzhavskiy mud volcano near Kerch city.

Fedorov (1939) listed the mud volcanoes on the Kerch Peninsula and emphasized the different origins of breccias, petroleum, and gases. He also underlined the confinement of mud volcanoes to anticlinal folds, and subsequently provided a compilation review of the Kerch-Taman mud volcanoes, together with some data on erupted gases that were based on literature.

Pokrovskiy (1933) described the Gora Gnilaya mud volcano and its boron sources. According to his exploration data, the reserves of iodine and borax were 146 t and 13,000 t, respectively. For

Fig. 1.8 Konstantin Andreevich Prokopov (1882–1972)

comparison, he estimated the reserves of borax in the Karabetova Gora mud volcano at 115,000 t.

Vladimir V. Belousov was an outstanding Soviet scientist-geologist, geotectonist, and the author of the radiomigration hypothesis of the development of the tectonosphere (Fig. 1.9).

Belousov and Yarotskiy (1936) summarized all published materials on the Kerch-Taman mud volcanoes, paying particular attention to the conditions of their occurrence, activities, structure, and the composition of their breccias, gases, and water. They proposed important ideas on the use of mud volcanic products in the building industry and in balneology. They supplemented these ideas with some data on the mining of boron by the Bulganakskiy mining-chemical enterprise.

Popov (1938) described the mineralogical composition of mud volcanoes, focusing particularly on the mineralogy of boron.

Fedorov (1939) devoted this study to proving a link between diapiric processes and the development of mud volcanoes.

Kudryashev (1939) described the geology of the Shugo mud volcano.

Avdusin (1939) provided new data on the petrography of ejecta emanating from mud volcanoes.

Sulin (1939) summarized the hydrological and geochemical studies of mud volcanoes conducted until that time.

Gulyaeva (1939) examined the geochemistry of boron and came to the conclusion that it derives from mud volcanoes in association with erupted sedimentary material.

Fig. 1.9 Vladimir Vladimirovich Belousov (1905–1990)

Itkina (1939) described the distribution of potassium, mainly in Caucasian mud volcanoes.

Levenson (1939) investigated the problem of mud volcanism and geochemical bituminosity.

Ginzburg-Karagicheva et al. (1939) provided some data on the microbiology of mud volcanoes.

Gubkin (1940), a long-time leader in the USSR's petroleum industry, produced a comprehensive listing of characteristics revealed by mud volcanoes of the USSR, including those from the Black Sea region. He described their stratigraphy and lithology; provided classification and petrology of volcanic products; identified more than 50 minerals; characterized their water geochemistry, rocks, and bitumen; and emphasized the connection between mud volcanoes and diapir tectonics. His data on the content of boron, iodine, gases, and the microbiology of mud volcanoes are of particular interest. He suggested using areas of mud volcano development in the Crimea-Caucasus geological province as priority fields for petroleum exploration and insisted that all mud volcanoes found there, especially those in Azerbaijan, must be drilled. The Kerch-Taman mud volcanoes were to be considered for deep drilling if the depth of potential oil-gas reservoirs was determined to be suitable for drilling from an industrial perspective.

Avdusin (1948) dedicated his research to petrographic studies of Azerbaijan mud volcanoes.

1.4 Post World War II

After the Second World War, the Kerch-Taman mud volcanoes were studied as part of a geological survey focused on oil and gas exploration. This survey revealed for the first time the geological complexity of the anticlinal structures within which mud volcanoes are confined (Lychagin 1952).

Ronov (1951) summarized his research on the composition and geochemistry of the Kerch-Taman and Azerbaijan mud volcanoes, while Gattenberg (1954) estimated the age of erupted gases in the former.

Alyaev (1947) and Maymin (1951) studied the oil and gas potential of mud volcanoes. The latter provided the first overview of mud breccias distributed in the Tertiary (from Maikop to Pontian) geological section of the Kerch Peninsula and compared them to those from the Taman Peninsula.

Nikitin (1955) provided general geological characteristics of the Shugo mud volcano.

Zenin (1955) and Zaytsev (1965) described the Taman mud volcanoes (Karabetova Gora, Dulcimer, Ahtanizovskiy, Blevaka, and some others) over the entire area of their development (the Sea of Azov) and confirmed their connection to diapir tectonics.

Al'bov (1956) found water enriched with boron on the Kerch Peninsula over the course of his hydrological investigations in Crimea. Together with Goryainov (1968), they drew attention to the nature of boron found in the mud and water of Chokrak Lake.

Shardanov et al. (1962) and Shardanov and Znamenskiy (1965) examined the origins and roots of mud volcanoes found on the Taman Peninsula. They concluded that their roots go down to Lower Cretaceous rocks and contained large reservoirs of oil and gas.

Kalinko (1964) confirmed the genetic connection of mud volcanoes with oil and gas reservoirs. He also considered the distribution of mud volcanoes around the globe, paying particular attention to the nature and origin of mud volcanoes found in the Alpine foothills and intermontane basins. He denied any connection between mud volcanoes and tectonic disturbances. In his opinion, the Kerch-Taman, Kobystan, and Azerbaijan mud volcanoes emerged in the Late Maikopian deposits during the formation of their uppermost part.

Kurishko et al. (1968) described the hydrogeology of the Kerch Peninsula's mud volcanoes.

Research on the Kerch-Taman mud volcanoes was renewed in the 1960s due to increased industrial interest in rare elements—mercury, lithium, arsenic, and boron. Investigations of mud volcano geochemistry revealed an elevated mercury concentration in gas, water, and mud breccia (Morozov 1965; Karasik and Morozov 1966; Shternov 1968).

Morozov (1965) provided general geological characteristics of the knoll fields of mud volcanoes on the Kerch Peninsula (Gornostayevskiy, Kayala-Sart, Dzhau-Tepinsksy, and Kostyrinskiy) and the Taman Peninsula (Gladkovskiy, Shuginskiy, and Semigorskiy). He also investigated the geochemistry of accessory chemical elements (e.g., boron, mercury, bromine, iodine, and lithium) in mud volcanism processes. Shternov (1968) described in detail the mud volcanism of the Kerch-Taman region. The formation of mud volcanoes was explained in accordance with the ideas of Kovalevskiy (1935) as being a result of "magma spike" intrusions into the sequence of sedimentary rocks (Fig. 1.10).

Another reason why interest in the Kerch-Taman mud volcanoes intensified at this time was connected to the successful use of erupted clays as raw materials for the production of steel pellets. An exploration was carried out on the Burulkayskaya, Korolevskaya, and Kayaly Sartskaya geological structures, with the main goal being to delineate the spatial distribution of mud breccia fields and provide a technological assessment of their exploration (Yuhanov 1968a, b).

Nesterov (1968) considered the possibility of using mud breccia for balneological purposes.

Fig. 1.10 Sergey Aleksandrovich Kovalevskiy (1889–1975)

1.5 Mud Volcanoes as Sources of Mineral Ores

The most important breakthrough in the study of the Kerch-Taman mud volcanoes relates to the discovery of ore-bearing deposits in compensated or compressed geosynclines that formed during the Kimmerian orogenic cycle. Their presence in some mud volcanic structures has been known since the days of Andrusov (1893), and they were later described by Konstantov et al. (1933), Gubanov (1961, 1963), and Malakhovskiy (1956), but only after the discovery of their uniqueness was a new type of iron ore deposit distinguished in mud volcanic structures with large-scale mineralization (Shnyukov and Naumenko 1964) (Fig. 1.11).

Pavel I. Naumenko was Chief Geologist of the Kamysh-Burunsky Iron Ore Plant and then Head of the Department of Mining Geology of the Ministry of Ferrous Metallurgy of the Ukrainian SSR. In close contact with the scientists of the National Academy of Sciences of Ukraine, P. I. Naumenko investigated the mud volcanoes of the Kerch Peninsula and discovered a new type of iron ore deposit associated with them. Together with the first author of this book, he located the large Novoselovskiy iron ore deposit, which contains more than 100 million tons of deposit as well as a number of small deposits of iron ore and one deposit of fluxes.

This breakthrough initiated serious geological exploration, which was carried out by the Department of Mining Geology of the USSR under the supervision of scientists from the Academy of Sciences of Ukraine and financed by the Kherson Economic Council, Ukraine. As a result of this work, a large field of oolitic iron ores was discovered within the Novoselovskiy mud volcanic hearth. Called the Novoselovskoe field, it contains 125 mln t of conditional iron ore, and together with non-conditional iron ores, up to 200 mln t were recognized by the USSR Commission of Reserves as being an important source of iron ore. In addition, several small fields (e.g., Rep'evskoe, Uzunlarskoye) containing these deposits were discovered (Shnyukov and Naumenko 1964). All obtained material was summarized in the fundamental monograph (Shnyukov et al. 1971).

Fig. 1.11 Pavel Ivanovich Naumenko (1928–1987)

After this discovery, the search for ore-bearing deposits in mud volcanic structures continued, and a number of others (e.g., Achinskoe Bulganakskoe, Arma-Elinskiy, Batalnenskoe, and Kamenskoe) were located (Shnyukov and Naumenko 1982; Shnyukov et al. 1985, 1987) and summarized in Shnyukov et al. (1992). The potential for ore-bearing fields existing within mud volcanic structures is still considered economically significant, and in 2016, important deposits were discovered within the Andreevskiy mud volcano (Shnyukov 2016).

1.6 Modern Geological Studies of Mud Volcanoes

Along with applied investigations into mud volcanoes, classical geological study of them has also been ongoing.

Gemp et al. (1970) conducted investigations of the CH_4 and CO_2 isotopic composition of gases from Kerch-Taman mud volcanoes.

Lagunova (1974) considered the origin of CO_2 and hydrochemical features (Lagunova and Gemp 1978) of the Kerch-Taman mud volcanoes and related them to endogenic processes.

Shnyukov et al. (1986) published an atlas titled *Mud Volcanoes of the Kerch-Taman Region*. It was later updated by a slightly different team of scientists and reprinted (Shnyukov et al. 2005).

Kholodov (2002a, b, 2012) provided general overviews of the world's mud volcanoes, placing special attention on those of the Kerch and Taman peninsulas.

A number of other publications have highlighted the geological structure and activities of the Taman mud volcanoes (Nesterovskiy 1990; Bogatikov et al. 2003).

Sobisevich et al. (2008) published new data on the deep structure of the Karabetova Gora mud volcano, while Tvertinova et al. (2015) described the structural position and specific features of its formation.

The lithology of mud breccia, including coarse material, from the Dzhau-Tepe mud volcano was described by Titova (2013) and Titova et al. (2013).

Shnyukov et al. (2013b) found gold in mud breccia from the Korolevskiy and Bulganakskiy mud volcanoes.

Gusakov (2004) identified a new mud volcano on the Taman Peninsula and called it Shkol'nyy.

In the Western Black Sea basin, Shnyukov and Netrebskaya (2013) traced the roots of mud volcanoes down to the surface of the Moho boundary.

Shnyukov (2016) studied minerals from knoll breccias deriving from mud volcanoes developed in Maikopian sediments of the Kerch-Taman region. He concluded that these minerals had been formed by hydrothermal formation processes. The book also contains new data on authigenic minerals in the breccias of northwestern Caucasian mud volcanoes that had developed in Lower Cretaceous sediments.

A full description of the characteristics of native minerals recovered within various types of mud volcanoes in the Black Sea region was provided by Shnyukov (2016). These minerals likely indicate the important role of deep fluids in the formation of the region's mud volcanoes (Shnyukov and Lukin (2011).

At the end of the twentieth century, the "center of gravity" for mud volcanism research moved to the Black Sea. Over the course of dozens of marine expeditions conducted aboard different research vessels (e.g., "Mikhail Lomonosov," "Kiev," "Professor Vodyanitskiy," "Vladimir Parshin," and some others), a substantial number of underwater mud volcanoes were discovered (Shnyukov et al. 2005). Their possible existence was predicted by other researchers much earlier (Zenkovich 1894; Andrusov 1918; Shepel' 1926), but of fundamental interest is the publication of Kovalevskiy (1960), in which he first expressed his opinion about the existence of mud volcanoes in the northeastern part of the Sea of Azov as well as in the deep waters of the Black Sea. He reckoned that mud and "pyrogenic" volcanoes should be located along the 38th meridian and should include the elevated seafloor banks of Zhelezinka (in the Sea of Azov) as well as Krasnov, Leonchevoy, and Shternov (in the Black Sea). Unfortunately, whether these are "pyrogenic" or mud volcanoes or just unrelated bottom elevations has yet not been confirmed. E.F. Shnyukov searched for them over a 24-h period in 1989 on board the R/V "Mikhail Lomonosov," but he could not find the 300-m-high underwater hill that was described by Kovalevskiy. Nevertheless, Kovalevskiy's idea about deepwater mud volcanoes in the Black Sea has proven to be correct.

In 1989, two mud volcanoes (MSU and Vassoevich) were discovered in the central part of the Black Sea basin at a water depth of 2000 m (Ivanov et al. 1989). Further investigations led to the discovery of dozens of similar morphological structures within the Black Sea, particularly in its central part (south of Crimea), in the Sorokin depression, in the Tuapse depression, and a few other places.

To summarize, study of Black Sea mud volcanoes has been conducted through multiyear programs pursued by many Russian and Ukrainian research teams from various organizations—such as "Yuzhmorgeologiya," Moscow State University, the National Academy of Sciences (Geological Institute, Department of Marine Geology and Sedimentary Ore-Formation, Institute of Biology of the Southern Seas, Geophysical Institute, and Marine Hydrophysical Institute), Odessa I.I. Mechnikov National University, and others. Significant contributions were made by Bulgarian, Romanian, German, and other European scientists, who worked together with Russian and Ukrainian specialists within the frameworks of international projects, such as (1) EC FR5 project METROL "Methane Flux Control in Ocean Margin Sediments," 2002–2005; (2) EC FR5 project CRIMEA "Contribution of high-intensity gas seeps in the Black Sea to Methane Emission to the atmosphere," 2003–2005; (3) the German National Geotechnologies program METRO "Methane and methane hydrates within the Black Sea: Structural analyses, quantification and impact of a dynamic methane reservoir," 2004–2007; (4) ESONET "The European Sea Floor Observatory Network," 2003–2004; and (6) EC FR6 project HERMES "Hot spot ecosystem research on margins of European Seas," 2006–2009. These and other projects—e.g., the UNESCO Training-Through-Research (TTR) Programme—have recorded a vast quantity of important data about mud volcanoes in the Pontic basin over the course of dozens of research cruises using various research vessels—e.g., "Le Suroît" (2001, 2002), "Mare Nigrum" (2005–2008), "Knorr," "Meteor" (2005, 2007), and "Le Marion Dufresne" (2004).

During the same period (1990–2015), the National Academy of Sciences of Ukraine carried out dozens of geological expeditions on the research vessels "Kiev," "Mikhail Lomonosov," "Akademik Vernadskiy," "Professor Vodyanitskiy," "Vladimir Parshin," and "Ichthyander." A few research cruises were conducted by scientists from Odessa I.I. Mechnikov National University. These expeditions focused on the study of mud volcanoes, specifically bottom topography in areas of volcanoes, volcanic activity, composition of clastic material and gases, etc. Several new mud volcanoes have been discovered, and presently, 68 mud volcanoes are known in the basin (Shnyukov et al. 2015). The most important outcome of the research can be found in a number of collective monographs (Shnyukov et al. 1993, 1997, 1998, 2005, 2006, 2007, 2013a).

Basov and Ivanov (1996) described Late Quaternary volcanism in the Black Sea. Meysner et al. (1996) highlighted mud volcanoes in the Western Black Sea basin. Konyukhov et al. (1990) described a number of newly discovered mud volcanoes in the deep part of the Black Sea. Bouriak (1994, 1995) published important results, one of which was a seismic and acoustic study of mud volcanoes together with their gas content in the deep part of the Black Sea. Limonov et al. (1999) compared mud volcanoes of the Black and Mediterranean seas. Shnyukov (1999) summarized the particular manifestations of mud volcanoes in the Black Sea. Bohrmann et al. (2003) provided new data on the Dvurechenskiy and Odessa mud volcanoes, as well as some others. Blinova et al. (2003) conducted a study of hydrocarbon gases from the mud volcanoes located in the Sorokin depression. Stadnitskaya and Belen'kaya (2000) studied hydrocarbon gases of mud volcanoes and their composition and isotopes. Minerals formed by mud volcanic fluids (gold, silver, copper, zinc, lead, and others) in the breccia of mud volcanoes within the Western Black Sea were described by Shnyukov et al. (2013a).

Along with mud volcanoes, about 4000 methane gas seeps and bogs have been discovered in the Black Sea. They clearly demonstrate enormous degassing activity, at least double that of any other basin in the world. This indicates an enormously high capacity to produce, hold, dissolve, and transform CH_4 into other chemical compounds (e.g., H_2S) or into another physical state (e.g., methane hydrates), and this process deserves close study. The gas seeps and bogs are plentiful on the shelf and continental slope. The most notable degassing occurs in the northwestern part of the Black Sea and in offshore areas of the Kerch-Taman, Bulgarian, and Georgian coasts. While degassing is moderately well documented there, the Turkish side (43% of the shelf) is largely unknown.

Egorov et al. (2011) present general characteristics of Black Sea degassing via gas seepages based on many years of research. Several mud volcanoes are also described, and the total volume of released methane is estimated to be 25×10^6 m^3 annually. The book also provides an overview of the physical nature of degassing together with detailed descriptions of the areas where gas seeps are located.

A summary of all known mud volcanoes of the Black Sea and its coastal areas was published by Shnyukov et al. (2013a). This book describes the mud volcanoes of the Black Sea region as a geological phenomenon. Particular attention is paid to manifestations of mud volcanism in different areas of the region, especially in its northeastern coastal zone. Numerous geophysical data on the deep roots of mud volcanoes and gas seeps are provided.

The catalogue of mud volcanoes of the Black Sea-Sea of Azov region was published by Shnyukov et al. (2015). It contains characteristics of the 68 known mud volcanoes that will undoubtedly facilitate their further study.

If gas seeps and mud volcanoes contribute to degassing of the Black Sea basin, gas (largely methane) hydrates bind methane beneath the seafloor. Gas hydrates take various shapes, such as small cakes, nodules, irregular crystals, veins, thickenings, etc. They were first discovered and described in the Black Sea by Efremova and Zhizhchenko (1974) as white crystals located in cavities within muddy sediments. Later, they were commonly found and repeatedly described by many other researchers (e.g., Korsakov et al. 1991; Ginsburg et al. 1990; Shnyukov et al. 1990; Byakov and Kruglyakova 2001). Extensive seismic data obtained by Romanian scientists (Popescu et al. 2007) enabled them to reveal the main characteristics of gas hydrates from bottom simulating reflections (BSRs) in the deepwater part of the Danube delta across an area of about 2900 km^2. In general, findings of gas hydrates are confined to the lower part of the continental slope (below a water depth of 700 m). The thickness of seismic facies containing gas hydrates ranges between 400 and 500 m, sometimes as much as 800 m. Their lower boundary is rather shallow (a few meters) and subparallel to the seafloor. Today, gas hydrates have been recovered in about 20 localities largely associated with three environment types: submarine fans, fractured zones, and mud volcanoes (suggesting their close genetic relationships). Age of the sediments hosting gas hydrates as well as their lithological properties (e.g., grain size) vary significantly. Relatively coarse-grained sediments make better hydrate reservoirs than fine-grained sediments. The area of the Black Sea suitable for gas hydrate formation is estimated at 288,100 km^2, representing about 68% of the total Black Sea, or almost 91% of the deepwater basin; the volume of gas hydrates has been estimated at 4.8 km^3 corresponding to $0.1–1 \cdot 10^{12}$ m^3 of free methane (Vasilev and Dimitrov 2002). Are these realistic numbers? Which gas hydrate layers are most suitable for exploration? What influences do climate and sea-level change have on the formation of gas hydrates in the basin? Many scientists are trying to answer these questions. It has been observed that methane hydrates are most often confined to mud diapir structures and river paleovalleys with seepage of freshwater at depths greater than 600–700 m.

In some places, a fivefold repetition of BSR boundaries was found in geological sections. This led to discussions on the nature of gas hydrates, e.g., the differences in their chemistry, and the influence of temperature variations on their development during glaciations (Popescu et al. 2007). At the request of the "Rosneft'" (Russian Oil) company, a survey for gas hydrates in the Tuapse depression was conducted by scientists from Moscow State University. As a result, a few areas of gas hydrate distribution were discovered.

Geophysical investigations of gas hydrates in the Ukrainian economic zone were carried out by European scientists during the course of biological studies to the west of Sevastopol in the mouth of the Paleo-Kalanchak (paleo-Dnieper) River. The area of gas hydrate development revealing thicknesses up to 100 m was estimated to be 805 km^2 at water depths of 700–1350 m (Lüdmann et al. 2004).

In 2010–2011, geophysical investigations of Black Sea gas hydrates in that area were completed by scientists from the Geophysical Institute of the Ukrainian National Academy of Sciences. Specialists from the Odessa Institute of Refrigeration put forward a proposal to create a special platform for extraction of gas hydrates, but no financial support was received.

In general, work on Black Sea gas hydrates holds great promise, both scientific and practical.

Mud volcano eruptions affect the geo- and ecosystems of the Black Sea (Shnyukov 2016). They play an active role in bottom degassing and may serve as high-efficiency, low-cost indicators for the investigation of gas hydrates, including those suitable for industrial exploration.

Gas outbursts from offshore mud volcanoes affect the hydrochemical regime of the sea, producing currents and acoustics. They also affect ecosystems and, most of all, routes of seasonal fish migrations. To a certain degree, the entire biota in the Black Sea are determined by fluctuations in the level of hydrogen sulfide exposure.

Mud volcanoes can cause great damage to the environment. Powerful eruptions cause ground subsidence in nearby areas, and this makes mud volcanoes a real threat when they occur within the limits of urban agglomerations, for example, Kerch in Crimea and Temryuk in Krasnodar Krai. Contamination of the air with mercury and other elements can also be hazardous to people.

Mud volcanoes can have a significant effect on maritime traffic, especially in narrow straits. For instance, seven mud volcanoes have been found offshore in Kerch Strait, through which nearly 10,000 ships cleave their way every year. There have been cases of ships running aground within the navigational channel (e.g., S/S "Caesar" in 1914, and some others). The shoal on which they were grounded proved to be mud volcanic in nature (Shnyukov and Yanko 2014).

At times, mud volcanoes within the strait have given rise to the formation of small islets. Some researchers propose methane outbursts as the cause of ship loss in the Bermuda Triangle. Accidents of that kind are likely to happen in the Black Sea as well. The probability of such accidents has been shown experimentally (Hueschen 2010).

Mud volcanoes can sometimes be used for precise localization of oil traps, as they are good indicators of oil- and gas-bearing provinces. This feature can also aid in the development of earthquake forecasting criteria in the future (Shnyukov et al. 2013a).

Another direction of research is the influence of released methane into the ecosystem. It appears that the influence of methane at different concentrations on benthic communities varies in a much more complicated way than was previously thought (Yanko et al. 2017).

In general, mud volcanism in the Black Sea region is an extremely interesting phenomenon of multidimensional importance, deserving in-depth study primarily as an indicator of the Earth's oil- and gas-bearing capacity. Today, mud volcano studies are largely focused on scientific, not applied, aspects. There have yet to be any surveys of mud volcanoes as indicators of gas hydrates, nor have there been any calculations performed to determine their contribution to the total degassing of the seafloor on a basin-wide scale. Likewise, no research has been conducted on the hazards they pose. At the same time, mud volcanoes are the seafloor's expression of endogenic processes and a "cheap window" by comparison to drilling into the deep geosphere; they may be considered valuable resources for industrial needs in the future. In this regard, wide spectrum investigation of mud volcanism should be continued.

References

Abich GV von (1873) Geologicheskiy obzor poluostrovov Kerchi i Tamani (Geological survey of the Kerch and Taman peninsulas). Zapiski Kavkazskogo otdeleniya russkogo geograficheskogo obshchestva 8:3–160 (in Russian)

Abryutskiy LP (1853) O deistvii gryazevogo vulkana na Tamanskom poluostrove v Svguste 1853 goda (On the activity of a mud volcano on the Taman Peninsula in August 1853). Gorny Zhurnal Book 10, Part 4:271–277 (in Russian)

Al'bov SV (1956) Mineral'nye istochniki Kryma (Mineral springs of Crimea). Krymizdat, Simferopol (in Russian)

Alekseev P (1880) Zametki o gryazevykh vulkanakh (Notes about mud volcanoes). Zapiski Kievskogo obschestva estestvoispytateley 6(1):137–163 (in Russian)

Alyaev SE (1947) Novye dannye o tektonike Kerchenskogo poluostrova (New data about tectonics of the Kerch Peninsula). Izvestya AN SSSR Seriya Geol 6:97–99 (in Russian)

Andrusov NI (1884) Geologicheskie issledovaniya na Kerchenskom poluostrove v 1862–1883 g. (Geological research on the Kerch Peninsula in 1862–1883). Zapiski Novorossiyskogo obshchestva estestvoispytateley X:1–76 (in Russian)

Andrusov NI (1893) Geotektonika Kerchenskogo poluostrova (Geotectonics of the Kerch Peninsula). Materialy dlya geologii Rossii 16:3–206 (in Russian)

Andrusov NI (1918) Geologicheskoe stroenie dna Kerchenskogo proliva (Geological structure of the bottom of the Kerch Strait). AN SSSR, Seriya 6, 12(1):23–28 (in Russian)

Ansted DT (1868) Gryazevye vulkany (Mud volcanoes). Russkiy Vestnik:73–78 (in Russian)

Arkhangel'skiy AD (1925) Neskol'ko slov o genezise gryazevykh vulkanov Apsheronskogo poluostrova i Kerchensko-Tamanskogo regiona (A few words about the genesis of the mud volcanoes of the Absheron Peninsula and the Kerch-Taman region). Bull MOIP Geological Branch 33(3/4):269–285 (in Russian)

Arkhangel'skiy AD, Blokhin AA, Menner VV et al (1930) Kratkiy ocherk geologicheskogo stroeniya neftyanikh mestorozhdenii Kerchenskogo poluostrova (A brief sketch of the geological structure of the Kerch Peninsula oil fields). Trudy Glavnogo Geologo-Razvedochnogo Upravleniya VSNKh SSSR 13:5–90 (in Russian)

Avdusin PP (1939) K petrografii produktov izverzheniya gryazevykh vulkanov Krymsko-Kavkazskoy geologicheskoy provintsii (On the petrography of the products of the eruption of mud volcanoes of the Crimean-Caucasian geological province). In: Results of a study of mud volcanoes of the Crimean-Caucasian geological province. AN SSSR, Moscow-Leningrad, pp 57–66 (in Russian)

Avdusin PP (1948) Gryazevye vulkany Krymsko-Kavkazskoy geologicheskoy provintsii (Mud volcanoes of the Crimean-Caucasian geological province). AN SSSR, Moscow (in Russian)

Basov EI, Ivanov MK (1996) Pozdnechetvertichny gryazevoy vulkanizm v Chernom more (Late Quaternary mud volcanism in the Black Sea). Litologiya i poleznye iskopaemye 2:215–222 (in Russian)

Beketov VN (1916) Yod, brom, bornaya kislota v okrestnostykh Kerchi i Tamanskom poluostrove (Iodine, bromine, boric acid in the vicinity of Kerch and on the Taman Peninsula). Izvestiya AN, seriya 6 (11) (in Russian)

Belousov VV, Yarotskiy LA (1936) Gryazevye sopki Kerchensko-Tamanskoy oblasti, usloviya ikh vozniknoveniya i deyatel'nosti (Mud hills of the Kerch-Taman region, the conditions of their occurrence and activity). ONTI, NKPT SSSR, Moscow (in Russian)

Blinova VN, Ivanov MV, Bohrmann G (2003) Hydrocarbon gases in deposits from mud volcanoes in the Sorokin Trough, north-eastern Black Sea. Geo-Mar Lett 23:250–257

Bogatikov OA, Voytov GI, Sobisevich LE et al (2003) O paroksizmal'nom izverzhenii gryazevogo vulkana gora Karabetova 6 maya 2001 goda. Tamanskaya gryazevulkanicheskaya provintsiya (On the paroxysmal eruption of the Mount Karabetova mud volcano on May 6, 2001. Taman mud volcanic province). Doklady RAN 6(390):805–808 (in Russian)

Bohrmann G, Ivanov M, Foucher J-P et al (2003) Mud volcanoes and gas hydrates in the Black Sea: new data from Dvurechenskii and Odessa mud volcanoes. Geo-Mar Lett 23(3–4):239–249

Borisyak AA (1907) Doklad po voprosu osmotra gryazevoy sopki bliz Vladislavovkiy (Report on the inspection of the mud hill near Vladislavovka). Izvestiya Geolkoma 26(3):34–36 (in Russian)

Bouriak S (1994) Mud volcanoes of the deepest part of the Black Sea: some special structures connected with mud volcanism of the region (according to seismic data of the Cruises of 1991 and 1993). Recent Marine Geological Research in the Mediterranean and Black Seas through the UNESCO Tredmar Programme and its "Floating University" Project. Abstracts. Free University, Amsterdam 31 January – 4 February 1994. Marine/94 UNESCO, June 1994, p 25

Bouriak S (1995) Black Sea deep water mud volcano area: seismic and acoustic images probably connected with gas charge: the evidence of gas responsibility for bright spot. Abstract, 3 postcruise meeting of UNESCO Tredmar "Floating University" Programme. Cardiff, 30 January – 3 February 1995, Marine/99 UNESCO 1995, pp 2–3

Byakov YuA, Kruglyakova RP (2001) Gazogydraty tolshchi Chernogo morya – uglevodorodnoe syr'e budushchego (Gas hydrates of the Black Sea stratum – the hydrocarbon material of the future). Razvedka i okhrana nedr 8:14–19 (in Russian)

Chirvinskiy PN (1908) Zametka o gryazevykh sopkakh Kerchenskogo poluostrova (A note on the mud hills of the Kerch Peninsula). Zapiski Kievskogo obshchestva estestvoispytateley 20(2):791–797 (in Russian)

Dubinevich V (1885) Analizy gryazi Enikal'skikh gryazevykh vulkanov (Mud analyses of Yenikal mud volcanoes). Izvestiya Varshavskogo universiteta 6 (in Russian)

Dubois de Montpéreux F (1837) Pis'mo o glavnykh geologicheskikh yavleniyakh na Kavkaze i Krymu Eli de Bomonu (Letter on the main geological phenomena in the Caucasus and the Crimea to Eli de Beaumont). Gornyy zhurnal 3, 4:345–394 (in Russian)

Dvoychenko PA (1914) Izverzhenie gryazevoy sopki Dzhau-Tepe (Eruption of the mud volcanic hill Dzhau-Tepe). Priroda 4:614–619 (in Russian)

Efremova AG, Zhizhchenko BP (1974) Obnaruzhenie kristallogidratov v osadkakh (Detection of crystalline hydrates in precipitation). Doklady AN SSSR 214(5):1179–1181 (in Russian)

Egorov VN, Artemov YuG, Gulin SB et al (2011) Metanovye sipy v Cherom more. Sredoobrasuyushchaya i ekologicheskaya rol'. Sevastopol' (Methane seeps in the Black Sea. Environmental and ecological role). Ekosagidrofisika, Sevastopol (in Russian)

Ergart AA (1916) Estestvennyy goryuchiy gas v g. Temryuke Kubanskoy oblasti (Natural combustible gas in Temryuk, Kuban region). Stavropol (in Russian)

Fedorov SF (1939) Gryazevye vulkany Krymsko-Kavkazskoy geologicheskoy provintsii i diapirizm (Mud volcanoes of the Crimean-Caucasian geological province and diapirism). Rezul'taty issledovaniya gryazevykh vulkanov Krymsko-Kavkazskoy geologicheskoy provintsii. AN SSSR, Moscow-Leningrad, pp 5–44 (in Russian)

Felitsyn EV (1902) Nekotorye svedeniya o gryazevykh vulkanakh Tamanskogo poluostrova (Some information about the mud volcanoes of the Taman Peninsula). Izvesiya obshchestva lyubiteley izucheniya Kubanskoy oblasti (in Russian)

Filippov N (1857) Poezdka po beregam Azovskogo morya letom 1856 goda (A trip along the shores of the Azov Sea in the summer of 1856). Morskoy sbornik 30(7):1–52 (in Russian)

Gattenberg YuP (1954) K voprosu o vozraste gazov gryazevykh vulkanov Kerchenkogo poluostrova (On the question of the age of gases of mud volcanoes of the Kerch Peninsula). Paper presented at the 8th scientific-practical conference of the scientific student society of the Oil Institute. Moscow, 1954 (in Russian)

Gembitskiy AA (1914) Izverzhenie Dzhau-Tepe (Eruption of Dzhau-Tepe). Yuzhnoe inzhenernoe obshchestvo 4:6 (in Russian)

Gemp SD, Dubrova NV, Nesmelova ZYA et al (1970) Izotopnyy sostav ugleroda uglerodsoderzhashchikh gazov (CH_4 i CO_2) gryazevykh vulkanov Kerchenko-Tamanskoy oblasti (The carbon isotope composition of carbon-containing gases CH_4 and CO_2 of mud volcanoes of the Kerch-Taman region). Geokhimiya 2:243–247 (in Russian)

Gerts K (1876) Istoricheskiy obzor archeologicheskikh issledovaniy i otkrytiy na Tamanskom poluostrove (Historical review of archaeological research and discoveries on the Taman Peninsula). Moscow (in Russian)

Ginsburg GD, Kremlev AN, Grigor'ev MN et al (1990) Fil'trogennye gazovye gidraty v Chernom more (Filtrogenic gas hydrates in the Black Sea). Geologiya i geofisika 3:10–20 (in Russian)

Ginzburg-Karagicheva TP, Veger RA et al (1939) Rezul'taty issledovaniya gryazevykh vulkanov Krymsko-Kavkazskoy geologicheskoy provintsii (Results of the study of the Crimean-Caucasian geological province). AN SSSR Moscow-Leningrad, pp 168–195 (in Russian)

Golovkinskiy NA (1889) Otchet gidrogeologa Tavricheskoy zemskoy upravy za 1889 god (Report of the hydrogeologist of the Taurian Zemstvo government for 1889). Simferopol (in Russian)

Goryainov EP (1968) Nekotorye osobennosti geologicheskogo stroeniya Chekush-Koyashskogo sernogo mestorozhdeniya i novye proyavleniya sery na Kerchenskom poluostrove (Some features of the geological structure of the Chekur-Koyashsky sulfur deposit and new manifestations of sulfur on the Kerch Peninsula). Paper presented at the IVth Scientific conference on the study of mineral resources of fossil sediments of the sedimentary complex of the south of Ukraine. Izdatel'stvo Kievskogo Universiteta, pp 55–58 (in Russian)

Grechishkin LA (1931) Geologicheskoe opisanie Dzhenalbatskoy kotloviny na Kerchenskom poluostrove (Geological description of the Genalbat basin on the Kerch Peninsula). Trudy glavnogo geologo-razvedovatel'nogo upravleniya VSNKH 38. book 1, Geolizdat, Moscow-Leningrad, pp 85–90 (in Russian)

Gubanov IG (1961) K geomorfologii Kerchnskogo poluostrova (On the geomorphology of the Kerch Peninsula). Izvestiya Krymskogo otdeleniya geograficheskogo obshchestva 6 (in Russian)

Gubanov IG (1963) Pro mineralogichyy sklad gryazovulkanichnykh brekcy v rimmeriyskikh vidkladakh Kerchens'kogo pivostrova (On the mineralogical composition of mud-alkaline breccia in the Cimmerian sediments of the Kerch Peninsula). Pamyati VI Vernadskogo. Bud. AN USSR, Kyiv (in Ukranian)

Gubkin IM (1913) Obzor geologicheskikh obrazovaniy Tamanskogo poluostrova (An overview of the geological formations of the Taman Peninsula). Izvesiya Geolkoma 32:8 (in Russian)

Gubkin IM (1940) Gryazevye vulkany Sovetskogo Soyuza i ikh svyaz's genezisom neftyanykh mestorozhdeniy Krymsko-Kavkazskoy geologicheskoy provintsii (Mud volcanoes of the Soviet Union and their connection with the genesis of oil fields of the Crimean-Caucasian geological province). Trudy 17 Mezhdunarodnogo geologicheskogo kongressa 4:33–66 (in Russian)

Gulyaeva LA (1939) Bor gryazevykh vulkanov (Boron of mud volcanoes). In: Results of the study of the Crimean-Caucasian geological province. AN SSSR Moscow-Leningrad, pp 103–134 (in Russian)

Gusakov IN (2004) Ritmika gryazevogo vulkanizma Tamanskogo poluostrova (Thc rhythm of mud volcanism on the Taman Peninsula). Azovskoe otdelenie "Kuban'heologiya," Temryuk, pp 61–62 (in Russian)

Helmersen GP (1864) Issledovanie gryazevykh vulkanov i neftyanykh istochnikov v Krymu i ne Tamanskom poluostrove (Study of mud volcanoes and oil sources in Crimea and on the Taman Peninsula). Zapiski Imperatorskogo Mineralogicheskogo Obshchestva Part 1: 294–295 (in Russian)

Hueschen MA (2010) Can bubbles sink ships? Am J Phys 78(2):139–141

Itkina ES (1939) Rasprostranenie kaliya v vodakh gryazevykh vulkanov i neftyanykh mestorozhdeniy Kavkaza (Distribution of potassium in waters of the Caucasus). In: Results of the study of the Crimean-Caucasian geological province. AN SSSR, Moscow-Leningrad, pp 125–145 (in Russian)

Ivanov MK, Konyukhov AI, Kul'chitsky LM et al (1989) V glubokovodnoy chφsti Chernogo morya (Mud volcanoes in the deep sea of the Black Sea). Vestnik Moskovskogo Universiteta, Seriya Geologicheskaya 3:48–54 (in Russian)

Izgaryshev NA, Sludskiy AF (1917) Gryazevye vulkany Kerchenskogo poluostrova i Temryuksko-Tamanskogo rayona (Mud volcanoes of the Kerch Peninsula and

Temryuk-Taman district). Rudnye vesti 2(3/4):97–112 (in Russian)
Kalinko MK (1964) Osnovnye zakonomernosti raspredeleniya nefti i gazov v zemnoy kore (Basic laws of distribution of oil and gas in the earth's crust). Nedra, Moscow (in Russian)
Karasik MA, Morozov VI (1966) Osobennosti rasprostraneniya rtuti v produktakh gryazevogo vulkanizma Kerchensko-Tamanskoy provintsii (Features of mercury distribution in mud volcanism products in the Kerch-Taman province). Geokhimiya 6:668–678 (in Russian)
Kessler JD, Reeburg WS, Seuthon J (2006) Basin-wide estimates of the input of methane from seeps and clathrates in the Black Sea. Earth Planet Sci Lett 243:366–375
Khlopin VG (1919) Bor i ego soedineniya (Boron and its compounds). Materialy glya izucheniya estestvemmykh proizvoditel'nykh sil Rossii 33 (in Russian)
Kholodov VN (2002a) Gryazevovulkanicheskiye provintsii i morfologiya gryazevykh vulkanov (Mud volcanic provinces and the morphology of mud volcanoes). Geologiya i poleznye iskopaemye Mirovogo okeana 3:227–241 (in Russian)
Kholodov VN (2002b) Geologo-geokhimicheskie osobennosti i model' firmirovaniya gryazevykh vulkanov. Geologiya i poleznye iskopaemye Mirovogo 4:339–358 (in Russian)
Kholodov VN (2012) Gryazevie vulkany: rasprostranenie i genesis (Mud volcanoes: distribution and genesis). Geologiya i poleznye iskopaemye Mirovogo 4:5–27 (in Russian)
Klepinin NN (1914) Gryazevye sopki Kerchenskogo poluostrova i izverzhenie sopki Dzhau-Tepe v 1914 godu (Mud hills of the Kerch Peninsula and eruption of the Dzhau-Tepe hill in 1914). Po Krymu No 2, Simferopol (in Russian)
Konstantov SV, Kechek GA, Krasil'nikov LK et al (1933) Kerchenskie zhelezorudnye mestorozhdeniya (Kerch iron ore deposits). Trudy VGRO NKGP SSSR 325:1–128 (in Russian)
Konyukhov AM, Ivanov MK, Kulnitskiy AM (1990) O gryazevykh vulkanakh i gazogidratakh v glubokovodnykh rayonakh Chernogo morya (About mud volcanoes and gas hydrates in the deep-sea areas of the Black Sea). Litologiya i poleznye iskopaemye 3:12–23 (in Russian)
Korsakov OD, Stupak SN, Byakov YuA (1991) Chernomorskie gazogidraty – netraditsionnyy vid uglerodnogo syr'a (Black Sea gas hydrates are an unconventional type of carbon raw material). Geological zhurnal 5:67–75 (in Russian)
Kovalevskiy SA (1935) Gazovyy vulkanizm (vulkany i vulkanoidy) (Gas volcanism (volcanoes and volcanoids)). Azerbaydzhanskoe neftyanoe khozyaystvo 1 (in Russian)
Kovalevskiy SA (1960) Geologicheskie cherty lineamenta 38 meridiana v rayone Chernogo morya (Geological features of the lineament 38 meridian in the Black Sea region). Doklady AN SSSR 130(6):1306–1309 (in Russian)
Kramarevskiy M (1866) Analizy gryazi Bulganakskikh sopok (Analysis of mud from the Bulganakskikh mud volcanoes). Izvestiya Varshavskogo universiteta 1 (in Russian)
Kudryashev YeV (1939) Geologicheskiy ocherk gryazevogo vulkana Shugo (Geological sketch of the Shugo mud volcano). Rezultaty issledovaniya gryazevykh vulkanov Krymsko-Kavkazskoy geologicheskoy provintsii. AN SSSR, Moscow-Leningrad, pp 45–56 (in Russian)
Kurishko VA, Mesyats IA, Tverdovidov AS (1968) Gidrogeologiya gryazevogo vulkanizma Kerchenskogo poluostrova (Hydrogeology of mud volcanism of the Kerch Peninsula). Geologicheskiy zhurnal 28(1):49–59 (in Russian)
Lagunova IA (1974) O genezise CO_2 v gazakh gryazevykh vulkanov Kerchensko-Tamanskoy oblasti (About the genesis of CO_2 in the gases of the mud volcanoes of the Kerch-Taman region). Geokhimiya 11:1711–1716 (in Russian)
Lagunova IA, Gemp SD (1978) Gidrokhimicheskie osobennosti gryazevykh vulkanov (Hydrochemical features of mud volcanoes). Soviet Geol 8:108–124 (in Russian)
Land F (1866) Neftyanye kolodtsy i gryazevye vulkany Tamanskogo poluostrova (Oil wells and mud volcanoes of the Taman peninsula). Meditsinskiy sbornik 1:14–24 (in Russian)
Levenson VE (1939) Problema gryazevogo vulkanizma i geokhimicheskaya bituminologiya. Rezultaty issledovaniya gryazevykh vulkanov Krymsko-Kavkazskoy geologicheskoy provintsii (The problem of mud volcanism and geochemical bituminology. The results of the study of mud volcanoes of the Crimean-Caucasian geological province). AN SSSR, Moscow-Leningrad, pp 145–167 (in Russian)
Limonov AF, Kozlova YeV, Meysner LB (1999) Struktura verkhney chasti osadochnogo chekhla v progibe Sorokina (Structure of the upper part of the sedimentary cover in the Sorokin trough). Geologiya i poleznye iskopaemye Chernogo morya. OMGOR NAN Ukrainy, Kiev, pp 167–172 (in Russian)
Lombardo S (2000) The essential Homer. Hackett Publishing, Indianapolis/Cambridge
Lüdmann T, Wang HK, Konerding P et al (2004) Heat flow and quantity of methane deduced from a gas hydrate field in the vicinity of the Dnieper Canyon, northwestern Black Sea. Geo-Mar Lett 24(3):82–193
Lychagin GA (1952) Iskopaemye gryazevye vulkany Kerchenskogo poluostrova (Fossil mud volcanoes of the Kerch Peninsula). Byulleten' moskovskogo obshchestva ispytateley prirody, Otdelenie Geologii 27(4):3–13 (in Russian)
Malakhovskiy VF (1956) Geologiya i geokhimiya kerchenskikh zheleznykh rud i ikh vazhneyshikh komponentov (Geology and geochemistry of Kerch iron ores and their major components). AN USSR, Kiev (in Russian)

Maymin ZL (1951) Tretichnye otlozheniya Kryma (Tertiary deposits of Crimea). Gostoptekhizdat, Moscow-Leningrad (in Russian)

Meysner LB, Tugolesov DA, Khakhalev YeM (1996) Zapadno-Chernomorskaya gryazevulkanicheskaya provintsiya (Western Black Sea mud volcanic province). Okeanologiya 35(1):119–127 (in Russian)

Morozevich V (1886) Analiz gryazi na Yenikalskoy spoke (Analysis of mud on Yenikalskiy mud volcano). Izvestiya Varshavskogo universiteta 3 (in Russian)

Morozov VI (1965) Rtut' v kaynozoyskikh otlozheniyakh Kerchenskogo poluostrova (Mercury in the Cenozoic sediments of the Kerch Peninsula). Doklady AN SSSR 163(1):209–211 (in Russian)

Murzaev PM (1928) Izverzhenie sopki Dzhau-Tepe v 1927 godu (Eruption of the Dzhau-Tepe mud volcano in 1927). Trudy Khymskogo NII 1:87–91 (in Russian)

Nesterov KV (1968) K voprosu o metodike issledovaniy mineralnykh vod (On the question of the method of research of mineral waters). Tezisy dokladov IV konferentsii po izucheniyu poleznykh iskopaemykh yuga Ukrainy, KGU, Kiev (in Russian)

Nesterovskiy VA (1990) Aktivizatsiya gryazevykh vulkanov Kerchensko-Tamanskoy oblasti (Activation of mud volcanoes of the Kerch-Taman region). Geologicheskiy zhurnal 1:138–143 (in Russian)

Nikitin PN (1955) Gryazevoy vulkan Shugo (Shugo mud volcano). Priroda 10:109–110 (in Russian)

Obruchev VA (1921) Mestorozhdeniya nefti i gaza Kerchenskogo poluostrova (Oil and gas fields of the Kerch Peninsula). Neftyanoe i slantsevoe khozyastvo 5/8:181–221 (in Russian)

Obruchev VA (1926) Novye tendentsii v tektonike (New trends in tectonics). Izvestiya geologicheskoy commissii 45:117–140 (in Russian)

Pallas PS (1795) Kratkoe fizicheskoe i topograficheskoe opisanie Tavricheskoy oblasti (Brief physical and topographical description of the Tauride region) St Petersburg (in Russian)

Pallas PS (1883) Poezdka vo vnutrennost Kryma vdol' Kerchenskogo poluostrova i na ostrov Taman (A trip to the interior of the Crimea along the Kerch Peninsula and the island of Taman). Zapiski imperatorskogo odesskogo obshestva istorii i drevnostey 13:35–108 (in Russian)

Pokrovskiy NF (1933) Razvedka na bor i yod na gryazevykh sopkakh Tamanskogo poluostrova (Exploration for boron and iodine on the mud hills of the Taman Peninsula). Geologiya na fronte industrializatsii 2:32–38 (in Russian)

Popescu I, Lericolais G, Panin N et al (2007) Seismic expression of gas and gas hydrates across the western Black Sea. Geo-Mar Lett 27:173–183

Popov SP (1938) Mineralogiya Kryma (Mineralogy of Crimea). AN SSSR, Moscow (in Russian)

Pototskiy FYE (1888) Gryazevye vulkany v Krymu i ikh svyaz' s istochnikami gornogo masla (Mud volcanoes in Crimea and their connection with sources of mountain oil). "Kolosya" 9 (in Russian)

Prokopov KA (1931) Geotektonicheskiy ocherk Kerchenskogo poluostrova i otnoshenie ego k Krymu i Tamani (Geotectonic essay on the Kerch Peninsula and its relation to Crimea and Taman). Trudy Geologo-Razvedochnogo Upravleniya 38:13–23 (in Russian)

Pyatnitskiy PP (1925) Tamanskiy poluostrov v gorno-geologicheskom otnoshenii (Taman Peninsula in geological terms). Izvestiya soveshchaniya po issledovaniyu i izucheniyu Kubanskogo kraya 3 (in Russian)

Ronov AB (1951) K voprosu o gryazevom vulkanizme yugo-vostochnogo Kavkaza (On the issue of mud volcanism in the southeastern Caucasus). Doklady AN SSSR 77(6):268–284 (in Russian)

Samoylov YA (1898) Yenikalskie gryazevye sopki (Yenikalskie mud volcanoes). Byulleten' impertorskogo obshchestva naturalistov 2/3:80–86 (in Russian)

Sedelshchikov VV, Kulchavov GK (1914) Izverzhenie sopki Dzhau-Tepe (Eruption of Dzhau-Tepe mud volcano). Materialy po estestvenno-istoricheskomu obsledovaniyu rayona deyatelnosti Dono-Kubano-Terskogo obshchestva selskogo khozyaystva 1:53–59 (in Russian)

Shardanov AN, Znamenskiy VA (1965) Gryazevoy vulkanizm i perspektivy neftenosnosti Tamanskogo poluostrova (Mud volcanism and oil potential of the Taman Peninsula). Geologiya nefti i gaza, pp 18–20 (in Russian)

Shardanov AN, Malyshok VG, Peklo VP (1962) O kornyakh gryazevykh vulkanov Tamani (On the roots of mud volcanoes of Taman'). Trudy Krasnodarskogo filiala VNIINeft 5(1):53–66 (in Russian)

Shepel' SA (1926) Kolebaniya dna Kerch-Yenikalskogo proliva (Oscillations of the Kerch-Yenikal Strait bottom). Dekadnyy byulleten' pogody i sostoyaniya morya po Chernomorskomu i Azovskomy poberezhyu 30:11–18 (in Russian)

Shnyukov EF (1999) Gryazevoy vulkanizm v Chernom more (Mud volcanism in the Black Sea). Geologicheskiy zhurnal 2:38–47 (in Russian)

Shnyukov EF (2016) Flyuidogennaya mineralizatsiya gryazevykh vulkanov Azovo-Chernomorskogo regiona (Fluidogenic mineralization of mud volcanoes of the Azov-Black Sea region). Logos, Kiev (in Russian)

Shnyukov EF, Lukin AE (2011) O samorodnykh elementakh v razlichnykh geoformatsiyakh Kryma i sopredelnykh regionov (About native elements in various geoformations of Crimea and adjacent regions). Geologiya i poleznye iskopaemye Mirovogo okeana 2:5–29 (in Russian)

Shnyukov EF, Naumenko PI (1964) Kimmeriyskie zheleznye rudy vdavlennykh sinklinaley Kerchenskogo poluostrova (Cimmerian iron ores of recessed synclines of the Kerch Peninsula). Krymizdat, Simferopol (in Russian)

Shnyukov EF, Naumenko PI (1982) Nakhodka Achinskoy rudonosnoy sinklinali na Kerchenskom poluostrove i ee paleogeograficheskoe znachenie (Discovery of the Achinskaya ore-bearing syncline on the Kerch Peninsula and its paleogeographic significance). Geologicheskiy zhurnal 5:51–57 (in Russian)

Shnyukov EF, Netrebskaya EYA (2013) Korni Chernomorskikh gryazevykh vulkanov (The roots of the Black Sea mud volcanoes). Geologiya i poleznye iskopaemye Mirovogo okeana 14:87–92 (in Russian)

Shnyukov EF, Yanko (also Yanko-Hombach) V (2014) Gazootdacha dna Chernogo morya: geologo-poiskovoe, ekologicheskoe i navigatsionnoe znachenie (Degassing of the Black Sea bottom: geological prospecting, environmental and navigational significance). Vestnik Odesskogo universiteta, Seriya geograficheskikh i geologicheskikh nauk 19, issue 3(23):225–241 (in Russian)

Shnyukov EF, Naumenko PI, Lebedev YuS et al (1971) Gryazevoy vulkanizm i rudoobrazovanie (Mud volcanism and ore formation). Naukova Dumka, Kiev (in Russian)

Shnyukov EF, Alenkin VM, Naumenko PI (1985) Rudonosnost' Armaelinskoy vdavlennoy sinklinali na Kerchenskom poluostrove (The ore-bearing capacity of the Armaelinskaya recessed syncline on the Kerch Peninsula). Doklady AN USSR, Seriya B, pp 23–25 (in Russian)

Shnyukov EF, Sobolevskiy YuV, Gnatenko GI et al (1986) Gryazevye vulkany Kerchensko-Tamanskoy oblasti. Atlas (Mud volcanoes of the Kerch-Taman region, Atlas). Naukova Dumka, Kiev (in Russian)

Shnyukov EF, Nesterovskiy VA, Alyonkin VM (1987) Geologichna budova i rudonosnist Bulganakskoï vdavlenoï sinklinali (Geological structure and ore deposits of Bulganak pushed syncline). Dopovidi AN USSR, Seriya B (11):26–29 (in Ukranian)

Shnyukov EF, Ivannikov AV, Bezborodov AA et al (1990) Rezultaty geologicheskikh issledovaniy 51-go reysa v Chernoe more NIS "Mikhail Lomonosov" (The results of geological studies of the 51st cruise of the R/V "Mikhail Lomonosov" to the Black Sea). Preprint IGN 90–8, Kiev (in Russian)

Shnyukov EF, Gnatenko GI, Nesterovskiy VA et al (1992) Gryazevye vulkanizm Kerchensko-Tamanskogo regiona (Mud volcanism of the Kerch-Taman region). Naukova Dumka, Kiev (in Russian)

Shnyukov EF, Kleshchenko SA, Avilov VI et al (1993) Gazovye anomalii v donnykh osadkakh severo-zapada Chernogo morya (Gas anomalies in the bottom sediments of the northwest of the Black Sea). Geologiya i geokhimiya goryuchikh kopalyn 4:7–9 (in Russian)

Shnyukov EF, Shcherbakov IB, Shnyukova YeYe (1997) Paleoostrovnaya duga severa Chernogo moray (Paleo-island arc of the north of the Black Sea). Chernobylinform, Kiev (in Russian)

Shnyukov EF, Ivannikov AV, Kobolev VP et al (1998) Geologiya, geofizika i gidrografiya severo-zapada Chernogo moray (Geology, geophysics and hydrography of the northwest of the Black Sea). OMGOR NAN Ukrainy, Kiev (in Russian)

Shnyukov EF, Starostenko VI, Ivannikov AV et al (2005) Gazovyy vulkanizm Chernogo morya (Black Sea gas volcanism). OMGOR NAN Ukrainy, Kiev (in Russian)

Shnyukov EF, Starostenko VI, Kobolev VP (2006) Gazogidratonosnost donnykh otlozheniy Chernogo morya (Gas hydrate potential of sediments of the Black Sea). Geofizicheskiy zhurnal 28(6):29–40 (in Russian

Shnyukov EF, Kutnyy VA, Naumenko SP et al (2007) Travertiny i drugie mineralnye obrazovaniya gazovodnykh istochnikov Kerchenskogo poluostrova (Travertine and other mineral formations of gas sources of the Kerch Peninsula). Geologiya i poleznye iskopaemye Mirovogo okeana 3:5–14 (in Russian)

Shnyukov EF, Kobolev VP, Pasynkov AA (2013a) Gazovyy vulkanizm Chernogo morya (Gas volcanism of the Black Sea). Logos, Kiev (in Russian)

Shnyukov EF, Sokol EV, Nigmatulina YeN et al (2013b) Zoloto v gryazevykh vulkanakh Kerchenskogo poluostrova kak pokazate' glubinnosti gryazevulkanicheskikh flyuidov (Gold in the mud volcanoes of the Kerch Peninsula as an indicator of the depth of the mud volcanic fluids). Geologiya i poleznye iskopaemye Mirovogo okeana 4:79–89 (in Russian)

Shnyukov EF, Deyak MA, Ivanchenko VV et al (2015) Neobychnaya mineralogiya gryazevykh vulkanov Kerchenskogo poluostrova (Unusual mineralogy of the mud volcanoes of the Kerch Peninsula). Geologiya i poleznye iskopaemye Mirovogo okeana 4:5–19 (in Russian)

Shternov AG (1968) Geologiya i genezis gryazevykh vulkanov Kerchensko-Tamanskogo rayona (Geology and genesis of mud volcanoes of the Kerch-Taman region). Dissertation, Kiev State University (in Russian)

Sivere L (1931) Izverzhenie Dzhardzhavskoy sopki (Eruption of the Dzhardzha mud volcano). Priroda 1:98–99 (in Russian)

Sobisevich AL, Gorbatikov AV, Ovsyuchenko AN (2008) Glubinnoe stroenie gryazevogo vulkana gory Karabetovoy (The deep structure of the mud volcano Karabetova mountains). Doklady RAN 422(4):542–546 (in Russian)

Stadnitskaya AN, Belen'kaya IYu (2000) Sostav i proiskhozhdenie uglevodorodnykh gazov i ikh vliyanie na diageneticheskoe karbonatoobrazovanie (progib Sorokina, zapadno-vostochnoy chasti Chernogo morya) (Composition and origin of hydrocarbon gases and their effect on diagenetic carbonate formation (Sorokin trough, west-east part of the Black Sea)). Geologiya Chernogo i Azovskogo morey, Gnosis, Moscow, pp 155–163 (in Russian)

Steber EA (1909/1910) Gryazevoy vulkan Karabetova gora bliz Tamani (Mud volcano Karabetova mountain near Taman). Izvestia Kavkazskogo otdeleniya Russkogo heograficheskogo obshchestva 20 (in Russian)

Steber EA (1915) Proiskhozhdenie nefti (Origin of oil). Priroda 4:1–30 (in Russian)

Stopnevich AD (1920) Mineral waters. Commission on the study of the natural productive forces of Russia of the RAS. The Wealth of Russia, Petrograd (in Russian)

Sulin VA (1939) Zadachi i nekotorye itogi gidrogeologicheskogo i geokhimicheskogo izucheniya gryazevykh vulkanov Krymsko-Kavkazskoy provintsii (Tasks and some results of hydrogeological and geochemical study of mud volcanoes of the Crimean-Caucasian province). Rezultaty issledovaniy gryazevykh vulkanov Krymsko-Kavkazskoy geologicheskoy provintsii, AN SSSR Moscow-Leningrad, pp 93–102 (in Russia)

Titova NO (2013) Litologichniy sklad gruboulamkovikh vikidiv gryazovogo vulkanu Dzhau-Tepe (Lithologic composition of coarse-volcanic emissions of the Dzhau-Tepe mud volcano). Geologiya i poleznye iskopaemye Mirovogo okeana 3:110–117 (in Ukranian)

Titova NO, Nesterovskiy VF, Deyak MA et al (2013) Sostav peschano-glinistoy fraktsii sopochnoy brekchii gryazevogo vulkana Dzhau-Tepe na Kerchenskom poluostrove (Composition of the sandy-clay fraction of the gryphon breccia of the mud volcano Dzhau-Tepe on the Kerch Peninsula). Geologiya i poleznye iskopaemye mirovogo okeana 4(34):90–94 (in Russian)

Turley GF (1930) Bulganakska grupa gryazovikh vulkaniv na Kerchenskomu pivostrovi ta produkti ïkhnogo vibukhu (Bulganaks'ka group of mud volcanoes on the Kerch peninsula and products of their explosion). Trudy Kharkivs'kogo tovaryustva doslidnikiv pryrody, pp 14–22 (in Ukranian)

Tvertinova TYu, Sobisevich AL, Sobisevich LYe et al (2015) Strukturnaya pozitsiya i osobennosti stroeniya gryazevogo vulkana gory Karabetovoy (Structural position and features of the structure of the mud volcano of Karabetova Mountain). Geologiya i poleznye iskopaemye Mirovogo okeana 2:106–122 (in Russian)

Vasilev A, Dimitrov L (2002) Spatial and quantitative evaluation of the Black Sea gas hydrates. Russ Geol Geophys 43(7):637–649

Vernadskiy VI, Popov SP (1899–1900) Yenikalskie gryazevye vulkany (Enikal mud volcanoes). Byulleten' moskovskogo obshchestva ispytateley prirody (Protokoly zasedaniy. Prilozheniya), pp 37–41 (in Russian)

Voskoboynikov NI, Gur'ev AV (1832) Geologicheskoe opisanie poluostrova Tamani, prinadlezhashchego k zemle voyska Chernomorskogo (Geological description of the Taman Peninsula belonging to the land of the Black Sea) 1 (in Russian)

Yanko (also Yanko-Hombach) V, Kravchuk A, Kulakova I (2017) Meiobentos metanovykh vykhodov Chernogo morya (Meiobenthos of methane outlets of the Black Sea). Phenix, Odessa (in Russian)

Yuhanov IS (1968a) Geologiya i veshchestvennyy sostav keramzitovogo syrya Kerchenskogo poluostrova (Geology and material composition of Keremzite raw materials of the Kerch Peninsula). Dissertation, Odessa (in Russian)

Yuhanov IS (1968b) O fiziko-khimicheskoy sushchnosti protsessov keramzitoobrazovaniya na primere kerchenskikh glin (On the physico-chemical nature of the claydite formation processes based on the example of Kerch clays). Tezisy dokladov IV nauchno-tekhnicheskoy konferentsii po izucheniyu poleznykh iskopaemykh otlozheniy osadochnogo kompleksa yuga Ukrainy, Kiev State University, Kiev (in Russian)

Yushkin EM (1909) O neftyanykh mestorozhdeniyakh Tamanskogo poluostrova, a takzhe o tamanskoy rude (On the oil fields of the Taman Peninsula, as well as on the Taman ore). Groznyy (in Russian)

Zaytsev AV (1965) Gryazevye vulkany Priazovya (Mud volcanoes of Azov). Dissertation, Rostov na Donu State University (in Russian)

Zenin GN (1955) Gryazevye vulkany Tamanskogo poluostrova (Mud volcanoes of the Taman Peninsula). Uchenye zapiski Krasnodarskogo pedagogicheskogo institute 14:54–59 (in Russian)

Zenin VN (2012) Arkheologicheskie obekty v landshaftakh gryazevogo vulkanizma. Problemy arkheologii, etnografii, antropologii Sibiri i sopredelnykh territoriy. Materialy itogovoy sessii instituta arkheologii i etnografii SO RAN 2012 g. Izd-vo In-ta arkheologii i etnografii SO RAN XVIII, Novosibirsk, pp 83–87

Zenkovich VKh (1894) Kerch v proshedshem i nastoyashchem: istoriko-arkheologicheskiy i geograficheskiy ocherk) (Kerch in past and present: historical-archaeological and geographical essay). Kerch (in Russian)

Zhivilo K (1909) Ekskursiya na Tamanskiy poluostrov (Excursion to the Taman Peninsula). Kubanskiy sbornik. Trudy publikatsiy obl. stat. otd. 14, Yekaterinodar, pp 3–17 (in Russian)

2 Study Area

Summary
The chapter provides an overview of the study area that includes the Black Sea, the Sea of Azov, the Kerch and Taman peninsulas, and adjacent northern Caucasus. The Black Sea considered as the easternmost of the seas of the Atlantic Ocean basin by itself is the largest meromictic basin in the world, with an area. The basin was formed in the Mesozoic as a back-arc structure above the northward subducting Tethyan oceanic lithosphere and is surrounded by Alpide fold belts. In terms of crustal structure, the megadepression comprises (1) the western and eastern deep basins separated from each other by the Andrusov Ridge and (2) a number of tectonic structures that also surround the basins on each side. The Kerch and Taman peninsulas' region lies at the junction of Crimean and Caucasian folding. It is potentially an oil- and gas-bearing region with a complex multitiered geological structure. The region is characterized by (1) the widest development of discontinuous tectonics of different levels that vary from deep and regional faults to small dislocations, (2) the presence of some folding systems in the upper structural floor, and (3) the widespread development of diapirism. The North Caucasus or Ciscaucasia is the northern part of the Caucasus region between the Sea of Azov and Black Sea on the west and the Caspian Sea on the east.

The study area includes the Black Sea, the Sea of Azov, the Kerch and Taman peninsulas, and the northern Caucasus.

The Black Sea is the easternmost of the seas of the Atlantic Ocean basin and the largest meromictic basin in the world, with an area of 423,000 km^2. If we take the ratio of the sea volume to the summary area of the cross sections of all its straits (which is 0.04 km^2 for the Bosporus and 0.02 km^2 for the Kerch Strait) as a measure of isolation of a sea basin, then the Black Sea can be considered the most isolated sea of the Global Ocean (Zubov 1956). Its maximum length (along 42° 29' N lat) and width are 1148 km and 611 km, respectively. Its surface area (excluding estuaries, such as the Dnieper-Bug liman—liman is a local term for ancient estuaries in the Black Sea and Sea of Azov) and its volume are about 416,790 km^2 and 535,430 km^3, respectively, and the maximum depth is 2212 m (Ivanov and Belokopytov 2013). The Sea of Azov is connected to the Black Sea via the Kerch Strait and has an area of 39,000 km^2 and a volume of 290 km^3. It is the shallowest sea in the world, with the depth varying between 0.9 and 14 m. There is a constant outflow of water from the Sea of Azov to the Black Sea. The sea bottom is also relatively flat with the depth gradually increasing from the coast to the center.

The Black Sea megadepression is a central structure of the region surrounded by a "necklace

E. Shnyukov, V. Yanko-Hombach, *Mud Volcanoes of the Black Sea Region and their Environmental Significance*, https://doi.org/10.1007/978-3-030-40316-4_2

of smaller depressions" and is referred by Patalakha et al. (2003) as a mysterious, deep, suboceanic depression—one of many peculiar depressions of the Western Tethys formed during different time (Tugolesov et al. 1985).

From tectonic point of view, the basin was formed in the Mesozoic as a back-arc structure above the northward subducting Tethyan oceanic lithosphere and is surrounded by Alpide fold belts. In terms of crustal structure, the megadepression comprises (1) the western and eastern deep basins separated from each other by the Andrusov Ridge and (2) a number of tectonic structures that also surround the basins on each side (Fig. 2.1).

The western deep basin is older than the eastern one; the former was rifted from the latter with the dissection of the Late Jurassic to Early Cretaceous carbonate platform that had been established on the southern margin (Moesian Platform) of the northern supercontinent, Laurasia (Nikishin et al. 2003). The western deep basin is floored by oceanic crust overlain by thick sediment units probably of Cretaceous and younger age; the latter has a thinned continental or oceanic crust overlain by sediments less than 10 km thick.

The bottom of the Black Sea reveals a compressional tectonic regime resulting in active faulting. Tectonic ruptures form weakened zones in the sedimentary cover and act as channels/chimneys/venting sites for upwardly migrating fluids; these sites indicate the presence of gas reservoirs beneath the sea floor that commonly relate to fluidized deformations, e.g., diapirism (Shnyukov and Ziborov 2004).

The Kerch and Taman peninsulas' region lies at the junction of Crimean and Caucasian folding. It is potentially an oil- and gas-bearing region with a complex multitiered geological structure (Ali-Zadeh and Tsimmel'zon 1966), frequent mismatching of structural plans belong-

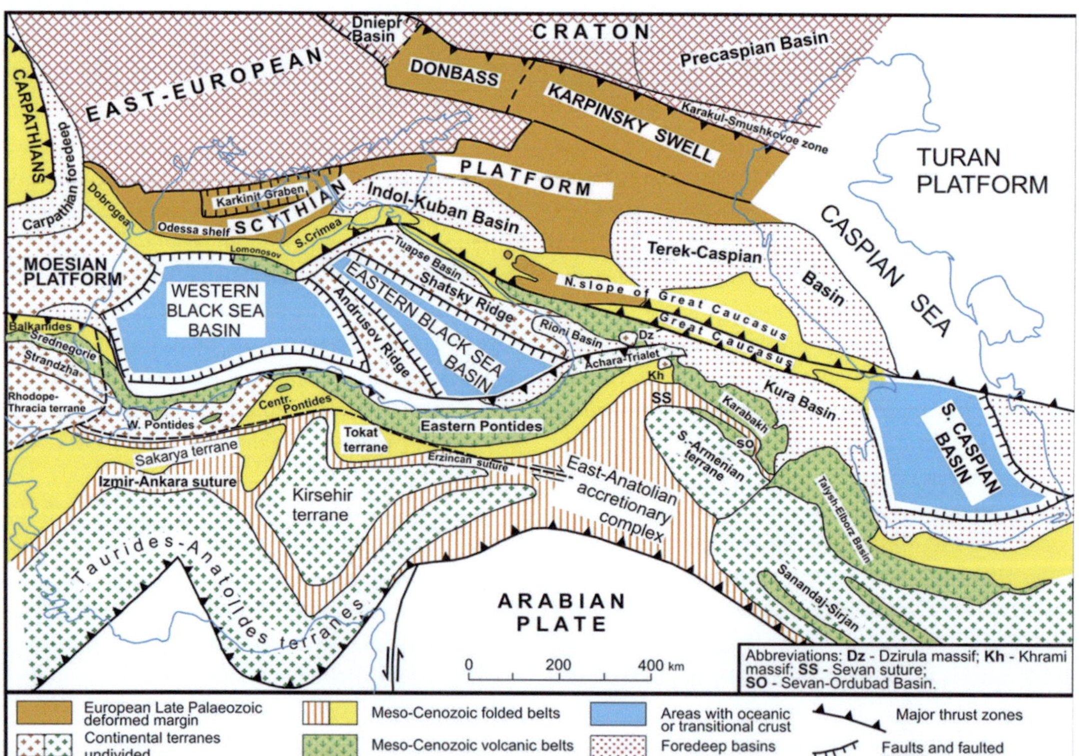

Fig. 2.1 Tectonic chart of the Azov–Black Sea region (Nikishin et al. 2011). Reproduced with permission from Turkish Journal from Earth Sciences

ing to different structural floors, and the presence of thick strata of Maikopian and Cretaceous clay deposits. The region is characterized by (1) the widest development of discontinuous tectonics of different levels that vary from deep and regional faults to small dislocations, (2) the presence of some folding systems in the upper structural floor, and (3) the widespread development of diapirism.

The North Caucasus or Ciscaucasia is the northern part of the Caucasus region between the Sea of Azov and Black Sea on the west and the Caspian Sea on the east, in Russia. Geographically, the term North Caucasus also refers to the northern slope and western extremity of the Caucasus Major mountain range, as well as a part of its southern slope to the West (until the Psou River in Abkhazia). Geologically, the Caucasus Mountains belong to a system that extends from southeastern Europe into Asia. The Greater Caucasus Mountains are mainly composed of Cretaceous and Jurassic rocks with the Paleozoic and Precambrian rocks in the higher regions. Some volcanic formations are found throughout the range.

References

Ali-Zadeh AA, Tsimmel'zon IO (1966) Glubinnoe stroenie Azerbaydzhana (The deep structure of Azerbaijan). Geotektonika 3:26–34 (in Russian)

Ivanov VA, Belokopytov VN (2013) Oceanography of the Black Sea. Hydrophysical Institute NAS of Ukraine, Sevastopol

Nikishin AM, Korotaev MV, Ershov AV (2003) The Black Sea basin: tectonic history and Neogene-Quaternary rapid subsidence modelling. Sediment Geol 156:149–168

Nikishin AM, Ziegler P, Bolotov SN et al (2011) Late Palaeozoic to Cenozoic evolution of the Black Sea-Southern Eastern Europe region: a view from the Russian Platform. Turk J Earth Sci 20:571–634

Patalakha EI, Gonchar VV, Senchenkov IK et al (2003) Indentornyy mekhanizm v geodinamike Krymsko-Chernomorskogo regiona (prognoz UV i seysoopasnosti) (The indenter mechanism in geodynamics of the Crimean-Black Sea region (forecast of hydrocarbons and seismic risks). EMKO, Kiev (in Russian)

Shnyukov EF, Ziborov AP (2004) Mineralnye bogatstva Chernogo morya (Mineral wealth of the Black Sea). Kiev (in Russain)

Tugolesov DA, Gorshkov AS, Meisner LB et al (1985) Tektonika mezo-kaynozoyskiy otlozheniy Chernomorskoy vpadiny (Tectonics of the Meso-Cenozoic sediments of the Black Sea Basin). Nedra, Moscow (in Russian)

Zubov NN (1956) Osnovy ucheniya o prolivakh Mirovogo okeana (Fundamentals of the studies of World Ocean straits). Geographgiz, Moscow (in Russian)

3 Materials and Methods

Summary

This chapter provides an overview of the materials and methods used in the study of the terrestrial and offshore mud volcanoes of the Black Sea region. The general strategy includes geomorphological, geological, geophysical, gas-geochemical, paleontological, and micropaleontological dimensions. The geomorphological dimension covers observations of the terrain/sea bottom relief in the search for mud volcanoes. As a rule, mud volcanoes on land are characterized by rounded positive forms of relief comprising hills up to 60 m in height. The hills are constituted largely of mud breccia. Some mud volcanoes look like small, rounded puddles filled with bubbling liquid mud. In the sea, the geomorphological dimension entails bathymetrical (morphometrical) analysis of bottom relief performed by multibeam echo sounder and side-scan sonar that enable mapping of local bottom elevations that could represent mud volcanoes. The geophysical dimension involves seismoacoustic profiling, multichannel seismic profiling, cross and ring seismic sounding (tomography) on reflected and refracted waves, in situ geothermal measurements, detailed gravimetric and magnetometric survey, sounding the formation of the electromagnetic field in order to determine the conductivity parameters of the bottom sediments and the layer of gas hydrates, and laboratory experimental thermodynamic studies of the physical properties of artificial samples of hydrate-containing bottom sediments. Seeps and gas emanations from mud volcanoes were studied by echo sounding with the help of modern digital echo sounders. The geological dimension applies to both terrestrial mud volcanoes and adjacent recessed synclines as well as offshore mud volcanoes. This dimension focuses on the study of sediments within mud volcanic structures, including identification of lithological, grain-size, and mineralogical properties of the sediments recovered by drilling, coring, and dredging. The gas-geochemical dimension covers geochemical investigation of gases emitted by mud volcanoes. The paleontological and micropaleontological dimensions investigated mollusks, foraminifers, ostracods, and nematodes.

The original version of this chapter was revised: Figure 3.6 has been updated now. The correction to this chapter is available at https://doi.org/10.1007/978-3-030-40316-4_12

Over the past 50 years, a multidisciplinary study of all known active and fossil mud volcanoes on the land surface has been conducted on the Kerch and Taman peninsulas, as well as the North Caucasus, and all cases of active eruptions were recorded (e.g., Shnyukov et al. 2005). Many of these terrestrial mud volcanoes and the recessed synclines associated with them were drilled (Fig. 3.1), yielding about 200 wells with lengths between 200 and 250 m each.

Drilling was performed by the Department of Mining Geology, the Ministry of Ferrous Metallurgy of Ukraine, the Crimean expedition, and some oil-producing organizations.

E. Shnyukov, V. Yanko-Hombach, *Mud Volcanoes of the Black Sea Region and their Environmental Significance*, https://doi.org/10.1007/978-3-030-40316-4_3

Fig. 3.1 Drilling of a well at the Andreevskiy mud volcano in May 2016

Recovered cores were initially described and sampled by the Department of Mining Geology (P.I. Naumenko and others) and then transferred for further study to the Department of Marine Geology and Mineral Resources of the Ukrainian Academy of Sciences.

Particular attention was paid to mud volcanic breccia and ores that were investigated lithologically, mineralogically, and paleontologically by all available methods (e.g., grain-size; binocular, polarized, and SEM microscopy; X-rays; thermal and spectral analyses; and wet chemistry). Isotopic methods were used as well for some carbonates and glauconites. A huge amount of data was collected and presented in several monographs, hundreds of articles, and in this book (see the extensive reference lists appended to each chapter).

Extensive studies of terrestrial mud volcanoes gradually declined, and in the late 1980s, exploration began to shift toward the study of offshore mud volcanoes located in the Sea of Azov and the Black Sea. As a first step, mud volcanoes of the Kerch Strait and southeastern Sea of Azov were investigated on board the small R/V "Geokhimik," which was equipped with a drilling rig (Fig. 3.2) that could drill a well up to 100 m in length within water up to 50 m in depth (Shnyukov et al. 1977).

In the mid-1990s, a powerful coordinating structure called the National Agency for Marine Research and Technology of Ukraine and the Government of Ukraine implemented the "National program for research and use of the resources of the Sea of Azov, Black Sea, and other parts of the World Ocean," for a period lasting until the year 2000. In the framework of this program, marine expeditionary research including the study of mud volcanoes was carried out annually in the Black Sea on the R/V "Kiev" (Shnyukov and Shchiptsov 1996; Shnyukov et al. 1997) (Fig. 3.3).

The R/V "Kiev" (formerly called "Academician Alexei Krylov") carried a displacement of 9920 t and was built to ice class specifications with unlimited navigation area at the Nikolaev Shipyard in 1982 to solve specific problems of hydro-exploration. In 1993, the R/V "Kiev" was re-equipped with a unique multifunctional laboratory for marine geological and geophysical research, including an advanced marine hydroacoustic complex. In the course of the vessel's movement, this complex enabled it to receive a continuous acoustic section of the upper layer of bottom sediments in real time with subsequent computer processing of the obtained information. The R/V "Kiev" was also equipped with an autonomous manned submarine vehicle capable of immersing to a depth of 550 m (Fig. 3.4). The six-person crew of the submarine vehicle was able to photograph, produce videos,

Fig. 3.2 R/V "Geokhimik" equipped with a drilling rig

Fig. 3.3 R/V "Kiev"

as well as deploy a telecontrolled manipulator to sample bottom sediments and lift pieces of rock and other objects weighing up to 300 kg.

Other Ukrainian R/Vs were also used in the study of offshore mud volcanoes. These include "Mikhail Lomonosov," "Ikhtiander" (Fig. 3.5), "Academician Vernadskiy" (Fig. 3.6), "Professor Vodyanitsky" (Fig. 3.7), and "Vladimir Parshin" (Fig. 3.8).

These R/Vs were equipped with advanced single- and multibeam echo sounders, landed automatic probe stations, and other instruments allowing remote detection, visual observation, and mapping of mud volcanoes and gas seeps on

Fig. 3.4 Autonomous submarine vehicle located on board the R/V "Kiev"

the sea bottom. The equipment and methods used are described elsewhere (e.g., Shnyukov et al. 2013, 2015; Shnyukov 2016; Yanko et al. 2017). Here, we provide our general strategy for the study of terrestrial and offshore mud volcanoes that includes geomorphological, geological, geophysical, gas-geochemical, paleontological, and micropaleontological dimensions.

The *geomorphological* dimension covers observations of the terrain/sea bottom relief used in the search of mud volcanoes. As a rule, mud volcanoes on land are characterized by rounded, positive forms of relief comprising hills up to 60 m in height. The hills are constituted largely of mud breccia (Figs. 1.1 and 4.1). Some mud volcanoes look like small, rounded puddles filled with bubbling liquid mud (Fig. 4.2). In the sea, this dimension entails bathymetrical (morphometrical) analysis of bottom relief performed by multibeam echo sounder and side-scan sonar that enable mapping of local bottom elevations that could represent mud volcanoes. Hills on the sea bottom are relatively small and quite rare. They have rounded cone shapes and generally rise about 30–50 m above the seafloor, occasionally 100–120 m. Sometimes, it was possible to observe calderas and breccia flow on their tops (e.g., thc Dvurcchenskiy, MSU, Yuzhmorgeologiya, and Parshin mud volcanoes).

Fig. 3.5 R/V "Ikhtiander"

Fig. 3.6 R/V "Academician Vernadskiy"

Fig. 3.7 R/V "Professor Vodyanitskiy"

On modern bathymetrical maps, all irregularities in bottom relief, including these hills, can be seen if their dimensions are greater than one tenth of the sea depth. Taking into account that the depth of the Black Sea exceeds 2000 m, mud volcanoes are not visible on regular bathymetrical maps. A

Fig. 3.8 R/V "Vladimir Parshin"

while ago, we obtained permission to examine the archives of the General Directorate of Navigation and Oceanography of the USSR. The archives contained working materials of the hydrographic survey that enabled us to draw a bathymetric map of the Black Sea bottom in order to contour the hills on it. We assume that at least some of the hills (coordinates of which we have) are, in fact, mud volcanoes, but unfortunately, we could not complete this work due the emerging geopolitical situation.

Additionally, in the search for cells of mud volcanic activity, hydroacoustic investigations of the near bottom water horizons were performed in the mode of two-frequency vertical sounding by the R/V as the ship moved with a speed of 4–10 knots, as well as at drift stations. This detailed survey was performed on selected polygons on a system of legs (rectangular grid, saw-like legs, etc.). In deep water (greater than 250 m), measurements were performed at a frequency of 38 kHz, and an RF channel (120 kHz) was used to observe the scattering of sound in the upper layer of the seawater. The duration of the emitted pulses (τ) and repetition period (***Tn***) varied depending on the depth of the place within ***Tn*** = 0.7–4.5, $\boldsymbol{\tau u}$ = 0.3–3 ms (38 kHz), and 0.3–1 ms (120 kHz). Data from the output of the EC 500 signal processor unit (an echogram and an array of the volume scattering force for each channel) were computerized and recorded together with the navigation data. The minimum interval between samples at depth is usually 0.1 to 0.03 m for the frequencies of 38 and 120 kHz, respectively.

During our expeditions, mud volcanoes described by colleagues were investigated as well.

The *geophysical* dimension involves:

- Seismoacoustic profiling, including that in the near bottom version with narrow-beam parametric and chirp profilographs (frequency ranges 0.3–1.5, 2–7, and 8–23 kHz, resolution 10–50 cm, penetration up to 50–200 m below the bottom, towing depth up to 600–1000 m)
- Multichannel seismic profiling, including the near bottom (hydropneumatic or vibrating radiator, 120/240-channel spit 600–1000 m long or 100–300 m in the bottom variant, frequency range 30–700 Hz, resolution 1–2 m, penetration under the bottom 0.5–1.0 km)
- Cross and ring seismic sounding (tomography) on reflected and refracted waves with the help of four-component bottom seismographs with determination of speeds and dynamic characteristics of longitudinal and transverse waves in bottom sediments
- In situ geothermal measurements by the "Geos" telemetric system of absolute temperature and thermal conductivity of bottom sediments for the purpose of calculating heat

fluxes and constructing thermo-barometric models for the zone of gas-hydrated formation
- Detailed gravimetric and magnetometric surveys to clarify the nature of the potential field anomalies and map tectonic disturbance zones
- Sounding the formation of the electromagnetic field in order to determine the conductivity parameters of the bottom sediments and the layer of gas hydrates
- Laboratory experimental thermodynamic studies of the physical properties of artificial samples of hydrate-containing bottom sediments

Seeps and gas emanations from mud volcanoes were studied by echo sounding with the help of modern digital echo sounders.

The majority of gas seeps were detected using the SIMRAD echo sounder that enables one to observe them and obtain their coordinates, depth, height, and other parameters during the passage of the tracks on the monitor screen, together with the storage of data in the computer's memory, and has the ability to create a database of the received data. SIMRAD sounder settings included an operating frequency of 38 kHz, a depth range of 10–110 m, and a 1.5 s period of signal repetition. The minimum signal level displayed on the monitor by gray color is 66 dB.

In recent years, the remote-controlled submersible Sophokl-1, developed by V.S. Blinov (National University of Shipbuilding, Nikolaev) has been used. This device has a system of underwater navigation (Fig. 3.9).

The vast majority of gas torches and mud volcanoes were studied with acoustic equipment. In this connection, when describing gas manifestations, we actually describe acoustic portraits of seeps and mud volcanoes.

The *geological dimension* applies to both terrestrial mud volcanoes and adjacent recessed synclines as well as offshore mud volcanoes. This dimension focuses on the study of sediments within mud volcanic structures, including identification of lithological, grain-size, and mineralogical properties of the sediments recovered by drilling, coring (about 800 hundred cores obtained by gravity core with lengths up to 7 m, Fig. 3.10), and dredging (Figs. 3.11 and 3.12).

On land, the wells were drilled by the Mine Geology Department of the Ministry of Ferrous Metallurgy of Ukraine and the Crimean expedition (Fig. 3.1). Some geological material was also provided by oil producing organizations and transferred to the authors.

Sedimentologically, fields of breccia are typical for terrestrial mud volcanoes but not for submarine ones as these are often buried in muddy sediments. Usually, it is not possible to pick up clastic material, especially if the study is carried out by coring. Over the course of lithological and mineralogical investigation, special attention was paid to the breccia and enclosed ores, which were studied by grain size, SEM, optical, X-ray, thermal, and spectral methods. In some cases, methods of wet chemistry were applied as well. Some carbonates and glauconite were studied by isotopic methods.

The *gas-geochemical* dimension covers geochemical investigation of gases emitted by mud volcanoes (for more details see Chap. 11).

The paleontological and *micropaleontological* dimensions investigated mollusks, foraminifers, ostracods, and nematodes using methods described in Nevesskaya (1963), Ilyina (1966), Yanko and Troitskaya (1987), Yanko and Gramova (1990), Yanko-Hombach (2007), Yanko et al. (2017), as well as in Chap. 11 of this volume.

In summary, the search criteria for terrestrial and offshore mud volcanoes of the Black Sea do not completely coincide with each other. Mud volcanoes on land are typically characterized by rounded positive forms of relief—hills up to 60 m above the surrounding terrain or small round liquid mud bodies, the presence of mud volcanic breccia, the activity of gas seeps, the presence of mud volcanic breccia, hydrogen sulfide springs, and geochemical signs (an increase in the concentration of Hg and a number of others elements, boron and borates, the presence of pyrite, etc.). Finally, the principal sign of a mud volcano is its activity. In subaqueous conditions, these signs are not always observed. Positive forms of underwater relief are relatively frequent

Fig. 3.9 Self-propelled remote-controlled underwater apparatus "Sophokl-1"

Fig. 3.10 Lifting of a gravity core aboard the R/V "Professor Vodyanitsky"

Fig. 3.11 Lifting a box dredge aboard the R/V "Professor Vodyanitsky"

Fig. 3.12 Lifting a mesh dredge aboard the R/V "Professor Vodyanitsky"

signs of mud volcanoes. It is impossible to identify submerged liquefied breccia by echo sounders. The presence of breccia is a valid criterion, but unfortunately, only for active volcanoes. The breccia and clastic material erupted by paleovolcanoes is normally hidden by younger sediments. Operation of gas seeps is a reliable indication, but again, only for active volcanoes. Paleovolcanoes often do not demonstrate this feature.

References

Ilyina LB (1966) Istoriia gastropod Chernogo moria (History of the gastropods of the Black Sea). Trudy Paleontologicheskogo Instituta Akademii Nauk SSSR 110, Moscow (in Russian)

Nevesskaya LA (1963) Opredelitel' dvustvorchatykh molliuskov morskikh chetvertichnikh otlozheniy Chernomorskogo basseina (Guide to the identification of bivalves from quaternary marine sediments of the Black Sea). Trudy Palentologicheskogo Instituta Academii Nauk SSSR 96. Nauka, Moscow (in Russian)

Shnyukov EF (2016) Flyuidogennaya mineralizatsiya gryazevykh vulkanov Azovo-Chernomrskogo regiona (Fluidogenic mineralization of mud volcanoes of the Azov-Black Sea region). Logos, Kiev (in Russian)

Shnyukov EF, Shchiptsov AA (1996) Geologicheskie issledovaniya NIS «Kiev» v Chernom more (6-y reys) (Geological research of the R/V "Kiev" in the Black Sea [6th voyage]). OMGOR NAMIT, Kiev (in Russian)

Shnyukov EF, Rybalko SI, Grigorev AB et al (1977) Nauchno-issledovatelskoe sudno «Geokhimik» i nekotorye itogi pervykh reysov v severo-zapadnuyu chasti Chernogo morya (Research vessel "Geochemist" and some results of the first voyages to the northwestern part of the Black Sea). Preprint. Institut Geokhimii s fisiky mineralov, Kiev (in Russian)

Shnyukov EF, Shchiptsov AA, Ivannikov AV et al (1997) Geologiya Chernogo morya (po rezultatam geologicheskikh i geofizicheskikh issledovaniy 5-go reysa NIS «Kiev») (Geology of the Black Sea (based on the results of geological and geophysical studies of the 5th cruise of the R/V "Kiev")). OMGOR, Kiev (In Russian)

Shnyukov EF, Sheremetyev VM, Maslakov NA et al (2005) Gryazevye vulkany Kerchensko-Tamanskogo regiona. (Mud volcanoes of the Kerch-Taman peninsula). Glavmedia, Krasnodar (in Russian)

Shnyukov EF, Kobolev VP, Pasynkov AA (2013) Gazovyy vulkanizm Chernogo morya (Gas volcanism of the Black Sea). Logos, Kiev (in Russian)

Shnyukov EF, Stupina LV, Rybak EN et al (2015) Gryazevye vulkany Chernogo morya. Katalog (Black Sea mud volcanoes. Catalogue). Logos, Kiev (in Russian)

Yanko VV, Gramova LV (1990) Stratigrafiya chetvertichnykh otlozheniy Kavkazskogo shelfa i kontinentalnogo sklona Chernogo morya po mikrofaune (Stratigraphy of quaternary deposits of the Caucasian shelf and the continental slope of the Black Sea based on the microfauna). Soviet Geol 2:60–72 (in Russian)

Yanko VV, Troitskaya TS (1987) Pozdnechetvertichnye foraminifery Chernogo moria (Late Quaternary foraminifera of the Black Sea). Nauka, Moscow (In Russian)

Yanko V (also Yanko-Hombach V), Kravchuk AO, Kulakova II (2017) Meiobentos metanovykh vykhodov Chernogo morya (Meiobenthos of methane outlets of the Black Sea). Phenix, Odessa (in Russian)

Yanko-Hombach V (2007) Controversy over Noah's flood in the Black Sea: geological and foraminiferal evidence from the shelf. In: Yanko-Hombach V, Gilbert AS, Panin N, Dolukhanov PM (eds) The Black Sea flood question: changes in coastline, climate and human settlement. Springer, Dordrecht, pp 149–204

4 Modern Ideas About Mud Volcanism

Summary

According to the authors' opinion, formed on the basis of numerous definitions proposed by many researchers as well as the recovery and analysis of new geological materials, mud volcanism represents one manifestation on the Earth's surface of local gas flows emanating from deep hydrocarbon sources. Accompanying these flows are eruptions of breccia, clastic rock material, and water, all contributing to the formation of specific relief features, such as conical hills with craters at their top and sometimes round lakes.

The eruptive channels of mud volcanoes are geodynamically traceable to great depths—up to 9000 m, eventually being lost at the base of the sedimentary cover, which in the Black Sea dates to the beginning of the Mesozoic. The root of the Mantiynyy mud volcano has been traced to the Moho surface, but it probably runs deeper.

The gases erupted by mud volcanoes are mainly represented by methane containing insignificant amounts of heavy hydrocarbons (ethane, etc.*), carbon dioxide, as well as nitrogen and rare gases.*

The chemical content of waters emitted from the Kerch-Taman mud volcanoes includes the following associations: sodium chloride-sodium hydrogencarbonate, *sodium* hydrogencarbonate *-sodium, sodium chloride-carbonate, sodium sulfate, sodium chloride, sodium chloride-carbonate-sulfate, and sodium chloride. The waters of the Black Sea mud volcanoes have not yet been studied.*

Mud volcanic breccia is a detrital rock, consisting in most cases of debris from various clayey rocks. It is cemented with clay and often porous. Sometimes it appears burned as a consequence of explosive eruptions.

The mineralogical composition of mud volcanic breccia is represented by hydromica, montmorillonite, and sometimes an admixture of kaolinite and metagalloisite in its finely dispersed part. The coarse clastic material comprises limestone, marl, and sandstone.

In several mud volcanoes on the Kerch Peninsula, the mud volcanic breccia is kaolinized and carbonatize, thereby acquiring a greenish-blue color. The numerous accessory clastic terrigenous minerals of mud volcanic breccia include zircon, garnet, ilmenite, pyroxene, amphibole, etc.*; these are similar to the terrigenous minerals of the breccia-forming Maikopian and Cretaceous rocks.*

4.1 General Remarks

The earliest Russian publications described mud volcanoes as abysses, salses, mud keys, bakings, splashes, blebs, and zalts (Voskoboynikov and Gur'iev 1832). The term "mud volcano" was introduced by Helmersen and Schrenck (1885) as a translation from the German *mudevulkan*

The original version of this chapter was revised: Caption for Fig. 4.4 has been updated now. The correction to this chapter is available at https://doi.org/10.1007/978-3-030-40316-4_12

E. Shnyukov, V. Yanko-Hombach, *Mud Volcanoes of the Black Sea Region and their Environmental Significance*, https://doi.org/10.1007/978-3-030-40316-4_4

previously used by Abich (1873). Within northern Eurasia (in the former USSR), small mud volcanoes are called mud hills (or gryazevaya sopka). Their shape depends on the density of the ejected liquid clay; when the material is very fluid, a cone does not form.

In the foreign literature, these formations were known as *pseudovolcanoes* or *volcanitis*. Steber (1914) proposed the term *volcanoids* and, for their morphological expression in relief, *mud salses*. All these terms reflected the geomorphological features of mud volcanoes. In fact, many authors describe them as simple topographical features complicating the surface topography (e.g., Jakubov et al. 1971; Hovland et al. 1997). They did not take into account that these landforms are expressions of a remarkable natural process initiated deep within the sedimentary succession.

In the Geological Dictionary (Paffengolts 1973, p. 122), these definitions are summarized as follows: "*Mud volcano* is a large flat-conical hill composed entirely or from the surface only by mud, possessing a funnel-shaped crater at the top, and extending to some depth by a channel from which gas, water, and mud are periodically or continuously released. Episodic violent changes are also noted." This seems to be a fair definition, but it is not complete, as it does not reflect the full nature of mud volcanism.

According to Gubkin (1937), the gas-oil manifestations and mud volcanism are both the result of similar phenomena, namely, the geological and tectonic formations called diapirs. Later, Avdusin (1948) would agree with this explanation. Bondarchuk (1959) expressed this idea in brief by pointing out that mud volcanism is a manifestation of diapir tectonics. Fedorov (1939) considered the presence of diapir structures to be an indispensable condition, even the rule, for the formation of mud volcanoes in the Crimean-Caucasian region. Kalinko (1964) agreed that there is a connection between mud volcanoes, diapirs, and cryptodiapir structures, but on a local scale. In his opinion, mud volcanoes result from the influence of hydrodynamic conditions typical for certain tectonic zones. This idea was supported by the presence of mud volcanoes in Azerbaijan restricted to zones of major tectonic disturbances/fractures (Yakubov et al. 1980).

We understand the term *mud volcano* to be an eruptive apparatus that ejects different materials (mud breccias, clastic debris, water, gases) from the Earth's interior and builds either positive forms of relief (hills) (Fig. 4.1a–c) or (rarely) depressions filled with liquid mud volcanic breccia (Fig. 4.2a–c).

The term mud volcanic field refers to the whole complex of mud volcanic phenomena (e.g., gryphons, breccia, hydrotherms) characteristic for a certain geographic area.

Mud volcanoes of the Black Sea region are quite complex and diverse in their morphology, composition, and internal structure. Small mud volcanoes are called salses (Fig. 4.2a–c), while small secondary vents are known as gryphons (Fig. 4.3a, b).

Some mud volcanoes are nonexplosive (Fig. 4.2), while others are eruptive (Fig. 4.4), often ejecting burning breccias (Fig. 4.5).

Gryphons are positive features with a conical shape where gas, water, oil, and mud are continuously expelled with variable density and volume. These structures normally gather in fields, cluster in the central part of the crater, or follow trends controlled by tectonic features (Mazzini and Etiope 2017).

Mud volcanoes can have a rounded, isometric, or elongated shape, with craters in the middle that can grow to several tens of meters in diameter. They can be spaced very close to each other, forming a chain of positive structures (Fig. 4.1b) comprising mud volcanic breccia that can generate massive, gradual, and progressive outflows of semiliquid material. Some authors call this a "diapiric mélange" (Barber et al. 1986; Barber 2013).

The basic structure and main elements of a mud volcano are shown in Fig. 4.6.

The entire mass of erupted mud breccia constitutes a modern or fossil mud field. Sometimes fossil mud deposits no longer demonstrate any active volcanism. Modern mud volcanic manifestations, however, represent only the top floor of what might be imagined as multistory buildings, their lower floors deeper within the Earth having been formed in the geological past.

Fig. 4.1 Series of mud volcano hills on (**a**) the Kerch Peninsula, (**b, c**) the Taman Peninsula

Fig. 4.2 The nonexplosive Bulganakskiy mud volcano in the vicinity of Bondarenkovo village looks like a depression (**a**) filled with liquid mud containing gas bubbles (**b, c**)

Fig. 4.3 Small gryphons on the Bulganakskiy mud volcano

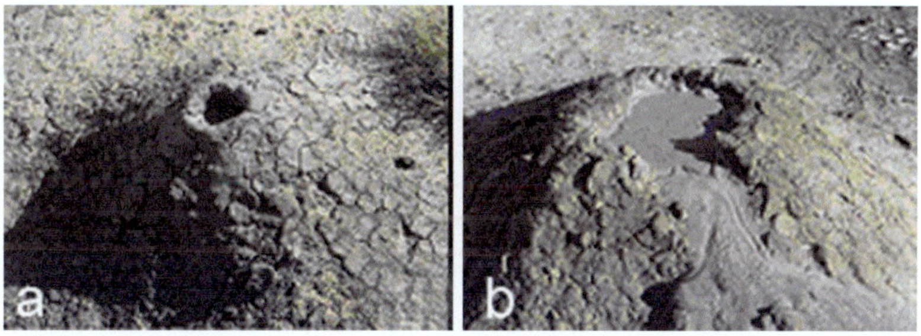

According to Avdusin (1948), a mud volcano's morphostructure depends largely on the physical properties of the sedimentary rocks involved in its formation, especially their ductility. The latter property depends on mineralogy, grain size, water saturation, and even composition. Avdusin admits to the presence of plastic clays in the lower part of the Cretaceous section of the Caucasus but considers the Maikopian deposits as exceptional in this regard, stating that they serve as the main "building" material of mud volcanoes. For example, the Ilanly mud volcano (Azerbaijan) takes the form of a regular cone composed of typical Maikopian mud breccia brought to the surface from a great depth (Fedorov 1939).

Time exerts considerable influence on the architecture of mud volcanic cones, modifying their initial form (Avdusin 1948). In addition to the natural weakening of mud volcanic activity over time, the peak of a mud volcano's cone can become weathered until it is completely smooth,

Fig. 4.4 Image of an explosion of the Golubitskiy underwater mud volcano in the Sea of Azov in 2008

Fig. 4.5 Gas flares at the final stage of a mud volcano eruption in Western Tsimbaly, NW of the Taman Peninsula (February 2002)

and the muzzle of the structure itself can be completely leveled out until it disappears. Currents of mud breccias can also distort the shape of the cone, and these changes usually persist for a long time. Sometimes, volcanoes or hills located separately from each other nevertheless possess joint foci and group together (Fig. 4.1b).

Mud volcanism is a fairly widespread phenomenon, but it is far from a universal one. There are certain conditions that must be met for their development:

1. *The presence of thick layers of plastic clayey rocks* (e.g., Avdusin 1948). In the Black Sea region, Maikopian (Oligocene to Lower Miocene) deposits with a thickness up to 3–5 km, sometimes more, and to a lesser extent the Lower Cretaceous clays are the main base strata that give rise to Ponto-Caspian region mud volcanoes. They have a monotonous structure, highly dispersed composition, and insignificant (2–3%) sand fractions not only in the Eastern Caucasus but Azerbaijan (Aliyev et al. 2015) and the Black Sea region (Shnyukov et al. 2013) as well. Their ductility significantly increases in the presence of interstitial water (Durmishyan 1961).

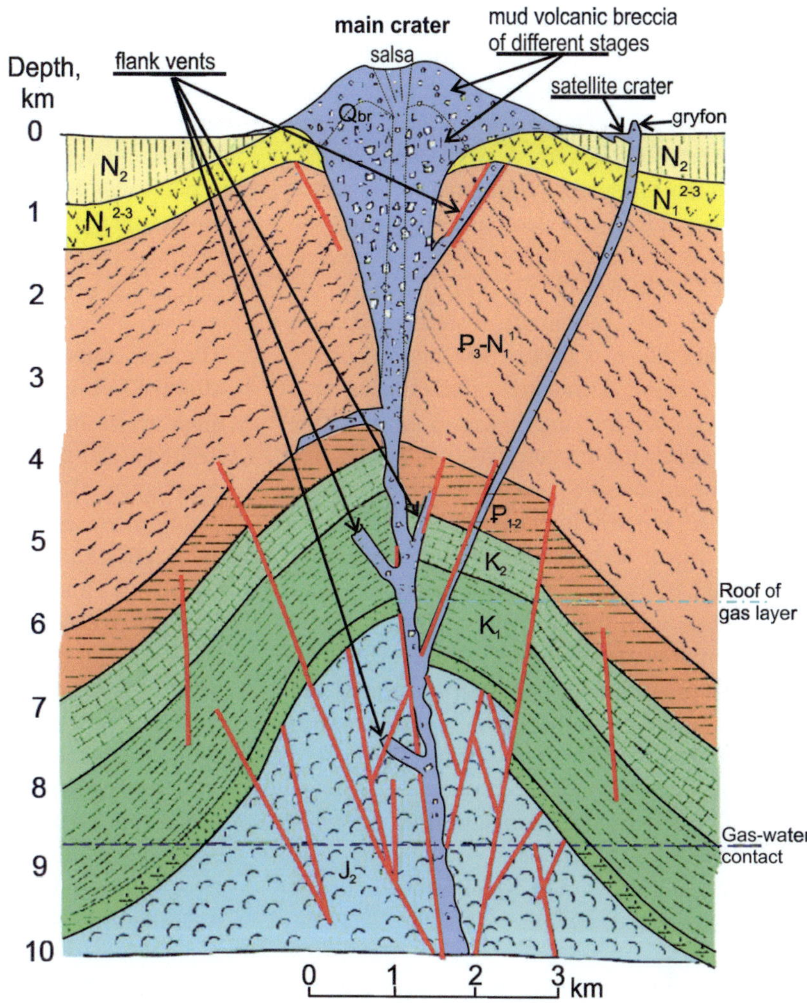

Fig. 4.6 Basic structure and main elements of a conical mud volcano. For legend, see Fig. 6.4

2. *Favorable tectonic conditions, such as accretion prisms arising in subduction zones* (Limonov 2004). Lateral tectonic compression or tension is also very important but not always a requirement. A key example of their power can be seen in the Black Sea's center, where lateral tension is particularly strong, and about ten large mud volcanoes exist.
3. *Tectonic fractures, especially in areas where they intersect* (Yakubov and Zeynalov 1964; Kalinko 1968). These fractures are important to the extent that no mud volcano can exist without tectonic structures. This is clearly demonstrated in the Black Sea region (Shnyukov et al. 2013), Azerbaijan (Aliyev et al. 2015), and Sakhalin (Siryk 1962). Some researchers emphasize the interrelation between mud volcanoes and deep faults in the Earth's crust (Gorin 1964, 1967; Gorin and Buniat-Zade 1968).
4. *Diapiric structures*. In fact, most, but not all, mud volcanoes are confined to diapirs. Diapiric structures are lacking in nearly all of the mud volcanoes of the eastern part of the Black Sea. Instead, their roots can be traced to the Paleozoic foundation (Ismagilov et al. 2006). With this specific case, one can conclude that diapirs are a frequent, but not a necessary factor responsible for the development of mud volcanoes.
5. *Gases from the Earth's interior*. Gases are constantly forming under contrasting tectonic

movements. They rush upward when a ratio of reservoir/hydrostatic pressure exceeding 2–2.2 becomes sufficient to break the continuity of the rocks (Durmishyan 1961). The gas flow, possibly together with deep water, breaks into the cracks with great force, carrying away interstitial water and fractured rock materials (Kalinko 1964). After gas escape reduces pressures in the gas reservoir, the eruption of a mud volcano ceases. Subsequently, new gases enter the main strata through isolated fractures, microcracks, and weakly permeable interlayers, a process that again raises the gas pressure. Although this increase in pressure does not reach the value it once had, it is still high enough to lift the gas-water mirror and encourage a re-eruption of the mud volcano. The process continues until the pressure in the gas reservoir drops below that required to overcome the resistance of the materials located in the eruptive channel. An eruption can also be halted if the strata were to be filled with condensate coming from gas or oil. Between the violent eruptions, a gryphon stage of activity can be manifested if permeable layers are saturated with water and located higher in the geological sequence (Shnyukov et al. 2013).

According to Kropotkin and Valyaev (1979), outbursts from mud volcanoes contain so much gas that no near surface gas reserves would be sufficient to feed them. Mud volcanoes must therefore be fueled by gases emanating from deep sources, and this means that only by recognizing the leading role of deep degassing along tectonic fractures can one obtain a satisfactory explanation of large-scale mud volcanism.

Mazzini and Etiope (2017, p. 83) highlighted four major points that are characteristic of mud volcanoes: "(a) The discharge of at least a three-phase system (gas, water, and sediment— and occasionally oil). (b) Gas and saline water related to a diagenetic or catagenetic hydrocarbon production system (accordingly gas is dominated by methane and subordinately C_2^+ hydrocarbons). (c) The involvement of sedimentary rocks with a gravitative instability resulting from rapid sedimentation, leading to the formation of mobile shales, diapirs or diatremes. (d) The (common) presence of breccia within the discharged material."

Specific features of the host formation and evolution over time can generally explain the variety in mud volcano morphology and activity. According to Kalinko (1964), three types of mud volcanoes can be distinguished.

1. The Lok-Batan type of mud volcano is named after the Lok-Batan Mud Cone, the largest and most active mud volcano in the world. It is located on the Absheron peninsula in Azerbaijan. This type is characterized by highly explosive activities commonly due to the ignition of emitted gases (Fig. 4.4). Short periods of violent activity alternate with long periods of inactivity. The extruded mud breccia has a highly viscosity, which serves to build well-formed, steep cones. Blockages of the eruptive channel by mud "corks," and the explosive breaking of these corks when the pore fluid pressure exceeds the retention force, explain the explosive character of this type of mud volcano. Lok-Batan-type mud volcanoes are found on the Apsheron Peninsula, Azerbaijan, and the Caspian Sea; Dzhau-Tepe, the biggest mud volcano of the Kerch Peninsula, Crimea, is also a Lok-Batan type.
2. The Chikishlyar type is named after the Chikishlyar mud volcano located in south-western Turkmenistan. It is characterized by calm, relatively weak, and continuous activity. Gas is permanently emitted in approximately uniform quantities through numerous vents that also spit out small amounts of water. The upper part of the sedimentary sequence contains water-saturated layers that form very low, flat, or bulging domes. These domes merge with the surrounding plain, or form plate-shaped depressions, often filled with water. This type of mud volcano is very common on the Kerch Peninsula.
3. The Shugin type is named after the Shugo mud volcano, located in western Caucasus. In places, its development is complicated by the appearance of large pockets in its porous

rocks. These pockets are filled with gas that moves upward, causing periodic emissions and primary eruptions. Eruptive periods alternate with periods of weak activity. This type comprises the majority of the known mud volcanoes worldwide and is characterized by a great variety of forms, mostly with composite craters.

No relationship has been noted between mud volcano types and their regional distribution. All three types can be present in any mud volcano belt, depending on the local lithological and tectonic properties of the host sediments. In a historical context, each type of mud volcano reflects only one of the stages of its activity; it may well be that in the geological past, mud volcanoes manifested different stages, varying from turbulent to relatively calm, and this would be because mud volcanic foci tend to function continuously for tens of millions of years.

One might wonder how clays of the Maikopian diapirs could be responsible for mud volcano development if their roots lie in the deeper Mesozoic sediments. In the light of new seismic data obtained from mud volcanoes of the Black and Caspian seas, Ismagilov et al. (2006) propose that mud volcanoes are much more complicated geological phenomena than has previously been thought.

In our opinion, a significant number of the Black Sea's mud volcanoes are still linked to clay diapirs, but this connection is the reverse of what has generally been assumed. The key to understanding this important relationship was recognized by Meisner and Tugolesov (1997) in their paper on fluidization deformations in the sedimentary cover of the Black Sea. They propose that the deep fluids penetrating through the sedimentary cover seem to facilitate the process of folding. Thus, the diapirs do not generate mud volcanoes, but on the contrary, flows of deep fluids generate mud volcanoes and create favorable conditions for the appearance of diapirs. Indications of these fluids can be seen as subvertical geological bodies or as a system of bulges that penetrate the surface, creating large discontinuities in the sedimentary cover.

If this is true, then a clarification of the mud volcano phenomenon is very much needed. In this respect, the definition of Kovalevskiy (1935, p. 22) can be quite helpful: "... Mud volcanoes are the brothers of an innumerable heteronymic family of water-and-gas sources, mud lakes and hills, mud and burnt mountains, infernal swamps, salses ... and many other native names that usually characterize different features of local manifestations of mud volcanism. The latter is only a partial form of manifestation of gas jets on the present-day surface that flow from the depths of the Earth's crust and give rise to a vast group of seemingly different phenomena united under the common name 'gas volcanism of the Earth'." Strictly speaking, the definition should be formulated more succinctly: "... mud volcanism is only one form of manifestation on the modern land surface of gas jets that flow from the depths of the Earth's crust." Updated to reflect the newest evidence, this definition should be modified to the following: "mud volcanism is created at the Earth's surface by local gas streams of deep hydrocarbons, and eruptions are accompanied by ejections of mud breccia, detrital material, and water that create peculiar forms of relief that look like cone-shaped hills topped by craters and often filled with bubbling liquid mud." Under marine conditions, mud volcanoes can be observed only when they create positive relief forms on the seafloor. They often appear to be flat truncated cones, crowned with a vent that can vary substantially in size and configuration. In general, both onshore and offshore mud volcanoes form cones of different sizes often with a complex appearance in panoramic view, but their base is always a round cone, with varying degrees of flatness (e.g., Fig. 4.7). As a rule, they are associated with the axial lines of anticlines and are thus useful for structural mapping of the seafloor.

Within the Black Sea region, onshore mud volcanoes of the Kerch and Taman peninsulas reach considerable sizes, but they are generally smaller than those of Azerbaijan. The peak of mud volcanic activity in the Black Sea region is confined to the Neogene, while in Azerbaijan, it is in the Quaternary.

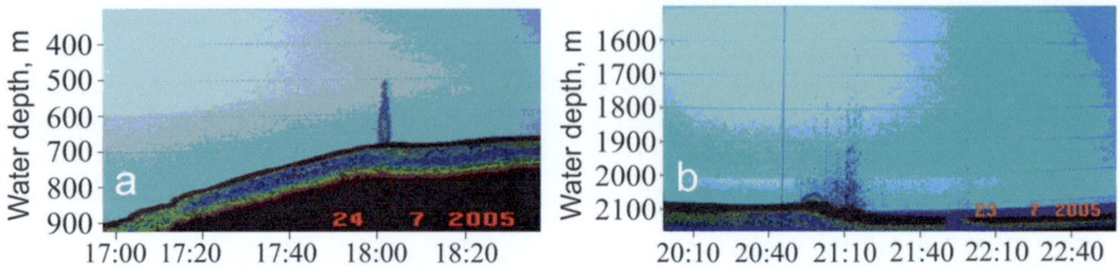

Fig. 4.7 Seismoacoustic profile across underwater mud volcanoes: (**a**) Mitin mud volcano: 44° 37′ 54″ N, 36° 01′ 03″ E; depth 687 m; height of high intensity seep is 200 m; (**b**) Sevastopol mud volcano: 44° 16′ 42″ N, 34° 52′ 43″ E; depth 2150 m; height of high intensity gas seep is 300–500 m (Shnyukov et al. 2006b)

4.2 Structure of the Eruptive Channel of Mud Volcanoes

Though mud volcanoes have long drawn the attention of people because of their uniqueness, little study has been focused on their structure.

One of the first ideas relating to the structure of the eruptive channel in the mud volcanoes of Azerbaijan was suggested by Avdusin (1948). He proposed that the morphology of mud volcanic structures is determined by three factors: mountain-forming (tectonic) forces, the volume and depth of mixing of ejected gases (largely methane) and liquids, and the lithologic properties of sediments involved in the eruptive process.

Excavation of the Ilanly mud volcano in Azerbaijan demonstrated that layers of breccia formed the cone. These layers overlapped with each other and folded as additional layers were accumulated. The breccia comprised several different types: tectogenic (friction, sliding, dragging, mylonites, etc.), sedimentogenic, and crystalline.

The feeder channel and the body of the mud volcano consisted of a ductile mass of crushed rock. The ductile deformation appeared to increase continuously toward its central part (Fig. 4.8).

In the periphery, the rocks were cracked but still preserved some continuity, while in the central part, they already revealed a clastic structure. Both breccia and gases moved at their greatest speed in the central part of the feeder channel, in the process crumpling the sediments. Clastic material of hard rocks (limestones, sandstones, etc.) made up 3–8% of the breccia. Clast sizes varied between 2 mm and 17 cm, but sometimes, these clasts attained the size of boulders, reaching 3–5 m^3 and even 10 m^3 in some cases. There were also huge (hundreds of m^3) lumps of clay covered by slickenlines. Sometimes, the breccia was completely disintegrated by gases and excess water, forming muddy pelites that represent a specific facies of the feeder channel. The thickness of the breccia in Azerbaijan varied between 0.5 and 3 m, and one can perhaps assume that these are the general sizes of breccia covers in the Black Sea region as well.

Each active volcano possesses a main crater (central vent) that varies in size and shape over the course of an eruption. The craters usually have a rounded form surrounded by a rim, and size can vary from meters to tens of meters.

The main crater of the volcano is linked to its subterranean depths through a system of vertically aligned chambers. These chambers are filled with semifluid, gas-saturated masses of mud or silt breccia. A specimen of such breccia taken from a fresh eruption of the Bulganakskiy mud volcano appears as a porous rock in Fig. 4.9.

Mud breccia is usually enriched with gaseous and liquid hydrocarbons. For example, mud breccia of the Lok Batan mud volcano contained ~1.5 cm^3/kg of heavy hydrocarbons and ~0.4 cm^3/kg of methane even 8 years after the eruption that produced it (Rakhmanov 1987).

An active mud volcano is usually characterized by pulsating eruptions, and the feeder channel consists of a long chain of these vertically linked gas-saturated chambers that convey erup-

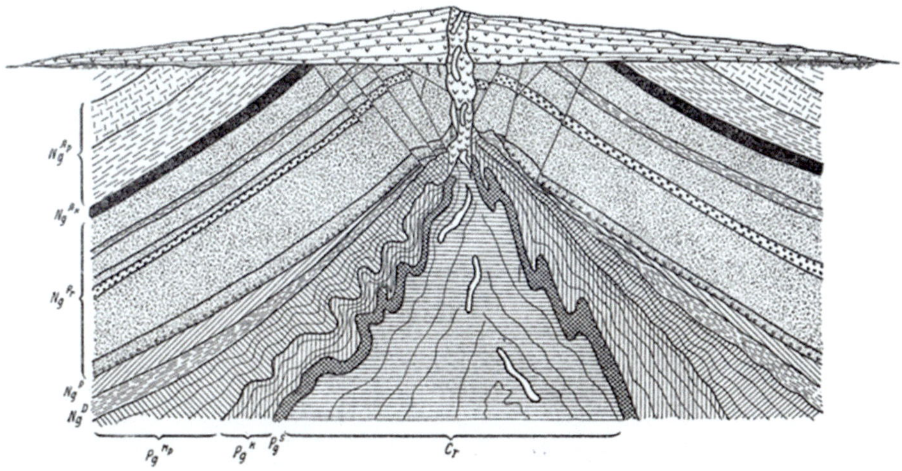

Fig. 4.8 Scheme of a feeder channel that illustrates the structure of a mud volcano in eastern Transcaucasia (Azerbaijan). From Avdusin (1948)

Fig. 4.9 Freshly erupted breccia with pores and cavities from the Bulganakskiy mud volcano

tive material upward from great depths (Fig. 4.10).

Some idea of the great depth of a feeder channel can be obtained from Alifan and Lysenkov (1968), who wrote of an incident that occurred during the drilling of well No. 63 in the Barca-Helmes mud volcano of Turkmenistan. A drilling tool was lost to free fall within the channel from a depth of 873 m; it came to rest at 1964 m, a drop of over 1000 m.

The structure of the upper parts of feeder channels has been studied via geophysical methods, and thus there is a wealth of observations to draw from. Alekseev et al. (2004) conducted studies of the Karabetova Gora mud volcano using a "LOZA" georadar instrument together with hydrogen flow sensors in order to investigate the subsurface structure of the channel to a depth of ~100 m. The study, conducted between 1998 and 2006, was based on the measurement of broadband electromagnetic pulses in a geological sequence with the registration of interfaces between layers. The researchers described the development of numerous gas chambers in the feeder channel to a depth of 17 m. These chambers were typically under high pressure and periodically erupted, releasing large amounts of breccia (Fig. 4.10). The authors were able to predict new eruptions of the mud volcano based on changes in the speed of hydrogen flows within the supply channels.

Another investigation performed within the Andrusov mud volcano by Olenchenko et al. (2015) was carried out using a modified method of vertical electrical sounding called electromotography, a technique based on the differences in electrical resistivity of rocks. Over the course of these studies, the researchers observed impulsive episodic eruptions of gases (predominantly methane) as well as water and mud breccias. The

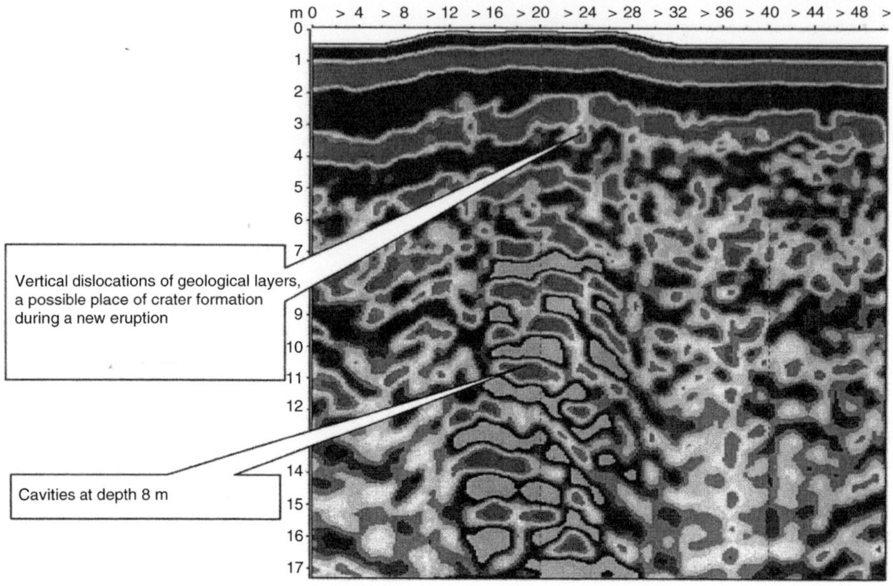

Fig. 4.10 Georadar section showing breccia structure in the area of the main crater of the Karabetova Gora mud volcano after its eruption in 2004 (Alekseev et al. 2004)

sounding profile was 115 m in length and passed through the central part of the Andrusov mud volcano, directly over the vent. During the course of an eruption, the dynamic change in the channel environment was analyzed in a series of electromotographic images that reflected the process whereby the gas-saturated fluid rose (t = 0–15 min). The feeder channel opened for 30 min, eruption ensued, and then tightening of the channel by mud breccia occurred during the next 60 min. This study provided key evidence that led to the discovery of the mechanism underlying mud volcano eruptions.

Sobisevich et al. (2015) described two zones of dilatancy (decompensation) in the Shugo mud volcano. The first zone was labeled "focal" because it corresponds to the foci of mud volcanoes where the forces of eruption occur. The second zone was labeled "boundary" due to its proximity to a free surface where a system of outflow channels forms over the course of an eruption. The sediments in this zone are characterized by multiscale, de-compacted, and fractured structures that are especially pronounced in the central part of the vent. They serve as passages for escaping gases and, afterward, water that carries away the breccia that sealed the eruptive channel, eventually widening the way for the discharge of sedimentary material (salse-gryphon stage).

In summary, studies of mud volcano feeder channels have revealed the presence of numerous chambers in the gas-saturated breccia of the mud cones. These can be traced to a depth of 200–500 m and possibly even deeper. Their presence is lithologically and tectonically conditioned, and beneath them, there is a single feeder channel formed of connected chambers or a column (Figs. 4.11 and 4.12).

The eruptive channels of Black Sea mud volcanoes were studied by Ismagilov et al. (2006) and Shnyukov et al. (2013) using seismic data. Significant contributions were provided by the numerous scientific expeditions of geologists from Yuzhmorgeologiya Enterprise.

Seismic profiles of feeder channels belonging to onshore and offshore Black Sea mud volcanoes are all traceable to a depth of 9 km; they become lost somewhere at the base of the Mesozoic. They are well expressed and quite diverse in appearance. In the upper part of the feeder channels, there is a system of isolated chambers, while in the main part, there is either a single large chamber or a system of large chambers connected to each other. Each chamber

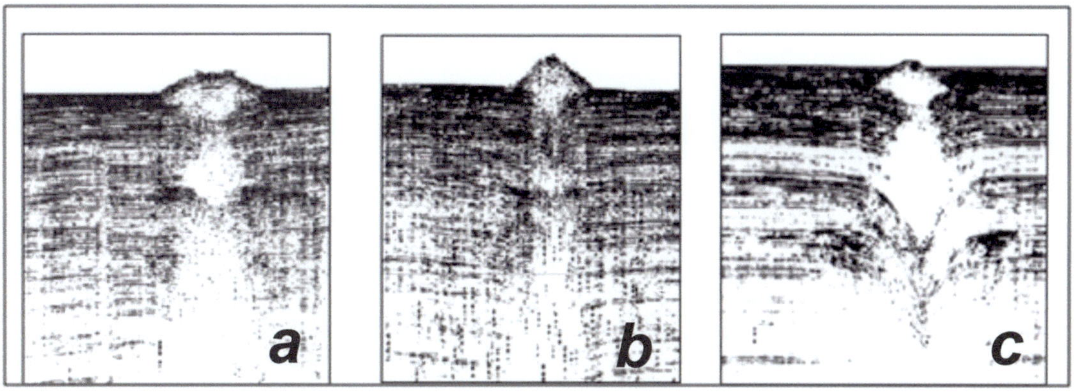

Fig. 4.11 Seismic records of Black Sea mud volcanoes: (**a**) MSU (profile PR-173); (**b**) Yuzhmorgeo (profile PR-173); (**c**) Strakhov (profile PR-172). The internal structure of the feeder channels is readily visible (Bouriak 1995)

Fig. 4.12 Time seismic section of the Vassoevich mud volcano in the western Black Sea depression (Shnyukov and Netrebskaya 2013)

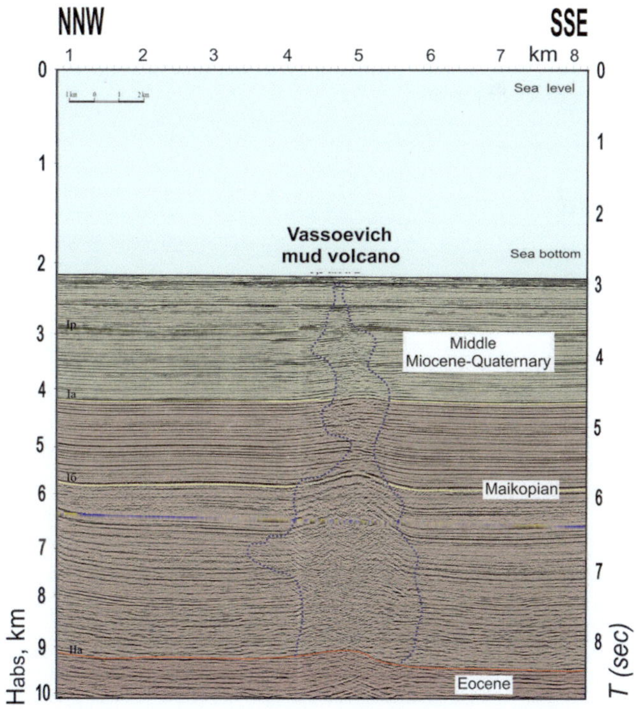

appears as a bulge in the feeder channel within tectonically prepared and lithologically favorable rocks (Fig. 4.12).

Not all offshore mud volcanoes have eruptive channels similar to those described above. In this regard, the Goncharov and Kovalevskiy mud volcanoes deserve special attention (Figs. 4.13 and 4.14).

These volcanoes do not possess intermediate gas chambers (for methane storage) in their feeder channel. Instead they incorporate a kind of columnar, subvertical channel that widens as it descends. Similar structures have been found in other mud volcanoes in the western Black Sea basin, where the Moho boundary is located closer to the surface, at a depth of 19 km.

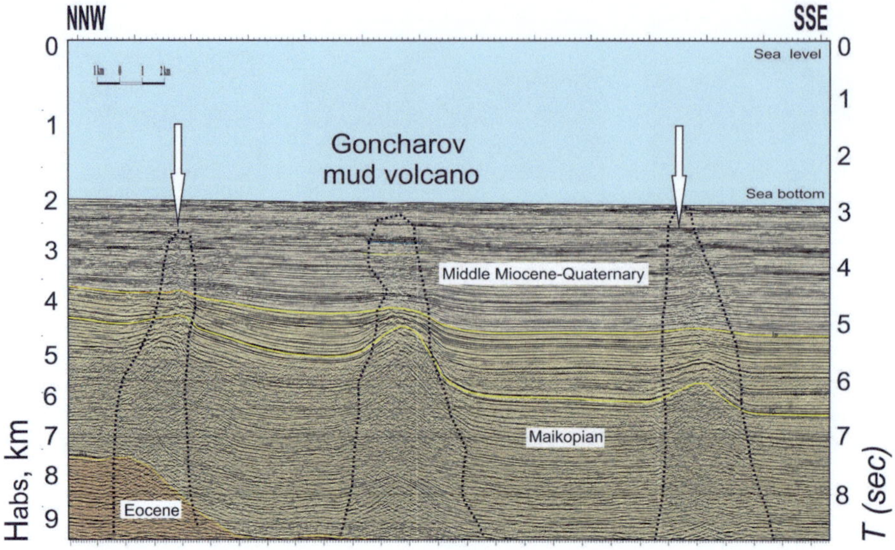

Fig. 4.13 "Blind bodies" similar to mud volcanoes (white arrows) located near the Goncharov mud volcano in the western Black Sea basin

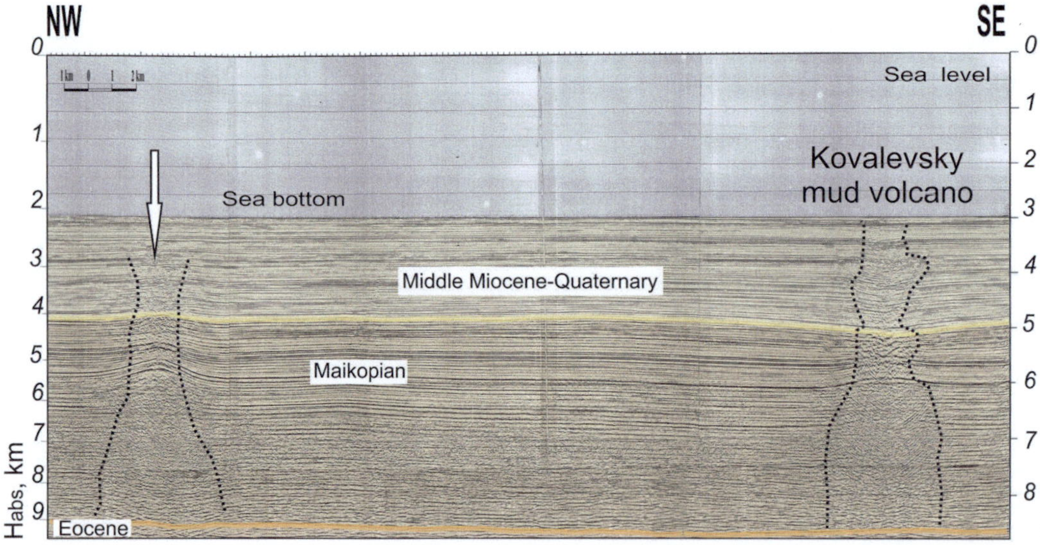

Fig. 4.14 "Blind body" (white arrow) similar to a mud volcano located near the Kovalevskiy mud volcano in the western Black Sea basin

Similar feeder channels devoid of intermediate chambers are recorded among the mud volcanoes of the eastern Black Sea basin, such as Yalta, Istanbul, and Kazakov in the Sorokin Trough) (Figs. 4.15, 4.16, and 4.17).

More often, however, the eruptive apparatus of mud volcanoes has a more complex structure in the form of a system of connected swells of indefinite shape (Fig. 4.12). Judging by their stratigraphic position in the geological section, these swells are most often motivated by the

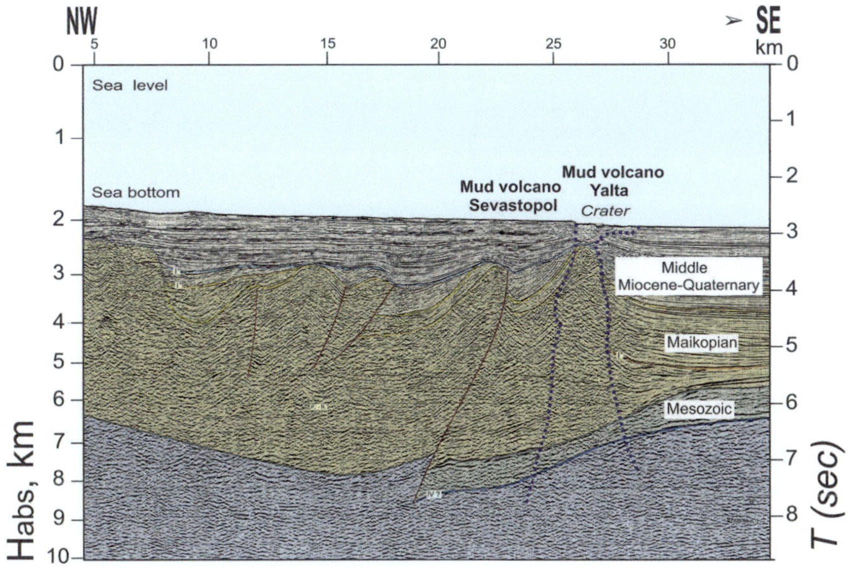

Fig. 4.15 Time seismic section of the zone including the Sevastopol and Yalta mud volcanoes located in the Sorokin Trough

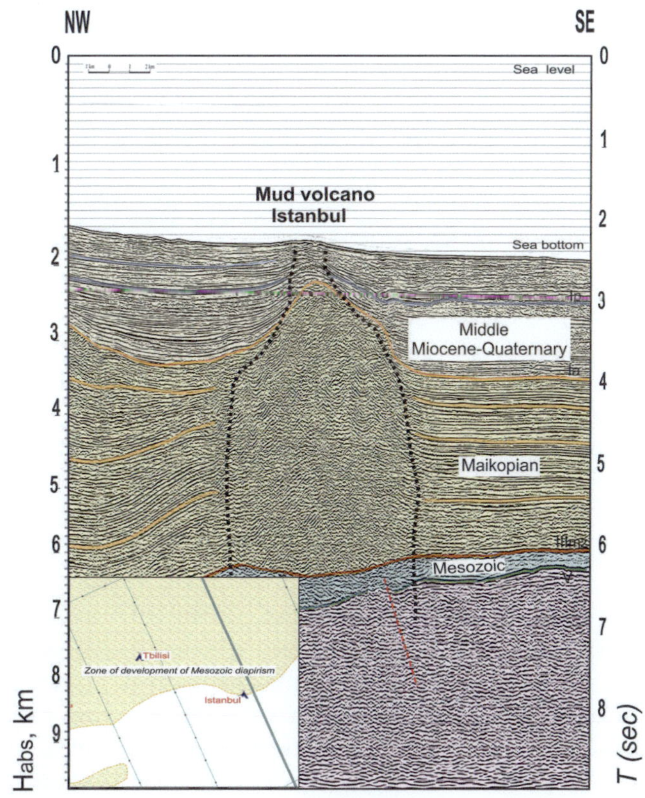

Fig. 4.16 Time seismic section of the Istanbul mud volcano zone (Sorokin Trough)

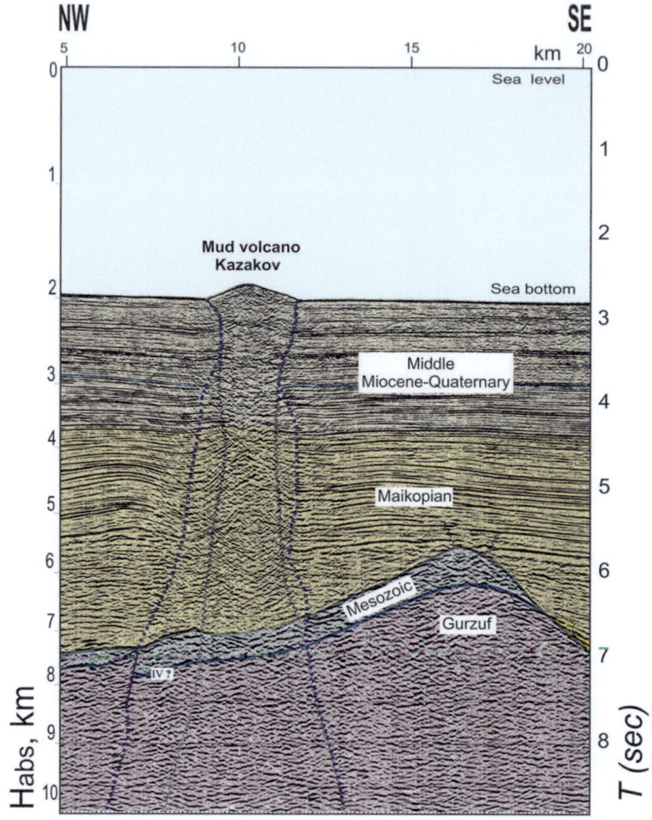

Fig. 4.17 Time seismic section of the Kazakov mud volcano (Sorokin Trough)

lithological features of the sedimentary complexes.

Subvertical "blind bodies" have been encountered that are similar to mud volcanoes but are not exposed at the surface. These features are not yet fully understood (Figs. 4.13 and 4.14) but may represent traces of fluid flow movements that escaped from the depths but do not have enough power to break through the sediment sequences to reach the surface. They are like malfunctioning mud volcanoes, and similarities can be drawn with subvertical geological structures described in the Caspian Sea (Mamedov and Guliyev 2003; Guliyev 2010; Khaustov 2011).

In general, studies of the eruptive channels of the Black Sea's mud volcanoes create an impression of gas flows moving from the interior and gradually fading as they approach the surface. Such evidence confirms the positions developed by Kropotkin and Valyaev (1979) and Dmitrievskiy and Valyaev (2002) about the deep degassing of the Earth.

Chapter 6 provides further descriptions of mud volcanoes that clearly illustrate the diversity of feeder channels.

4.3 The Scale and Composition of Gases Released by Mud Volcanoes

The release of gases from mud volcanoes is often the first thing that draws attention to them. The scale of this phenomenon varies immensely, as has been shown by Rakhmanov (1987) and Aliyev (2016) in the case of the mud volcanoes in Azerbaijan. Another example of this variability can be seen in the Black Sea region, where gas release tends to be relatively modest but occasionally can be quite large. Onshore mud volcanoes are often accompanied by dangerous methane ignitions, as was seen after the Crimean earthquake of September 1927, when the seabed released gases that spontaneously combusted upon contact with the air (Nikonov 2002).

A number of authors have proposed estimates of daily gas emissions from the Kerch-Taman mud volcanoes. Amounts proposed during quiet stages have been 400 m^3 (Belousov and Yarotskiy 1936), 350 m^3 (Zaitsev 1965), and 520 m^3 (Shternov 1968). During erupting stages, they emit 1500–2000 m^3 (Shternov 1968). The last author also calculated that approximately 11–15 trillion m^3 of gases have been released over the 20 million years in which mud volcanoes are known to have been active. There are currently no such calculations for offshore Black Sea mud volcanoes, but we must assume that those figures are also quite large. For example, powerful fountains of gas that can achieve a height of up to 800–1000 m and a diameter of ~400 m have been observed in the Sorokin Trough. Their collective release has been calculated at 2.6×10^5 m^3 (Egorov et al. 2011). These data also include degassing from numerous gas seeps. The highest estimates of total degassing from the floor of the Black Sea are between 5.0×10^9 and 6.0×10^9 m^3 per year (Kessler et al. 2006) and 2.2×10^9 m^3 per year (Egorov et al. 2003).

Of course, all calculations made to date are arbitrary and show an order of magnitude rather than actual numbers. In addition, the variability of the volume of ejected gases should be taken into account in connection with the instability of the geological environment, e.g., tectonic activity, lithological properties of sediments, etc. The gases released by mud volcanoes, along with their discharged water, serve as the main indicators for determining mud volcano activity in the region.

The chemistry and isotopic composition of emitted gases have been studied by many specialists over the years (e.g., Steber 1915; Lukashchuk 1933; Burkser and Bronshtein 1933; Belousov and Yarotskiy 1936; Kudryavtsev 1963; Subbota 1964; Zaitsev 1965; Kurishko et al. 1968; Shternov 1968; Gemp et al. 1970; Shnyukov et al. 1971; Lagunova 1974; Gemp and Lagunova 1978; Voytov 2001; Herbin et al. 2008; Egorov et al. 2011; Lein and Ivanov 2018). Gases emerging from mud volcanoes in the Black Sea region contain methane and its homologues, carbon dioxide, nitrogen, argon, helium, hydrogen, hydrogen sulfide, and carbon monoxide. The main components are methane and carbon dioxide, with the methane content in most cases reaching 80–99%; in some cases, however, methane content has been found to be as low as 0.5%, and in even fewer cases, it has been recorded as zero. Carbon dioxide varies from fractions of a percent to 91.7% (Bolshetarkhanskiy mud volcano, Crimea), but in the vast majority of cases, it averages ~10%. Nitrogen varies from 0.2% to 5.0% in a majority of mud volcanoes; however, it can reach 11.5%, 23.6%, and even 82.7% in emissions from the Dhzau-Tepe, Kayaly-Sart, and Bulganakskiy mud volcanoes, respectively. The highest concentrations of argon vary between 0.051% in outbursts of the Gladkovskiy mud volcano and 0.53% in those of the Nasyrskiy mud volcano. Helium is present everywhere, amounting to thousandths, sometimes hundredths, of a percent, e.g., 0.011% and 0.014% in outbursts of the Oldenburgskiy and Bulganakskiy mud volcanoes, respectively. Substantial amounts of hydrocarbons are found sporadically, their content varying significantly from tenths of a percent (in most cases) to 77.28% (Bulganakskiy and Pavlov mud volcanoes), and in one rare instance 83.31% (hydrogen sulfide spring associated with the Dzhau-Tepe mud volcano).

The ratios of gas components can also vary significantly at different times in individual mud volcanoes and even in gryphons. Thus, according to long-term observations of the Bulganakskiy mud volcano, the main gas components have changed as follows (Table 4.1).

The distribution of different gases released by mud volcanoes across the Kerch-Taman area and the changes in their chemical composition over time are illustrated in Fig. 4.18.

Judging from the ratio of the gas components released by terrestrial mud volcanoes, six types of gas emissions are distinguished: methane, methane-carbon dioxide, carbon dioxide, nitrogen, heavy hydrocarbon, and carbon dioxide-nitrogen-methane.

Table 4.1 Change over time in the % composition of gases in the Bulganakskiy mud volcano

Main components	Date of sampling				
	24.09.1925 (Lukashchuk 1933)	1932 (Belousov and Yarotskiy 1936)	18.08.1962 (Kurishko et al. 1968)	10.07.1965 (Kurishko et al. 1968)	June 2001 (Herbin et al. 2008)
CH_4	56.7	56.7	94.81	95.67	77.1[a]
CO_2	22.3	42.2	4.10	2.31	15.8
N_2 + rare elements	15.0	0.1	0.71	1.87	7.1

[a]0.1% C_2H_6 (ethane) was also detected

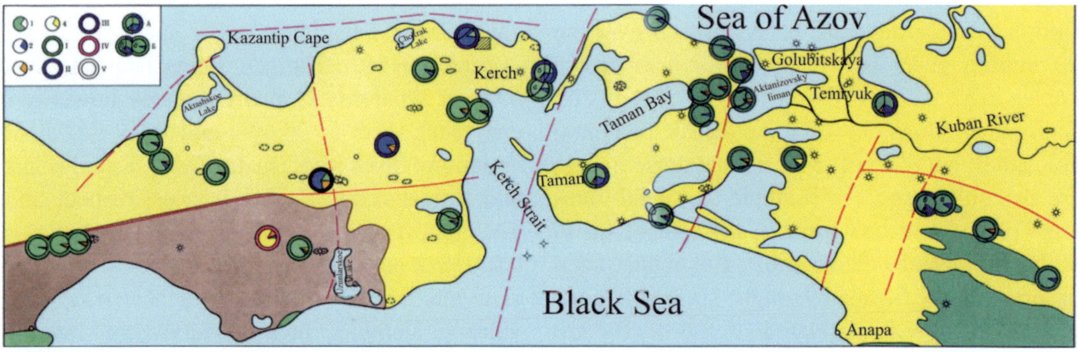

Fig. 4.18 Mud volcanic gases of the Kerch-Taman region (Shnyukov et al. 2005). Chemical content of gases: (*1*) methane, (*2*) carbon dioxide, (*3*) nitrogen, (*4*) heavy hydrocarbons. Types of mud volcanic gases: (*I*) methane, (*II*) methane-carbon dioxide, (*III*) carbon dioxide, (*IV*) nitrogen, (*V*) heavy hydrocarbons. (*A*) Change of gas content over time, (*B*) mud volcanoes with different chemical content of gases. Age of sediments is indicated by color—for legend here and in Fig. 4.19, see Fig. 6.4

The *first type* of mud volcanic gases—methane—is typical for the majority of volcanoes on the Kerch Peninsula: Vladislavovskiy, Eastern and Western Hyrtsyz-Shibanskiy, Korolyevskiy, Nasyrskiy, Kamenskiy, Boruch-Oba, Burashskiy, Chongelekskiy, Soldatsko-Slobodskiy, Voskhodovskiy, Yenikalskiy, and the majority of mud volcanoes of the Bulganakskiy and Malotarkhanskiy hearths. Among the gases that this type of volcano releases, methane is dominant; a methane content of about 80–99% is characteristic, while small quantities of carbon dioxide, nitrogen, and some heavy hydrocarbons are also present. Most of the volcanoes of the Taman Peninsula fall within this type, with the exception of the volcanoes of Karabetova Gora, Gora Gnilaya, and western gryphons of the Shugo mud volcano.

The *second type*—methane-carbon dioxide—is typical for the Trubetskoy mud volcano, which releases gases comprising 67.8% methane and 32% carbon dioxide.

The *third type*—carbon dioxide—is characteristic of gases from the the Gornostaevskaya recessed syncline and the Bolshetarkhanskiy mud volcano, which release gases comprising 89.6% and 91.7% carbon dioxide, as well as 9.1% and 8.3% methane, respectively. The gas emitted by the Gornostaevskaya recessed syncline also contains 1.3% heavy hydrocarbons.

The *fourth type*—nitrogen—is characteristic of the Abih mud volcano (Bulganakskiy mud volcanic hearth), which releases gases composed of 82.7% nitrogen.

The *fifth type*—heavy hydrocarbon—is characteristic of only one mud volcano, Dzhau-Tepe, which releases gases containing 83.3% heavy hydrocarbons, 11.5% nitrogen, and 5.2% carbon dioxide.

The *sixth type*—carbon dioxide-nitrogen-methane—is typical for gases of the Kayaly-Sartskaya recessed syncline. The main gas components here include 61.2% carbon dioxide, 23.6% nitrogen, and 14.8% methane.

All types of mud volcanic gases demonstrate a relatively consistent gas chemistry when studied at different times. There are exceptions, however. The Tsentralny (Bulganakskiy mud volcanic hearth), Enikalskiy, Karabetova Gora, and Gora Gnilaya mud volcanoes, as well as the western gryphons of the Shugo mud volcano, are distinguished from the others by sharp changes in the ratio of gas components over time. Released gases of the Enikalskiy mud volcano, for example, belong to the second type, yet its carbon dioxide content varies from 28.5 to 48.96% (Kurishko et al. 1968).

For the other mud volcanoes showing inconsistent gas chemistry, changes in the ratio of erupted gases have been noted over time, so that they appear to shift between methane and methane-carbon dioxide types. The gases erupted by the Tsentralny mud volcano can serve as an example: in 1962, methane was the dominant component (94.81%), while in 1968, the methane concentration dropped to 56.5%, and the rest was made up of carbon dioxide (42.8%) with lesser components (0.7%) of other gases (Kurishko et al. 1968). A similar phenomenon was observed in the same years for the Karabetova Gora and Gora Gnilaya mud volcanoes, and the western gryphons of the Shugo mud volcano (Lagunova 1974). No regularities have been observed among the mud volcanoes showing different gas compositions that regard their geographic distribution within the structures of the Neogene-Quaternary structural floor. This can be related to the structural-tectonic peculiarities of the floor lying at considerable depth as well as the duration, activity, and intensity of mud volcanoes having different depths of origin. If differences in gas composition occur within mud volcanoes of the same field of distribution (e.g., the Malotrakhanskiy and Bulganakskiy hearths), this variability could be connected to the confinement of feeder channels (i.e., roots of mud volcanoes) to fault zones located at different depths.

The study of carbon isotope composition in gases is of particular importance in understanding isotope-geochemical features of the mud volcanic process. Such analyses have been performed on mud volcanic gases of the Kerch-Taman region (Gemp et al. 1970; Gemp and Lagunova 1978; Valyaev et al. 1980), and results have indicated that for methane, $\delta^{13}C$ varies from −3.33‰ to −4.84‰, corresponding to the values of gas in oil and gas fields. The ^{13}C of the carbon dioxide varies between −0.99 and + 1.17‰, most often between −0.57 and +0.74‰. This is close to the values of $\delta^{13}C$ from volcanic gases of the Kuril volcanoes (−0.71 to +0.38‰), and it is typical for thermal springs, gases within closed pores of igneous rocks, and volcanic gases (Gemp et al. 1970). Characteristically, the gases of neighboring volcanoes and even of different gryphons of one volcano can give a rather variegated picture of isotopic values.

The research conducted so far suggests that analyses of carbon isotopes from erupted gases could serve as a reliable indicator of methane and carbon dioxide origin points (Valyaev et al. 1980). Valyaev et al. determined the isotopic composition of carbon dioxide for 17 mud volcanoes. On the Taman Peninsula, the concentration of CO_2 in the composition of gases is 20–50%. The $\delta^{13}C$ of carbon dioxide varies from −6.5 to +14.6‰, and the $\delta^{13}C$ of methane reaches +67.3‰.

On the Kerch Peninsula, the concentration of CO_2 in the composition of gases is also 20–50%, yet in one case (Bolshetarkhanskiy mud volcano), it is 91.7% (Lagunova and Gemp 1978). The $\delta^{13}C$ of carbon dioxide varies between 10.1 and 8.5‰; for the Tarkhanskiy mud volcano, it is 8.5‰, while for the Tischenko mud volcano, it is 7.3‰. One of the conclusions of Lagunova and Gemp (1978) is that the carbon dioxide found in gases released by mud volcanoes possesses a complex nature—showing a deep-metamorphic and even a magmatic origin—meaning that it is coming from the Earth's interior.

Changes in the carbon dioxide content of spontaneously released gases may be due to neotectonic movements or an increase in seismic activity. The question of the nature of methane in the composition of mud volcanoes remains controversial.

Data on the isotopic character of gases from mud volcanoes of the Kerch-Taman region undoubtedly testify to the fact that the degassing products of the deep Earth vary in their composition. In general, the composition of mud volcano gases is determined by a number of factors, such as features of the tectonics of various structural floors, depth of the mud volcano roots, the activity of fault zones (including seismicity), the thickness of the sedimentary cover, the presence of oil and gas deposits at various stratigraphic levels, and the activities of these fluids.

Lately, researchers have been paying special attention to instability in the isotopic composition of gases, for example, in the gryphon gases of the Kipyaschy Bugor (southwestern Turkmenistan), Dashgil (the Shemakhino-Kobystan mud volcanic region of Azerbaijan), Shugo (the Western Kuban foredeep), and Bugazskiy (the Kerch-Taman foredeep) mud volcanoes. Features of their chemical and carbon isotope instabilities have been considered (Voytov 2001), and it has been demonstrated that this instability is related to the geodynamic seismic activity of the Alpine folding belt, as well as the mechanical mixing of the gases from individual strata that were exposed by mud volcano channels. The roots of the channels extend to great depths and are characterized by critical and supercritical temperatures (more than 373 °C), where apparently, the foci of tectonic earthquakes and deep fluidization of the interior are localized.

In May 2001, a paroxysmal explosion of the Karabetova Gora mud volcano occurred on the Taman Peninsula (Bogatikov et al. 2003). Geological surveys were conducted immediately afterward (according to eyewitness accounts, the paroxysm was accompanied by a strong roar, flames, and smoke and dust columns up to 100 m high). At the site of the explosion's center, a rounded array of mud breccias up to 500 m^2 and a volume of up to 800 m^3 together with spots of brick-red slags were present. Later, the emitted gases were sampled to determine their chemical and carbon isotope compositions. The studied gases were represented by mixtures of hydrocarbons, a series of methane and its homologues up to iso- and normal pentanes and hexanes, as well as nitrogen, carbon dioxide, helium, and in single samples, molecular hydrogen. Based on their helium content, the ejected gases were natural mixtures, the genesis of which could be connected with the complex of rocks of the conditional basement. This permitted the assumption that the feeder region of the gases is localized under the volcano in Paleozoic deposits or even in subcrustal depths.

Throughout 2001–2002, joint research by the staff of the Department of Marine Geology and Sedimentary Ore Formation of the National Academy of Sciences of Ukraine and the French Institute of Petroleum was conducted to study and monitor the mud volcanoes of the Kerch Peninsula. During the year, weekly gas sampling was carried out within the Bulganakskiy field (the Oldenburgskiy, Tishchenko, Andrusov, Obruchev, Pavlov, and Tsentralnoe Ozero mud volcanoes). The French Institute of Petroleum analyzed the gases, and the data obtained indicated once again that there was significant variability over time in the percentages of gas components. For each of the investigated mud volcanoes, a peculiar variation in the composition of their gases was detected. During certain active periods, either methane, carbon dioxide, or nitrogen was predominant.

According to Voytov (2001), the dynamics of variability in the composition of gases through time is even more complicated. He found that the percentage of gas constituents (CH_4, CO_2, N_2, etc.) in the samples varied at the sampling points over the course of days or even mere hours! In the opinion of Kurishko et al. (1968), estimation of gas chemistry must be conducted differentially, as composition is constantly changing, even within the same field.

Several important studies were carried out by Gutsalo and Plotnikov (1981) on the ratio of carbon isotopes in carbon dioxide-methane systems as an indicator of their origin. Within these systems, three different regions can be distinguished according to this ratio that reflect the main genetic varieties of natural gases from the Earth's crust: (1) region of variation in the isotopic composition of mantle methane and carbon dioxide, (2) region of variation in the isotopic composition of biogenic methane and carbon dioxide, and (3) region of variation in the isotopic composition of methane and carbon dioxide formed during thermometamorphism (pyrolysis) of organic matter in argillaceous shales and coal. The established values and limits in the $\delta^{13}C$ variations of methane and carbon dioxide in gases of the Kerch-Taman mud volcanoes lie between regions 1 and 2. This suggests that there is mixing of gases from two sources in the feeder channels of volcanoes.

The most recent study of the chemical composition and isotopic content of mud volcanic gases from the Kerch Peninsula was performed by Etiope et al. (2004) (Table 4.2).

The data obtained by this study confirm the previously noted general trends in the composition of gases.

The number of studies of erupted gases from offshore mud volcanoes is incomparably smaller than those of onshore mud volcanoes. Generalized data on offshore mud volcanoes have been provided by Lein and Ivanov (2018), while data for those onshore have been published by Stadnitskaya and Belenkaya (2007), Kruglyakova et al. (2009), Sahling et al. (2009), and some others.

Table 4.2 Chemical composition and isotopic content of mud volcanic gases from the Kerch Peninsula (Etiope et al. 2004)

Mud volcano	$\delta^{13}C$ CH_4, ‰	$\delta^{13}C$ CO_2, ‰	CH_4, %	CO_2, %
Armaelinskiy	−43.7	20.7	91.64	8.25
Tishchenko	−39.3	5.8	95.39	3.93
Tishchenko	−42.3	7	80.26	18.64
Bulganakskiy	−38.5	7	94.72	4.62
Obruchev (western)	−48.5	−10.1	83.18	13.4
Obruchev (central)	−50.2	0.3	35.13	64.22
Shilov	−46.2	−3.2	84.83	14.8
Travertine	−43.0	2.6	47.33	11.65
Soldatsko-Slobodskiy	−49.3	−19.4	96.72	1.72
Krasnopolskiy	−36.8	−14.7	97.99	0.86
Nasyrskiy	−53.6	−16.7	87.28	8.53

In the overwhelming majority of underwater mud volcanoes, methane dominates. Among the erupted gases of the Dvurechenskiy mud volcano, for example, the concentration of methane is 99.56%; for Vodyanitskiy, it is 99.77%; for Odessa, it is 99.1%; and for Heligoland, it is 99.78%. There are admixtures of carbon dioxide (0.04–0.38%) and traces of ethane, isobutane, n-butane, and n-pentane. In our opinion, as on land, the composition of the gases from offshore mud volcanoes changes quite noticeably over time. Data of this kind are not yet available, but it seems to us that similar findings are inevitable.

4.4 Water from Mud Volcanoes

Hydrochemical features of the volcanic waters of the Kerch-Taman mud volcanoes have been studied by many scientists over many years (e.g., Stopnevich 1920; Belousov and Yarotskiy 1936; Popov 1938; Sulin 1939; Al'bov 1956, 1971; Kurishko et al. 1968; Shternov 1971; Gemp et al. 1970; Gemp and Lagunova 1978; Nesterov 1979; Lagunova and Gemp 1978; Uchiteleva 1979).

The flow rate of mud volcanoes is generally small. Sulin (1939) considers that the water richness of mud volcanoes is determined mainly by the number of gryphons, their size, absolute elevation of the vent, the difference between the elevation of aquifers on the surface and the elevations of their contact with the volcanic breccia, the filtering capacity of breccia, and the climatic features of the area.

The flow rate is a function of the tectonic activity of the mud volcanic region as a whole and of individual parts within the region in particular.

The most complete analytical data on the hydrochemical characteristics of volcanic water, as well as its geological interpretation, have been provided by Kurishko et al. (1968) and Gemp and Lagunova (1978). Several analyses performed on

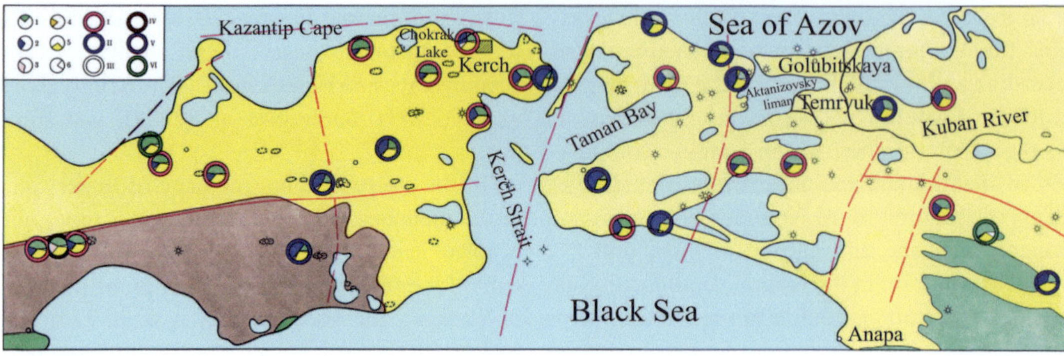

Fig. 4.19 Volcanic water of the Kerch-Taman mud volcanoes. Chemical content: (*1*) chlorine ion, (*2*) bicarbonate ion, (*3*) carbonate ion, (*4*) sulfate ion, (*5*) sodium ion, (*6*) other ions. Type: (*I*) chloride-hydrocarbon-sodium, (*II*) hydrocarbonate-chloride-sodium, (*III*) chloride-carbonate-sodium, (*IV*) sulfate-chloride-sodium, (*V*) chloride-hydrocarbonate-sulfate-sodium, (*VI*) chloride-sodium (Lagunova and Gemp 1978)

the volcanic water of onshore Kerch-Taman mud volcanoes during different years show many similarities in their hydrochemistry, distribution, and isotopic composition. In general, they reveal relatively low mineralization, ranging from 3.84 g/l (Borukh-Oba mud volcano) to 23.36 g/l (Bulganakskiy and Vernadskiy mud volcanoes). Their composition is rather uniform.

Three hydrodynamic zones can be distinguished (Lagunova and Gemp 1978): (1) the western zone (the southwestern part of the Kerch Peninsula), (2) the central zone (the northeastern part of the Kerch Peninsula and the western part of the Taman Peninsula), and (3) the eastern zone (the remainder of the Taman Peninsula). The distinctiveness of these zones is conditional, as different characteristics can quite often be observed within a single zone. Even volcanic water found in breccias and gryphons may display different characteristics (Fig. 4.19).

Cation/anion ratios suggest the existence of six types of volcanic water (see Fig. 4.19).

The first type, chloride-hydrocarbonate-sodium, is characteristic of volcanic water from the western (Soldatsko-Slobodskiy, Burashskiy, Bolshetarkhanskiy, Malotarkhanskiy, or Trubetskoy mud volcanoes) and part of the eastern (Blue Gulch and Shugo mud volcanoes) hydrodynamic zones (Fig. 4.19, I). For this type, a prevalence of chlorine over hydrogencarbonate is typical. Cations of K, Mg, and Sr are present permanently but in insignificant quantities (Table 4.3).

The second type (II), hydrogencarbonate-chloride-sodium, is typical of mud volcanoes in the central zone (Table 4.3).

The third type (III), chloride-carbonate-sodium, is recorded only for the western Hyrtsyz-Shiban mud volcano (Table 4.3).

The fourth type (IV), sulfate-chloride-sodium, is found in the Shilov (Malo-Tarkhanskiy) mud volcano (Table 4.3).

The fifth type (V), chloride-hydrogencarbonate-sulfate-sodium, has been detected within the Bulganakskiy mud volcano (Table 4.3).

The sixth type (VI), sodium-chloride, is known only from the westernmost Kamenskiy mud volcano and the eastern Gladkovskiy mud volcano. It typically shows an exceptionally high chlorine content and a minor admixture of the hydrogencarbonate ion.

It is very likely that the waters of mud volcanoes in the region possess deep origins. First, Cs dominates over Rb. Rb occurs only during the process of magmatic differentiation. Otherwise, the migration capacity of Cs is much lower compared to that of Rb during rock leaching and the attendant high temperatures and pressure. Such conditions are inherent for sodium chloride flowing solutions (Khitrov and Kolonin 1962). Second, there is a similarity in the distribution of both rare elements (Cs and Rb) in the volcanic

Table 4.3 Content of selected chemical elements and compounds in the water ejected by Kerch-Taman mud volcanoes (Lagunova and Gemp 1978)

Type	Chemical element/compound	Min, mg/l	Max, mg/l	Mud volcano
I	Cl	3894		Soldatsko-Slobodskiy
			8350	Shugo
	CHO^{3-}	1506		Korolevskiy
			4409	Shugo
	B	18.26	634.0	Trubetskoy
	Li	1.0		Soldatsko-Slobodskiy
			19.4	Korolevskiy
	As (sporadically)		0.05	Soldatsko-Slobodskiy Korolevskiy Trubetskoy
	Hg	1×10^{-3}	1.5×10^{-3}	Soldatsko-Slobodskiy Korolevskiy Trubetskoy
	CO_2		115.8	Vladislavovskiy
	H_2S	7.82		Vladislavovskiy
			9.12	Eastern Hyrtsyz-Shiban
II	CHO^{3-}	4026.0		Borukh-Oba
			8956	Bugazskiy
	Cl	1310.0		Borukh-Oba
			5896.83	Vernadskiy
	Na	2725		Borukh-Oba
			4342	Gnilaya Gora
	Ca		100	Borukh-Oba, Gnilaya Gora
	Mg		632.55 (exception)	Vernadskiy
	Li	1.4		Karabetova Gora
			9.0	Tsentralnoe Ozero
	Hg		1×10^{-3}	Borukh-Oba
			2.5×10^{-3} (exception)	Oldenburgskiy
	B	298.0		Enikalskiy
			632.0	Karabetova Gora
	H_2S	7.48		Borukh-Oba
			7.8	Vernadskiy
III	Cl		6108.55	Western Hyrtsyz-Shiban
	Na		5094.58	Western Hyrtsyz-Shiban
IV	SO_4^{2-}		2839.4	Shilov
	Cl		2702.15	
	HCO^{3-}		244.0	
	CO_3^{-2}		360	
	Na		3021.45	
	B		398.72	
	Li		1.0	
	Hg	Not detected		
	As	Not detected		
	H_2S		1.14	
V	Cl		7801.16	Andrusov
	HCO^{3-}		5448	
	SO_4^{2-}		1939.0	
	Na		7191.0	

(continued)

Table 4.3 (continued)

Type	Chemical element/compound	Min, mg/l	Max, mg/l	Mud volcano
	Ca	Insignificant		
	Mg	Insignificant		
	B		928.0	
	Li		2.4	
	As		0.02	
	Hg		1×10^{-3}	
	H_2S		2.7	
VI	Cl		12170.0	Gladkovskiy
	HCO^{3-}		270.0	
	Na		6095.0	
	Ca		1297.0	
	Mg		282.0	
	K		210.0	
	B		250.82	
	Li		14.4	
	Hg		1×10^{-3}	
	Rb	0.36	4.0	Obruchev Trubetskoy Karabetova Gora
	Cs	0.36	4.0	
	P		7.0	
	F		1.0	

and thermal water of volcanic regions (Gemp and Lagunova 1978). Third, the high content of boron, which sometimes exceeds 1000 mg/l during the most active period of a mud volcano's activities (Shnyukov et al. 1992), is similar to volcanic water of the Alpine folding regions (Lagunova and Gemp 1978). Fourth, the closeness of the chemical composition of the water to productive oil- and gas-bearing structures speaks in favor of a similar genesis. Fifth, elevated concentrations of B, Li, and Hg indicate an additional source of these elements that could come from great depths along the faults (Kurishko and Sivan 1971).

There is an alternative opinion that mud volcanic water represents a product of the biochemical transformation of the elusive waters from Maikopian deposits that penetrate upward into the overlying sedimentary rocks. This would mean that the appearance of carbon dioxide in them is evidence of active thermometamorphic processes occurring in the interior, for example, the Indolo-Kuban Trough (Nesterov 1979).

Volcanic water strongly affects mud breccia. Its influence can be readily seen in the wide distribution of kaolinite, which can present itself either as a uniform fine impregnation, or in various forms of carbonatized zones in which breccia is enriched by veins of kaolinite. The breccia is usually heavily crystallized, such as in the Kayala-Sartskiy mud volcano field (Shnyukov et al. 1971). The veins of kaolinite formed under the influence of a carbonate solution circulating during the development of the breccia at great depth within the mud volcanic foci. Kaolinized zones engulf the carbonatized sections and transgress their boundaries, involving both the cement and debris of the breccia. The degree of carbonatization can vary; sometimes, the brecciated structure of the original rock is preserved. With continuous carbonatization, the rock acquires a microgranular structure, and only minerals in the aleuritic size range remain unchanged.

The $\delta^{18}O$ values of volcanic water fluctuate with respect to the SMOW (Standard Mean Ocean Water) standard, ranging from 3.8 (Baksinskiy mud volcano) to 6.0 (Bulganakskiy mud volcano). This shows their similarity to oil and fumarolic water (Shnyukov et al. 1986).

Unfortunately, the study of volcanic water in offshore mud volcanoes is not yet possible.

4.5 Mud Volcanic Breccia

Along with mud volcanic gases and waters, the main product of mud volcanic activity is mud breccia—a specific rock that has no analogues among sediments arising from other well-known geological processes. Mud breccia is a type of rock formed by mud volcanic eruptions. Proceeding from the generally accepted classification of rocks, the term mud breccia is to a certain extent conditional. In fact, the classification of mud breccia is itself controversial (Arkhangel'skiy 1925; Shatskiy et al. 1929; Avdusin 1939, 1948; Khain 1953; Khain et al. 1953; Shnyukov 1999, 2000; Shnyukov et al. 2006a, 2013). Most researchers take tectonic breccia (or tectonites) as their primary referent, from which mud volcanic breccia is derivative; however, while the former is produced by brittle deformation, often in fault zones, the latter is formed through layer-by-layer rubbing, disharmonic crushing, the bulging actions caused by diapiric processes, and finally reworking by water and gases erupted from mud volcanoes. Mud breccia is often found flamed and burned (Figs. 4.20, 4.21, and 4.22).

Breccia formed in the vents of mud volcanoes, known as vent breccia, is usually represented by argillaceous pelite. The stratified sequence of fossil mud breccia forms specific compensatory and subsidence structures (Shnyukov et al. 2013).

Initial sediments that produce mud volcanic breccia are themselves created through the processes of layer-by-layer grinding, disharmonic crumpling, and diapiric bulging and then reworked in the course of their eruption as they are mobilized by water and gas. Collapse breccia develops from the collapse of rock overlying an opening, which thereby introduces fragments of rock of various lithological types into the breccia. Ventral facies, as a rule, are characterized by argillaceous pelites, and stratified mud volcanic breccia includes colossal deposits of fossil breccias within mud volcanic compensatory structures, called recessed synclines.

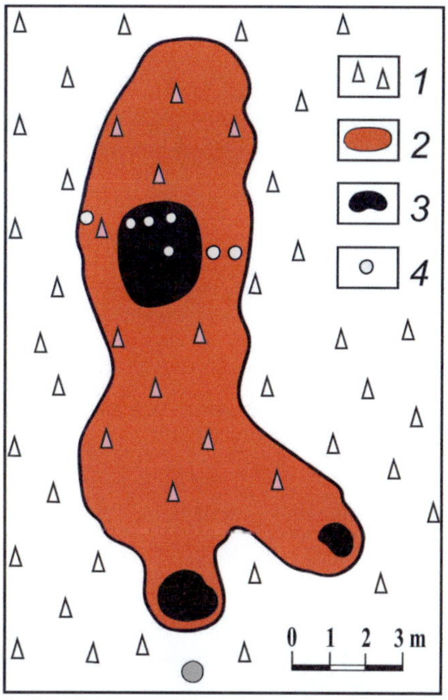

Fig. 4.20 Schema of the development of mud breccias for the Karabetova Gora mud volcano, 1982. (*1*) Unaltered clay breccia, (*2*) red burned breccia, (*3*) black slag-like mud breccia, (*4*) sampling points

Fig. 4.21 Field of burned mud volcanic breccias on the Karabetova Gora mud volcano (photographed in 2003)

Lithologically, mud volcanic breccia is a clastic rock consisting predominantly of argillaceous sediments cemented by clay; it is often porous. Proceeding from the formal and generally accepted classifications, the term mud volcanic breccia is to a certain extent conditional, since it includes not only fragments of pebble-gravel

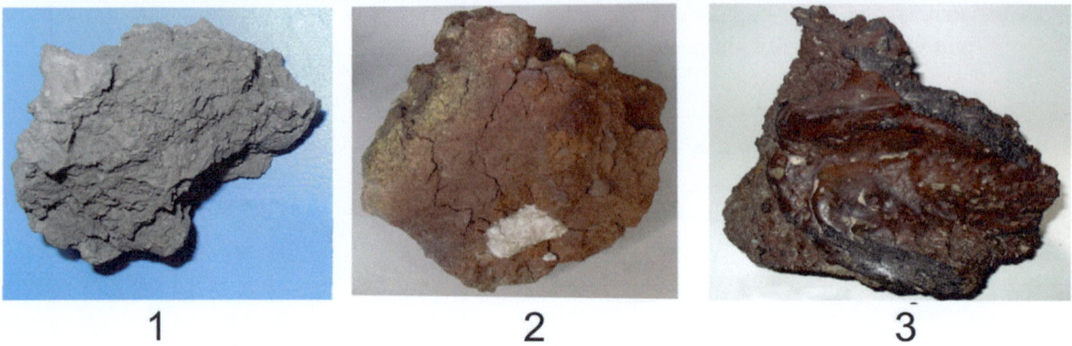

Fig. 4.22 Different stages of burned mud volcanic breccia. (*1*) Initial stage, (*2*) brick-red stage (900–950 °C), (*3*) black, slag-like stage (above 1050 °C)

dimensions but also gravel, sand, and silty fragments of varying degrees of roundness, from angular to well-rounded. Therefore, among the co-deposits based on particle size, there are pelites, silts, sandstones (sand), gravelites, breccias, and conglobreccia.

For example, the mud breccias of the Bulganakskiy, Dzhau-Tepe, and Burul'kayskiy mud volcanoes are all largely composed of argillaceous pelites, rock debris, and gravelstones, respectively. These lithologic types are represented by normal sedimentary rocks that were erupted and then processed and redeposited in an aquatic environment enriched with gases. The spatial and vertical distribution of mud breccia, even within a single mud volcanic field, is complex and heterogeneous. This distribution can best be illustrated by a 100 m geological section through Chokrak-Karagan sediments that was recovered by drilling into a marginal part of the Korolyovskiy mud volcano (Fig. 4.23).

In this section, the alternation of different lithological sediment types has been recorded with conglomerate breccia on the bottom and sandy-pelite breccia at the top. The sequence was formed by rocks collapsing into the localized water body of a compensatory syncline, which then filled with diluted mud breccia. Under an active hydrodynamic regime, with time-varying intensity, these deposits were repeatedly washed and redeposited. The presence of glauconite sands without fragments of clays and other rocks corresponds to a period of attenuation in the mud volcano's activity and the accumulation of normal sedimentary rocks. Similar interstratification of the Sarmatian normal sedimentary clays with mud breccia is recorded in geological sections of the Kayala-Sarta compensatory syncline on the Kerch Peninsula.

Fresh mud volcanic breccia is usually black or gray, brecciated, or homogeneous clay. Sometimes, the unusual bluish-greenish-gray of strongly carbonatized breccia can be observed, for example, in the Novoselovskaya, Kayaly-Sartskaya, and Burulkayskaya recessed synclines of the Kerch Peninsula. Along with carbonatization, which causes discoloring of the mud breccia, the processes of kaolinization are widespread. Zones of kaolinization engulf the carbonatized places and extend beyond them.

The relationship between unchanged and changed (or discolored) breccia is complicated. Sometimes, a gradual transition from changed bluish-greenish-gray and unchanged dark gray and black varieties can be observed, although the structural and textural features of the rocks remain homogenous throughout the entire section. At the same time, however, fragments of these breccia types can intrude into each other, especially near their contacts. In other cases, an alternation of unchanged and changed breccia layers can be observed. The former is usually composed of pelitic particles that often have a stream-flaked structure, constituted by randomly oriented flakes of clay minerals. Individual stripes and micro-lenses of clays are distinguishable from the main mass of sediments by saturated coloration. The main pelite mass contains

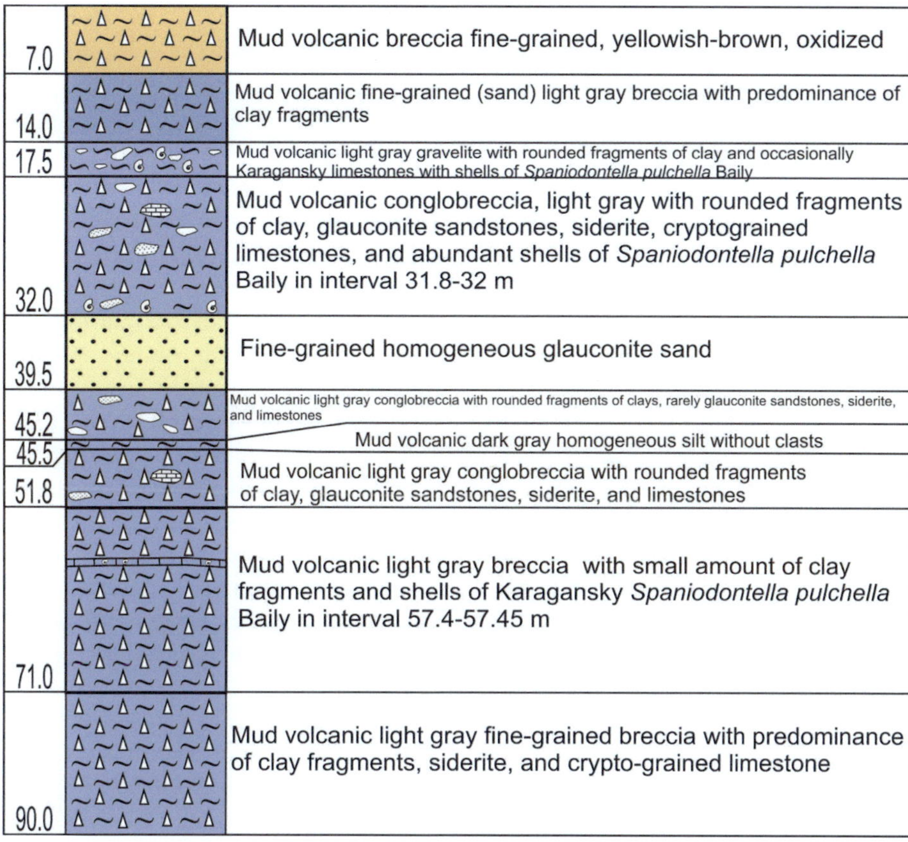

Fig. 4.23 Geological section from the Korolyovskiy mud volcano

an insignificant (up to 1%) admixture of acute-angled fragments of quartz, ore minerals, or less often muscovite, plagioclase, and microcline—all of aleuritic size. A small amount (up to 10% of the rock volume) of semi-rounded isometric fragments of differing clays, between 0.5–1.5 mm in size, tends to be submerged within the bulk rock, being distributed irregularly. In both composition and structure, these fragments are indistinguishable from unchanged breccia but can sometimes contain more silty material. Varying shades of gray (up to 5–10%) also tend to be present within unchanged breccias; they are close to black ones but differ from them in their large content of silty material.

Carbonatization processes tend to develop in equal measure for unchanged and changed breccias. There are two types of carbonatization: solid and vein. The former is the most common and happens when the entire breccia mass (debris and cement) undergoes carbonatization. Carbonate develops in the form of clusters of microgranular masses, as well as single fine (0.01 mm) grains. The degree of carbonatization tends to be uneven. The primary breccia-like texture of the rock is preserved, and the rock acquires a micrognanular structure. Only minerals of aleuritic size remain unchanged.

Sometimes larger (0.05–0.1 mm) pellets develop among the pelitomorphic carbonate. They are transparent and are not dusty with pelite particles. Their shape is tapered or sometimes rhombic. In some cases, small areas of recrystallization of a more coarse-grained carbonate can be observed, and small concentric-to-radial spherulites are sometimes seen. Inside, they are composed of pelitomorphic aggregates of carbonate, while outside they are surrounded by a thin single-crystal shell. Between the two extreme cases described above, intermediate cases are also observed, which can range from single carbonate grains to continuous carbonate aggregates that completely obscure the primary brecciated rock texture. Sometimes in thin sections of carbonatized breccias, small foraminiferal shells may be present.

The vein type of breccia is less widely distributed, usually developing along cracks. The original textural-structural breccia pattern is often preserved. As a rule, the veins are formed in the clayey cement of the breccia without affecting the material of the rock fragments.

The processes of kaolinization are expressed in the formation of an abundant white powdery coating and in the development of veins of kaolinite. The latter is represented by an aggregate of tiny grains. Occasionally, kaolinite fills the interior of foraminiferal tests, while their walls are composed of carbonate. Based on the character of precipitation, the kaolinite formed seems to be secondary in nature. Almost ideal hexahedral crystals of kaolinite exclude an allochthonous genesis, suggesting that they were formed in situ.

Carbonatized and kaolinitized mud volcanic breccias tend to contain some rare mineralogical findings, such as yarosite, realgar, and pyrite. Yarosite in the form of yellow powder was discovered in the Novoselovskiy mud volcano; the powder contained fine gypsum crystals along the cracks of the changed breccia (Borehole No. 64, depth 11.0–11.2 m, 45° 14′ 58.32″ N; 36° 06′ 09.72″ E). The smallest orange-red grains of realgar were also present (Borehole No. 60, depth 80 m, 45° 14′ 55.22″ N; 36° 06′ 11.36″ E), as well as small (up to 1.0–1.5 cm) joints and intergrowths of crystals (Boreholes No. 60, depth 62, 64 m), as well as infrequent, thin veins of pyrite (Borehole No. 60, depth 203.4 and 208.0 m). In one of the boreholes, the pyrite crystals were coated with very small deposited particles of realgar.

The heating curve of the bluish-gray carbonatized breccia is similar to that of a mixture of calcite and a high-temperature modification of alumina in the proportion of 25% $CaCO_3$ and 75% Al_2O_3 (Cuthbert and Rowland 1947).

The carbonate mineral in the changed breccias is calcite, which has been confirmed by chemical analyses of both types of breccia. The contents of the main chemical components in both unchanged and changed breccia are presented in Table 4.4, which shows that, in general, the chemical contents are higher in unchanged breccia, except for CO_2.

The unchanged gray breccia from Novoselovskiy mud volcano contained Mn (0.06%), Ni (0.06%), Co (0.003%), Ti (0.3%), V (0.02%), Cr (0.01%), Zr (0.02%), Cu (0.003%), Pb (0.003%), Ga (0.001%), Be (0.0001%), Sc (0.001%), La (0.006%), Y (0.003%), Sr (0.03%), and Yb (0.0003%).

The changed breccia contained the same amount of Ni, Ti, V, Cr, Zr, Cu, Ga, Sc, Y, and Yb, but less Mn (0.01%) and more Pb (0.02%), Co (0.005%), and Sr (0.1%), and Ba (0.03%) also appears. In the white powdery kaolinite coating, the content of Cr (0.05%) and Pb (0.004%)

Table 4.4 Chemical content of unchanged[a] and changed[b] breccia (%)

Chemical component	Unchanged	Changed	Measured component	Unchanged	Changed
SiO_2	57.04	53.35	Na_2O	0.88	0.55
TiO_2	0.72	0.62	K_2O	2.19	2.27
$A1_2O_3$	14.77	10.74	P_2O_5	0.18	0.10
Fe_2O_3	3.14	2.77	H_2O hygroscopic	2.09	1.53
FeO	4.31	2.87	Loss on ignition	5.52	3.34
MnO	0.12	0.08	S	0.93	0.51
MgO	2.66	2.37	CO_2	3.0	9.69
CaO	1.68	9.94	C	1.68	–

[a]Unchanged breccia of gray color (Borehole No. 64, depth. 103.0 m; Novoselovskiy mud volcano)
[b]Changed breccia of greenish-bluish color, carbonatized and kaolinitized (Borehole No. 64, depth 107.5–108.0 m, Novoselovskiy mud volcano)

increased, while the content of Mn (0.01%), Ni (0.001%), Ti (0.1%), Zr (0.001%), Cu (0.001%), and Sr (0.03%) decreased. At the same time, As (0.0002%) appeared, while Co, Be, Sc, La, Y, Ba, and Yb disappeared. The content of Cr and Pb in the white powdery coating exceeded their Clarke value. Compared to the Clarke value, the content of V was elevated in both the unchanged and changed breccia; the content of Sr and Pb was elevated in the unchanged and changed breccia, respectively (Vinogradov 1962).

Changed breccia is confined mainly to areas of tectonic disturbances. The processes of kaolinization and subsequent carbonatization develop in local geological sections, rather than gravitating toward the deep zones of the mud volcanic strata; they do not manifest themselves in the upper horizons.

There are gradual transitions between the bluish-greenish changed and dark gray or black unchanged breccia. Sometimes fragments of changed breccia can be present in unchanged breccia (and vice versa). Sometimes interbedding of changed and unchanged breccia can be observed.

In the case of post-mudflow volcanic transformations, there would likely not be any inclusions of bluish-greenish breccia in the unchanged varieties. It appears that the solutions causing these changes come from mud volcanic foci along certain tectonically weakened zones, and this would be due to the clays acting as impermeable rocks, leaving tectonic disturbances to serve as the easiest way for the solutions to circulate.

Judging by the nature of these changes, the circulating water would have to be saturated with CO_2, which is common for Kerch mud volcanoes. At present, travertines and the powdery carbonate coating found on rocks are widely known in places with modern seepage of mud volcanic waters. Under the influence of carbonate waters, changes would occur during the initial composition of the clay breccia. The essence of this process appears to be saturation of the rocks with calcite and removal of alumina and silicic acid from the clay, which are the main constituents of the accompanying carbonatization of kaolinite. That is why kaolinization develops more widely than carbonatization.

The noted complex of changes—carbonatization and kaolinization—is quite remarkable. It is similar in many respects to the phenomena recorded in the mercury deposits of Krasnodar Territory (dikkitization, carbonatization) as well as the mercury deposits of Transcarpathia (kaolinization, carbonatization).

4.6 Clastic Material

The autochthonous components of the clastic material and cement found in mud volcanic breccia are represented by clays of the Maikopian (late Paleogene to early Neogene), as well as the Paleogene, Cretaceous, and occasionally Jurassic ages—with the Maikopian being predominant. Allochthonous clastic material makes up no more than 3–8% of all erupted breccia.

The clastic material found in mud volcanic breccia has been studied by many investigators (e.g., Steber 1912; Klepinin 1914; Prokopov 1931; Grechishkin 1936; Belousov and Yarotskiy 1936; Popov 1938; Avdusin 1948). General works exploring this matter for the Kerch Peninsula were published by Shnyukov and Kulichenko (1969), Shnyukov et al. (1971), Nesterovskiy and Titova (2012), and Titova et al. (2013) and for the Taman Peninsula by Shardanov et al. (1962). The results of our research, together with the published data, form the basis of this chapter.

The size of the debris present in mud volcanic breccia varies greatly, although very large pieces show up much less frequently. For example, a piece of Chokrak limestone with a volume of up to 1.3 m^3 was found in the Bulganakskiy mud volcano, a clump of glauconite quartzite sandstone that measured up to 2.5 m in size was found within the Voskhodskiy mud volcano, and large siderite nodules (<50 cm) were discovered within the Bulganakskiy, Nasyrskiy, and Karabetova Gora mud volcanoes. Karabetova Gora also contained huge (1–2 m^3) blocks of limestone that were observed during its last eruption. Large erupted clasts tend to be very diverse and include mainly conglomerates, conglobreccia, gravelites, sandstones, siltstones, limestones, dolomites, and siderites. Sporadically, they contain fragments of phosphorites, iron ores, marls, and in very rare cases, igneous rocks. They have been found within the boundaries of the Bulganakskiy and Malo-Tarkhanskiy mud volcanic hearth as well as within the Novoselovskiy, Shugo, and Karabetova Gora mud volcanoes. They differ in the composition of their detrital material and cementitious matrix, and this difference is primarily determined by the lithological content of the rocks erupted by the mud volcano. Thus, among the coarse clastic rocks of the Kerch mud volcanic breccia, Miocene and Pliocene limestones typically predominate. They are represented by organogenic-detrital, reef-like (formed by Bryzoa), oolite-pseudo-oolitic rocks as well as siderite, glauconite, quartzite sandstones (silicites) and less frequently by phosphorites. The cement is basal; it comprises mainly calcite and occasionally clay-hydroghetite.

The coarse clastic fragments found in mud volcanic breccia are mainly represented by siderites, recrystallized limestones, siliceous rocks, and occasionally by basalts, quartz porphyries, quartzite-like glauconite sandstones, and phosphorites. The sandy component of the fragments includes feldspars, pyroxenes, and amphiboles, and the cement of these rocks is clay-hydroghetite. In the hydroghetite mass, relicts of fine-grained pyrite are often observed. A peculiar bone breccia (a deposit of fragments of bones of vertebrates mixed with sediments) cemented by calcite has been encountered within the material erupted by the Shugo mud volcano. The composition and age of clastic material are determined by its proximity to the Greater Caucasus, where the Lower Cretaceous rocks erupted by mud volcanoes lie at shallow depths.

The contents of mud volcanic breccia can be exemplified by the largest Bulganakskiy mud volcano. Its breccia is composed of sandstone (40%), limestone (35%), marl (10%), clay siderite (7–9%), and siltstone (5–7%) (Nesterovskiy and Titova 2012).

Lithologically, sandstones are divided into two types: oligomictic quartz-glauconite and polymictic. Oligomictic sandstones are found within the Bulganakskiy, Voskhodovskiy, Malo-Tarkhanskiy, Ak-Tube, Dzhau-Tepe, Bolshe-Tarkhanskiy, and Korolevskiy (Kerch Peninsula) mud volcanoes as well as most of the studied volcanoes of the Taman Peninsula. They are dense, sometimes thin-layered rocks. Clastic material is represented mainly by quartz and glauconite, which comprise from 15% to 30% of the rock. Redeposited and authigenic glauconite are clearly present. Cement consists of regenerated quartz and sometimes clay-glauconite. Often, these sandstones are pyritized. Under surface conditions, pyrite is oxidized, turning into hydroghetite, which impregnates the primary cement. In individual cases, single rhombohedral crystals of secondary calcite and dolomite have been observed.

Polymictic sandstones have been encountered only within the Bulganakskiy and Novoselovskiy mud volcanic breccia. As a rule, constituent sandstones are thinly bedded with calcite, and a clay-siderite cement of basal, contact-pore type, or sometimes substitution type. In the sandstones of the Novoselovskiy mud volcano, cement revealing a fontainebleau structure—poikilitic structure, in which the cement is represented by large (spathic) crystals of calcite or gypsum with numerous sand particles that are typical of sandstone—is present. The sandstones of the Ak-Tube mud volcano possess poikilitic secondary gypsum cement. The clastic material in this sandstone is represented by quartz, plagioclase, clay fragments, clayey siderites, siliceous rocks, muscovite, and chlorite. Glauconite has also been found to occur sporadically in small amounts. The glauconite-bearing sandstones of the Taman Peninsula have developed in strata dating from the Early Cretaceous to the Eocene, but not younger than Oligocene (Shardanov et al. 1962).

According to Shnyukov et al. (1971), the isotopic ages of the described sandstones are 32–42 Ma (Early Maikopian) for the Kerch samples and 120 Ma (Late Jurassic to Early Cretaceous) for the Taman samples. These dates are in good agreement with geological data.

A significant amount of debris from glauconite-bearing sandstones is typically linked to the mud volcanoes of the Vneparpachskoy zone located within the Bulganakskiy and Malo-Tarkhanskiy hearths, as well as the Voskhodovskiy and Korolyovskiy mud volcanoes. At the same time, within the mud volcanoes of the Southwestern Plain, where Maikopian deposits have developed, these rocks are found in the form of individual fragments.

Carbonate rocks are present in virtually all mud volcanoes in the form of debris within the breccia.

Thermographic study of this carbonate rock debris has shown the presence of limestone, dolomitic limestone, and lime-dolomitic marl in their composition (Shnyukov et al. 1971).

Carbonate rocks are extremely diverse in their lithological features. They are represented by organogenic-detrital, organogenic-clastic, reef-like crypto-grained, clayey crypto-grained, pseudo-oolite, partially or completely recrystallized, and other varieties.

This diversity is defined not just by the lithologic composition of the rocks erupted by mud volcanoes but also by the facies from which they emerge.

Based on the constituent rock-forming organisms, there are foraminiferal, foraminiferal-pelecypod, foraminiferal-gastropod, bryozoan, stromatolitic, and other varieties of carbonate rocks. Carbonate rocks often contain sand, silt, and clay admixtures, evenly distributed or in the form of thin layers. Lamination is sometimes due to the segregation of biogenic material. Cryptograined carbonate rocks have been formed mostly by chemogenesis, and because of this, identifying their age is problematic, as everything is mixed.

Stratigraphically, the carbonate rock debris is of Late Miocene to Pliocene (Tarkhanian, Chokrakian, Karaganian, Sarmatian, Meotian, and Pontian) age. For the Taman mud volcanoes, clastic material of Upper (Turonian, Coniacian) and even Lower Cretaceous age is present (Shardanov et al. 1962).

Recrystallized marble-like limestones have been found in the mud volcanic breccia of some mud volcanoes of the Taman Peninsula (Shugo, Gladkovskiy, and Karabetova Gora). They may belong to Upper Jurassic formations.

Marl-based debris is extremely rare in mud volcanic breccia. The bluish-gray marl of the Dzau-Tepe mud volcano contains Early Eocene fauna, while similar marl of the Ak-Tube mud volcanoes has been found to contain poorly preserved fauna, presumably of Upper Cretaceous date.

Siderite is widespread in the breccia of a majority of mud volcanoes. Its abundance is approximately the same within the Southwestern Plain—where the Maikopian series is widely distributed—and the Vneparpachskaya area of the Kerch Peninsula.

In the overwhelming majority of cases, siderite forms nodular concretions with a dense cryptogranular structure; less often, fragments of slabby siderite are present. The mineralogical and chemical features of the concretions have enabled them to be dated as late as the Maikopian (Shnyukov et al. 1968). Slabby siderites have an even greater age (as in the Shugo, Akhtanizovskiy, and Eastern Tsymbaly mud volcanoes). The two varieties of siderite possess a different chemistry. The content of MnO_2 in the concretion ranges between 1.06% (Voskhodovskiy mud volcano) and 3.19% (Enikalskiy mud volcano), while in the slabby variety, it varies from 5.06% (Eastern Tsymbaly mud volcano) to 9% (Akhtanizovskiy mud volcano).

Carbonate bundles with cone-shaped structures were found in the debris of the Dzhau-Tepe, Ak-Tube, Nasyrskiy, Bulganakskiy, Bolshoy-Tarkhanskiy, Malo-Tarkhanskiy, Shugo, and Gladkovskiy mud volcanoes. Usually, the size of the bundles does not exceed a few centimeters. Within the Dzau-Tepe mud volcano, large cones are present in that large carbonate cones aggregate in cone-like formations and can be found in sizes of up to several dozen centimeters in length; they are supposedly of Early Cretaceous age.

Fragments of igneous rocks found in mud volcanic breccia are extremely rare, and the interpretation of their genesis is ambiguous and highly contradictory. For example, fragments of quartz porphyries and porphyrites encountered in discharges of the Shugo and Gladkovskiy mud volcanoes are considered to be coarsely fragmented elements of the conglomerates from Cretaceous formations lying at depth (Avdusin 1948). In Avdusin's opinion, the rounded and angular fragments of olivine andesite and olivine dolerite found within volcanic mud breccia of the Shopurskiy mud volcano were collected by people from Phanagoria, a town of the Bosporan Kingdom located in the immediate vicinity of the mud volcano. Shardanov et al. (1962) encountered pebbles of effusive rocks as well as fragments of trachydacite (Karabetova Gora mud volcano) and a black mesh-type rock of obsidian (Sopka mud volcano) among the conglobreccia of the Early Cretaceous age (Dzhiginskiy mud volcano).

We discovered fragments of greenish-gray fine-grained rocks in the breccia of the Semigorskiy mud volcano that were likely previously attributed to sandstones. They range in size from 1 to 3 cm and are sometimes slightly rounded. Their thin sections demonstrate heterogeneity duc to the presence of larger inclusions of white crystalline aggregates in the fine-grained matrix. Detailed study of these fragments by A.A. Yushin (Shnyukov et al. 2015) revealed that, in most cases, the white inclusions are xenomorphic, possibly representing altered crystallites of feldspars. But in sample No. 3 (Table 4.5), the white inclusions are represented by spherulites of a siliceous substance, and the structure of the rock matrix bears obvious traces of metasomatic transformations, including also sulfide mineralization.

Structural and textural differences in rock fragments are also reflected in their chemical composition (Table 4.6).

Some fragments are low in SiO_2 but high in Fe, Mg, Ti, Al, Ca, Ni, Ti, Cr, and Zn. This corresponds to basic rocks. An increase in SiO_2, together with a noticeable decrease in Fe, Ni, Cr, and a slight increase in Zn, indicate metasomatic processing of the original mafic rocks. The low levels of Pb in these rocks indicate minimal accessory mineralization of Pb in the matrix of the mud volcanic breccia. It appears that the fragments of fine-grained greenish-gray rocks could have formed due to the brecciation of rocks of basic composition that were altered to varying degrees (probably tuffs). The determination of their initial geological composition requires more detailed studies.

We also found in the solid ejecta of the Shugo and Gladkovskiy mud volcanoes angular fragments of hyalobasalts, basaltic porphyrites, chloritized diabases, and oligoclase that most likely belong to coarse clastic sediments of the Cretaceous age.

Even taking all of this information into consideration, we do not exclude the possibility that mud volcanoes emerging at considerable depths broke through volcanogenic dike rocks that were older than the Cretaceous age. Therefore, the age of some erupted igneous rocks may be Jurassic and perhaps even older. On the other hand, a local manifestation of young Neogene volcanism should not be excluded as well.

Complex studies of clay minerals (thermographic, X-ray, etc.) that appear as fragmentary clay clumps and clayey cement in mud volcanic breccia have revealed a hydromica-montmorillonite composition, sometimes with a slight admixture of kaolinite and metagalloisite. The composition of clay minerals is quite stable spatially and suggests that similar formation conditions exist across a vast territory.

Peculiar fragments have been found in the solid ejecta of the Karabetova Gora, Gora Gorelaya, and Azovskoe Peklo mud volcanoes. They are represented by dense, dark gray lithified breccia, which in all likelihood are ancient, and were ejected by these mud volcanoes from strata

Table 4.5 Chemical composition of fragments of chloritized rocks in the breccia of the Semigorskiy mud volcano; results from X-ray fluorescence analysis; analyst V.V. Zagorodny

Content, g/t															
Sample no.	SiO_2	TiO_2	$Al2O_3$	Fe_2O_3	MnO	MgO	CaO	Na_2O	K_2O	P_2O_5	S	Cl	H_2O-	Loss on ignition	Σ
1	24.83	1.63	16.40	24.45	0.06	9.28	5.10	0.15	0.12	0.16	0.03	0.17	0.29	10.71	93.37
3	44.69	1.54	13.81	10.60	0.06	8.56	5.10	2.62	0.12	0.29	0.20	0.16	0.29	7.18	95.21

Table 4.6 Elements-admixtures in the fragments of chloritized rocks in breccia of the Semigorskiy mud volcano; results from X-ray fluorescence analysis; analyst A.V. Andreev

	Content, g/t													
Sample no.	Mn	Ni	Ti	V	Cr	Mo	Cu	Pb	Zn	As	Zr	Y	Sr	Ba
1	4416	222	>1	–	478	3	23	<2	148	5	141	25	94	53
3	2626	85	>1	–	5	5	23	1	210	13	99	25	91	86

Notes:
Clarke values: Nb, Bi, Sn, Ge, Ga, Sc, Be, Yb, Li, Ag, Co
Below detection limit: W, Ta, Hf, Sb, Cd, As, Th, Ce, La

of considerable depth. Study of these fragments in thin sections has revealed that they consist of fine-grained siderite, in some parts recrystallized, with fragments of clays and sandstones. In this case, the clay cement of the breccia as well as some clay fragments are completely replaced by siderite.

Solid fragments found in the breccias of marine mud volcanoes have received far less study. This is due to the difficulty of obtaining such material, as well as the relatively shorter time that has been spent studying offshore mud volcanoes compared with onshore ones. In the breccia of offshore mud volcanoes, fragments of limestone, marl, and shale are often found, but detailed studies have usually not been carried out. Only the MSU mud volcano contained both angular and rounded sandstone, siltstone, and Late Eocene limestone, while no other volcano has revealed this (Konyukhov et al. 1990). A piece of Chokrak limestone (15 × 20 cm) and a small fragment of travertine have been identified by Shuraev (2016).

In the Malyshev mud volcano, dolomite from the Late Miocene was recorded (Konyukhov et al. 1990). Debris from Maikopian argillites and Pliocene limestones was observed in the Dvurechenskiy and Bezymyannyy mud volcanos (Sorokin Trough), respectively. Limestones of uncertain age have been found in the Tbilisi mud volcano. Debris of gray calcareous clay was observed in the Dolgovskiy mud volcano. In the volcanoes of the Tuapse Trough, Sarmatian foraminifera were discovered (Konyukhov et al. 1990). An SEM and microprobe study of solid rock microdebris (with particle sizes ranging from a few μm to a few mm) shows as a first approximation the presence of magnetite quartzite (?), hard coal, limestone, shale with sulfides, ultrabasic tuff (?), and graphite (?). In SEM images, elongated grains of magnetite interspersed among quartz grains have been observed (Fig. 4.24) (Table 4.7).

Fig. 4.24 Supposed magnetite quartzite (MSU mud volcano) in reflected light. Numbers 1 (magnetite) and 2 (quartz) indicate places sampled, and their chemical spectrum is represented in Table 4.7 as spectrum 1 and 2

On the Ukrainian Shield, similar magnetite quartzites of Precambrian age were encountered during the drilling of the Zelenovskaya Magnetic Anomaly (one of the right bank anomalies north of the Krivoy Rog iron ore basin and the Azov Sea shore). They were also found in breccia from the offshore MGU mud volcano that is located in close proximity to the magnetic anomaly in the Black Sea. Its assumed age is Precambrian.

The fact that fragments of coal were found in the MGU, Yuzhmorgeo, and Tredmar mud volcanoes is very interesting. The size of the fragments ranges up to 1 × 2 mm, and the grains are angular in shape (Fig. 4.25).

Table 4.7 X-ray fluorescent spectra of magnetite quartzite

Spectrum 1 (magnetite)			Spectrum 2 (quartz)		
Element	Weight, %	Atomic, %	Element	Weight, %	Atomic, %
O	29.72	59.36	O	55.03	68.89
Si	0.76	0.86	Si	42.27	30.15
Fe	69.52	39.78	Fe	2.70	0.97
Total	100.00		Total	100.00	

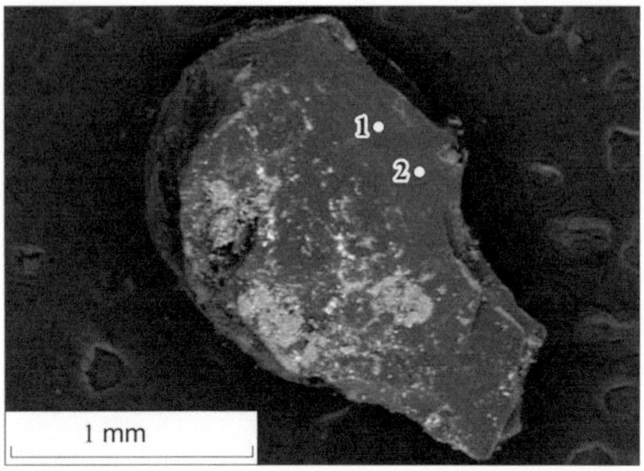

Fig. 4.25 Fragment of coal with impregnation of sulfides (Tredmar mud volcano). Numbers 1 and 2 indicate places sampled and represented in Table 4.8 as spectrum 1 and 2

Their chemical content is shown in Table 4.8.

The area of coal development closest to the Tredmar mud volcano is Beshuyskie Kopi in Crimea; the coal found there is of Middle Jurassic age. Less exposed coals of the Middle Jurassic are also known in the area of Balaklava. Perhaps, Jurassic coal-bearing strata have developed under the bottom of the Black Sea.

The chemistry of coals has not yet been compared, but sulfur has been fixed both in Beshuyskiy coal (up to 2–3%) and in coal debris from the Tredmar mud volcano (0.2–2.49%).

Fluids brought upward to the surface apparently broke through Middle Jurassic layers and carried fine clastic material, including coal fragments. Jet, which is a dense variety of hard coal, has already been found in Quaternary sediments on the shelf near Sevastopol, though it originates in deposits of the Middle Jurassic (Shnyukov and Shchiptsov 1986). Tugolesov et al. (1985) noted the presence of Middle Jurassic deposits on the Andrusov Ridge, i.e., near mud volcanoes. The accidental introduction of coal into mud volcanic breccia as a result of anthropogenic activities is unlikely.

The least informative discovery was a fragment of carbonaceous limestone. Establishing at least its approximate age is not yet possible. Other clastic fragments found in mud volcanic breccia are largely controversial, and as such they carry too much uncertainty to be considered here.

4.7 Accessory Mineralization

Accessory minerals are often present in mud volcanic breccia of the Black Sea region. In reference books, accessory (from the Latin "accessorius") minerals are treated as constituent minerals that are present within rocks in small quantities (>1%); they are useful for establishing qualitative characteristics of the rock as well as determining the conditions of mineral formation (Small Soviet Encyclopedia, v. 1, p. 212). The Geological Dictionary (Paffengolts 1973) interprets accessory minerals as insignificant parts of a rock, but they are important in many respects. By origin, accessory minerals can be allogenic and authigenic. The former are widely used for the determination of drift-feeding provinces and for the stratigraphic correlation of geological sections. The latter are significant indicators of sedimentary conditions and their subsequent changes.

Formally, accessory minerals include not only terrigenous clastics but newly forming minerals developed in the mud volcanic breccias due to chemical reactions under fluid currents. These minerals will be described in Chap. 8.

In lithological manuals (Frolov 1992), the content of accessory minerals present in a rock is insignificant (1–2%). It must be noted that all minerals can be accessory, but only 60 of them

Table 4.8 X-ray fluorescent spectrum of coal from Fig. 4.25

Spectrum 1			Spectrum 2		
Element	Weight, %	Atomic, %	Element	Weight, %	Atomic, %
C	92.29	94.19	C	89.67	92.18
O	7.46	5.71	O	9.90	7.64
S	0.25	0.10	Al	0.09	0.04
Total	100.00		Si	0.13	0.06
			S	0.21	0.08
			Total	100.00	

are most commonly found, and these include gold, platinum, and silver. A number of these minerals come from igneous rocks (e.g., pyroxenes, olivine, amphiboles, biotite, magnetite, ilmenite, sphene, rutile, anatase, bruxite, leucoxene, zircon, apatite, tourmaline, garnet, chromite, picotite, monazite, cassiterite, xenotime, spinel, etc.). Others derive from metamorphic rocks (e.g., amphiboles, magnetite, chlorites, epidote, zoisite, kyanite, staurolite, sillimanite, andalusite, garnets, corundum, chloritoid, muscovite, biotite, topaz, etc.).

The first mineralogical study of mud volcanic breccia was conducted by Abich (1840). Swedish geologist Hjalmar Sjögren (1888) studied the muddy clays, feldspars, quartz, amphiboles, calcareous spar, and pyrite present in breccia from the Lok-Batan mud volcano (Azerbaijan). Gümbel (1889) conducted a microscopic investigation of breccia from Italian mud volcanoes and discovered diatoms, radiolarians, as well as softened claycy shales that had underlain the mud volcanoes. He did not find the real products of volcanic activities, such as ash and lava. In his research, undertaken in 1909–1913, Steber (1913) proved that mud volcanic breccia contains thick rubbing debris saturated with water. He found quartz, orthoclase, olivine, chlorite, and calcite in the Karabetova Gora mud volcano. In general, there has been insufficient study of the mineralogy of mud volcanic breccia from the Kerch-Taman and Azerbaijan mud volcanoes.

Avdusin (1939) found 70 minerals within the solid fractions of mud volcanic ejecta, and he divided them into three groups (A, B, and C) according their genesis (Table 4.9).

The first group (A) includes minerals representing the products of mechanical disintegration of sedimentary rocks. These include feldspars, quartz, amphiboles, pyroxenes, ilmenite, beidellite, halloysite, mica, etc. It is of interest that approximately the same composition of minerals was observed by us in the mud volcanoes of the Black Sea region. The second group (B) includes minerals associated with the processes occurring in the breccia. The third group (C) includes minerals from the thermal metamorphism of breccia, formed when exposed to high temperatures caused by the hot gases.

Mud volcanic breccia differs in its accessory mineral composition based upon the underlying geology, e.g., whether the mud volcano is rooted in Maikopian or Cretaceous strata (Shnyukov 2016).

Breccia erupted from mud volcanoes deriving from Maikopian strata (3–5 km thickness) is represented by clays. Their Maikopian age is supported by a complex of benthic, e.g., *Tschokrakella caucasica* (Bogdanovich), *Bolivina budensis* (Hantken) (dominant), *Chiloguembelina cubensis* (Palmer) (subdominant), *Guembelina gracillima* Andrae, *Nodosaria spinescens*, *Robulus* aff. *crassus* (d'Orbigny), *Cibicides amphisyliensis* Andrae, and *Uvigerinella majcopica* Kraeva, as well as planktonic, e.g., *Globigerina brevispira* Subbotina and *Globorotalia* cf. *hexacamerata* Subbotina, foraminifera, often with pyritized tests. The Maikopian strata cover huge areas, including the southeastern part of Ukraine, northern and eastern Crimea, the North Caucasus, and the Black Sea.

Breccia is represented by a crushed clay clastic material that was carried upward by powerful hydrocarbon fluids. These fluids broke through the Paleozoic-Mesozoic deposits and penetrated into the tectonically crumpled strata along cracks.

Table 4.9 Minerals from mud volcanic breccia

A		B	C
Relic minerals from underlying sedimentary strata		Minerals of mud volcanoes	Minerals of thermal metamorphism
Quartz	Serpentine	Calcium carbonates	Alite $3CaO \times SiO_2 + CaO \times Al_2O_3$
Feldspars	Tremolite	Dolomite	Belite $2CaO \times Al_2O_3$
Analcime	Actinolite	Siderite	Lime-soda feldspars
Beidellite	Common hornblende	Sulfur	Alpha iron (ά-Fe)
Clay minerals	Glaucophane	Pyrite	Amorphous silica
Opal	Augite	Marcasite	Calcium oxide
Chalcedony	Diopside	Hydrotroilite $FeS \bullet nH_2O$	Brown and green glass
Volcanic glass	Epidote	Gypsum	
Coal formations	Zoisite	Anhydrite	
Garnet	Disthene (Kyanite)	Celestine	
Rutile	Sillimanite	Jarosite $(K, Na)_2Fe_6(OH)_{12}(SO_4)_4$	
Brown spinel	Andalusite	Saltpeter	
Zircon	Titanite	Halite	
Anatase	Glauconite	Chalcopyrite	
Brookite	Olivine	Botallackite $CuCl_2 \times 3Cu(OH)_3$	
Apatite	Staurolite	Aragonite	
Collophane	Tourmaline	Anthracolite	
Delite	Ilmenite	Thermonatrite	
Colorless mica	Leucoxene	Epsomite	
Biotite	Magnetite	Carbonate	
Green Mica	Red iron oxides	Borax	

This explains why breccia contains abundant foraminifera, as well as accessory minerals typical of Maikopian sediments. Accessory mineralization in Maikopian strata is insignificant in content but varies significantly regionally.

On the Kerch Peninsula, Maikopian sequences contain aluminosilicates, kyanite, sillimanite, and staurolite among the terrigenous minerals, as well as garnet, zircon, magnetite, and ilmenite among the heavy fraction (Pashchenko 1960). There are also interlayers of fine- and medium-grained siltstones enriched with quartz, clayey minerals (e.g., montmorillonite), glauconite (up to 10%), oligoclase (up to 10%), rutile, ilmenite, leucoxene, zircon, tourmaline, andalusite, and hornblende, occasionally barite, and in minor quantities, native aluminum, red lead, and malachite that derive from the coastal zone of the Sea of Azov (Skorik and Bayrakov 2007). The bulk of terrigenous mineral ejecta also derive from areas adjacent to the Sea of Azov.

Oligocene deposits of the northwestern Black Sea corresponding in age to part of the Maikopian strata were studied by Tsikhotskaya et al. (1986) and Klyushina (2006). Their work was based on the study of 35 boreholes recovered from the Arkhangel'skiy, Golitsynskiy, Southern Golitsynskiy, Krymskiy, and Olimpiyskiy underwater uplifts. The following minerals have been discovered in a light fraction of clays interlayered with sandstones: quartz (up to 80%), feldspars (albite, orthoclase, microcline), muscovite, flint fragments, graphite, and chabazite. In the heavy fractions, the most common minerals are ordinary hornblende (mostly of aleuritic size), kyanite, ilmenite, tourmaline, leucoxene, garnet, zircon, rutile, epidote, sphene, and apatite. Other minerals were found less often, and these include biotite, clinozoisite, anatase, sillimanite, staurolite hematite, rhombic pyroxenes, chromite, spinel, corundum, and single grains of moissonite. Authigenic minerals include pyrite, siderite, calcite, glauconite, silica minerals, limonite, colophane, and zeolite. Clay minerals include montmorillonite, hydromica, kaolinite, and chlorite.

The mineralogical systematics of the terrigenous deposits of the Cenozoic and Mesozoic in the northern Caucasus and Ciscaucasia was established by Grossgeim (1961). He focused most attention on the accessory mineralization of the Oligocene deposits of Ciscaucasia. According to his data, the region contained the same set of terrigenous minerals as the Kerch Peninsula (Table. 4.10).

Pyrite, siderite, barite, celestite, glauconite, and phosphates in the heavy fraction as well as opal, chalcedony, and clay minerals in the light fraction most likely have an authigenic origin.

Our study of mud volcanic breccia enables us to identify approximately the same terrigenous minerals typical for Maikopian strata in various regions. As noted by Skorik and Bayrakov (2007), we also did not find native aluminum, red lead, and some silicates, but it is not necessary in this case to exclude any minerals from the list of those possibly present since some are rare due to being occasionally discharged from mud volcanoes.

The black ore minerals are widely distributed (Grossgeim 1961). Most of them are well-rounded, although the general outline of crystals is often preserved, indicating their terrigenous genesis. At the same time, grains of completely unabraded minerals are also present, which suggests a fluidogenic origin (Figs. 4.26, 4.27, and 4.28).

Zircon and rutile grains are very clearly rounded (Figs. 4.29 and 4.30).

Garnets are represented by several varieties, e.g., andradite and almandine-spessartine. The latter contains up to 16% MnO (Fig. 4.31).

One sharp-angled fragment of andradite was observed in breccia from the Andrusov mud volcano. There are also slightly rounded pyroxenes and clinopyroxene with inclusions of titanomagnetite (Fig. 4.32).

Amphiboles are usually well-rounded (Fig. 4.33), but sometimes they retain facet features (Fig. 4.34).

Table 4.10 Maximum concentration of accessory minerals in the Lower, Middle, and Upper Oligocene sediments of the Ciscaucasian terrigenous-mineralogical province (Grossgeim 1961)

Fraction	Minerals	Ciscaucasian terrigenous-mineralogical province	
		Lower Oligocene (max %)	Middle and Upper Oligocene (max %)
Heavy	Black ore (magnetite, ilmenite, chromite)	32	51
	Zircon	11	5.6
	Garnet	11.2	4.5
	Tourmaline	10.7	9.2
	Rutile	7.5	3.8
	Leucoxene	30.7	28
	Sphene	4.6	2.5
	Anatase	<1	<1
	Brookite	<1	<1
	Pyroxenes	2	<1
	Amphiboles	<1	<1–12
	Disthene (kyanite)	<1	1
	Staurolite	<1	2
	Andalusite	–	–
	Sillimanite	–	–
	Mica	27	30
	Chloritoid	6.6	<1
	Epidote group	15	27
	Chromium spinels	<1	62
	Apatite	<1	35
	Corundum	<1	–
	Perovskite	–	–
	Pyrite	100	100
	Limonite	4	34.4
	Siderite	78	95
	Barite	<1	–
	Celestine	–	–
	Glauconite	32	5
	Phosphates	–	Minor
	Quartz	83	78
Light	Fragments of rock and weathered minerals	17	45–98
	Mica	+	+
	Glauconite	+	45–90
	Volcanic glass	–	<1
	Opal	–	–
	Chalcedony	–	–
	Clay minerals	–	–

Fig. 4.26 Unabraded grain of magnetite from the Soldatsko-Slobodskiy mud volcano

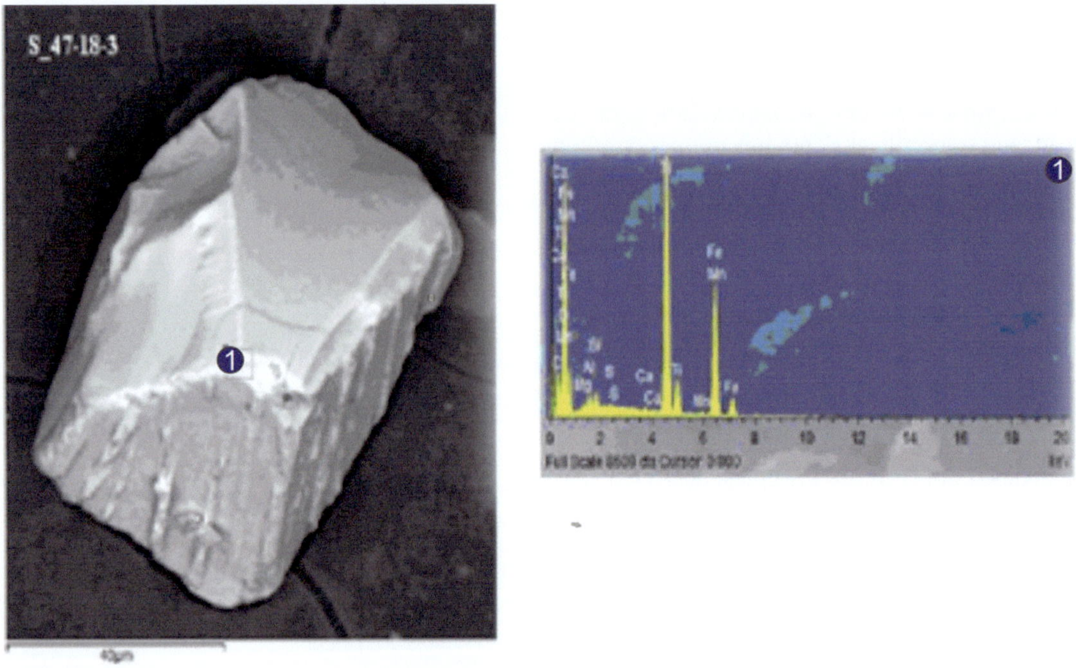

Fig. 4.27 Unabraded grain of ilmenite and its geochemical spectrum from the Korolyovskiy mud volcano

Fig. 4.28 Unabraded grain of ferriferous chromite and its geochemical spectrum from the Shilov mud volcano

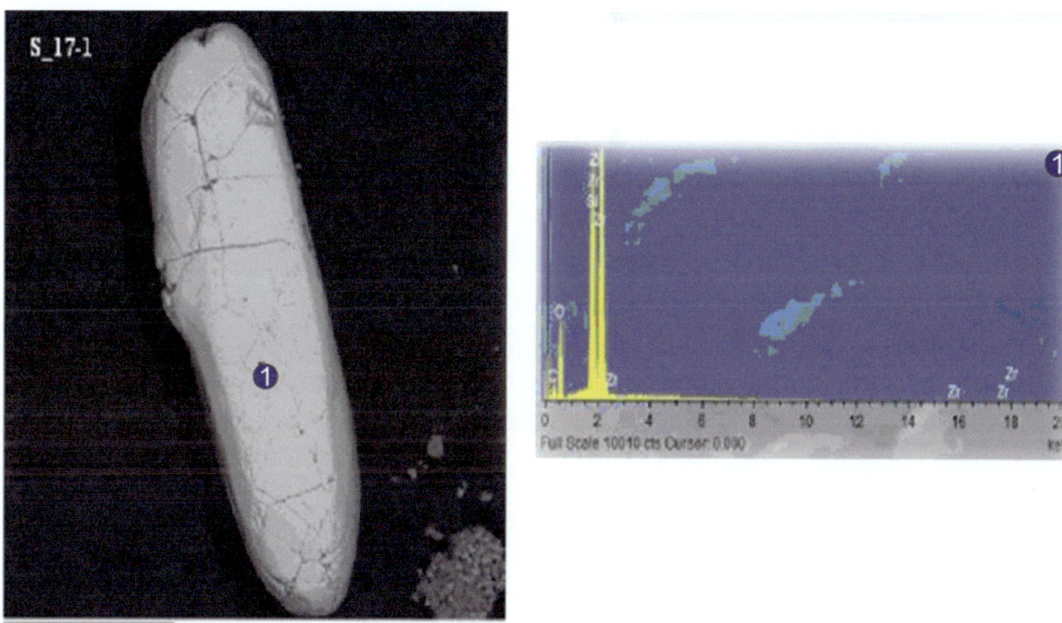

Fig. 4.29 Rounded grain of zircon and its geochemical spectrum from the Trubetskoy mud volcano

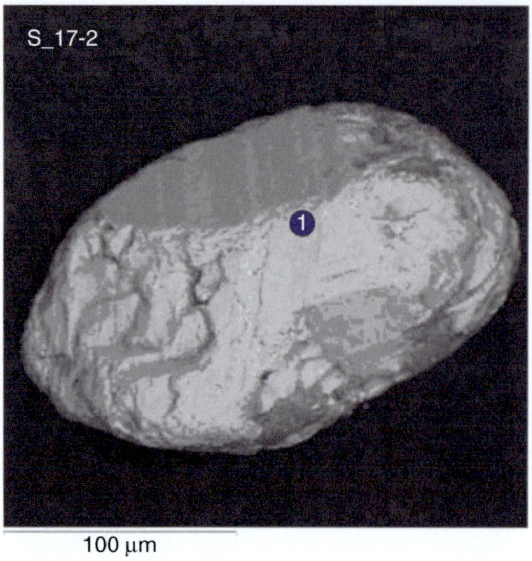

Fig. 4.30 Semirounded grain of rutile from breccia of the Andrusov mud volcano; only the upper side in the image preserves an undamaged crystal facet

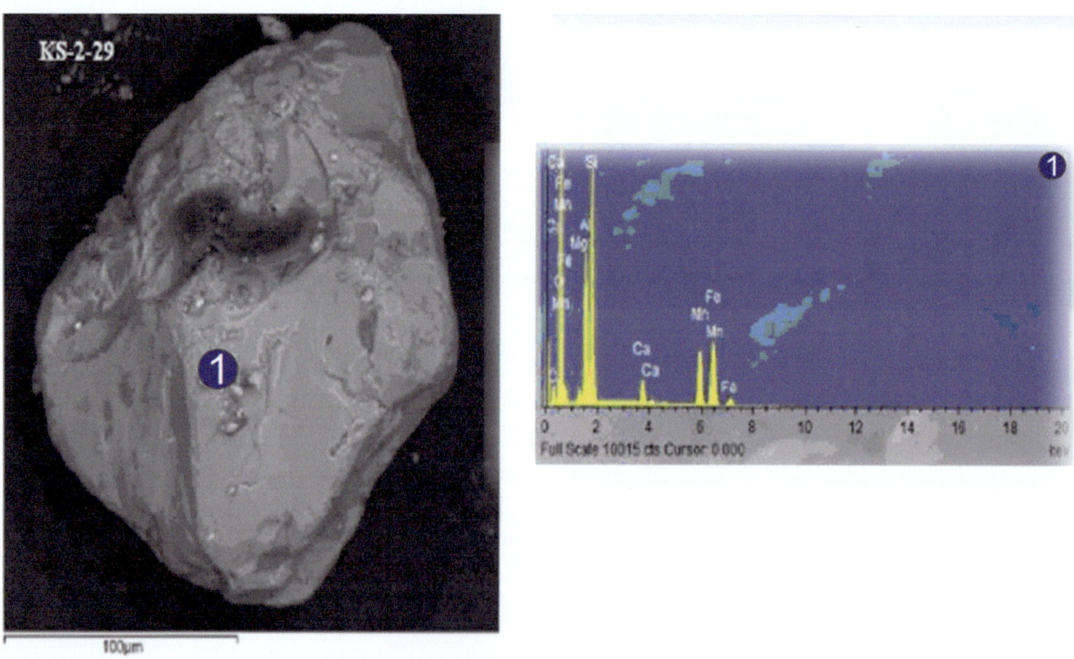

Fig. 4.31 Rounded grain of almandine-spessartine with principal elements assayed at 1 from breccia of the Andrusov mud volcano

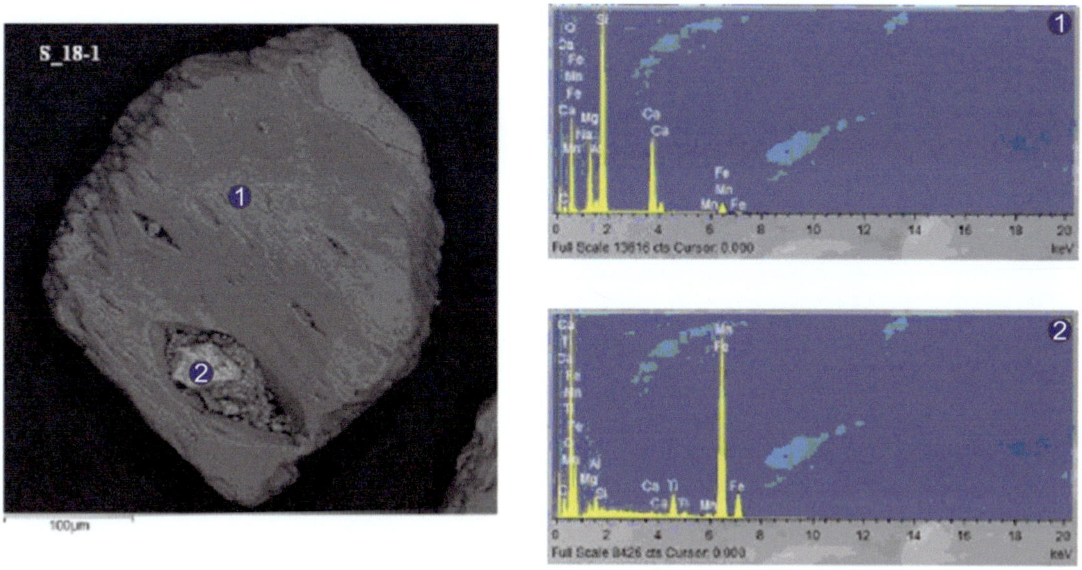

Fig. 4.32 Rounded grain of clinopyroxene with inclusions of titanomagnetite and with principal elements assayed at *1* and *2* from the Andrusov mud volcano

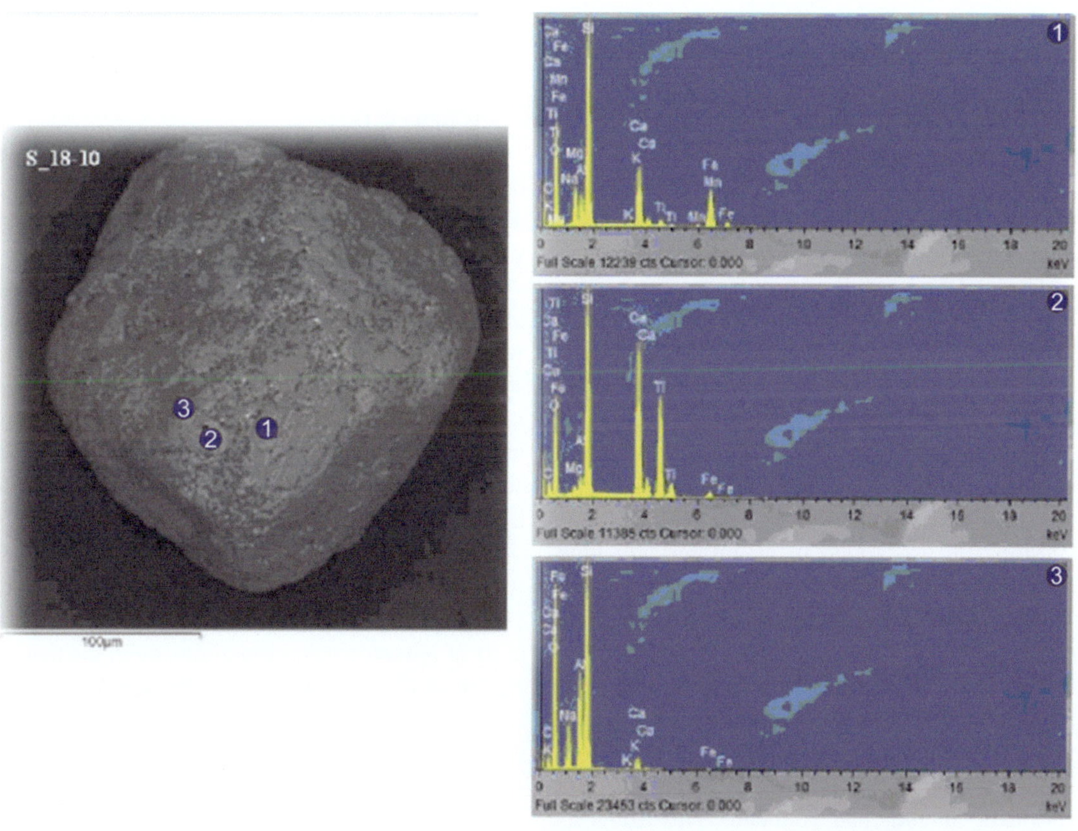

Fig. 4.33 Well-rounded grain of amphibole with principal elements assayed at *1*-amphibole, *2*- titanite, and *3*-plagioclase (top to bottom) from the Korolyovskiy mud volcano

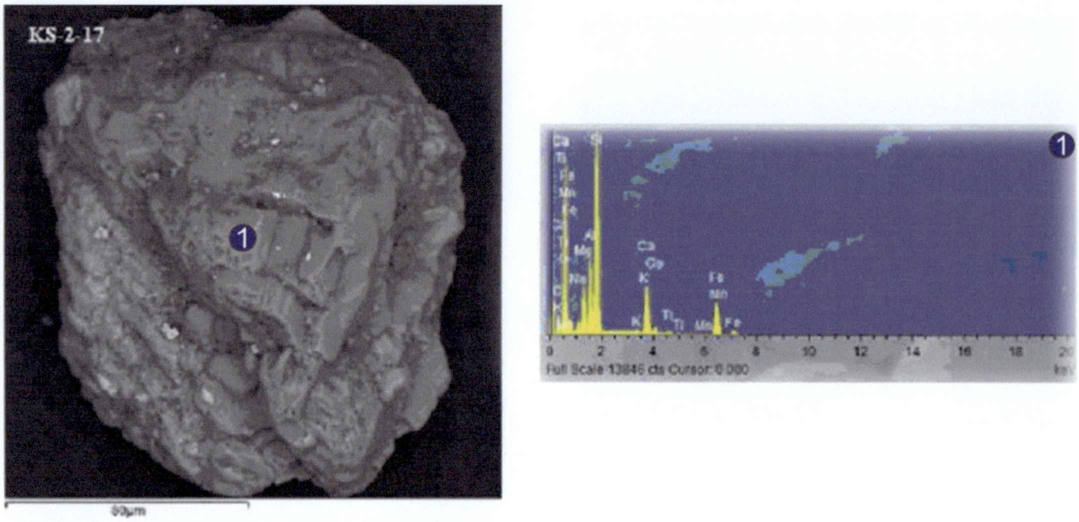

Fig. 4.34 Amphibole grain that retains facet features with principal elements assayed at 1 from breccia of the Andrusov mud volcano

A high degree of roundness is typical for andalusite (Fig. 4.35), sillimanite (Fig. 4.36), and staurolite (Fig. 4.37).

As a rule, epidote grains are well rounded (Fig. 4.38), although partly rounded individuals can also be present.

Glauconite, which is newly formed in the sedimentary process, forms fairly typical small nodules (Fig. 4.39) that indicate its growth within mud volcanoes.

Chalcedony as well as tiny fragments of clay minerals and some other rocks are also present.

The origin of phosphates is controversial. Some of them (e.g., apatite) have been found in various mud volcanoes (Fig. 4.40).

Apatite can be represented by an intergrowth of crystals (Fig. 4.41) as well by chamfered crystals (Fig. 4.42).

Monazite was found in an intergrowth with chlorite (Fig. 4.43).

Accessory minerals from mud volcanoes emerging from Lower Cretaceous strata have rather peculiar terrigenous-mineralogical features in both heavy and light fractions. These minerals are often overgrown with iron hydroxides and carbonate crusts, and they are pyritized, corroded, altered, and rounded. This suggests that mineral-forming processes continue throughout the duration of mud volcanic eruptions.

The widest area of mud volcano distribution is located within the Western Kuban (North Caucasus) terrigenous-mineralogical province (Grossgeim 1961) where the Lower Cretaceous transgression reached its maximum during Albian time (Table 4.11).

This region is characterized by a high content of stable minerals, especially garnet and zircon, as well as a relative diversity of mineral composition and polymicticity within the light fraction.

Terrigenous minerals listed in Table 4.11 are also present in other Lower Cretaceous stages, as well as in Jurassic deposits. In addition to them, some other minerals (e.g., topaz, dumortierite, vesuvianite, melilite, etc.) are found in Lias sediments, while sphalerite is not found in Malm sediments, but rather in a different terrigenous-mineralogical province, known as Baksanskaya.

In summary, the list of accessory minerals found in breccia that was compiled by Grossgeim (1961) can be divided into three groups: (1) minerals of igneous rocks (found mostly in the heavy fraction), (2) minerals formed by fluid processes (found mostly in the light fraction), and (3) minerals formed in breccia under conditions of weathering.

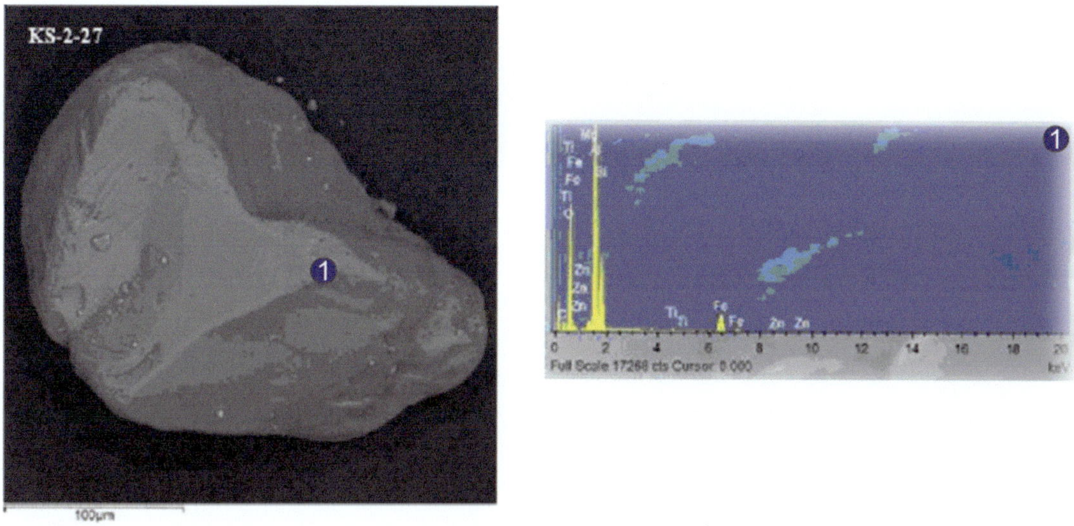

Fig. 4.35 Well-rounded andalusite grain with principal elements assayed at *1* and *2* from breccia of the Shilov mud volcano

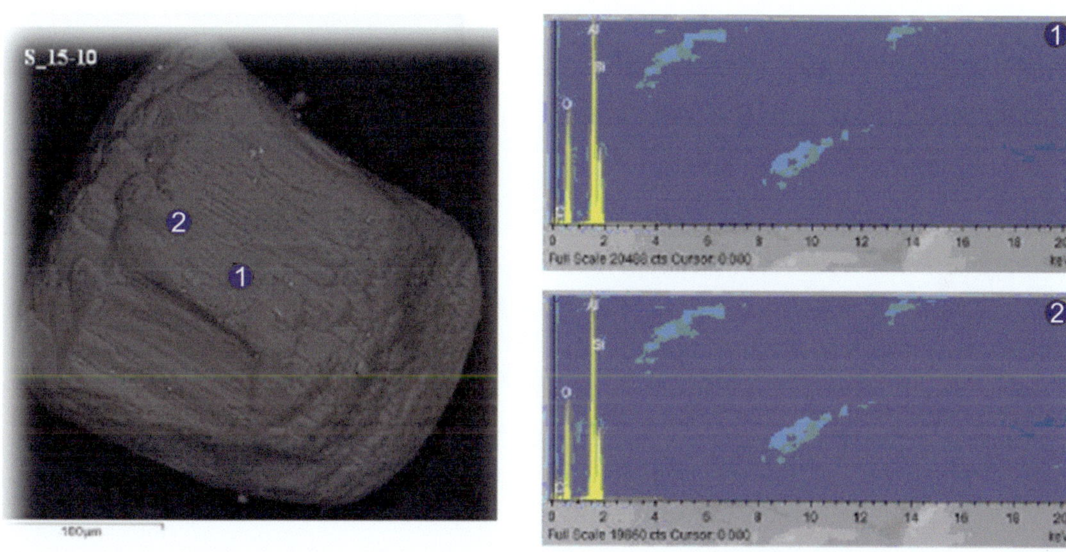

Fig. 4.36 Well-rounded sillimanite grain with principal elements assayed at 1 and 2 from breccia of the Trubetskoy mud volcano

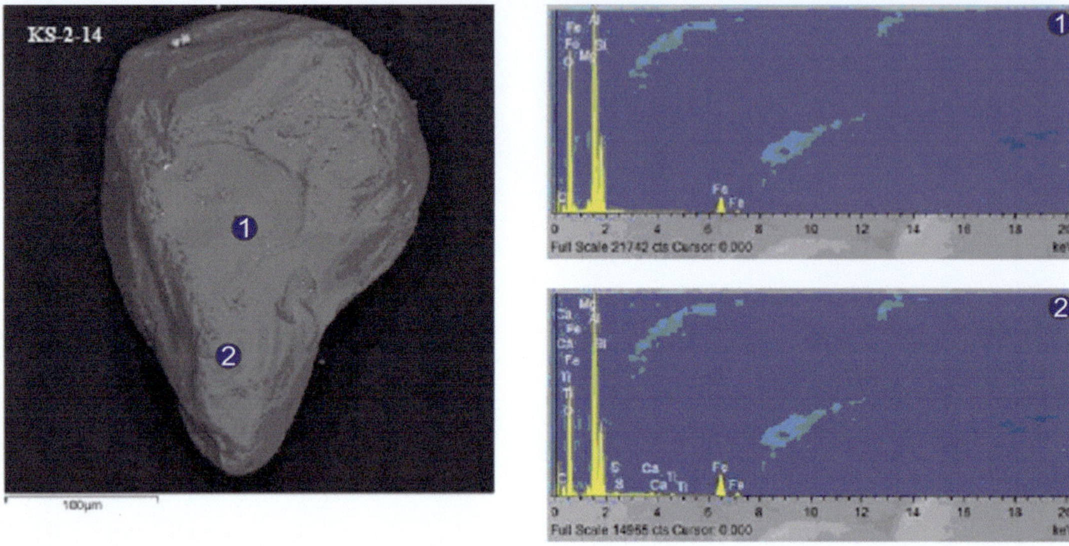

Fig. 4.37 Well-rounded staurolite grain with principal elements assayed at *1* and *2* from breccia of the Trubetskoy mud volcano

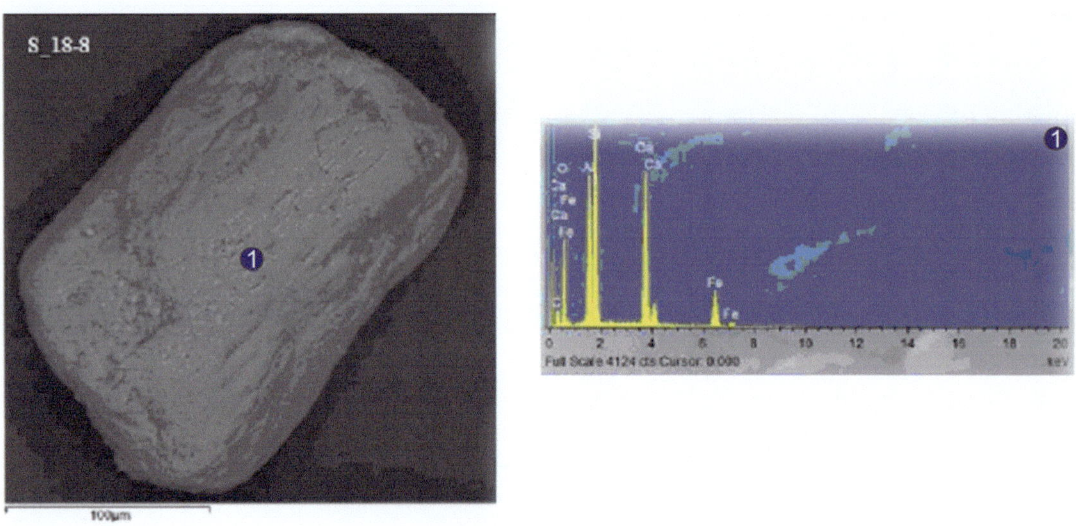

Fig. 4.38 Well-rounded grain of epidote with principal elements assayed at *1* and *2* from breccia of the Korolyovskiy mud volcano

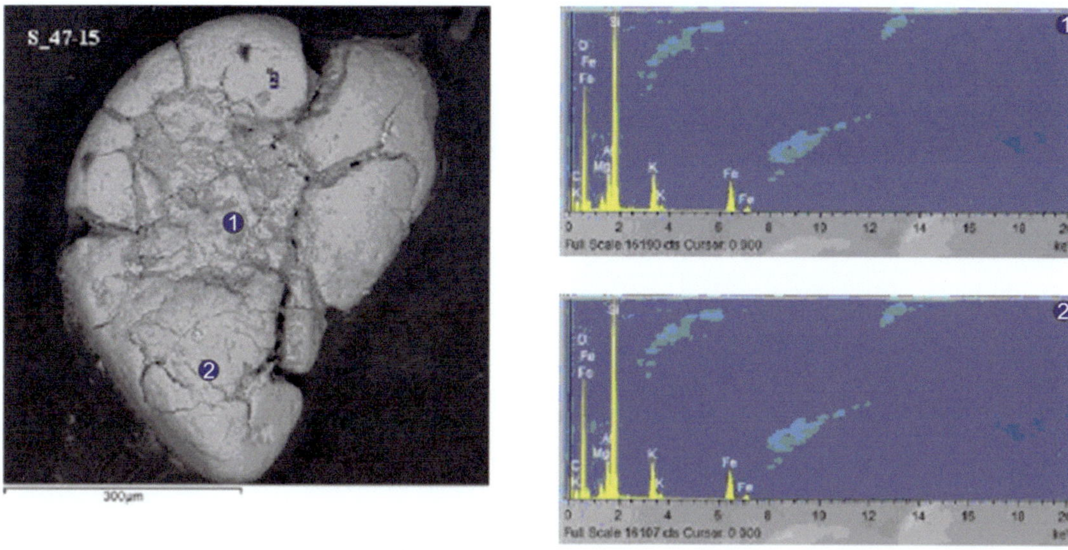

Fig. 4.39 Glauconite grain with principal elements assayed at *1* and *2* from breccia of the Korolyovskiy mud volcano

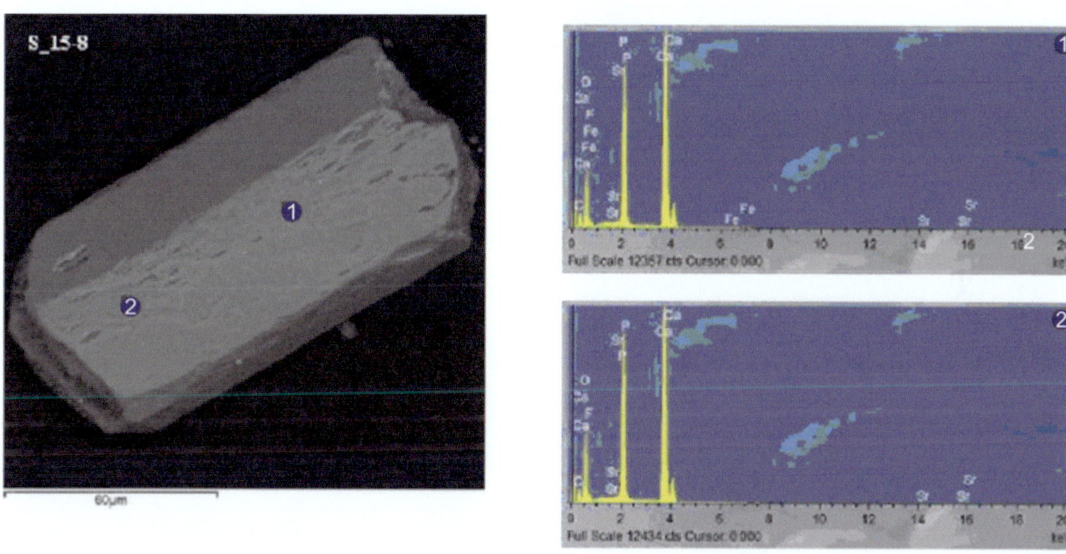

Fig. 4.40 Apatite grain with principal elements assayed at *1* and *2* from the Trubetskoy mud volcano

Fig. 4.41 Glomercrystalline cluster of fluorapatite crystals with inclusions of very fine scheelite grains with principal elements assayed at *1* from breccia of the MGU mud volcano

Fig. 4.42 Chamfered crystal of apatite with principal elements assayed at *1* from breccia of the Tredmar mud volcano

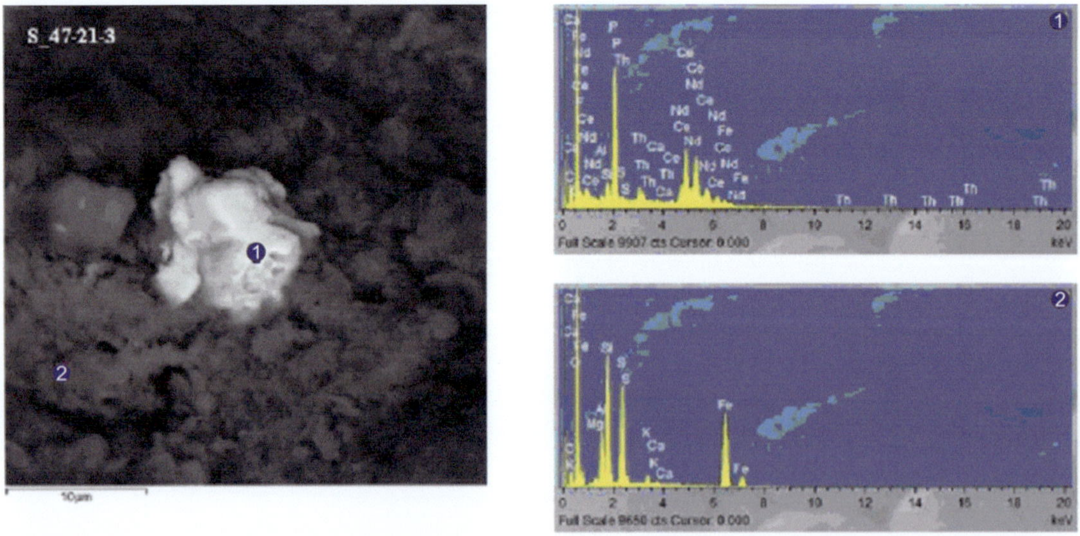

Fig. 4.43 Monazite (*1*) in chlorite (*2*) with principal elements assayed at *1* and *2* from breccia of the Trubetskoy mud volcano

Table 4.11 Mineralogical content of terrigenous deposits of the Western Kuban province of the North Caucasus region; from Grossgeim (1961)

Mineral		Max content, %
Heavy fraction	Black ore (magnetite, ilmenite, chromite)	15
	Zircon	31
	Garnet	26
	Tourmaline	10.8
	Rutile	31
	Leucoxene	36.9
	Sphene	1
	Anatase	1
	Brookite	1
	Pyroxene	1
	Amphiboles	1
	Disthene (kyanite)	–
	Staurolite	1
	Andalusite	Insignificant
	Sillimanite	–
	Mica	11.3
	Chloritoid	1
	Epidote group	1.8
	Chromium spinels	1
	Apatite	4
	Corundum	1
	Anhydrite	1.8
	Dolomite	1
	Pyrite	4
	Limonite	–
	Hematite	51.6
	Siderite	100
	Calcite	–
	Barite	57.6
	Celestite	–
	Glauconite	16
	Phosphates	–
Light fraction	Quartz	72–99
	Feldspars	48
	Fragments of weathered rocks and minerals	1.2–80
	Mica	1
	Glauconite	1
	Opal	1–65.8
	Chalcedony	–
	Volcanic glass	1
	Spicules of sponges	–
	Clay minerals	7.8

References

Abich GV (1840) O poyavivshemsya na Kaspiyskom more ostrove (On the appearance of a Caspian Sea island. Materials concerning knowledge of the mud volcanoes of the Caspian region). Trudy Geologicheskogo instituta Azerbaidjanskogo filiala AN nauk SSSR, Baku (in Russian, translation from German) Reissue

Abich GV (1873) Geologicheskiy obzor poluostrovov Kerchi i Tamani (Geological survey of the Kerch and Taman peninsulas). Zapiski Kavkazskogo Otdeleniya Russkogo Geograficheskogo Obshchestva 8:3–160 (in Russian)

Al'bov SV (1956) Mineral'nye istochniki Kryma (mineral springs of Crimea). Krymizdat, Simferopol (in Russian)

Al'bov SV (1971) O Kerchensko-Tamanskoy gidrogeokhimicheskoy i gryazevulkanicheskoy oblasti (On the Kerch-Taman hydrogeochemical and mud volcanic region). Doklady AN SSSR 197(1):175–177 (in Russian)

Alekseev AS, Geza NI, Glinskiy BM (2004) Aktivnaya seysmologiya moshchnymi vibratsionnymi istochnikami (Active seismology with powerful vibration sources). Sibirskoe otdeleneie RAN (in Russian)

Alifan II, Lysenkov SP (1968) Redkiy sluchay proyavleniya gryazevogo vulkanizma na neftyanom mestorozhdenii Barsa-Gelmes (A rare case of mud volcanism in the Barsa-Helmes oil field). Geologiya nefti i gaza 8:53–64 (in Russian)

Aliyev AA (2016) Zakonomernosti prostranstvenno-vremennogo raspredeleniya gryazevykh vulkanov Dzheyrankechmez-Yuzhnokaspiyskoy vpadiny (Patterns of spatio-temporal distribution of mud volcanoes in the Jeyrankechmez-South Caspian depression). Geologiya i poleznye iskopaemye mirovogo okeana 4:67–80 (in Russian)

Aliyev ADA, Guliev KS, Rakhmanov RR (2015) Sravnitelnyy analiz gryazevogo vulkanizma v Chernomorskom i Kaspiyskom regionakh (Comparative analysis of mud volcanism in the Black Sea and Caspian regions). Geologiya i poleznye iskopaemye mirovogo okeana 2:92–105 (in Russian)

Arkhangel'skiy AD (1925) Neskol'ko slov o genezise gryazevykh vulkanov Apsheronskogo poluostrova i Kerchensko-Tamanskogo regiona (A few words about the genesis of the mud volcanoes of the Absheron Peninsula and the Kerch-Taman region). Bull MOIP Geol Branch 33(3/4):269–285 (in Russian)

Avdusin PP (1939) K petrografii produktov izverzheniya gryazevykh vulkanov Krymsko-Kavkazskoy geologicheskoy provintsii (On the petrography of the products of the eruption of mud volcanoes of the Crimean-Caucasian geological province). In: Results of a study of mud volcanoes of the Crimean-Caucasian geological province. AN SSSR, Moscow-Leningrad, pp 57–66 (in Russian)

Avdusin PP (1948) Gryazevye vulkany Krymsko-Kavkazskoy geologicheskoy provintsii (Mud volcanoes of the Crimean-Caucasian geological province). AN SSSR, Moscow (in Russian)

Barber AJ (2013) The origin of mélanges: cautionary tales from Indonesia. J Asian Earth Sci 76:428–438

Barber AJ, Tjokrosapoetro S, Charlton TR (1986) Mud volcanoes, shale diapirs, wrench faults, and melanges in accretionary complexes, Eastern Indonesia. AAPG Bull 70:1729–1741

Belousov VV, Yarotskiy LA (1936) Gryazevye sopki Kerchensko-Tamanskoy oblasti, usloviya ikh vozniknoveniya i deyatel'nosti (Mud hills of the Kerch-Taman region, the conditions of their occurrence and activity). ONTI, NKPT SSSR, Moscow (in Russian)

Bogatikov OA, Voytov GI, Sobisevich LE (2003) O paroksizmal'nom izverzhenii gryazevogo vulkana gora Karabetova 6 maya 2001 goda. Tamanskaya gryazevulkanicheskaya provintsiya (On the paroxysmal eruption of the Mount Karabetova mud volcano on May 6, 2001. Taman mud volcanic province). Doklady RAN 6(390):805–808 (in Russian)

Bondarchuk VG (1959) Geologiya Ukrainy (Geology of Ukraine). AN USSR (in Ukrainian)

Bouriak S (1995) Black Sea deep water mud volcano area: seismic and acoustic images probably connected with gas charge: the evidence of gas responsibility for bright spots (data from the 1st and 3rd TTR cruises). In: Third Post-cruise Meeting of the UNESCO/TREDMAR "Floating University" Programme, Cardiff, 30 January–3 February 1995, pp 2–3

Burkser YS, Bronshtein KG (1933) Gazy Kryma i Kerchenskogo poluostrova (Gases of Crimea and the Kerch Peninsula). Prirodnye gazy SSSR 13:1–11 (in Russian)

Cuthbert FL, Rowland RA (1947) Differential thermal analysis of some carbonate minerals. Am Mineral 32:111–116

Dmitrievskiy AN, Valyaev BM (2002) Uglevodorodnaya degazatsiya cherez dno okeana: lokalizovannye proyavleniya, masshtaby, znachimost' (Hydrocarbon degassing through the ocean floor: localized manifestations, extent, significance). Degazatsiya Zemli i genezis uglevodorodnykh flyuidov i mestorozhdeniy. GEOS, Moscow (in Russian)

Durmishyan AG (1961) O genezise gryazevykh vulkanov v svete novykh dannykh (On the genesis of mud volcanoes in the light of new data). Izvestiya vysshikh uchebnykh zavedeniy "Neft' i gaz" 12:21–26 (in Russian)

Egorov VN, Polikarpov GG, Gulin SB et al (2003) Sovremennye predstavleniya o sredoobrazuyushchey i ekologicheskoy roli struynykh metanovykh gazovydeleniy so dna Chernogo morya (Modern ideas about the environmental and ecological role of jet methane gas emissions from the bottom of the Black Sea). Morskoy ekologicheskiy zhurnal 2(3):5–26 (in Russian)

Egorov VN, Artemov YuG, Gulin SB (2011) Metanovye sipy v Cherom more. Sredoobrasuyushchaya i ekologicheskaya rol' (Methane seeps in the Black Sea. Environmental and ecological role). Ekosa-gidrofisika, Sevastopol (in Russian)

Etiope G, Feyzullaev A, Baciu CH et al (2004) Methane emission from mud volcanoes in eastern Azerbaijan. Geology 32(6):465–468

Fedorov SF (1939) Gryazevye vulkany Krymsko-Kavkazskoy geologicheskoy provintsii i diapirizm (Mud volcanoes of the Crimean-Caucasian geological province and diapirism). In: Rezul'taty issledovaniya gryazevykh vulkanov Krymsko-Kavkazskoy geologicheskoy provintsii. AN SSSR, Moscow-Leningrad, pp 5–44 (in Russian)

Frolov VT (1992) Litologiya (Lithology). MGU, Moscow (in Russian)

Gemp CD, Lagunova IA (1978) Svyaz' gryazevogo vulkanizma s endogennymi protsessami (The connection of mud volcanism with endogenous processes). Trudy VNIGRI – Vliyanie endogennykh faktorov na formirovanie zalezhey nefti i gaza, pp 75–97 (in Russian)

Gemp SD, Dubrova NV, ZYa N et al (1970) Izotopnyy sostav ugleroda uglerodsoderzhashchikh gazov (CH_4 i CO_2) gryazevykh vulkanov Kerchenko-Tamanskoy oblasti (The carbon isotope composition of carbon-containing gases CH_4 and CO_2 in mud volcanoes of the Kerch-Taman region). Geokhimiya 2:243–247 (in Russian)

Gorin VA (1964) Strukturnaya migratsiya uglevodorodov i ee rol v formirovanii neftyanykh i gazovykh mestorozhdeniy Azerbaydzhana (Structural migration of hydrocarbons and its role in the formation of oil and gas fields in Azerbaijan). Izvestiya AN Azerbajanskoy SSR, Seriya geol-geograf nauk 3:34–48 (in Russian)

Gorin VA (1967) Yevraziyskiy (Uralo-Kavkazsko-Araviyskiy) neftegazonosnyy poyas razlomov zemnoy kory (Eurasian [Ural-Caucasian-Arabian] oil and gas belt of faults in the Earth's crust). Doklady AN SSSR 172(4):909–912 (in Russian)

Gorin VA, Buniat-Zade ZA (1968) O zakonomernostyakh proyavleniya gazo-neftyanogo vulkanizma zapadnogo borta Yuzhno-Kaspiyskoy vpadiny i migratsii uglevodorodov (On the patterns of gas-oil volcanism on the western side of the South Caspian depression and the migration of hydrocarbons). In: Sokolov VA (ed) Genezis nefti i gaza. Nauka, Moscow, pp 175–184 (in Russian)

Grechishkin LA (1936) Geologicheskie issledovaniya v Khinalugskom i Shakhdagskom rayonakh (Geological research in Khinalug and Shahdag districts). ONTI (in Russian)

Grossgeim VA (1961) Istoriya terrigennykh mineralov v mezozoe i kaynozoe Severnogo Kavkaza i Predkavkazya (History of terrigenous minerals in the Mesozoic and Cenozoic of the Northern Caucasus and Ciscaucasia). Gostotekhizdat, Leningrad (in Russian)

Gubkin IM (1937) Neft' v SSSR (Oil in the USSR). Energeticheskie resursy SSSR, Akademiya Nauk, Moscow-Leningrad, v. 1, pp 169–194 (in Russian)
Guliyev IS (2010) Subvertikalnye geologicheskie tela—novye obekty poiskov mestorozhdeniy uglevodorodov (Subvertical geological bodies—new objects of hydrocarbon exploration). In: Vserossiyskaya konferentsiya "Degazatsiya Zemli: geodinamika, geoflyuidy, neft', gaz i ikh parogenezy.". GEOS, Moscow, pp 140–145 (in Russian)
Gümbel CW (1889) Geologische Bemerkungen über die warmen Quellen von Gastein und ihre Umgebung. Bayerische Akademie der Wissenschaften (München). Mathematisch-Physikalische Klasse: Sitzungsberichte
Gutsalo LK, Plotnikov AM (1981) Izotopnyy sostav ugleroda v sisteme CO_2-CH_4 kak kriteriy proiskhozhdeniya metana i dioksida ugleroda v prirodnykh gazakh Zemli (carbon isotopic composition in the CO_2-CH_4 system as a criterion for the origin of methane and carbon dioxide in the Earth's natural gases). Dokl Akad Nauk SSSR 259:470–473 (in Russian)
Helmersen G, Schrenck L (1885) BAND VIII. Gemischten Inhalts. Beiträge zur Kenntnis des russischen Reiches und der angränzenden Länder Asiens; Folge 2, mit 15 lithographischen Tafeln. Kaiserlichen Akademie der Wissenschaften, St-Petersburg, Riga, Leipzig
Herbin JP, Saint-Germes M, Maslakov NA et al (2008) Oil seeps from the "Boulganack" mud volcano in the Kerch Peninsula (Ukraine–Crimea), study of the mud and gas: inferences for the petroleum potential. Oil Gas Sci Technol 63(5):609–628
Hovland M, Hill A, Stokes D (1997) The structure and geomorphology of the Dashgil mud volcano, Azerbaijan. Geomorphology 21(1):1–15
Ismagilov DF, Krasikova TA, Terekhov AA et al (2006) Osobennosti formirovaniya osnovnykh strukturno-tektonicheskikh elementov rossiyskoy chasti Chernogo morya (Features of the formation of the main structural-tectonic elements of the Russian part of the Black Sea). Geologiya, geofizika i razrabotka neftyanykh mestorozhdeniy 2:4–11 (in Russian)
Jakubov AA, Alizadeh AA, Zeinalov MM (1971) Mud volcanoes of the Azerbaijan SSR Atlas. Publishing House of Academy of Sciences of AzSSR, Baku
Kalinko MK (1964) Osnovnye zakonomernosti raspredeleniya nefti i gazov v zemnoy kore (Basic laws of distribution of oil and gas in the Earth's crust). Nedra, Moscow (in Russian)
Kalinko MK (1968) Neorganicheskoe proiskhozhdenie nefti v svete sovremennykh dannykh (Inorganic origin of oil in the light of current data). Nedra, Moscow (in Russian)
Kessler JD, Reeburgh WS, Southon J et al (2006) Basin-wide estimates of the input of methane from seeps and clathrates to the Black Sea. Earth Planet Sci Lett 243:366–375
Khain VE (1953) Bolshoy Kavkaz. Geologiya Azerbaydzhana, tom "Tektonika, gryazevoy vulkanizm, istoriya geologicheskogo razvitiya" (Greater Caucasus. Geology of Azerbaijan, volume "Tectonics, mud volcanism, history of geological development"). AN Azerb. SSSR, Baku, pp 9–91 (in Russian)
Khain VE, Zhabrev IP, Shapiro AR (1953) Tipy tektonicheskikh i gryazevulkanicheskikh brekchiy Kobystana (Types of tectonic and mud volcanic breccias of Kobystan). Azerbaijankoe neftyanoe khozyastvo 6:5–7 (in Russian)
Khaustov VV (2011) O glubinnykh vodakh Yuzhno-Kaspiyskoy vpadiny (on the deep waters of the South Caspian depression). Uchenye zapiski: elektronnyy nauchnyy zhurnal Kurskogo gosudarstvennogo universiteta 2(18) (in Russian)
Khitrov NI, Kolonin RG (1962) O perekhode redkikh shchelochnykh elementov iz albitizirovannogo mikroklina v rastvor v gidrotermalnykh usloviyakh (On the transition of rare alkaline elements from an albitized microcline to a solution under hydrothermal conditions). Eksperimentalnye issledovaniya v oblasti glubinnykh protsessov. Nauka, Moscow, pp 132–145 (in Russian)
Klepinin NN (1914) Gryazevye sopki Kerchenskogo poluostrova i izverzhenie sopki Dzhau-Tepe v 1914 godu (Mud hills of the Kerch Peninsula and eruption of the Dzhau-Tepe hill in 1914). Po Krymu No. 2, Simferopol (in Russian)
Klyushina GKh (2006) Litologiya i fatsialni osoblivosti oligotsenovikh vidkladiv pivnichno-zakhidnogo shelfu Chornogo morya (Lithology and facies features of the Oligocene deposits of the northwestern shelf of the Black Sea). Dissertation, IGN, Kiev (in Ukrainian)
Konyukhov AM, Ivanov MK, Kulnitskiy AM (1990) O gryazevykh vulkanakh i gazogidratakh v glubokovodnykh rayonakh Chernogo morya. Litol Pol Iskop 3:12–23 (in Russian)
Kovalevskiy SA (1935) Gazovyy vulkanizm (vulkany i vulkanoidy) (Gas volcanism [volcanoes and volcanoids]). Azerbaydzhanskoe neftyanoe khozyaystvo 1, 40 p (in Russian)
Kropotkin PN, Valyaev BM (1979) Glubinnye razlomy i degazatsiya Zemli (Deep faults and degassing of the earth). In: Pucharovskiy YM, Yanshin AL (eds) Tektonicheskoe razvitie zemnoy kory i razlomy. Nauka, Moscow, pp 257–267 (in Russian)
Kruglyakova RP, Kruglyakova MV, Shevtsova NT (2009) Geologo-geokhimicheskaya kharakteristika estestennykh proyavleniy uglevodorodov v Chernom more (Geological and geochemical characterization of natural manifestations of hydrocarbons in the Black Sea). Geologiya i poleznye iskopaemye mirovogo okeana 1:27–51 (in Russian)
Kudryavtsev NA (1963) O kolichestvennykh sootnosheniyakh mezhdu uglevodorodami osadochnykh porod i nefti (The quantitative relationships between hydrocarbons of sedimentary rocks and oil). Geologiya nefti i gaza 9:45–49 (in Russian)
Kurishko VA, Mesyats IA, Tverdovidov AS (1968) Gidrogeologiya gryazevogo vulkanizma Kerchenskogo poluostrova (Hydrogeology of mud volcanism of the Kerch Peninsula). Geologicheskiy zhurnal 28(1):49–59 (in Ukrainian)

Kurishko VA, Sivan GP (1971) Genetichne spivvidnoshennya vod naftogazonosnikh struktur i gryazovikh vulkaniv Kerchensko-Tamans'koï zoni (Genetic correlation of waters of oil and gas structures and mud volcanoes of the Kerch-Taman zone). Geologiya i geokhimiya goryuchikh kopalyn 25:46–51 (in Ukrainian)

Lagunova IA (1974) O genezise CO_2 v gazakh gryazevykh vulkanov Kerchensko-Tamanskoy oblasti (On the genesis of CO_2 in the gases of the mud volcanoes of the Kerch-Taman region). Geokhimiya 11:1711–1716 (in Russian)

Lagunova IA, Gemp SD (1978) Gidrokhimicheskie osobennosti gryazevykh vulkanov (Hydrochemical features of mud volcanoes). Soviet Geol 8:108–124 (in Russian)

Lein AY, Ivanov M (2018) Interaction of biogeochemical sulfur and carbon cycles in marine basins by the example of the Black Sea. Dokl Earth Sci 481(1):939–942

Limonov AF (2004) Gryazevye vulkany (Mud volcanoes). Sorosovskiy obrazovatelnyy zhurnal 8(1):63–69 (in Russian)

Lukashchuk AI (1933) Gazy kerchenskikh sopok. Mineralnyy istochnik Adzhi-Su. Rezultaty analizov (Gases of Kerch hills. Mineral spring Adzhi-Su. Results of analyses). Prirodnye gazy SSSR 13:11–18 (in Russian)

Mamedov IS, Guliyev IS (2003) Subvertikalnye geologicheskie tela v osadochnom chekhle Yuzhno-Kaspiyskoy vpadiny (Subvertical geological bodies in the sedimentary cover of the South Caspian basin). Izvestiya Azerbaydzhanskoy Akademii Nauk o Zemle 3:139–146 (in Russian)

Mazzini A, Etiope G (2017) Mud volcanism: an updated review. Earth Sci Rev 168:81–112

Meisner LB, Tugolesov DA (1997) Flyuidogennye deformatsii v osadochnom vypolnenii Chernomorskoy vpadiny (Fluidogenic deformations in the sedimentary performance of the Black Sea depression). Razvedka i okhrana nedr 7:18–21 (in Russian)

Nesterov KV (1979) Geokhimiya mineralnykh vod zapadnoy chasti Indolo-Kubanskogo progiba (Geochemistry of mineral waters of the western part of the Indo-Kuban deflection). Geokhimiya landshaftov i podzemnykh vod. Severo-Kavkazsky nauchnyy tsentr, pp 80–98 (in Russian)

Nesterovskiy VA, Titova NO (2012) Litologichniy sklad tverdikh vikidiv sopok Bulganatskogo gryazovogo vulkanu (Lithologic composition of solid emissions of the Bulganatskiy mud volcano). Zbirnik naukovikh prats Institutu geologichnykh nauk NAN Ukraïni 5:60–64 (in Ukrainian)

Nikonov AA (2002) Krymskie zemletryaseniya 1927 goda. Neizvestnye yavleniya na more (Crimean earthquakes of 1927. Unknown phenomena at sea). Priroda 9:13–20 (in Russian)

Olenchenko VV, Shnyukov EF, Gaskova OL et al (2015) Dinamika izverzheniya gryazevogo vulkana na primere sopki Andrusova (Bulganakskiy gryazevulkanicheskiy ochag, Kerchenskiy poluostrov) (Dynamics of the eruption of a mud volcano based on the example of the Andrusov hill [Bulganakskiy mud-volcanic hearth, Kerch Peninsula]). Dokl Akad Nauk 464(2):214–219 (in Russian)

Paffengolts KN (ed) (1973) Geologicheskiy slovar' (Geological Dictionary). V I, Issue 2

Pashchenko YE (1960) Paleogeografiya maykopskogo baseynu Krimu (Paleogeography of the Maikop basin of Crimea). AN USSR, Kiev (in Ukrainian)

Popov SP (1938) Mineralogiya Kryma (Mineralogy of Crimea). AN SSSR, Moscow (in Russian)

Prokopov KA (1931) Geotektonicheskiy ocherk Kerchenskogo poluostrova i otnoshenie ego k Krymu i Tamani (Geotectonic essay on the Kerch Peninsula and its relation to Crimea and Taman). Trudy Geologo-Razvedochnogo Upravleniya 38:13–23 (in Russian)

Rakhmanov RR (1987) Gryazevye vulkany i ikh znachenie v prognozirovanii gazoneftenosnosti (mud volcanoes and their significance in forecasting of oil and gas). Nedra, Moscow (in Russian)

Sahling H, Bohrmann G, Brüning M et al (2009) Vodyanitskii mud volcano, Sorokin Trough, Black Sea: geological characterization and quantification of gas bubble streams. Mar Pet Geol 26(9):1799–1811

Shardanov AN, Malyshok VG, Peklo VP (1962) O kornyakh gryazevykh vulkanov Tamani. Trudy Krasnodarskogo filiala VNII Neft 5(1):53–66 (in Russian)

Shatskiy NS, Zhukov MM, Milanovskiy YV et al (1929) Dislokatsionnye brekchii i gryazevye vulkany v Azerbaydzhane (Dislocation breccias and mud volcanoes in Azerbaijan). Byull Mosk Obshch Ispyt Prir Otd Geol 7(1–2):97–161 (in Russian)

Shnyukov EF (1999) Gryazevoy vulkanizm v Chernom more (Mud volcanism in the Black Sea). Geologichekiy zhurnal 2:38–47 (in Russian)

Shnyukov EF (2000) Gryazevye vulkany Chernomorskogo regiona. Putevoditel' ekskursii po gryazevym sopkam Kerchenskogo poluostrova s 6 po 15 iyulya 2000 goda (Mud volcanoes of the Black Sea region. Narrative of an excursion among the mud volcanoes of the Kerch Peninsula from July 6 to 15, 2000). OMGOR NNPM NAN Ukrainy, Kiev (in Russian)

Shnyukov EF (2016) Flyuidogennaya mineralizatsiya gryazevykh vulkanov Azovo-Chernomrskogo regiona (Fluidogenic mineralization of mud volcanoes of the Azov-Chernomr region). Logos, Kiev (in Russian)

Shnyukov EF, Kulichenko VG (1969) Vikoviy spektr krupnoulamkovogo materialu u vikidakh gryazovykh vulkaniv Kerchenskogo pivostrova (Age range of large-scale material in the emission of mud volcanoes of the Kerch Peninsula). Dopovidi AN USSR (seriya B) 11:796–798 (in Ukrainian)

Shnyukov EF, Netrebskaya EY (2013) Korni Chernomorskikh gryazevykh vulkanov (The roots of the Black Sea mud volcanoes). Geologiya i poleznye iskopaemye mirovogo okeana 2:66–80 (in Russian)

Shnyukov EF, Shchiptsov AA (1986) Geologicheskie issledovaniya NIS "Kiev" v Chernom more (6-y

reys) (Geological research of the R/V "Kiev" in the Black Sea [6th voyage]). OMGOR NAMIT, Kiev (in Russian)
Shnyukov EF, Naumenko PI, Lebedev YS (1968) Mineralnyy sostav karbonatnykh porod vdavlennykh sinklinaley Kerchenskogo poluostrova (Mineral composition of carbonate rocks in depressed synclines of the Kerch Peninsula). In: Tezisy dokladov IV nauchnteknicheskoy konferentsii po izucheniyu poleznykh iskopaemykh otlozheniy osadochnogo chekhla yuga Ukrainy. KGU, Kiev, pp 44–46 (in Russian)
Shnyukov EF, Naumenko PI, Lebedev YS et al (1971) Gryazevoy vulkanizm i rudoobrazovanie (Mud volcanism and ore formation). Naukova Dumka, Kiev (in Russian)
Shnyukov EF, Sobolevskiy YV, Gnatenko GI et al (1986) Gryazevye vulkany Kerchensko-Tamanskoy oblasti (Mud volcanoes of the Kerch-Taman province) atlas. Naukova Dumka, Kiev (in Russian)
Shnyukov EF, Gnatenko GI, Nesterovskiy VA et al (1992) Gryazevyy vulkanizm Kerchensko-Tamanskogo regiona (Mud volcanism of the Kerch-Taman region). Naukova Dumka, Kiev (in Russian)
Shnyukov EF, Starostenko AV, Rusakov OM et al (2005) Glubinnaya priroda gazovykh fakelov zapadnoy chasti Chernogo morya po rezultatam geofizicheskikh issledovaniy. Geologiya i poleznye iskopaemye mirovogo okeana 5:70–82 (in Russian)
Shnyukov EF, Sheremet'ev VM, Maslakov NA et al (2006a) Gryazevye vulkany Kerchensko-Tamanskogo regiona (Mud volcanoes of the Kerch-Taman region). Glavmedia, Krasnodar (in Russian)
Shnyukov EF, Starostenko AV, Ivannikov AV et al (2006b) Geologicheskie issledovaniya Chernogo morya (po rezultatam 62-go reysa NIS "Professor Vodyanitskiy") (Geological studies of the Black Sea based on the results of the 62nd cruise of the R/V "Professor Vodyanitsky"). OMGOR NNPM NAN Ukrainy, Kiev (in Russian)
Shnyukov EF, Sokol EV, Nigmatulina YN et al (2013) Zoloto v gryazevykh vulkanakh Kerchenskogo poluostrova kak pokazate' glubinnosti gryazevulkanicheskikh flyuidov (Gold in the mud volcanoes of the Kerch Peninsula as an indicator of the depth of the mud volcanic fluids). Geologiya i poleznye iskopaemye mirovogo okeana 4:79–89 (in Russian)
Shnyukov EF, Deyak MA, Ivanchenko VV et al (2015) Nalozhennaya mineralizatsiya gryazevykh vulkanov Kerchenskogo poluostrova (Imposed mineralization of the mud volcanoes of the Kerch Peninsula). Geologiya i poleznye iskopaemye mirovogo okeana 4:5–19 (in Russian)
Shternov AG (1968) Geologiya i genezis gryazevykh vulkanov Kerchensko-Tamanskogo rayona (Geology and genesis of mud volcanoes of the Kerch-Taman region). Dissertation, KGU, Kiev (in Russian)
Shternov AG (1971) Pro formuvannya vod gryazovikh vulkaniv Kerchenskogo pivostrova (On the formation of waters of the mud volcanoes of the Kerch Peninsula). In: Geologiya i geokhimiya goryuchikh kopalyn. Naukova Dumka, pp 51–55 (in Ukrainian)
Shuraev IN (2016) Oblomochnyy material sopochnoy brekchii gryazevogo vulkana MGU (Clastic material of the mud breccia of the Moscow State University mud volcano). Geologiya i poleznye iskopaemye mirovogo okeana 4:82–86 (in Russian)
Siryk IM (1962) Gryazevye vulkany Yuzhnogo Sakhalina – veroyatnye sputniki neftyanykh i gazovykh mestorozhdeniy (Mud volcanoes of South Sakhalin – probable satellites of oil and gas fields). 7:66–75 (in Russian)
Sjögren H (1888) Om bildningen af Kaspiska hafvets bäcken (On the formation of the Caspian Sea hill). Geologiska Föreningen i Stockholm Förhandlingar 10:49–74 (in Swedish)
Skorik IM, Bayrakov VV (2007) Geologogeokhimicheskie osobennosti alevrolitov Maykopa Kerchenskogo poluostrova (Geological and geochemical features of Makopian alevirite Kerch Peninsula). Doklady AN USSR 10:112–117 (in Russian)
Sobisevich AL, Tveritinova TY, Likhodeev DV et al (2015) Glubinnoe stroenie gryazevogo vulkana Dzhardzhava v predelakh Yuzhno-Kerchenskoy antiklinalnoy struktury (Depth structure of the Dzhardzhava mud volcano within the South Kerch anticlinal structure). Voprosy inzhenernoy seysmologii 42(2):73–80 (in Russian)
Stadnitskaya AN, Belenkaya IYu (2007) Sostav i proiskhozhdenie uglevodorodnykh gazov i ikh vliyanie na diageneticheskie karbonatoobrazovaniya (progib Sorokina, SV chast Chernogo morya) (Composition and origin of hydrocarbon gases and their influence on diagenetic carbonate formations [Sorokin deflection, the north part of the Black Sea]). Geologiya Chernogo i Azovskogo morey. Gnosis, Kiev, p 155–163 (in Russian)
Steber EA (1912) Iezheuasa – gryazevoy vulkan severozapadnoy okonechnosti Kavkaza (Yezheuasa – the mud volcano of the northwestern tip of the Caucasus). Kavkazskogo Otdeleniya Russkogo Geograficheskogo Obshchestva 21(1):51–63 (in Russian)
Steber EA (1913) Cheredovanie burnykh i spokoynykh izverzheniy gryazevykh vulkanov (The alternation of turbulent and calm eruptions of mud volcanoes). Izvestiya NTO pri Yekaterinoslavskom gornom institute (in Russian)
Steber EA (1914) Gryazevye vulkany i proiskhozhdenie nefti (Mud volcanoes and the origin of oil). Yuzh. Inzhener 4:120–128, 6:144–153 (in Russian)
Steber EA (1915) Proiskhozhdenie nefti (Origin of oil). Priroda 4:1–30 (in Russian)
Stopnevich AD (1920) Mineralnye vody. Komissiya po izucheniyu estestvennykh proizvoditelnykh sil Rossii, sostoyashchaya pri Rossiyskoy AN, Petrograd (Mineral water). Commission for the Study of the Natural Productive Forces of Russia at the Russian Academy of Sciences, Petrograd (in Russian)
Subbota MI (1964) Novye dannye o gazakh i tipakh gryazevykh vulkanov rayona Zapadnoy Kubani (New data on gases and types of mud volcanoes in the area of the Western Kuban). Trudy Vsesoyuznogo

nauchno-issledovatel'skogo geologo-razvedochnogo neftyaogo instituta nedr 41:156–178 (in Russian)
Sulin VA (1939) Zadachi i nekotorye itogi gidrogeologicheskogo i geokhimicheskogo izucheniya gryazevykh vulkanov Krymsko-Kavkazskoy provintsii (Tasks and some results of hydrogeological and geochemical study of mud volcanoes of the Crimean-Caucasian province). In: Rezultaty issledovaniy gryazevykh vulkanov Krymsko-Kavkazskoy geologicheskoy provintsii. AN SSSR, Moscow, pp 93–102 (in Russian)
Titova NO, Nesterovskiy VF, Deyak MA et al (2013) Sostav peschano-glinistoy fraktsii sopochnoy brekchii gryazevogo vulkana Dzhau-Tepe na Kerchenskom poluostrove (Composition of the sandy-clay fraction of the gryphon breccia of the mud volcano Dzhau-Tepe on the Kerch Peninsula). Geologiya i poleznye iskopaemye mirovogo okeana 4(34):90–94 (in Russian)
Tsikhotskaya NN, Pustovoyt II, Krochak MD (1986) Litologo-fatsialnaya kharakteristika porod maykopskoy serii severo-zapadnogo shelfa Chernogo morya (Lithologic-facies characteristics of rocks of the Maikop series of the northwestern shelf of the Black Sea). Dopovidi AN URSR (Ser B) 12:21–24 (in Ukrainian)
Tugolesov DA, Gorshkov AS, Meisner LB et al (1985) Tektonika mezo-kaynozoyskiy otlozheniy Chernomorskoy vpadiny (Tectonics of the Meso-Cenozoic sediments of the Black Sea Basin). Nedra, Moscow (in Russian)
Uchiteleva LG (1979) Metallonosnost' subtermal'nykh s termal'nykh vod Vostochnykh Karpat i Kryma (Metallurgy of subthermal and thermal waters of the Eastern Carpathians and Crimea). Trudy VSEGEI 260:46–59 (in Russian)
Valyaev BN, Grinchenko YUI, Prokhorov VS (1980) Izotopnyy sostav ugleroda gazov gryazevykh vulkanov Sovetskogo Soyuza (The carbon isotope composition of the gases of mud volcanoes in the Soviet Union). In: Vsesoyuznyy simpozium po stabilnym izotopam v geokhimii (11–14 noyabrya 1980 goda), Moscow, pp 240–241 (in Russian)
Vinogradov AP (1962) Srednie soderzhaniya khimicheskikh elementov v glavnykh tipakh izverzhennykh gornykh porod (The average content of chemical elements in the main types of igneous rocks). Geokhimiya 7:555–571 (in Russian)
Voskoboynikov NI, Gur'iev AV (1832) Geologicheskoe opisanie poluostrova Tamani, prinadlezhashchego k zemle voyska Chernomorskogo (Geological description of the Taman Peninsula belonging to the land of the Black Sea army). Gornyy zhurnal 1 (in Russian)
Voytov GI (2001) O khimicheskikh i izotopnouglerodnykh nestabilnostyakh grifonnykh gazov gryazevykh vulkanov (na primere Yuzhno-Kaspiyskoy i Tamanskoy gryazevulkanicheskikh provintsiy) (On the chemical and carbon isotope instabilities of gryphon gases of mud volcanoes [based on the example of the South Caspian and Taman mud volcanic provinces]). Geokhimiya 4:422–433 (in Russian)
Yakubov AA, Zeynalov MM (1964) Gryazevoy vulkanizm Azerbaydzhana (Mud volcanism of Azerbaijan). Ocherki po geologii Azerbaydzhana. AN Azerbaijanskoy SSR, Baku (in Russian)
Yakubov AA, Grigoryants BV, Aliev AD et al (1980) Gryazevoy vulkanizm Sovetskogo Soyuza i ego svyaz s neftegazonosnostyu (Mud volcanism of the Soviet Union and its relationship with petroleum potential). ELM, Baku (In Russian)
Zaitsev AV (1965) Gryazevye vulkany Priazovya (Mud volcanoes of Azov). Dissertation, RGU Rostov na Donu (in Russian)

5 Accompanying Mud Volcanic Structures

Summary

This chapter describes the structures associated with mud volcanoes. Mud volcanic activity is accompanied by the eruption of huge volumes of gases, mud breccia, detrital rock material, and water. This has led to the creation of weakened zones in the depths near mud volcanoes due to the removal of sedimentary material. The dimensions of these weakened zones are proportional to the volume of ejected masses. Weakened zones express themselves in the surface relief of the area as peculiar lowlands that usually take the shape of oval depressions. The depressions of mud volcanoes that developed in the Neogene seas filled with sediments accumulated in accordance with the position of the mud volcano within the marine basin—silts, sands, as well as carbonate and ferrous sediments. Over time, the sediments turned into clays, limestones, sands, iron ores, etc. These rocks imposed a heavy loading that exacerbated the downward deflection, and in the process of subsidence, numerous discontinuities in the rocks occurred, which often merged into ring disturbances.

The lowering of the Earth's surface as a result of the removal of underlying sedimentary material is not a property unique to mud volcanoes. The same phenomenon occurs in real volcanoes that eject lava.

*Such depressions are clearly manifested on the Kerch Peninsula where they have received the local name "recessed synclines." The first genetically correct interpretation of these structures belongs to Golovkinskiy (*1889*). Andrusov (*1893*) and Arkhangel'skiy (*1925*) challenged the views of Golovkinskiy, while subsequent researchers (e.g., Prokopov* 1931*; Alyaev* 1947*; Maymin* 1951*; Lychagin* 1952*; Shnyukov and Naumenko* 1964*; Shnyukov et al.* 1971*) confirmed that Golovkinskiy's description was correct.*

The total number of known recessed synclines on the Kerch and Taman peninsulas is 39. All of them are located in the arched parts of anticlines composed of Maikopian clay rocks. Many of them contain layers of Kimmerian iron ore deposits in their section.

In their structure, recessed synclines can be divided into two main types. The first type includes those recessed synclines in which erupted material accumulates in the center of the structure, i.e., where the mud volcano itself is located. The calderas, in fact, represent the compensatory troughs filled with Neogene sediments that delineate its boundaries. The rocks composing the recessed synclines themselves form the frame of the caldera. The second type includes those recessed synclines in which erupted material accumulates in the vicinity of the mud volcano, not in the center of the recessed syncline. In this case, the development of compensatory troughs occurs to the side, and accumulated sediments do not outline the calderas, which apparently are eventually destroyed by weathering processes.

E. Shnyukov, V. Yanko-Hombach, *Mud Volcanoes of the Black Sea Region and their Environmental Significance*, https://doi.org/10.1007/978-3-030-40316-4_5

Undoubtedly, mud volcanoes on the floor of the Black Sea are also accompanied by recessed synclines. Their sedimentary thickness will undoubtedly differ from that of analogues on land. Geologists have yet to figure out the structural features of mud volcanoes on the sea bottom.

Mud volcanic activity produces huge emissions of gases, breccia, and water. According to approximate calculations, mud volcanoes of the Kerch Peninsula alone have ejected at least 25 billion tons of solid products during their geological history (Lychagin 1952). To expel such an enormous amount of earth materials, many trillions of m^3 of gases would be required to propel such matter upward.

Based upon data obtained from 150 boreholes drilled into mud volcanoes of this region, the amount of ejected matter might be even higher, perhaps 40 billion tons (Naumenko 1966). The average volume of gases released by mud volcanoes of the Kerch Peninsula is about 70,000 m^3 per year, or 25×10^{12} m^3 during their geological history, at least from Maikopian time (Plotnikov 1967). This estimate is in good agreement with the present level of activity seen in some mud volcanoes. For example, the Dzhau-Tepe mud volcano ejected approximately 50 thousand m^3 of breccia over the course of 20–30 min on March 18, 1914. The eruption was accompanied by fires, a cloud of white vapor and smoke, a buzzing sound, and the formation of a caldera on top of the volcano possessing a depth of 4.5 m and a diameter of 40–50 m (Dvoychenko 1914). The Gora Gorelaya mud volcano ejected about 1 million m^3 of breccia during the 4 days of February 24–27, 1794. It was accompanied by a crackling thunder-like sound, white steam, black smoke, pillars of flame reaching twice the height of the mountain, and fire that lasted for 25 min (Alekseev 1880). Much greater eruptions of breccia have been observed in Azerbaijan. The Boz-Dag mud volcano ejected up to 5 million m^3 of breccia during a single eruption (Rakhmanov 1987), and the Touragay, Dawali-Dag, Otman-Boz-Dag, and Boz-Dag mud volcanoes each ejected several million m^3 of breccia during a single eruption cycle (Avdusin 1948). These figures pertain to the largest mud volcanoes, but smaller ones also release significant amounts due to their abundance and frequent manifestations (Shnyukov et al. 2006).

The eruption of huge masses of breccia supplemented by gas and water from underlying eroded crustal strata and aquifers creates mass deficits deep underground, which become, as a result, weakened zones. The mass deficits grow with the intensification of mud volcanic activity. In this way, the dimensions of the weakened zones are proportional to the volume of ejected material. These weakened zones create peculiar depressions in the surface topography surrounding the mud volcanoes. For example, the Bulganakskiy mud volcano is located within a depression of considerable size. On closer examination, similar depressions can be seen around other Kerch and Taman mud volcanoes, but such depressions are quite pronounced in the southwestern lowland of Turkmenistan (the Pribalkhanskaya and Chikishlyarskaya zones) where they have developed at the tops of anticlinal folds. These depressions can reach up to 5–10 km in diameter, and the amplitude of their concavity varies between 50 and 500 m in recent relief. For example, the top of the Chelekenskiy uplift is complicated by the activity of large mud volcanoes, such as Pink Porsukel, Western Porsukel, and Aligul (Semenovich 1956; Sungurov 1958), each of which lies at the center of a depression (Fig. 5.1).

In such depressions, the layers are inclined downward toward the center, where the vent of the volcano is located. Kosygin (1935) indicates that the mud volcanoes of the Keymiro-Chikishlyarskiy district contain flexures and depressions similar to those of Chelekenskiy. They cannot be seen from the surface but are visible on geological sections and seismic profiles at depths from 1.3 km to 7 km. The Bakinian (early Pleistocene) beds that overlie them are not sufficiently bent to mirror the flexures at the surface. Sometimes, the mud volcano vent can be displaced toward the side of a depression. The formation of depressions is accompanied by the emergence of small "secondary folds" up to 2.5 × 1 km in size (Fig. 5.2).

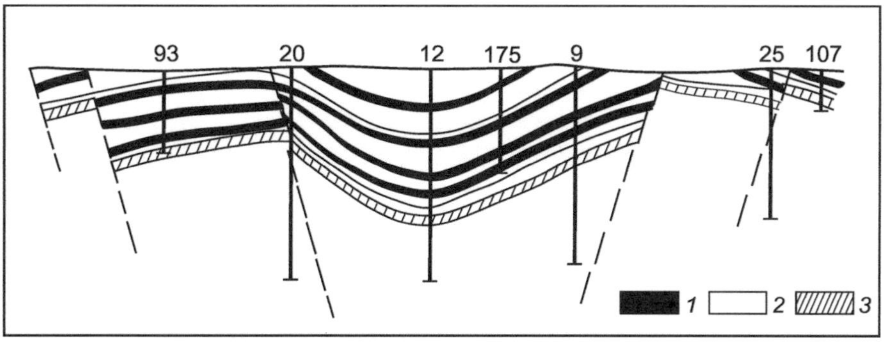

Fig. 5.1 Tectonic diagram of the Porsukel mud volcano in southwestern Turkmenistan. Key: black layers, sandstones; white layers, clays; hatched layers, carbonate rocks (Sungurov 1958)

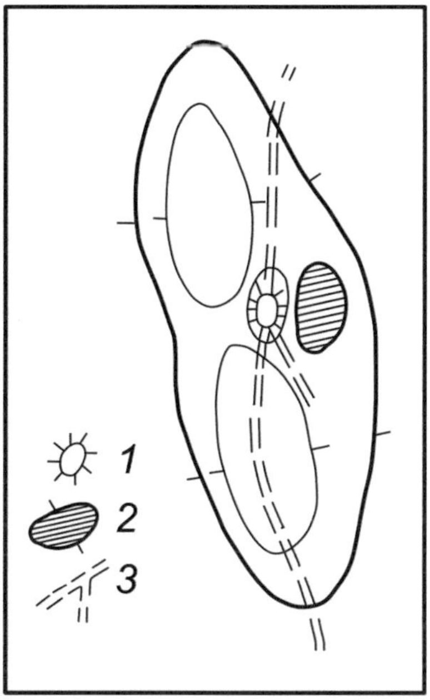

Fig. 5.2 Generalized scheme of the Kotur-Tepinskaya and Khan-Guli-Chekurkuinskaya structures (southwestern Turkmenistan). Key: (*1*) supposed buried mud volcano, (*2*) secondary fold, (*3*) tectonic rupture (Sungurov 1958)

Both depressions and "secondary folds" are broken by a system of radial, semicircular, and ring downthrows. As to their shape, the depressions most often represent oval grabens fragmented into separate blocks. Mud volcanic activity, which leads to the onset of the depressions, sometimes causes a kind of the brachianticline bifurcation (as it does in the Porsimskiy uplift). As an example of secondary folds, Sungurov (1958) describes a fairly large (5 × 4 km) trough in the Keimiro-Okaremskaya anticlinal structure with a flexure amplitude of 300 m and 600 m on the western and eastern slopes, respectively. Based on seismic data, this trough is hidden beneath ~1.5 km of Apsheronian (Eopleistocene) and older strata.

On terrestrial surfaces under subaerial conditions, mud volcanic activity produces depressions of different sizes, which gradually fill with sediments and typically lead to the alignment of relief. In a subaqueous environment, it is different. Such underwater geological processes took place during the Neogene on the Kerch Peninsula, in the Azov-Black Sea basin, and partly on the Taman Peninsula, and they continue to take place in the Azov-Black Sea basin today.

Within the depressions near mud volcanoes of the Neogene seas, the accumulation of sediments (muds, carbonate and ferruginous sediments, sands, etc.) occurred in accordance with local conditions and the position of the volcanoes on the sea bottom. Over time, the sediments turned into clays, limestones, iron ores, sands, etc., and in so doing, they created a heavy weight pressing down upon the depression, accelerating its downward deflection, and sometimes creating bent structures at depths exceeding 100 m.

Strictly speaking, a lowering of the Earth's surface as a result of deep sediment removal is not a feature unique to mud volcanoes. Collapse of real (magmatic) volcanoes subsequent to eruption is a widely known phenomenon described in the geological literature (e.g., Kovalevskiy 1935a, b).

These depressions, which are in fact compensatory synclines found near mud volcanoes, are manifested most clearly on the Kerch Peninsula. They are locally called "recessed synclines" and are broadly distributed at various depths in areas where accumulation of marine sediments occurred simultaneously with mud volcanic activity during the entire duration of the Neogene. This explains why all the disputes over the origin of these recessed synclines are based on studies of material sampled on the Kerch Peninsula. In fact, any recessed syncline is evidence for the activity of a given mud volcano (Naumenko 1967). The correct genetic evaluation of the recessed synclines of the Kerch Peninsula was provided by Golovkinskiy (1889), who drew attention to isolated bedrock exposures of limestones in the Maikopian clays near the Dzhau-Tepe mud volcano and in many anticlinal valleys (e.g., Sartskaya, Aspar-Belukuyskaya, Tarkhanskay, Baksinskaya, and Dzhardzhavskaya) and related these finds to mud volcanic processes. He was able to explain the confinement of mud volcanic cones to the anticlinal valleys due to the destruction and erosion of rocks in the underlying fold zones that facilitated the ejection of deep-seated gases and water causing the eruption of breccia. This, in turn, decreased sediment mass at depth and led to subsidence of the neighboring terrain. Subsidence could occur gradually if the underlying rocks were plastic or if they were non-plastic, with the accompaniment of landslides. He wrote that gradual subsidence can be expected in neighboring synclines, while landslides occur in the immediate vicinity of the vent, where the continuity of the rocks has already been partially broken. As we see, the recessed synclines were interpreted as dips formed by the removal of masses of breccia causing the appearance of holes in the interior of the Earth into which limestones and other hard rocks of these structures descended. The principal aspect of this explanation, namely, the interpretation of recessed synclines as deriving from the mud volcanic process, is accepted by most modern researchers; it is not, of course, about filling the subterranean holes with limestone but about the sedimentary process that results in the development of depressions in the surface relief.

Strangely, however, geologists who studied the Kerch Peninsula in the late nineteenth and early twentieth centuries rejected this thesis. Andrusov (1884) argued that the explanation offered by Golovkinskiy (1889) did not fully coincide with the nature of the phenomenon or, at least, did not have so significant an impact. His opinion was that, on the Kerch Peninsula, there were no significant traces of mud volcanic activity; the amount of mud carried out by eruptions was insufficient for any significant lowering of the terrain; the mud originated, most likely, not at great depths but along the path of the gas migration. Further, there was no reason to assume that previous geological epochs witnessed a greater scale of mud volcanic activity compared to the present. He explained the occurrence of recessed synclines and their composition differently: (1) they were a result of more rapid erosion of clayey-marly rocks that underlie the stable Sarmatian limestones of the anticlinal arch, after which the remnants of the Sarmatian limestones descended and survived in the center of the anticlinal valley; (2) they were a result of local lowering of the anticlinal arch due to folding processes, etc. In his opinion, the normal bedding of rocks in recessed synclines did not fit the hypothesis that they descended in an area of mud volcanic eruptions. These synclines were inserted within the anticline, which possessed considerable width. Finally, in the development of anticlinal valleys, considerable numbers of landslides could have occurred down their sides. If dense rocks (e.g., Sarmatian or Chokrakian limestones) had slipped, they could have remained inside the anticlinal valley in the form of separate hills, isolated from the eroded crests.

Obruchev (1926) noted that some anticlines have depressions in their vaulted part with the formation of small anti- and synclines within the depressions. He treated their emergence as secondary bends in the wide anticlinal folds.

A tectonic origin for recessed synclines was considered by Arkhangel'skiy (1925) and some others. In their opinion, the depressions on the vaults of the anticlines developed at the expense of dense Chokrakian or Sarmatian limestones, which, when bent, exerted greater resistance to bending forces than the clays containing them, thereby forming synclinal deflections. With further destruction, the soft clays were squeezed out and transformed into breccias with crushed limestones preserved as the remnants.

Prokopov (1931) pointed out that observed cup-shaped forms (i.e., recessed synclines) filled with many meters of mud volcanic breccia (e.g., near the village of Mayak) and possessing isolated cliffs were not caused by tectonics (e.g., compressional stress), and as such, strictly speaking, they cannot be called synclines or troughs. Instead, they were formed by smoothly descending sections of various parts of the anticlines. As facts accumulated, it became obvious.

The study by Alyaev (1947) of the geological features of recessed synclines significantly contributed to a better understanding of their origin. Maymin (1951) stressed the preservation of typical Kerch Peninsula bedding in the sediments of recessed synclines, and he interpreted them as surviving remnants but not as a result of tectonic complications. Their formation as a kind of specific structure—a recessed syncline—was associated with the eruptive activity of mud volcanoes accompanied by the removal of a significant amount of plastic clayey brecciated mass and the formation of hollows with subsequent accumulation within them of younger sediments. These structures, folded along with more solid rocks in the future (Maymin 1951), resisted the folding forces and deformed in rather complicated ways, not keeping up with the general pattern of regional tectonics. Maymin emphasized the different ages of the recessed synclines and the anticlines underlying them, and he generally agreed with the views of Golovkinskiy (1889).

Lychagin (1952) clearly formulated that recessed synclines of the Kerch Peninsula were formed by mud volcanic activities under certain paleogeographical conditions.

Shnyukov and Naumenko (1964, 1966) and Shnyukov et al. (1971) developed the main views about the mud volcanic origin of recessed synclines on the Kerch Peninsula and revealed their significance as ore-bearing geological structures.

At present, the mechanism by which recessed synclines develop is clear.

According to Prokopov (1931), once formed, the anticlinal fold becomes the arena for intensive manifestation of mud volcanism and, in certain areas, begins to bend as a result of the release of significant amounts of mud onto the surface. The trough develops very slowly and smoothly, which causes plastic bending of the accumulating sediment layers and their cup-shaped structure (Fig. 5.3).

The initial structure is represented by clay overlain by limestone (Fig. 5.3a). At a certain point in the development of the structure in its

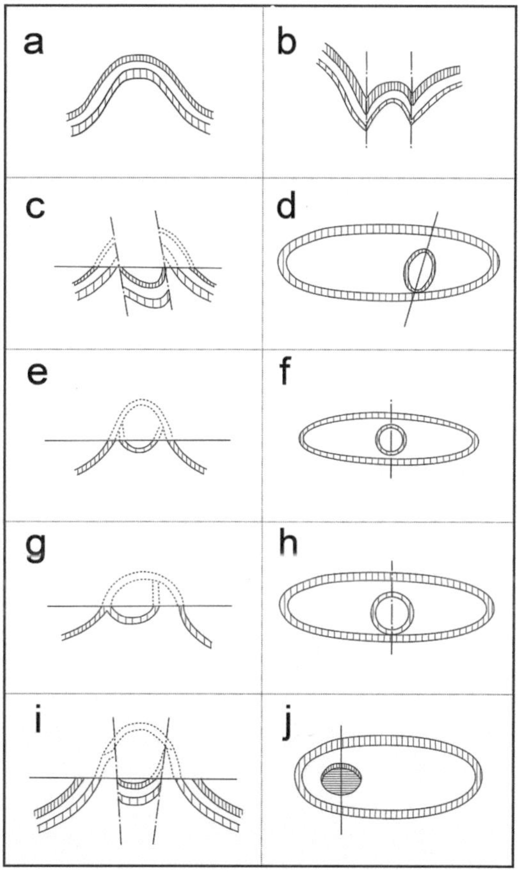

Fig. 5.3 Stages in the development of a recessed syncline (Prokopov 1931). The wider hatched layer is clay, and the tightly hatched layer is limestone. (**a**) Initial anticline represented by clay overlain by limestone, (**b**) certain sections of the anticline bent by mud volcanic activity forming a cuplike deflection, (**c**, **d**) further bending leads to the formation of a ring graben, (**e**–**j**) differences in the development of recessed synclines

arch or in its weakened part, certain sections of the anticline are bent by mud volcanic activity forming a cuplike deflection (Fig. 5.3b). Further bending leads to the formation of a ring graben (Fig. 5.3c, d). Variations can lead to differences in the development of recessed synclines (Figs. 5.3e–j). The Korolevskaya recessed syncline, which is located symmetrically in the center of the anticline along its axis, corresponds to Fig. 5.3e, f. The Burulkayskaya recessed syncline and other similar structures show one of the anticlinal wings connected to the cuvette by flexural inflection, and the common features are shown in Fig. 5.3g, h. Finally, some asymmetry of the ring fault with respect to the deflection explains the appearance of semi-moon shapes in the cases of the Kayaly-Sartskaya, Kara-Sidzheutskaya, Kashikskaya, and some other recessed synclines (Figs. 5.3i, j).

In our opinion, the hypothesis of recessed syncline development proposed by Prokopov (1931) is correct in general; however, new and extensive studies have revealed additional details. Final formation of the anticlinal structure and the manifestation of mud volcanism create the initial prerequisites for the onset of a recessed syncline. The activity of the mud volcano inevitably produces a depression under all paleogeographic conditions, marine or subaerial (Shnyukov et al. 2013), and this leads to bending of the rocks in the arched part of the anticline. Though Lychagin (1952) expressed his opinion that the occurrence of depressions under subaerial conditions was impossible, evidence from actual materials refutes his view and shows that even in terrestrial settings, pronounced depressions can occur due to mud volcanic activities.

Further development of depressions and their transformation into recessed synclines are determined by many circumstances.

During submarine eruptions of a mud volcano within a sea basin, ejected breccia together with normally accumulating marine sediments and the weight of the overlying water column— which is not present in onshore mud volcanoes—creates a heavy loading, which further accelerates the settling and lowering of surfaces on anticlinal structures that overlie mud volcanic foci.

On land, erupted breccia is eroded, largely washed away, and spread around, thereby slowing the accumulation of sediments and decreasing the scale and amplitude of depression formation. This process can nevertheless occur on a considerable scale, as in the case of the Bulganakskiy depression. It is reasonable to assume that similar structures were also formed in this manner in southwestern Turkmenistan with subsequent burial of some depressions.

Under a continuous marine regime, development of a recessed syncline proceeds further: as natural sedimentation and breccia accumulate, the beds in the arch of the anticlines are lowered and their continuity is disturbed. Since the form of the descending structure is most often a rounded compensation funnel, the shape of the corresponding disruption is more or less a pronounced ring dip-slip fault produced by a system of smaller tectonic fractures. A peculiar rounded graben filled with sediments and breccia is the result. In some places, the mud volcanic eruption occurs directly within the recessed syncline, sometimes along the tectonic fractures framing it, and occasionally, the center of the volcano is shifted outside the recessed syncline. Mud volcanic deposits are present in geological sequences of the overwhelming majority of recessed synclines, indicating clearly that they are interrelated. When marine recessed synclines are exposed subaerially by sea retreat, they disintegrate due to weathering processes, often to the extent that they are transformed into positive relief forms—hills, low cliffs, etc. This occurs due to a difference in lithological composition and hardness of the rocks that constitute the recessed synclines and the structures enclosing them. The former comprise dense dolomitized limestones, while the latter consist predominantly of Maikopian clays. If recessed synclines are composed exclusively of Maikopian clays, they are not expressed in the relief.

The development of recessed synclines proceeds in a far more complicated way, as the vast majority of them represent an extremely complex layering of various sedimentary rocks and mud volcanic breccias. On their periphery, reef limestones are usually developed and fall toward the center under the influence of continuous lowering. Under terrestrial conditions, the further activity of mud volcanoes leads to continued subsidence and leveling of the terrain. Finally, with further sea-level rise, deepening of the sea, and cessation of mud volcanic activity, recessed synclines become buried (e.g., the Ansatasievsko-Troitskaya recessed syncline on the Taman Peninsula). On the Kerch Peninsula, most recessed synclines ceased their development in the third stage because they were exposed above sea level by marine regression. With the persistence of a marine environment, recessed synclines are either buried or they continue to develop.

The recessed synclines of the Kerch Peninsula deserve special attention because they contain the Kimmerian iron ore deposits as well as limestones, dolomites, and mud breccias that attract interest due to chemical (e.g., boron), building (e.g., limestones, dolomites), and balneological (e.g., mud) raw materials. The breccias are also used as a raw material in the metallurgical industry. Only the recessed synclines of the Kerch Peninsula contain such a wide range of mineral resources, as no others in the Black Sea region provide a similar wealth of useful materials. Even on the Taman Peninsula, mud volcanic clays are sandier and therefore less suitable as building materials and pellets for metal manufacture.

The total number of recessed synclines on the Kerch Peninsula is 39 (Fig. 5.4).

Some of these structures are not completely understood, and referring to them as recessed synclines is debatable. For example, Gubanov (1961) interprets Gora Opukskaya as a large recessed syncline, while many other authors do not agree. Also controversial is the origin of the Ossovinskaya structure.

Some general patterns can be noted in the geographic situation of recessed synclines. They are localized along the lines of large anticlinal structures of latitudinal and northeast-southwest strike, often in places where the vaults are most elevated. The morphology of the recessed synclines was repeatedly considered in the scientific literature as an inalienable part of the Kerch Peninsula (Blagovolin 1960; Gubanov 1961). Relief in the territories of their occurrence is conditioned by tectonics, lithologic properties of the anticlinal structures, intensity of mud volcanic activity, genetic type and age of the rocks involved in their structures, as well as erosion processes (Blagovolin 1960; Gubanov 1961).

The different lithologic composition of the rocks that form recessed synclines causes a certain variation in their relief. Recessed synclines located in the arched parts of anticlines composed of Maikopian clayey rocks are usually eroded and expressed in the terrain as anticlinal valley hollows. Those composed of durable Miocene carbonate rocks would be preserved as uplifts, ridges, cliffs, steep banks, etc. The remnants of the rocks are particularly complicated if recessed synclines are composed of an alternation of hard (limestones) and soft (clays) rocks, leading sometimes to the formation of peculiar structures, such as "table elevations," half-moons, or sickles (e.g., Kashikskaya, Kayaly-Sartskaya, Eastern Illichevskaya), ovals (e.g., Baksinskaya, Kostenkova, Ossovinskaya, Opukskaya), and, less often, circles (e.g., Burulkayskaya, Korolevskaya). The rock remnants are usually elevated above the terrain by less than several tens of meters.

The Ak-Tube recessed syncline displays a mysterious morphology. Despite its carbonate sandstone and mud volcanic breccia composition, it takes a positive form of relief in the southwestern plain of the Kerch Peninsula.

Elevation of outliers above the terrain usually does not exceed several tens of meters. The terrain appears quite different if the recessed synclines are composed of easily destroyed sandy-argillaceous deposits, most often of Pliocene age. Such structures are usually not positively expressed in the relief, not even weakly. In areas of recent mud volcanic activity, they form peculiar recessed depressions, as in the field of the Bulganakskiy mud volcano. Mud volcanic cones are often located within recessed synclines (e.g., the Burashskiy mud volcano), and sometimes they attach to their wings (e.g., the Dzhau-Tepe mud volcano attached to the Dzhau-Tepe recessed syncline).

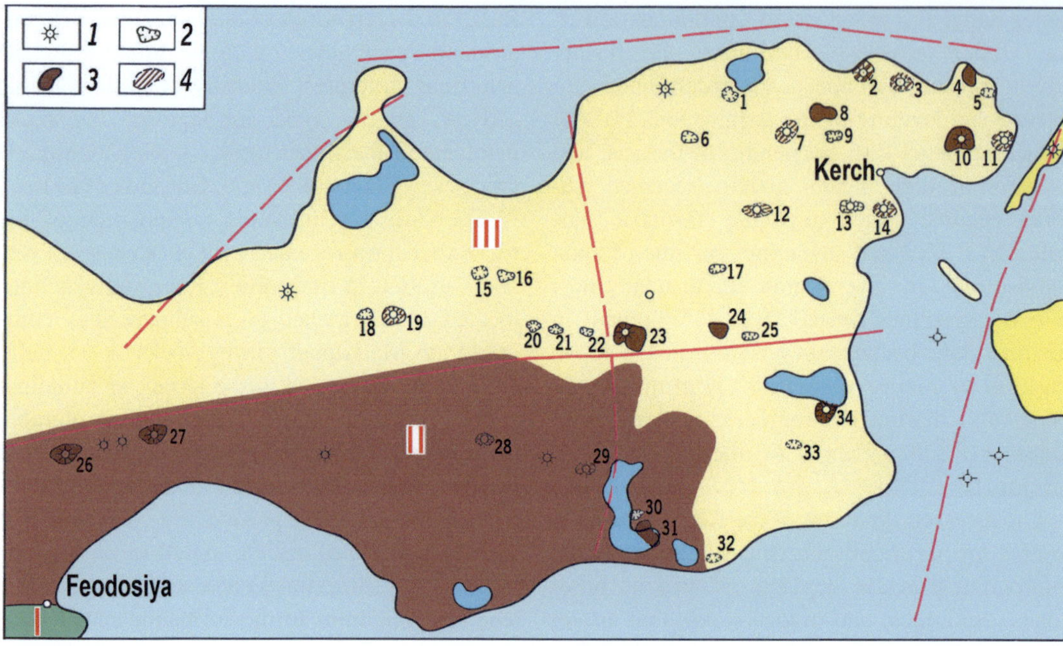

Fig. 5.4 Location of recessed synclines on the Kerch Peninsula. (*1*) Mud volcanoes, (*2*) recessed synclines, (*3*) iron ore bearing recessed synclines, (*4*) presumably iron ore-bearing recessed synclines. I–III, the main structural elements of the region. Red dotted lines, faults. Recessed synclines: (*1*) Chokrakskaya, (*2*) Bolshetarkhanskaya, (3) Bulganakskaya, (*4*) Ossovinskaya, (*5*) Borzovskaya, (*6*) Chumnaya Balka, (*7*) Burashskaya, (*8*) Kezenskaya, (*9*) Malobabchikskaya, (*10*) Baksinskaya, (*11*) Enikalskaya, (*12*) Andreevskaya, (*13*) Voskhodovskaya, (*14*) Soldatsko-Slobodskaya, (*15*) Kashikskaya, (*16*) Burulkayskaya, (*17*) Gornostaevskaya, (*18*) Illichevskaya, (*19*) Kenegezskaya, (*20*) Western Kara-Sidzheutskaya, (*21*) Eastern Kara-Sidzheutskaya, (*22*) Kayaly-Sartskaya, (*23*) Novoselovskaya, (*24*) Rep'evskaya, (*25*) Sokolskaya, (*26*) Achinskaya, (*27*) Arma-Elinskaya, (*28*) Dzhau-Tepe, (*29*) Ak-Tube, (*30*) Konchekskaya, (*31*) Uzunlarskaya, (*32*) Opukskaya, (*33*) Chorelekskaya, (*34*) Chongelekskaya

Outcrops of rocks in recessed synclines of the Kerch Peninsula were investigated by many geologists (e.g., Andrusov 1884, 1893; Obruchev 1921; Prokopov 1931; Ilyin 1931; Grechishkin 1931; Davitashvili 1932; Ebersin 1933, 1940; Zhizhchenko 1940; Kolesnikov 1940; Lychagin 1952). Generalization of the stratigraphic data was provided by Alyaev (1947, 1953) and Maymin (1951). Materials on the stratigraphy of ore-bearing deposits are given by Shnyukov and Naumenko (1964).

Almost the entire complex of Tertiary sediments can be found within the composition of recessed synclines on the Kerch Peninsula. The main stratigraphic peculiarity, first noted by Alyaev (1947), is the loss of certain units from their geological sections without any regularity. Maymin (1951) describes the deposition of recessed synclines directly upon Maikopian strata ranging from Lower and Middle Maikopian (e.g., Dzhau-Tepe and Ak-Tube recessed synclines) to Upper Maikopian (almost all other recessed synclines).

As a rule, the geological sequence of recessed synclines begins from the Chokrakian sediments, which are cracked and overlie Maikopian strata with a steep angular unconformity; they reveal thicknesses up to 50 m near the village of Korolyovo. Andrusov (1884) first described the Chokrakian sediments of the recessed syncline near Lake Chokrak. In the carbonate-sandy sediments, he identified shells of numerous mollusks, e.g., *Angulus fuchsi* (Toula), *Donax tarchanensis* Andrusov, and *Lucina dujardini* Deshayes; in the detrital limestones, he found *Cerithium cattleyae* Baily, etc.

Ebersin (1968) identified small gastropods from the genera *Bulla*, *Trochus*, and *Cerithium* in dense, recrystallized, fine-grained, and algal-*Spirorbis* limestones that underlie ore deposits in the Novoselovskaya recessed syncline. This enabled him to identify the age of these sediments as Miocene (probably Chokrakian). However, other researchers consider these sediments to be Middle Sarmatian (Kulichenko, personal communication in mid-1970). Fauna from outcrops and borings of other recessed synclines of the Kerch Peninsula have been insufficiently studied. In some recessed synclines (e.g., Gornostaevskaya), a substantial thickness of mud volcanic breccia is present in the Chokrakian beds.

The Karaganian stage—first described by Andrusov (1965) as "Spaniodont layers"—has a more limited distribution in some recessed synclines compared to the Chokrakian stage. This stage is represented by carbonate facies. Maymin (1951), referring to Andrusov (1893), mentioned the finding of *Spaniodontella*, *Mohrensternia grandis*, *Spirorbis*, and *Serpula* in the Karaganian sediments of Konchekskaya and Chumnaya Balka. Zhizhchenko (1940) discovered shells of *Spaniodontella* in stromatolite and brecciated limestones of the Burashskaya as well as limestones, sands, and clays in the Opukskaya structures, which we consider to be recessed synclines after Alferov (1931) and Gubanov (1961). The upper Karaganian marls contain folads, some of which were identified as *Pholas dactylus* L. var. *callosa* Cuv., as well as elongated ostracoda. The thickness of the Karaganian sediments in recessed synclines reaches 25 m near the village of Korolevo.

The Konkian beds enriched with folads have a restricted spatial distribution. They are found in the Konchekskaya, Uzunlarskaya, Opukskaya, Gornostayevskaya, and Malobabchikskaya recessed synclines. Their fauna have not been studied in detail. Only in the Malobabchikskaya anticline, where the recessed syncline of the same name formed, did Arkhangel'skiy et al. (1930) identify some other mollusks (e.g., *Corbula gibba* O., *Syndesmia alba* Wood var. *scythica* Sok., *Cardium* ex gr. *vindobonense* Pt., *Spirialis andrussovi* Kittl) in addition to folads. The thickness of the Konkian deposits does not exceed 10 m.

The Sarmatian stage is the most widely distributed within recessed synclines. It is subdivided into three substages: the Lower, Middle, and Upper Sarmatian. Carbonate-clay facies dominate, and their total thickness exceeds 100 m.

None of the studied recessed synclines revealed undisturbed sections of Sarmatian sediments. As a rule, only one substage can be found, with the exception of the eastern and central parts of the Kerch Peninsula, where Middle and Upper Sarmatian deposits were observed. Also, Sarmatian beds are not generally underlain by Konkian ones, as there is no uniform bedding in the sequence of the Sarmatian substages within recessed synclines; outside the recessed synclines, uniform bedding does exist.

The Lower Sarmatian substage has a limited distribution. Sediments of this age were encountered in the Slyusarevskaya anticline as well as in the northern and southwestern parts of the Kerch Peninsula. Paleontologically, they remain insufficiently studied.

The Middle Sarmatian substage shows a wider distribution. Its sediments were observed in the Sartskaya, Slyusarevskaya, and Korolevskaya anticlines. Its greatest presence is revealed in the Ossovinskaya recessed syncline, where it is represented by deepwater clays with *Cryptomactra pesanseris* Mayer-Eimar and *Cardium barboti* R. Hoern (Kolesnikov, 1940). Andrusov (1893) described similar sediments in the upper part of the Kezenskoe mineral field as limestones containing the mollusks *Mactra vitaliana* Orb. var. *ponderosa* Eichw., *Tapes gregarius* Partsch., *Donax hornesi* Sinz., *Solen subfragilis* Eichw., *Cardium fittoni* Orb., *C. donginski* Nordm., *C.* cf. *karreri* Fuchs., *Nassa duplicata* Orb., and the foraminifera *Ammonia beccarii* (Orb.), *Nonion depressulum* (Walker & Jacob), and *Elphidium striatopunctatum* (Orb).

Marls of the upper Middle Sarmatian substage in the Ossovinskaya recessed syncline as well as the nubecularian limestones that replace them to the west and are found in the Burashskaya, Kayaly-Sartskaya, and Western

Kara-Sidzheutskaya recessed synclines contain the following fossils: mollusks, *Tapes vitalianus* Orb., *Modiola volhynica* Eichw., *Modiola clenysiana* Orb., *Cryptomactra* (nov. g.) *pes anseris* C. Mayer, *Cardium barboti* R. Hoern., *C. papyraceum* Sinz., *C. loveni* Nordm., *C. pseudosemisulcatum* Andrus., *Nassa verneuili* Orb., *N. subsriatula* Sinz., *Phasianella kischeneviala* Orb., *Delphinula squamosospinosa* Sinz., *Acmaea angulata* Orb., *A. compressiuseula* Eichw.; bryozoans, *Schizoporella sanguinea* Nordm., *Diastopora primigenia* Eichw.; and foraminifera, *Vertebralina sarmatica* Karr. The nubecularian limestones of the Dzhau-Tepe and Kashikskaya recessed synclines contain abundant *Nubecularia* and rare *Mactra caspia* Eichw. (Andrusov 1893).

Sediments of the Upper Sarmatian substage are rarely present in recessed synclines in the center and north of the Kerch Peninsula. They are represented by carbonate sandstones with *Mactra caspia* Eichw. (e.g., in the Opukskaya recessed syncline) and clays with diatoms *Grammatophora*, *Achnanthes*, *Cocconeis*, *Navicula*, and *Epithemia* that give them the appearance of diatom shapes (trepels). Sometimes, terrestrial *Helix* and *Pupa* are present (e.g., Babchikskoe Gorge).

Meotian stage sediments reveal a limited distribution in recessed synclines. They have been identified in the Kezenskaya structure where they are represented by Lower Meotian reef limestones and clays with *Potamides* aff. *novorossica* (Sinz.), *Membranipora*, *Hydrobia*, *Modiola*, and *Helix* (Kolesnikov 1940). Their thickness varies from 50 m (Kezenskaya) to 100 m (Opukskaya).

Pontian beds overlie Meotian and older geological formations with an angular unconformity and are developed mainly in the recessed synclines of the Vneparpachskiy region of the Kerch Peninsula. Carbonate facies dominate. In the Burashkaya recessed syncline, they contain *Didacna subcrassatellata* Andrus. (Andrusov 1918) and bones of whales (Seninskiy 1904); in the Ossowinskaya, they contain *Paradacna abichi* Hoern (Kolesnikov 1940); and in the Rep'evskaya—*Sopgeria subcarinata* Desh., which is typical of the Upper Pontian (Ebersin 1940).

The transition from the Pontian to the Kimmerian stage deserves special attention due to its high variability from place to place. On the Kerch Peninsula, the latter overlie the former transgressively, sometimes with an unconformity as can be observed in the fields of mineral deposits of the Kerch Peninsula (Shnyukov 1965) and in the coastal zone of the Sea of Azov (Semenenko 1965). The relationship between the Kimmerian and Pontian stages in Krasnodar territory seems to be different, as the former overlie the latter concordantly. According to Andrusov (1903), there was underwater folding of the bottom sediments at the Pontian-Kimmerian boundary. This caused a visible continuity in their transition despite the small angular unconformity.

The Pontian-Kimmerian boundary can be traced in the Baksinskaya and Rep'evskaya recessed synclines only and is characterized there by complex intertransitions. At Baksinskaya, the contact between the Pontian and Kimmerian sediments—composed of loose shell limestones that Andrusov (1918) called phalena—is typical for the Kerch mineral fields. These sediments differ lithologically, and there is a sharp angular unconformity because phalena have steeper angles of inclination.

The Kimmerian stage is represented in recessed synclines by ore deposits of the Middle Kimmerian substage. They can be underlain by either Pontian or Meotian strata.

The Rep'evskaya recessed syncline is different. Here, the Pontian sediments (of unknown thickness) are represented by brownish-yellow clayey sands and sandstones with remains of mollusk shells and occasionally clay pebbles. In the central part, they are gradually overlain by Kimmerian sediments (up to 40 m in thickness), while a noticeable angular unconformity can be seen on its periphery. Toward the bottom, the ore layer is markedly impoverished in iron, the content of which gradually approaches the Clarke of concentration for the sandy clay-detrital rock.

The thickness of the Pontian sediments ranges from 70 m to 100 m in the Andreevskaya and Burashskaya recessed synclines, respectively (Naumenko 1966).

The Lower Kimmerian substage is reliably established in the Rep'evskaya, Uzunlarskaya, Baksinskaya, Novoselovskaya, Ossovinskaya, and Kesenskaya recessed synclines and presumably in Soldatskaya and Mayakskaya as well. In the first two, the Lower Kimmerian formations were discovered by Ebersin (1933, 1940) and in Novoselovskaya by Shnyukov and Naumenko (1964) and Shnyukov et al. (1971). In the Lower Kimmerian ferruginous sandy coquina from the Rep'evka and Konchek mountains, Ebersin (1933) identified *Arcicardium subacardo* Andrus., *Didacnomya corbuloides* Desh. var. *major* Andrus., and others. In the Lower Kimmerian clayey conglomerates of Novoselovka village, Semenenko (1965) found numerous shells of *Dreissensia iniquivalvis* Desh., *D. rostriformis akmanaica* Andrus., *Plagiodacna carinata* Desh., *Prosodacna* cf. *semisulcata* Reuss, *Paradacna stratonis* Andrus., and *Arcicardium* sp. (drillhole 47, internal 150–160 m), while Ebersin (1968) identified in tobacco clays *Didacnomya major* Andrus., *Stenodacna praeangusticostata* Ebers., poorly preserved *Plagiodacna* (?) sp., and *Viviparus duboisi* Mayer. The maximum thickness of the Lower Kimmerian deposits in well 41 drilled into the Novoselovskaya recessed syncline is 56 m.

In the Middle Kimmerian ore layers of recessed synclines, there are fairly abundant molluscan fauna reminiscent of the typical Kerch mineral deposits. Dreissenids and cardiids predominate. Freshwater gastropods occur in a subordinate amount. In the iron ore deposits of the Novoselovskoe mineral field, Semenenko (1965) identified numerous shells of *Dreissena iniquivalvis* Desh., *D. angusta* Reuss, *D. polymorpha* Pall., *Didactia* ex gr. *multistriata* (Reuss), *Arcicardium* sp., *Oraphocardium* sp., *Prosodacna* sp., *Stenodacna* sp., *Pseudocatillus* sp., *Viviparus* sp., *Bithynia* sp., and *Micromelania* sp. Ebersin (1968) also found there *Dreissena angusta* Reuss, *D. iniquivalvis* Desh., *D. rostriformis* Desh., *Arcicardium acardo* Desh., *Didacna gourieffii* Desh., *D. romanensis* R. Hoern., *D. crassatellata* Desh., *L. (Tauricardium) squamulosum* Desh., *Oraphocardium* sp., *O. alatoplanum* Andrus., *Paradacna stratonis* Andrus., *Plagiodacna modiolaris* Reuss, *Prosodacna* ex gr. *macrodon* Desh., *Pseudocatillus donacoides* Andrus., *P. zlatarskii* Andrus., *Pterodacna edentula* Desh., *Stenodacna angusticostata* Reuss, *Bithynia cyclostoma* Reuss, and *Hydrobia* (?) sp.

In the Kimmerian deposits of recessed synclines as well as typical Kerch troughs, the bones of whales and seals have been found. Seninskiy (1904) discovered bones of *Cetotherium* and *Phoca* in the iron ore layers at the periphery of the Baksinskaya and Uzunlarskaya recessed synclines. Opinions differ on the origin of these whale and seal bones, but most researchers support the idea of Andrusov (1918) about their redeposition from Sarmatian deposits as evidenced by their consistent roundness. Pidoplichko (1966) and Bryuzgin (1968), however, think that seals inhabited Kimmerian water reservoirs of the recessed synclines. In fact, fish bones have been discovered in the iron ore deposits (drillhole 47) of the Novoselovskaya recessed syncline. There are also rare remains of terrestrial vertebrates, which are absent outside of these structures. In the ore deposit of the Uzunlarskoe mineral field in the debris of the coastal scarp, Shnyukov and Naumenko (1964) discovered bones of deer *Megaloceras* ex gr. *verticornis* and horse *Equus* sp. (small individual, identified by A.I. Shevchenko) and bones of Stenon zebra (Bryuzgin, personal communication July 24, 1984). In the Rep'evskaya recessed syncline, at the Pontian-Kimmerian boundary, Pavlova (1902) discovered bones of *Mastodon borsoni* Hays. Apparently, the recessed synclines possessed the most favorable conditions for burial and preservation of vertebrates that died in the viscous environment of rapidly accumulating clay and ore deposits. In such settings within the Novoselovskaya and Rep'evskaya structures, half-charred tree trunks were found. The thickness of the Middle Kimmerian deposits in the Novoselovskaya depression varies from 40 to 99 m.

Upper Kimmerian deposits are tentatively distinguished in the Kezenskaya and Novoselovskaya recessed synclines. Faunistically, they are characterized only in the latter where their maximum thickness reaches 57 m. Ebersin (1968) discovered there *Pachydacna* ex gr. *cimmerica* Andrus., *Prosodacna* ex gr. *macrodoti* Desh., *Didacna tamanensis* R. Hoern., *Dreissena supracimmeria* David., *D. rostriformis* Desh., and fragments of *Pseudocatillus* sp.

The Kuyalnitskian stage has limited distribution in recessed synclines. It is represented by sands, clayey silts, and clays, all without organic remains. They overlie unconformably the older strata only in the peripheral parts of the Baksinskaya, Kesenskaya, Novoselovskaya, and some other structures. Their greatest thickness is 60 m in the Baksinskaya recessed syncline.

Quaternary deposits are represented by brownish-gray clays and loams, and they are developed almost everywhere within recessed synclines.

The structure of the recessed synclines of the Kerch Peninsula has been long discussed in the geological literature. It was noted that they are confined to localities on the axes of anticlinal structures showing the greatest swelling (Lychagin 1952) but not necessarily in their central parts. For example, the Rep'evskaya, Alagolskaya, Kayaly-Sartskaya, western and eastern Kara-Sidzheutskaya, and Novoselovskaya recessed synclines are located within the Sartskaya anticline. They can also be located on the anticline's periphery, especially in the case of anticlinal asymmetry. For example, the Ilyichevskaya recessed syncline is localized at the extreme western part of the Korolevskaya anticline. The Burashskaya recessed syncline lies at the periphery of a large anticline where the maximum uplifting of folding is confined.

Recessed synclines, it seems, can be traced only in the upper structural floor of the Maikopian sequence that encloses them; in the lower structural floor, they fade out. Usually, they are located in close proximity to the mud volcanic foci and are confined to the axes of anticlinal structures. The latter, according to Belousov (1960), are situated above peculiar tectonic seams, such as the junctions between blocks of the Paleozoic basement. In other words, for the Kerch Peninsula, one can assume a relationship between mud volcanism (with its derivatives, such as recessed synclines) and deep faults.

There is a direct relationship between the size of the anticlines and the number of recessed synclines or, more precisely, with their total area. The larger the anticlinal structure, the greater the area encompassed by the recessed synclines that develop within it. This relationship can be very clearly seen in the Vulkanovskaya, Ak-Tubinskaya, Korolevskaya, Slyusarevskaya, and Sartskaya recessed synclines, as well as some others. Generally, recessed synclines occupy between 3% and 17.5% of the area occupied by their anticlinal structure.

The recessed synclines of the Kerch Peninsula reveal an isometric shape. They can be rounded (from circular to an elongated oval), bowl-shaped, half-moon, etc. (Lychagin 1952), and their sizes can vary from hundreds of meters to 4 km in diameter. The largest recessed syncline is Kezenskaya (27 km^2), and the smallest is Kara-Sidzheutskaya (0.1 km^2). Their general features include (1) steep inclinations in the layers at the periphery of the structures and (2) greater steepness in the underlying beds compared to those above indicating a continuous bending of the recessed synclines from the Maikopian to the Quaternary. The rate of this bending can be calculated for different moments in geological time using the thickness of the sediments deposited during different ages.

The steep angles of inclination along the sides of recessed synclines are caused by vertical movements. When certain angles of inclination are reached, a variety of bedding discontinuities occur. The bending of recessed synclines results jointly from continuous mud volcanic activities and the accumulation of erupted material that places increasing pressure on the bottom of the structures.

Disturbances in the continuity of the strata in recessed synclines are, however, somewhat unusual. They represent a kind of ring or semiring fault, along which the concave bowls of the depressions drop, forming small grabens. This was noted for the first time by Prokopov (1931), who described such ring faults, for example, in the Rep'evskaya recessed syncline.

In some cases, the amplitude of ring faults reaches hundreds of meters. The peripheral areas of recessed synclines that remain exposed are almost universally eroded. While the descending central part of the structure is often characterized by almost horizontal bedding, the layers on its periphery are steeply inclined (e.g., in the Novoselovskaya). Ring faults are not always present, however. Instead, almost all structures framed by bryozoan or shell limestone rings are marked by radially directed transverse disturbances.

Lychagin (1952) exploring the mechanism of formation of recessed synclines explained this as follows. Bryozoan or molluscan limestone strata around depressions incline toward the outside in early stages of sinking, but later, with greater descent of the trough bottom, they begin to slope down into the depression and crack parallel to the rim of the syncline (transverse faulting?) (Table 5.1).

In their structure, the recessed synclines of the Kerch Peninsula can be divided into two main types. The first type includes those recessed synclines in which erupted material accumulates in the center of the structure, where the mud volcano itself is located. The calderas, in fact, represent the compensatory troughs filled with Neogene sediments that delineate its boundaries. The rocks composing the recessed synclines themselves form the frame of the caldera. The second type includes those recessed synclines in which erupted material accumulates in the vicinity of the mud volcano, not in the center of the recessed syncline. In this case, the development of compensatory troughs occurs to the side, and accumulated sediments do not outline the calderas, which apparently are eventually destroyed by weathering processes.

On the Taman Peninsula, recessed synclines are far rarer than on the Kerch Peninsula. They have been discovered only in a few places: the Polivadina Balka (Buryak, personal communication, autumn 1974); on the shore of the Bugazskiy Liman; and in the geological section of the Makotra mountain, where shallow Pontian rocks form a small bowl with steep angles of inclination. The Makotra mountain bowl looks as if it had been inserted into the mud volcanic breccia. Recessed synclines have also been found near the Kapustinskaya gully, where Vassoevich (1928) recorded a synclinal complication on the axis of the large diapiric structure.

Of particular interest is the recessed syncline hidden under thick Pliocene-Quaternary deposits within the Anastasievsko-Troitskaya anticline (Shaulov 1967, 1968). The latter is an integral part of the Anastasievsko-Krasnodarskaya anticlinal zone, stretching from the Taman Peninsula in the west to the city of Krasnodar in the east (Fig. 5.5).

The Anastasievsko-Troitskaya anticline is composed of Lower-Middle Pliocene sediments and consists of three arches separated by shallow saddles lying on the axis of the northwest-southeast strike. The northwestern arch is called the Anastasievskiy uplift; the two southeastern parts belong to the Trotskiy uplift.

Deep wells drilled into the northwestern periclinal fold of the Anastasievsko-Troitskaya anticline have revealed a buried syncline at depths of 1300–2200 m in a kind of oval graben. The latter is filled with a thick layer (up to 300 m) of mud volcanic breccia underlain by Upper Sarmatian and Meotian and overlapped by Pontian deposits (Fig. 5.6).

From the surface, the entire graben is covered by Pliocene (Kimmerian, Kuyalnitskian) and

Table 5.1 Correlation between the angles of rock inclination and faults

Recessed syncline	Wing	Angle of inclination	Tectonic disturbance
Rep'evskaya	Southern	10	Absent
	Northern	50	Fault
Novoselovskaya	Western	45	Fault
	Eastern	20	Absent
Burashskaya	Western	35	Fault
	Eastern	15	Absent

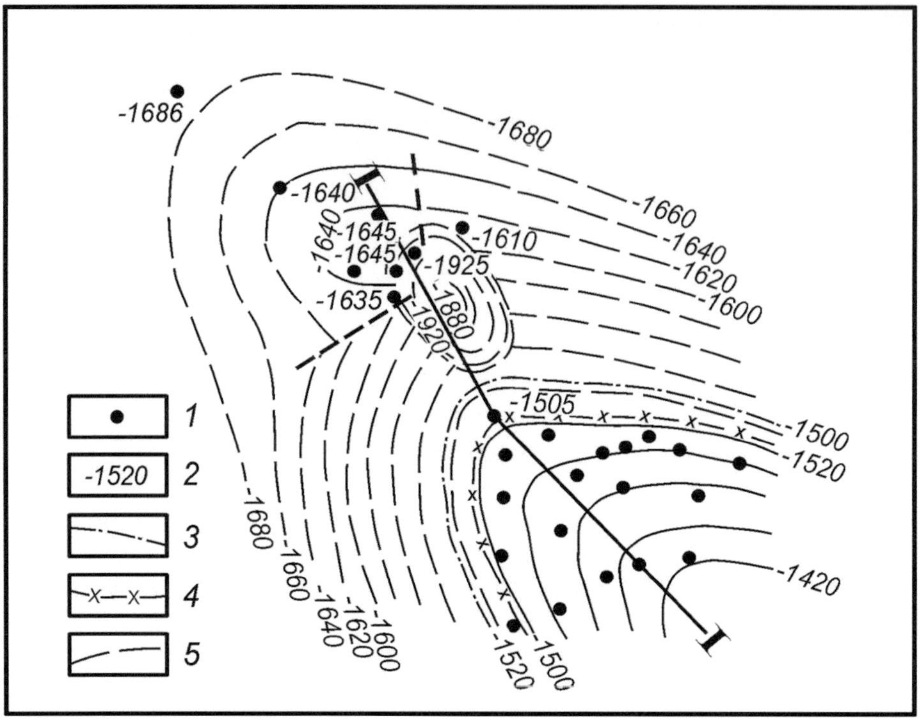

Fig. 5.5 Structural diagram of the northwestern periclinal fold of the Anastasievsko-Troitskaya anticline (Krasnodar Territory): (*1*) drilled wells, (*2*) isohypses of the roof of the IV meiotic horizon (see Fig. 6.6), (*3*) the outer contour of the oil-bearing capacity of the main horizon deposit, (*4*) the outer contour of gas-bearing capacity, (*5*) tectonic fracture (Shaulov 1967)

Quaternary sediments with a total thickness of ~1300 m. This graben reveals a number of common features that it shares with recessed synclines of the Kerch Peninsula, such as the ringlike shape of the surrounding tectonic ruptures, the presence of mud volcanic breccia, and the loss of stratigraphic units (e.g., part of the Meotian) (Shaulov 1967, 1968). In this area, mud volcanic activities ceased in the early Pontian; however, subsidence continued subsequently. Despite the great thickness of the overlapping Pliocene-Quaternary sediments, a slight depression can be seen even in the modern relief due to the compacting of the breccia. The formation of the Anastasievsko-Troitskaya recessed syncline can be explained by the turbulent mud volcanic activity on the Taman Peninsula at the Meotian-Pontian boundary, as well as by favorable paleogeographic conditions that preserved this syncline. New findings of similar structures on the Taman Peninsula, as well as in northwestern Ciscaucasia, can be expected.

Recessed synclines are almost unknown at the southeastern tip of the Caucasus. This region is characterized by intense modern mud volcanic activity. An analogue of the Kerch recessed synclines was recorded only at the Malaya Kyaniza mud volcano, which is currently inactive (Fedorov 1939). This volcano is located 4 km west of the village of Duvanny. In shape, it looks like a truncated cone 304 m high and 1.5 km in diameter at its base. Initially, it was supposed that Malaya Kyaniza was a mud volcano confined to a syncline. This impression was created by the Bakinian limestones inclined toward the center of the volcano at an angle of 5–10°. In fact, the Duvanny dome is a typical arch of the anticline composed of rocks from the Productive Formation of the Akchagylian, Apsheronian, and Bakinian stages. In their northwestern plunge, the

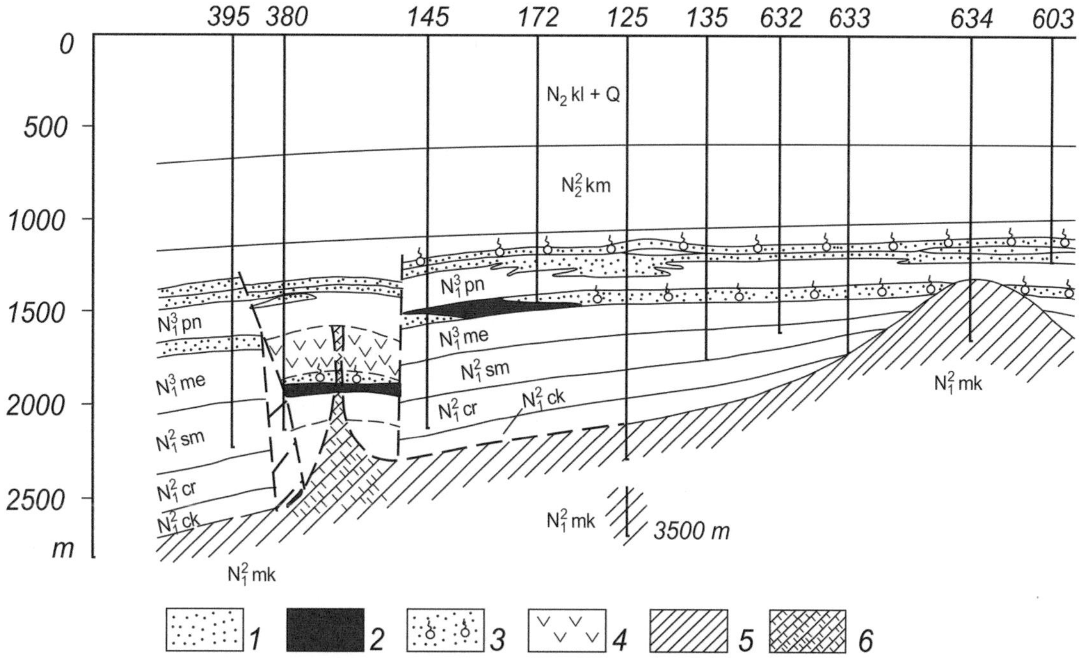

Fig. 5.6 Longitudinal geological section through the northwestern periclinal fold of the Anastasievsko-Troitskaya anticline: (*1*) sand, sandstone, and siltstone, (*2*) oil, (*3*) gas, (*4*) fossil breccia, (*5*) crumpled Maikopian clay, (*6*) breccia in the mud volcano caldera (Shaulov 1967). For stratigraphic indexes see Fig. 6.4

Akchagylian rocks sloped periclinally, and the overlying Bakinian deposits produce a bowl-like structure in the center of which this mud volcano is located (Fig. 5.7).

According to Fedorov (1939), the bending of the Bakinian limestones was caused by the complications of brachianticline formation due to the activity of mud volcanic processes as in the forming of the Burashskaya recessed syncline on the Kerch Peninsula.

As we can see, almost all authors emphasize the mud volcanic origin of compensatory troughs and the synclinal arrangement of Tertiary sediments and breccia that produces them. In this connection, these structures should be called "compensatory synclines" of mud volcanic origin. This name reflects both the structural and genetic features of all landforms of this type in different regions. As far back as the early 1970s, Shnyukov et al. (1971) noted that the term "recessed syncline," widely used and applied by many researchers, is insufficient, as it does not imply features reflecting genesis. Similarly, the definition of these structures as complications is not entirely successful either. This term has been used for many years, however, and has become familiar to a wide range of geologists. In this connection, it seems to us, it is possible to admit its use as a local, Kerch designation, reserving the use of "compensatory syncline" as the more universal and precise term, generally characteristic of structures of this type from different regions.

Analysis of geological data shows that the most important condition for the development of recessed (compensatory) synclines is the combination of favorable paleogeographic conditions with intensive mud volcanic activity. The Kerch Peninsula is now a region of damped mud volcanism. Here, the period of intensive mud volcanism was marked by the existence of a marine regime over the greater part of the Kerch Peninsula during the Neogene, as a result of which the peninsula can be considered a unique region for the development of these peculiar structures. Southwestern Turkmenistan is now also characterized by damped mud volcanic processes (Sungurov 1958).

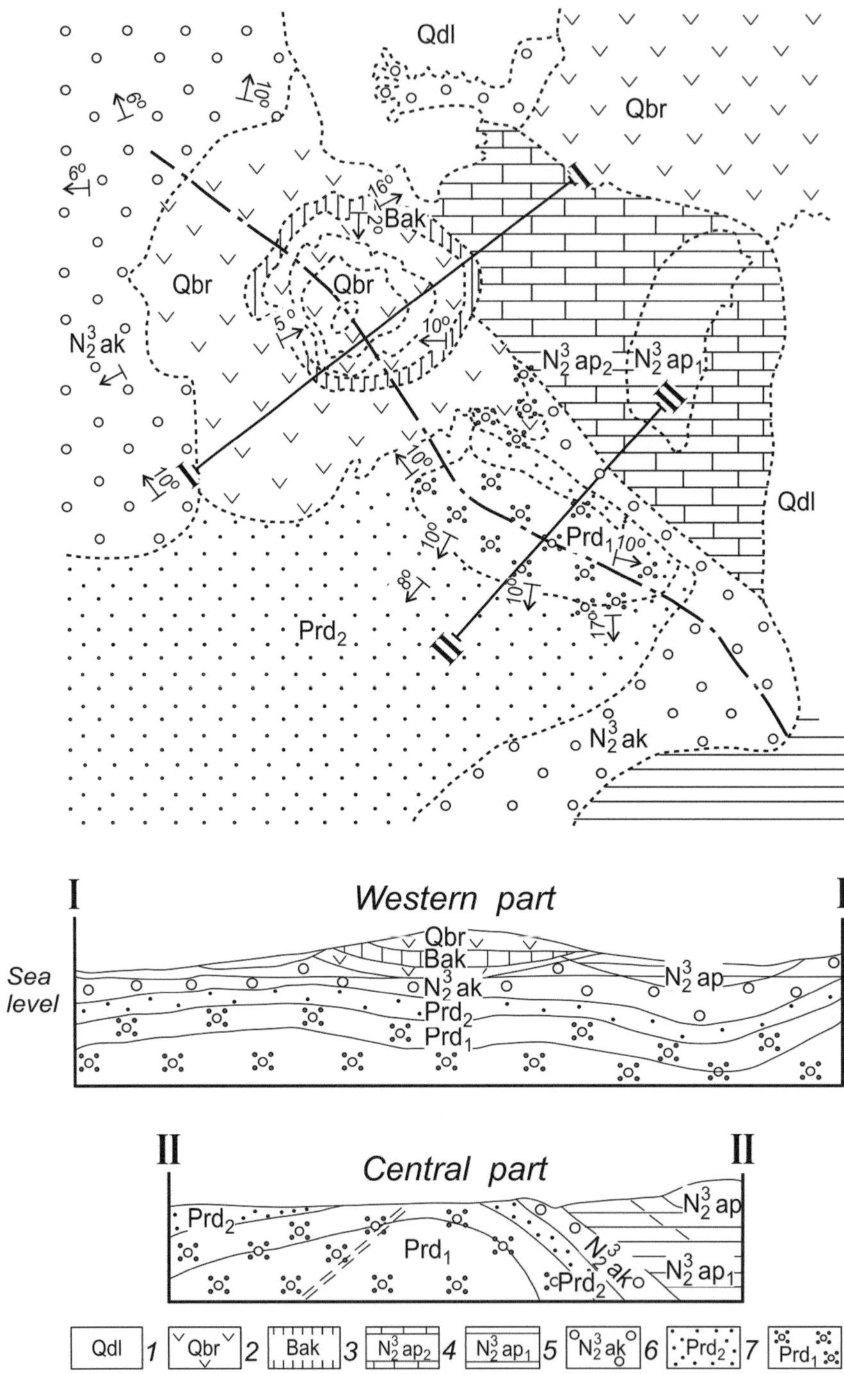

Fig. 5.7 Geological scheme and sections of the area surrounding the recessed syncline of the Malaya Kyaniza mud volcano, village of Duvanny (Azerbaijan): (*1*) modern sediments, (*2*) mud volcanic breccia, (*3*) Bakinian, (*4*) Middle Apsheronian, (*5*) Lower Apsheronian, (*6*) Upper Akchagylian, (*7*) the upper 100 m of the Productive Series, (*8*) Productive Series (Fedorov 1939)

Undoubtedly, the mud volcanoes of the Black Sea are also accompanied by compensatory synclines. Presumably, the mechanism of their formation is similar to that of the Kerch structures. One can assume a fairly long period of activity for the Black Sea volcanoes, which are, moreover, located in an aquatic environment, and especially in deep-sea locations. Most definitely, these would have resulted in the development of compensatory synclines, although the sediments they comprise would be different from subaerial analogues. Their study is one of the most important tasks of research into mud volcanic processes, because these specific structures can be a "storehouse" of varied mineral resources.

References

Alekseev P (1880) Zametki o gryazevykh vulkanakh (Notes about mud volcanoes). Zapiski Kievskogo Obschestva Estestvoispytateley 6(1):137–163 (in Russian)

Alferov VA (1931) Geologorazvedochnye raboty na neft v yugo-vostochnoy chasti Kerchenskogo poluostrova (Geological exploration for oil in the southeastern part of the Kerch Peninsula). Trudy GGRU 39:10 (in Russian)

Alyaev SE (1947) Novye dannye o tektonike Kerchenskogo poluostrova (New data about tectonics of the Kerch Peninsula). Izvestya AN SSSR Seriya Geol 6:97–99 (in Russian)

Alyaev SE (1953) Ob osobennostyakh stroeniya struktur Kerchenskogo poluostrova (On the peculiarities of the structures of the Kerch Peninsula). Byulleten' moskovskogo obshchestva ispytateley prirody, Otdelenie geologii XXVIII(2):70–72 (in Russian)

Andrusov NI (1884) Geologicheskie issledovaniya na Kerchenskom poluostrove v 1862–1883 g. (Geological research on the Kerch Peninsula in 1862–1883). Zapiski Novorossiyskogo obshchestva estestvoispytateley X:1–76 (in Russian)

Andrusov NI (1893) O sostoyanii Chernomorskogo basseyna v pliotsenovuyu epokhu (On the state of the Black Sea basin in the Pliocene epoch). Mélanges géologiques et paléontologiques tirés du Bulletin de l'Académie impériale des sciences de St.-Pétersbourg I:165–178 (in Russian)

Andrusov NI (1903) Geologicheskie issledovaniya na Tamanskom poluostrove (Geological research on the Taman peninsula). Materialy po geologii Rossii 21(2):257–383 (in Russian)

Andrusov NI (1918) Geologicheskoe stroenie dna Kerchenskogo proliva (Geological structure of the bottom of the Kerch Strait). AN SSSR Seriya 6 12(1):23–28 (in Russian)

Andrusov NI (1965) O vozraste morskikh posletretichnykh terras Kerchenskogo poluostrova (On the age of the marine post-tertiary terraces of the Kerch Peninsula). Izbrannye trudy, Nauka, Moscow, IV, pp 143–163 (Reprint, in Russian)

Arkhangel'skiy AD (1925) Neskol'ko slov o genezise gryazevykh vulkanov Apsheronskogo poluostrova i Kerchensko-Tamanskogo regiona (A few words about the genesis of the mud volcanoes of the Absheron Peninsula and the Kerch-Taman region). Bul MOIP Geological Branch 33(3/4):269–285 (in Russian)

Arkhangel'skiy AD, Blokhin AA, Menner VV et al (1930) Kratkiy ocherk geologicheskogo stroeniya neftyanikh mestorozhdenii Kerchenskogo poluostrova (A brief sketch of the geological structure of the Kerch Peninsula oil fields). Trudy Glavnogo Geologo-Razvedochnogo Upravleniya VSNKH SSSR 13:5–97 (in Russian)

Avdusin PP (1948) Gryazevye vulkany Krymsko-Kavkazskoy geologicheskoy provintsii (Mud volcanoes of the Crimean-Caucasian geological province). AN SSSR, Moscow (in Russian)

Belousov VV (1960) Nekotorye obshchie voprosy tektoniki oblasti sopryazheniya Kryma i Kavkaza (v svyazi s problemoy proiskhozhdeniya skladchatosti) (Some common questions of tectonics in the region of conjugation of Crimea and the Caucasus [in connection with the problem of the origin of folding]). Problemy tektonofiziki. Gosgeoltekhizdat, Moscow, pp 261–275 (in Russian)

Blagovolin NS (1960) Proiskhozhdenie i istoriya razvitiya Kerchenskogo proliva (The origin and history of the Kerch Strait). Izvestiya AN SSSR, seriya geograficheskaya 2:105–109 (in Russian)

Bryuzgin K (1968) K voprosu izucheniya iskopaemykh kitoobraznykh i tyuleney Kerchenskogo poluostrova (On the study of fossil cetaceans and seals of the Kerch Peninsula). Tezisy dokladov III nauchnoy konferentsii po izucheniyu Kerchenskikh osadochnykh zheleznykh rud, Kiev, p 37 (in Russian)

Davitashvili LSh (1932) O stratigraficheskom podrazdelenii kimmeriyskogo yarusa (On the stratigraphic division of the Kimmerian stage). Byulleten' moskovskogo obshchestva ispytateley prirody, Otdelenie geologii XI(4):268–383 (in Russian)

Dvoychenko PA (1914) Izverzhenie gryazevoy sopki Dzhau-Tepe (Eruption of mud volcanic hill Dzhau-Tepe). Priroda 4:614–619 (in Russian)

Ebersin AG (1933) Geologicheskie issledovaniya zhelezorudnykh mestorozhdeniy Kerchenskogo poluostrova (Materialy k stratigrafii pliotsena Kerchenskogo poluostrova) (Geological studies of the iron ore deposits of the Kerch Peninsula [Materials for the Pliocene stratigraphy of the Kerch Peninsula]). Trudy Vsesoyuznogo geologo-razvedochnogo obedineniya NKTP SSSR 325:129–167 (in Russian)

Ebersin AG (1940) Sredniy i verkhniy pliotsen Chernomorskoy oblasti. Kimmeriyskiy yarus (Middle and Upper Pliocene of the Black Sea. Kimmerian stage). Stratigrafiya SSSR, Neogen, pp 477–527 (in Russian)

Ebersin AG (1968) O kimmeriyskikh otlozheniyakh Novoselovki (Kerchenskiy p-ov) (On the Kimmerian deposits of Novoselovka). Tezisy dokladov chetvertoy nauchno-tekhnicheskoy konfererentsii po izucheniyu poleznykh iskopaemykh otlozheniy osadochnogo kompleksa yuga Ukrainy. KGU, Kiev, pp 30–31 (in Russian)

Fedorov SF (1939) Gryazevye vulkany Krymsko-Kavkazskoy geologicheskoy provintsii i diapirizm (Mud volcanoes of the Crimean-Caucasian geological province and diapirism). Rezul'taty issledovaniya gryazevykh vulkanov Krymsko-Kavkazskoy geologicheskoy provintsii. AN SSSR, Moscow-Leningrad, pp 5–44 (in Russian)

Golovkinskiy NA (1889) Otchet gidrogeologiya Tavricheskoy zemskoy upravy za 1989 god (Report of the hydrogeologist of the Taurian Zemstvo government for 1889). Simferopol (in Russian)

Grechishkin LA (1931) Issledovanie neftyanykh mestorozhdeniy v severnoy i tsentralnoy chastyakh Kerchenskogo poluostrova (Study of oil fields in the northern and central parts of the Kerch Peninsula). Geologo-razvedochnye issledovaniya v neftenosnykh rayonakh Kerchenskogo poluostrova v 1926 gody 2:49–71. Geologizdat, Moscow-Leningrad (in Russian)

Gubanov IG (1961) K geomorfologii Kerchenskogo poluostrova (On the geomorphology of the Kerch Peninsula). Izvestiya Krymskogo otdeleniya geograficheskogo obshchestva 6 (in Russian)

Ilyin SN (1931) Issledovanie neftyanykh mestorozhdeniy vostochnoy chasti Kerchenskogo poluostrova (Investigation of oil fields in the eastern part of the Kerch Peninsula). Trudy Geologo-Razvedochnogo Upravleniya 39:37–48 (in Russian)

Kolesnikov VP (1940) Verkhniy miotsen. Nizhniy pliotsen (Upper Miocene. Lower Pliocene). Stratigrafiya SSSR. AN SSSR, Moscow (in Russian)

Kosygin AI (1935) Melitopolskiy gazonosnyy rayon (Melitopol gas region). Prirodnye gazy SSSR. ONTI (in Russian)

Kovalevskiy SA (1935a) Gazovyy vulkanizm (vulkany i vulkanoidy) (Gas volcanism [volcanoes and volcanoids]). Azerbaydzhanskoe neftyanoe khozyaystvo 1 (in Russian)

Kovalevskiy SA (1935b) Yeshche o gazovom vulkanizme (More about gas volcanism). Azerbaydzhanskoe neftyanoe khozyaystvo 9 (in Russian)

Lychagin GA (1952) Iskopaemye gryazevye vulkany Kerchenskogo poluostrova (Fossil mud volcanoes of the Kerch Peninsula). Byulleten' moskovskogo obshchestva ispytateley prirody, Otdelenie geologii 27(4):3–13 (in Russian)

Maymin ZL (1951) Tretichnye otlozheniya Kryma (tertiary deposits of Crimea). Gostoptekhizdat, Moscow-Leningrad (in Russian)

Naumenko PI (1966) Novye dannye po geologii i metallogenii vdavlennykh sinklinaley Kerchenskogo poluostrova (New data on geology and metallogeny of depressed synclines of the Kerch Peninsula). Tezisy dokladov III nauchno-tekhnicheskoy konferentsii po izucheniyu kerchenskikh i drugikh osadochnykh zheleznykh rud USSR, Naukova Dumka, Kiev (in Russian)

Naumenko PI (1967) Geologiya vdavlennykh sinklinaley Kerchenskogo poluostrova (Geology of recessed synclines of the Kerch Peninsula). Avtoref. dis. kand. Geol. nauk, Odessa I.I. Mechnikov State University (in Russian)

Obruchev VA (1921) Mestorozhdeniya nefti i gaza Kerchenskogo poluostrova (Oil and gas fields of the Kerch Peninsula). Neftyanoe i slantsevoe khozyastvo 5/8:181–221 (in Russian)

Obruchev VA (1926) Novye tendentsii v tektonike (New trends in tectonics). Izvestiya Geologicheskoy Commissii 45:117–140 (in Russian)

Pavlova MV (1902) Novye nakhodki *Mastodon borsoni dart* na yuge Rossii (New finds of *Mastodon borsoni dart* in the south of Russia). Yezhegodnik po geologii i mineralogii Rossii 5(1):27–33 (in Russian)

Pidoplichko IG (1966) O nekotorykh voprosakh paleogeograficheskikh issledovaniy (On some issues of paleogeographic studies). Tezisy dokladov III nauchnoy konferentsii po izucheniyu Kerchenskikh osadochnykh zheleznykh rud, Kiev, pp 4–5 (in Russian)

Plotnikov AM (1967) Poteri uglevodorodov na gryazevykh vulkanakh Kerchenskogo poluostrova (Loss of hydrocarbons in mud volcanoes of the Kerch Peninsula). Geologiya i neftegazonosnost Prichernomorskoy vpadiny. Naukova Dumka, Kiev (in Russian)

Prokopov KA (1931) Geotektonicheskiy ocherk Kerchenskogo poluostrova i otnoshenie ego k Krymu i Tamani (Geotectonic essay on the Kerch Peninsula and its relation to Crimea and Taman). Trudy Geologo-Razvedochnogo Upravleniya 38:13–23 (in Russian)

Rakhmanov RR (1987) Gryazevye vulkany i ikh znachenie v prognozirovanii gazoneftenosnosti (Mud volcanoes and their significance in the prospecting of oil and gas). Nedra, Moscow (in Russian)

Semenenko VN (1965) Oriktotsenozy mollyuskovykh faun v nizhnekimmeriyskikh otlozheniyakh Severnogo Priazovya (Orictocenoses of mollusk faunas in the Lower Kimmerian sediments of the Northern Azov region). Tezisy dokladov seminara-soveshchaniya po metodike izucheniya kimmeriyskikh zheleznykh rud Azovo-Chernomorskoy provintsii, Kiev, pp 22–25 (in Russian)

Semenovich VV (1956) Tektonika Chelekena i gryazevoy vulkanizm (Cheleken tectonics and mud volcanism). Trudy instituta geologii AN TSSR 1:268–279 (in Russian)

Seninskiy KI (1904) Zametka o pliotsenovykh otlozheniyakh okrestnostey Kerchi i Tamani (A note on the Pliocene deposits of the vicinity of Kerch and Taman).

Protokol zasedaniya obshchestva estestvoispytateley prirody pri Yurevskom universiteti 14(1) (in Russian)
Shaulov MA (1967) Gazoneftenosnost' severo-zapadnoy periklinali Anastisievsko-Troitskoy skladki v svyazi s drevnim gryazevym vulkanizmom (Gas-oil-bearing capacity of the northwest pericline Anastisievsko-Troitskaya fold in connection with ancient mud volcanism). Izvestiya vuzov Neft' i gaz 7:5–8 (in Russian)
Shaulov MA (1968) K voprosu o proiskhozhdenii vdavlennostey na svodakh antiklinalnykh skladok (Toward the question of the origin of depressions on the arches of anticlinal folds). Tezisy dokladov IV nauchnteknicheskoy konferentsii po izucheniyu poleznykh iskopaemykh otlozheniy osadochnogo chekhla yuga Ukrainy. KGU, Kiev, pp 30–31 (in Russian)
Shnyukov EF (1965) Genezis kimmeriyskikh zheleznykh rud Azovo-Chernomorskoy rudnoy provintsii (Genesis of Kimmerian iron ores of the Azov-Black Sea ore province). Naukova Dumka, Kiev (in Russian)
Shnyukov EF, Naumenko PI (1964) Kimmeriyskie zheleznye rudy vdavlennykh sinklinaley Kerchenskogo poluostrova (Kimmerian iron ores of recessed synclines of the Kerch Peninsula). Krymizdat, Simferopol (in Russian)
Shnyukov EF, Naumenko PI (1966) Kimmeriyskie zheleznye rudy vo vdavlennykh sinklinalyakh Kerchenskogo poluostrova (Kimmerian iron ores of recessed synclines of the Kerch Peninsula). Soviet Geol 4:142–143 (in Russian)
Shnyukov EF, Naumenko PI, Lebedev YuS et al (1971) Gryazevoy vulkanizm i rudoobrazovanie (Mud volcanism and ore formation). Naukova Dumka, Kiev (in Russian)
Shnyukov EF, Starostenko VI, Kobolev VP (2006) Gazogidratonosnost donnykh otlozheniy Chernogo morya (Gas hydrate potential of sediments of the Black Sea). Geofizicheskiy zhurnal 28(6):29–40 (in Russian)
Shnyukov EF, Kobolev VP, Pasynkov AA (2013) Gazovyy vulkanizm Chernogo morya (Gas volcanism of the Black Sea). Logos, Kiev (in Russian)
Sungurov AM (1958) Osobennosti stroeniya gryazevykh vulkanov Yugo-Zapadnoy Turkmenii (Features of the structure of mud volcanoes in Southwest Turkmenistan). Novosti neftyanoy tekhniki, seriya geologicheskaya 9:12–14 (in Russian)
Vassoevich NV (1928) Materialy po geologii Tamanskogo poluostrova i Anapskogo rayona Steblievskiy mys (Materials on the geology of the Taman Peninsula and Anapa region, Steblievskiy cape). Neftyanoe khozyaystvo (in Russian)
Zhizhchenko VP (1940) Sredniy miotsen (Middle Miocene). Stratigrafiya SSSR 12:51–227. AN SSSR, Moscow-Leningrad (in Russian)

6 Mud Volcanoes of the Black Sea Region

Summary

This chapter covers the entire catalogue of mud volcanoes known so far in the Black Sea region, in particular the Kerch and Taman peninsulas, the northwestern Caucasus, the Black Sea itself, and the southeastern part of the Sea of Azov.

Based mostly on the authors' research data, characteristics of mud volcanoes are provided region by region. Descriptions involve the geological configurations of mud volcanoes, the nature of the associated compensatory subsiding structures (called "recessed synclines"), and contextual details about the larger geological structures within which the mud volcanoes are confined.

The total number of mud volcanoes in the Black Sea region is currently 180. About 100 of them occur on land.

The information offered here is very extensive and not yet covered in the English language literature. Despite their random distribution at the local level, all discovered mud volcanoes are confined to areas where the underlying sediments contain the Maikopian geological formation. The Maikopian formation is almost homogeneous and represented by clays with variable thicknesses ranging up to 5000 m. The pattern of mud volcano distribution does not depend upon the thickness of the Maikopian clays, however, and mud volcanoes are often absent in areas where Maikopian sediments are the thickest. Presumably, mud volcanoes occur most often in the zones of deep tectonic dislocation within the Maikopian sequence and through which deep fluid flows break through. They contribute to the appearance of diapir folding and individual diapirs.

The Kerch-Taman mud volcanic region extends far beyond the Kerch and Taman peninsulas. The largest modern mud volcanic foci, such as Dzhau-Tepe, Bulganakskiy, Gora Gnilaya, Karabetova Gora, and Shugo, are located not along the periphery of the region, but in its axial part. Perhaps there was a different situation in the geological past, or the boundaries of the region were different, but now this trend can be traced quite clearly.

At present, more than 50 mud volcanoes are known on the Kerch Peninsula. At least 50 mud volcanoes have been discovered on the Taman Peninsula and in northwestern Caucasus. Many of them are still active, manifesting with relatively quiet eruptions and constant release of gases, mud breccia, water, and sometimes petroleum products. There are also mud volcanoes that explode violently. According to current knowledge, 67 strong mud volcanic explosions have occurred based on the literature and human recollections. Of course, this figure is understated because it is impossible to count all eruptions in the absence of systematic observations carried out by a special surveillance service. Out of the 68 known strong eruptions, 16 occurred on the Kerch Peninsula, 13 occurred in the Kerch Strait and the Sea of Azov, and 39 were located on the Taman Peninsula and in northwestern Caucasus.

The original version of this chapter was revised: Figure 6.27 has been updated now. The correction to this chapter is available at https://doi.org/10.1007/978-3-030-40316-4_12

E. Shnyukov, V. Yanko-Hombach, *Mud Volcanoes of the Black Sea Region and their Environmental Significance*, https://doi.org/10.1007/978-3-030-40316-4_6

About 80 mud volcanoes are known today in the Black Sea. Most of them are located in deep water, but some volcanoes have been found at relatively shallow depths. Conditionally, they are separated by us into ten groups.

Group I is located in the northwestern part of the Black Sea and includes the Parshin and Vodyanitskiy-2 mud volcanoes.

Group II is localized in the Western Black Sea depression and consists of 11 mud volcanoes: Grigor'ev, MGU, Yuzhmorgeo, Malyshev, Kornev, Goncharov, Mantiynyy, Strakhov, Tredmar, Vassoevich, and Kovalevskiy. As can be seen, the Eastern Black Sea depression is far poorer in mud volcanoes compared to the Western Black Sea basin. This impression may result from insufficient study of the region, but geological reasons are also quite possible.

Group III is localized within the Sorokin Trough that was formed in the Oligocene with the Crimean-Alpine folding and is an extremely interesting area for mud volcano development. Twenty-six mud volcanoes are known within the Sorokin Trough and have been studied in varying degrees of detail. They are grouped into four types. Compared to mud volcanoes in the Western Black Sea depression, those of the Sorokin Trough are smaller, but their gas flares are much more powerful. Such flares are unknown elsewhere in the Black Sea. All mud volcanoes discovered so far in the Sorokin Trough are confined to the same sublatitudinal fault zone and are characterized by the release of gases. Most of these mud volcanoes are active at present, and breccia overlies recent mud sediments. Almost all cores recovered from these mud volcanoes are enriched with gases, and in five of them, gas hydrates were found.

Group IV is localized near Kerch within the Kerch-Taman shelf and continental slope that are located largely within the Kerch-Taman Trough that is composed of Neogene and Paleogene sedimentary rocks transgressively overlapping Upper Cretaceous sediments. The sedimentary strata are folded and dislocated. Geological and geophysical studies have identified 16 morphostructures resembling mud volcanoes. So far, only three mud volcanoes have been confirmed by geological research. Others are proposed based on geophysical methods, but geological investigation has been hampered by thick shell banks that have formed over them.

Group V contains 11 mud volcanoes. Eight of them are localized within the Kerch Strait, and three within the Sea of Azov. One of the most interesting is the underwater Golubitskiy mud volcano that is located about 500 m seaward from the shore in front of the village of Golubitskaya. Numerous eruptions of this mud volcano were almost always accompanied by the emergence of an island that was quickly destroyed by the waves. The places at sea where mud volcanoes are developing may be dangerous for navigation, and for this reason, they are indicated in sailing directions, such as Temryukskaya and Golubitskaya banks.

Group VI includes 16 mud volcanoes located in the Tuapse Trough and Shatskiy Ridge. The mechanism of mud volcanism within the Tuapse Trough is related to abnormal pressure buildup in clay sediments; this leads to the development of mud volcanoes through which deep fluids can outflow to the surface.

Groups VII–X are localized in the Georgian, Turkish, Bulgarian, and Central sectors of the Black Sea. There is not much information available on them and they are not described in this chapter.

6.1 Geological Characteristics of Mud Volcanism in the Black Sea Region

Mud volcanoes were discovered within the Black Sea not long ago, in the twentieth century (e.g., Kovalevskiy 1935, 1960; Ivanov 1999; Shnyukov et al. 1999, 2013; Egorov et al. 2011), although one of them—the Golubitskiy mud volcano in the Azov Sea—was described as early as 1795 by Pallas, who linked its catastrophic explosion to an earthquake. The mud volcanoes on the sea bottom are largely confined to the Sorokin, Kerch-Taman, and Tuapse troughs, all in the Eastern Black Sea basin, and to the abyssal part of the Western Black Sea basin. In addition, there are also about 4000 gas seeps (Shnyukov 2016), which are sometimes difficult to distinguish from mud volcanoes. In the Sea of Azov, mud volcanoes are present in its southern part and the adjacent coast, where they are localized on the Kerch and Taman peninsulas as well as in the northwestern Caucasus (geologically speaking, in the Kerch-Taman Trough) (Fig. 6.1).

Despite their random distribution at the local level, all discovered mud volcanoes are confined to areas where the underlying sediments contain the Maikopian geological formation (Fig. 6.2).

The Maikopian formation is almost homogeneous and represented by clays with variable thicknesses ranging up to 5000 m. The pattern of mud volcano distribution does not depend upon the thickness of the Maikopian clays, however, and mud volcanoes are often absent in areas where Maikopian sediments are the thickest. For example, the Indolo-Kuban Trough contains a great thickness of Maikopian sediments everywhere, yet mud volcanoes are located on its southern flank while they are absent in the center. In the Western Black Sea basin, a swarm of mud volcanoes is concentrated in the northeast, while the greatest thickness of Maikopian deposits occurs in the southwest (Fig. 6.2).

Presumably, mud volcanoes occur most often in the zones of deep tectonic dislocation within the Maikopian sequence and through which deep fluid flows break through. They contribute to the appearance of diapir folding and individual diapirs.

The Kerch-Taman mud volcanic region extends far beyond the Kerch and Taman peninsulas. The largest modern mud volcanic foci, such as Dzhau-Tepe, Bulganakskiy, Gora Gnilaya, Karabetova Gora, and Shugo, are located not along the periphery of the region, but in its axial part. Perhaps there was a different situation in the geological past, or the boundaries of the region were different, but now this trend can be traced quite clearly.

The oldest exposed rocks on the surface or overlain by Quaternary sediments are Cretaceous in age. They are developed in the southeast (megan-

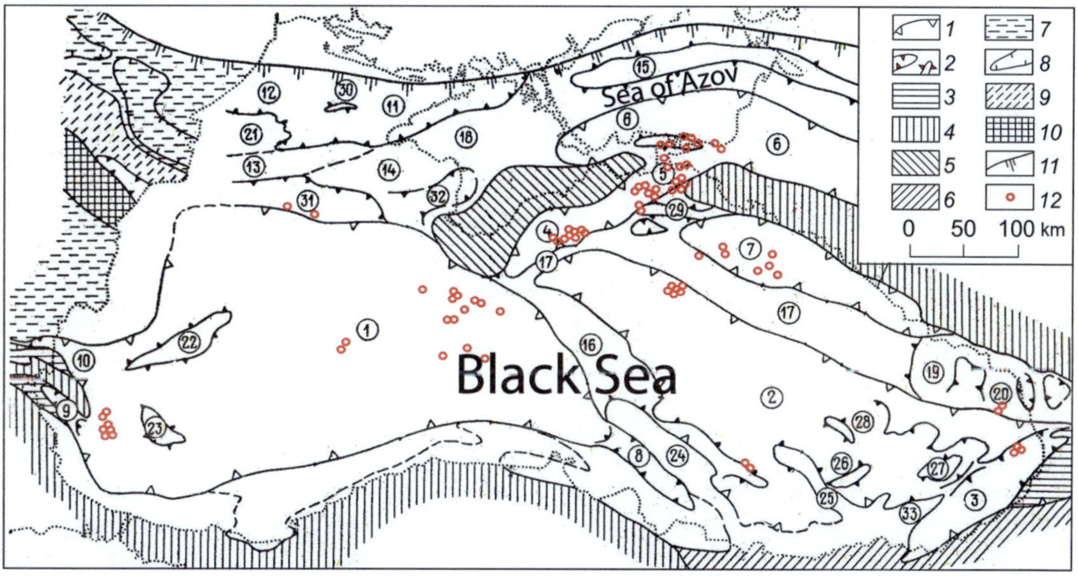

Fig. 6.1 Map of the Black Sea tectonic structures with the locations of mud volcanoes: (I) Contours of large Paleogene depressions and troughs; (II) contours of minor uplifts and depressions; (III) Paleocene-Eocene troughs of the Adzharo-Trialetskaya structure and the Balkans; (IV) folded structures of the Western Caucasus, Western Black Sea, and the Balkans; (V) Crimean meganticlinorium; (VI) Eastern Black Sea; (VII) Paleozoic-Mesozoic sedimentary cover of the Moesian platform and Pre-Dobrogea trough; (VIII) contours of Jurassic depressions within the Pre-Dobrogea trough; (IX) folded Paleozoic and Triassic deposits of the North Dobrogea and Priprutskiy protrusion; (X) Riphean horst of Central Dobrogea; (XI) southern border of the ancient East European platform; (XII) mud volcanoes. Depressions (numbers in circles): (1) Western Black Sea; (2) Eastern Black Sea. Troughs: (3) Gurian; (4) Sorokin; (5) Kerch-Taman; (6) Indolo-Kuban; (7) Tuapse; (8) Sinopskiy; (9) Burgaskiy; (10) Lower Kamchiyskiy; (11) Karkinitskiy; (12) Krylovskiy. Elevations: (13) Gubkin; (14) Kalamitskiy; (15) Azovskiy; (17) Shatskiy. Ridges: (16) Andrusov. Arches: (18) Crimean; (19) Gudautskiy; (20) Ochamchirskiy; (32) Alminskiy. Uplifts: (21) Kiliyskiy; (22) Polshkov; (23) Druzhba; (24) Archangel'skiy; (25) Muratov; (26) Chikhachev; (27) Dzhanelidze; (28) Strakhov; (29) Barrier; (30) Golitsyn; (31) Shelf edge; (33) Trabzon protrusion. Geological base is taken from Tugolesov et al. (1985)

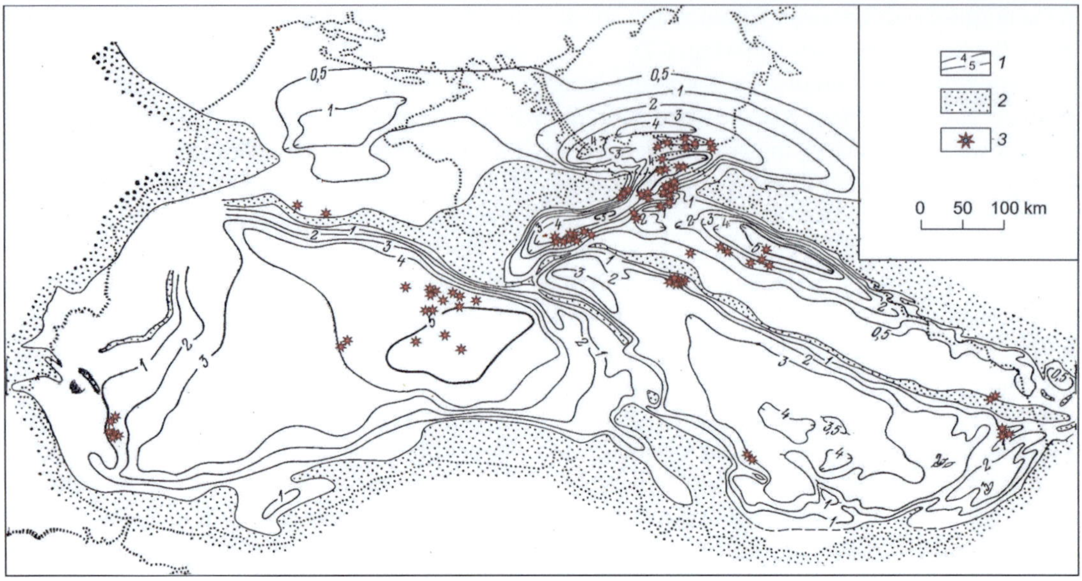

Fig. 6.2 Map of the distribution and thickness of Maikopian clays with mud volcano locations in the Black Sea region: (1) isopachytes (km); (2) areas lacking Maikopian sediments; (3) locations of mud volcanoes. Geological base is taken from Tugolesov et al. (1985)

ticlinorium of the Caucasus), the southwest (the area of Feodosiya town—the meganticlinorium of the Crimean Mountains), and the core of the Dyurmenskaya anticline on the south of the Kerch Peninsula. Eocene deposits are also localized here. Oligocene (Lower and Middle Maikopian) and Lower Miocene (Upper Maikopian) deposits are particularly widespread within the Southwestern Plain of the Kerch Peninsula, and they make up the core of the anticlines in the Vneparpachskaya area of the Kerch-Taman Cross-Interpericlinal Trough and the northern part of the positive structures of the Taman Peninsula. The Middle and Upper Miocene, as well as the Pliocene deposits, are widely distributed and form the wings of the anticlinal folds, producing synclinal depressions and recessed synclines as a result of mud volcanic activity. In some parts of the geological map (e.g., the Aktashskaya lowland on the Kerch Peninsula, and the territory adjacent to the Kuban valley in the West Kuban trough, etc.), the development of Quaternary sediments is evident due to their great thickness and the lack of information on sediments that underlie them (Fig. 6.3).

The deep geological structure of the region was studied by V.A. Vereskun using the geophysical and deep drilling data of Yu.S. Krotenko, V.I. Chernov, I.L. Sviridenko, and V.S. Popovich (Shnyukov et al. 2013).

On the sub-latitudinal and sub-meridional geological transects through the Kerch and Taman peninsulas (Figs. 6.4, 6.5, and 6.6), the relationships of all five structural floors of the region (Neogene, Maikopian, Eocene-Upper Cretaceous, Lower Cretaceous, and Jurassic) are clearly visible.

Most of the profiles record the coincidence of structural plans of the Maikopian, Eocene-Upper Cretaceous, and Lower Cretaceous-Upper Jurassic floors. As a rule, the vaults of anticlines either coincide or are somewhat displaced. The steep subsidence of the Maikopian structural floor in the vicinity of the Kerch Strait has been noted.

Based on the deep tectonic data, the vast majority of mud volcanoes are confined to anticlinal structures of the Maikopian structural floor and partially overlap the deeper structures of the Jurassic-Lower Cretaceous and even Triassic structural floors (Figs. 6.4, 6.5, and 6.6). Moreover, mud volcanoes are confined to tectonic dislocations. Among them, there are crosscutting dislocations developed in the Triassic-Jurassic and filled in with sediments of the same age. They extinguish in the Late Jurassic strata. There are also younger crosscutting dislocations developed

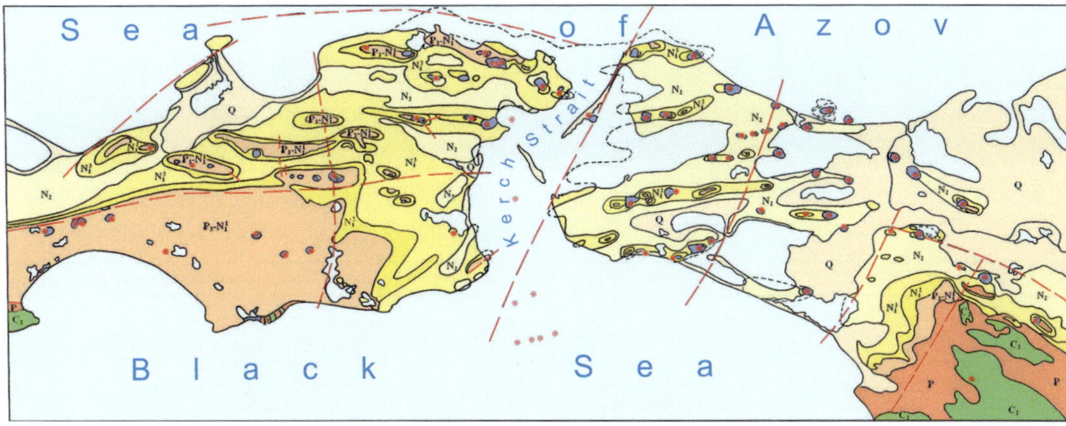

Fig. 6.3 Geological map of the Kerch and Taman peninsulas. The map does not include the names of the stages to simplify the drawing; these names are provided in the legend of the geological transect in Fig. 6.4

in the Jurassic-Cretaceous and filled in with sediments of the same age. They extinguish in the Lower Cretaceous strata. Both kinds of tectonic dislocations are called "blind faults."

From west to east, the thickness of the Neogene structural floor increases, while that of the Maikopian decreases. The intense bending of the Taman Peninsula is reflected in the thick cover of Quaternary sediments.

The Kerch-Taman mud volcanic region is geologically and historically peculiar compared to other areas of mud volcanism. Its maximum activity occurred most likely during Chokrakian and Sarmatian times, while at present, it is much less dynamic. This can be seen in both the smaller sizes of recent mud volcanoes and the decreased scale of their manifestations.

At present, more than 50 mud volcanoes are known on the Kerch Peninsula. At least 50 mud volcanoes have been discovered on the Taman Peninsula and in northwestern Caucasus. Many of them are still active, manifesting with relatively quiet eruptions and constant release of gases, mud breccia, water, and sometimes petroleum products. There are also mud volcanoes that explode violently. According to current knowledge, 67 strong mud volcanic explosions have occurred based on the literature and human recollections. Of course, this figure is understated because it is impossible to count all eruptions in the absence of systematic observations carried out by a special surveillance service. Out of the 68 known strong eruptions, 16 occurred on the Kerch Peninsula, 13 occurred in the Kerch Strait and the Sea of Azov, and 39 were located on the Taman Peninsula and in northwestern Caucasus (Table 6.1).

The largest explosions were recorded from the Gora Gorelaya mud volcano, where about one million m^3 of breccia were thrown out during a single eruption. Quite often, large eruptions of breccia have been accompanied by rumbling, shaking, fire, smoke, and sometimes steam, and debris can be thrown up to a height of 50–100 m. If explosions of breccia are accompanied by fire, burned annular or elongated zones of folded black and red slag-like formations are usually formed around the mud volcanic foci, and the breccia reveals various stages of burning.

The western boundary of the Kerch-Taman region is marked by the Vladislavovskiy mud volcano, located near the hill of Tush-Oba and Lake Achi. The eastern boundary of the region is less certain, and the boundaries are quite mobile, probably due to the abnormally high pressure of gas deposits in the interior that is at least partially responsible for mud volcanic processes. This can be demonstrated by the fact that the largest presently active mud volcanoes (e.g., Dzhau-Tepe, Bulganakskiy, Gnilaya Gora, Karabetova Gora, and Shugo) are located in the axial part of the region but not on its periphery. This may be related to a decrease in gas pressure toward the periphery and the fading of mud volcanic processes.

A wide array of mud volcanoes is located within the Taman Peninsula and the northwestern end of the Caucasian anticlinorium. This series

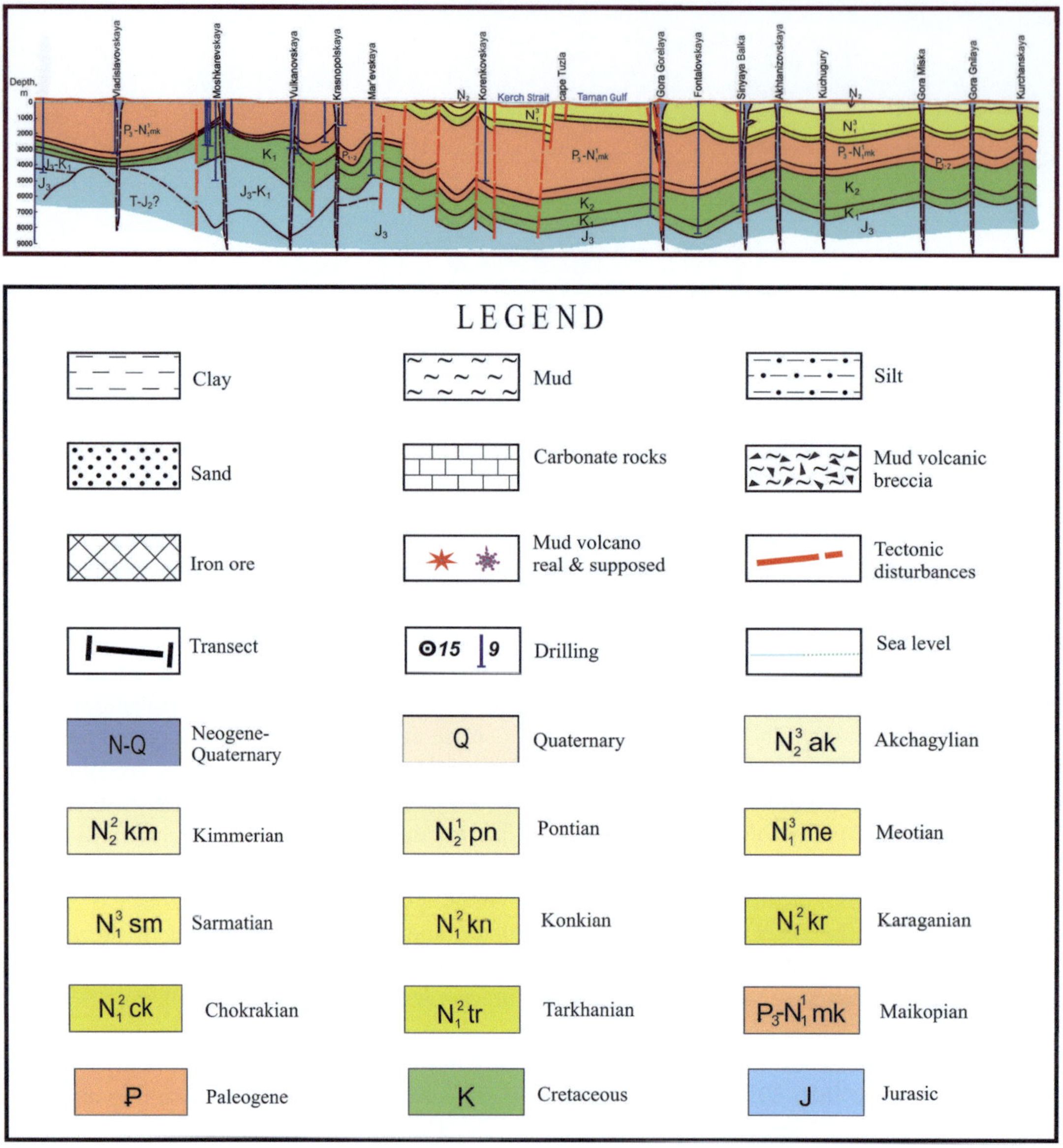

Fig. 6.4 Sub-meridional transect extending from Lake Uzunlar to the Bulganakskiy mud volcanic hearth. For legend location and names of anticlines and mud volcanoes (the names at the modern surface) in Figs. 6.4, 6.5, and 6.6), refer to Fig. 6.7 and Table 6.2

begins with the Raznokolskiy mud volcano and continues with those of Shugo, Gladkovskiy, Kievskiy, Semenovskiy, Shapsugskiy, and Semigorskiy. It is very likely that Semigorskiy, located on the southern slope of the Caucasian anticlinorium in the vicinity of the Natukhayevskaya village near Anapa, is not the last in a row. Other mud volcanoes to the east are quite possible (Slavyanov 1928). The information published in the literature on local lore about the discovery of a mud volcano near the city of Stavropol (popular literature) is of interest because this volcano is currently flooded beneath a water storage reservoir. Some mud volcanoes (e.g., Gladkovskiy, Semigorskiy, Shkol'nyy, and Shapsugskiy) are "sitting" not on Maikopian but on Lower Cretaceous sediments.

Locations of mud volcanoes and recessed synclines are shown in Fig. 6.7 and described in Table 6.2.

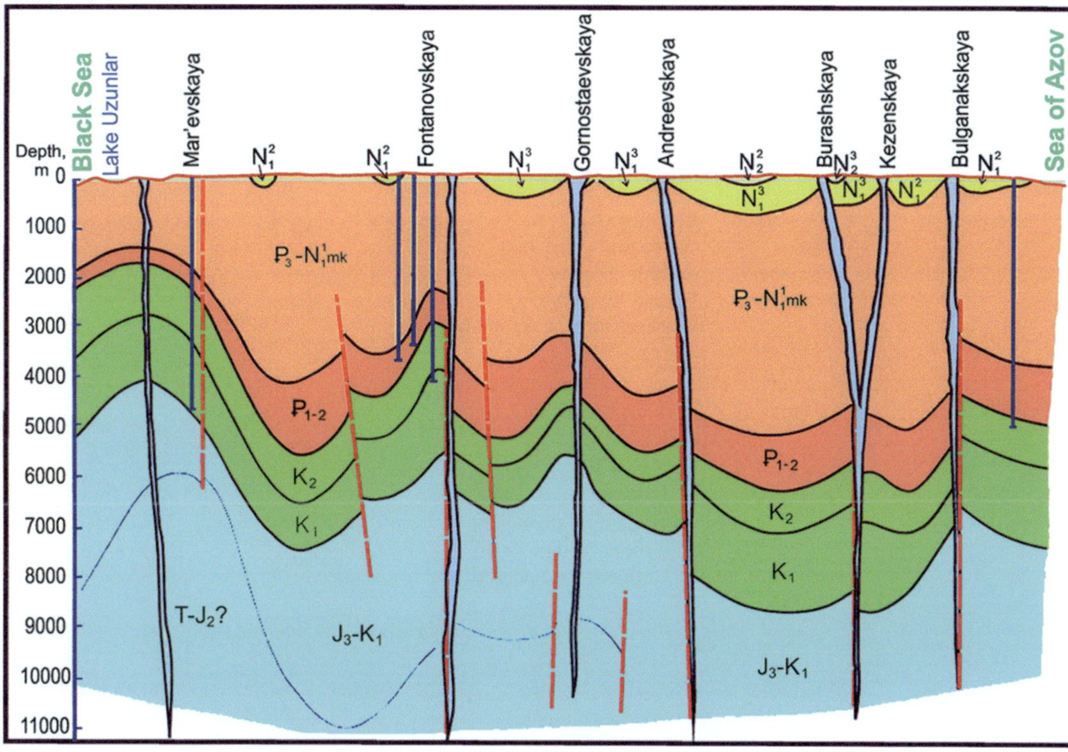

Fig. 6.5 Sub-latitudinal transect extended from Lake Achi to the Taman Peninsula

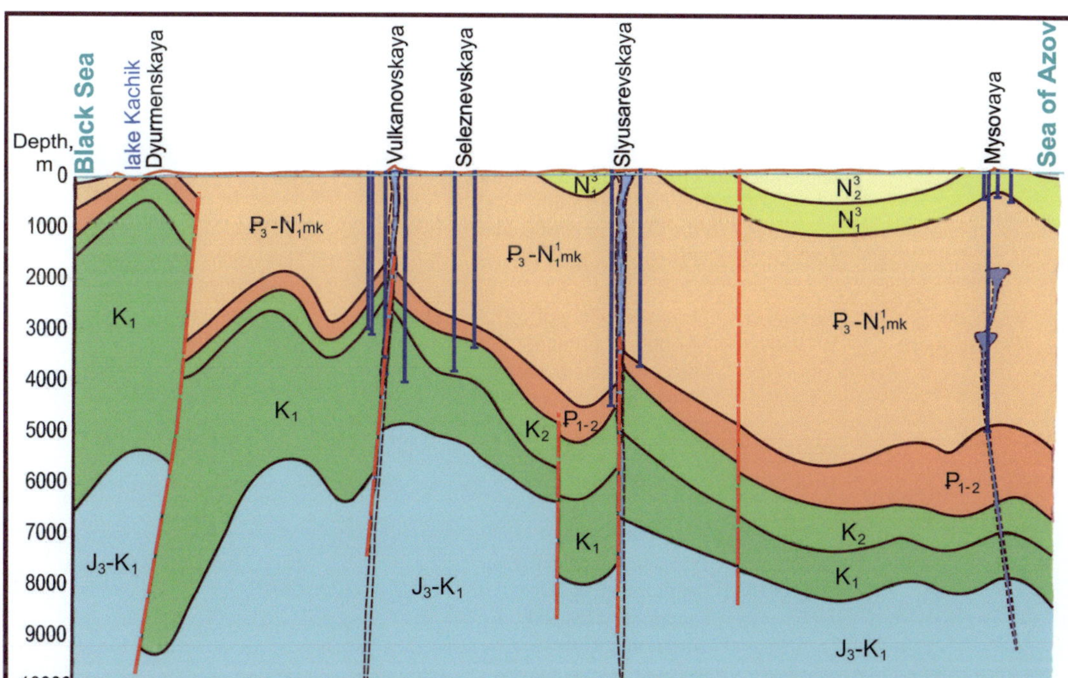

Fig. 6.6 Sub-meridional transect extending from the Kerch Peninsula to Durmen town and the coast of the Sea of Azov

Table 6.1 Recorded eruptions of mud volcanoes in the Kerch-Taman region and in northwestern Caucasus. For the location of mud volcanoes and recessed synclines mentioned throughout this book, see Fig. 6.7 supplemented by Table 6.2

No.	Mud volcano	Date of eruption	Type of eruption	Duration and character of activity	References
Kerch Peninsula					
1	Dzhau-Tepe	Beginning of the seventeenth century	Eruption of breccia, which destroyed the ancient village of Dzhau-Tepe on the western slope of the mud volcano	No data	Dvoychenko (1914)
		Eighteenth century	Eruption of mud volcanic breccia	"	Pallas (1795)
		1884	"	"	Dvoychenko (1914)
		March 16–17, 1909	On the first day, the peak of the mud volcano was broken up; on the second day, about 64 thousand m^3 of liquid mud poured out	48 h	Klepinin (1914)
		August 17–18, 1909	No data	No data	Dvoychenko (1914)
		March 18, 1914	Eruption of about 51,000 m^3 of mud breccia accompanied by fire, a cloud of white vapor, smoke, and a roar; formation of a caldera on the mud volcano peak with a depth of 4.5 m and diameter of 40–50 m	20–30 min	Dvoychenko (1914), Klepinin (1914), Gembitskiy (1914), Sedelshchikov and Kul'chavov (1914)
		1920	Grandiosc cruption of mud breccias	No data	Popov (1928)
		1925	Eruption of mud breccia	"	Murzaev (1928)
		September 12–13, 1927	Quiet extrusion of liquid breccia coinciding with the Crimean earthquake	"	Murzaev (1928)
		1942	Eruption of mud breccia	"	Naumenko (verbal communication)
2	Voskhodovskiy (Dzhardzhava)	November 8, 1930	Eruption of about 37,000 m^3 of mud breccia accompanied by explosion, smoke, and fire	10–15 min	Sivere (1931)
		March 16–17, 1982	Eruption of about 100,000 m^3 of mud breccia accompanied by vapor and roar	72 h	Authors' observation
		December 3–4, 2015	Opening of calderas, an appearance of large gaping cracks, and degassing of formed salses	2 h	Pasynkov and Vakhrushev (2016)

(continued)

Table 6.1 (continued)

No.	Mud volcano	Date of eruption	Type of eruption	Duration and character of activity	References
3	Bulganakskiy	Spring 1923	Eruption of mud breccia from Andrusov mud volcano	No data	Turley (1930)
		September 1926	"	"	"
		September 1, 1986	Large eruption of the Andrusov mud volcano, ejection to a height of 40 m, the volume of the ejected breccia about 500 m^3	10–15 min	Shnyukov et al. (1992)
		September 10, 2001	Large eruption of the Andrusov mud volcano	"	A.A. Abashin (verbal communication)
4	Nasyrskiy	1988–1990	Noticeable revival of activity with the appearance of new eruptions of liquid mud with an admixture of oil	Repeated emissions	Shnyukov et al. (1992)
5	Novoselovskiy	December 3–4, 2015	Powerful eruption from 4 am on December 3, 2015, to 8 pm the next day. A huge amount of mud breccia was ejected from the mud volcano caldera that formed two tongue-like streams covering about 2.5 hectares	44 h, pulsating	Pasynkov and Vakhrushev (2016)
6	Soldatsko-Slobodskoy (Nikitskiy)	December 3–4, 2015	Opening of a caldera, appearance of large gaping cracks, and degassing of formed salsas	No data	Pasynkov and Vakhrushev (2016)
7	Small mud volcanoes located near residential area of Solnechny (absent in Table 6.2)	December 3–4, 2015	Opening of a caldera, appearance of large gaping cracks, and degassing of formed salsas at a distance from the residential area that was not dangerous for inhabitants	"	Pasynkov and Vakhrushev (2016)
Taman Peninsula and Northwestern Caucasus					
8	Karabetova Gora	1818	Eruption of mud breccia	No data	Abryutskiy (1853); Belousov and Yarotskiy (1936)
		1819	"	"	"
		1820	"	"	"
		1821	"	"	"
		1822	"	"	Voskoboynikov and Gur'iev (1832)
		1832	"	"	Belousov and Yarotskiy (1936)
		1833	"	"	Abryutskiy (1853)

(continued)

Table 6.1 (continued)

No.	Mud volcano	Date of eruption	Type of eruption	Duration and character of activity	References
		1835	Turbulent eruption of mud breccias	"	Abikh (1873)
		August 1853	Turbulent eruption of mud breccia accompanied by a column of flame, thick black smoke, and release of breccia into the air. The area of the overflowed breccia was 340 m in diameter and 1.5 m in height	~3 h	Abryutskiy (1853)
		August 6, 1856	Eruption of mud breccia	No data	Gerts (1876)
		1868	"	"	"
		1882	"	"	Steber (1909/1910)
		1907	"	"	Belousov and Yarotskiy (1936)
		1908	Ejection of 500,000 cubic sazhens of mud	"	"
		1927	Column of flame with black smoke and small eruption of breccias	"	Accounts of local residents
		1947	Turbulent eruption of mud breccias	"	"
		1952	"	"	"
		1963	"	"	"
		1968	Explosive eruption, turbulent outpouring of mud and gases, appearance of a small cone on the southeastern slope; its height reached 3.5 m in 10 years	"	I.D. Lozovoy (verbal communication)
		1974	"	"	"
		August 1980	Eruption of mud breccia with a volume up to 5000 m^3	"	Authors' observations
		August 2, 1981	Eruption of mud breccia with volume up to 8–10,000 m^3 followed by roar and ejecting breccia but without flame	3 h	Herdsman I.I. Chaly (verbal communication)
		1982	Eruption of mud breccia	No data	Authors' observation
		August 19, 1984	Explosion with flame and release of large clastic fragments over a considerable distance, significant amount of mud breccia and gases	1 h	I. Lyut (verbal communication)

(continued)

Table 6.1 (continued)

No.	Mud volcano	Date of eruption	Type of eruption	Duration and character of activity	References
		May 6, 2001	Large eruption of mud breccia	No data	"
		April 20, 2002	"	"	I.N. Gusakov (verbal communication)
		Summer 2003	No data	No data	"
		June 19, 2004	Turbulent eruption of mud breccia with volume up to 47,000 m^3	"	I. Lyut (verbal communication)
	Malaya Karabetka	Summer 2006	Turbulent eruption of mud breccias	"	I.N. Gusakov (verbal communication)
		Summer 2010	No data	"	"
		Summer 2014–2016	"	"	"
9	Gora Gorelaya	February 24, 1794	Turbulent eruption of mud breccia with volume more than 100,000 cubic fathoms (one million m^3) accompanied by a thunder-like noise, white steam, black smoke, columns of flame reaching double the height of the mountain, and 25 min of fire	72 h	Alekseev (1880), Pallas (1883), Voskoboynikov and Gur'iev (1832)
10	Chushka Spit (Blevaka)	1986	Eruption of mud breccia	Still active	Gusakov (2004)
		August 29, 2002	Turbulent explosion	"	"
11	Sinyaya Balka (Tizdarskiy)	December 1987, October 1988	Eruption of mud breccia enriched with oil	"	Shnyukov et al. (1992)
12	Gora Sopka	Beginning of 1988	Eruption of mud breccia and gas	No data	"
13	Akhtanizovskiy	1818	Explosive eruption of mud breccia	"	Alekseev (1880)
		August 6, 1853	Eruption of mud breccia (on the same day as an eruption of Karabetova Gora)	"	Abryutskiy (1853)
		Spring 1988	Explosions of water, mud, and gases in the islands that appeared a year earlier in the Akhtanizovskiy estuary	By autumn, the islets were blurred and no traces of activity were observed	Shnyukov et al. (1992)

(continued)

Table 6.1 (continued)

No.	Mud volcano	Date of eruption	Type of eruption	Duration and character of activity	References
		2001	Intense outflow of mud	6 days	
		March 24, 2002	Large eruption of mud breccia	"	Gusakov (2004)
14	Tsymbaly Western	First half of the nineteenth century	Powerful outflow of mud breccia with length of tongue about 300 m	No data	Belousov and Yarotskiy (1936)
		March 4, 1977	Powerful outflow of mud breccia	"	Gusakov (2004)
		February 2002	A powerful outflow of breccia, repeated ignition, large mud tongue with a width of 100 m and a length of about 2 km	"	"
15	Gora Gnilaya	1810	Several intense eruptions of mud breccia	"	Land (1866)
		October 1978	Explosion of the crater, ejection of lumps of soil, stone, and mud	Several hours	I.N. Gusakov (verbal communication)
		2001	Eruptions of mud breccia, and gas (presumably methane) outlets	"	Gusakov (2004)
		February 26, 2002	Powerful (up to1000 m^3) outburst of breccia	"	"
		Summer 2003	No data	"	"
		June 2004	Colossal eruptions of mud breccia	"	"
		March 2008	Large eruption of mud breccia	"	"
		Summer 2010	No data	"	"
		September 2012	Large eruption (1000 m^3) of mud breccia	"	"
		Summer 2014–2016	Large eruption (400 m^3) of mud breccia	"	"
16	Gora Miska	1812	Catastrophic eruption with release of vapor, smoke, and flame	"	Zenin (1955)
		1843	Turbulent eruption of mud breccia	No data	Belousov and Yarotskiy (1936)
		1844	"	"	"
		1905	Catastrophic eruption with formation of huge cracks	"	"

(continued)

Table 6.1 (continued)

No.	Mud volcano	Date of eruption	Type of eruption	Duration and character of activity	References
		September 4, 1988	Intense ejecting of breccia, appearance of cracks	"	Shnyukov et al. (1992)
17	Gora Neftyanaya (southern)	June 6, 2002	Turbulent outburst of oil	"	Gusakov (2004)
		March 28, 2002	No data	"	I.N. Gusakov (verbal communication)
		2014	No data	"	"
18	Gora Boris and Gleb	April 26, 1818	Explosion resulted in the formation of a mud volcano	"	Dubois de Montpéreux (1837)
		1844	Catastrophic eruption with release of vapor, smoke, and flame	"	Zenin (1955)
		1905	Column of flame, outburst of mud breccia	"	"
		Beginning of 1980	Explosion of gases, which created a burned focus of mud volcanic breccia of round shape with a diameter of 3–5 m	"	Evidence of local residents
19	Shugo	Summer 1903	Turbulent eruption of mud breccia accompanied by noise and a high flame column	More than 24 h	Steber (1912)
		1907	Turbulent eruption of mud breccia accompanied by noise and a high flame column	No data	"
		1908	Small eruption accompanied by light explosions, strong noise, and release of mud to a height of 5–7 m	Few minutes	Grechishkin (1931)
		1909	No data	No data	"
		June 24, 1959	Turbulent eruption of 1500 m^3 of mud to a height of ~5 m	~2 h	Gusakov (2004)
20	Peklo Azovskoe	Summer 2006	Medium strength eruption of mud breccia	"	I.N. Gusakov (verbal communication)
		Summer 2009	"	"	"
21	Krasny Oktyabr	Summer 1991	Bulging of the ground over an area of 250 m^2	"	"

(continued)

Table 6.1 (continued)

No.	Mud volcano	Date of eruption	Type of eruption	Duration and character of activity	References
		August 2002	Bulging of the ground up to 3 m high over an area of 250 m^2	"	"
		Summer 2006	Bulging of the ground up to 4 m high over an area of 180 m^2	"	"
22	Polivadina	Summer 1997	Bulging of the ground over an area of 400 m^2	"	"
23	Gora Zelenskiy	March 20, 2002	Large eruption with release of gas and collapse of the shore ledge	"	"
24	Bogatyr	August 2002	No data	"	"
		Summer 2010	No data	"	"
25	Mys Kamenny	May–June 2010	Bulging of the ground up to 3.5 m over an area of 3 hectares	"	"
26	Gora Girka	Summer 2016	Oil spots	"	"

Within the Kerch part of the Kerch-Taman region, the following large tectonic structures of the first order can be recognized: (I) the Southwestern Plain, (II) the Vneparpachskaya area, and (III) the Kerch-Taman Cross-Interpericlinal Trough. Within the Taman part of the region, they are (IV) the northern Taman zone of positive structures and (V) the spurs of the northwestern Caucasus (Fig. 6.7).

The Southwestern Plain represents the eastern periclinal depression of the Crimean Mountain meganticlinorium. If one considers the relationship between the Plain and structures of the meganticlinorium that developed in its mountainous part, it becomes obvious that the Plain represents the eastern periclinal closure of the Tuakskiy anticlinorium.

From the north, the Plain is contoured by the Parpachskiy crest, which is composed of Middle Miocene rocks that monoclinally submerge into the Vneparpachskiy area. From the east, the Plain is contoured by the branch of the Parpachskiy crest called the Parabolic Ridge, which has the same geological characteristics and extends submeridionally toward the Black Sea and Koyashskiy Lake.

The Plain is underlain predominantly by Maikopian deposits. Lower Cretaceous sediments are exposed only within the core of the Durmenskaya anticline, and the recessed synclines are the downward bending and filled in by Middle and Upper Miocene deposits as well as those of the Pliocene.

Morphologically, the Plain is weakly dissected. On its flat background, there are wavelike, cone-shaped, or oval (in plan) elevations up to 150 m corresponding in some cases to mud volcanoes (e.g., Dzhau-Tepe, Ak-Tube, Konchekskiy, etc.), recessed synclines, or exposed anticlinal cores.

There are several zones of anticlines. Some of them (e.g., Arma-Elinskaya, Moshkarevskaya, Vulkanovskaya, Krasnopolskaya, Sadovaya, and Uzunlarskaya) contain mud volcanoes and recessed synclines. As a rule, the anticlines are oriented in a northeastern direction except for the westernmost ones (e.g., Arma-Elinskaya and Andreevskaya Western) that possess an almost sub-latitudinal alignment. Except for the Durmenskaya anticline, the cores of all others are composed of Middle and occasionally Lower Maikopian rocks, and the recessed synclines are composed of Upper Maikopian deposits.

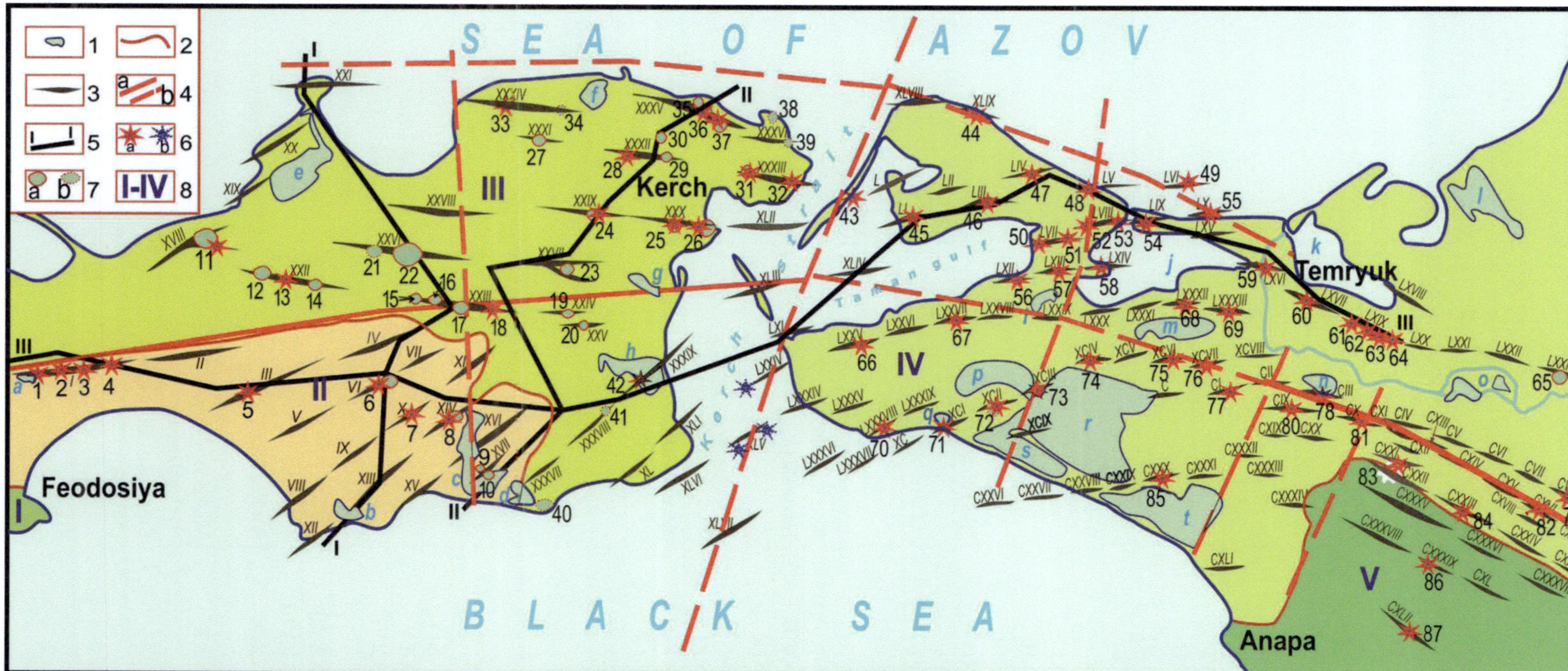

Fig. 6.7 The main structural elements of the Kerch-Taman region (the numbers in the map correspond to those in Table 6.2). 1. Lakes: (a) Achi, (b) Kachik, (c) Uzunlar, (d) Koyash, (e) Aktash, (f) Chokrak, (g) Churbash, (h) Tobechik; Limans: (i) Yanovskiy, (j) Akhtanizovskiy, (k) Kurchanskiy, (l) Vostochnyy, (m) Starotitarovskiy, (n) Bolshoy Raznokol, (o) Kurkiy, (p) Tsokur, (q) Solenoe, (r) Kiziltashskiy, (s) Bugazskiy, (t) Vityazevskiy. 2. Parpachskiy crest. 3. Anticlines. 4. Faults (a, reliable; b, assumed). 5. Geological transects. 6. Mud volcanoes (a, on land; b, prospective underwater mud volcanoes based on geophysical data. 7. Recessed synclines (a, reliable; b, assumed). 8. Main structural elements: (I) Eastern closure of the meganticlinorium of the Crimean Mountains, (II) Southwestern Plain, (III) Vneparpachskaya area (the southern side of the Indol-Kuban Trough), (IV) Northern Taman zone of positive structures, (V) spurs of the northwestern Caucasus (see Table 6.2)

Table 6.2 The main structural elements of the Kerch-Taman region (recessed synclines having the same name as the mud volcanoes from which they originated are not separately numbered)

Anticlines		Mud volcanoes		Recessed synclines	
Kerch Peninsula, Southwestern Plain					
Arma-Elinskaya (Batalnenskaya)	I	Vladislavovskiy	1	Achinskaya	1
		Khyrtsyz-Shiban Western	2		
		Khyrtsyz-Shiban Eastern	3		
		Arma-Eli	4	Arma-Elinskaya	4
Andreevskaya Western	II				
Moshkarevskaya	III	Kerleutskiy	5		
Seleznevskaya	IV				
Ulyanovskaya	V				
Vulkanovskaya	VI	Dzhau-Tepe	6	Dzhau-Tepe	6
Vostochnaya Eastern	VII				
Beloborodskaya	VIII				
Gavrilenkovskaya	IX				
Krasnopolskaya	X	Borukh-Oba	7		
Marfovskaya	XI				
Chaudinskaya	XII				
Dyurmenskaya	XIII				
Sadovaya	XIV	Ak-Tube	8	Ak-Tube	8
Karangatskaya	XV				
Mar'evskaya	XVI				
Uzunlarskaya	XVII		9	Konchekskaya	9
			10	Uzunlarskaya	10
Vneparpachskaya county, Northern part of the Kerch Peninsula					
Kamenskaya	XVIII	Nasyrskiy	11	Kamenskaya	11
Krasnokutskaya	XIX				
Belokamenskaya	XX				
Mysovaya	XXI				
Korolyovskaya	XXII	Korolyovskiy	13	Illichevskaya	12
				Kenegezskaya	14
Sartskaya (Fontanovskaya)	XXIII			Western Kara-Sidzheutskaya	15
				Eastern Kara-Sidzheutskaya	16
				Kayaly-Sartskaya	17
		Novoselovskiy	18	Novoselovskaya	18
Rep'evskaya	XXIV			Rep'evskaya	19
Sokolskaya	XXV			Sokolskaya	20
Slyusarevskaya	XXVI			Kashikskaya	21
				Burulkayskaya	22
Gornostaevskaya	XXVII			Gornostaevskaya	23
Novonikolaevskaya	XXVIII				
Andreevskaya (Chanlugarskaya)	XXIX	Andreevskiy	24	Andreevskaya	24
Voskhodovskaya	XXX	Voskhodovskiy (Dzhardzhava)	25	Voskhodovskaya	25
		Soldatsko-Slobodskoy (Nikitskiy)	26	Soldatsko-Slobodskaya	26
Chumnaya Balka	XXXI			Chumnaya Balka	27

(continued)

Table 6.2 (continued)

Anticlines		Mud volcanoes		Recessed synclines	
Malobabchikskaya	XXXII	Burashskiy	28	Burashskaya	28
				Malobabchikskaya	29
				Kezenskaya	30
Glazovskaya	XXXIII	Baksinskiy	31	Baksinskaya	31
		Enikalskiy	32		
Karalarskaya	XXXIV	Syuyurtashskiy	33	Chokrakskaya	34
Bulganakskaya	XXXV	Bolshetarkhanskiy	35	Bolshetarkhanskaya	35
		Malotarkhanskiy	36	Malotarkhanskaya	36[a]
		Bulganakskiy	37	Bulganakskaya	37[a]
Borzovskaya	XXXVI			Borzovskaya	38[a]
				Ossovinskaya	39
Opukskaya	XXXVII			Opukskaya	40
Pogranichnaya	XXXVIII			Chorelekskaya	41
Priozernaya	XXXIX	Chongelekskiy	42	Chongelekskaya	42
Yakovenkovskaya	XL				
Korenkovskaya	XLI				
Kerch Strait					
Anticlines[a]	XLII–XLVII	Chushka	43		
Taman Peninsula					
Mys Kamenny	XLVIII				
Mys Peklo	XLIX	Peklo Azovskoe	44		
Negiereevskaya	L				
Gorelaya	LI	Gora Gorelaya	45	Gorelaya	45[a]
Beregovaya	LII				
Fontalovskaya	LIII	Fontalovskiy	46		
Kuchugurskaya	LIV	Kuchugurskiy	47		
Tizdarskaya	LV	Tizdar (Sinyaya Balka)	48		
Temryukskaya	LVI	Temryukskiy (Kazbek) marine	49		
Tsymbaly	LVII	Tsymbaly Western	50		
		Tsymbaly Eastern	51		
Akhtanizovskaya	LVIII	Akhtanizovskiy	52		
		Akhtanizovskiy Northern	53		
Peresypskaya	LIX	Gora Sopka	54		
Golubitskaya	LX	Golubitskiy (marine)	55		
Tamanskaya	LXI				
Phanagoriyskaya	LXII	Shopurskiy	56		
Yanovskaya	LXIII	Sennoy	57		
Borisoglebskaya	LXIV	Gora Borisa and Gleba	58		
Limannaya	LXV				
Gora Miska	LXVI	Gora Miska	59		
Gora Figura	LXVII	Gora Gnilaya	60		
Kurchanskaya Northern	LXVIII				
Kurchanskaya Southern	LIX	Kurchanskiy	61		
		Kurchanskiy 2	62		
		Kurchanskiy Dachnyy	63		
		Krasnooktyabrskiy	64		
Andreevskaya Eastern	LXX				

(continued)

Table 6.2 (continued)

Anticlines		Mud volcanoes		Recessed synclines	
Vostochnaya	LXXI				
Anastasievskaya Western	LXXII				
Anastasievsko-Troitskaya	LXXIII	Anastasievsko-Troitskaya	65		
Krotkov	LXXIV				
Karabetovskaya	LXXV	Karabetova Gora	66		
Komendantskaya	LXXVI				
Chirkovskaya	LXXVII	Gora Chirkova	67		
Bliznetskaya	LXXVIII				
Girkskaya	LXXIX				
Kopantskaya	LXXX				
Staro-Titarovskaya	LXXXI				
Neftyanaya Northern	LXXXII	Neftyanoy Northern	68		
Dubovyi Rynok	LXXXIII	Dubovyi Rynok	69		
Gora Zelenskogo	LXXXIV				
Kostenkova	LXXXV				
Aksenovskaya	LXXXVI				
Banki Andreeva	LXXXVII				
Mys Zheleznyi Rog	LXXXVIII	Peklo Chernomorskoye	70		
Kruglaya Gora	LXXXIX				
Bugazskaya marine	XC				
Bugazskaya	XCI	Bugazskiy	71		
Polivadina	XCII	Polivadina	72		
Kiziltashskaya Northern	XCIII	Kiziltashskiy Northern	73		
Vishesteblievskaya	XCIV	Vyshesteblievskiy	74		
Kapustina Balka	XCV				
Neftyanaya Southern	XCVI	Neftyanoy Southern	75		
Kamyshevataya	XCVII	Kamyshevatskiy	76		
Strelchanskaya	XCVIII				
Kiziltashskaya	XCIX				
Girlyanaya Gora	C				
Belyi Khutor	CI	Belyi Khutor	77		
Ereminskaya	CII				
Raznokolskaya	CIII	Raznokolskiy	78		
Varenikovskaya	CIV				
Adagirskaya Northern	CV				
Pavlovskaya	CVI				
Keslerovskaya	CVII				
Kievskaya	CVIII	Kievskiy	79		
Dzhiginskaya	CIX	Dzhiginskiy	80		
Ust-Chekupskaya	CX	Ust-Chekupskiy	81		
Varenikovskaya Western	CXI				
Varenikovskaya Southern	CXII				
Varenikovskaya Eastern	CXIII				
Adagumskaya	CXIV				
Medovskaya	CXV				
Novo-Mikhailovskaya	CXVI	Semenovskiy	82		
Psifskaya Blagoveshchenskiy	CXVII				
Arnautskaya	CXVIII				

(continued)

Table 6.2 (continued)

Anticlines		Mud volcanoes		Recessed synclines	
Utashskaya	CXIX				
Yurovskaya	CXX				
Shuginskaya	CXXI	Shugo	83		
Verkhne-Chekupskaya	CXXII				
Gladkovskaya	CXXIII	Gladkovskiy	84		
Damanskaya	CXXIV				
Moldavanskaya	CXXV				
Banka Mariya Magdalena	CXXVI				
Blagoveshchenskaya marine	CXXVII				
Bugazskaya Eastern	CXXVIII				
Blagoveshchenskaya Western	CXXIX				
Blagoveshchenskaya	CXXX	Blagoveshchenskiy	85		
Blagoveshchenskaya Eastern	CXXXI				
Suvorovo-Cherkesskaya	CXXXII				
Levitskaya	CXXXIII				
Gostagaevskaya Western	CXXXIV				
Shumayskaya Northern	CXXXV				
Psebepskaya	CXXXVI				
Krasny Molot	CXXXVII				
Gostagaevskaya	CXXXVIII				
Nikolaevskaya	CXXXIX	Shkol'nyy	86		
Sibzirskaya	CXL				
Tsybanova	CXLI				
Semigorskaya	CXLII	Semigorskiy	87		

Note: [a]Assumed mud volcanoes identified based on published and seismoacoustic data

The Vneparpachskaya area is located outside the Parpachskiy region and represents the southern flank of the Indol-Kuban Trough. It is characterized by thick (up to 1000 m) Neogene deposits. The zones of anticlinal uplift are not significant in length, but there are about 24 anticlines that are arranged in an echelon formation. They do not strike uniformly. In general, the anticlines of the southern part of the Vneparpachskaya area are oriented sub-latitudinally, while those in the northern and central part run in a northwestern or near sub-latitudinal direction. The linearity of the anticlines is very often complicated by the presence of mud volcanoes in their axial parts. Maikopian deposits form the cores of anticlines, sometimes without being exposed to the surface. Both the wings of the anticlines and the recessed synclines are filled with Middle and Upper Miocene deposits together with those of the Pliocene.

In comparison with the Southwestern Plain, the relief of the Vneparpachskaya area is very fragmented. This can be explained in two ways. First, large synclinal troughs (e.g., iron ore troughs) are expressed in the form of flat trough-shaped basins. Second, the so-called inverted valleys that arise in the axial parts of the anticlinal folds are characteristic of this territory. They are composed of ductile, easily eroded Maikopian clays, while their wings rise in the form of ridges composed of carbonate rocks. The inverted valleys are exemplified by the valley of the Voskhodovsko-Soldatskaya anticline, which is contoured along the northern periphery by the Mithridatskiy Ridge and along the south by the Yuz-Oba Ridge. Other examples are the peculiar ring structures that are well expressed in relief and form anticlinal wings composed of Sarmatian and Meiotic reefogenic limestones. In the center of the ring structure, there is a closed cavity that fixes the core of the brachianticlinal fold and that is composed of partly eroded Sarmatian clays (e.g., the Mysovaya anticline confined to Cape Kazantip). The same type of relief is also characteristic of numerous recessed synclines (e.g., Kayaly-Sartskaya, Burulkayskaya, Rep'yevskaya, etc.).

Mud volcanoes and recessed synclines are confined to the maximum uplift of anticline hinges. Several mud volcanic hearths and recessed synclines can be confined to a single anticline (e.g., the Sartskaya anticline with four recessed synclines; Bulganakskaya with three recessed synclines; Korolyovskaya and Voskhodskaya with two recessed synclines). In this zone, there are identifiable bifurcating anticlines (e.g., Kamenskaya, Gornostaevskaya) between the branches of which large recessed synclines have developed.

The Kerch-Taman Cross-Interpericlinal Trough (hereafter, the Trough) within the Kerch Peninsula is bordered on the west by the Parabolic branch of the Parpachskiy Ridge and on the north by a band of sub-latitudinal anticlinal uplifts. To the east, it continues into the Kerch Strait and, further on, into the Taman part of the region. In all likelihood, the northern boundary of the Trough should be drawn along the line of the Parpachskiy deep fault between the Kamysh-Burun syncline oriented sub-latitudinally and the Eltigen-Ortelskaya syncline of northeasterly strike, which is typical for the majority of anticlinal and synclinal structures of this zone. In the northwestern part of the Trough, there are three dome-shaped anticlinal uplifts—the Rep'yevskaya, Sokolskaya, and Alagolskaya. Their arches are composed of Maikopian sediments, while their wings are made up of Middle Miocene deposits. The same-named recessed synclines are confined to the cores of these structures. In the Rep'yevskaya anticline, Kimmerian iron ores are developed.

In the southeastern part of the Trough (on land), two anticlinal zones are present that possess confined recessed synclines (Opukskaya, Chongelekskaya, and Chorelekskaya).

The general pattern of geostructural zoning consists of a continuation of the Kerch Peninsula structures under the Kerch Strait and onto the Taman Peninsula. Thus, the Chokrak-Jurakov-Kutskaya anticlinal zone located in the north of the Kerch Peninsula persists along the bottom of the Kerch Strait and connects with the anticline of the Mys (Cape) Kamenny on the Taman Peninsula.

The Baksinskaya anticline is located in the northeastern corner of the Kerch Peninsula and partially enters the Kerch Strait up to the Chushka Spit. The Caterlezskaya syncline (or the Kimmerian iron ore trough) is clearly recorded in the Kerch Strait in the form of a syncline filled with Kuyalnitskian deposits.

The Chushkinskaya anticline is located in the Kerch Strait. It seems to be building a bridge between the Voskhodovsko-Soldatskaya anticline on the Kerch Peninsula and the Fontalovskiy anticline on the Taman Peninsula. The Blevaka mud volcano is confined to this anticline, within the spit of the same name. To the south of these structures, the sub-latitudinal strike of the folding is not clearly expressed. The southernmost sub-latitudinal syncline, the Kamysh-Burunskiya iron ore trough, does not currently have a synclinal continuation within the Kerch Strait despite the continuation of this structure within the Taman Peninsula (where the large syncline of Taman Bay is located). These two synclines are separated by the sub-latitudinally oriented bimodal Tuzlinskaya anticline, which is composed of Lower and Middle Sarmatian rocks. The junction of the Kerch and Taman folds of the northeastern stretch within the Kerch Strait is quite complex and, unfortunately, difficult to understand due to the lack of drill sampling. The most reliably traced structure is the complex Kiyz-Aulskaya synclinal zone located to the southeast of the Kerch Peninsula. It strikes across the Kerch Strait toward Cape Panagia on the Taman Peninsula. South of this zone is a large anticline composed of Sarmatian deposits. Geologists of "Yuzhmorgeologiya" mapped a structure resembling a recessed syncline within the core of this anticline. Farther to the south, there is a little-studied area composed of Kuyalnitskian-Kimmerian and Pontian-Kimmerian deposits.

The Taman Peninsula is bounded on the north by the shoreline of Taman Bay (Sea of Azov), on the west by the Kerch Strait, and on the south by the Black Sea. It is characterized by low-lying flat relief, a considerable ruggedness of the coastline, and the presence of numerous bays, shorelines, estuaries, and mud volcanoes. The Taman Peninsula shows subsidence rates of 0.4 m/kyr to 1.6 m/kyr (Brückner et al. 2010).

The beginning of geological studies into the Taman Peninsula dates back to the nineteenth century and is closely linked to the names of G.V. Abikh, A.M. Konshin, P.S. Pallas, A.D. Arkhangel'skiy, V.I. Vernadskiy, N.I. Andrusov, and other scientists. Systematic study was initiated by the Russian Geological Committee in 1912 under the leadership of I.M. Gubkin. A geological survey of the entire peninsula was conducted between 1915 and 1937 in order to study its geotectonics, deep structure, and mud volcanism. The survey was continued after the Second World War. Over the years, magnetic and electrical prospecting and surveying by gravimetric and seismic methods were carried out. To date, most of the Taman Peninsula has been covered by seismic MOV-OGT surveys, and more than 3400 km of transects have been performed. In addition, 500 wells have been drilled within the peninsula, more than 400 of which were for structural and mapping purposes, and more than 100 of which were for prospecting and parametric investigation (Engibaryan 2006).

It was discovered that clay diapirs on the Taman Peninsula are grouped into shaft-like uplifts forming the anticlinal zones of a sub-latitudinal expanse separated by synclinal deflections. Some zones have a continuation into the sea and some onto the Kerch Peninsula. In the west of the peninsula, there is a southern divergence of zones that is generally consistent with the orientation of the folded systems of the Caucasus and Crimean Mountains.

The Fontalovskaya anticlinal zone includes several isolated anticlines located at the junction of the land and the Sea of Azov, and it contains many active mud volcanoes.

According to the tectonic scheme of folding suggested by Shardanov et al. (1962), 14 anticlinal zones with 85 anticlinal uplifts (including some that lie underwater) are distinguished within the Taman part of the region. Their orientation, disposition, and morphology are not uniform throughout the territory except for the Cenozoic structures. The latter are characterized by fairly uniform folding conditions expressed in the development of linearly extended anticlinal zones, in which brachiform folds are present.

Striking of these zones is not the same. For the folds of the northwestern subsiding of the Greater Caucasus and the West-Kuban Trough, the general Caucasian northwestern strike is characteristic, while for the greater part of the Taman Peninsula and the adjacent part of the Black Sea, the orientation of the folds is sub-latitudinal, shifting to a northeastern-southwestern direction. The Dzhiginskiy fault of the sub-meridional strike can be considered to be the conditional boundary of the change in the strike of the anticlinal zones.

The anticlinal zones are separated by large, extended synclines filled with Pliocene deposits. In many cases, deposits of the Maikopian series are not exposed in cores taken from the anticlines, and sometimes they were submerged to a considerable depth.

Most of the mud volcanoes lie to the west of the Dziginskiy fault. The anticlines to which the mud volcanoes are confined are characterized by a fairly monotonous stratigraphy. As a rule, their cores contain sharply uplifted Maikopian deposits contoured on their wings and periclinal by submerged Miocene rocks.

Mud volcanoes of the southeastern part of the territory are confined either to Cretaceous deposits (e.g., the Semigorskiy mud volcano) or to areas of shallow bedding (e.g., the Shugo and Gladkovskiy mud volcanoes). A distinctive feature of the Taman mud volcanoes is the almost complete absence of recessed synclines. Their presence is more or less reliably established within the Anastasievsko-Troitskaya anticline, while in other structures (e.g., Polyvadina Gora, Gora Gorelaya), their presence is only assumed.

Although the tectonic structures of the Kerch and Taman parts of the region reveal much in common and form a uniform zone of Cenozoic folding, they are different in some aspects. The Kerch part is characterized by the development of brachisynclinal structures separated by fairly wide anticlinal zones that contain mud volcanoes and recessed synclines. The Taman part is characterized by the development of wide and deep synclinal troughs separated by narrow anticlinal zones that also show an intensive development of mud volcanoes, but almost no recessed synclines. This probably can be explained by differences in paleogeographic conditions, e.g., paleorelief that existed in the Neogene within the region. The Cenozoic structure of the Kerch Strait includes tectonic features common to both parts of the region, and this allows one to consider the Kerch Strait as a connecting structural link between the two parts of the Kerch-Taman region.

6.2 Mud Volcanoes of the Kerch Peninsula

The extreme western positive structure of the Kerch Peninsula is the Arma-Elinskaya (Batalienskaya) anticline (#I at the extreme left in Fig. 6.7). The anticline is surmounted by mud volcanoes in its arched part forming the so-called Vladislavovskiy mud volcanic hearth. The latter includes a cluster of large, medium, and small mud volcanoes, as well as recessed synclines and breccias of different ages.

The westernmost mud volcano of the cluster is the Vladislavovskiy or Tush-Oba (#I in Fig. 6.7) located near the hill of Tush-Oba and Lake Achi (Fig. 6.8).

For some reason, this mud volcano was omitted on the detailed maps produced by Andrusov (1893). It was described for the first time by Borisyak (1907), and according to him, it is an ancient mud volcano situated 10.7 km east of Feodosiya City. It stands 4.26 m in height, and its flat top is dotted with fragments of spherosiderite and sandstone. In our opinion, Borisyak mistakenly described the entire Tush-Oba hill as a mud volcano when, in fact, the mud volcanic manifestations were small parasitic gryphons that he described to the north, west, and south of the Tush-Oba hill. They appear to be flat cones with a height of 0.7–1.4 m and craters with diameters of 2.5–5 m. Today, there are about ten active and a few inactive gryphons. Some of the active ones whistle and hiss, while others throw out a small amount of greasy mud at a rate of 0.1–0.5 m^3/day. The released gas burns as well.

While investigating the Vladislavovskiy mud volcanic field, Belousov and Yarotskiy (1936) found three small mud volcanoes around the Tush-Oba hill. On its slopes, they noted visible evidence of alkali soils (solonchaks), and to the east of the hill, there were many insignificant and partially weathered craters that created a well-defined picture of a dried stream bed possibly related to tectonic rupture. The Tush-Oba hill itself did not attract much attention to these researchers. They expressed only a casual assumption concerning its mud volcanic origin and composition based on the evidence of its Chokrakian and Sarmatian limestone composition. No attention was paid to the Tush-Oba hill during subsequent geological surveys as well.

Shnyukov and Naumenko (1982) clarified the issue surrounding the Tush-Oba hill. They found that the hill is not composed of mud volcanic breccia, but instead, it comprises Chokrakian and Sarmatian limestones and a thick sequence of Kimmerian deposits that is lithologically complex and variegated in composition. Yet, the hill and the 2.3 km-wide plain adjacent to it on the east are covered by a thick layer of mud volcanic breccia. In its eastern part, this layer reveals a high concentration of fragments made up of Sarmatian limestones and sporadically ferruginous sandy clays of Kimmerian age. In its central part, the layer contains rare fragments of limestone and poor Kimmerian ores. Closer to the Tush-Oba hill, the fragments become more abundant, varying from numerous pieces of ferruginous clays and sandstones to the rare appearance of small fragments of pseudo-oolitic ores that are typical of recessed synclines.

Within the mud volcanic hearth, there are numerous small salses located mainly along the lines of temporary watercourses. In this way, the whole area to the east of Lake Achi emerges as an extremely complicated geological structure, which makes understanding this part of the Arma-Elinskaya anticline difficult due to the mud volcanic breccia that covers it (Fig. 6.9).

Fig. 6.8 View of the Vladislavovskiy (Tush-Oba) mud volcano from the north of Lake Achi. The Achinskaya recessed syncline is marked by an arrow. For the location of this mud volcano and those in the following figures, refer to Fig. 6.7 and Table 6.2

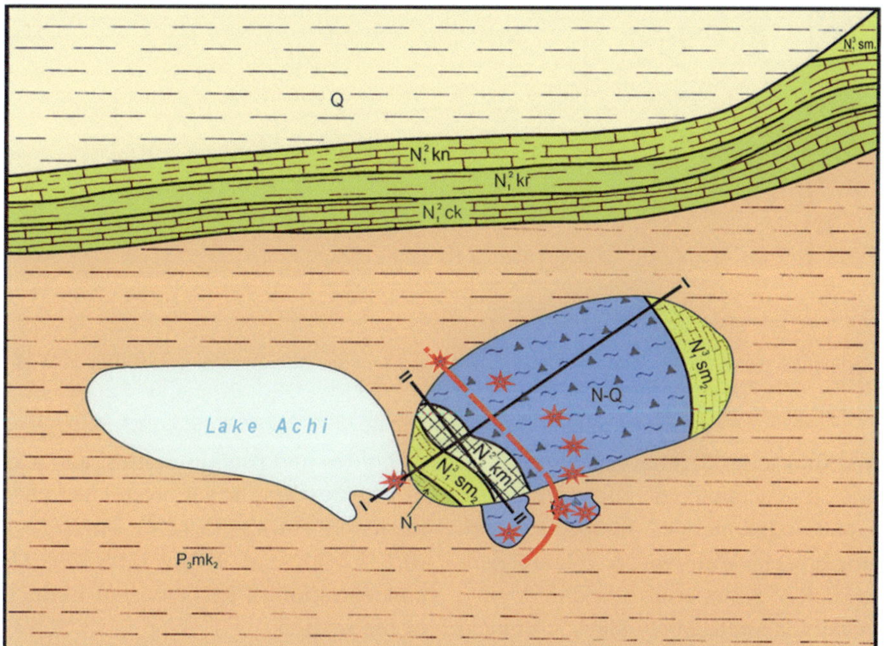

Fig. 6.9 Diagram of the structure of the Achinskaya recessed syncline. (For legend see Fig. 6.4)

Academy of Sciences and the Office of Mining Geology of the Ministry of Ferrous Metallurgy of Ukraine led to the discovery of sandy, ore-bearing Kimmerian deposits that reveal thicknesses up to 75–100 m (Figs. 6.10 and 6.11).

Well No. 1 (for the location of this and other wells described below, see Figs. 6.10 and 6.11) was drilled into the eastern slope of Tush-Oba hill and recovered Kimmerian ferruginous sandstones, conglomerates, and occasionally ore

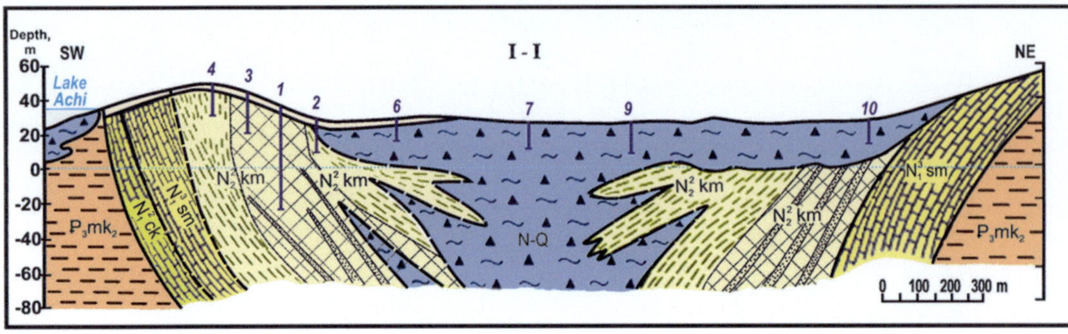

Fig. 6.10 Geological cross section through the Achinskaya recessed syncline along transect I–I. Location of the transect can be seen in Fig. 6.9. (For legend see Fig. 6.4)

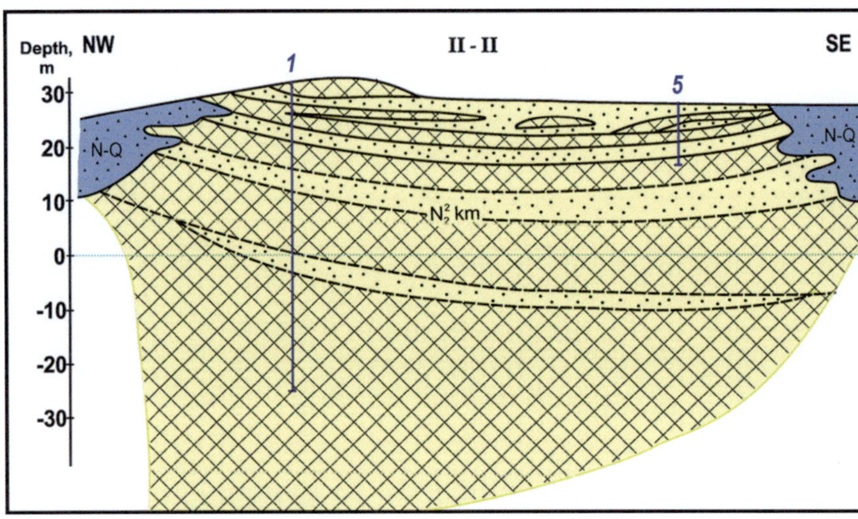

Fig. 6.11 Geological section through the Achinskaya recessed syncline along transect II–II. Location of the transect can be seen in Fig. 6.9. (For legend see Fig. 6.4)

sandstones with a total thickness of 58 m. Well Nos. 2 and 5 encountered Kimmerian ore sandstones and conglomerates interbedded with mud volcanic breccia. Well Nos. 6, 7, and 9 penetrated into 50 m of Kimmerian black, dark gray, and occasionally layered, sandy clays, crumpled within some depth intervals.

The recovered ore-bearing Kimmerian section has a structure typical for recessed synclines that is characterized by interbedded sediments enriched to varying degrees by iron. The upper 10 m of the section are represented by varieties of oxidized iron sediments that are dominated by a limonitized clayey component, an increased sand content, insignificant content of oolites and pseudo-oolites, and a wide range of the most important ore-related chemical elements: Fe (14.82–35.42%), Mn (0.33–2.51%), P (0.1–1.5%), and V (0.2–0.057%). The highest concentration of Fe is present in the upper part of the section.

The study of these rocks in thin section is difficult due to their total limonitization. Most often, the affected rocks are sandstones with different types of cemented terrigenous components and ooids. In Well No. 1, unoxidized tobacco iron ores were recovered and represented by lithologically different varieties, such as sandy iron ores, sandstones with pebbles (up to 1 cm in size), and siderite cement as well as tobacco conglomerates. The concentration of Fe varies between 9.7 and 43.41%, Mn between 0.18 and 2.5%, P between 0.1 and 0.2%, and V up to 0.021%. Spectral analysis of ores has also revealed Ba (from traces to 0.08%, or in some intervals—e.g., 8.2 m in Well No. 1—from 0.5 to 1.0%); Ti (from 0.05 to 0.4%); Zr (from 0.01 to 0.02%); Ni (from 0.002 to 0.01%); Co (from 0.001 to 0.005%); Cr (from 0.001 to 0.008%); and Cu, Pb, Be, Sc, Y, Yb, and La (from one thousandth to ten thousandths of a percent).

The clay pebble fraction (including sizes up to 1 cm) in the ores reaches 20%, and the rock bears an overall resemblance to tobacco ore conglomerate of a dark green or more rarely a gray-green color that is typical for mud volcanic structures. Sandy conglomerates and sandstones contain numerous remains of Kimmerian fossils. The rock is sideritized and compacted, while inclusions of sulfides and veins of carbonates are also visible. In thin sections, these are sandy iron ores that contain rounded fragments of clay or sandstone with basal and/or contact carbonate-chlorite-clay cement. Carbonate is represented by siderite. Among the terrigenous minerals, quartz predominates with rare plagioclase. Microcline, magnetite, biotite, amphibole, glauconite, zircon, sphene, etc. are also present. The grains of minerals with sizes around 0.1–0.4 mm are slightly rounded. The concentration of oolites ranges between 5 and 10%; as a rule, the oolites are developed around cores and, in terms of their size, they are similar to the dominant terrigenous particles. In polished sections of breccia, one's attention is drawn to the widespread development of sulfide mineralization represented by pyrite, which forms numerous xenomorphic or spherulite-like particles. In the breccia itself, sulfide mineralization is represented by realgar.

Two hypotheses regarding the structure of the Tush-Oba mud volcano can be considered. The first one assumes that Kimmerian deposits developed only within the Tush-Oba hill as well as a narrow zone adjoining the hill on the east that is limited by a fault.

The second and more realistic hypothesis supposes an extensive development of Kimmerian deposits beneath the entire mud volcanic field, the periphery of which is contoured by different size pieces of limestone and remains of Kimmerian mollusks. If this is the case, it is quite possible that a large iron ore deposit could be discovered in the recessed syncline located on the western extremity of the Tush-Oba mud volcano. A well drilled 2 km to the east of the hill recovered 10 m of limonitized mud volcanic breccia containing realgar.

Near the well, thick-walled shells of freshwater Kimmerian cardiids were found together with fragments of limonitized tobacco clays but without the ore. According to the paleontologist V.N. Semenenko (verbal communication), the presence among the mollusk shells of *Congeria* speaks in favor of prodelta conditions. The elevated sandiness of the ores does not contradict this. Despite the fact that ore deposits were not recovered, their presence is still possible. Judging from the lithological properties of the recovered section of Kimmerian deposits, including the composition of terrigenous material, the ore strata would be obviously enriched with sands. Apparently, there were freshwater flows that carried sand and chemogenic material into the Kimmerian lagoon formed by mud volcanic activity. This sedimentary material could have been brought from the low hills represented by the Botegechskiy and Egetskiy ridges that stretch from Vladislavovka station to the Crimean Mountains. In the geological section of these ridges, there are 70 m of sandy-argillaceous sediments of uncertain age overlain by pebbles. It would be desirable to determine the influence of alluvial processes on their formation.

The discovery of iron ores in mud volcanic deposits on the well-studied Kerch Peninsula is unique and deserves a certain amount of drilling and sampling to assess its industrial significance. It should also be taken into account in further paleogeographic reconstructions. This applies to other mud volcanoes of the Arma-Elinskaya anticline that run parallel to the Parpachskiy Ridge. The geological structure of the Parpachskiy Ridge is quite complicated because it is composed of three brachianticlinal swellings. On one of them, far to the west, the Vladislavovskiy mud field is situated. Another one, located far to the east, contains the Khyrtsyz-Shiban mud volcanic field.

The Khyrtsyz-Shiban mud volcanic hearth consists of two small mud volcanoes: a western (#2 in Fig. 6.7) and an eastern (#3 in Fig. 6.7) one, located about 900 m from each other and rising to a height of about 5 m. They are crowned with two or three gryphons and mud puddles that release gas (mostly methane). The Western Khartsyz-Shibanskiy mud volcano is currently located within a water reservoir. It forms a small islet dotted with large "flat cakes" of kaolinitized siderite as well as smears and stains of oil and gases that burn easily.

The Eastern Khyrtsyz-Shiban mud volcano is paleogeographically interesting. It is situated on the Arma-Elinskaya anticline and located 1.5 km to the west of the village of Batalnoye (#I in Fig. 6.7). Similar to Tush-Oba, it is also characterized by the exposure of Chokrakian limestones. The mud volcano takes the shape of an oval with a widest diameter of up to 200 m, and it rises above the terrain to a maximum height of 4 m. In relief, the volcano is poorly expressed, but it is clearly visible due to a change in vegetation type and the presence of a few mud puddles that release gas (mostly methane). Already in 1936, Belousov and Yarotskiy had proposed the presence of a recessed syncline here. In our opinion, they were right, and this recessed syncline, similar to that of Achinskaya, which was established in Chokrakian time, should contain ore deposits. In 1984, this hypothesis was confirmed. In the area of Batalnoye village, wells discovered for the first time two recessed synclines, one of which contained ore deposits (Shnyukov et al. 1985).

As early as 1981, it was suggested that Kimmerian deposits should be present in the Eastern Khyrtsyz-Shiban mud volcano by analogy with the Achinskaya recessed syncline (#3 in Fig. 6.7), where Chokrakian and Kimmerian sediments are present. In 1984, the first author found pieces of clastic ore material 1.5 km to the west of Batalnoye village. This encouraged the drilling of a few wells, and at a depth of 8–10 m under grayish-brown loam, caviar-like, oolitic-pseudo-oolitic brown and tobacco ores were recovered with a total thickness up to 7 m (Shnyukov et al. 1985). The area of the ore deposit was 200 × 250 m^2. It appeared typical of recessed synclines and contained lots of clastic ore material, with its Fe content ranging between 33.05 and 42.24% and its Mn varying between 1.34 and 7.77%.

A typical section of the ore deposit encountered in Well #5 with depth intervals is as follows:

1. Yellow ferruginous loam with manganese inclusions over cracks (0–10.1 m).
2. Fine-grained pseudo-oolitic, caviar-like ore, enriched with Mn, dark brown in color, loose, with thin horizontal layering parallel to the axis of the core stratification (10.1–10.5 m). Tobacco ore, oolitic-pseudo-oolitic with chlorite cement, carbonatized, solid, green in color, with compacted clay pebbles oriented to the most developed plane perpendicular to the core axis (10.5–13.1 m).
3. Tobacco clay, sandy, significantly oxidized, with gravel admixture (13.1–13.4 m). Large gravel oolitic-pseudo-oolitic with chlorite cement, intensely green in color, sometimes weakly cemented to loose gravelly sandstone with small (up to 5 mm) shells of pelecypods, and a layer of green clayey sand with gravel at the foot of the interval (13.4–13.7 m).
4. In contact with the ore layer, there is a 2 cm layer of strong clayey siderite, below which there is mud volcanic breccia, carbonatized, gray in color, bleached in places (13.7–14.1 m).
5. Mud volcanic breccia, gray, finely fragmented, sometimes conglobreccia (14.1–19.2 m).
6. Organogenic-detrital limestone from light to dark gray in color with typical Chokrakian fauna (19.2–23.1 m).

An unexpected feature of the ore deposit is its almost horizontal bedding and very rapid facies wedging. The ore layer, or facies, dissipates into the thickness of the mud volcanic breccia that fills the recessed synclines bordered by Chokrakian sediments. In the future, this area should be investigated with the help of drilling to clarify details of its composition.

Dzhau-Tepe (#6 in Fig. 6.7) is the largest mud volcano of the Kerch Peninsula. It is located in the arch of the Vulkanovskaya anticline (#VI in Table 6.2), 10 km south of the village of Leninskiy (Fig. 6.12).

This volcano is confined to the axial line of the anticline (Belousov and Yarotskiy 1936). The core of the anticline is composed of Lower Maikopian clay, while the slopes are Middle and Upper Maikopian sediments (Alyaev 1947).

The mud volcano forms a large hill with a height of 116 m asl. It rises 60 m above the level of the terrain, creating a majestic sight. Shape-wise, it is an asymmetric cone constituted by a few layers of mud volcanic breccia. Its slopes are cut by ravines (Fig. 6.13a, b), and the top is cracked (Fig. 6.14a, b).

Several deep wells were drilled into the Dzhau-Tepe mud volcanic field. In 1982, one of them located to the north of the volcano ejected gas from the Cretaceous sediments. To the east of the volcano, there is a hydrogen sulfide spring with a significant stream discharge. Against the background of a monotonous brown terrain, the bright greenery and iridescent shades of the spring water look like a beautiful and alien spot.

The area of erupted breccia and its volume reach 1.5 km^2 and 55 million m^3, respectively (Kalinko 1960). The recessed syncline is filled with Chokrakian, Karaganian, and Sarmatian sediments (exposed by excavation) indicating an explosion of this mud volcano in the geological past (Fig. 6.15a, b).

Modern activity of the Dzhau-Tepe mud volcano is recorded in traditional stories since the Middle Ages. Actually, the name of the mud volcano is translated from Tatar into Russian as "enemy mountain" or "enemy top," testifying to its active nature. It can be also translated as "the flowing peak" (Klepinin 1914). The explosive character of the Dzhau-Tepe mud volcano and its profuse emission of gases and breccia are shown in Fig. 6.16.

Some stages of its activity are provided in Table 6.1. Here, we offer more information about the catastrophic explosions of the Dzhau-Tepe mud volcano as obtained from different sources. In the beginning of the seventeenth century, erupted mud volcanic breccia destroyed the village of Dzhau-Tepe on the western slope of the mountain (Kurishko et al. 1968). The modern cone of the mud volcano was formed as a result of a large eruption in the eighteenth century (Pallas 1883). Andrusov (1884), who visited the mud volcano on September 4, 1884, pointed out some signs of recent activity but did not describe its extent. The beginning of the twentieth century was marked by a sharp increase in activity. The first eruption occurred on March 16–18, 1909, although local residents had noted the appearance of cracks on the hill a month earlier, in February 1909. Here is a description of the eruption from the words of eyewitnesses collected by P.A. Dvoychenko (1926), who wrote that the top of the mountain swelled up one sazhen (1

Fig. 6.12 General view of the Dzhau-Tepe mud volcano

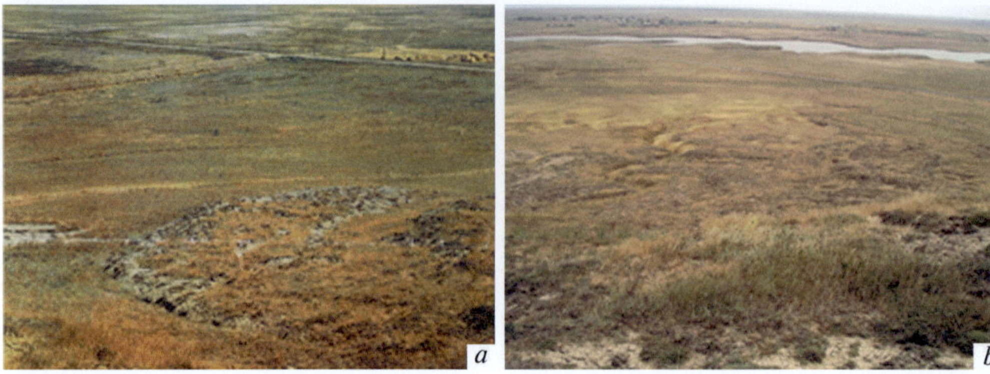

Fig. 6.13 Currents of mud volcanic breccia descending from the top of the Dzhau-Tepe mud volcano: (**a**) pictured in 1982, (**b**) pictured in 2004

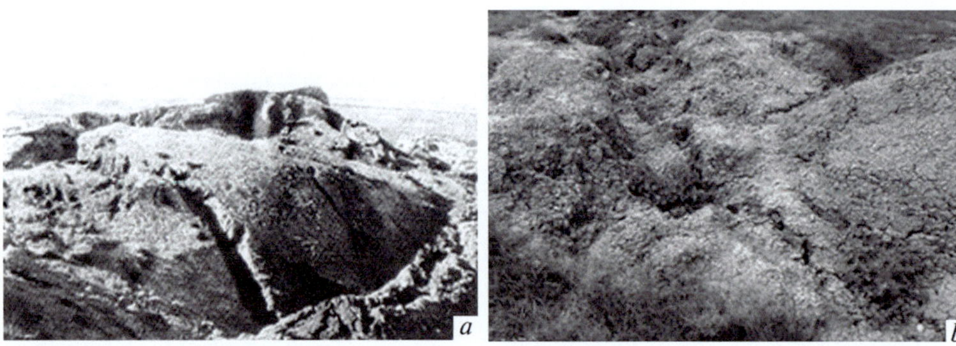

Fig. 6.14 Top of the Dzhau-Tepe mud volcano: (**a**) pictured in 1909 (Klepinin 1914), (**b**) pictured in 2004

sazhen = 2.133 m) and then sank two fathoms (1 fathom = 1.8288 m) below its original position, forming a system of cracks around the central depression. Subsequently, the outer shaft on the south side broke through to the surface and a stream of thick mud five sazhens wide began slowly to descend the slope. On the second day, a considerable amount of liquid mud with a weak hydrogen sulfide smell poured out, forming a stream of 160 sazhens in length, 1–3 sazhens in thickness, and 20–30 sazhens in width. On the third day, the thick mud moved slowly and soon stopped. Klepinin (1914) estimated that during the first eruption, the amount of discharged mud reached eight million pudi (64,000 m^3).

On August 17–18, 1909, another large eruption occurred. A stream of thick mud poured out in the same south-southeastern direction and blocked the first stream at a distance of up to 213 m, creating a ledge at the end of the flow. Dvoychenko (1926), who visited the peak of Dzhau-Tepe in 1912, observed a volcanic crater with a diameter of 53 m and a depth of up to 4 m, in the center of which there were three active gryphons each up to 1 m high. In the south, the crater was torn, and from the vent (10 m in diameter), a cylindrical stream of fresh mud was spewing and traveling downward along the slope after dividing into a number of parts.

On March 18, 1914, at 7 am, there was another eruption. Dvoychenko (1926) reported that within a radius of 10 km, the inhabitants of the surrounding villages reported slight oscillations of the soil. The eruption was short-lived: only 20 minutes. Residents of the village of Vulkanovka were awakened by the strong sound of thunder. A column of black smoke rose toward the clouds, but there was no fire. During all 20 minutes, there was a strong noise and fine earthen particles were thrown up to a height of

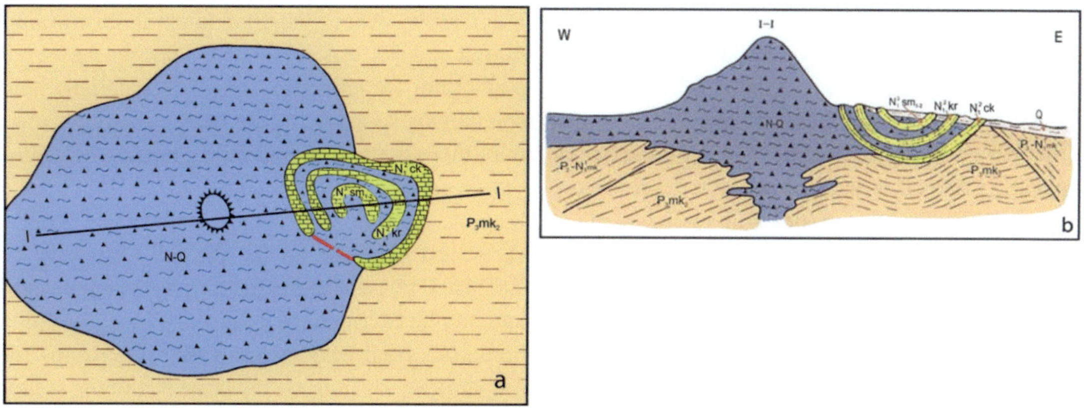

Fig. 6.15 (**a**) Geological structure of the Dzhau-Tepe mud volcano and the Dzhau-Tepe recessed syncline. (**b**) Geological section of the Dzhau-Tepe recessed syncline along the transect I–I. (For legend see Fig. 6.4)

Fig. 6.16 Explosion of the Dzhau-Tepe mud volcano. (Modified by N.A. Maslakov after Steber 1915)

40–60 m, while large stones the size of a grown man were tossed several meters into the air. A powerful mud stream poured southward, cutting into the old stream and carving out a new channel 4–6 m deep and 10–13 m wide. During this eruption, 50,000 m^3 of mud were ejected (Sedelshchikov and Kul'chavov 1914), which yielded the largest mud flow in terms of covered area compared to other volcanic events. The crater at the peak of Dzhau-Tepe survived, and the mud was less viscous compared to previous eruptions. The volume of discharged gases was the greatest compared with previous ones. The temperature of the mud was 12–12.5 ° C, while the air temperature was 16 ° C. The mud volcano exploded again in 1925 and 1927 (Murzaev 1928). In 1925, the total length of the flow reached 60 m and crossed over the stream of 1914. On September 13, 1927, a very strong outpouring of mud again emerged from the volcano. A stream of lighter, ashen-gray mud filled in the crater and moved toward the southwest. Its length was 41 m, its width 6–12 m, and its thickness 1.0–1.5 m. The volume of ejected mud was relatively small: 1500 m^3. This quiet extrusion of mud from beneath the Earth was unusual for this mud volcano; it was caused most likely by the Crimean earthquake of 1927.

In the early 1930s, Belousov and Yarotskiy (1936) observed three small gryphons at the top of the volcano that released gases. Eyewitnesses reported a small volcanic eruption in 1942. When the authors of this book visited the Dzhau-Tepe mud volcano within the past few years, no signs of modern activity were observed. As it currently appears, the mud volcanic breccia is represented by weathered brown varieties with abundant clastic material composed of siderite, limestone, calcite, sandstone, iron sulfides, etc.

To the east of the Dzhau-Tepe mud volcano, at the top of the Krasnopol'skaya anticline and oriented southwest-northeast, there is dried or watered (depending on the season) solonchak on the surface of Maikopian sediments experiencing poor aeration. There is also a small clay pit with weak rudimentary gryphons. Together they form the fossil Borukh-Oba (#7 in Fig. 6.7) mud volcano that is only weakly expressed in the relief. During our observation, this volcano represented a dry lake with an area of 55 × 150 m and stretching from north to south. The thickness of the dry crust under the lake reached 20 cm. The lake bottom contained 11 very small and slightly fizzy puddles, the largest of which had a diameter of about 1 m. The 70 m-deep well drilled into the east of the dry lake recovered a soil-vegetation layer (0.3 m), brown loam (0.3–6.8 m), gray-brown coarse breccia (6.8–9.6 m), dark olive mud volcanic breccia (9.6–12.6 m), and layered black clays (12.6–15.3 m) that were crumpled in the interval 13.3–14.0 m. Benthic foraminifera *Saccammina grzybowski* (Schubert) and fish teeth were discovered within the interval 12.8–15 m, identifying its age as Upper Maikopian (L.A. Digas, verbal communication). Above the 10 m interval, this foraminiferal species was replaced by poorly preserved planktonic foraminifera, single radiolarians, and spicules of sponges. In the heavy sediment fraction, there were abundant ferruginous concretions that resemble oolites. The age of the breccia remains unclear. The authors expected to find a depressed syncline in the Borukh-Oba area, as suggested by two small hills near the southwestern and southeastern edges of the dry lake and quite numerous fragments of Sarmatian limestones. To our surprise, drilling on the large hill to the southwest did not confirm this assumption; however, there are still insufficient data for a final conclusion as the well recovered only 5 m of sediments and stopped in the loam.

Yet farther to the east, within the Sadovaya anticline, lies the fossil Gora Sadovaya (Ak-Tube) mud volcano (#8 in Fig. 6.7) accompanied by a recessed syncline of the same name, presumably of Sarmatian age (Fig. 6.17).

This mud volcano is located subparallel to the Vulkanovskaya and Krasnopolskaya anticlines. The core of the Sadovaya anticline is composed of Lower Maikopian clays. At present, no signs of activity can be observed although the thickness of the fossil breccia is substantial—over 35 m (drilled wells did not fully penetrate to its bottom and stopped at this depth). There is also a small, bitterly salty spring.

Finally, in the extreme eastern part of the Southwestern Plain, at the peak of the Uzunlarskaya anticline, two recessed synclines have been discovered and studied: Konchekskaya

Fig. 6.17 General view of the Gora Sadovaya (Ak-Tube) mud volcano. (Pictured in 2004)

is located to the north, and Uzunlarskaya is located to the south. Together they make up the picturesque hill called Gora Konchek (Trekhgornaya), where Chokrakian, Karaganian, Sarmatian, and Kimmerian sediments are exposed in the middle of a plain that is covered by Maikopian deposits (Fig. 6.18).

The Konchekskaya recessed syncline is filled with Chokrakian, Karaganian, and Sarmatian sediments. The Uzunlarskaya recessed syncline contains the same composition with the addition of Kimmerian sediments, and it is enriched with iron ore. In both cases, the sediments are interbedded with mud volcanic breccia (Figs. 6.19 and 6.20).

As determined by wells, the thickness of the mud volcanic breccia in the Kimmerian ore deposits reaches 3–4 m. No sign of modern mud volcanic activity could be observed.

The Kimmerian iron ore deposit in the southern part of the Uzunlarskaya recessed syncline cannot be explored because it is covered by Lake Uzunlar. Structurally, the Uzunlarskaya recessed syncline represents a deep, probably elongated bowl of as yet undetermined dimensions. The greatest thickness of the ore deposit reaches 35 m in the explored part with a steep drop of the Kimmerian ore beds up to 45°. The ore bed sharply wedges toward the sides of the syncline. It is recorded by the steep angles of incidence of the ore-bearing structure (Fig. 6.21).

In the section of the iron ore deposit, there are two layers of clay facies that pass into mud volcanic breccia. The ore deposit has a pebbly-ooidal appearance, and the content of ooids reaches almost 70%. Among the ooids, there are mainly pseudo-oolites (rounded pieces of rocks reminiscent of oolites), while oolites are observed in insignificant amounts. Pebbles and small boulders of clay are quite abundant.

The iron ore deposit is represented mainly by tobacco ore with an average content of Fe (27.3%) and Mn (0.85%). The upper part of the ore deposit reveals a thickness of about 5 m and is represented by brown ores with Fe and Mn content of 33.8% and 1.28%, respectively. Approximate reserves of terrestrial (non-lake) ore deposit is 16.5 million tons.

Presumably, Lake Koyash and the mountain Opuk are also mud volcanic structures. All mud volcanoes described above are located on the Southwestern Plain, within the Parpachskiy region (Fig. 6.7). They were formed during Chokrakian time and are characterized by fading activity (except for the Dzhau-Tepe mud volcano). Some hold opinions that these mud volcanoes were formed during Middle Maikopian time, but there are few data to support this position.

The Vneparpachskaya area is much richer in modern mud volcanoes than the Southwestern Plain. The extreme western mud volcano, Nasyrskiy (#11 in Fig. 6.7), is confined to the core of the Nasyrskaya anticline, which is composed of Maikopian clay. The anticline is surrounded by Chokrakian-Karaganian-Konkian rocks. In the terrain relief, the Maikopian core is expressed by a closed depression at the bottom of which there is a relatively small, oval (250 × 175 m) hill with a height up to 12 m. On

Fig. 6.18 General view of the Gora Konchek mud volcano and Uzunlar Lake. (Photographed in 2004)

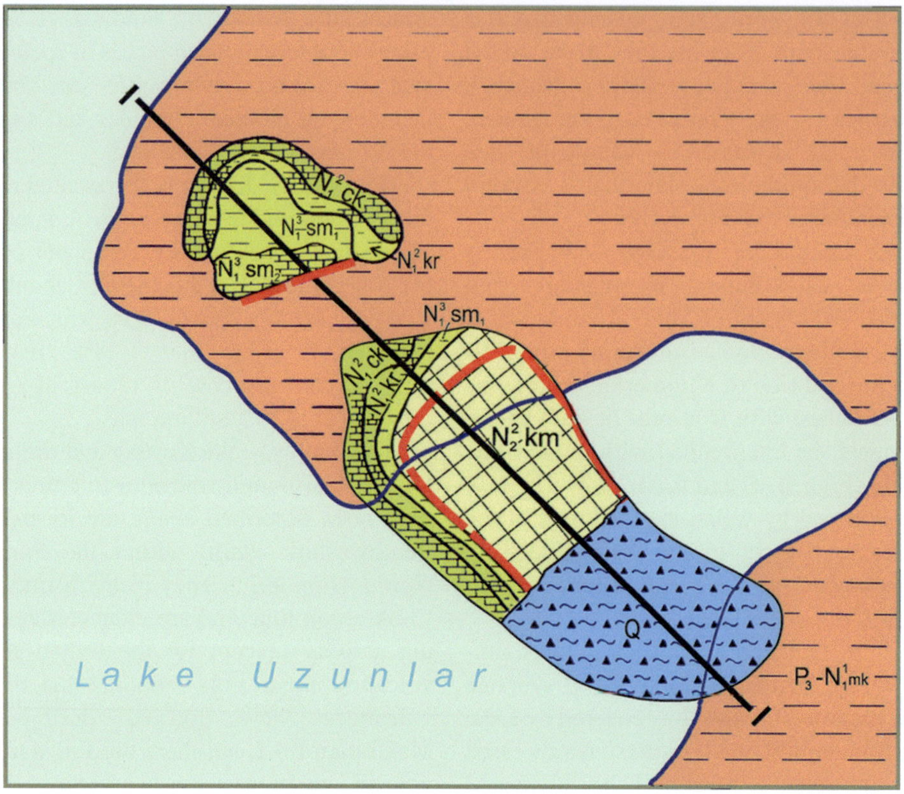

Fig. 6.19 Geological structure of the Uzunlarskaya recessed syncline. (For legend see Fig. 6.4)

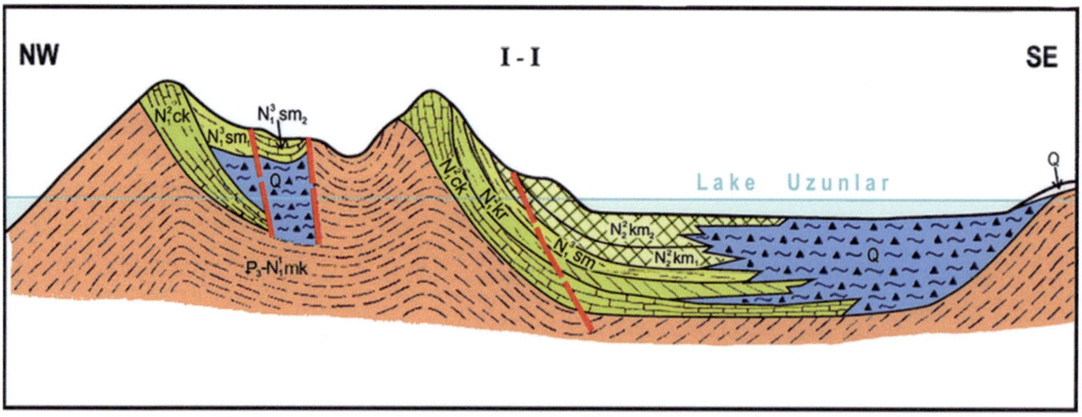

Fig. 6.20 Geological section of the Konchekskaya and Uzunlarskaya recessed synclines. For the location of this transect, see Fig. 6.19. (For legend see Fig. 6.4)

its flat top, there are several small gryphons and slight gas emanations from the mud puddles. The most characteristic feature of the Nasyrskiy mud volcano is that all eruptions have been impregnated with oil. In addition to the hill, there are three small lakes within the Nasyrskaya anticlinal valley, and their origin is somehow related to mud volcanic activity. The Nasyrskaya anticline is a southeastern branch of the large Kamenskaya anticlinal structure (Fig. 6.22).

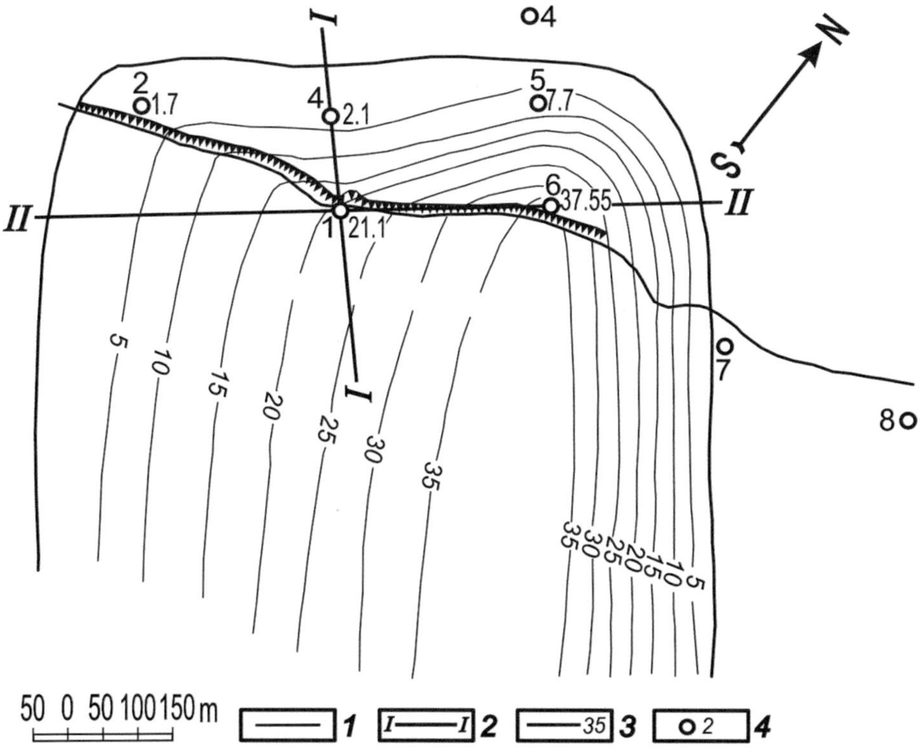

Fig. 6.21 Schematic diagram of the total thickness of ore deposits in the Uzunlarskaya recessed syncline: (1) contours of iron ore deposit, (2) line of transect, (3) isolines of thickness, (4) wells (Shnyukov and Naumenko 1964)

The latter has a southwest-northeast strike and bifurcates into two branches in its southeastern part. Its southeastern branch—the Nasyrskaya anticline—contains in its center a Maikopian core framed by Sarmatian bryozoic limestones. The bifurcation of anticlines on the Kerch Peninsula is a rare phenomenon.

There is only one known anticline that is bifurcated at its eastern end: Gornostaevskaya. Here, bifurcation is caused by mud volcanic activity and the appearance of a recessed syncline at the very edge of the anticline. This occurrence enabled us to assume by analogy that a similar reason likely accounts for the bifurcation of the Kamenskaya anticline. We believe such a bifurcation is caused by the ore-bearing recessed syncline with its eastern boundary that runs along a meridional fault, which therefore opens the syncline to the east and connects it to the iron ore-bearing Chegercinskiy graben-syncline (Figs. 6.23 and 6.24).

The data recovered from geological explorations carried out by different organizations in the 1950s, 1960s, and 1970s seemed to confirm the opinion of Konstantov et al. (1933) on the limited development of iron ore here and its confinement to the sides of the large Chegercinskiy graben-syncline.

Upon our request, ten deep wells were drilled on the Kamenskaya anticline by the Crimean expedition and the Management of Mining Geology of the Ministry of Ferrous Metallurgy at the end of the last century. The material thus obtained confirmed the presence of an ore-bearing recessed syncline as the main reason for bifurcation of the Kamenskaya anticline on its eastern end (Shnyukov et al. 1987). The Kamenskaya recessed syncline is delineated by the northeastern wing of the Kamenskaya anticline. From the surface, it is composed of Sarmatian limestones and framed by Meotian and Pontian limestones. On the southeast, where

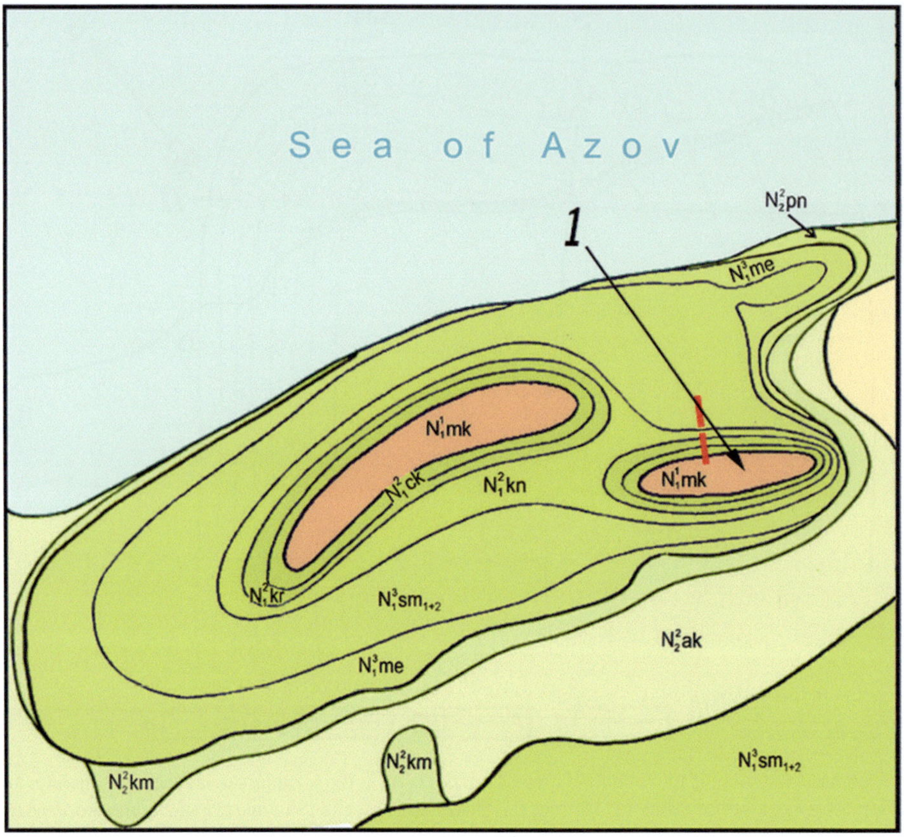

Fig. 6.22 Diagram of the Kamenskaya anticline's geological structure: 1—Nasyrskaya anticline. (For legend see Fig. 6.4)

the Nasyrskiy mud volcano described above is located, it is framed by the Nasyrskaya anticline, which is composed of Maikopian and framed by Chokrakian, Karaganian, and Konkian deposits. The boundaries of the Kamenskaya recessed syncline have been determined by drilling. Its dimensions are 1500–1700 m in width and 1800–1850 m in length. The total area is 3 km^2. At its base, the recessed syncline is composed of Meotian limestones, and Pontian sediments are almost absent. Kimmerian deposits are represented by three horizons that lie directly on the Meotian limestones and are overlain by Kuyalnitskian and Quaternary sediments. The deposit of Kimmerian iron ore reaches a thickness of 20 m and is represented exclusively by unoxidized ores, such as tobacco oolitic, tobacco oolitic-pseudo-oolitic, and cementing tobacco ores and tobacco clay varieties (Fig. 6.25).

Ore deposits are composed predominantly of oolites (up to 25%), pseudo-oolites, and pelletized ore fragments (35–37%). There are also clay pebbles. Hydroferriclorite acid cement makes up 20–30% of the deposit.

Iron content is highest in the oolitic ores (37–45%) making such deposits of great interest industrially (Shnyukov et al. 1992). The distribution of Fe, Mn, and P is shown in Figs. 6.26 and 6.27.

To the east of the Kamenskaya anticline, the orientation of anticlinal structures changes. Most (except for those located to the southeast of the Kerch Peninsula) strike in a sub-latitudinal direction and are often complicated by several mud volcanoes and recessed synclines in their arched parts. Within the large Korolyovskaya anticline, there is a small mud volcanic structure—the Ilyichevskaya recessed syncline—as well as the

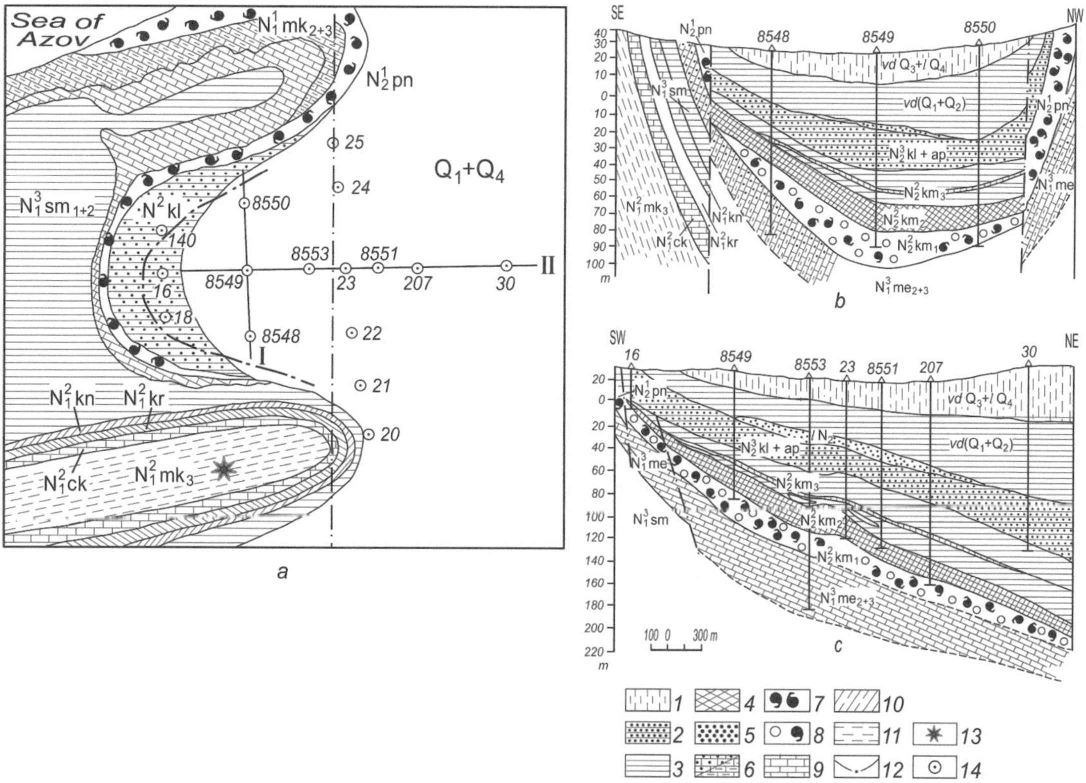

Fig. 6.23 Kamenskaya recessed syncline: (**a**) geological diagram; (**b**), (**c**) geological transects along the lines I–I and II–II, respectively. Key: (1) loam, loess, fossil soil, loess-like loam; (2) interbedding of sands and clays; (3) clays; (4) iron ore; (5) sand, clayey sand, aleurite; (6) calcareous sandstone (numerator), clays with interlayers of limestones and calcareous clays (denominator); (7) organogenic detrital limestones; (8) ferruginous shells; (9) limestones; (10) clays with interbeds of marls; (11) the upper part of the Maikopian deposits, and non-carbonate clays with interbeds of siderites; (12) tectonic disturbance; (13) mud volcano; (14) well

Korolyovskiy mud volcano with a recessed syncline of the same name on the top (Fig. 6.28).

The Korolyovskiy (#13 in Fig. 6.7) mud volcano looks like a large sloping mountain composed of mud volcanic breccia. There is also the Korolyovskaya recessed syncline that drilling has shown to be made up of Chokrakian and Karaganian sediments. The latter is redeposited fossil mud volcanic breccia represented by clayey conglomerates with a relatively high content of cinnabar and graphite (Shnyukov et al. 1971), as well as green clays that remind one of the Kimmerian tobacco clays (Naumenko 1977). Adjacent on the west and east of the Korolyovskiy mud volcano are cup-shaped limestone structures, which are probably also recessed synclines. The eastern one is almost completely covered with human-generated trash. Though it was very powerful in the geological past, at present, the Korolyovskiy mud volcano is inactive, and the minor gryphons on its southern slope are the last manifestations of its former activity.

To the east, the Korolyovskaya anticline is replaced by the Sartskaya anticline, which includes the Kara-Sidzheutskaya (Western and Eastern), Kayaly-Sartskaya, and Novoselovskaya recessed synclines, and the Novoselovskiy mud volcano (Fig. 6.29).

It is quite possible that another mud volcano also exists near the Adzimende cavity, in the southwest of the Sartskaya anticline (Shnyukov et al. 1971). The Kara-Sidzheutskaya (Western and Eastern) and the Kayaly-Sartskaya recessed synclines are composed of Sarmatian limestones only (Figs. 6.30 and 6.31).

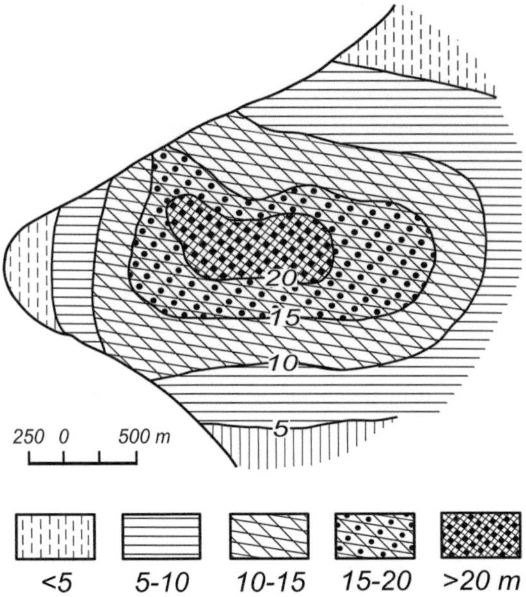

Fig. 6.24 Thickness of the ore deposit in the Kamenskaya recessed syncline

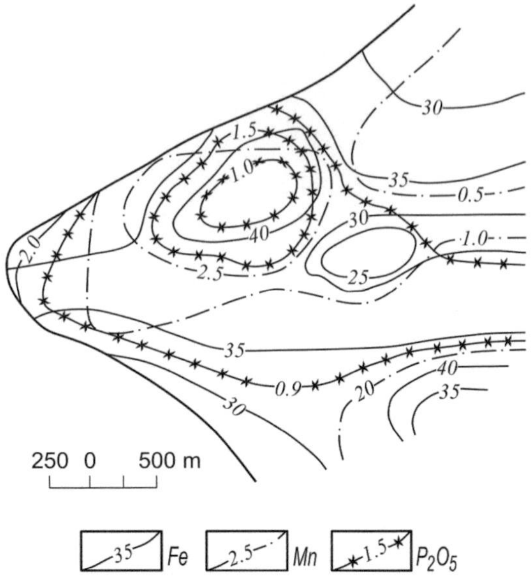

Fig. 6.26 Diagram showing the distribution of Fe, Mn, and P (weighted average values, %) in Kimmerian deposits of the Kamenskaya recessed syncline

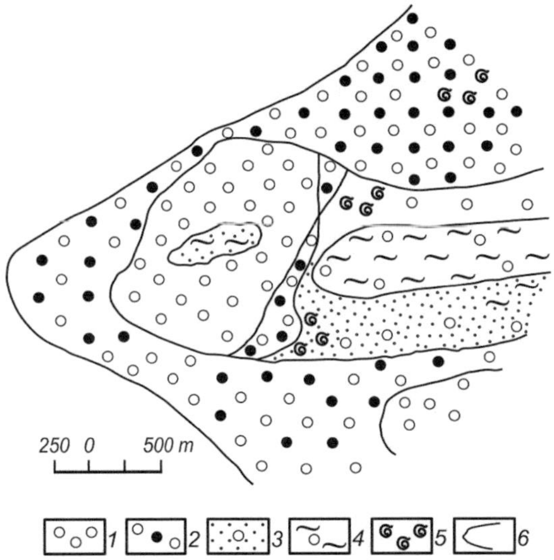

Fig. 6.25 Lithofacies scheme of the ore deposit in the Kamenskaya recessed syncline. Key: (1) tobacco oolitic ore; (2) tobacco oolitic-pseudo-oolitic ore; (3) tobacco sandstone; (4) poor oolitic tobacco ores and tobacco clays; (5) places of fossil accumulation; (6) contour of the ore deposit

Wells drilled into these structures recovered mud volcanic breccia, locally kaolinitized and carbonatized.

The Novoselovskiy (#18 in Fig. 6.7) mud volcano is now small in size, and signs of its activity are sporadic. However, the scale of its activity in the geological past was grandiose. The amount of erupted breccia accumulated since the beginning of the mud volcano's activity (presumably from as early as the Upper Maikopian) is estimated to be several billion m^3. A large outburst occurred just recently in 2016. Due to the long activity of this mud volcano, the large Novoselovskaya recessed syncline developed with an iron ore deposit, which was discovered and explored by the Institute of Geological Sciences of the Academy of Sciences of the Ukrainian SSR and the Management of Mining Geology of the Ministry of Ferrous Metallurgy of Ukraine in 1963–1970 (Shnyukov and Naumenko 1964, 1966, 1982).

The geological sequence of the Novoselovskaya recessed syncline preserves the general patterns of bedding that are typical for other similar structures of the Kerch Peninsula. In particular, there are no Chokrakian, Karaganian, Konkian, Lower and Upper Sarmatian, and possibly Pontian sediments. The

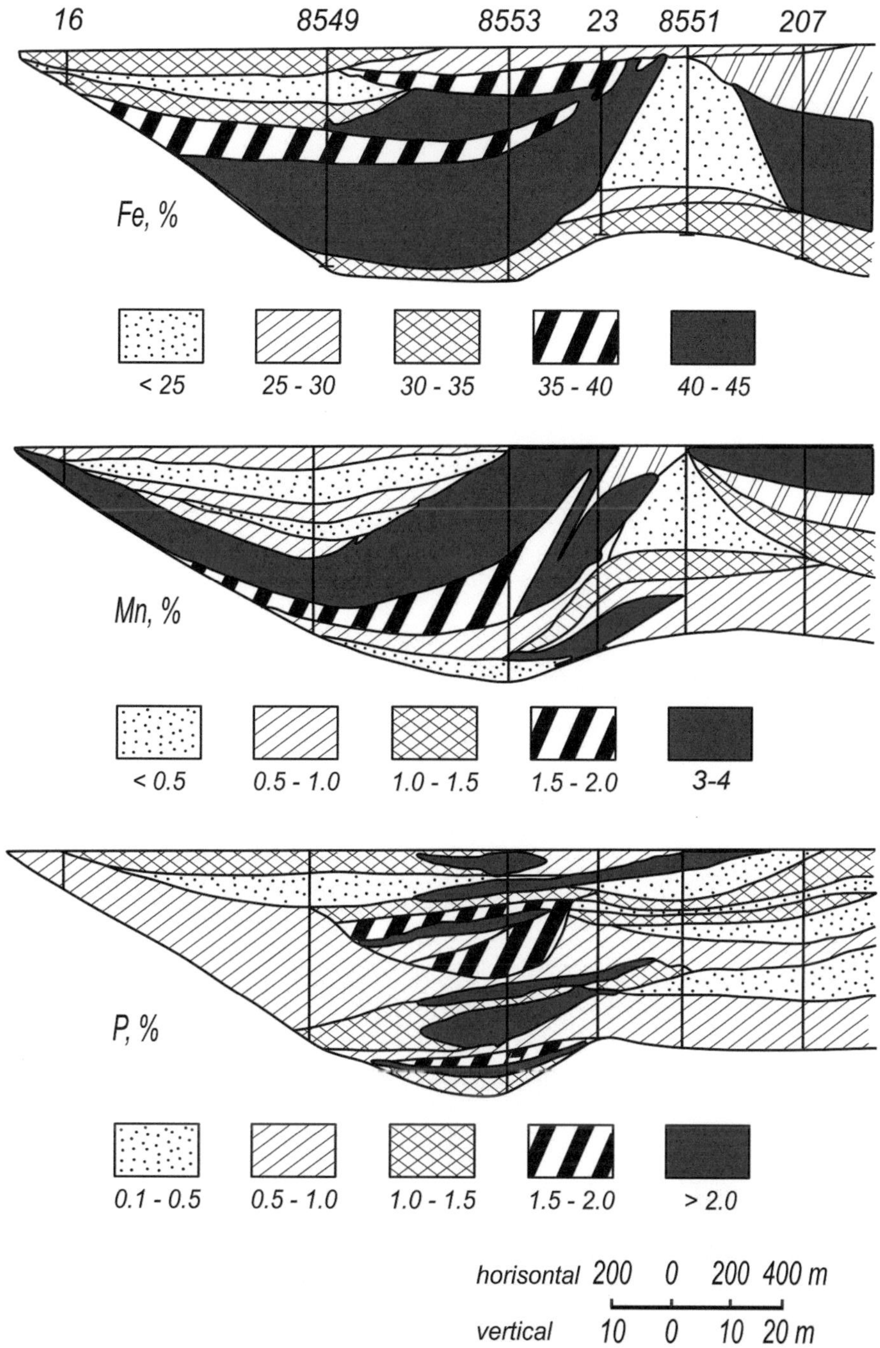

Fig. 6.27 Content of Fe, Mn, and P in the Kimmerian ores of the Kamenskaya recessed syncline (transect II–II, weighted average values, %). For location of the geological transect, see Fig. 6.23)

oldest deposits are represented by the Upper Maikopian devoid of fossils dark gray shale clays, which are overlain by Middle Sarmatian limestones with *Cardium obsoletum* Eichw., *Modiola navicula* Dub., and *Trochus* cf. *papilla* (Ilyin 1931). The latter are exposed on the top and slopes of the hill. Their true thickness presumably reaches 70 m. Adjacent to them are sup-

Fig. 6.28 General view of the Korolyovskiy mud volcano

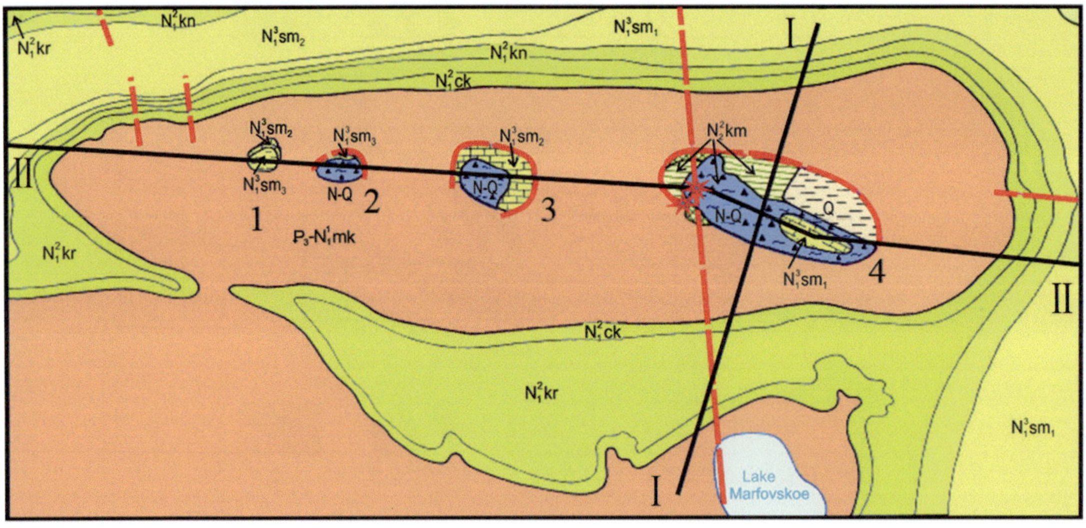

Fig. 6.29 Diagram of the geological structure of the Sartskaya anticline with the Western Kara-Sizheutskaya (1), Eastern Kara-Sizheutskaya (2), Kayaly-Sartskaya (3), and Novoselovskaya recessed synclines and Novoselovskiy mud volcano (4). For legend, see Fig. 6.7

posedly Meotic limestones discovered in one of the boreholes at a depth of 234 m (Kulichenko and Semenenko, verbal communication) as well as clayey coquina and barren clays of the lower (Azovian) horizon of the Kimmerian stage.

The lower Kimmerian molluscan fauna is concentrated in the marginal eastern and northern zones of the ore deposit. In the deeper central part of the recessed syncline, it disappears, and clays become barren. Judging from the available mate-

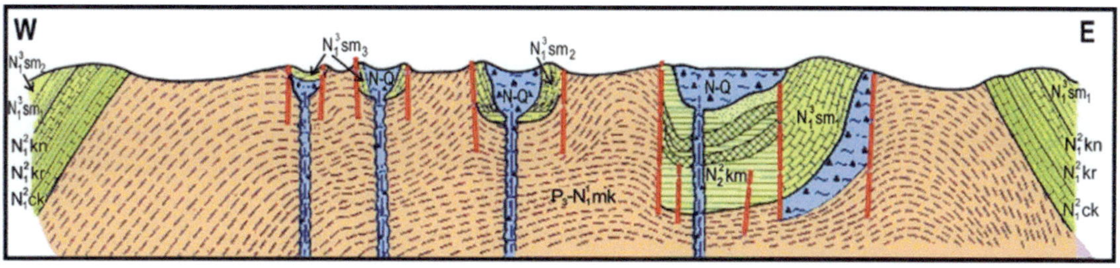

Fig. 6.30 Diagram and sub-latitudinal geological transect through the Sartskaya anticline along the line I–I. For location of the transect, see Fig. 6.29. For legend, see Fig. 6.4

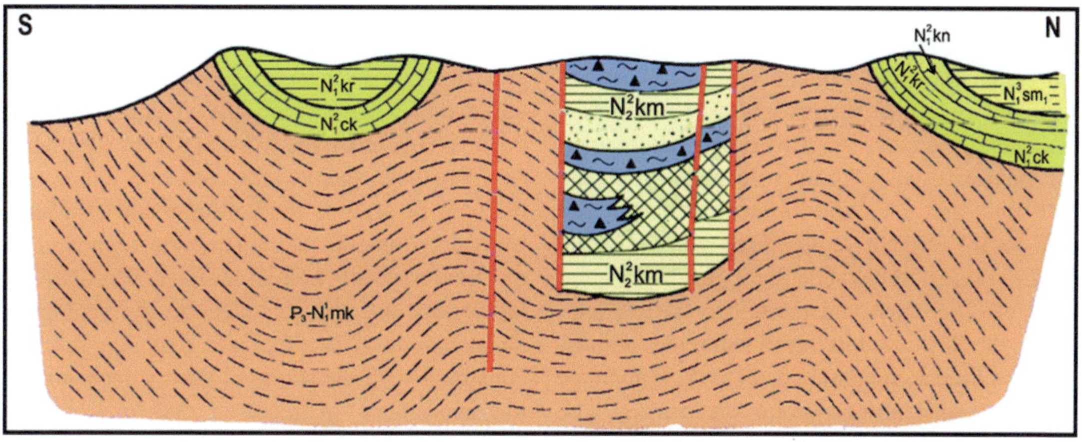

Fig. 6.31 Sub-meridional geological transect through the Sartskaya anticline on the traverse of the Novoselovskaya recessed syncline along the line II–II. For location of the transect, see Fig. 6.29. For legend, see Fig. 6.4

rials, one can assume a stratigraphic unconformity at the Lower Kimmerian-Maikopian boundary.

The Middle Kimmerian (Kamysh-Burunian) horizon is composed of oolitic-pseudo-oolitic iron ores interbedded with tobacco and brecciated clays, ferruginous conglomerates, and clayey pebbles. They conformably overlie Lower Kimmerian beds. Transition from the Lower to Middle Kimmerian beds is gradual, even subtle, and the beds can be distinguished from each other only by their iron content. As a rule, the foot of the ore deposit is enriched with mollusk shells and is gradually transformed into coquina or clays of the Azovian horizon. Neither the Azovian nor the Kamysh-Burunian horizon is exposed at the surface.

The Upper Kimmerian beds are often represented by barren sandy-aleuritic-brecciated clays with interlayers of redeposited iron ores and tobacco clays. They overlie the Middle Kimmerian beds without unconformity and cover the entire iron ore area of the recessed syncline.

Barren, slightly clayey sands and sandstones of these beds with ferruginous interlayers were exposed in places within an artificial exploration near the pump house in the village of Novoselovka. Andrusov (1893) and Ilyin (1931) referred them to the Pontian, Eberzin (1940) assigned them to the Upper Kimmerian, and Maymin (1951) linked them to the ore layers of the Kimmerian stage. Following Eberzin (1940), we refer these deposits to the Upper Kimmerian beds based on their lithological properties, in particular the presence of ferruginous interlayers.

The total thickness of the Kimmerian deposits exceeds 200 m; the lower horizon is greater than 57 m, the middle 99 m, and the upper 66 m.

Kuyalnitskian deposits in the Novoselovskaya recessed syncline are indicated conditionally based on analogy with those in the Kamysh-Burunskiy Trough.

Quaternary red-brown loams can be traced in the form of a narrow strip along the northern wing of the Novoselovskaya recessed syncline.

Structurally, the Novoselovskaya recessed syncline represents an oval-shaped deep graben located in the arch of the anticline. It is contoured from the north, south, and west by a semi-ring fault (Figs. 6.29 and 6.30). The graben stretches from northwest to southeast at an angle of 45° to the axis of the Sartskaya anticline over a distance of 3.2 km. The width of the graben is 1.6 km. The fault character of the graben is fixed by (1) the steep angles of incidence (45–90°) of the enclosing Maikopian clays toward the iron ore deposit, (2) manifestations of mud volcanism along the fault line, and (3) sulfide mineralization of the clay rocks adjacent to the fault zone. On the southern side of the graben, light gray Maikopian clays are crumpled and rotated to the vertical; within the clays are imprints of fish scales aligned to the stratification plane as well as an abundance of pyrite concentrated along the cracks. The angles of incidence of the layers composing the graben are different. On the wings, they incline at 10–25°, while in the center, the beds lie horizontally. The younger strata have smaller angles of incidence than the older ones.

Within the Novoselovskaya recessed syncline-graben, three structural blocks can be distinguished (Fig. 6.32).

The first block is confined to the Middle Sarmatian limestones, which form the eastern part of the recessed syncline-graben; the second block is limited to the area of the ore deposit; and the third block is located to the north of the first one. Geomorphologically, they are expressed according to their elevation, gentle spur, and closed lowland. In general, the Kimmerian deposits level out the unevenness of the Sarmatian basement. The structure of the ore deposit enables one to characterize the western part of the graben as a half bowl open to the east, and the eastern part can be seen as the northern side of a large bowl open on the south. They are separated by the mud volcano and its erupted breccia. The block structure reflects stages in the geological development of the recessed syncline that are consistent with the stages of mud volcanism and the appearance of recessed synclines on the Kerch Peninsula.

The ore deposit reaches a substantial thickness, up to 92.7 m, but the deposit's shape is

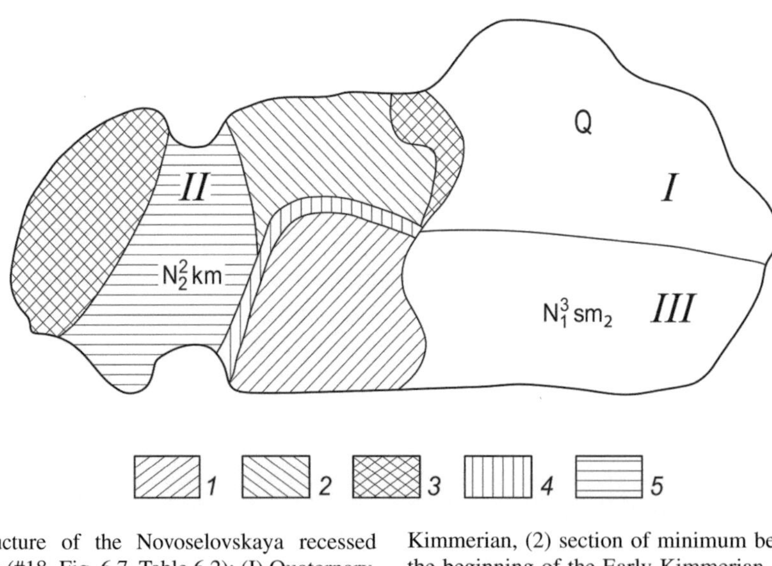

Fig. 6.32 Structure of the Novoselovskaya recessed syncline-graben (#18, Fig. 6.7, Table 6.2): (I) Quaternary, (II) Middle Kimmerian, (III) Middle Sarmatian. Structure of the Kimmerian block broken in parts: (1) section of maximum bending lowered to the beginning of the Early Kimmerian, (2) section of minimum bending lowered to the beginning of the Early Kimmerian, (3) sections lowered to the beginning of the Middle Kimmerian, (4) locus of bending, (5) mud volcano

complicated. It stretches 1.7 km from northwest to southeast, and its width is 1.2 km, giving an area of 2.0 km^2. Iron ores are represented by two main types: brown and tobacco. The tobacco ores consist mainly of hydroferrichlorite as well as iron and manganese hydroxides and carbonates. Hydroferrichlorites formed mainly as the cementitious component of the ores, and to a lesser extent of the oolites and pseudo-oolites. In the overwhelming majority of cases, both the brown and tobacco ores give the appearance of fish eggs. Their main components are cemented pseudo-oolites, while oolites play a subordinate role (Fig. 6.33).

A characteristic feature of all varieties of iron ore is the constant presence of clay pebbles in varying quantities. In the upper part of the section, they are almost absent. In the central part of the iron ore deposits, the ore layers are interbedded with tobacco clays, clayey conglomerates, and greenish-gray and gray clays with thicknesses ranging from 0.6 m to 2.0 m. To the north, they wedge into a solid ore deposit that forms three zones (Fig. 6.34).

Zone 1 (pseudo-oolitic-clastic ores) encircles the western and northern wings of the deposit in the form of an arcuate strip 250 m wide, penetrated by a mud volcano in one place (the narrowest part in Fig. 6.34). To the immediate south lies Zone 2 (oolitic-pseudo-oolitic ores), and this is replaced by Zone 3 of the same ores interbedded with tobacco clays and conglomerates yet farther south.

Some specific features characterize the geological section of the Novoselovskoe ore deposit. The bed of the deposit is composed of banded brown and tobacco "fish egg" pseudo-oolitic iron ores. It is followed by strata of gray clays interbedded with tobacco clays, ferruginous clay conglomerates, mud volcanic breccia, and poor (low iron content) tobacco ores. The lower bed contains tobacco oolitic-pseudo-oolitic-clastic ores with lenses of brown ore. Among the ores of the upper bed, there are two interlayers of ferruginous clay conglomerates with thicknesses ranging from 0.6 to 1.3 m. The number of interlayers increases southward. On the strike of the ore deposit, no significant reduction in thickness has been observed. At the junction of the ore deposit with the Sarmatian limestones on the east, the former wedge out. Friable brown iron ores extend for 1.8 km in the form of a strip along the outer frame of the structure; their width ranges between 60 and 200 m, and their thickness varies from 6 to 32 m. Their angles of inclination vary from 2° in the east to 40° in the west. The total thickness of

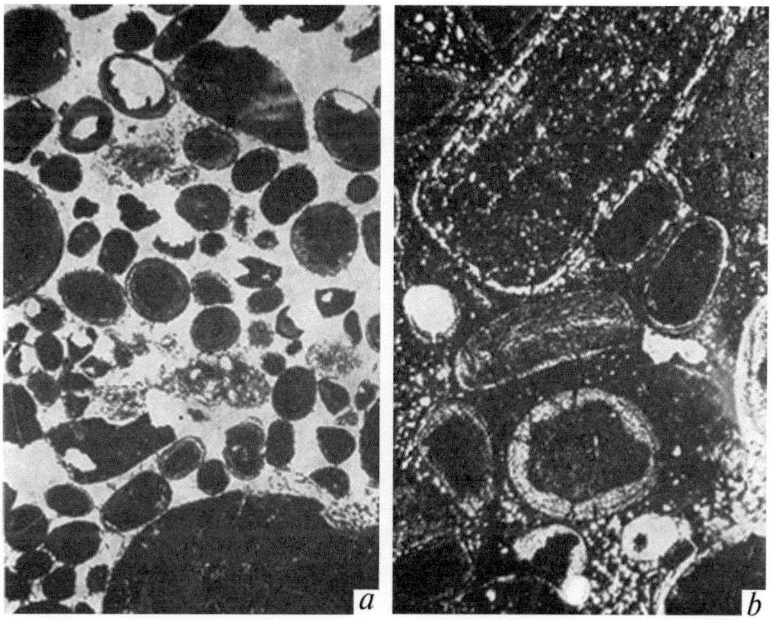

Fig. 6.33 Pseudo-oolitic (**a**) and pseudo-oolitic-oolitic (**b**) iron ores of the Novoselovskoe ore deposit. Oolites and pseudo-oolites are hydrogoethite. Most of the pseudo-oolites are well rounded. X 20 in plane polarized light

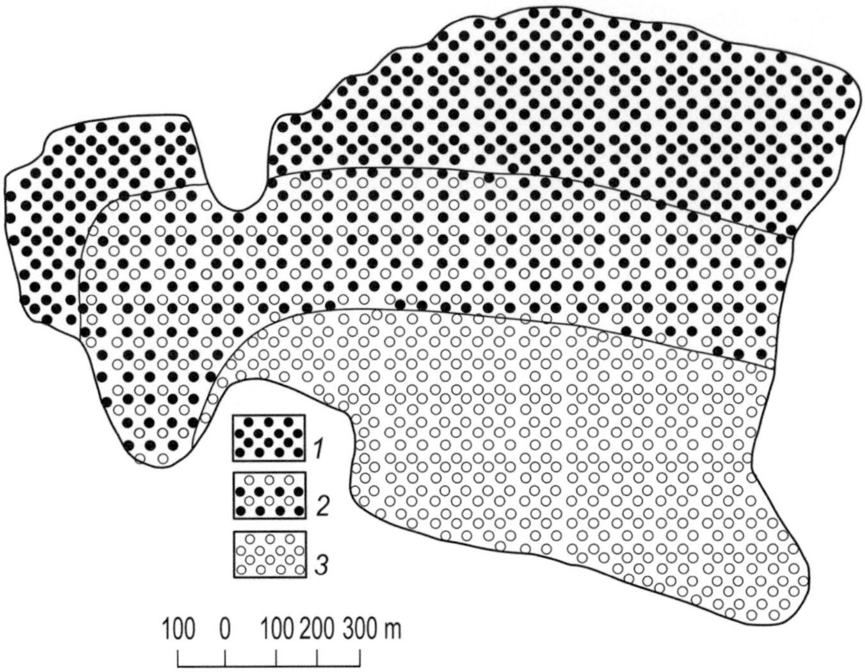

Fig. 6.34 Three zones of the Middle Kimmerian Novoselovskoe iron ore deposit. Key: (1) pseudo-oolitic-clastic ores; (2) oolitic-pseudo-oolitic ores; (3) tobacco clays and conglomerates interbedding with iron ores

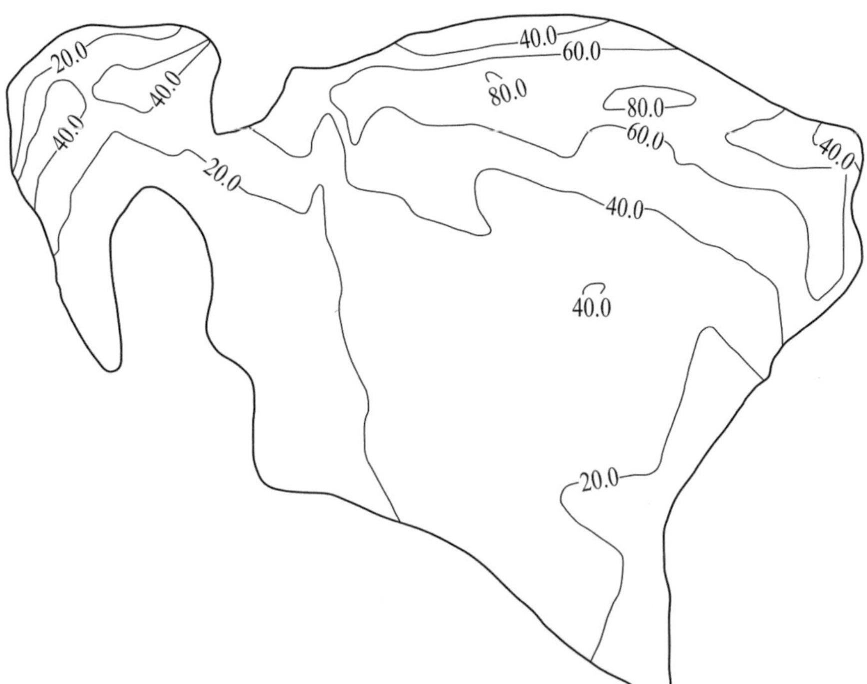

Fig. 6.35 Diagram of the Middle Kimmerian Novoselovskiy iron ore deposit in isopachytes (m)

the conditional (suitable for industrial exploration) beds of iron ores within the ore deposit varies from 22 to 92.7 m, reaching its anomalously greatest thickness in the northeastern and western parts of the deposit (Fig. 6.35).

Zoning of the ore body has been determined not only by the distribution of ores but also by the distribution patterns of chemical element concentrations within the ore deposit. According to sampling data, the chemical composition of ores varies considerably. Thus, the content of iron varies from 8–9% (tobacco clays) to 44.2% (average 36.5%), manganese varies from 0.64% to 11.3% (average 2.1%), phosphorus from 0.18% to 1.0% (average 0.86%), and arsenic from 0.0% to 0.153% (average 0.073%). As a rule, the highest concentrations of iron coincide with the highest concentrations of pseudo-oolitic and oolitic-pseudo-oolitic ores.

Isolines connecting the highest concentrations of iron (39–40%) map out a narrow zone in the northern third of the ore deposit. Areas enriched with iron and manganese do not coincide (Fig. 6.36).

The highest concentration of manganese (as high as 2.5%) in the west and north decreases to 1.0% in the central part of the ore deposit (Fig. 6.37).

The concentration of phosphorus is 0.8–0.9% over the entire area of the ore deposit. The content of boron decreases from the Middle to the Lower Kimmerian and Maikopian as well as the Middle Sarmatian and Pliocene strata of the Novoselovskoe ore deposit, thereby indicating that this element is closely related to mud volcanic processes (Table 6.3).

The total reserve of the Novoselovskiy iron deposit is estimated at 100–125 million tons of ore, with iron content over 30% and 30 million tons of ore with poorer content. The ore deposit is overlain by sandy-argillaceous sediments that contain mud volcanic breccia with a volume of about 20 million m^3 (40 million tons).

Geophysical and geological work carried out at the end of the twentieth century revealed that the geological structure of the eastern part of the Novoselovskaya recessed syncline is

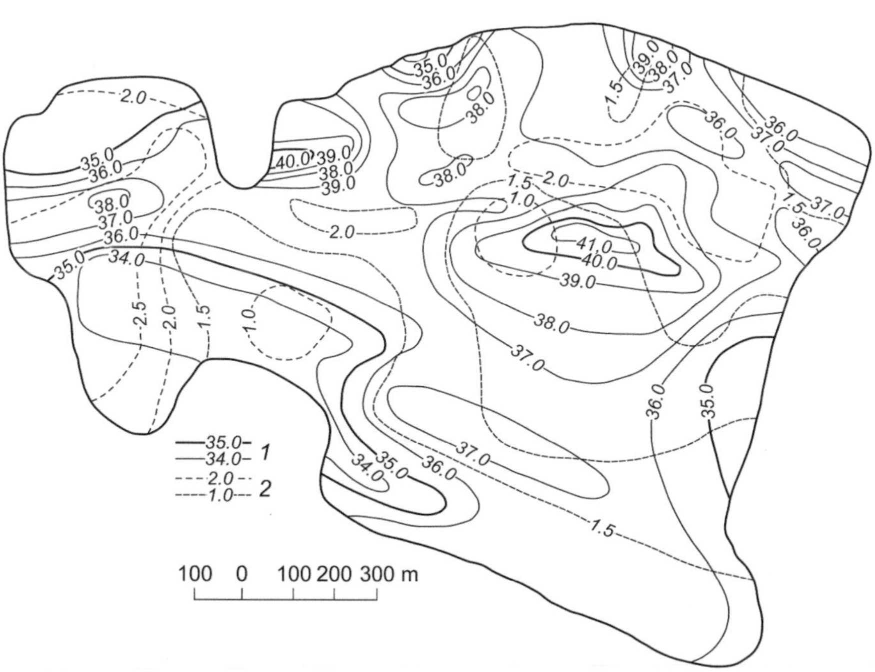

Fig. 6.36 Diagram of iron and manganese content in the Middle Kimmerian ore deposit of the Novoselovskiy ore body (isolines, m): (1) iron content; (2) manganese content

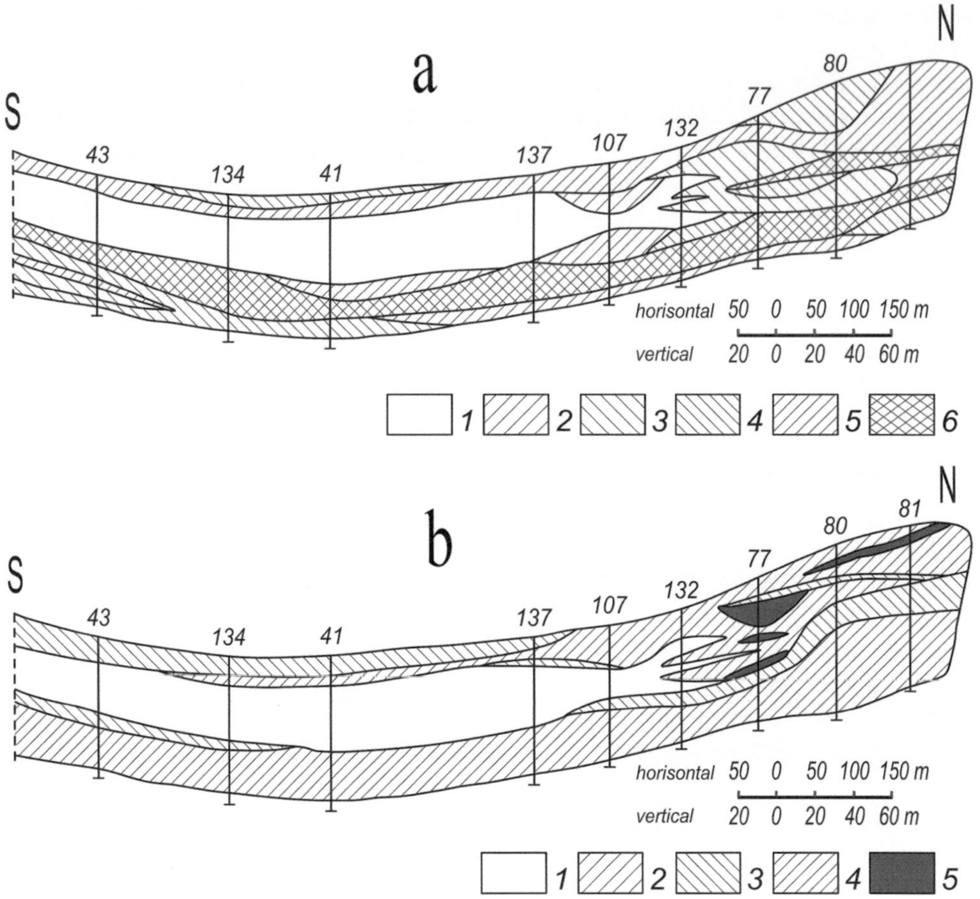

Fig. 6.37 Content of Fe (**a**) and Mn (**b**) in the Middle Kimmerian deposit of the Novoselovskiy ore deposit along the transect I–I. For Fe (%): 1 = 10–20; 2 = 20–25; 3 = 25–30; 4 = 30–35; 5 = 35–40; 6 = 40–45. For Mn (%): 1 = up to 0.5; 2 = 0.5–1.3; 3 = 1–2.4; 4 = 2–4.5; 5 = 4–6. For location of transect, see Fig. 6.29

somewhat more complicated than previously thought, and the size of the structure is larger. It is still unclear where the eastern and southern wings of the ore deposit lie. Perhaps, they have not eroded due to uplift, but on the contrary, they were submerged and might be explored in the course of further investigations.

The so-called Sokolskie domes are located to the east of the Sartskaya anticline. They represent brachianticlinal swellings crowned by the Rep'yevskaya and Alagolskaya recessed synclines. The former is particularly interesting as an ore-bearing structure (Ilyin 1931; Eberzin 1933, 1940; Shnyukov and Naumenko 1964).

The Rep'yevskaya recessed syncline has an oval shape, slightly elongated in the latitudinal direction, with long and short axes of about 500 and 300 m, respectively. To the west, it is adjoined by a small area of limestone (Pontian?) exposed by two abandoned quarries. Obviously, erosional processes have separated it from the main body of the recessed syncline. The recessed syncline is located in the Maikopian core of the anticline and is composed of Pontian and Kimmerian rocks. The former are represented by clayey sand and limestone with a thickness of 15 m. The latter are represented by a 37 m-thick iron ore deposit of lenticular shape and pebbly-oolitic-pseudo-oolitic structure. The average iron content is 37%; that of

Table 6.3 Auxiliary admixture of boron in the Novoselovskoe iron ore deposit.[a] Some wells are shown in Fig. 6.37

	Well																
Stage Horizon	3	5	9	23	29	31	40	41	42	43	44	45	46	47	49	52	Avg
Q	–	–	–	0.03 5	–	–	0.03 1	–	–	–	–	–	–	0.018 1	–	–	0.03 7
N_2^3ku				–	–	–	–	–	–	–	–	–	–	0.018 1	–	–	0.018 1
$N_2^2km_3$	–	–	–	–	0.064 18	0.066 15	0.025 23	0.014 13	0.056 7	0.019 10	0.021 1	–	0.038 4	0.062 2	–	0.04 1	0.04 94
$N_2^2km_2$	0.44 22	0.22 24	0.25 18	–	–	–	0.026 22	0.043 13	0.059 13	0.053 17	–	–	0.05 21	0.092 24	0.09 21	0.81 18	0.133 213
$N_1^3km_1$	–	–	–	–	–	–	0.03 10	0.027 11	0.021 3	0.028 4	–	–	–	–	0.076 3	0.078 2	0.035 33
N_1^3sm	–	–	–	–	–	–	–	–	0.0–0.0005 2	–	–	–	–	–	–	–	0.0025 2
$N_1^1 + P_3$		–	–	–	–	–.	–	–	–	–	–	Traces 2	–	–	–	–	Traces

[a]Numerator is the content of B_2O_3 (%), denominator represents the number of analyses

manganese is 1.46%; and that of phosphorus is 0.81%. The bulk of the deposit contains tobacco ore. The upper 4–6 m of the ore deposit are oxidized and composed of brown ore. The reserve of the ore deposit is estimated at three million tons. The main peculiarity of the Rep'yevskaya recessed syncline is that no mud volcanic breccia is present.

The more northern zone contains the Slyusarevskaya and Gornostaevskaya anticlinal structures. Both are devoid of modern mud volcanoes, although several fossil mud volcanic centers and the recessed synclines created by them have been recorded. One of them is the Gornostaevskaya recessed syncline, which is confined to the eastern extremity of the anticline of the same name. The mud volcano that created this recessed syncline also caused its bifurcation (Figs. 6.38 and 6.39).

The Novo-Nikolaevskaya, Chanlugarskaya, and Voskhodskaya anticlines form the next anticline zone located to the north.

The arch of the Chanlugarskaya anticline is crowned by the large Andreevskiy (Chanlugarskiy) mud volcano (#24 in Fig. 6.7); it shows signs of modern activity (Fig. 6.40).

In the 1930s, another mud volcanic structure was discovered in the western part of the Voskhodskaya anticline along with the Andreevskiy mud volcano. This new structure was called the Eastern Andreevskaya recessed syncline; it is prominently expressed in relief within an oval composed of Sarmatian limestone. Inside, there is a patch of rusty-brown soil with

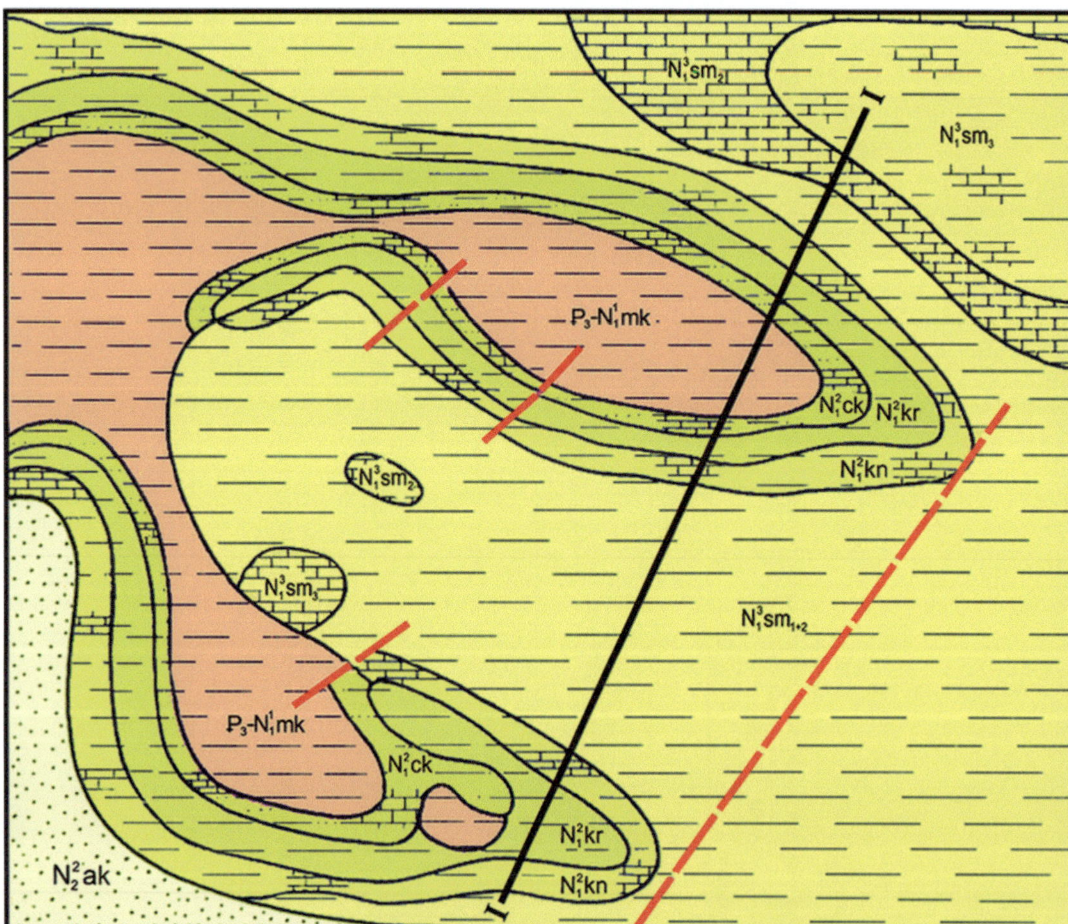

Fig. 6.38 Diagram of the geological structure of the Gornostaevskaya recessed syncline

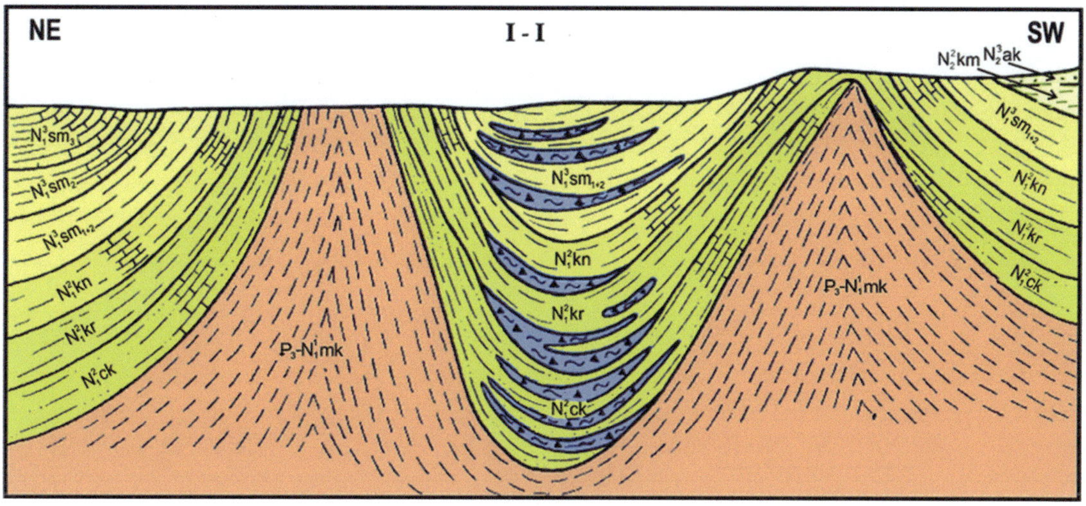

Fig. 6.39 Diagram of the geological section of the Gornostaevskaya recessed syncline along transect I–I. For the location of the transect, see Fig. 6.38

Fig. 6.40 General view of the Andreevskiy (Chanlugarskiy) mud volcano (indicated by the red arrow). (Photographed in 2003)

large shells of Kimmerian (?) mollusks together with very small and rare fragments of iron ore. Earlier, Shnyukov et al. (1971) mistakenly considered the sediments inside the oval to be of Pontian age.

Studies of breccia from the Andreevskiy mud volcano show the constant presence of ooids. Their composition suggested the existence of an ore deposit within the recessed syncline, created by the activity of this mud volcano, and the prediction was later confirmed by a well drilled in 2016 that, before stopping, passed through 7 m of iron ore deposit typical for such structures (Fig. 6.41).

It will be necessary to conduct additional exploration in order to assess the size, quality, and total reserve of the iron ore deposit (Shnyukov 2016). The Andreevskiy (Changlugarskiy) volcano has rarely erupted in recorded history.

The Vozhodovskaya anticline carries two mud volcanoes in its arched part:

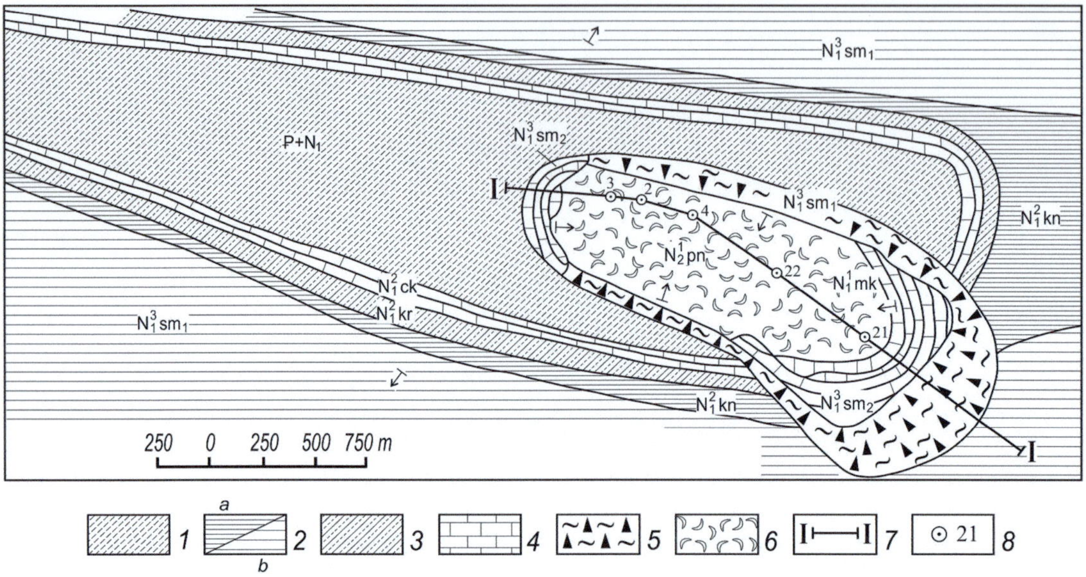

Fig. 6.41 Geological diagram of the Andreevskiy (Changlugarskiy) mud volcanic field. Key: (1) shelly clay; (2) shelly clay of different age (a, Konkian; b, Sarmatian); (3) clay interbedded with marl; (4) limestone; (5) mud volcanic breccia, (6) coquina and accumulated mollusk shells; (7) transect line; (8) number of drilled well

Fig. 6.42 General view of the Voskhodovskiy (Dzhardzhava) mud volcano. (Photographed in 1982)

Fig. 6.43 Mud flow from the eruption of the Voskhodovskiy (Dzhardzhava) mud volcano. (Photographed from a helicopter in 1982)

Voskhodovskiy (Dzhardzhava) and Soldatsko-Slobodskoy (Nikitskiy).

The Voskhodovskiy (#25 in Fig. 6.7) mud volcano (Figs. 6.42, 6.43, and 6.44) and the recessed syncline of the same name are located in the western outskirts of the town of Kerch.

The core of the anticline is composed of Maikopian sediments and complicated by a diapiric structure. The recessed syncline is composed of Chokrakian-Sarmatian limestones (Figs. 6.45 and 6.46).

On the eastern slope of the Voskhodovskaya (Dzhardzhava) recessed syncline, there are three small mud volcanoes named Vysokiy, Kizyachnyy, and Melnichnyy. They form a hill rising 20–25 m above the level of the terrain with its longest diameter reaching 300 m. Helmersen (1864) described one of these volcanoes for the first time, and according to his description, it was a small cone with a height of 1.0 m and a diameter of 3.0 m that released gas and liquid mud. It is difficult to know which of the three small mud volcanoes was described, but most likely, it was Vysokiy because Kizyachnyy was at the time only a rounded depression with a mud puddle up to 1 m in diameter in its center. All researchers (e.g., Belousov and Yarotskiy 1936) noted a calm regime for these volcanoes that were also characterized by small oil spills.

Fig. 6.44 Cracks in the breccia erupted from the Voskhodovskiy (Dzhardzhava) mud volcano. (Photographed in 1982)

At 5.10 pm on November 8, 1930, the Vysokiy mud volcano suddenly exploded. The eruption lasted only 10–15 minutes, but a large amount of mud breccia was blasted into the air (Sivere 1931). A stream of liquid mud ran down the slope

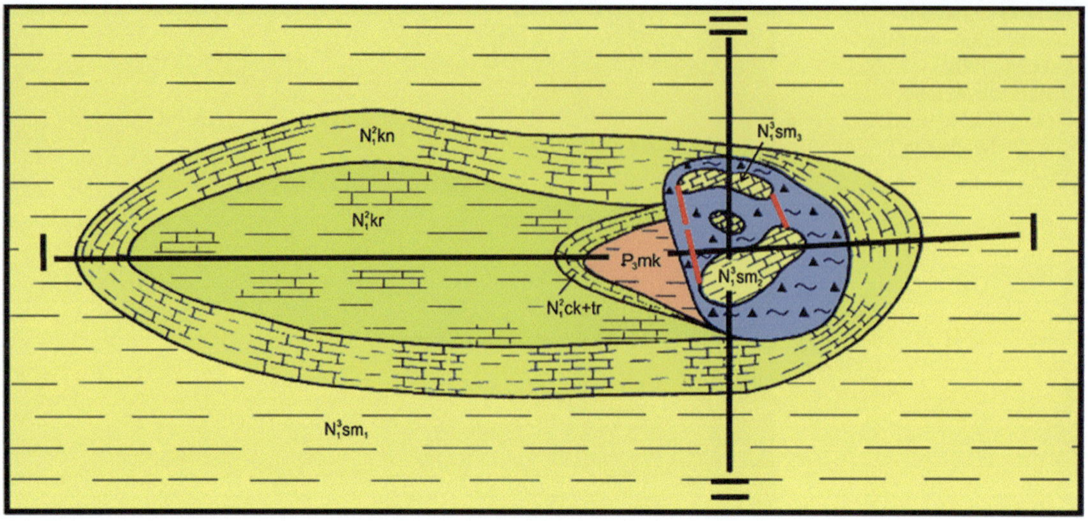

Fig. 6.45 Diagram of the geological structure of the Voskhodovskaya (Dzhardzhava) recessed syncline. For its location, see Fig. 6.7

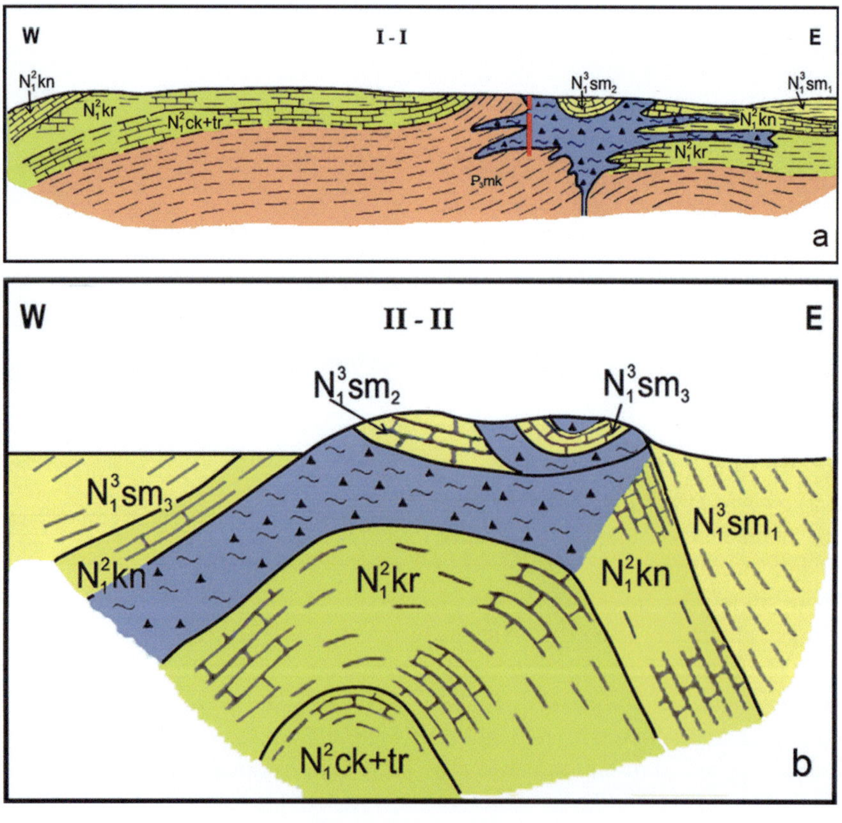

Fig. 6.46 Geological cross sections of the Voskhodovskiy (Dzhardzhava) mud volcano located in the recessed syncline of the same name along transects I–I and II–II. For the location of these transects, see Fig. 6.45

Fig. 6.47 Consequences of the explosion and flow of the Vysokiy mud volcano in 1930 (Belousov and Yarotskiy 1936)

of the hill toward the Kerch Strait. Its length reached 500 m, its width 75 m, and its thickness up to 1 m. One of the stream's channels attained a thickness of up to 3 m and moved westward toward the village. Two houses located along its path were half flooded (Fig. 6.47).

The main mud stream displaced light carriages, haystacks, and many other objects. As a result of the explosion, large and radially diverging cracks appeared that were up to 2 m wide. According to eyewitnesses, the crater emitted smoke and fire, and the coarse mud retained a strong smell of oil even a few days later. The total volume of erupted breccia reached 37.5 thousand m^3. The local inhabitants did not remember similar explosions in the past, but the old name of this mud volcano and the village, Dzhardzhava (translated as "Enemy Ravine"), speaks volumes. Indeed, Belousov and Yarotskiy (1936) recorded at least two fossil streams leading from the mud volcano that preceded the explosion of 1930.

After this eruption, no mud volcanic activity was observed for many years, and minor eruptions did not attract much attention. City buildings eventually were constructed closer and closer to the Voskhodovskiy (Dzhardzhava) mud volcano. Neither humans nor animals felt signs of impending catastrophe, but the calm was deceptive. In the depths of the Earth, a slow accumulation of energy was taking place, and on the night of March 16 to 17, 1982, a new strong explosion of the Voskhodovskiy (Dzhardzhava) mud volcano suddenly occurred. Naumenko and Blazhko in the newspaper *Crimean Truth* of April 17, 1982, referred to eyewitness testimony and wrote that the explosion "... was accompanied by a roar resembling the sound of jet engines during take-off of a passenger liner." In the morning, they watched outbursts of rock in pieces up to 2 m in diameter that blasted from the vent up to 10 m into the air together with gases and vapor in the form of a cloud as well as liquid mud that ran continuously down the slope. This explosion caused a small earthquake. Around the Vysokiy mud volcano, the Earth's surface subsided 0.2 m or more within a radius of about 400 m. Residential buildings and city water supplies were disturbed by this subsidence and broke apart. The size of the vent was 15 × 20 m, and the width of the cracks in it reached 0.5 m. The emissions (gas and breccia) started at 23.00 on March 18, 1982, and continued for almost 2 days, eventually attaining a volume of 100 thousand m^3. Mud streams threatened the Voskhodovskoe village, poured into the valley, and approached the railway as they moved toward the sea.

Based on the descriptions of eyewitnesses, the mud volcanic explosion was captured by V.I. Kantemirov (Fig. 6.48).

Since the eruption of 1982, almost no mud volcanic activity has been observed.

Geophysical study of the deep architecture of the Voskhodovskiy (Dzhardzhava) mud volcano was conducted using low-frequency microseismic sounding (Sobisevich et al. 2015). This permitted the tracing of subvertical fluid-supply structures in the upper and middle parts

Fig. 6.48 Explosion of the Voskhodovskiy (Dzhardzhava) mud volcano on March 17, 1982. (Painting by V.I. Kantemirov)

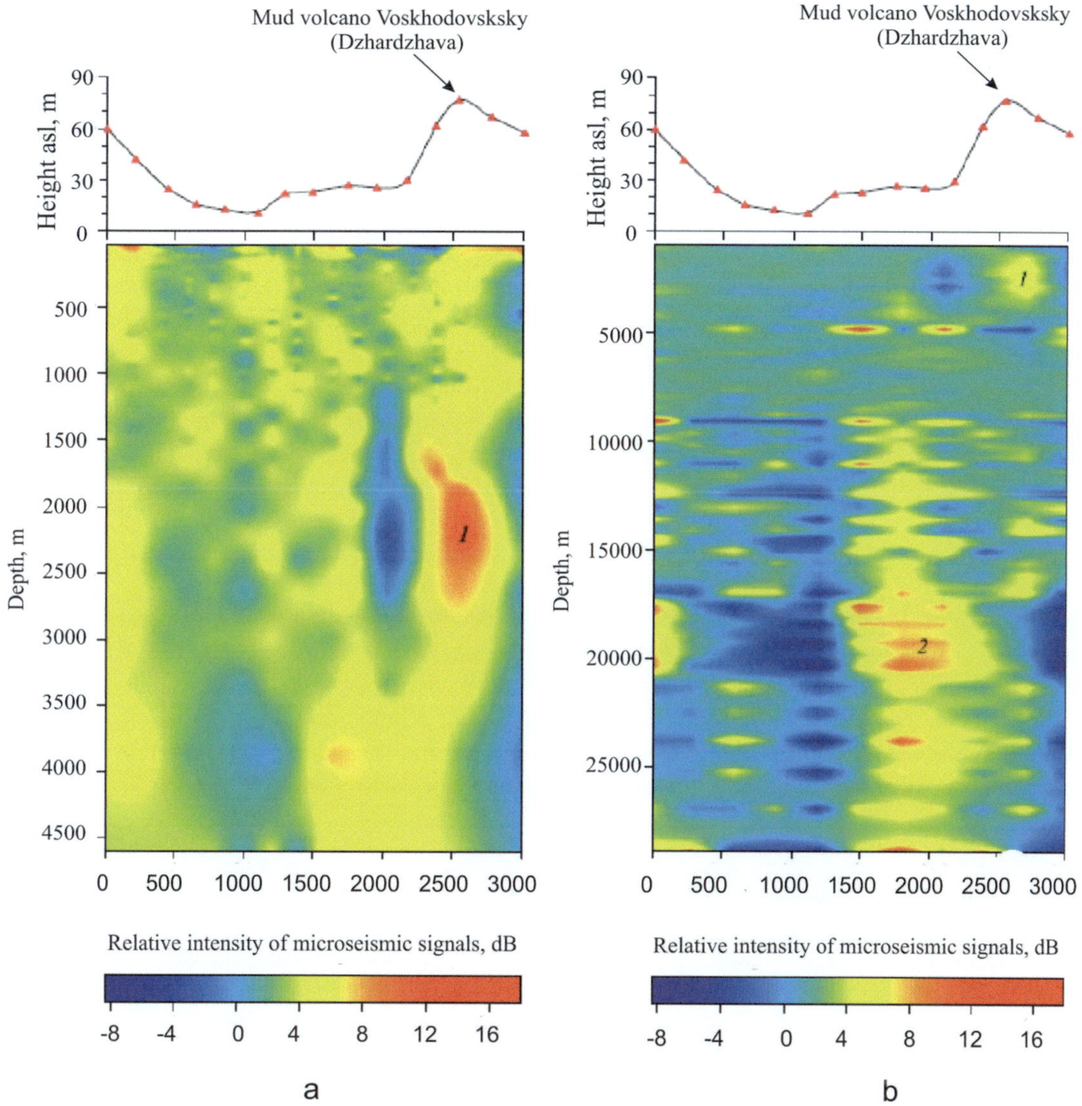

Fig. 6.49 The Voskhodovskiy (Dzhardzhava) mud volcano: (**a**) geophysical section measuring points along the height transect (above) to a depth of 4.5 km, (**b**) the same to a depth of 28 km; accumulations of mud volcanic products: (1) near surface, (2) deep (Sobisevich et al. 2015)

of the crust along a 3 km-long transect that crossed the mud volcano and the eastern pericline of the Voskhodovskaya anticline from north to south (Fig. 6.49).

The geophysical sounding conducted to a depth of 4.5 km (Fig. 6.49a) is comparable to the total thickness of the Maikopian and younger deposits. There are three steep zones showing lower velocity values that correspond to areas of fluid-permeable structures with activated (softened by the sea) Maikopian clays in the core of the anticline. The section probably corresponds to the near-surface mud volcanic center and is located directly under the mud volcanic focus at a depth of 1700–2700 m (zone 1 in Fig. 6.49a). Between the steeply dipping zones of increased permeability, there are relatively dense blocks. In addition, there are inclined heterogeneities that probably correspond to the layered structure of sedimentary Cenozoic strata. To the south of the mud volcanic cone (in the central part of the transect), there is a decrease in heterogeneity, which most likely reflects the fall of rocks on the southern wing of the diapir anticline.

The geophysical sounding conducted to 25 km (Fig. 6.49b) is comparable to the entire sequence of the Alpine sedimentary complex and goes deep into the pre-Alpine foundation. There, a large subvertical inhomogeneity (zone 2) with reduced velocities is located to the south of the mud volcanic construction along the axis of the anticlinal structure. Parallel to it, in the central part of the section, there are less significant steeply dipping subvertical fluid-permeable zones that create a complex picture. The contrast in homogeneities at depths of about 5 km and 9 km corresponds, probably, to the foot of the Maikopian and the Alpine-Mesozoic-Eocene sedimentary complex, respectively; these overlie the pre-Mesozoic geological formations. At depths of 17 and 21 km, there are subhorizontal inhomogeneities corresponding probably to a deep accumulation of gases. The interpretation of these imaging results illuminates the deep deposition of the Voskhodovskiy (Dzhardzhava) mud volcano.

To the east and within the same anticline valley, there is a large mud volcanic field represented by manifestations in the Soldatskaya-Slobodkoy (Nikitskiy) mud volcano of the town of Kerch. Previously, these manifestations included only a small mud volcano, Nikitskiy, located near the hospital. It is a low, flat hill crowned by a puddle that slowly releases gas. In the 1980s, P.I. Naumenko (1976) succeeded in proving that the Soldatsko-Slobodskaya mud volcanic field (Fig. 6.50) is very extensive and includes, in addition to the abovementioned Nikitskiy mud volcano, a large recessed syncline with a number of small gryphons and a mud puddle that releases gas (mostly methane).

The Soldatsko-Slobodskaya recessed syncline is composed of Sarmatian, Pontian- Kimmerian, and recent blocks of sediments. A minor manifestation of Kimmerian iron ore was discovered in one of the blocks of the syncline near the Container Factory on a hill (Figs. 6.51 and 6.52).

Fig. 6.50 General view of the Soldatsko-Slobodskoy mud volcanic field near the town of Kerch

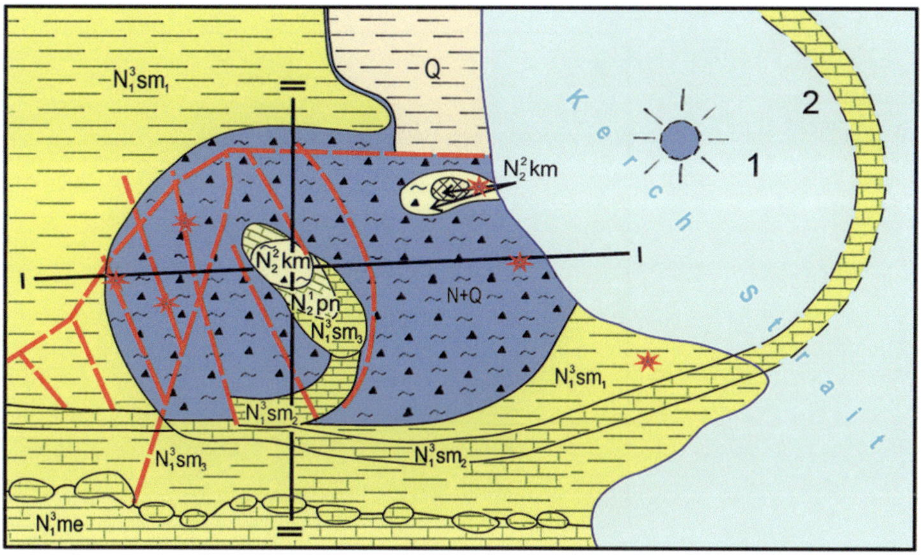

Fig. 6.51 Diagram of the geological structure of the Soldatsko-Slobodskoy mud volcanic field: (1) an "island" that was formed by mud volcanic eruption and washed out in 1880; (2) a presumed extension of the Yuz-Oba Ridge constituted by Sarmatian limestones

In 1880, a mud volcanic island appeared in the Kerch Strait (Zenkovich 1894). Its exact location unfortunately was not established because it quickly eroded away, but one can assume it was located within sight of the shore. In 2016, a small volcano in the form of a mud puddle releasing gas appeared on Cape Ak-Burun in the coastal zone of the Kerch Strait. Here, on the shore, there are also remains of a recessed syncline containing Kimmerian rock, apparently a remnant fragment of a large eroded mud volcanic structure. Geologically speaking, the appearance of all these phenomena would be possible only within the closure of the Voskhodovskaya anticline (Fig. 6.45), across the southern part of which runs the construction route for the Kerch bridge linking Crimea and the Caucasus. There is no certainty that the engineers designing the bridge took into full account the danger of reoccurring mud volcanic activity in this area.

The northernmost anticlinal zone of the Kerch Peninsula also contains a number of large mud volcanoes. A medium-size mud volcano, Syuyurtashskiy (#33 in Fig. 6.7), is located within the western pericline of the Karalarskaya anticline (Fig. 6.53).

At the top of the hill, there are a few bubbling puddles, a weak outpouring of mud, a hydrogen sulfide spring with a large production rate (Fig. 6.54), and a recessed syncline.

Not all researchers recognize this morphostructure as a mud volcano, however. For example, Belousov and Yarotskiy (1936) consider the hill to be a system of mud avalanches of Maikopian clays.

The eastern extremity of the Karalarskaya anticline is thought to be a mud volcano with its caldera filled by Lake Chokrak. On the eastern coast of the lake, there are several hydrogen sulfide springs with significant production rates. The favorable climate, therapeutic mud of the lake formed by the mud volcano, hydrogen sulfide springs, and proximity to the sea and sandy beaches have contributed to the establishment of a medical and resort complex.

The chain of anticlines represented by the Chumnaya Balka, Malobabchikskaya, and Glazovskaya (Baksinskaya) structures (#XXXI, XXXII, and XXXIII in Fig. 6.7) runs sub-latitudinally and has mud volcanic manifestations at the top of each structure. The Chumnaya Balka with its recessed syncline of the same name derives from a fossil mud volcano; it is located atop the extreme western

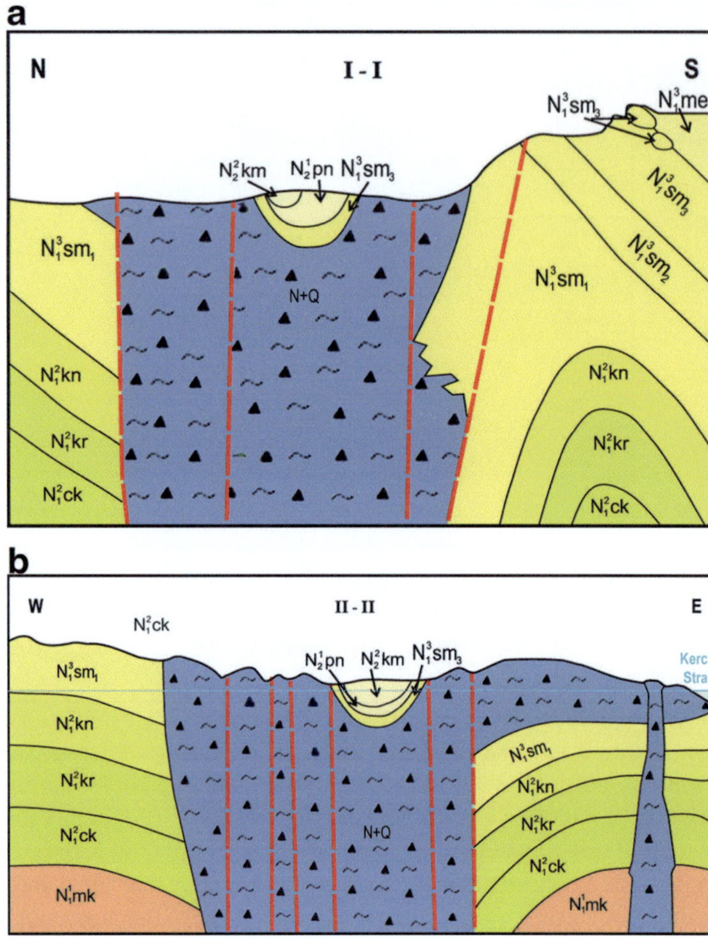

Fig. 6.52 Geological sections of the Soldatsko-Slobodskoy mud volcanic field and the recessed syncline of same name along transects I–I and II–II (for location, see Fig. 6.51)

anticline and will be described later as an iron ore-bearing structure.

The Malobabchikskaya anticline is also crowned by a recessed syncline bearing the same name. In the center, there is a 35 m-thick layer of mud volcanic breccia interbedded with Sarmatian limestone. There is also the Burashskaya recessed syncline, one third of which is occupied by mud volcanic breccia (up to 600,000 m^3 in volume); it contains two small mud volcanoes with five or six small gryphons.

To the north of this structure, there is the large Kezenskaya recessed syncline. This structure has long been interpreted as an ordinary trough devoid of modern volcanism. The syncline takes the shape of a fairly large oval with an area of 15.5 km^2 contoured by Sarmatian limestone with an iron ore deposit within. On the east, the ore deposit reveals a thickness of 6.1 m, on the southwest 3.5 m, in the center 0.5 m (average thickness is 2–2.5 m). It contains a high content of clastic ore material, is quite sandy, and is overlapped by Upper Pliocene sands. The latter is typical for the entire anticlinal zone. The finding of mud volcanic breccia in the Sarmatian limestone, the absence of some strata (e.g., Pontian), and sharp changes of elevation at the base of the ore deposit (from +22 m to +120 m) suggest that the Kezenskaya recessed syncline should be attributed to a mud volcanic structure formed during the Sarmatian. The ore deposit contains mainly brown varieties with a 25% concentration of oolites of predominantly

Fig. 6.53 General view of the Syuyurtashskiy mud volcano. (Photographed in 2003)

Fig. 6.54 Hydrogen sulfide spring within the Syuyurtashskiy mud volcano

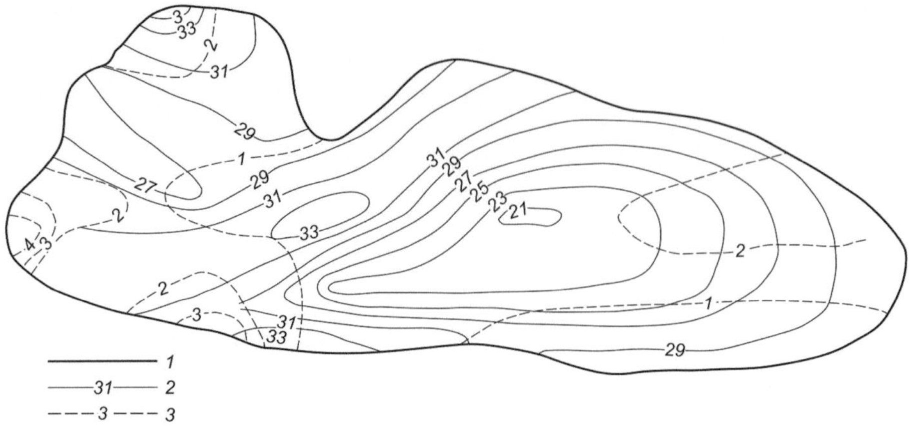

Fig. 6.55 Diagram of the iron and manganese content in the Middle Kimmerian ore deposit of the Kezenskaya recessed syncline. Key: (1) boundary of the ore deposit, (2 and 3) isolines of iron and manganese content, respectively

0.10–0.25 mm in diameter. The content of oolites and pseudo-oolites is approximately equal. The average iron content in the ores is 27.6% (from 17.45 to 37.12%), manganese 1.8%, phosphorus 0.66%, and arsenic 0.036% (Fig. 6.55).

The iron ore reserve is estimated to be 26 million tons (Arbuzov et al. 1967), but it is significantly poorer by comparison to the overall basic iron ore reserves of the Kerch Peninsula.

The Glazovskaya (Baksinskaya) anticline with an area of about 3 km^2 contains the Baksinskiy and Enikalskiy mud volcanoes.

The Baksinskiy mud volcano is located within a recessed syncline of the same name. In the modern terrain, this is a circus depression within which a hill exists without any sign of modern mud volcanic activity. The hill is 2 m in height and 50 m in diameter, and a hydrogen sulfide spring lies about 250 m away. The estimated volume of mud volcanic breccia is 125 million m^3 (dated to the Kimmerian), 35–40 million m^3 (Kuyalnitskian), and 35 million m^3 (Quaternary), but in fact the true figures may in reality be much higher (Shnyukov et al. 1971). The geological structure of the Baksinskaya recessed syncline (Fig. 6.56) was studied by Andrusov (1893), Prokopov (1931), Konstantov (1929), Konstantov et al. (1933), Gubanov (1959), Shnyukov and Fesyunov (1965), and Shnyukov and Naumenko (1964).

The Baksinskaya recessed syncline is filled with Meotian, Pontian, Kimmerian, and Kuyalnitskian sediments interbedded with mud volcanic breccia (Fig. 6.57).

The ore deposit forms a wide strip that transgressively covers Pontian sediments and frames the syncline from the inside, wedging toward the center of the breccia field like a horse's tail. It represents a single iron ore deposit with a thickness of 10–16 m, existing only along the periphery of the syncline. In the center, it submerges to 100 m and wedges out. The ores have an oolitic-clastic-pseudo-oolitic composition typical of the Kerch region. The ores are represented by tobacco (predominant) and brown varieties. On average, the tobacco ore contains 37.63% Fe, 0.38% Mn, and 0.841% P; the brown ore contains 38.61% Fe, 1.76% Mn, and 0.88% P. The total reserve is estimated to be 42 million tons (Arbuzov et al. 1967).

The Enikalskiy mud volcano (#32 in Fig. 6.7) is located in the extreme eastern part of the Glazovskaya anticline on the outskirts of the village of Podmayachnoe (Fig. 6.58).

It was first observed by Pallas (1795) and then by Samoylov (1898), Vernadskiy and Popov (1899–1900), Chirvinskiy (1908), Izgaryshev and Sludskiy (1917), Belousov and Yarotskiy (1936), and others.

The description by Pallas (1795) is of special interest. He wrote that about a mile to the north of

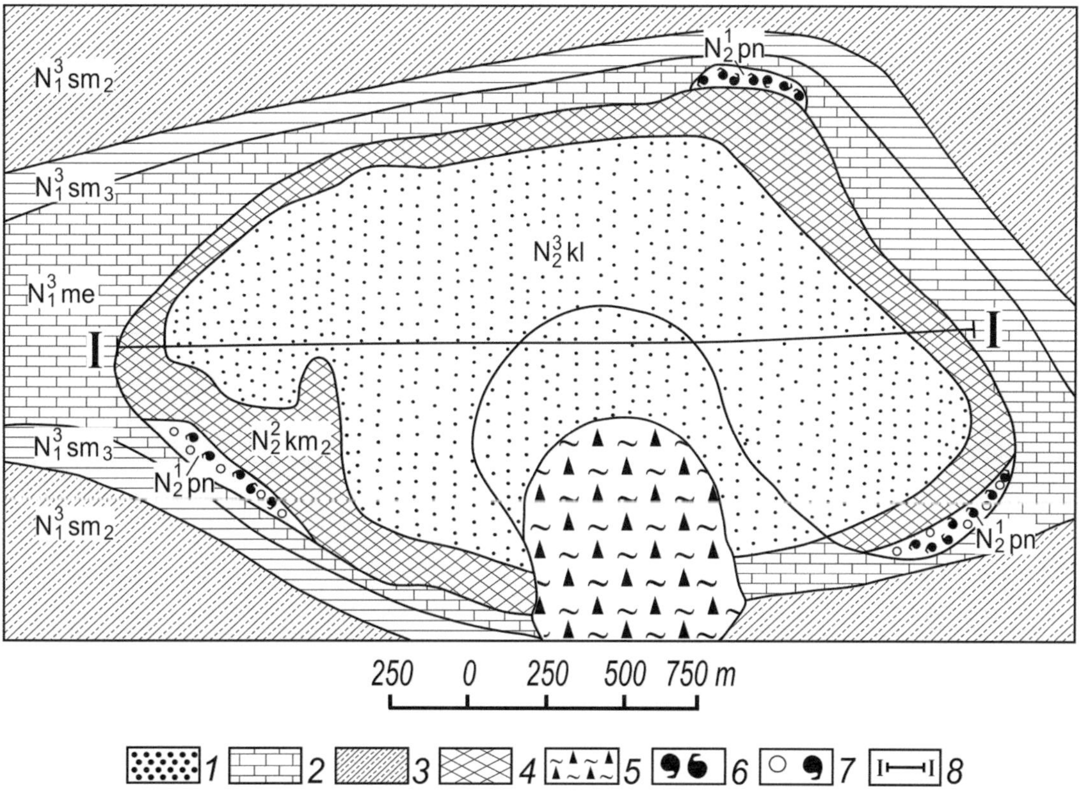

Fig. 6.56 Diagram of the Baksinskiy mud volcanic field and the Baksinskaya recessed syncline (Shnyukov and Naumenko 1964). Key: (1) sands, clayey sands; (2) limestone, dolomitic limestone; (3) clay with interlayers of marl; (4) brown iron ore; (5) mud volcanic breccia; (6) clayey coquina; (7) granular coquina; (8) line of transect

Enikale village, there were salt springs at the top of the mountain, which emerge like boiling water from the ground together with a black oil that he referred to as "mountain oil." The oil floated on the surface of the water in excavations conducted nearby. Then, it was collected and used for smearing on wheels as a lubricant and burning in lamps instead of candles. It was also used as a medicine, especially in the treatment of chilled body members. After purification by distillation, it was used in pharmacies as well as for some artillery solutions. In the circle of the aforementioned springs, the surface of the clayey soil is covered by this mountain oil, and therefore the soil appears black as resin and burns easily, yielding visible flames and a strong, unpleasant smell.

The most detailed description of this volcano was made by Samoylov (1898). He wrote that the volcano is located northwest of the village of Mayak (Enikale), in the middle of an elliptical kettlehole (recessed syncline?), and it is outlined by a chain of Meotian reef limestones forming a small hill on which gryphons and small mud volcanic holes are visible. Our observations of the area in 2002, 2003, and 2010 did not locate any gryphon activity. In the mud volcanic breccia, there are shells representing typical Kimmerian mollusks. This indicates that the mud volcanic eruptions and the formation of ore-bearing sediments occurred underwater. This fact was not previously taken into account in the reconstruction of the northern anticlinal zone of the Kerch Peninsula.

The Enikalskiy mud volcano is typical of the Kerch Peninsula. It is located near the settlement and convenient for observation, but unfortunately, it cannot be recommended for tourism because it is covered by garbage. An oil seepage exists one and a half km to the north-northwest of this volcano (Fig. 6.59).

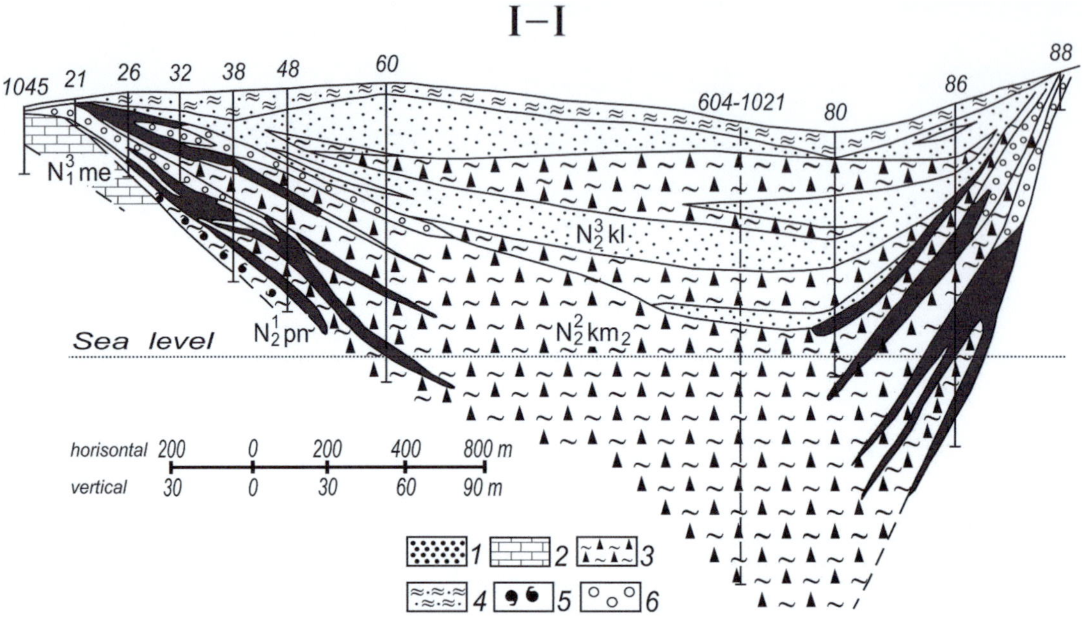

Fig. 6.57 Geological cross section through the Baksinskiy mud volcano field and the Baksinskaya recessed syncline along transect line I–I (Shnyukov and Naumenko 1964). Key: (1) sands, clayey sands; (2) limestone, dolomitic limestone; (3) mud volcanic breccia; (4) reworked mud volcanic breccia; (5) clayey coquina; (6) iron ore. For the transect location, see Fig. 6.56

Fig. 6.58 General view of the Enikalskiy mud volcano field on the outskirts of the village of Podmayachnoe. (Photographed in 2003)

Fig. 6.59 Oil seepage along the slope of the ravine near the Enikalskiy mud volcano. (Photographed in 2003)

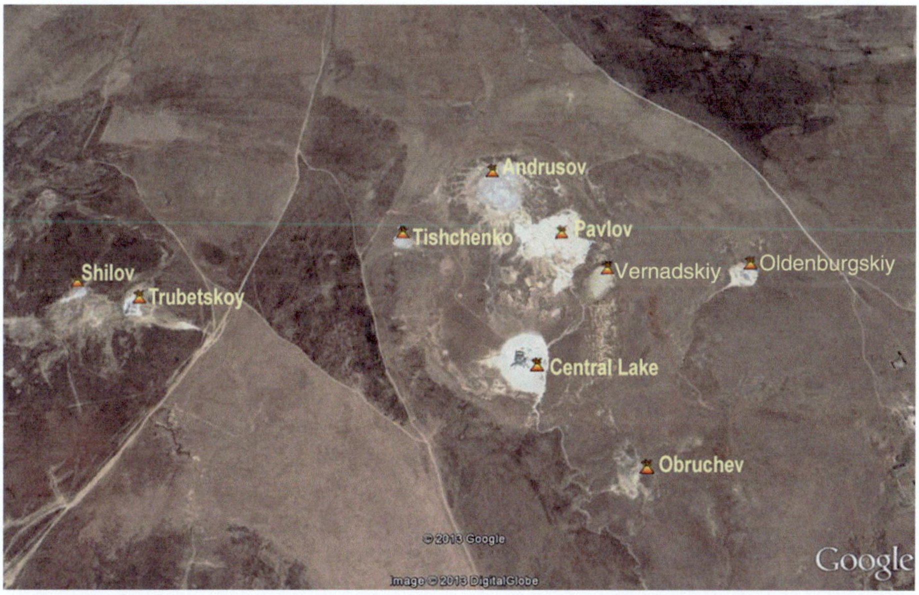

Fig. 6.60 The Bulganakskiy mud volcanic hearth seen from above

Fig. 6.61 Aerial view of the Bulganakskiy mud volcanic hearth. (Photographed from a helicopter in 1982)

The Bulganakskiy mud volcanic hearth (#XXXV, Table 6.2) is one of the most peculiar of the Kerch Peninsula (Fig. 6.60).

It has a round shape and contains a number of mud lakes and mud volcano cones (Fig. 6.61).

It has been studied for more than a century by many famous Russian geologists (e.g., Helmersen 1864; Abich 1873; Golovkinskiy 1889; Andrusov 1918; Vernadskiy and Popov 1899–1900; Popov 1928; Klepinin 1914; Izgaryshev and Sludskiy 1917; Prokopov 1931; Belousov and Yarotskiy 1936). Their descriptions have been very useful in estimating mud volcanic activity over a prolonged time interval.

The central part of the Bulganakskiy mud volcanic hearth is covered by solonchaks with a depth of subsidence up to 25 m below the modern ground level. The southwestern edge of the solonchaks is occupied by the Tsentralnoe Ozero (Central Lake), which is filled with liquid mud. The lake's diameter ranges from 50 m (in autumn) to 60 m (in spring). On a daily basis, it releases about 100 m^3 of gas and 5000 liters of liquid mud (Belousov and Yarotskiy 1936) that slowly flows out of the lake. Large mud bubbles are constantly swelling and gurgling over the lake surface (Figs. 6.62, 6.63, and 6.64).

The main gases produced by the Bulganakskiy mud volcano are methane and carbon dioxide; the variation depends upon the particular vent, but over time there has been little variation, either in rate or content of outflow (Herbin et al. 2008).

In the northern part of the solonchaks, there are five mud volcanoes: Andrusov, Abikh, Pavlov, Vernadskiy, and Tishchenko (Fig. 6.64).

The Andrusov mud volcano is located in the northeastern part of the solonchaks. It is the largest of the five and forms a truncated cone about 7 m in height; its upper platform is 150 m in diameter, and at the base, it is about 300 m across. Its slopes are incised by ravines and covered by mud flows of different ages. In the ravines, mud volcanic deposits are exposed. Their color varies from dark brown to light gray, indicating different degrees of weathering. This mud volcano has shown unstable activity. In 1936, there were three groups of active gryphons, each up to 2 m high (Belousov and Yarotskiy 1936). In 1969–1970, we observed up to ten gryphons; in 1982–2004,

Fig. 6.62 Bulganakskiy mud volcanic hearth with the Tsentralnoe Ozero mud volcano in the center. (Photographed from a helicopter in 1982)

Fig. 6.63 General view of the Bulganakskiy mud volcanic hearth with the Tsentralnoe Ozero mud volcano in the center. (Photographed in 2003)

their number decreased to five and fewer. Presently, there is only one gryphon (Fig. 6.65).

Turley (1930) described a strong explosion from this mud volcano in spring of 1923 and then again in September of 1926, when approximately one million poods (16,000 tons) of breccia were ejected. He wrote that the explosions were accompanied by outbursts of flame that were explained by the presence of hydrogen fluoride among the gases that were flammable in

Fig. 6.64 General view of the Bulganakskiy mud volcano hearth; the Tsentralnoe Ozero mud volcano is visible in the center, and the Andrusov mud volcano can be seen in the upper right corner. (Photographed from a helicopter in 1982)

air. He discovered in the breccia reddish-pink rock and rightly interpreted it as burned products of the explosion, thereby confirming the existence of explosive events on the hill in the geological past. Similar slag-shaped, reddish-pink fragments were also repeatedly found by the first author of this book, who observed an explosion of this mud volcano on September 1, 1986, together with other geologists. It occurred 1 day subsequent to an earthquake, the epicenter of which lay in the Vrancea Mountains of Romania. A few hours before the explosion occurred, shepherds heard an incomprehensible rumbling. All the sheep suddenly gathered in one place and then left the pasture. The first explosion was very strong. A massive explosion of mud not less than 50 m in diameter was thrown into the air about 30–40 m in height and then collapsed in front of the geologists. In the turmoil and panic, this first ejection could not be photographed. The second explosion involved a mud outburst 15–20 m in diameter that rose 5–10 m in height. The third one also released water, gas, and debris in addition to ejected mud. The solid components of the blast flew away from the source of the explosion

Fig. 6.65 General view of a gryphon on the southern slope of the Andrusov mud volcano. (Photographed in 2004)

Fig. 6.66 Photographs of the explosion of the Andrusov mud volcano on September 1, 1986

landing between 15 and 50 m away. In total, all three explosions occurred over the course of 10–15 min separated by 1–2 min breaks between them. The second and third explosions were photographed (Fig. 6.66).

According to local residents, similar explosions occurred in 2001. Careful examination of the mud volcanic hill shows that its central part is composed of gray, slightly dried, but not weathered breccia, i.e., relatively fresh (Fig. 6.67).

During our annual visits once a year over the last 40 years, there have been continuous appearances of fresh breccia in the center of the hill. This indicates that eruptions of the mud volcano are quite regular and that we simply miss them because no systematic observations have been conducted. The eruptions appear quiet in character and lacking in violent explosions.

The Abikh mud volcano is located to the south-southeast of the Andrusov mud volcano. It is mostly a steep-sided structure (Fig. 6.68a), rounded (Fig. 6.68b) at the top, with a height of up to 4 m and a diameter up to 9 m.

Located nearby is the Pavlov mud volcano, represented by a round lake of mud with a diameter up to 25 m. The lake constantly releases gas and feeds a flowing stream leaking outward (Fig. 6.69).

The daily emission of gas and mud from this volcano in summer is estimated to be 10–12 m^3 and 4000 liters, respectively (Belousov and Yarotskiy 1936), but in winter, these numbers are higher. This mud volcano can be easily traced on photographs by mud flows from the cone.

About 150 m to the southeast of the Pavlov mud volcano lies the Vernadskiy mud volcano. It is represented by a small (2 × 3.5 m) lake of liquid mud (Fig. 6.70).

The Tishchenko mud volcano is located on the western slope of the main depression. It is represented by a cone with gently bubbling muddy water at the top (Fig. 6.71).

Fig. 6.67 General view of freshly erupted breccia at the top of the Andrusov mud volcano. (Photographed in 2002)

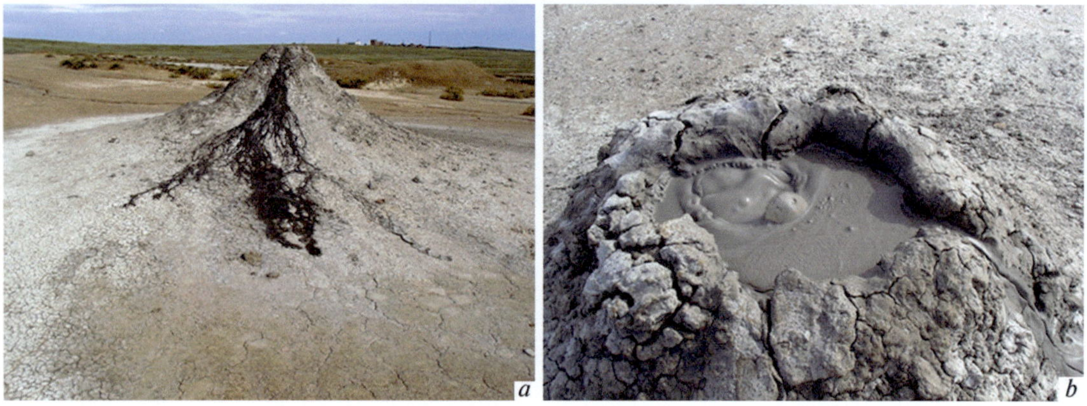

Fig. 6.68 (**a**) General view of the Abikh mud volcano. (Photographed in 2003). (**b**) Top of the Abikh mud volcano showing a bubbling mud puddle that releases gas (nitrogen) in the center. (Photographed in 2003)

In the south, the Bulganakskiy mud volcanic hearth is bounded by the Obruchev mud volcano (Fig. 6.72).

It rises 20 m above the level of the central solonchak, stretches from north to south, and has been modified by construction and mining work. On the top, there are a few small gryphons that release slight amounts of gas.

To the east of the central solonchak lies the Oldenburgskiy mud volcano (Fig. 6.73).

It is a depression with a diameter of 70–150 m and a depth up to 4 m that is filled by a lake of liquid mud 10 m in diameter on the top. Sometimes, mud from the lake flows into the nearby ravine and the flow almost dries up, leading to the appearance of small, and very small, gryphons that release about 50–60 m^3 of gas and 2000 liters of liquid mud per day (Khain et al. 1953). The western border of the Bulganakskiy mud volcanic hearth has been fixed at about 1 km to the east of the main depression. The former represents a slightly aerated lake of liquid mud up to 5–6 m in diameter, and the latter appears as a field with mud volcanoes Trubetsko and Shilov, each up to 6 m tall,

Fig. 6.69 General view of the Pavlov mud volcano. (Photographed in 2003)

Fig. 6.70 General view of the Vernadskiy mud volcano. (Photographed in 2003)

with gryphons on the top and a salt spring on its southern slope (Figs. 6.74, 6.75, and 6.76).

Together, these mud volcanoes form the Malotarkhanskiy mud volcanic field, which is, in fact, a continuation of the Bulganakskiy mud volcanic hearth.

In the last century, the commercial firm "Krymneftegeologiya" explored the ideas of

Fig. 6.71 General view of the Tishchenko mud volcano. (Photographed in 2003)

Fig. 6.72 General view of the Obruchev mud volcano. (Photographed in 2003)

Fig. 6.73 General view of the Oldenburgskiy mud volcano. (Photographed in 2003)

Fig. 6.74 General view of the Trubetskoy mud volcano. (Photographed in 2005)

Fig. 6.75 General view of the Shilov mud volcano. (Photographed in 2005)

Fig. 6.76 Chain of gryphons on the Shilov mud volcano. (Photographed in 2005)

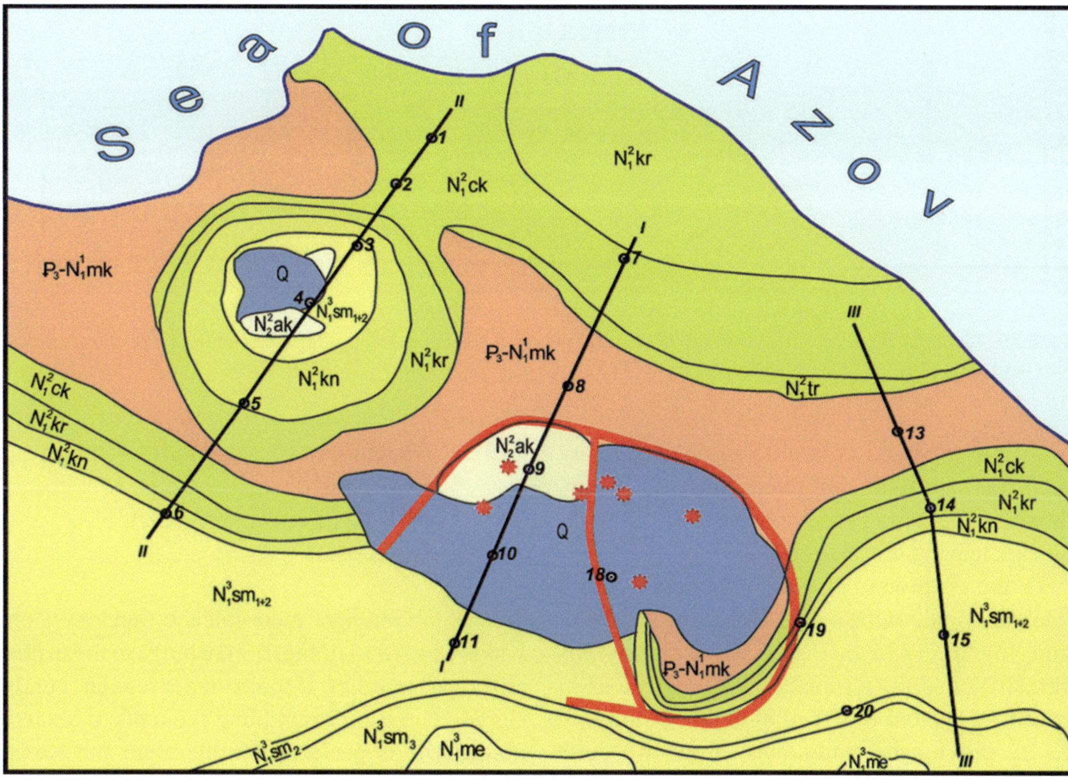

Fig. 6.77 Geological map of the Bulganakskiy mud volcanic hearth

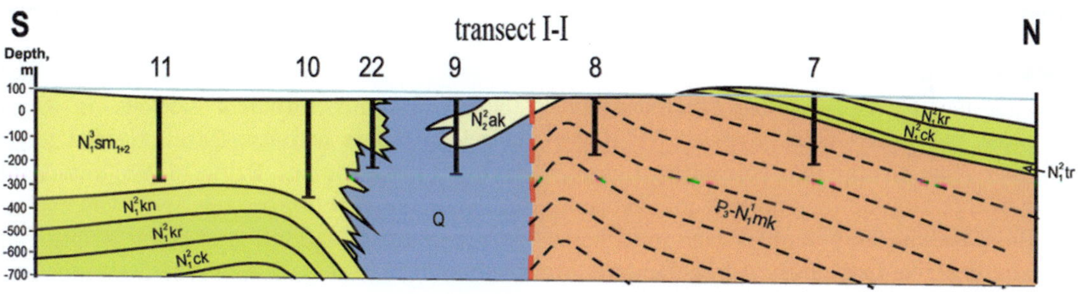

Fig. 6.78 Geological transect I–I through the western part of the Bulganakskiy mud volcanic field (Bolshetarkhanskiy mud volcano). For transect locations here and in Figs. 6.79 and 6.80, see Fig. 6.77

I.M. Gubkin on the oil and gas potential of mud volcanoes and, in so doing, drilled 20 wells that contributed significantly to a better understanding of the Bulganakskiy mud volcanic hearth (Figs. 6.77 and 6.78).

The area where the Andrusov, Tishchenko, Oldenburgskiy, Tsentralnoe Ozero, and Obruchev mud volcanoes, as well as other smaller volcanic manifestations, are concentrated has been shown to be only the central part of a much larger mud volcanic field (Fig. 6.60).

To the east of this area, Well #15 was drilled. It penetrated directly into the vent of a previously unknown mud volcano, which is buried now under a sequence of Quaternary sediments, located most probably within a recessed syncline (Fig. 6.79).

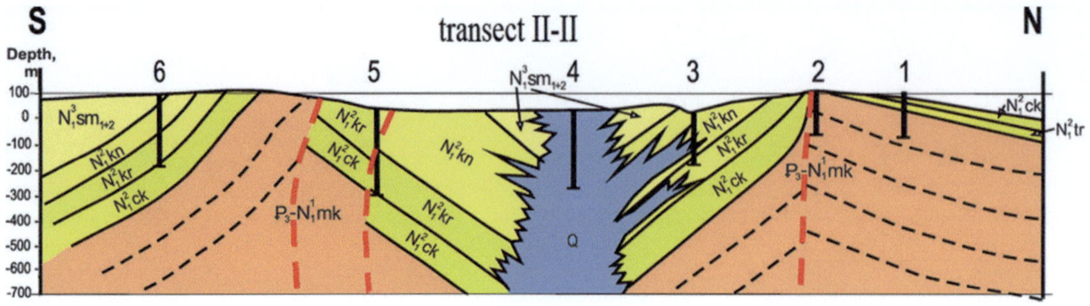

Fig. 6.79 Geological transect II–II through the central part of the Bulganakskiy mud volcanic field. For transect locations here and in Figs. 6.78 and 6.80, see Fig. 6.77

This well passed through 5 m of Quaternary sediments and then continued down to 400 m, always encountering mud volcanic breccia without ever leaving this material.

To the northwest of the Bulganakskiy mud volcanic field, within the Bolshetarkhanskiy mud volcano (#35 in Fig. 6.7), several wells were drilled along a transect showing the existence of a typical recessed syncline belonging to a previously unknown mud volcano (Fig. 6.80).

The Bolshetarkhanskiy and Malotarkhanskiy mud volcanoes represent a large, now inactive, group of small mud volcanic cones in the center of a rounded recessed syncline of Chokrakian deposition. A number of small mud puddles release gas (mostly methane) and gryphons in the Tarkhanskiy ravine forming a line that separates the two volcanoes. Judging by the large volume of mud volcanic breccia, the Bolshetarkhanskiy mud volcano must have been quite active in the past. A large amount of clastic material around the hill speaks in favor of strong explosions.

In summary, the Bulganakskiy and Malotarkhanskiy mud volcanic fields represent a complex geological structure of significant size comprising a number of large recessed synclines superimposed one on top of another and permeated by several mud volcanoes. The entire area is covered by mud volcanic breccia, making it difficult to understand the local tectonics (Prokopov 1931; Shnyukov et al. 2013).

6.3 Mud Volcanoes of the Taman Peninsula and the Northwestern Caucasus

The Gora Gorelaya mud volcano (ancient name Kuku-Oba) (#45 in Fig. 6.7) belongs to the explosive type. In relief, it looks like a regular, gently sloping elevation resembling the tents of ancient nomads; it is perfectly visible from the Kerch Peninsula in clear weather (Fig. 6.81).

This mud volcano is located in the core of an anticline composed of Maikopian deposits and framed by Sarmatian, Meotian, Pontian, and Kimmerian rocks (Figs. 6.82 and 6.83).

History has recorded some information about numerous turbulent periods in Gora Gorelaya's past activity. Based on the testimony of eyewitnesses, Pallas (1795) described in detail the eruption of this mud volcano in March of 1794. According to his description, a column of black smoke first rose from the middle of the hill, and then a pillar of fire appeared, which seemed from afar to have risen to 50 sazhens in height and girth (a sazhen is an Old Russian unit of distance equal to 2.16 m). The flames were visible for about 3 h. Over the course of 3 days, mud blasted from the vent reaching double the height of a human (about 2 m). Thunderous rumblings accompanied the eruption. One surveyor at the time found a hole on the top of the Kuku-Oba hill that was 10–12 arsheens in size (1 arsheen = 71.26 cm) with steam and earthen matter spewing from the hole

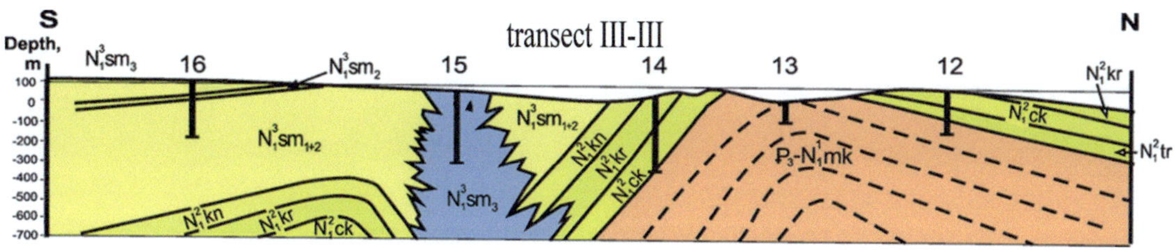

Fig. 6.80 Geological transect III–III through the western part of the Bulganakskiy mud volcanic field

Fig. 6.81 General view of the Gora Gorelaya (Kuku-Oba) mud volcano looking north across Taman Bay. (Photographed in 2002)

together with oil. Alekseev (1880) described the great astonishment of archaeologists who were working there. By following the most precise instructions of Strabo, they found the Kuku-Oba mud volcano instead of the tomb of King Satyr. During the eruption of this volcano in 1794, fragments of an ancient statue were ejected. Apparently, the Gora Gorelaya mud volcano had attracted ancient people to the location, but at the same time, it was surely hazardous to be in close proximity because its catastrophic eruptions probably destroyed human settlements.

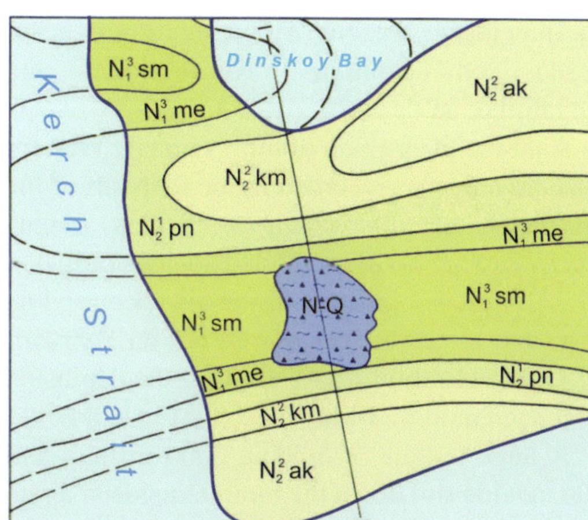

Fig. 6.82 Diagram of the geological structure of the Gora Gorelaya mud volcano

To the east, the Fontalovskiy mud volcano (#46 in Fig. 6.7) is confined to the anticline of the same name, and farther to the east—the Tsymbalskaya anticlinal zone. This zone consists of the Tsymbalskaya, Akhtanizovskaya, Peresypskaya, and Golubitskaya anticlines and includes six mud volcanoes: Tsymbaly Western (#50 in Fig. 6.7), Tsymbaly Eastern (#51 in Fig. 6.7), Akhtanizovskiy (#52 in Fig. 6.7), Akhtanizovskiy Northern (#53 in Fig. 6.7), Gora Sopka (#54 in Fig. 6.7), and the underwater Golubitskiy mud volcano (#55 in Fig. 6.7).

The Tsymbaly Western (also called Shumukoy or Puchina), Tsymbaly Eastern, and Akhtanizovskiy mud volcanoes form a com-

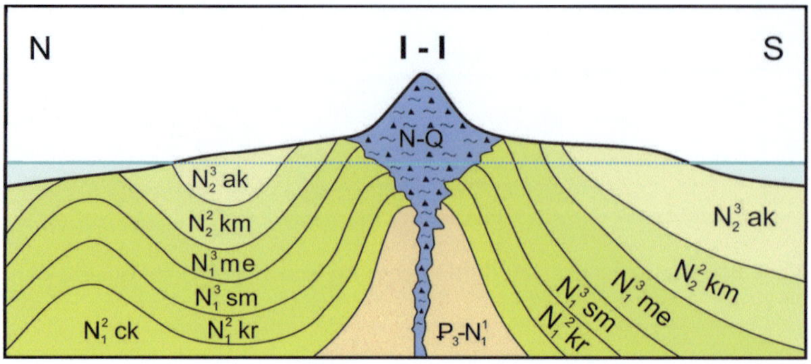

Fig. 6.83 Geological cross section of the Gora Gorelaya mud volcano. For the location of the transect I–I, see Fig. 6.82

mon elevation stretching 6.5 km in a northeastern direction. This elevation has steep slopes that descend directly to the side of the motorway near the village of Zaporozhskaya. The three mud volcanoes form gently sloping hills, each rising up to 50 m in height. In recent years, only the Tsymbaly Western mud volcano has shown activity, and the other volcanoes remain inactive. At their tops, small fields of mud volcanic breccias (0.5–1.5 km in diameter) are exposed. At the top of the Tsymbaly Eastern mud volcano, the breccia is impregnated with oil, and the remnants of drilling equipment are present. This elevation is composed of Pontian, Kimmerian, and Kuyalnitskian deposits.

A powerful eruption of the Tsymbaly Western mud volcano was recorded in the first half of the nineteenth century, when a breccia tongue reached 300 m in length (Belousov and Yarotskiy 1934). A catastrophic explosion also occurred on February 13, 2002, and reoccurred on February 17, 2002. The mud flow following the explosion continued until mid-May of 2002 (Fig. 6.84).

A huge tongue of breccia 2000 m long and 100 m wide slid down the mountainside at a rate of 1 m per day, blocked the road, and disrupted communication lines. A huge caldera of subsidence formed that encompassed a total area of 45 hectares. The total mass of erupted material reached one million m^3 (I.N. Gusakov, personal communication). There were dozens of episodically flaring torches in the breccia of the caldera (Fig. 6.85) that left behind stains of burned slags, black and brownish-red in color, and 1–3 m in diameter.

Three mud volcanoes are known within the Phanagorian anticlinal zone. They are Shopurskiy (#56 in Fig. 6.7), Sennoy (#57 in Fig. 6.7), and Gora Boris and Gleb (#58 in Fig. 6.7), each located within the Phanagoriyskaya, Yanovskaya, and Borisoglebskaya anticlines, respectively.

The Shopurskiy (Fig. 6.86) and Sennoy mud volcanoes are located in the immediate vicinity of the southern coast of Taman Bay, within a single anticlinal ridge.

The two flat-topped hills with heights up to 50–60 m are divided by a saddle. They are composed of Sarmatian, Meotian, Pontian, Kimmerian, and Kuyalnitskian strata. The mud volcanic breccia occupies a small area (from 400 to 500 m in diameter). In the immediate vicinity of the Shopurskiy mud volcano, in the coastal cliff, the great ancient city of Phanagoria was founded and is now being excavated (Fig. 6.87).

Scientific excavations began at Phanagoria in 1936 under the direction of Professor V.D. Blavatskiy. Research from that time to the present has determined the city's boundaries, thickness of the cultural layer, chronological range of Phanagoria's occupation, and the location of the necropolis. Excavations have also revealed additional pages of the city's history and material culture (Yanko-Hombach et al. 2007).

At the top of the Shopurskiy mud volcano, fragments and shards of amphoras have been found. It may be that periodic explosions and/or smoking of the mud volcano caused a sense of fear among the ancient inhabitants who worshiped it. Avdusin (1948) believed that the fragments of crystalline rocks found within the mud volcanic breccia on both the Shopurskiy and

Fig. 6.84 Mud flow after the explosion of the Tsymbaly Western mud volcano in 2002

Fig. 6.85 Flaring torches in the breccia of the Tsymbaly Western mud volcano after its explosion in 2002

Sennoy hills were brought there by ancient inhabitants together with the cultural objects for cultic rites.

A wide mountain with a gentle top rises above the narrow cape near the western shore of the Akhtanizovskiy liman. This mountain has been eroded by water from the liman on three sides, and only in the western part is it connected by a narrow isthmus with the shore. This mud volcano, called Gora Boris and Gleb, is

Fig. 6.86 Shopurskiy mud volcano: (**a**) general view, (**b**) top of the hill with its puddle of mud. Photographed in 2002

one of the most picturesque of the Taman Peninsula. It has an oval flat-topped shape and shows evidence of numerous landslides and collapses on its slopes (Figs. 6.88 and 6.89).

Almost the entire cone of this mud volcano is composed of mud volcanic breccia. Only on its northern slope are Pliocene sands interlayered with conglomerates exposed. Currently, no activity has been observed for this mud volcano; however, evidence of its violent behavior in the past exists in the literature. Dubois de Montpéreux (1837) described an explosion that occurred on April 26, 1818, that resulted in the formation of a volcanic cone. According to him, stones from the foundation of the Temple of Diana Agrotera together with remains of a monument with Greek inscriptions were ejected from the earth. In all probability, the volcano was at rest during the time of the Bosporan Kingdom; otherwise the Greeks could not have interred bodies of the deceased there. In 2000, the authors found round fired patches of mud volcanic breccia 3–5 m in

Fig. 6.87 Archaeological site of Phanagoria: (**a**) general view, (**b**) amphorae. Photographed in 2007

diameter. Given the instability of such burned products under surface conditions, it was assumed that the explosion generating these patches had occurred no more than 5 years previously.

A seismoacoustic transect of the bottom of Akhtanizovskiy liman, near the Boris and Gleb mud volcano, showed that its feeder channel is enormous in size (Fig. 6.90).

Southeast of Akhtanizovskiy liman is the Karabetovskaya anticlinal zone with the Karabetova Gora (#66 in Fig. 6.7), Gora Chirkova (#67 in Fig. 6.7), Neftyanoy Northern (#68 in Fig. 6.7), and Dubovyi Rynok (#69 in Fig. 6.7) mud volcanoes. The Karabetova Gora mud volcano (first described by Pallas in 1795) is the largest and most exotic of the four (Fig. 6.91).

Fig. 6.88 Aerial view of the Gora Boris and Gleb mud volcano. (Photographed from a helicopter in 1982)

Fig. 6.89 General view of the Gora Boris and Gleb mud volcano photographed from above the village of Akhtanizovskiy in 2003

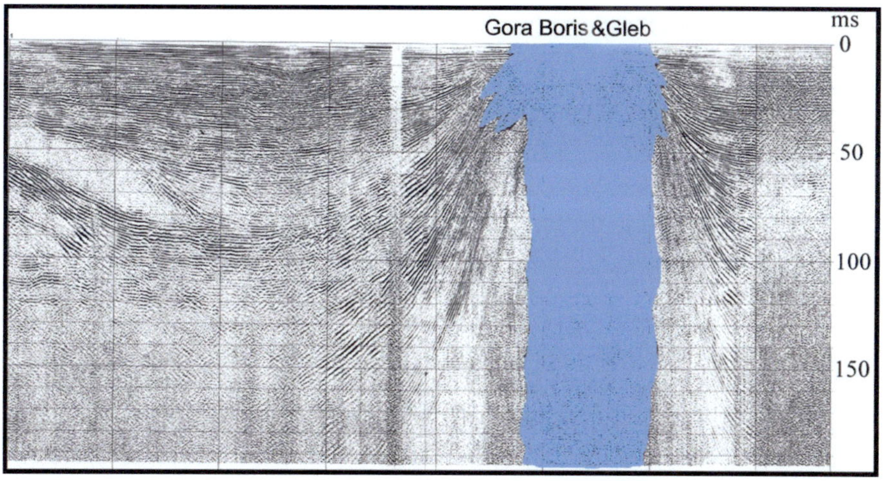

Fig. 6.90 Seismoacoustic transect conducted by "Yuzhmorgeologiya" across the bottom of Akhtanizovskiy liman near the Boris and Gleb mud volcano

Fig. 6.91 View of the Karabetova Gora mud volcano. (Photographed from a helicopter in 1982)

The mud volcano is located 5 km east of the village of Taman and forms a plateau rising up to 142 m asl. It includes a number of separate hills, salsas, and small crater lakes, and it is composed of mud volcanic breccia. The relief of the plateau is extremely heterogeneous and resembles the surface of the Moon due to repeated violent eruptions in various parts of the mud volcano that have been recorded many times throughout the nineteenth, twentieth, and twenty-first centuries. One of the most famous and significant explosions occurred on August 6, 1856, at 6:30 am. It was described by Herz (1876), who wrote that on the previous day, the neighborhood residents heard thunderous underground shocks. The eruption began with a colossal explosion, during which huge masses of breccia were blasted from the crater to heights of 30–40 m. The explosion was accompanied by flames and clouds of black smoke, and the eruption continued for 3 h. Afterward, the top of the mountain was covered with a thick (up to 2 m) layer of black mud. In 1869, on the western slope of the volcano, two large cracks formed. The surface of the mountain was rugged and elevated by 1 m in places (Land 1866). Renewal of mud volcanic activity has occurred many times at least since the early 1880s. All explosions were of the same type, each representing a one-time ejection of large masses of breccia (5–10 thousand m^3) accompanied by a rumble and rapid ground movements (Fig. 6.92).

On May 6, 2001, a large eruption of breccia occurred at the top near the location of previous outbursts. On June 19, 2004, the explosion was repeated, and about 47 thousand m^3 of breccia were ejected. Both explosions were accompanied by gas combustion yielding fresh reddish-brown and black slag-like breccia (Fig. 6.93).

Among the scattered clumps of breccia are huge (1 m^3 and larger) boulders of Chokrakian and Sarmatian limestones (Fig. 6.94).

An aerial view of the Karabetova Gora mud volcano illustrates its complexity and diversity (Fig. 6.95).

At the top of the Karabetova Gora mud volcano, there are numerous lakes situated within the mud volcanic breccia. The lakes (dark blue in Fig. 6.95) have bitter-salty water with green, exotic vegetation in some of them.

The powerful driving forces of mud volcanic processes generate landscapes that are similar to those of real lava volcanoes. Ruptured beds and crumbling erupted breccia create a picture of primordial chaos (Fig. 6.96).

Geologically, the Karabetovskaya anticline is composed of Maikopian, Chokrakian, Karaganian, Sarmatian, Meotian, Pontian, and Kimmerian strata. The Karabetova Gora mud volcano is confined to the Maikopian core of the anticline, forming a large and thick blob of breccia, which is in turn broken up by new supplies of mud volcanic products (Figs. 6.97, 6.98, and 6.99).

A microseismic sounding of the Karabetova Gora mud volcano is illustrated in Fig. 6.99.

A relatively narrow vertical zone associated with saturated fluids brought by supply channels was identified under the Karabetova Gora mud volcano. According to experimental data, the

Fig. 6.92 Mud volcanic breccia at the top of the Karabetova Gora mud volcano after the eruption of 2004

Fig. 6.93 Burned mud volcanic breccia formed by large-scale gas combustion on the top of the Karabetova Gora mud volcano. (Photographed in 2004)

Fig. 6.94 Lumps of limestone and large clods of breccia (in gray color to the left of the reddish object) on the surface of the Karabetova Gora mud volcano. (Photographed in 2004)

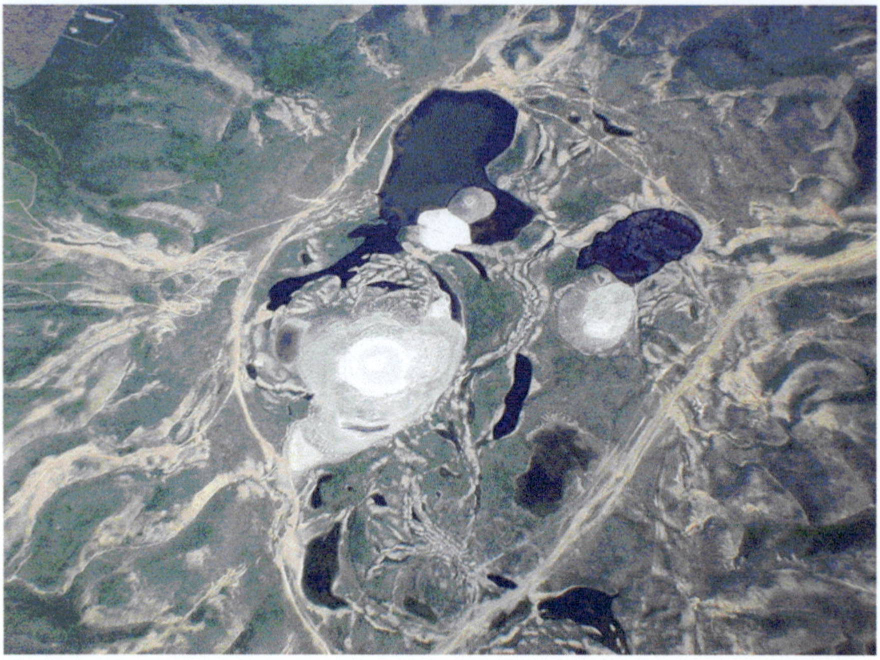

Fig. 6.95 Partial aerial photograph of the Karabetova Gora mud volcano. Photographed in 2005. Currently inactive mud volcanoes appear in the lower center of the figure; somewhat to the east (right) can be seen fresh outpourings of breccia (?) with well-defined concentric shafts fixing different stages of eruptions. The peak of the Karabetova mountain is completely formed by mud volcanic breccia. The plateau seems to be built from numerous eruptions, and in this way, the mountain grew upward

Fig. 6.96 Deep cracks on the surface of the Karabetova Gora mud volcano after the eruption of 2001. (Photographed in 2004)

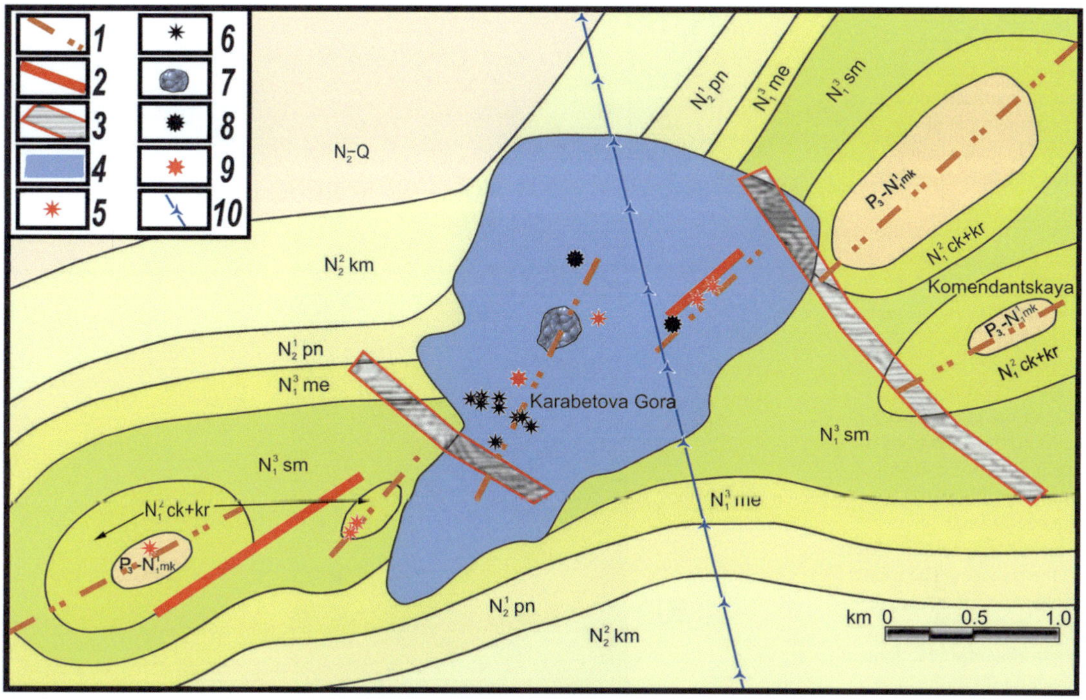

Fig. 6.97 Geological scheme of the Karabetova Gora mud volcano area (Sobiesievich et al. 2008). The relief contours are spaced 5 m apart. Legend: (1) axes of anticlinal folds and their numbers; (2) reliably established faults; (3) faults identified by structural geomorphological features; (4) breccia cover; (5) active gryphons; (6) inactive gryphons; (7) salsa; (8) large cone-shaped gryphons several meters high; (9) center of the explosion on May 6, 2001; (10) points along the microseismic survey transect. P_3-N_1 = Maikopian clay; N_1ck + kr = Chokrakian and Karaganian dark gray clays with interlayers of marls and limestones; N_1sm = Sarmatian stage, greenish-gray clays with numerous interbeds of porous marls and limestones; N_1me = Meotian dark clay with interlayers of marls and limestones; N_2pn = Pontian dark gray and black clay with interlayers of loose limestones, marls, and sandstones; N_2km = Kimmerian clay, sand with interlayers and lenses of oolitic iron ores in the lower part; N_2-Q = clay sandy loam and sand in the upper part of the horizon of red-brown clay (Tveritinova et al. 2015)

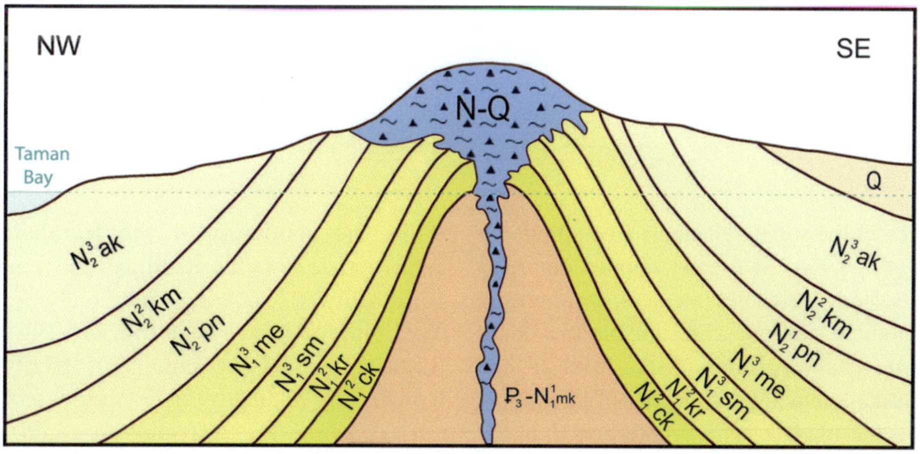

Fig. 6.98 Simplified geological section of the Karabetova Gora mud volcano

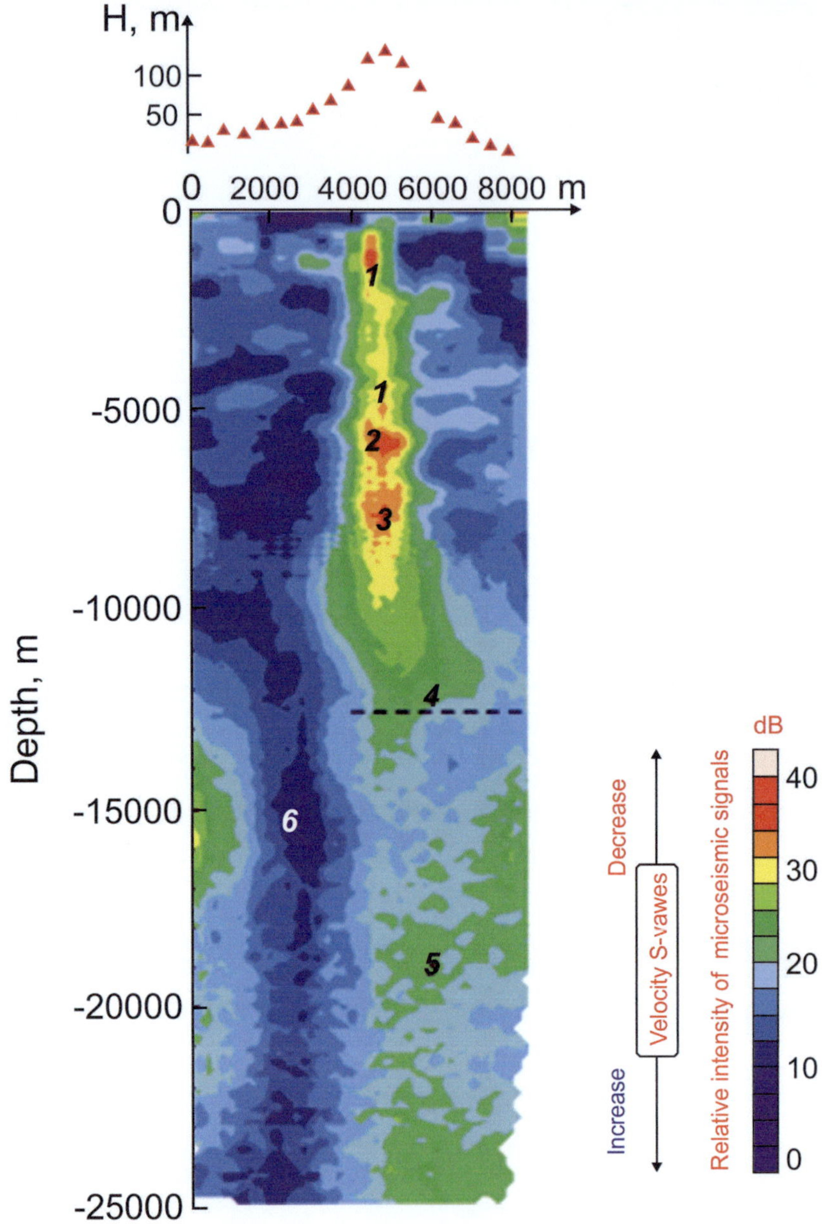

Fig. 6.99 Vertical geophysical section along the transect of the microseismic sounding (Tveritinova et al. 2015). Key: (1) near-surface mud volcanic chamber; (2, 3) deep mud volcanic reservoirs; (4) transition area to consolidated foundation; (5) deep fluid-permeable structures; (6) impermeable consolidated rocks. Above the geophysical section, the elevations of the transect measuring points are indicated, and the conditional coordinate system is tied to the transect. The tone scale shows the relative intensity of environmental reactions, reaching up to the generation of Rayleigh microseismic waves with different penetration depths. An increase in the intensity of the reaction within the environment corresponds to reduced S-wave velocities and vice versa

feeder area of the supply channels was found at a depth of 4.5–9 km. At greater depths, contrasts between seismic velocities are less pronounced, but the anomaly continues, presumably to a depth of 15 km. The formation of the core of the Karabetovskaya anticline, composed of Maikopian clays, and the mud volcano is interpreted as the result of deep processes determined by dilatant structures (Tveritinova et al. 2015).

To the southeast of the Karabetovskaya anticlinal zone is the Kiziltashskaya anticlinal zone with the following mud volcanoes: Peklo Chernomorskoye (#70 in Fig. 6.7; Fig. 6.100), Bugazskiy (#71 in Fig. 6.7), Kiziltashskiy Northern (#73 in Fig. 6.7), Vyshesteblievskiy (#74 in Fig. 6.7), Neftyanoy Southern (#75 in Fig. 6.7), and Kamyshevatskiy (#76 in Fig. 6.7).

Fig. 6.100 Cropped photo of the Peklo Chernomorskoye mud volcano. (Photographed in 2004)

The cores of the anticlines, within which the mud volcanoes are confined, are composed of Maikopian series, while their peripheries are made up of Chokrakian, Karagantian, Sarmatian, Meotian, Kimmerian, and Kuyalnitskian deposits. The anticlines with sediments fringing them as islands are observed among the vast area of Quaternary deposits. The latter have high thicknesses, accumulated in many places due to the Kuban River, which has changed its riverbed countless times.

The most significant and exotic mud volcano of this zone is the Bugazskiy (Fig. 6.101) mud volcano, which is located on the western shore of the liman of the same name.

This volcano consists of several groups of small gryphons totaling 98 in all. One of the groups, located on the western shore of the Bugazskiy estuary and 1400 meters from the village of Veselovka, has been known for a long time.

On the western slope of the estuary, Pallas (1795) discovered a small cone from which muddy gray silt had erupted. On the eastern slope, he found two similarly acting gryphons. He probably meant the Bugazskiy mud volcano when he wrote (p. 45): "In general it is filled with a cloudy salty water, on the surface of which the mountain oil rises, particularly in those springs which are located close to the road to the liman, and which are covered by oil about a foot thick, and this oil is much thicker and blacker than the oil under Enikale town."

Zenin (1955), who examined this mud volcano 150 years after Pallas, indicated that the picture had changed somewhat. On the western slope of the liman, there were no traces of mud volcanic activity, while on the eastern slope, there were three groups of active gryphons. The first group included two large gryphons that stood several meters apart, both with heights of 1 m and diameters at the base of 1.5 m. They were surrounded by twelve smaller gryphons erupting gray mud with oil films and a strong smell of hydrogen sulfide. Local people use this mud for medicinal purposes and washing their clothes. The second group contained eight small gryphons between the Bugazskiy liman and the Salt Lake on both slopes of the ravine (Fig. 6.7). They act in the same manner as the previous ones. The third group contained 18 small, active gryphons and was located 1 km to the northeast of the village of Veselovka, on the slopes of the gully. Gray mud erupting from these gryphons all contained oil films. The easternmost gryphon erupted clean oil. Our recent

Fig. 6.101 General view of the Bugazskiy mud volcano with ejected breccia. (Photographed in 2004)

observations show that this mud volcano still preserves the same grouping of gryphons and hills described by Zenin (1955), and although the number of gryphons and hills has decreased, the intensity of the activity endures.

Most of the mud volcanoes described above appear in the form of peaked mountains and plateaus, but the Bugazskiy mud volcano is like a combination of the Bulganakskiy and Karabetova Gora mud volcanoes, both in its morphology and in the character of its activity, though it is somewhat smaller.

The Neftyanoy Southern mud volcano is located to the east of the road linking the villages of Vyshesteblyevskaya and Dzhiginka; it is unusual and does not quite correspond to the accepted concept of a mud volcano (Fig. 6.102).

It is represented by an oil-soaked area of 15×30 m^2 with two large mud puddles that intensively release gas (mostly methane). The gas-oil-water seepages are quite weak and almost do not flow. There are numerous oil clots that simply spin in the water during aeration. No clastic material or ejected breccia is present around the puddles, but there are efflorescence blossoms of borates and salt and signs of burning to a brick-red color here and there indicating recent flaming outbursts. On June 6, 2002, we observed a rapid release of oil from this mud volcano.

The Kamyshevatskiy mud volcano has a conical shape typical of these geological structures; it is completely devoid of any signs of mud volcanic activity.

To the southeast of the Bugazskaya anticline, there are two anticlinal zones of northeastern strike: Ereminskaya and Blagoveshchenskaya. The Ereminskaya anticlinal zone is composed of Sarmatian, Meotian, Pontian, Kimmerian, and Kuyalnitskian deposits. The mud volcanic field is small and adjoins directly onto Maikopian deposits. The Blagoveshchenskaya anticlinal zone is similar to that of Ereminskaya.

Within the Ereminskaya anticlinal zone, there is an anticline called Belyi Khutor which possesses a mud volcano of the same name (#77 in Fig. 6.7). Within the Blagoveshchenskaya anticlinal zone, there is also a Blagoveshchenskiy mud volcano (#85 in Fig. 6.7).

Fig. 6.102 General view of the Neftyanoy Southern mud volcano. (Photographed in 2004)

The mud volcanoes described above are confined to anticlinal folds of northeastern strike that have been grouped into anticlinal zones of the Kerch-Taman transverse trough. One of them, located on the west, is the Anastasievsko-Troitskaya anticline, which is the largest one in the Anastasievsko-Krasnodarskaya anticlinal zone (Shaulov 1967, 1969; Shnyukov et al. 1971).

The Anastasievsko-Troitskaya anticline consists of three arches separated by saddles lying on the axis of its northwest-southeast strike. It is composed of Lower Middle Pliocene sediments (Fig. 6.103).

Wells drilled into its northwestern periclinal section revealed a buried negative structure in the form of an oval graben at depths of 1300–2200 m filled with a 300 m-thick layer of mud volcanic breccia. The breccia is underlain by Meotian and Upper Sarmatian sediments containing oil, and it is overlain by Pontian sediments. This negative structure is hidden by a thick (up to 1300 m), gently sloping, almost horizontal sequence of Kimmerian, Kuyalnitskian, and Quaternary sediments. This negative structure was first described by Shaulov (1967), who considered it to be a graben that complicated the anticline. In his later work (Shaulov 1969), he admitted its closeness to recessed synclines of the Kerch Peninsula. Shnyukov et al. (1971) called this structure the Anastasievsko-Troitskaya recessed syncline because it possesses a number of specific features characteristic of this type of geological structure. It has the almost round form of structure-contorting tectonic ruptures expressed as a series of discontinuities; it contains mud volcanic breccia; and it lacks some stratigraphic horizons in its geological section. Mud volcanic activity had ceased already in early Pontian time, while the process of subsidence continued even later. This is evidenced by the development of a small depression in the modern relief.

Within the same Anastasievsko-Krasnodarskaya anticlinal zone to the northwest of the Anastasievo-Troitskaya anticline, the Gora Gnilaya mud volcano is confined to the Gora Figura anticline.

The Gora Gnilaya mud volcano (#60 in Fig. 6.7) was described by Pokrovskiy (1933). It is located

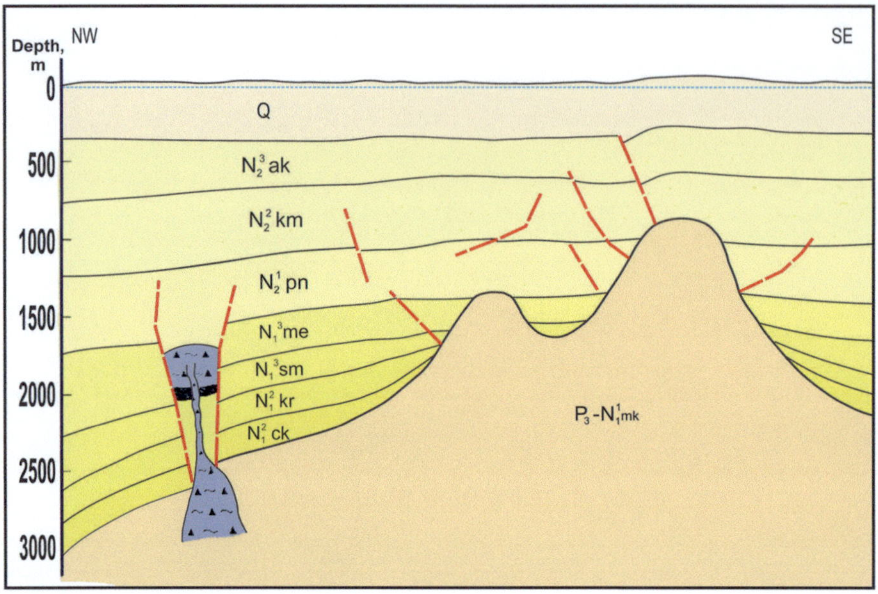

Fig. 6.103 Geological section through the Anastasievo-Troitskaya anticline

5 km southeast of the eastern outskirts of the town of Temryuk. It assumes the shape of a truncated cone with long and short axes of 550 and 500 m, respectively, and its peak is elevated 13–14 m above the ground (and 32 m above the level of the Sea of Azov). On its slightly hilly surface, there is recently erupted mud volcanic breccia of a dark gray color modified by light spots of eolian sand with numerous small gryphons, temporarily inactive and active. The gryphons are clustered into separate groups with heights sometimes up to 2 m (Fig. 6.104).

One of the gryphons located in the marginal southeastern part of the mud volcanic breccia field continuously releases gas and mud volcanic breccia that flows down the slopes. Against the background of the dried mud volcanic breccia, there are absolutely fresh tongues of mud breccia emerging from the vents of active gryphons (Figs. 6.105 and 6.106).

Numerous microgryphons are widely distributed within the area of the fresh but already weathered mud volcanic breccia, where numerous desiccation cracks form very quickly.

In the past, the Gnilaya Gora mud volcano was occasionally very active. Several intense eruptions occurred in the first half of the nineteenth century, but observations of events during the next century and a half are not reflected in the literature. A quite powerful explosion of breccia and gas was recorded in 2001, and this was repeated on February 26, 2002, when the amount of erupted breccia reached approximately 1000 m^3. It is quite interesting that this explosion occurred synchronously with that of the Tsymbaly Western mud volcano.

The Gora Miska (#59 in Fig. 6.7) mud volcano is confined to the northwestern closure of the Anastasievsko-Krasnodarskaya anticlinal zone. It is located in the eastern part of the town of Temryuk, reaches 74 m in height, and has a large round caldera at its top. The caldera is surrounded by a ring shaft that rises 10–12 m above the caldera. On the shaft, there is an operating Museum of Military Technology ("Military Hill") with a monument represented by a military tank. In the past, the volcano was 5 km away from the eastern outskirts of the town, but today, it is located inside the town and does not show any activity. An asphalt road crosses the volcano, leading to the town from the east. The segment of the road that passes alongside the caldera is continuously under repair because the covering asphalt bursts due to subsidence of the mud volcanic breccia. Not far from the road, there is a small salsa filled with liquid mud that periodically dries out.

Fig. 6.104 Line of gryphons on the mud volcanic breccia associated with the Gnilaya Gora mud volcano. (Photographed in 2003)

Fig. 6.105 An active gryphon on the mud volcanic breccia field associated with the Gnilaya Gora mud volcano. The halophytic vegetation is red. (Photographed in 2003)

In the past, this mud volcano was quite active. Referring to numerous literary and oral reports, Zenin (1955) described strong explosions with eruption of breccia, fumes, smoke, and flames in 1812, in 1843, and less intensely in 1860. The most severe explosion occurred in 1843. It was accompanied by thunderous rumblings and the outburst of a large column of flames. After the explosion, the crater filled with liquid mud and sagged, forming a depression. In 1905, in the northern part of the depression, an

Fig. 6.106 A group of gryphons seen against the backdrop of the town of Temryuk. City buildings are gradually encroaching upon the Gnilaya Gora mud volcano. (Photographed in 2003)

intensive release of mud breccia occurred, and cracks 0.75 m deep and 0.5 m wide were formed on the ground. In the 1950s, the middle part of the caldera was uplifted according to local eyewitnesses. Intense tremors and the appearance of cracks were noted on the volcano on September 4, 1988.

Despite its stable state at present, the Miska mud volcano poses a certain threat to nearby buildings and their inhabitants (Fig. 6.107).

The Raznokolskiy mud volcano (#78 in Fig. 6.7) is located in the North Caucasus near the village of Yurievo, 750 m northeast of the mountain called Raznokol. Its peculiar appearance is noteworthy: a huge, rounded protrusion of gray clayey breccia in the form of a "sausage" standing about 2 m above the surrounding terrain. It emerges directly from soil overgrown with grass on the southern high coastal cliff of the Old Kuban River. The length of the "sausage" reaches 1.5 km, and its mud breccia slides down the slope forming a kind of delta projecting into the river valley.

The sequence of erupted mud flow from its beginning to end is clearly noticeable (Fig. 6.108).

The rate of breccia protrusion is uneven. According to Fedorov (1931), it is 2.5–8.0 cm per day. During our short-term observations in 1982–1983, almost no daily flow was visible. At the same time, the authors had no doubt that the flow was moving. To prove it, a simple experiment was carried out. In August 1984, six stakes were inserted across the base of the extruding body of breccia: two of them were benchmarks introduced into the soil along the sides of the flow and the other four were set into the clay breccia, all on the same line. A year later, a return visit revealed that the stakes on the right side of the flow had moved forward 9–10 m while those on the left had moved only 1.5 m. On average, the speed of this recorded movement ranged between 0.4 and 2.5–3.0 cm per day. In the middle of the flow, numerous glide mirrors oriented diagonally in the direction of the movement were observed, and the breccia had a fresh appearance. By comparison with the data of Fedorov (1931), it can be seen that the flow rate was different depending on the point of measurement, and this was also confirmed by the observations of local residents, who reported that in different years, the flow of mud breccia moves at varying speeds. Within the stream, there is a variety of clastic material, in places a weak seepage of water, and whitish mineral

Fig. 6.107 General view of the Miska mud volcano. There is an operating museum of military technology sitting atop the volcano. (Photographed in 2003)

Fig. 6.108 Aerial view of the Rasnokolskiy mud volcano. A tongue of mud volcanic breccia (center of the photograph) is visible descending into the valley of the Old Kuban River in the distance. Its final part (3) is overgrown with bushes and trees; the middle part (2) is covered by a layer of weathered brownish breccia and overgrown by grass; and the initial part (1) in the right foreground appears as a fresh bluish-gray mud volcanic breccia. (Photographed from a helicopter in 1982)

Fig. 6.109 Protrusion of mud breccia from the Raznokolskiy mud volcano. (Photographed in 1982)

Fig. 6.110 Cracked, dry crust on the surface of the mud flow from the Raznokolskiy mud volcano. (Photographed in 1982)

incrustations. Geologically, the mud volcano is confined to the large Raznokolskaya anticline, which is a typical diapiric structure with a broken Maikopian core out of which the mud volcanic breccia has been squeezed (Figs. 6.109 and 6.110).

While visiting the Raznokolskiy mud volcano in 2002, the authors noticed that the stream had shrunk and become less spectacular. Yet, the geological significance of the Raznokolskiy mud volcano is substantial, as this unique phenomenon clearly demonstrates the mechanism of the mud volcanic process, which is shown schematically in Figs. 6.111 and 6.112.

One of the largest mud volcanoes of the northwestern Caucasus is Shugo (#83 in Fig. 6.7); it is the most picturesque mud volcano of the Kerch-Taman mud volcanic region (Figs. 6.113 and 6.114). The Shugo mud volcano is located on the right bank of the Shugo River in mountainous woodland, 6 km south of the village of Varenikovskaya in the territory of Krasnodar.

This volcano has several names: Jezeuasa ("ash mountain" in Circassian), Divnaya Gora ("wonder mountain" as it was called by V.I. Vinda, a mining engineer who first described this volcano in 1902), and Shugo (the name used by the Kuban Cossacks) (Steber 1912).

The Shugo mud volcano is confined to the top of the Shuginskaya anticline, which is composed mainly of Cretaceous and, to a lesser extent, Lower Tertiary deposits (Kudryashev 1939). Judging from descriptions in the literature (e.g., Steber 1912; Kudryashev 1939; Nikitin 1955; Shnyukov et al. 1971), this mud volcano has not changed much over the last century. At the top of a large cone-shaped mountain up to 2 km in diameter, there is a hollow representing the caldera of the mud volcano. Framed by a sea of green forest on all sides (Fig. 6.114), it is especially beautiful from a bird's-eye view. Within the caldera, there is an irregularly rounded basin, or cup, with a diameter from 300 to 500 m filled with gray mud volcanic breccia and devoid of vegetation.

The ring shaft framing the caldera is overgrown with trees and falls steeply toward the center of the mud volcano. In the lower part, the shaft is torn by a large gulch, which serves as a channel for the escape of the mud breccia. Between the ring shaft and the central swollen

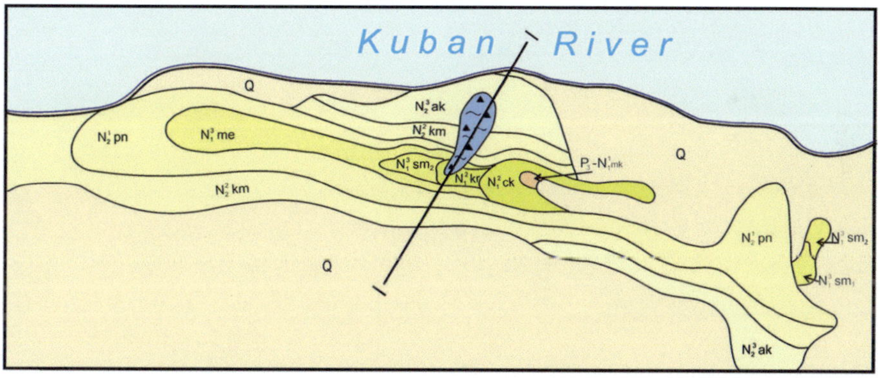

Fig. 6.111 Geological diagram of the Raznokolskiy mud volcano

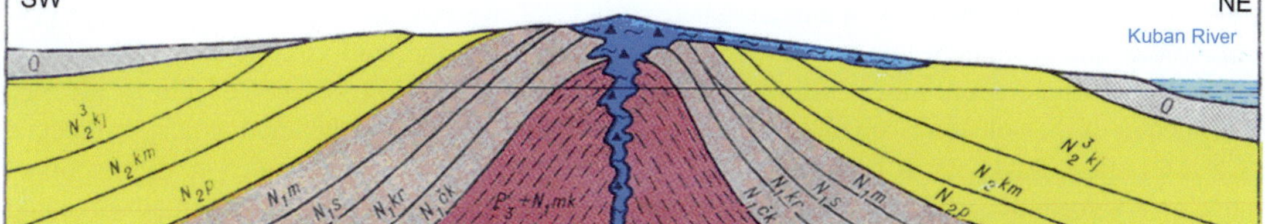

Fig. 6.112 Geological section through the Raznokolskiy mud volcano indicating how mud volcanic breccia is being squeezed out from the broken Maikopian core

Fig. 6.113 Aerial view of the Shugo mud volcano. (Photographed from a helicopter in 1982)

Fig. 6.114 General view from the west of the Shugo mud volcano. (Photographed in 2003)

part of the volcano, there is a small circular depression of about 100 m in diameter where atmospheric and volcanic water accumulate. The cover of dried and cracked mud volcanic breccia in the center of the volcano is small and does not exceed 60 cm in thickness. It is similar to other active mud volcanoes of the Kerch-Taman region where the greatest thickness of dried breccia crust is 55–70 cm (Kudryashev 1939). In the excavated ditch, the capping of breccia is underlain by a layer of semiliquid mud. In the summer of 1982, we recorded 33 small and 5 large gryphons within the mud volcano's cup. They were quite diverse. Small gryphons seem to ulcerate the gray surface of the mud volcanic breccia. Sometimes they are represented by large spots of fresh, wet mud, or they appear as small spreading gryphons that are obviously produced by mud of a less viscous consistency. Small gryphons are usually arranged in clusters of five to ten units, their localization being due to the existence of small semicircular or radial cracks along which they are aligned. Gryphons that appear in the periphery of the cup often yield small flows of liquid mud. Large gryphons gather in groups along the periphery of the mud volcano cup. They release gas, water, and mud that flow down the slopes of the cones. Mud bubbles appear when gases are released. On the surface of the puddles within the caldera, spots and films of oil are present (Figs. 6.115, 6.116, and 6.117).

The latest distribution of gryphons within the Shugo mud volcano was compiled by Sobisevich et al. (2014); it can be seen that they are concentrated in its southwestern part (Fig. 6.118).

Outpouring mud often spreads not just in one direction but in many directions about the cone of the gryphon. Watery gryphons emit a semiliquid mud that, when dried, forms peeling crusts. Streams of water and mud from large gryphons form efflorescences of salts and borates at the end of their flow when they dry. Puddles produced at the tops of large gryphons reach several meters in diameter. Along with large active gryphons that rise a few meters in height, a number of other large ones appear dried and residual. Numerous groups of small gryphons periodically bubble due to the release of gas. The flow of drying mud creates unique micro-landscapes reminiscent of lava streams from real volcanoes, but in miniature. Thickets of a very interesting plant called Kermek (*Statice limonioum* L.) grow on the perimeter of the Shugo mud volcano, indicating the presence of boron. Thus, Kermek bushes penetrate into the central part of the volcano, and their lilac flowers look very beautiful on the light gray background of the breccia, which often has a whitish color created by numerous salt efflorescences (Fig. 6.119).

The Shugo mud volcano ejects dozens of cubic meters of gas, mostly methane, and a lot of mud daily. Clastic material in the mud is most often represented by Cretaceous rocks, siderite, and crystalline rocks. It is generally suspected that Shugo is a calmly developing mud volcano that is experiencing a gryphon stage of development. Meanwhile, local residents retain memories of the volcano's violent explosions, one of which allegedly occurred about 150 years ago. There are echoes of these statements in the literature. Thus, Steber (1912) wrote about the "last turbulent eruption" in July or August of 1903, which began in the daytime with a terrible noise. At night, a high column of flame could be seen from the neighboring mountains emanating from the center of the crater. Steber also wrote about a

Fig. 6.115 Central part of the breccia field in the Shugo mud volcano. (Photographed in 2003)

Fig. 6.116 An active gryphon intensely spewing out gas and associated with the Shugo mud volcano. (Photographed in 2003)

Fig. 6.117 An active gryphon associated with the Shugo mud volcano. (Photographed in 2003)

large stream of mud in 1907 and a small, almost instantaneous (15–20 s) eruption in 1908, accompanied by light explosions, ejection of mud to a height of 5–7 m, and a strong noise. He concluded that small, turbulent eruptions of the Shugo mud volcano are not uncommon. In other literary sources, there are even more explicit statements. Thus, Land (1866) cited a legend according to which the Shugo mud volcano once housed a Circassian village that "for the sins of its inhabitants fell through the earth, and only one righteous widow was saved."

A rapid eruption of the Shugo mud volcano occurred at 9 am on June 24, 1959, when approx-

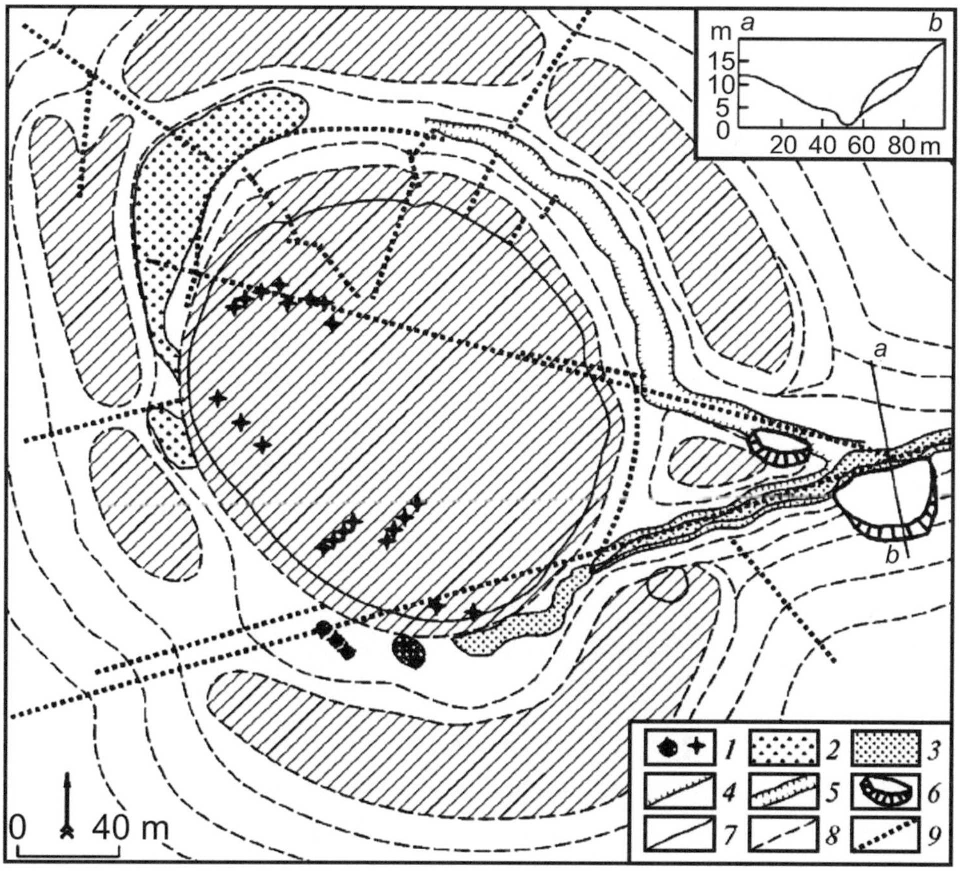

Fig. 6.118 Diagram of the apical part of the Shugo mud volcano. Key: (1) active gryphons and salsa (cross), (2) separate mud hills, (3) fresh eruptions of mud breccia, (4) renewed ring faults expressed by fresh ledges, (5) steep sides of gully with active erosion cut, (6) fresh landslides, (7) dry riverbeds, (8) provisional relief contours, (9) active tectonic ruptures expressed in relief (Sobisevich et al. 2014)

imately 1500 m^3 of mud breccia was ejected, and the height of the outflows reached 5 m. There were also weak eruptions in subsequent years.

Today, the mud volcano has lost its beauty in many respects and is being used for mud baths.

Geological and geophysical studies conducted by the Institute of Earth Physics of the Russian Academy of Sciences have revealed the deep structure of the Shugo mud volcano and that of the Akhtyrskaya flexure-discontinuous zone as a whole. The mud volcano chamber is located at a depth ranging between 15 and 20 km, while the root structures were observed at 25 km deep and more (Fig. 6.120).

The data obtained suggest a new approach to the study of deep processes in areas where mud volcanism has developed in the late Alpine folded structure of the northwestern Caucasus.

The Gladkovskiy mud volcano (#84 in Fig. 6.7, Table 6.2) is located near the village of the same name and is somewhat reminiscent of the Shugo mud volcano. The panorama of the Gladkovskiy mud volcano looks quite picturesque: a vast gray field of breccia amid the greenery of forests and vineyards (Fig. 6.121).

In its structure, the Gladkovskiy mud volcano and those of Shkol'nyy (#86 in Fig. 6.7), Semigorskiy (#87 in Fig. 6.7), and Shapsugskiy (absent from Fig. 6.7 because it is located outside of the territory covered by the map) are developed not on Maikopian but on Lower Cretaceous deposits. The Gladkovskiy mud volcano is con-

Fig. 6.119 Efflorescences of borates on the surface of the breccia at the Shugo mud volcano. (Photographed in 2003)

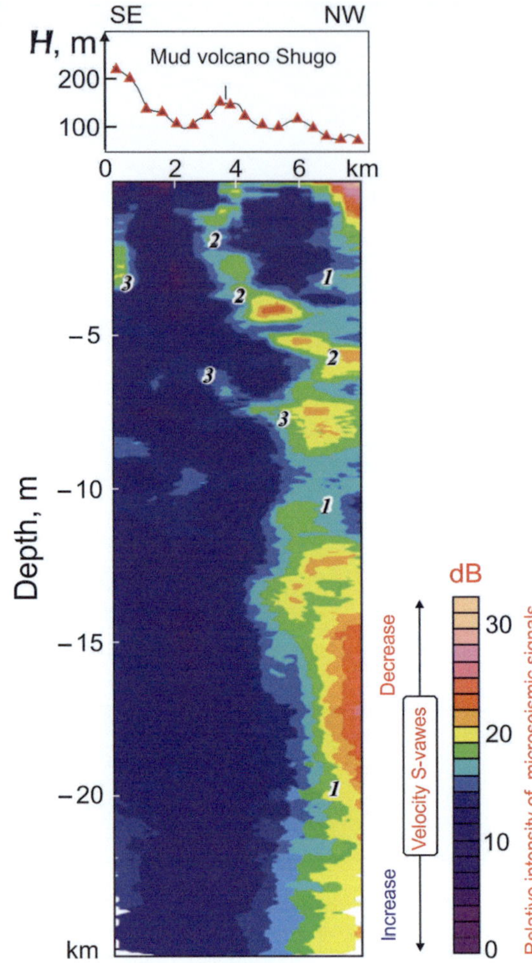

Fig. 6.120 Vertical geophysical section showing the contrasts of seismic velocities through the Shugo mud volcano. An increase in the relative intensity of the response of the medium corresponds to decrease in seismic velocities and vice versa. Key: (1) zone of reduced velocities associated with the Akhtyrskiy fault, (2) zone of reduced velocities associated with the fault that feeds the Shugo mud volcano, (3) zone of reduced velocities associated with the Gladkovskiy fault (Sobisevich et al. 2014)

fined to the western part of a small, elongated diapiric fold in the same anticlinal zone as the Shugo volcano, and it is broken by longitudinal tectonic ruptures.

The activity of the Gladkovskiy mud volcano is insignificant, but the number of gryphons and microgryphons is very high: at least 100. The most recent geological survey of this mud volcano corresponded closely to the previous survey performed by Kudryashev (1939). One slight difference is the presence of four small gryphons in the far south (Fig. 6.122).

The drying of some gryphons and the appearance of new ones nearby is typical for the area. The total number of relatively large gryphons is 35. They look like small cones containing mud puddles of 1–2 m in diameter that constantly release gas. The rest of the gryphons are small and very small, with diameters not exceeding 0.1–0.3 m. These gryphons release water, mud, and gas. As a whole, the image of the Gladkovskiy mud volcano is quite chaotic and complex. In a number of nearby geological structures, other mud volcanoes are found—Malye Blevaki (salsa of the mud volcano Gladkovskiy), Kiev, Semenovskiy, and Shapsugskiy.

The Semigorskiy mud volcano (#87 in Fig. 6.7) is located on the western extension of the Semigorskaya anticline near the village of

Fig. 6.121 Aerial view of the Gladkovskiy mud volcano. (Photographed from a helicopter in 1982)

Natukhayevskaya, 400 meters from the Anapa-Novorossiysk highway. The anticline has a northwest-southeast strike and is asymmetric because its southern wing is steeper. The mud volcano is sitting on a major tectonic disturbance (Slavyanov 1928).

According to our observations in 2014, the mud volcano is relatively large in size and represents an elliptical hill of 35–40 m in diameter rising 2–3 m above the surrounding terrain (Fig. 6.123).

The hill is composed mainly of mud volcanic breccia represented by gray clay with abundant debris (not more than 4–5 mm in size) of a harder clay. The western third of the hill is overgrown with grass; otherwise, the appearance of the breccia field is typical.

In 1964, 39 small active gryphons were found, and no volcanic activity was present (Subbota 1964). In 1983, the number of gryphons decreased to 27, and fewer were present in 2014 (Shnyukov 2016). Gas emitted by the gryphons is mostly methane and partially carbon dioxide. One gryphon emitted hydrogen sulfide (Subbota 1964).

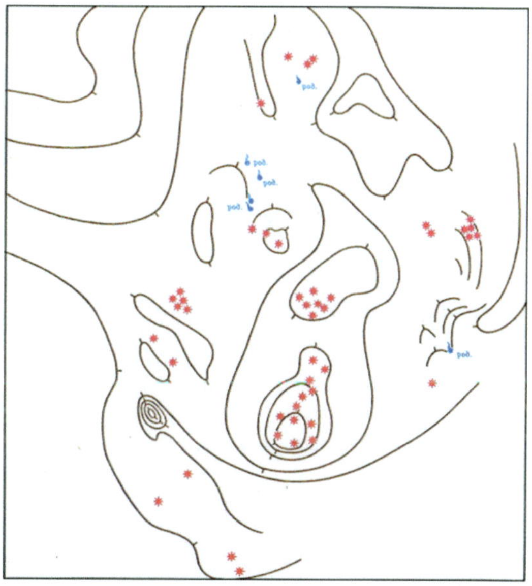

Fig. 6.122 Diagram of the location of active gryphons, salsas (in red), and springs (in blue) of the Gladkovskiy mud volcano

The Semigorskiy mud volcano sits upon Cretaceous sediments and demonstrates no activity. Clastic material is represented by solid fragments of Lower Cretaceous gray marl and gray sandstone. In 2014, numerous angular pieces of yellow and pink marls, calcite veins with pyrite, and greenish-gray rocks were investigated in detail. Study revealed that they belong to altered intermediate basic rocks (Shnyukov 2016).

Fig. 6.123 General view of the Semigorskiy mud volcano

There is a widely known Semigorskiy mineral spring located 1.5 km from the hill. Its water is similar in chemical composition to that of the Semigorskiy mud volcano. It contains iodide-salt-alkaline water with a high boric acid content bubbling with methane. There are traces of lithium, strontium, and barium within the spring flow, which yields 200 m^3 per day (Slavyanov 1928).

There are presumably few ore mud volcanic phenomena in the northern Caucasus. Thus, during his geological survey in the valley of the Chepsy River, V.M. Andreev (verbal communication) described a small mud hill sitting on Cretaceous sediments. Unfortunately, we could not find it.

There are also mineral springs (two to six) that resemble those of the Shapsugskiy mud volcano (Fig. 6.124).

They are located about 3 km to the southeast of the village of Shapsugskaya, between the rivers Abin and Shabari. Slavyanov (1928) wrote that there are no typical mud volcanic cones, but in their general appearance, conditions of outflow, and physicochemical properties of the mud and water eruptions, these springs are very reminiscent of mud volcanoes. Geologically, they are confined to clays of the upper siderite series of the Aptian stage of the Cretaceous.

There is another supposedly mud volcanic area located northwest of the village of Shapsugskaya, near the farm of Glubokiy Yar, on an elevation to the northeast of the creek near the crossroad. According to P.I. Ivchenko (verbal communication), there are several cones that remind one of mud volcanoes with springs releasing gas as well as concretions of calcite and siderite indicating that the cones lie above Cretaceous rocks. The survey performed by the authors discovered that these geological phenomena are reminiscent of mud volcanic structures though there are some differences. The springs are located in a widening (30 × 15 m^2) of the creek valley. There are seven springs, and they emit slightly bluish-gray mud and a limited outflow. Sometimes, the mud smells of hydrogen sulfide. The nature of this phenomenon remains wholly unclear (Fig. 6.125).

Fig. 6.124 General view of the Shapsugskiy mud volcano. (Photographed in 1982)

Fig. 6.125 Hydrogen sulfide spring of the Shapsugskiy mud volcano. (Photographed in 1982)

6.4 Mud Volcanoes of the Black Sea

About 80 mud volcanoes are known today in the Black Sea. Most of them are located in deep water, but some volcanoes have been found at relatively shallow depths. Conditionally, they are separated by us into ten groups: I, Northwestern; II, Western; III, Sorokin Trough; IV, near Kerch (Kerch-Taman

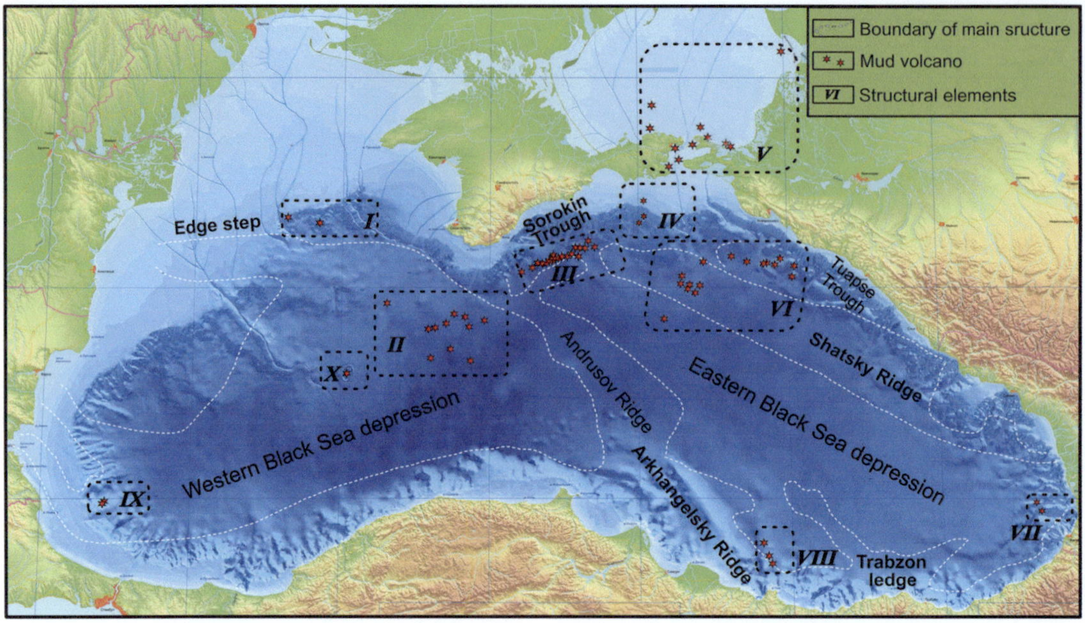

Fig. 6.126 Map of the Black Sea showing the major geological morphostructures and the groups of mud volcanoes: (I) Northwestern, (II) Western, (III) Sorokin Trough, (IV) near Kerch (Kerch-Taman shelf and continental slope), (V) Kerch Strait and the Sea of Azov, (VI) Tuapse Trough, (VII) Georgian, (VIII) Turkish, (IX) Bulgarian, (X) Central

shelf and continental slope); V, Kerch Strait and the Sea of Azov; VI, Tuapse Trough; VII, Georgian; VIII, Turkish; IX, Bulgarian; and X, Central (Fig. 6.126).

Groups I–VI are described below. Descriptions for groups VII–X are not provided here because little information is available. Moreover, they seem to possess features that are similar to those of other mud volcanoes.

6.5 Mud Volcanoes of the Northwestern Part of the Black Sea

Group I comprises the Parshin and Vodyanitskiy-2 mud volcanoes. The former lies at coordinates 44° 37′ 58″ N and 31° 13.21′ E and was discovered during the 27th voyage of the R/V "Vladimir Parshin" that was carried out in 2006 (Shnyukov et al. 2007). The mud volcano is located on a gentle slope of the Paleo-Dniester River tributary at a water depth slightly exceeding 200 m. It is a relatively large (440 × 240 m), plateau-like morphostructure that rises about 30 m above the seabed (Figs. 6.126, 6.127 and 6.128).

The plateau has an irregular, paw-like shape, somewhat elongated in the southwest-northeast direction, and it is crowned by a few conical craters, some of which release gas flares (Figs. 6.129 and 6.130a, b).

The gas flares differ in their height and size (Table 6.4).

An attempt by us to extract a core from the crater of the Parshin mud volcano did not succeed due to stormy conditions. Gravity core 2/27 (44° 37.70′ N; 31° 13.17′ E) was recovered at a water depth of 219 m, a few hundred meters away from the mud volcano. The core is 2.3 m long and includes the Neochernomorian (Late Holocene) (0–10 cm), which overlies with unconformity Late Neoeuxinian (Late Pleistocene) sediments. Both layers are represented by silt enriched with molluscan shells. In the Neochernomorian, it is *Mytilus galloprovincialis*, and in the Late Neoeuxinian, it is *Dreissena polymorpha* and *D. rostriformis*. The

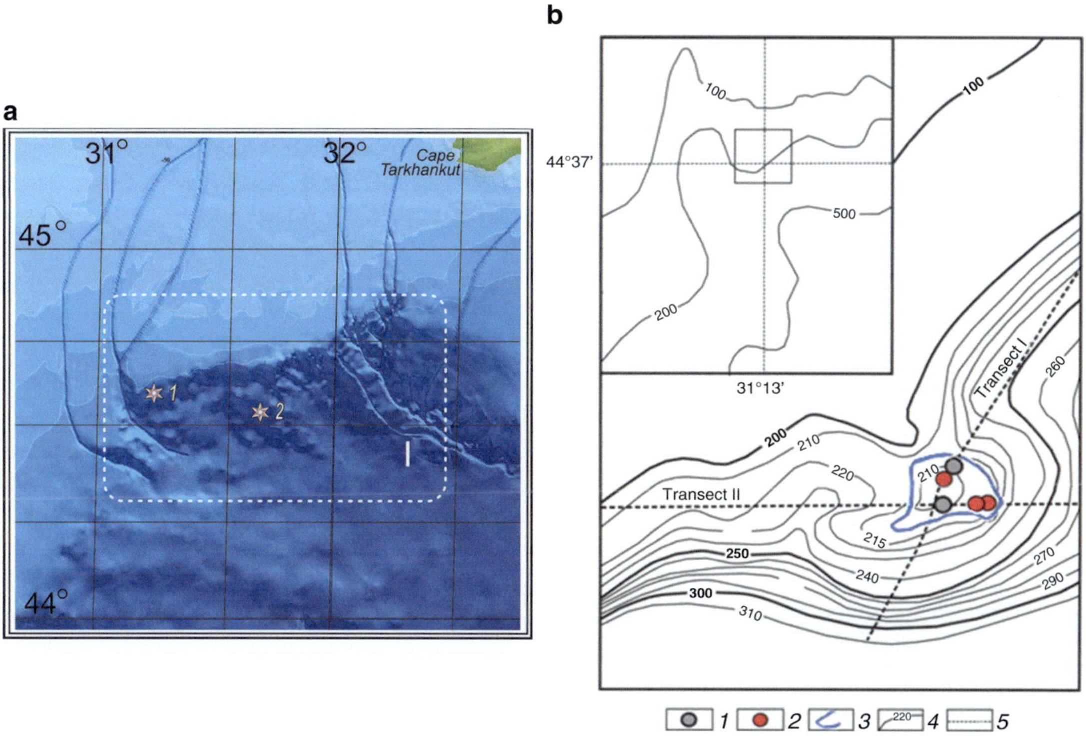

Fig. 6.127 (**a**) Location of the mud volcanoes Parshin (1) and Vodyanitskiy-2 (2) in the northwestern part of the Black Sea; for the location of group I, see Fig. 6.126. (**b**) Diagram of the mud volcanic plateau showing the location of the Parshin mud volcano, including non-active (1) and active (2) gas flares, contours of the Parshin mud volcano (3), isobaths (4), and echo sounder transects (5)

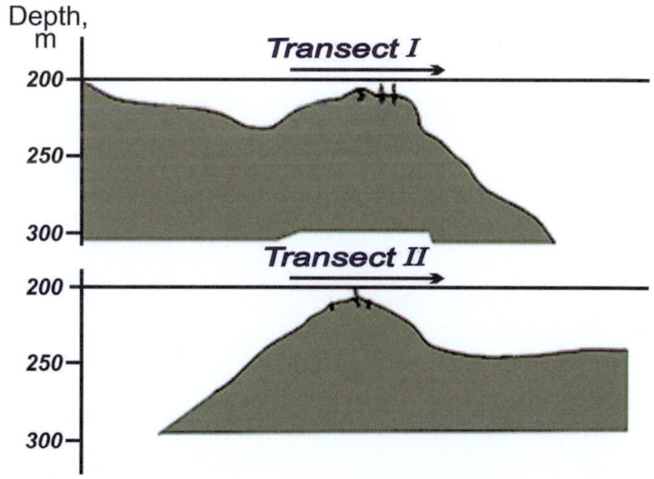

Fig. 6.128 Cross-transects of the seabed relief in the area of the Parshin mud volcanic manifestations (for the location of the profiles, see Fig. 6.127)

former is typical of marine environments, while the latter indicate the brackish conditions that existed at the time when the Black Sea is called the Neoeuxinian Lake (Yanko-Hombach 2007). There is also microbreccia represented by fine (mm to μm in size) clay particles in the silt. The clayey suspension ejected from the volcano became the microbreccia. Together with *Dreissena* shells, this indicates that the mud volcano has been active from at least the end of the Pleistocene.

The deep geological section along transect I–I was obtained by using the new geophysical method of geopolariton probing; it was performed by Yu.A. Bogdanov (Shnyukov et al. 2007) during the passage of the R/V "Vladimir Parshin" over the Parshin mud volcano. The results enabled us to trace the location of a powerful channel ("chimney") that delivers gas to the top of the anticline structure, which is composed presumably of Cretaceous, Paleogene, Neogene, and Quaternary sediments.

The section reveals the presence of gas pockets at depths of 600 and 2000 m. Their unloading determines the activity of the mud volcano and its gas flow from deeper horizons within the Earth's interior (Fig. 6.131).

The following geological horizons are distinguishable in the section presented above (outside the anticlinal domes): foot of the Quaternary and Neogene (1950 m depth); roof of the Eocene (2900 m depth); roof of the Cretaceous (about 3200 m depth, but in many places, it cannot be traced with certainty); roof of the Lower Cretaceous (3900 m depth on the left, and 3800 m depth on the right); foot of the Cretaceous (5300–4900 m depth); roof of the basalts/or granites? (9600–8900 m depth in the form of weakly sloping blocks with faults); and surface of the consolidated part of the basalts (10,650–9900 m depth). Several other horizons have also been discovered between the Cretaceous sediments and the foundation, one of which is perhaps wedged out.

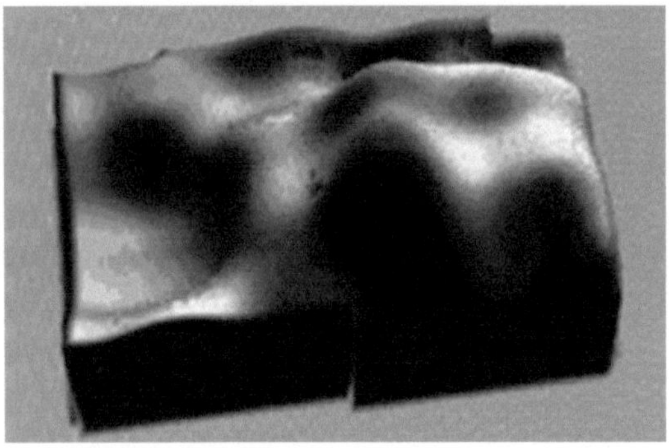

Fig. 6.129 3D image of the Parshin mud volcanic plateau crowned by a few conical craters. The image was obtained by echo sounder during the 27th voyage of the R/V "Vladimir Parshin" in 2006

Table 6.4 Coordinates, depth, and height of gas flares in the area of the Parshin mud volcano

No	Coordinates: Latitude, N	Coordinates: Longitude, E	Water depth, m	Height of gas flares, m
1	44° 37′ 58″	31° 13′ 21″	213.6	160.0
2	44° 37′ 69″	31° 13′ 14″	211.0	100.0
3	44° 37′ 64″	31° 13′ 09″	206.6	110.0
4	44° 36′ 32″	31° 11′ 54″	219.6	<100

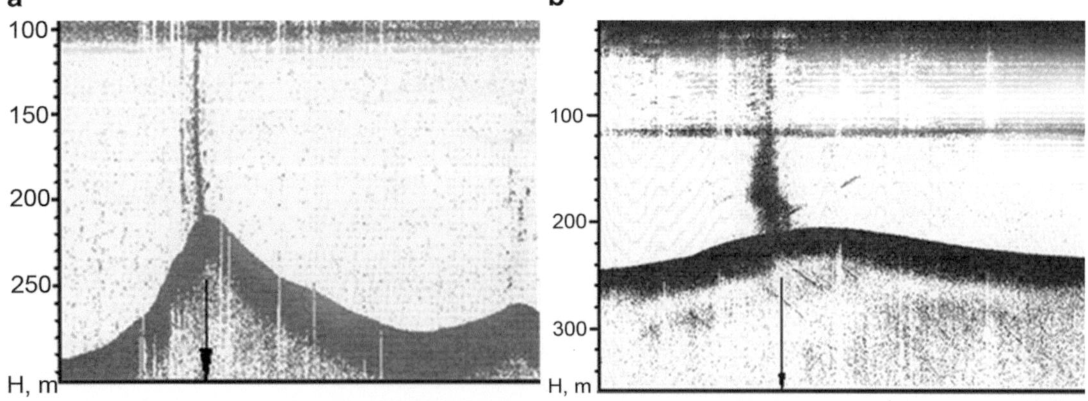

Fig. 6.130 Echograms of gas flares in the area of the Parshin mud volcano; for the schematic location and coordinates, see Fig. 6.126 and Table 6.4

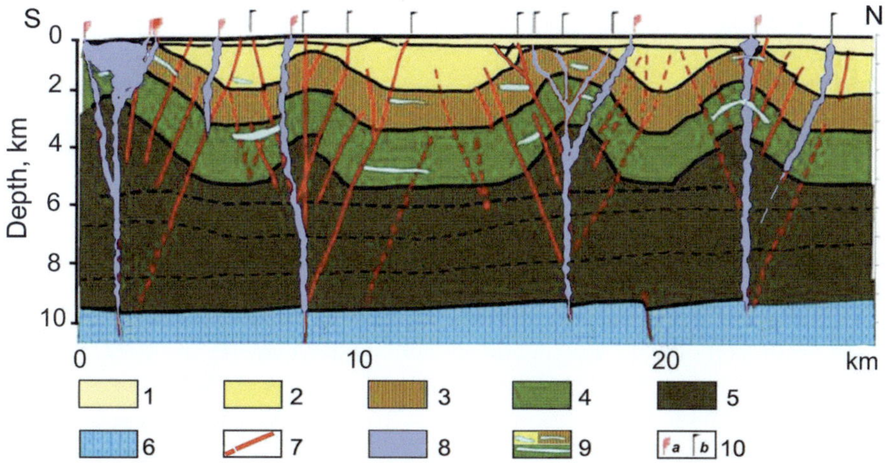

Fig. 6.131 Geological interpretation of the geophysical section obtained by the method of spontaneous electromagnetic emission of the Earth (Shnyukov et al. 2007). Key: (1) Quaternary and Neogene; (2) Maikopian; (3) Eocene; (4) Upper Cretaceous; (5) Lower Cretaceous; (6) crystalline basement; (7) tectonic disturbances and stressed zones; (8) gas transmission channels; (9) pockets of gases and water; (10) gas flares (a, recorded; b, presumed). There are more diagonal fractures (like the red lines in the figure) that go deeper into the crystalline basement rocks

Each recorded flare corresponds to a gas delivery channel (Fig. 6.131) that can be traced to a depth of 10 km. In the sedimentary layers of the upper part of the section (up to 5 km), anticlinal-like geological structures have been mapped within which gas-fluid accumulations may be concentrated.

According to Shnyukov et al. (2007), four more mud volcanoes have been recorded on the profile (Fig. 6.131).

The geological interpretation of the section displayed in Fig. 6.131 leads to the following conclusions. The roots of the mud volcanoes are located above the basement ruptures from which gas is released. This is supported by the presence of a certain uplifting by the action of escaping gas in the layers between the basement and the beginning of the Cretaceous. At the same time, conditions favorable for the development of mud volcanoes are created at a depth of no more than 6–7 km (Bogdanov et al. 2006).

In the area of the volcanoes, there is a series of diagonal stress zones or disturbances traced to a depth of at least 16 km. There are more diagonal fractures (like the red lines in the figure) that go deeper into the crystalline basement rocks.

In the arches of the anticlines, where the volcanoes are developed, there is a series of cracks enabling the upward migration of hydrocarbons to the surface of the seabed. As a consequence, the diagonal fractures allow gas emanations to travel some distance (up to 4 km) from the volcano's vent. A series of inclined disturbances in the intervolcanic space can create favorable conditions for the formation of subhorizontal traps for gas and water. In this case, traps have been identified between the second and third volcanoes (44° 49′ 54″ to 44° 49′ 57″ N).

Most of the accumulations of gas and water in mud volcanic collectors occur at depth intervals of 1170, 1750–1900, 3200, 3500, and 4500–4700 m. Less confident indicators suggest also depths of 750, 1000, and 2200–2300 m. In Fig. 6.131, these pockets are marked by a blue color (map key no. 9). At the top of the figure, existing and supposed flares are flagged in red and black color, respectively (map key no. 10).

In general, this is a large mud volcanic field, but it is probably one of many that make up an entire mud volcanic region.

6.6 Mud Volcanoes of the Western Black Sea Depression

Group II includes mud volcanoes that are distributed in the Western Black Sea depression. This depression comprises oceanic to suboceanic crust overlain by sediment cover with a combined thickness of about 19 km. Here, the mantle is located at the shallowest depth within the Black Sea basin, and only 17 km separates the seabed surface from the Moho boundary. The western basin appears to be undergoing an overall upward state of flexure (with an undercompensated basin center and overcompensated basin flanks), and the eastern basin is experiencing a downward state of flexure (an overcompensated basin center and undercompensated basin flanks). This suggests that the level of necking involved in the deformation is deep in the west but shallow in the east (Spadini et al. 1996).

The Group II contains 11 mud volcanoes (Fig. 6.132).

It was studied largely at the end of the twentieth century, especially in the 1990s, during the voyages of research vessels "Gelendzhik" (1991–1996), "Ikhtiandr" (1993), "Professor Vodyanitskiy" (1994), "Kiev" (1994–1997), and others. The work was carried out by the "Yuzhmorgeologiya," Moscow State University, the National Agency for Marine Research and Technology of Ukraine, the organizations of the National Academy of Sciences of Ukraine, and partially within the framework of UNESCO projects (Limonov et al. 1992, 1994, 1995; Meisner et al. 1996; Ivanov et al. 1997; Basov and Ivanov 1996; Shnyukov et al. 1994a, b, 1995, 2005; Kruglyakova et al. 2004).

To Group II belong some of the largest mud volcanoes of the Black Sea (Table 6.5).

These mud volcanoes are characterized by (1) morphology, represented mainly by shallow cones sometimes rounded by fresh eruptions that demonstrate their modern activity, and (2) the structure of their eruptive channels, which usually have a straight, pillar-like shape but sometimes may be complicated by swellings or "chambers" at varying depths (e.g., MSU, Strakhov). These swellings are lithologically or tectonically caused and related to different stages in the growth of the mud volcano (e.g., MSU has a wide expansion in the channel in the Late Maikopian) (Fig. 6.133).

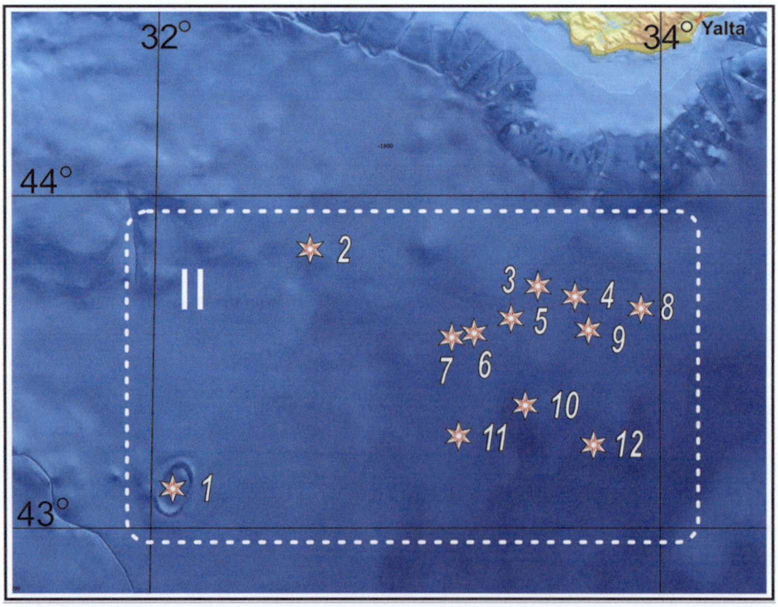

Fig. 6.132 Location of ring morphostructure (1) and mud volcanoes in the Western Black Sea basin: (2) Grigor'ev; (3) MGU; (4) Yuzhmorgeo; (5) Malyshev; (6) Kornev; (7) Goncharov; (8) Mantiynyy; (9) Strakhov; (10) Tredmar; (11) Vassoevich; (12) Kovalevskiy

Table 6.5 Coordinates and some characteristics of the mud volcanoes in the western part of the Black Sea

Mud volcano	Coordinates		Depth of the sea above the top of the cone, m	Cone height above the seabed, m	Diameter, km^2	Volume, km^3
	N	E				
Goncharov	43° 41.4′	33° 39.6′	2010	95	1 × 1	0.02
Grigor'ev	43° 49.2′	32° 37.4′	1825	60	2.0	
Kornev	43° 43.2′	33° 29.4′	1985	95	1 × 1	0.02
Kovalevskiy	43° 14.4′	33° 42.0′	2140	50	1.2	0.04
Malyshev	43° 38.0′	33° 21.9′	2027	75	1.5–2.5	0.08
Mantiynyy (Mantle)[a]	43° 39.6′	33° 09.2′	2100	?	?	
MSU	43° 31.8′	33° 07.2′	2100	80–90	4.0	0.001
Strakhov	43° 32.0′	33° 52.8′	2140	25–30	0.2	
Tredmar	43° 15.6′	33° 06.0′	2157	60	1.5	
Vassoevich	43° 53′	33° 12′	2156	40	1 × 1	0.01
Yuzhmorgeo	43° 31.8′	33° 12.0′	2066	110	2.5–4	

[a]Not expressed in the seabed relief
Modified after Meisner et al. (1996)

Part of the eruptive channel located between 7 and 5 km depth differs because of a configuration that is inversed compared to the general structure of horizons: in the arc there is a sagging of horizons accompanied by a decrease in frequencies and a change in polarity. The width of the zone in the section is 4–5 km, and the sagging reaches 150 ms, which is quite high for a velocity anomaly.

There are also "blind bodies" that resemble subvertical eruptive channels that do not reach the surface of the seafloor, e.g., near the Goncharov (Fig. 4.13) and Kovalevskiy (4.14) mud volcanoes described in Chap. 4.2.

The Goncharov mud volcano (Table 6.5) was discovered and sampled during a geological and geophysical survey conducted by scientists from the Geological Faculty of Moscow State University and "Yuzhmorgologiya" (Ivanov et al. 1989; Konyukhov et al. 1990). The regional profile passes the mud volcano at a considerable distance (Fig. 4.13). The cross section incorporates 9 s of signal registration and characterizes the Cenozoic part of the geological section up to the foot of the Maikopian series. An elevation of small dimensions (~4 km) with an amplitude of about 800 m is visible along the Maikopian sediment. The southeastern wing is steeper, suggesting the presence of a deep fault in the Mesozoic-Paleocene sediments. In the bottom relief, the Goncharov mud volcano has a weakly expressed caldera that is not visible in Fig. 4.13.

At a depth of −2.7 km, there is a "bright spot" (a bluish subhorizontal line in the image) accompanied by a shadow (screening) representing an anomaly of the reservoir type. At approximately −3.8 and − 4.2 km, there is a visible decrease in the frequency of the reflected waves (expressed by a wavy line in Fig. 4.13) typical for mineral deposits. Seismic data clearly show that the Goncharov mud volcano is confined to a tectonically active zone with signs of a fluidogenic channel of mantle hydrocarbons. In the core obtained at that mud volcano, gas hydrate is present (Kruglyakova et al. 2009).

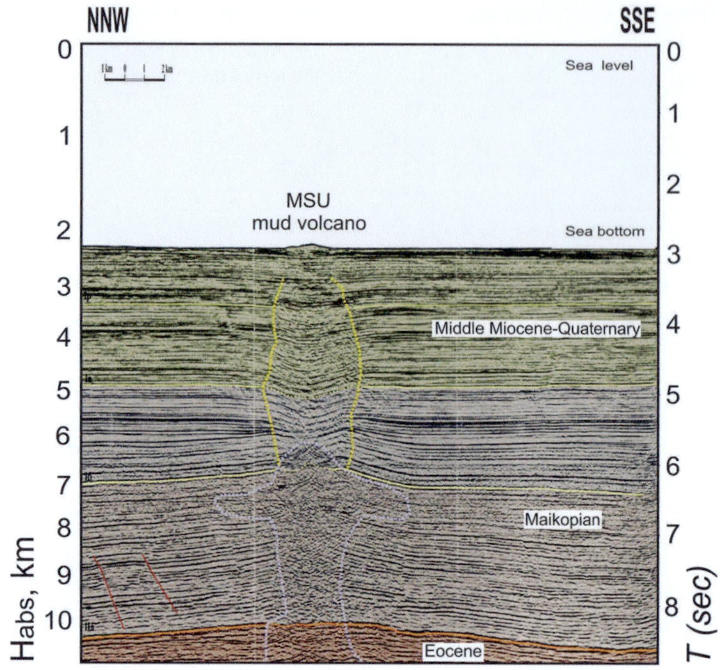

Fig. 6.133 Time seismic section near the MSU mud volcano. Swellings along the length of the eruptive channel are visible (Shnyukov and Netrebskaya 2016)

The Grigor'ev mud volcano (Table 6.5) was discovered in 1994 on board the R/V "Professor Vodyanitskiy." Its presence was indicated by an unevenness on the seafloor as well as the lithological content of cores and some geophysical anomalies (Shnyukov et al. 1995). The mud volcano was named after the now deceased participant in all our expeditions, A.V. Grigor'ev, and it is located on the northern slope of the Western Black Sea basin, appearing on the sea bottom as a depression with a diameter of 4–5 km with several cones up to 60 m tall and diameter about 2 × 2 km each.

Also observed were two gas flares (presumably) up to 60 m high. It was not possible to test the gases from the flares; therefore, complete confidence is lacking in the nature of the gas in the torches, although their images on the echograms differ sharply from those representing unevenness in the seabed.

Two gravity cores at St. 40–94 and St. 45–94 with coordinates 43° 49′ 15″ N; 32° 37′ 28″ E and 43° 49′ 20″ N; 32° 37′ 40″ E, respectively, were obtained within the Grigor'ev mud volcano. Their lithological logs are shown in Fig. 6.134. Down the cores, they are represented by Novochernomorian (New Black Sea) deepwater coccolithic mud, then by Drevnechernomorian (Old Black Sea) sapropelic mud, underlain by Neoeuxinian aleuritic-pelitic terrigenous mud, and then by Pre-Neoeuxinian mud volcanic breccia greater than 1 m in thickness.

The mud volcanic breccia is unique in having debris of clay and fragments of white marl up to 2 cm in size, as well as being gas-saturated (Core 45–94) and containing hydrotroilite in its lower part. The volcanic mud with its interlayers of hydrotroilite is black in color, thereby suggesting intense H_2S flow from an underbottom source during eruptions. The rock content within the breccia reaches 15% and is represented by Maikopian clays, occasionally argillite, glauconitic-quartz sandstones, and varied limestones. Pieces of hard Cretaceous clays and marls are present as well. The mud volcanic breccia has been recovered at many sampling stations around the mud volcano, having been scattered by bottom currents. The cores show that the Grigor'ev mud volcano was active prior to the deposition of Drevnechernomorian sediments, i.e., the Late

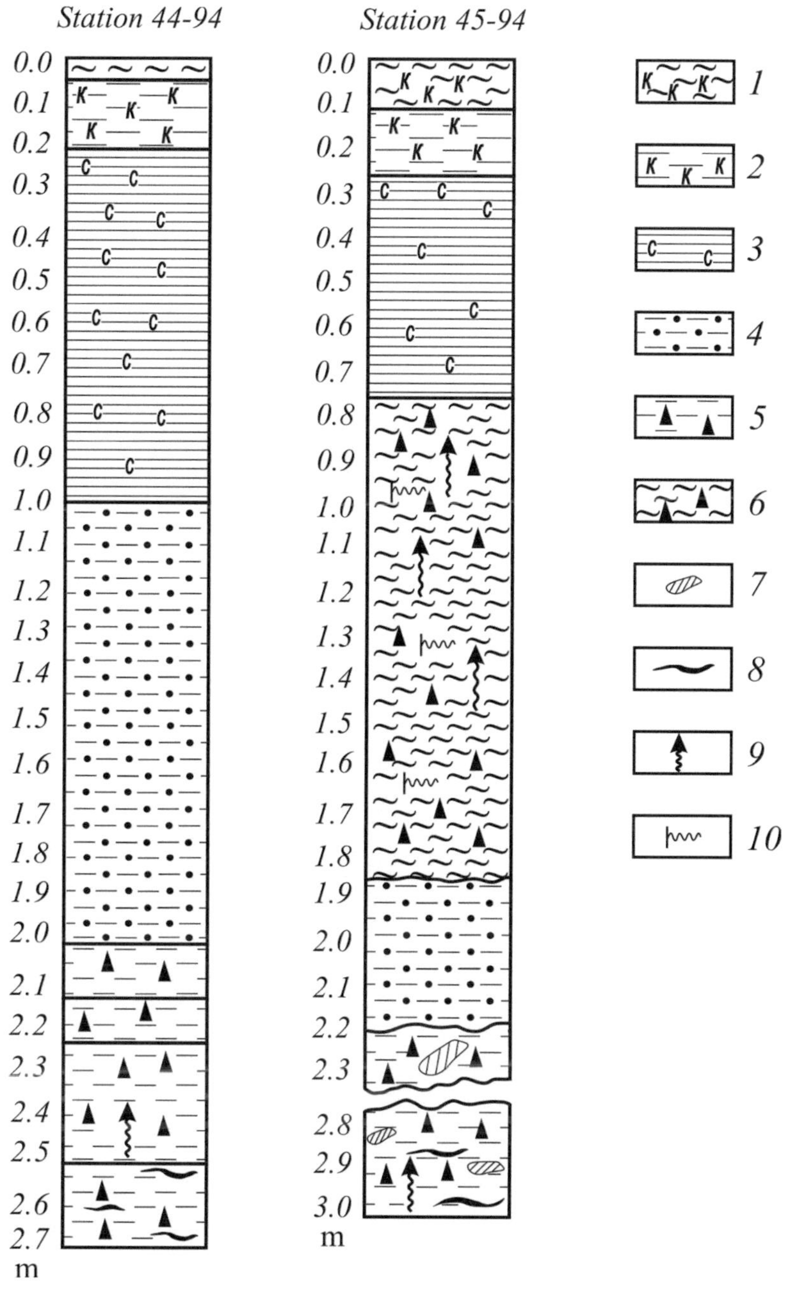

Fig. 6.134 Lithologic logs of two cores (44–94 and 45–94) recovered at the Grigor'ev mud volcano. Novochernomorian: (1) coccolithic liquid mud, (2) coccolithic thin-layered consolidated mud; Drevnechernomorian: (3) sapropelic mud; Neoeuxinian: (4) aleuritic-pelitic terrigenous mud; Pre-Neoeuxinian: (5) consolidated and schistose volcanic mud and breccia, (6) volcanic unconsolidated mud, (7) rock debris, (8) volcanic mud with interlayers of hydrotroilite, (9) gas-saturated areas, (10) sites of carbonization

Pleistocene. At present, its activity is insignificant and consists of small seepages.

The Kornev mud volcano (Table 6.5; Fig. 6.135) was first identified on the flat sea bottom south of Crimea by the work of "Yuzhmorgeologiya" in 1973–1974, and it was repeatedly confirmed between 1978 and 1983 (Tugolesov et al. 1985; Ivanov et al. 1989).

The time section (Fig. 6.135) reflects the entire sedimentary complex from the Mesozoic to the Quaternary, and there are no complications in the bottom relief. A "bright spot" anomaly at a time of 3200 ms (with subsequent shielding of the energy signal) can be interpreted as a hydrocarbon pool. The Kornev mud volcano corresponds to a local uplift with visi-

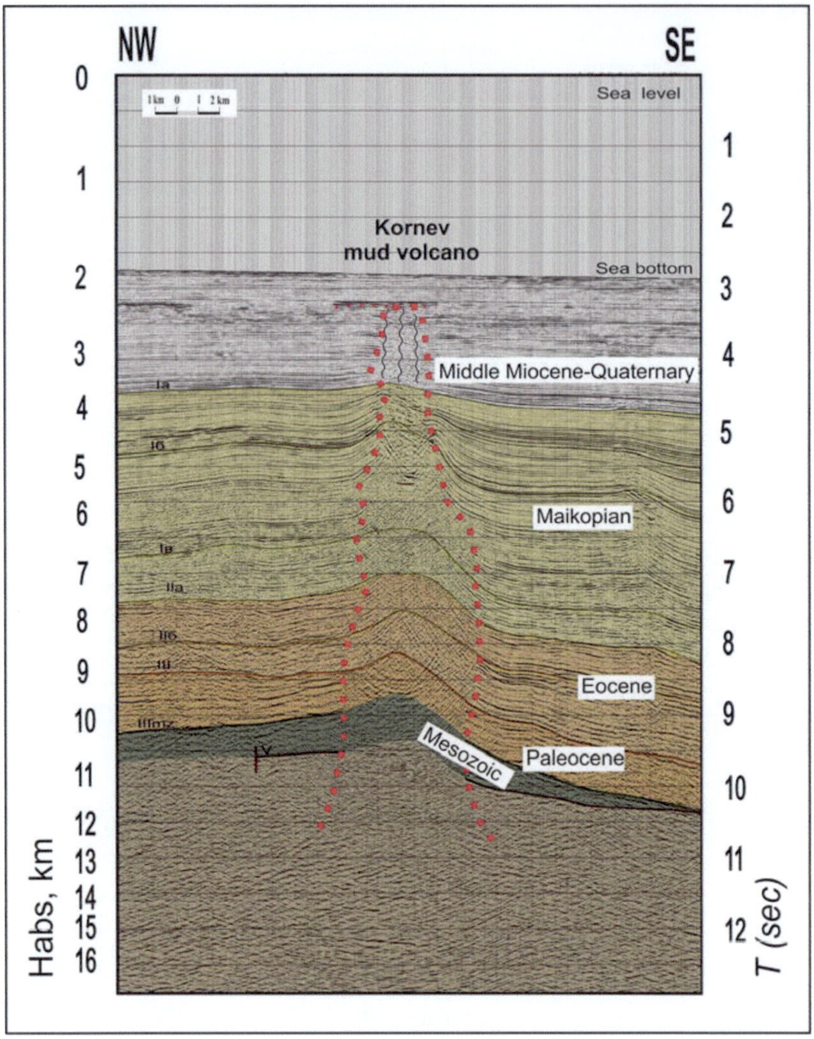

Fig. 6.135 Stratification of seismic horizons obtained by the common mid-point method (CMP) at a distance of 2.5 km from the Kornev mud volcano (Shnyukov et al. 2013)

ble dimensions of about 5 km; it was formed largely in the Early Miocene. Its southeastern wing is steeper, possibly because the process of folding was followed by a shift in the SE direction. The amplitude of the fold at the roof of the Oligocene (approximately 9 km in depth) is about 600 m. The termination of dynamic expressiveness of almost all reflecting boundaries of the sedimentary complex in the subvertical zone corresponding to the core of the fold suggests the likelihood that the uplifted geological section is strongly gas saturated due to the breakthrough of mantle fluids.

The sedimentary section within the Kornev mud volcano consists of two layers. The lower part in the cores is composed of mud volcanic breccia with a known thickness of 75 cm (Core TTR11-BS-313G, Shnyukov et al. 2013). It is overlain unconformably by thinly layered nanoplanktonic mud of Novochernomorian age. The breccia ranges from a light brown to a dark gray color and consists of non-rounded fragments of various sizes (2 cm maximum). The methane content of the sediments reaches 13 cm^3/kg.

Thus, this mud volcano is characterized by only one stage of activity, which terminated most likely in the Early Holocene. This stage of mud volcanism within a deep basin is unique and unknown so far beyond the Kornev mud volcano (Meisner et al. 1996).

The Kovalevskiy mud volcano (Table 6.5; Fig. 4.14) was discovered by "Yuzhmorgeologiya" in 1987–1990 together with the MSU, Vassoevich, and Yuzhmorgeo mud volcanoes. It is located in the deepest part of the Western Black Sea depression. Based on the time section obtained by CDP 200 km from the mud volcano, there appear to be no complications in the bottom relief. In the sedimentary section, an upwarping is visible within the channel at the 5.5 km level within the Maikopian sediment layer. The highest visible amplitude is about 250 m fixed at the Oligocene roof. There is a sagging of the horizons of the Upper Maikopian and above, probably due to an anomaly of low frequencies likely caused by gas saturation of the sediments, and as such, there is a typical pseudosyncline (Fig. 6.136).

Seismic data suggest that the Kovalevskiy mud volcano is a promising location for finding hydrocarbons within the geological morphostructure of the Cenozoic age; it is genetically related to tectonic movements and an inflow of hydrocarbons from the Earth's interior. This is confirmed by the pieces of gas hydrate that have appeared in cores recovered at this mud volcano.

Confirmation of present activity can be seen in the radial flows of mud volcanic breccia that have widened the structure up to 1200 m but are not overlain by recent chemopelagic sediments.

The Malyshev mud volcano (Table 6.5) was discovered in 1993 (Limonov et al. 1994). A sonar image and stratification of seismic horizons of the volcano is provided in Figs. 6.13 and 6.137, respectively.

The size of the Maikopian dislocated structures in the Western Black Sea depression rarely exceeds 8 km. Bottom relief on the regional profile appears without complications, and the upper part of the sedimentary complex up to 3.3 s does not reflect any lateral changes. Below 3.3 s, a subvertical zone of anomalous seismic reflection is visible. Above the Maikopian interval, an anomaly suggesting a reservoir is visible

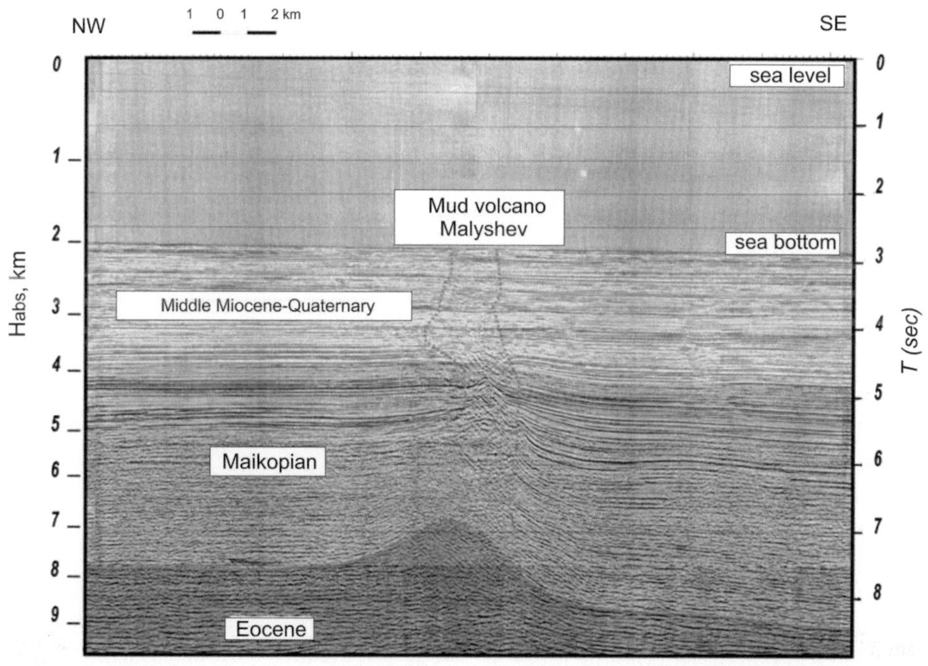

Fig. 6.136 Stratification of seismic horizons obtained by the CMP method in the area of the Malyshev mud volcano. The lateral zone of sharp change (in the center of the volcanic channel) in the characteristics of the wave field is highlighted and expressed by a loss of coherence, polarity change, and weakening of the wave dynamics (Shnyukov and Netrebskaya 2016)

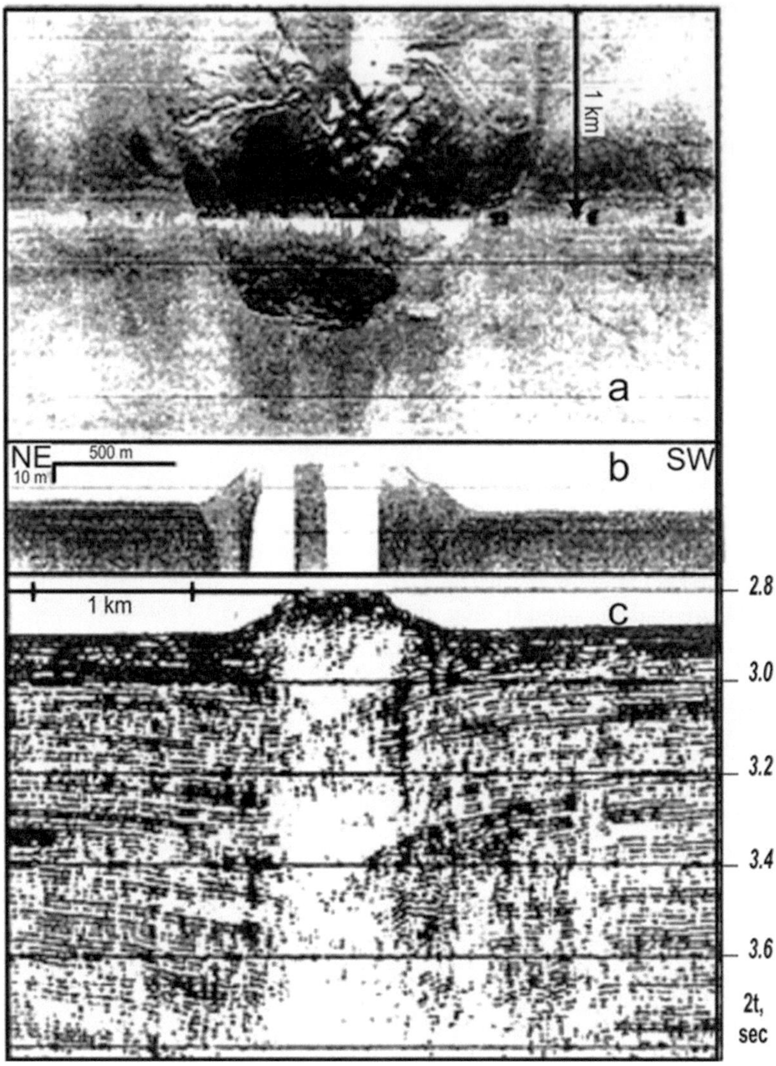

Fig. 6.137 The Malyshev mud volcano (Modified after Limonov et al. 1994). (**a**) MAK-1 deep-towed sidescan sonar image with its corresponding sub-bottom profilograph line (**b**) and seismic section (**c**) parallel to the MAK-1 track (Dimitrov 2002, see his Fig. 7 for location). Clearly seen are the dome-like structure of the mud volcano, numerous mud flow patches on the sonogram, and several bright spots around the well-delineated feeder channel on the seismic section

at around 3 km; the lower phase of the wave sags and registers with some delay. Structural complications in the behavior of the layers are observed only near the roof of the Maikopian sediments and older. The section reflected in the CDP (common depth point) profile is characteristic of positive structural forms complicated by a long-lived fault-based disturbance. Its maximal amplitude at the foot of the Maikopian series is estimated at 600 m. Alternating renewal of the fault, especially at the end of the Paleogene-beginning of the Neogene, formed a local uplift, most likely a block structure with a lowered southeastern wing. Signs of diapirism in the Maikopian interval are not evident, and the thickness changes regionally and gradually, without a noticeable increase in the core of the fold. The structure of the Eocene-Paleocene and Mesozoic intervals within the sedimentary section relates to tectonic disturbances most likely of a deep origin.

The sedimentary section within the Malyshev mud volcano consists of three layers (Shnyukov et al. 2013, Fig. 2.29). The lower part within core BS-260 (220–70 cm) begins with a roof composed of mud volcanic breccia, then Neoeuxinian (?) hydrotroilite mud (0.7–0.15 cm), and Novochernomorian coccolithic mud (0.00–

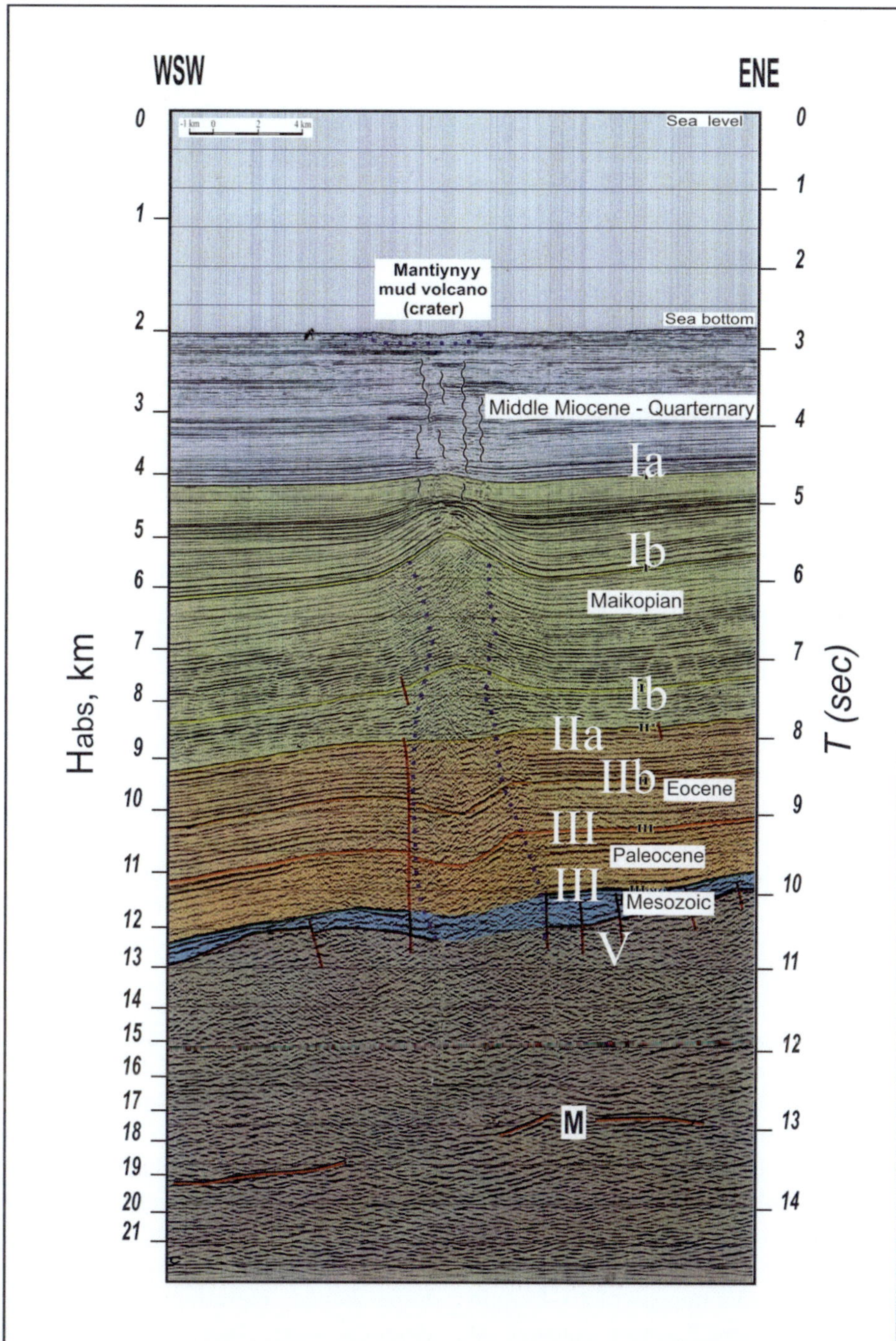

Fig. 6.138 Stratification of seismic horizons obtained by CDP within the Mantiynyy mud volcano

0.15 cm) with a clear unconformity at its foot. The presence of Holocene sediments at the top of the breccia may indicate the end of, or a pause in, mud volcanic activity during the Late Pleistocene. The content of methane in the sediments reaches 3.5 cm^3/kg.

The Mantiynyy (Mantle) mud volcano (Table 6.5, Fig. 6.138) was previously designated on the seismic profile of Meisner and Tugolesov (2003) as being buried. Its appearance on the sea bottom has not yet been described.

In-depth study of the CDP seismic section in an attempt to illustrate the wave field stratification has been performed by L.A. Pobedash (Shnyukov and Netrebskaya 2013). A geomorphologic anomaly appears in the bottom relief, presumably representing the mud volcano's crater. In the interval of the Middle and Upper Neogene, the structural complications of the reflecting horizons show a visible amplitude not exceeding 50 m, but within the mud volcano, there is a subvertical zone of wave field anomalies that sharply distinguishes this area from lateral ones.

The Mantiynyy mud volcano is extremely interesting as a kind of fluid plume, piercing almost 10 km through the sedimentary cover, including Quaternary-Miocene (3 km), Maikopian (5 km), Eocene (2 km), Paleocene (1.2 km), and Mesozoic (0.25–0.75 km), and then penetrating deep into the top of the basement, supposedly up to the Moho surface (Starostenko et al. 2010). This discovery figured prominently in the naming of the mud volcano.

It is quite possible that this mud volcano, as well as other similar structures, serves as a channel for the intrusion of hydrocarbons associated with deep faults or injection zones (Dmitrievskiy and Valyaev 2010).

The idea of the connection of mud volcanoes with the mantle and deeper horizons of the crust is sometimes expressed by specialists engaged in geochemistry, but, unfortunately, they are usually only assumptions or general provisions.

The MSU mud volcano (Table 6.5; Fig. 6.133) has a regularly rounded shape and gently sloping (2–3°, sometimes up to 6°) cone with a flat top complicated by a circular rampart of 10–15 m in height and a crater about 2 km in diameter. At least three eruptions of different ages and extent are clearly visible on MAK-1 high-resolution echograms. They are noted from the center of the crater to a distance of about 2 km to the north of the mud volcano, where there is a relief structure expressed in the form of a hill. This structure is interpreted as a young gryphon genetically related to the hearth of the MSU mud volcano. This mud volcano is complicated by disorders and roughness created by stiffened and dense mud volcanic breccia as well as by the presence of small gryphons (Shnyukov et al. 2015).

A high-resolution echogram performed at a frequency of 100 kHz revealed that the structure of the crater is complex; it contains two smaller craters with diameters of 500 and 350 m, respectively, each corresponding probably to younger eruptions. These craters are located in the main structure partially overlapping each other (denoted by numbers 3 and 4 in Fig. 6.139).

In its structural plan, the MSU mud volcano is confined to a local structure of 5 × 8 km at the foot of the Middle Oligocene (horizon at the level of ~8 s), which in turn is an element of the inclined fold elongated in the sub-meridional direction with the submerged hinge to the south (Fig. 6.133).

The Strakhov mud volcano (Table 6.5; Fig. 6.140) is located approximately 30 miles northeast of the MSU mud volcano. It was discovered in 1989 by scientists from Moscow State University during a cruise on the R/V "Feodosiya." The name of this mud volcano was given in honor of academician N.M. Strakhov for his very significant contributions to the geological and geochemical investigation of Black Sea bottom sediments.

In the CDP time series (Fig. 6.141), passing from the Strakhov mud volcano in the western periclinal at a distance of 4 km, there is a very shallow elevation in the bottom relief.

In the upper 500 m of the geological section, no anomalies have been identified in the mid-range of the wave field. From a depth of 2.7 km up within the structure of the geological section, there is a bending that is typical of a dome-shaped elevation with a subvertical anomaly zone; here, the waves lose their expressiveness and coherence until they acquire a completely chaotic character. Approaching the apical part of the volcano, the wave pattern seems to be more contrastive and spread over the entire section toward the bottom. The local Maikopian uplifts of 4 km in height are divided by a calm field with a length

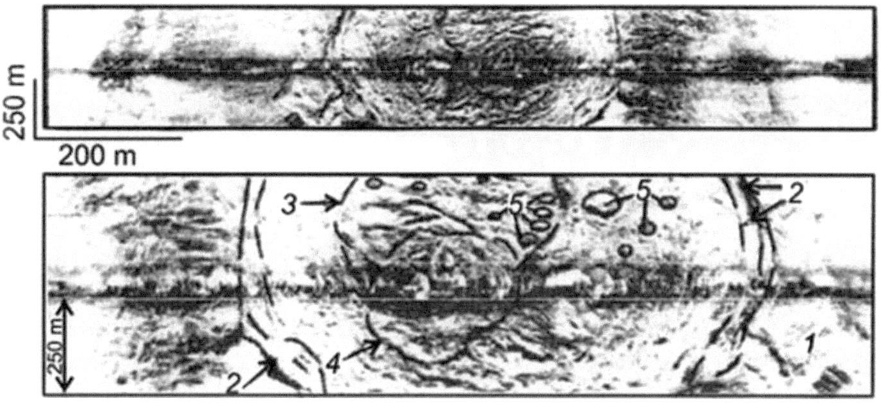

Fig. 6.139 Structure of the MSU mud volcano's crater (Ivanov et al. 1997). Key: (1) flows of mud-volcanic breccia; (2) annular rim of the main (more ancient) crater; (3) remains of the annular rim of a younger crater partially destroyed by eruption; (4) annular rim of the youngest crater; (5) ring structures that include local mud-volcanic buildups and places where gas was released to the surface

of 8–10 km. For all structural anomalies, no subverticality and displacement of inflection points in plan for different levels have been observed. The amplitude of the inflection is the highest at the roof of the Oligocene, while the Lower and Middle Maikopian reveal a greater thickness in the cores of the structures, which is a sign of clay diapirism.

The upper 96 cm beneath the sea bottom are represented by horizontal Holocene and Upper Pleistocene (Neoeuxinian) turbidites, which are typical of the Western Black Sea basin (Konyukhov et al. 1990). These sediments are underlain by mud volcanic breccia. The methane content of the bottom sediments reaches 93.3 volume % (Kruglyakova et al. 2009).

The Tredmar mud volcano (Table 6.5; Fig. 6.142) was discovered in 1991 and studied in greater detail in 1993 by two geological-geophysical expeditions on board the R/V "Gelendzhik" within the framework of the UNESCO international project "Training Through Research."

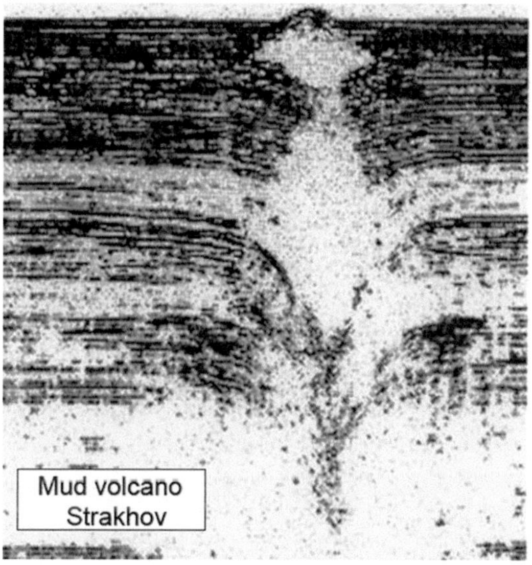

Fig. 6.140 Seismic acoustic recording of the Strakhov mud volcano providing a good view of the internal structure of the supply channels, which are characterized by the presence of large-scale inhomogeneous formations of the dilatant type in which viscosity increases with the rate of shear strain (Gonchar et al. 2004)

According to Kenyon et al. (2002, 20, 21, Fig. 18), "It is a mound 60 m high, and 1.5 km wide at the base. A small depression is visible at the surface of the eastern side of the mud volcano corresponding to the rim. Around the feeder channel depletion of reflectors can also be observed in the same way as around the Kovalevskiy mud volcano. The feeder channel is about 2 km wide and shows small variations in width with depth, a slight widening down section is also observed. This seems to be a common feature for all mud volcanoes studied in this area. A bright spot related to a fault system was observed at the depth of 3.3 s on the eastern side."

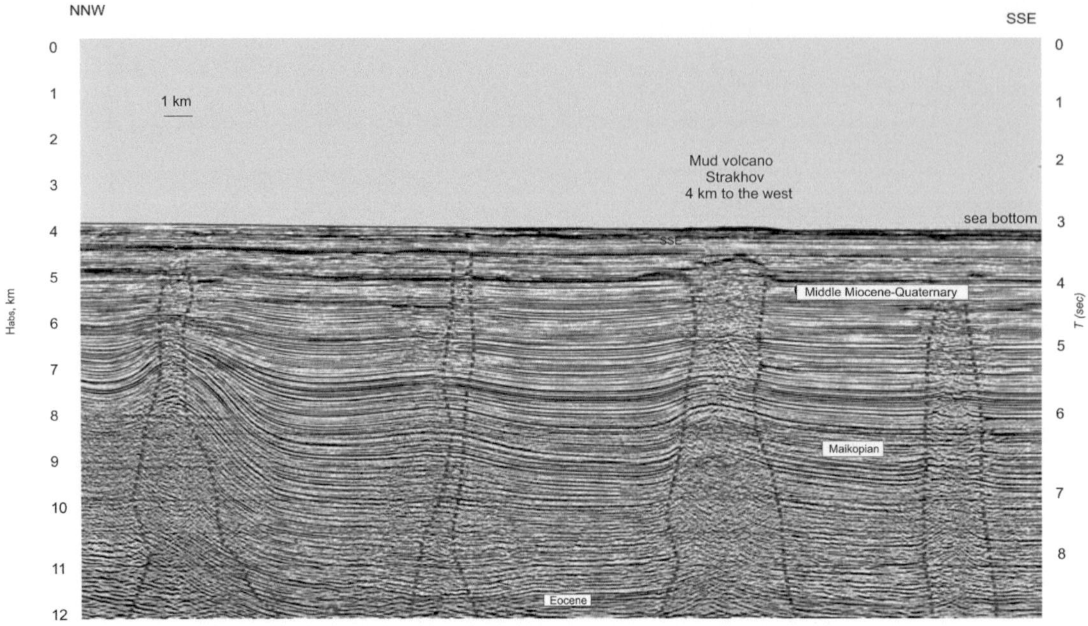

Fig. 6.141 Characteristics of the Maikopian diapirism within the Strakhov mud volcano on the CDP time series (Shnyukov and Netrebskaya 2016)

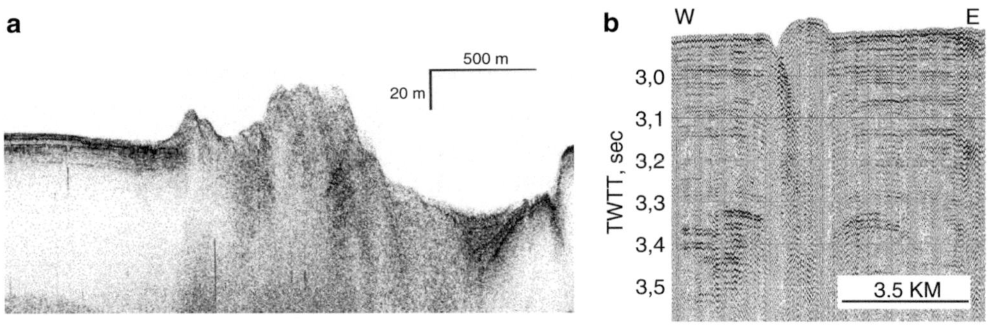

Fig. 6.142 (**a**) Profilogram of the Tredmar mud volcano (Gonchar et al. 2004) and (**b**) fragment of seismic line PS-294BS across the Tredmar mud volcano (Kenyon et al. 2002). The depressions in both images appear to correspond with each other

The Tredmar mud volcano differs morphologically from all the others. In its bottom relief, it is partially expressed as a negative structure surrounded by an annular rim indicating that breccia is highly saturated with water and has an elevated liquescency (Ivanov et al. 1997). The rim possibly indicates that the volcanic eruption had an explosive character. The Tredmar volcano is in all probability active at the present time.

The northern and northeastern parts of the mud volcano have collapsed and subsided due to surface outflow of sedimentary material from the Earth's interior. This has led to the development of a depression (recessed syncline) that is now filled with a well-stratified alternation of normal pelagic and mud volcanic sediments.

The Vassoevich mud volcano was discovered in 1988 during geological-geophysical student training on board the R/V "Feodosiya." It was named after the famous petroleum geologist N.B. Vassoevich and is located 30 km south of the MSU mud volcano.

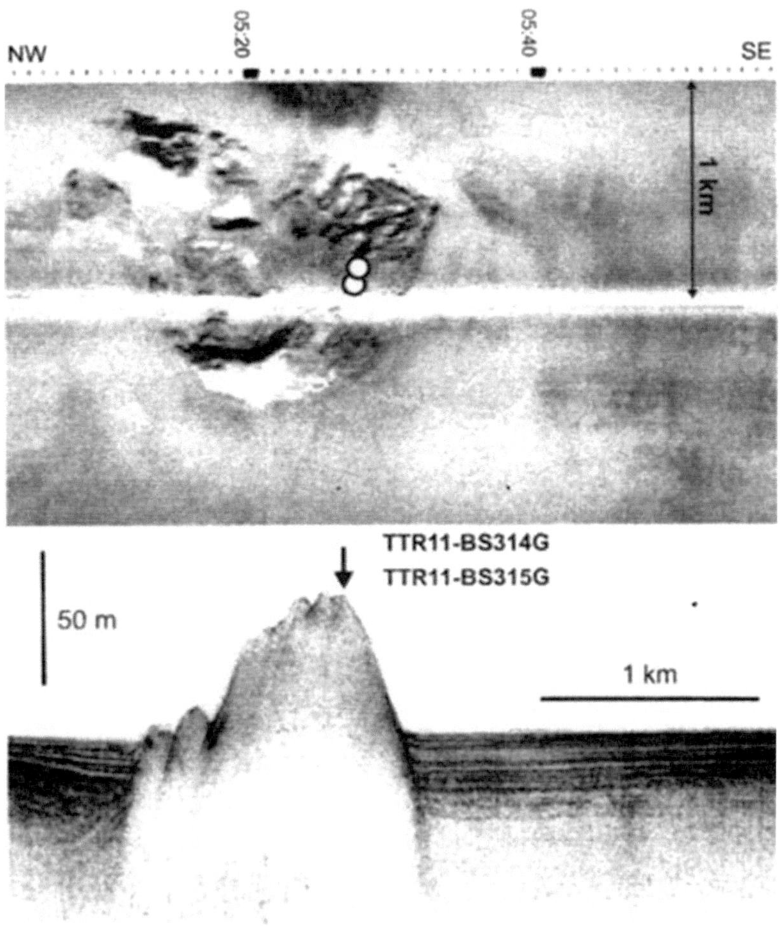

Fig. 6.143 A fragment of the sonogram and bottom profile through the Vassoevich mud volcano (Kenyon et al. 2002)

The Vassoevich mud volcano (Fig. 4.12) can be seen on the sonograph as an area of variable backscatter with an irregular outline (Fig. 6.143).

Kenyon et al. (2002, 24) provide the following description: "On the southwestern side of the volcano an escarpment is seen as a high backscattering semicircular feature; 415 m wide with a talus of collapsed sediments expanding for about 1 km to the north. On the subbottom profiler the mud volcano is represented by typical acoustically transparent facies, which is 2 km wide and 60 m high. The contact with surrounding hemipelagic sequence is sharp and on the eastern side layers are bent upwards at the contact, suggesting that extrusive processes took place during the mud volcano's formation. On the western side the hemipelagic sequence dips toward the mud volcano as a result of subsidence. The morphology of the edifice is affected by the collapse of its western flank with the remaining eastern flank being 20 m higher. The undisturbed slope of the eastern flank is 9.5° and the collapsed slope is 4.1°."

Cores have recovered mud breccia overlain by hemipelagic sediments with fragments of lithified carbonate crust in them.

The Yuzhmorgeo mud volcano was first identified by "Yuzhmorgeologiya" during seismic profiling in 1973–1974. It is located 10 km east of the MSU mud volcano, and both volcanoes are comparable in their physical dimensions (Table 6.5). The Yuzhmorgeo mud volcano reveals an asymmetric structure: the eastern slope is shallower than the western one probably because mud flows have been propagated mainly in the eastern direction (Fig. 6.144).

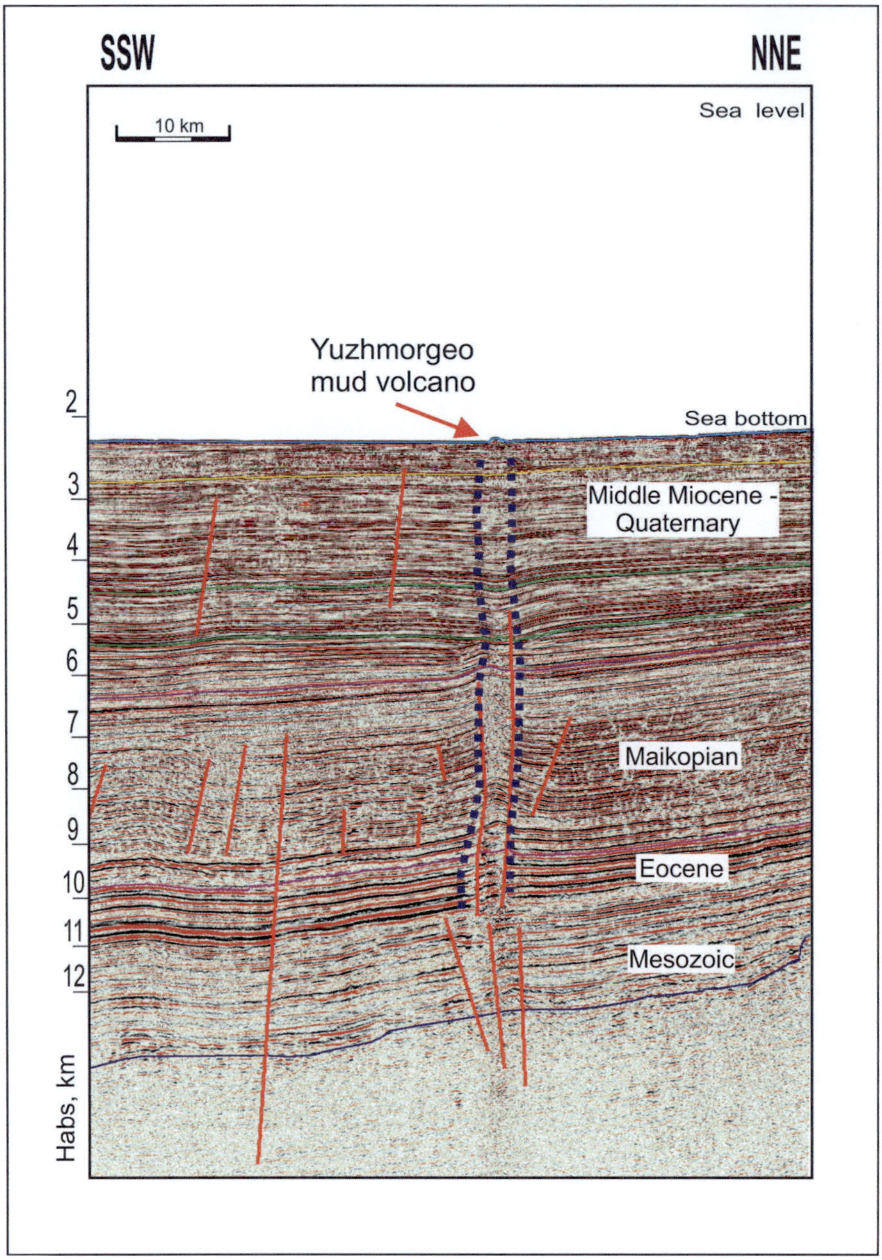

Fig. 6.144 CDP time section in the area of the Yuzhmorgeo mud volcano (Shnyukov and Netrebskaya 2013). Note that the horizontal scale is five times smaller relative to the vertical scale

The younger generations of flows (darker lineations of the high-resolution CDP time section shown in Fig. 6.144) have spread in a northeastern and southwestern direction. In the central part of the volcano, the structure of the flow is clearly visible with characteristic excrescences reflecting the outlines of the flow.

In its deep structure, this mud volcano is quite unusual. It has a subvertical, tectonically active eruptive channel created by a local breakthrough of fluid from great depths (Fig. 6.144).

The Yuzhmorgeo mud volcano is controlled by two factors: (1) a subvertical zone of tectonic complications in the Cenozoic horizons that is 2 km in width and (2) the block structure of the Mesozoic. In the Paleogene range, there is a flexure that flattens out in the upper Maikopian. Above, it manifests itself as a typical "pseudosyncline" expressed as a sharp, high-speed anomaly. A decrease in velocities is due to the high gas saturation of the sediments (Shnyukov, Netrebskaya 2013).

In the 1990s, during the course of a research cruise on board the R/V "Kiev," a ring morphostructure was discovered at a water depth of 2050–2080 m in the center of the Western Black Sea basin (Shnyukov, Schiptsov 1996). The ring morphostructure was recorded by two crossing profiles with initial and final coordinates 43° 08′ N; 32° 16′ E and 43° 12′ N; 32° 06′ E, respectively (Fig. 6.145).

In the 62nd voyage of the R/V "Professor Vodyanitskiy," it was established that earlier exploratory work by the R/V "Kiev" recorded the southwestern and northeastern marginal ledges of this morphostructure, which had a diameter of 6.5 miles (Shnyukov et al. 2005) (Fig. 6.146).

The morphostructure was measured along a series of multidirectional profiles that crossed it from southwest to northeast (Traverse 15), east to west (Traverse 16), northwest to southeast (Traverse 17), and the central part of the basin to the east (Traverse 18).

In addition, a number of small-scale spatial measurements were performed along the morphostructure's edges. It was discovered that the marginal ledges of the morphostructure are in fact sharp ledges of the seabed. They rise above

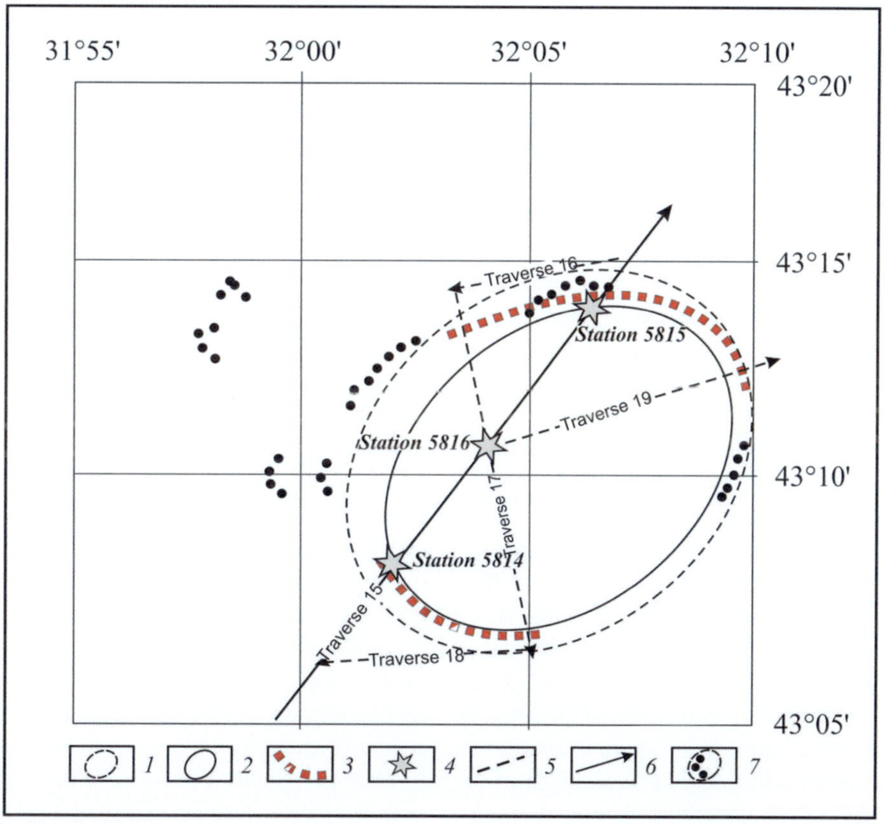

Fig. 6.145 Location and diagram of the ring morphostructure at the bottom of the Western Black Sea basin. Key: (1) outer boundaries; (2) supposed inner boundaries; (3) defined boundaries; (4) sampling stations; (5) survey routes; (6) hydroacoustic profile line; (7) revealed ledges

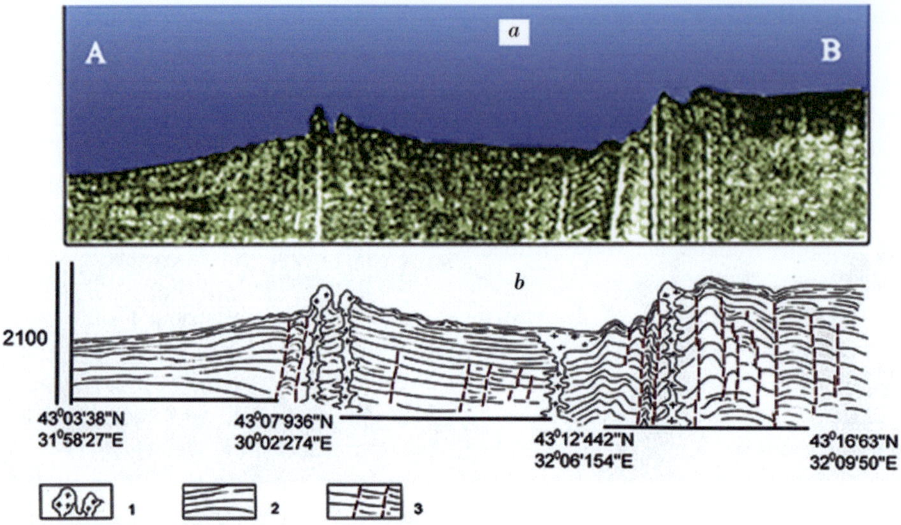

Fig. 6.146 Acoustic profile through the ring morphostructure: (a) profile, (b) interpretation of the profile. Key: (1) mud volcanic breccia, (2) layered bottom sediments, (3) tectonic dislocations

the abyssal plain, on average 30 m, from a depth of 2091 to 2060 m. The morphostructure takes the shape of an ellipse elongated from southwest to northeast, and its long and short axes measure 6.5 and 6 miles, respectively (Fig. 6.145). There are ledges that ascend gradually and stepwise toward the central part of the morphostructure from 2075 to 2045 m in water depth.

The southwestern ledge of the morphostructure resembles a mud volcanic hill with a clearly expressed caldera. Its foot manifests hilly relief that continues into the sea plain.

On the northeastern boundary (43° 12.07′ N; 32° 10.322′ E) of the morphostructure, there are five clearly expressed ledges with subvertical walls of disjunctive displacement traced in the section up to 70 m. The northwestern boundary has similar ledges that ascend stepwise from a depth of 2089 m to 2072 m in water depth. The relief of the lower central part (43° 13.184′ N; 32° 03.3469′ E) of the morphostructure is complicated by a series of low (up to 5–10 m) hills located at a depth of 2086 m from which they rise to 2072 m. In general, the morphostructure is one of many similar structures of the diapiric zone that have developed here.

Three 9 m gravity cores were collected on the morphostructure for the purpose of lithological study and a search for mud volcanic products. Two of them were recovered on the edges and one in the center of the morphostructure (Fig. 6.147).

The upper layer of terrigenous mud is present everywhere and reveals no specific features. The lower layer of hydrotroilitic mud recovered at St. 5814 wedges out toward St. 5816. This peculiarity of the core recovered at St. 5814 was possibly created by the intermittent and sufficiently powerful local accumulation of fine particulate material at the time of hydrotroilite deposition and, probably, by the additional intake of hydrogen sulfide bound from the hydrotroilite deriving from sources at the bottom. Fine clayey silt in the sections is layered, sometimes lumpy, but typical mud volcanic breccia is not found. We probably did recover mud from the volcano's vent facies, but these are difficult to distinguish from ordinary muds, and thus there can be no certainty about this. Cores from St. 5814 and St. 5815 are characterized by very high gas saturation. Similar sediments enriched with hydrotroilite were encountered by us in some large mud volcanoes.

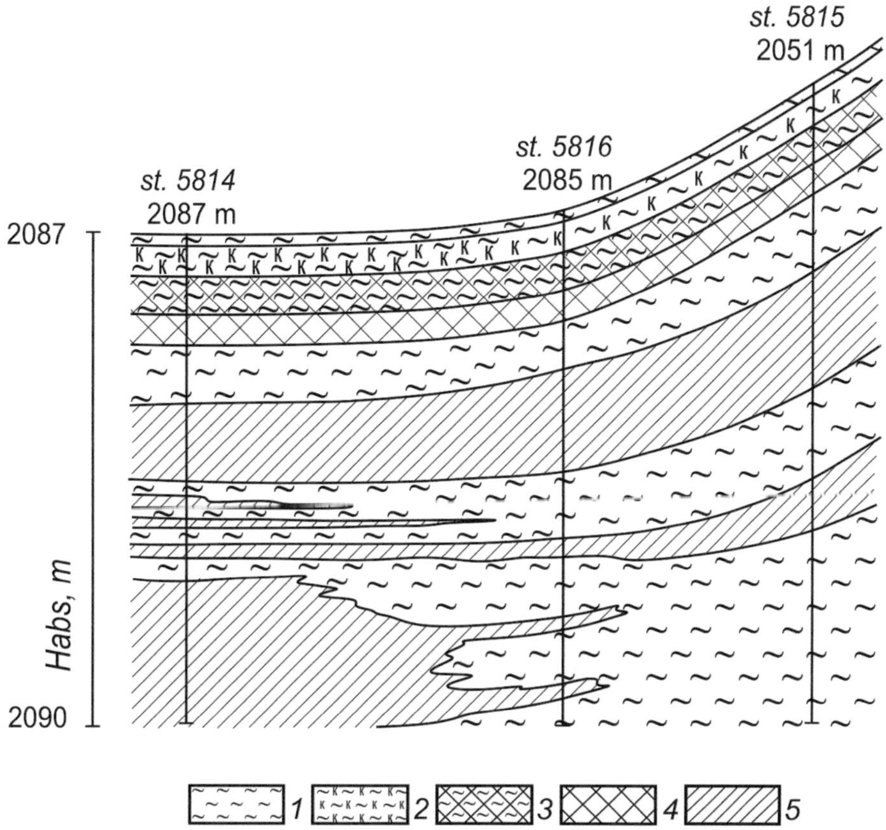

Fig. 6.147 Lithological section through the ring morphostructure. Key: (1) terrigenous mud; (2) coccolithic mud; (3) sapropelic mud; (4) sapropel; (5) hydrotroilitic mud

In general, a number of indirect signs enable us to assume a mud volcanic origin for this morphostructure. They include a similarity of relief with the Bulganakskiy mud volcanoes (Shnyukov et al. 1986), specific stratification and dislocation of the sedimentary strata, the presence of hydrotroilite, and a high gas saturation of the mud.

It is very likely that the morphostructure was formed as a result of repeated manifestations of powerful mud volcanoes scattered over the area and along its marginal parts (ledges). Intensive degassing from recovered sediments also indicates that the bottom of the morphostructure represents a so-called gas swamp, which is typical for mud volcanic fields and gas-saturated muds of the Black Sea. In such areas, the seabed is composed mainly of liquid, highly gas-saturated mud, and this can probably explain the bent character of the central part of the morphostructure, which is comparable to the huge Bulganakskaya recessed syncline (Fig. 6.148).

To the northwest of the ring morphostructure, a new satellite morphostructure of smaller size was detected that is connected to the elliptical structure from the northwest.

Clearly, the study of mud volcanoes in the Western Black Sea basin has been insufficient. It is very probable that many new mud volcanoes will be discovered there.

6.7 Mud Volcanoes of the Eastern Black Sea Depression

This region is virtually devoid of mud volcanoes, if we use only the existing information obtained during previous research.

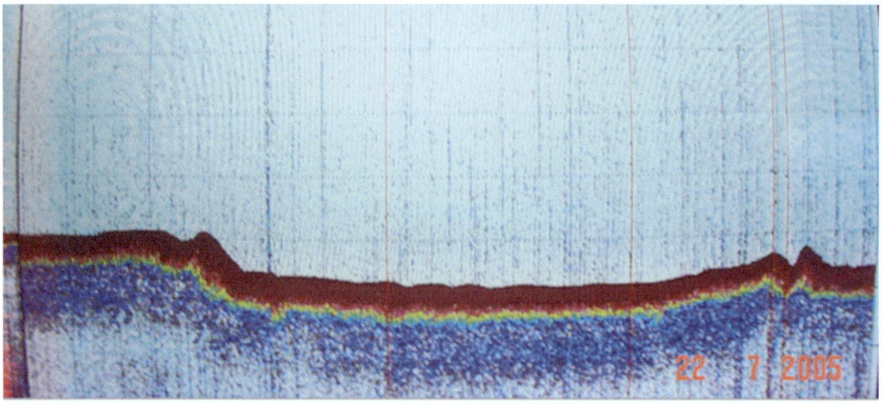

Fig. 6.148 Profile of measurements across the ring morphostructure (Traverse 17)

The existence of mud volcanoes within the Eastern Black Sea depression was proposed by Kovalevskiy (1960). During hydrographic work in the 1930s, a conical elevation with gentle slope rising 340 m above the flat seabed was discovered at a water depth of 1833 m; it lay along the line of the 38th meridian, 175 km south of the Shugo mud volcano on the Taman Peninsula. Near this elevation, the water depth was 2170 m. This elevation was defined by Kovalevskiy as a volcanoid, i.e., a mud volcano. However, the views of Kovalevskiy were somewhat vague because he combined mud volcanoes with pyrogenic ones, i.e., real volcanoes.

In 1989, Kovalevskiy's data were checked by an expedition on board the R/V "Mikhail Lomonosov." A multi-hour polygon with a square tract 10 miles on a side was executed. It would have been simply impossible to miss an elevation of 300 m on the seabed, but no hill could be found on the sea bottom. The search was led by an experienced navigator, L.I. Mitin. Two variants of an explanation were discussed: either poor navigation (because in the 1930s, geolocation was less accurate) or erosion of the hill (if it was composed of comparatively semiliquid breccia).

More recently, two presumed mud volcanoes have been described by geologists of "Yuzhmorgeologiya" along the "Blue Stream" gas pipeline, but due to problems related to work in Turkish waters, further study of these manifestations was not carried out.

According to Andreev et al. (2006), who refer to V.V. Gubenkov, an elevation with a diameter of 8 km is located within the Eastern Black Sea basin on the meridian of the Kerch Strait. It is interpreted as an "elevation of gas swelling" with a 1.2 km amplitude of rise. A mud volcano named Dwarf is confined to the center of this elevation.

As can be seen, the Eastern Black Sea depression is far poorer in mud volcanoes compared to the Western Black Sea basin. This impression may result from insufficient study of the region, but geological reasons are also possible.

6.8 Mud Volcanoes of the Sorokin Trough

The Sorokin Trough is about 150 km long and 50 km wide, and it represents a structural depression along the southeastern margin of the Crimean Peninsula in the Eastern Black Sea Basin at water depths of 800–2200 m. Structurally, the Sorokin Trough is considered to be a foredeep of the Crimean Mountains and belongs to a system of Oligocene-Miocene troughs, such as the Kerch-Taman Trough and the Indolo-Kuban Basin. It occupies the lower part of the Crimean continental slope and the adjacent lower part of the Crimean continental slope and its foot. In the south and southeast, the Sorokin Trough is bordered by the buried Cretaceous-Eocene Shatskiy Ridge and Tetyaev Rise (Fig. 6.149).

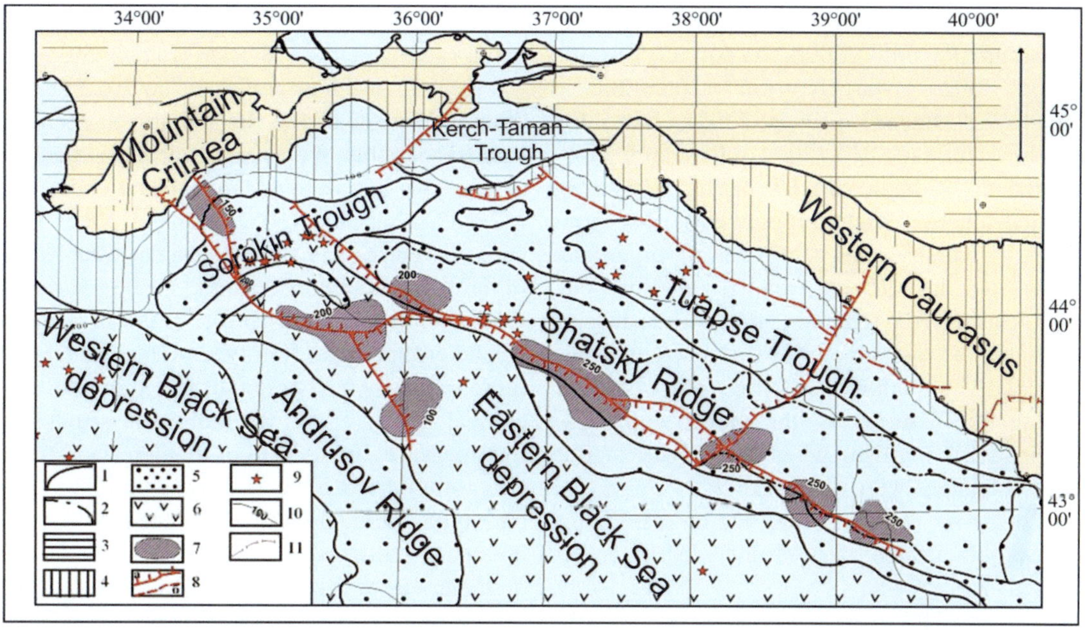

Fig. 6.149 Diagram of the northeastern part of the Black Sea region. Key: (1) borders of tectonic structures; (2) boundaries of the arch of the Shatskiy Ridge; (3) Epihercine Scythian Plate; (4) Kimmerian and Alpine mountain-folded structures; (5) Epihercine Transcaucasian-Pontian Plate; (6) post-rift Cenozoic depressions; (7) magnetic anomalies (values in Teslas); (8) faults; (9) mud volcanoes; (10) isobaths; (11) state border of Russia. (Compiled by Almendinger 2011)

The Sorokin Trough was formed in the Oligocene with the Crimean-Alpine folding and is an extremely interesting area for mud volcano development.

In terms of its relief, the Sorokin Trough is a shallow abyssal plain. Its flat bottom undergoes insignificant dismemberment in places where it intersects with continuations of underwater canyons on the continental slope and in areas manifesting gas volcanism.

In the Sorokin Trough, gas hydrates have been recovered from several mud volcanoes (Ginsburg et al. 1988, 1990; Bouriak and Akhmetzhanov 1998; Ivanov et al. 1996; Kenyon et al. 2002; Bohrmann et al. 2003). The gas hydrates are thought to form local accumulations; they are controlled by mud volcanoes and can be related to zones of intensive fluid flow upward toward the seafloor (Ginsburg et al. 1988, 1990; Bouriak and Akhmetzhanov 1998). Evidence for extensive fluid flow to the seafloor comes from acoustic anomalies in the seismic data, the formation of mud volcanoes, the geochemical composition of high gas-saturated sediments, and the light carbon isotope content of carbonates (Ivanov et al. 1996). In the Black Sea, gas hydrates are stable below a water depth of 700 m (Ginsburg et al. 1988, 1990).

There are two main sedimentary units in the Sorokin Trough. The lower unit comprises the upper part of the Maikopian and Pliocene sediments. The upper unit consists of Quaternary sediments (Woodside et al. 1997). The Maikopian deposits are particularly thick, up to 4–5 km. They are crumpled into brachyanticlinal linear folds with signs of tipping toward the south, i.e., toward the deep Black Sea basin (Tugolesov et al. 1985).

The area of the Sorokin Trough has been studied during dozens of research cruises using various R/Vs—e.g., "Mikhail Lomonosov" (1989), "Akademik Vernadskiy" (1989, 1992), "Ikhtiandr" (1992–1993), "Gelendzhik" (1991–1996), "Kiev" (1995–1997), "Professor Vodyanitskiy" (1994, 2000–

2004), "Professor Logachev" (2001, 2005), "Vladimir Parshin" (2006, 2008), "Le Suroît" (2001, 2002), "Mare Nigrum" (2005–2008), "Knorr," "Meteor" (2005, 2007), and "Le Marion Dufresne II" (2004). Research conducted on Russian and Ukrainian R/Vs was carried out by "Yuzhmorgeologiya," Moscow State University, and a number of organizations from the National Academy of Sciences of Ukraine. All together, these expeditions collected a vast amount of important data. Especially valuable were the data obtained during the course of the multi-year R/V "Gelendzhik" cruises when a majority of the mud volcanoes were discovered.

Twenty-six mud volcanoes are known within the Sorokin Trough; they are included in Group III (Fig. 6.126). Twenty of them have been studied in varying degrees of detail (Fig. 6.150).

Acoustic profiling performed in 2010 during the 15th voyage of the R/V "Maria S. Merian" located a number of new mud volcanic structures (Fig. 6.150) that were assigned the numbers M1, M2, M3, M4, and M10 (Bohrmann 2011). Gas emission during the voyage was not established for any of these mud volcanoes.

The mud volcanoes in the Sorokin Trough can be grouped into four types. The first type is represented by fairly regular, round in plan and conical in cross section, geological structures that measure 700–800 m in diameter and from tens to hundreds of meters in height. The second type includes mud volcanoes with pronounced calderas of various shapes and diameters up to 1 km formed as a result of col-

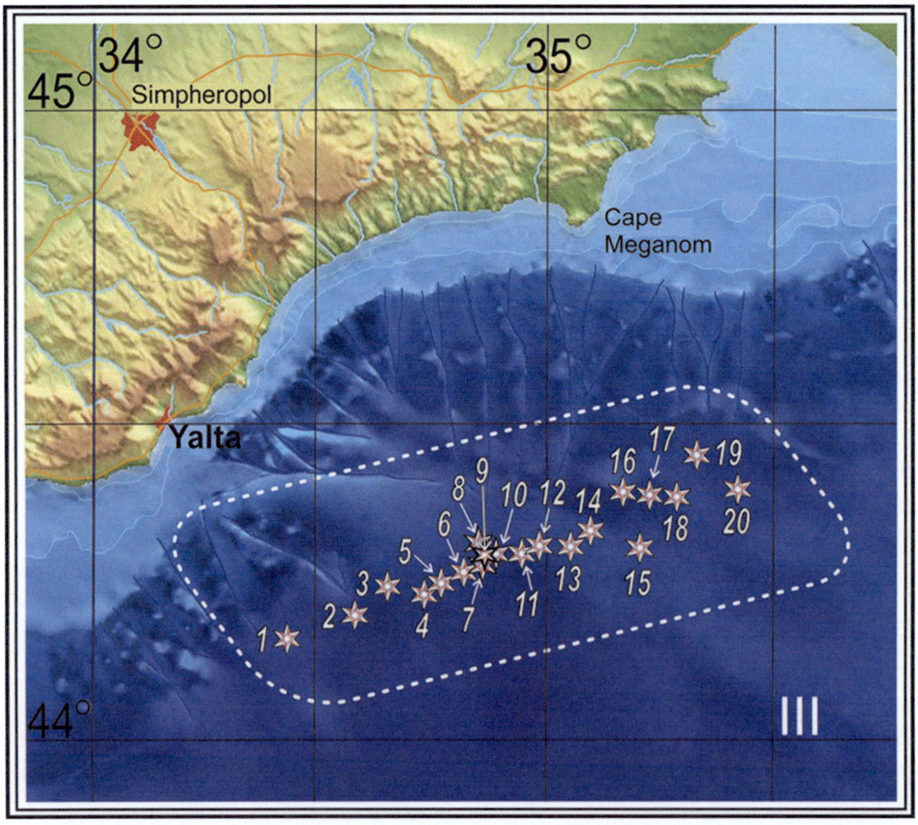

Fig. 6.150 Location of mud volcanoes in the Sorokin Trough. Key: (1) M10; (2) M4; (3) M3; (4) Sevastopol; (5) Yalta; (6) Karadag; (7) Massandra; (8) Bezymyannyy; (9) M2; (10) M1; (11) Dvurechenskiy; (12) Helgoland; (13) Vodyanitskiy-1; (14) NIOZ; (15) Kazakov; (16) Odessa; (17) M12; (18) Balaklava; (19) Tbilisi; and (20) Istanbul. For the location and coordinates of Group III, see Table 6.6 and Fig. 6.126

lapse through a system of concentric downthrows. The third type is represented by a single mud volcano, Dvurechenskiy, that is close to the "Barbados type" (Ivanov 1999) of volcano. It has a diameter of 1100 m, is round in shape, is flat at the arch, and ejects heavily diluted eruptive products. The fourth type is not actually a mud volcano but fissured outpourings of mud breccia that flow down the continental slope in the central part of the Sorokin Trough. The width of these mud flows is up to 1.5 km, and their length can be 4 km. The mudflows move down the continental slope from Yalta in an easterly direction along a system of parallel cracks oriented in the sublatitudinal direction (Ivanov et al. 1996).

Compared to mud volcanoes in the Western Black Sea depression, those of the Sorokin Trough are smaller, but their gas flares are much more powerful. Such flares are unknown elsewhere in the Black Sea. The volcanoes vary by height as can be seen in Table 6.6, and their character can be described using as examples the largest of them: Kazakov (Figs. 4.17, 6.151, and 6.152), NIOZ (Fig. 6.152), and Dvurechenskiy (Figs. 6.153 and 6.154).

The NIOZ mud volcano is quite similar to that of Kazakov.

According to seismic data, the Kazakov mud volcano lies within an anomalous sedimentary zone interpreted by wave dynamics and propagation velocity as an eruptive channel for the release of gaseous hydrocarbons (Fig. 6.152).

The Dvurechenskiy mud volcano has a very flat-topped central high (Krastel et al. 2003) that is 1.3 by 0.95 km in area and situated within a larger caldera-like depression measuring 3.2 km in diameter (Fig. 6.153).

The steep northern and western flanks of this central "mud pie" rise only 18 m above the bottom of the depression, whereas the eastern and southern flanks descend from the top of the mud volcano at 2062 m down to 2130 m (Greinert et al. 2006).

This volcano was studied in 2002 by a German expedition using R/V "Meteor" as well as by the

Table 6.6 Coordinates and some characteristics of mud volcanoes in the Sorokin Trough

Mud volcano	Coordinates N	Coordinates E	Depth of the sea above the top of the cone, m	Cone height above the seabed, m	Area, km²	Volume, m³
M10	44° 07.933′	34° 26.300′	–	–		No data
M4	44° 10.067′	34° 37.683′	–	–		No data
M3	44° 12.983′	34° 42.483′	–	–		No data
Sevastopol	44° 17′ 11″	34° 52′ 25″	2085	25	1	No data
Yalta	44° 17′ 00″	34° 54′ 03″	2024	60	1	No data
Karadag	44° 16′ 41″	34° 50′ 22″	2054.5	–	–	No data
Massandra	44° 16.81′	34° 52.43′	2038.4	–	–	No data
Bezymyannyy	44° 14.58′	34° 47.15′	2100	50	1	No data
M2	44° 17.250′	34° 52.033′	–	–		No data
M1	44° 07.933′	34° 26.300′	–	–		No data
Dvurechenskiy	44° 17.00′	34° 58.57′	2060	80	0.8	No data
Helgoland	44° 17.28′	35° 00.015′	2089	–	–	No data
Vodyanitskiy	44° 17.67′	34° 01.97′	2070	20	0.5	No data
NIOZ	44° 18.79′	35° 04.39′	2015–2020	70	1.2	No data
Kazakov	44° 17.54′	35° 10.50′	1920	135	2.5	No data
Odessa	44° 23.01′	35° 09.23′	1836	–	–	No data
M12	44° 23.750′	35° 12.067′	1850	–	–	No data
Balaklava	44° 22.65′	35° 14.91′	1841.1	–	–	No data
Tbilisi	44° 25.30′	35° 16.20′	1700	50	0.5	No data
Istanbul	44° 23.10′	35° 24.00′	1700	50	0.7	No data

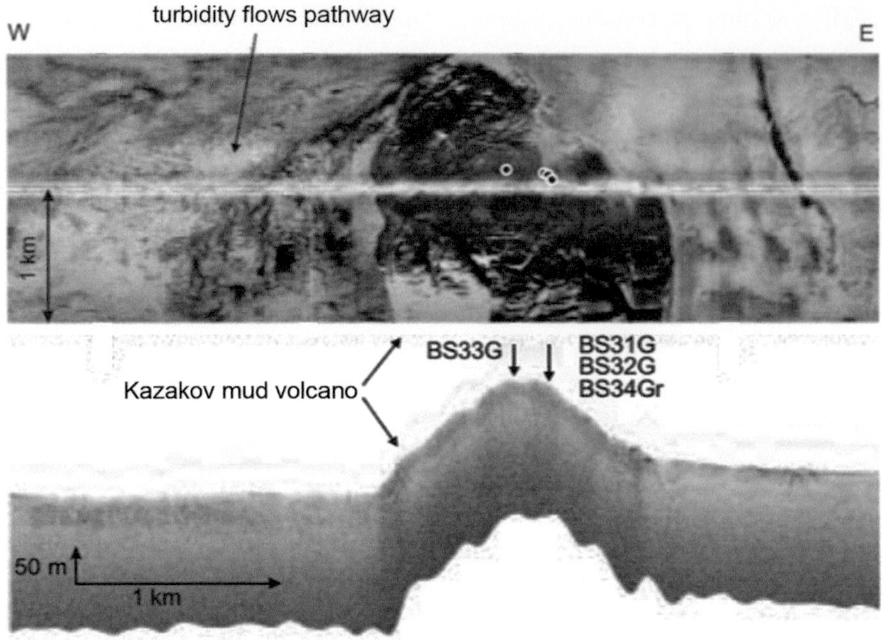

Fig. 6.151 Fragment of 30 kHz MAK-54BS sidescan sonograph and sub-bottom profiler record across the Kazakov mud volcano (Kenyon et al. 2002)

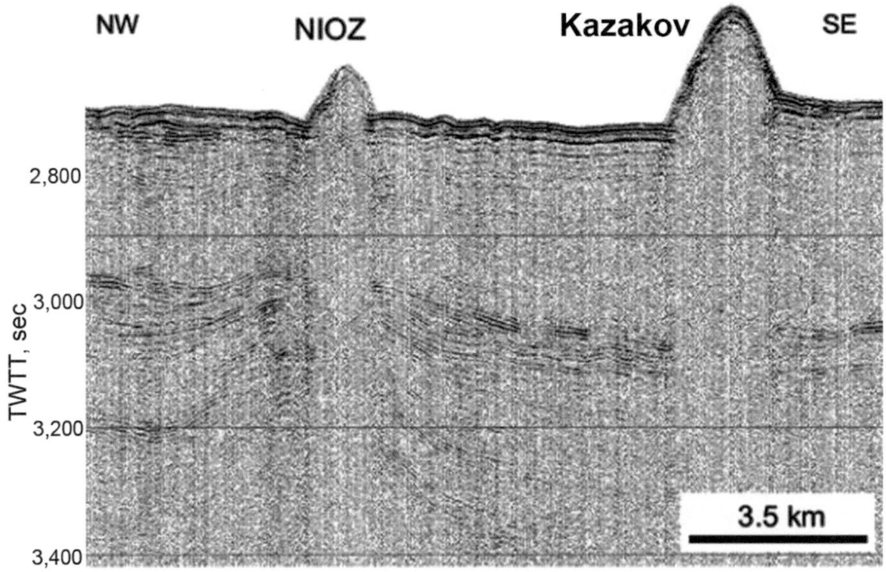

Fig. 6.152 Fragment of seismic line PS-297BS across the NIOZ and Kazakov mud volcanoes (Kenyon et al. 2002)

National Academy of Science of Ukraine using R/V "Professor Vodyanitskiy" (Shnyukov et al. 2003). Two powerful gas flares were observed (Fig. 6.154).

In the summer of 2002, the flares confined to the mud volcano's caldera were observed at a water depth of about 2000 m. Their diameter and height reached 400 m and 850 m, respectively.

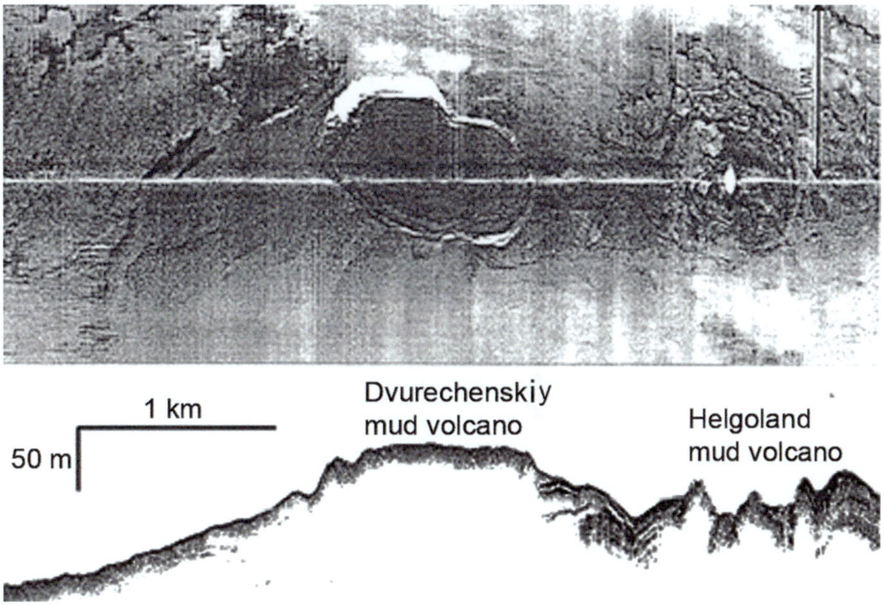

Fig. 6.153 Sidescan sonar image of the Dvurechenskiy mud volcano obtained during TTR-6 cruise by the deep-towed MAK-1 (Woodside et al. 1997). To the east (right) lies the Helgoland mud volcano

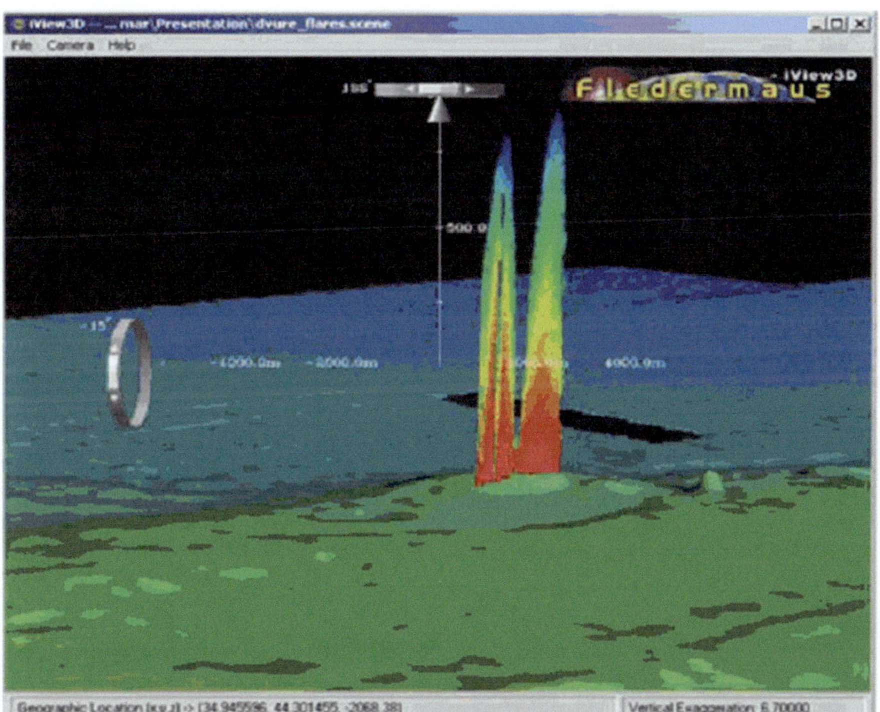

Fig. 6.154 3D image of gas flares emerging from the Dvurechenskiy mud volcano. Acoustic observations were made during the 57th cruise of the R/V "Professor Vodyanitskiy" in 2002. Bathymetry data were obtained with the swath bathymetry system HYDROSWEEP in the R/V "Meteor" M52/1 (MARGASCH) cruise. The image was created by Dr. Jens Greinert (Geomar, Germany) with the use of the program Fledermaus (Artemov 2006)

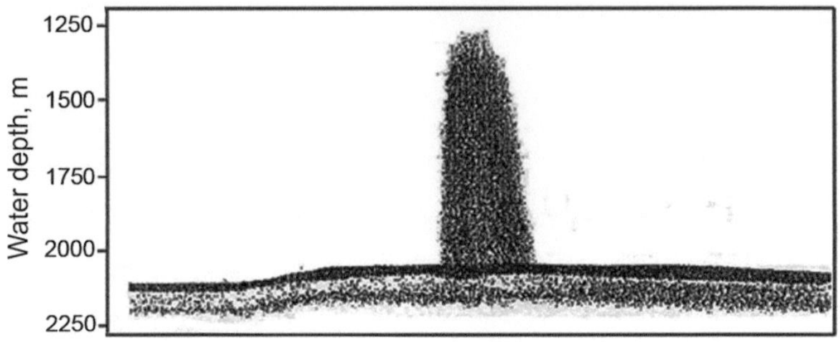

Fig. 6.155 Echogram of a gas flare about 800 m high above the seabed of the Dvurechenskiy mud volcano; image obtained during the 59th cruise of the R/V "Professor Vodyanitskiy"

None of the researchers who studied the Dvurechenskiy volcano before 2002 reported the existence of such powerful gas flares. The European expedition in May 2003 (Cruise #58-1 of R/V "Professor Vodyanitskiy") recorded no flares on this mud volcano.

During our cruise No. 59 on R/V "Professor Vodyanitskiy" in July of 2003, we observed once again a powerful gas flare rising about 800 m above the seabed (Fig. 6.155) that was confined to the same structure but somewhat offset from the two flares shown in Fig. 6.154.

Our repeated observations of this mud volcano during the years 2002–2011 have shown that gas emissions follow a pulsating character, and therefore this volcano can be related to the Lokbatan type (Kalinko 1960).

A lithological section of the area surrounding the Dvurechenskiy mud volcano is shown in Fig. 6.156.

The area is covered with carbonate crusts pierced through by holes and methane smokers. The radiocarbon age of the crusts is 5200 ± 120 years BP; the smokers yield ages of 3500 ± 140 years BP and 4300 ± 140 years BP. The bacterial mats have ages of 320 ± 90 years. Recovered pieces of breccia contained gas hydrates characterized by elevated values of heavy $\delta^{13}C$.

Homogeneous and layered clay fragments predominate in the debris. Light-colored limestones and argillites are also present. Geologically, this site is located in the fault zone of the intensively dislocated diapir folding of the central part of the Sorokin Trough. In the structure of the folds, Pliocene deposits with thicknesses of 0.2 and 1 km and Miocene deposits with thicknesses of a few hundred meters are present. The total thickness of the Maikopian deposits in the Sorokin Trough reaches 5–6 km.

The Vodyanitskiy-1 mud volcano (44° 34′ 25″ N; 31° 42′ 88″ E) is located at a water depth of 829 m. It is situated on a gentle slope resulting in an asymmetric shape that is quite complex (Fig. 6.157).

According to Sahling et al. (2009), the Vodyanitskiy-1 mud volcano consists of a cone about 500 m in diameter and 20 m in height surrounded by a depression (which the authors propose might be a recessed syncline); it is most pronounced on its eastern side. A sidescan sonar image shows the characteristic configuration of the mud volcano. The central crater-like crest is most likely the site where the mud flows erupted. It is surrounded by small-scale ridges arranged radially and a concentric pattern of anomalous backscatter that was interpreted as mudflows of different generations.

About 500 m northeast of the Vodyanitskiy-1 mud volcano, another structure with the appearance of a mud volcano can be distinguished, though it is much less evident. It has a positive morphology of about 10 m with very slightly higher backscatter compared to its surroundings. These are probably seafloor features of an old and now inactive mud volcano, which has been buried by regular hemipelagic sedimentation, leading to an indistinct morphology and weak backscatter anomaly.

Sediment samples were taken at three gravity corer stations at the summit of the Vodyanitskiy-1 mud volcano. One core contained mud clasts, whereas two cores recovered clay or sandy clay. Gas hydrates were found in two cores at a sedi-

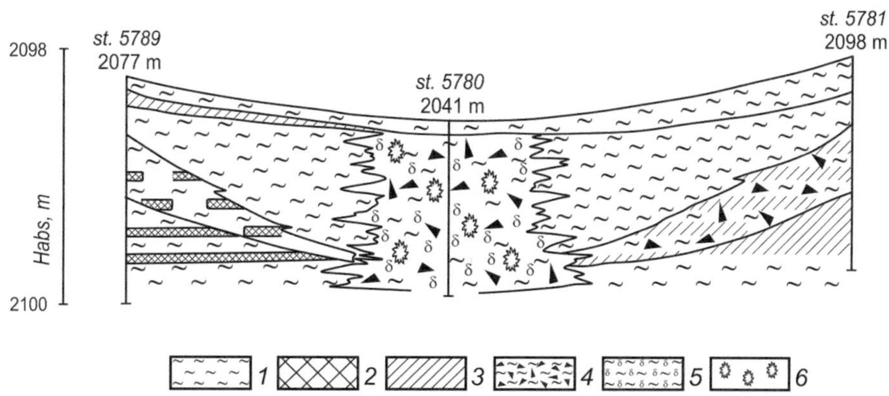

Fig. 6.156 Lithological section of the area surrounding the Dvurechenskiy mud volcano (Shnyukov et al. 2013). Key: (1) silt, (2) sapropel, (3) hydrotroilite silt, (4) mud volcanic silt, (5) gas-saturated silt; (6) inclusions of gas hydrates

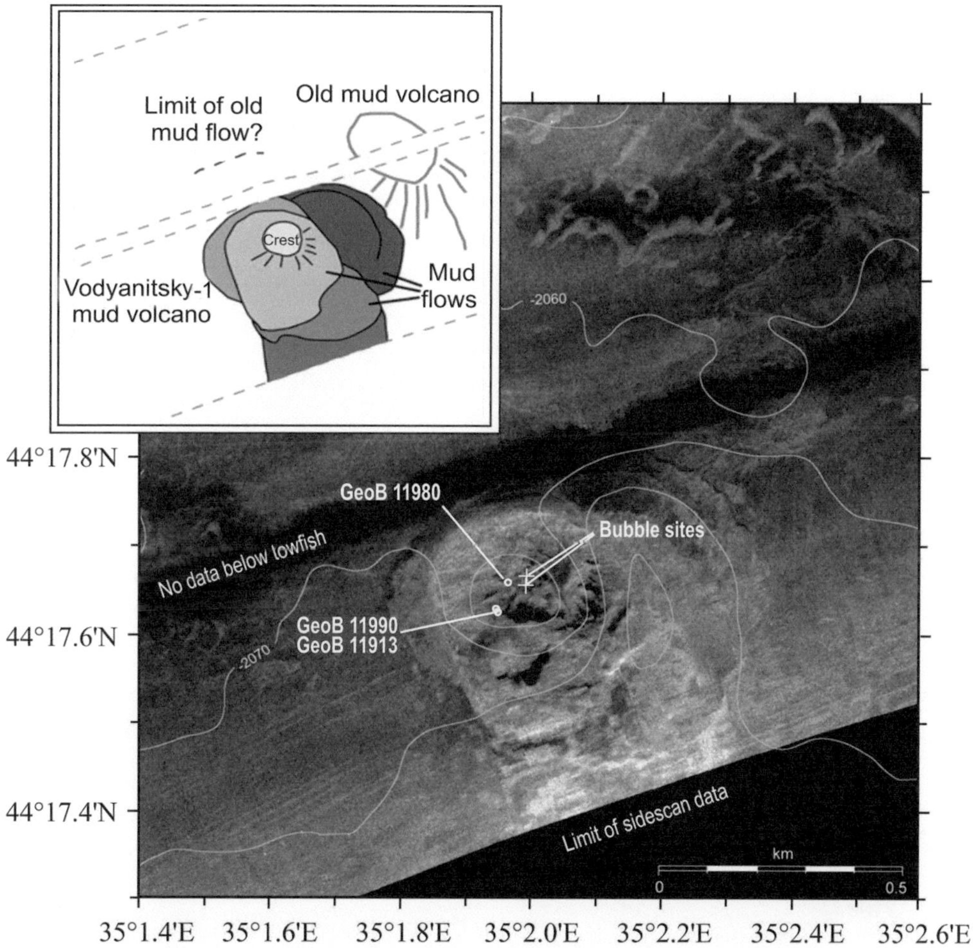

Fig. 6.157 Bathymetric contours overlaid onto a 75 kHz sidescan sonar profile (high backscatter in white) at the Vodyanitskiy-1 mud volcano and another mud volcano that is less distinct and probably old. The inset shows the interpretation of the backscatter image. The locations of the two bubble streams detected by ROV QUEST and the gravity corer-sampling sites are indicated (Sahling et al. 2009)

ment depth of about 1.5 m, which was generally the maximum penetration depth of the cores. This suggests that a layer of gas hydrate may have prevented the cores from penetrating any deeper. Carbonates occurred in the upper 35 cm of two of the cores. The carbonates were rounded in shape, and their size reached 9 cm, which is typical for authigenic carbonates probably related to anaerobic oxidation of methane. In addition, the core material had a mousse-like texture below a depth of about 40 cm, indicating a high gas content that expanded during the core recovery procedure and revealed increasing chlorinity with depth.

The Vodyanitskiy-1 mud volcano has one of the largest gas flares related to mud volcanic processes. Against the background of a hilly slope composed of turbidites, the gas is released from a small hill 4 m tall. The flare reaches 1300 m above the seabed (Greinert et al. 2006).

Three more mud volcanoes—Yalta, Sevastopol, and Odessa—with gas flares in their areas (Figs. 6.158, 6.159, and 6.160) were dis-

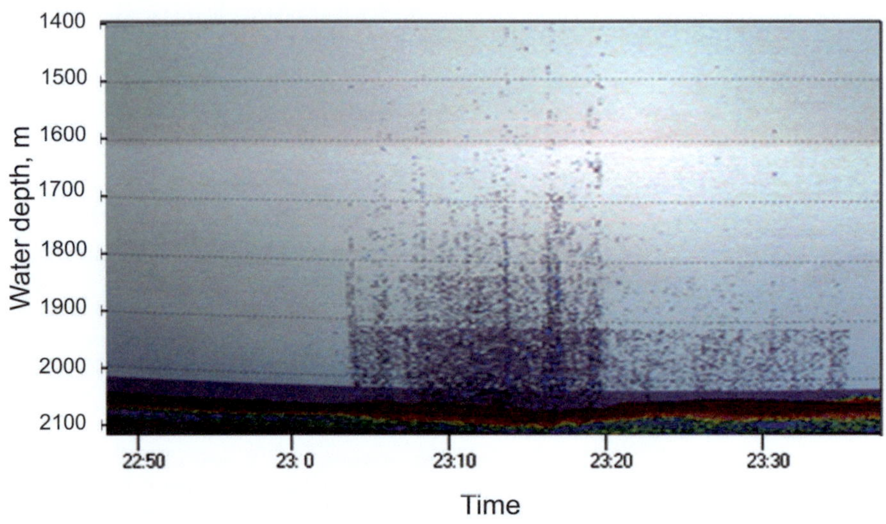

Fig. 6.158 Gas flares in the area of the Yalta mud volcano

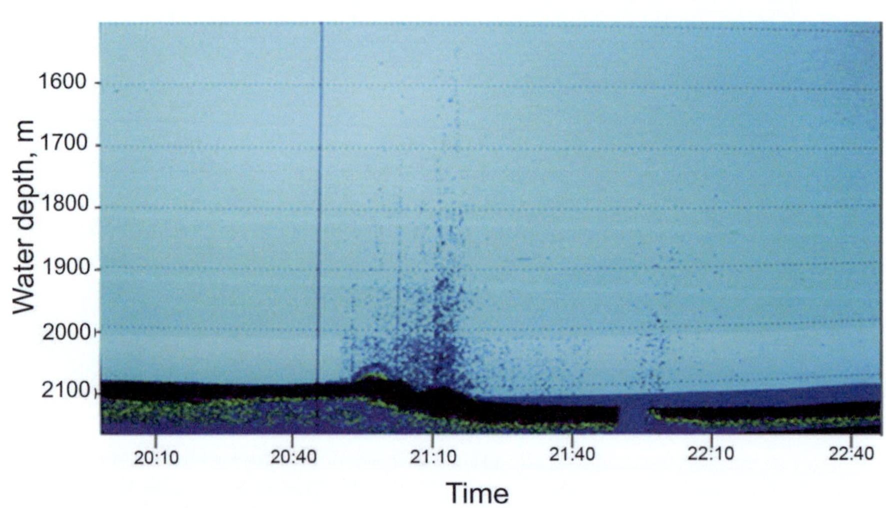

Fig. 6.159 Gas flares in the area of the Sevastopol mud volcano

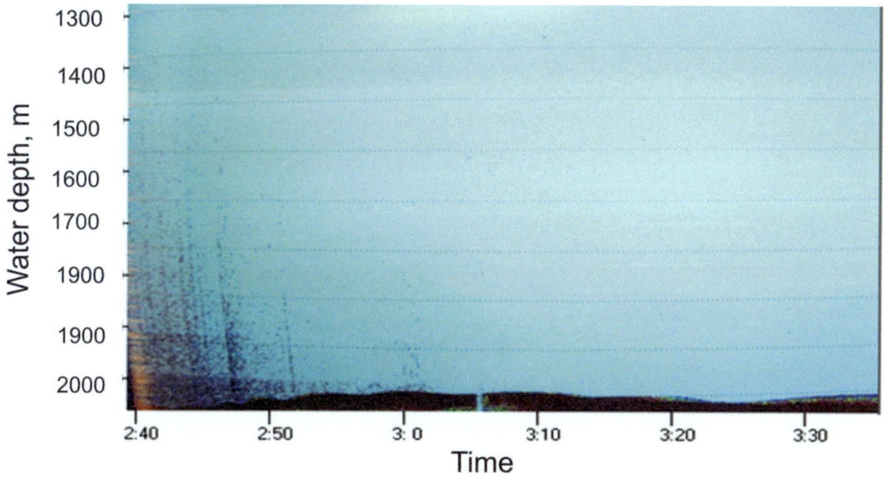

Fig. 6.160 Gas flares in the area of the Odessa mud volcano

covered in 2005 during the research cruise of R/V "Professor Vodyanitskiy."

All mud volcanoes discovered so far in the Sorokin Trough are confined to the same sublatitudinal fault zone and are characterized by the release of gases. Most of these mud volcanoes are active at present, and breccia overlies recent mud sediments. Almost all cores recovered from these mud volcanoes are enriched with gases, and in five of them, gas hydrates were found. This indicates the presence of large gas reservoirs located beneath the seafloor on the shelf and continental slope of the Black Sea.

6.9 Mud Volcanoes of the Kerch-Taman Shelf and Continental Slope

The Kerch-Taman shelf and continental slope are located largely within the Kerch-Taman Trough. The Trough is composed of Neogene and Paleogene sedimentary rocks transgressively overlapping Upper Cretaceous sediments. The sedimentary strata are folded and dislocated. Geological and geophysical studies have identified 16 morphostructures resembling mud volcanoes (Fig. 6.161).

With all the varieties of mud volcanic manifestations, the search criteria in this area of the Black Sea are also forms of relief—mostly flat, cone-shaped hills, often with subsidence calderas or small, funnel-shaped flutes. The heights of positive relief forms created by mud volcanoes do not exceed 50 m, rarely 100–110 m. Signs indicative of the existence of mud volcanoes include the presence of breccia, gas outlets, and characteristic lithological features from recovered cores.

Sixteen deep morphostructures with signs of mud volcanoes were identified using seismic profiles during the 65th voyage of R/V "Professor Vodyanitskiy." Seismic research was carried out by the association "Ukrgeofizika," and the interpretation was presented by the researcher M. Pobedash.

Hydroacoustic studies have shown that the deep-seated (and most likely fossilized) morphostructures identified on seismic profiles are not expressed on the seabed surface. Gas torches have been found emerging from many of them or next to them. Coring of the morphostructures for the purpose of clarifying the lithological composition of their sediments was effective only at water depths exceeding 70–100 m. At shallower depths, the coring was prevented by the presence of a hard bottom consisting of modern shell banks.

So far, only three mud volcanoes—Mitin, Naumenko, and OMGOR—have been confirmed by geological research (Fig. 6.162). Others are proposed based on geophysical methods, but

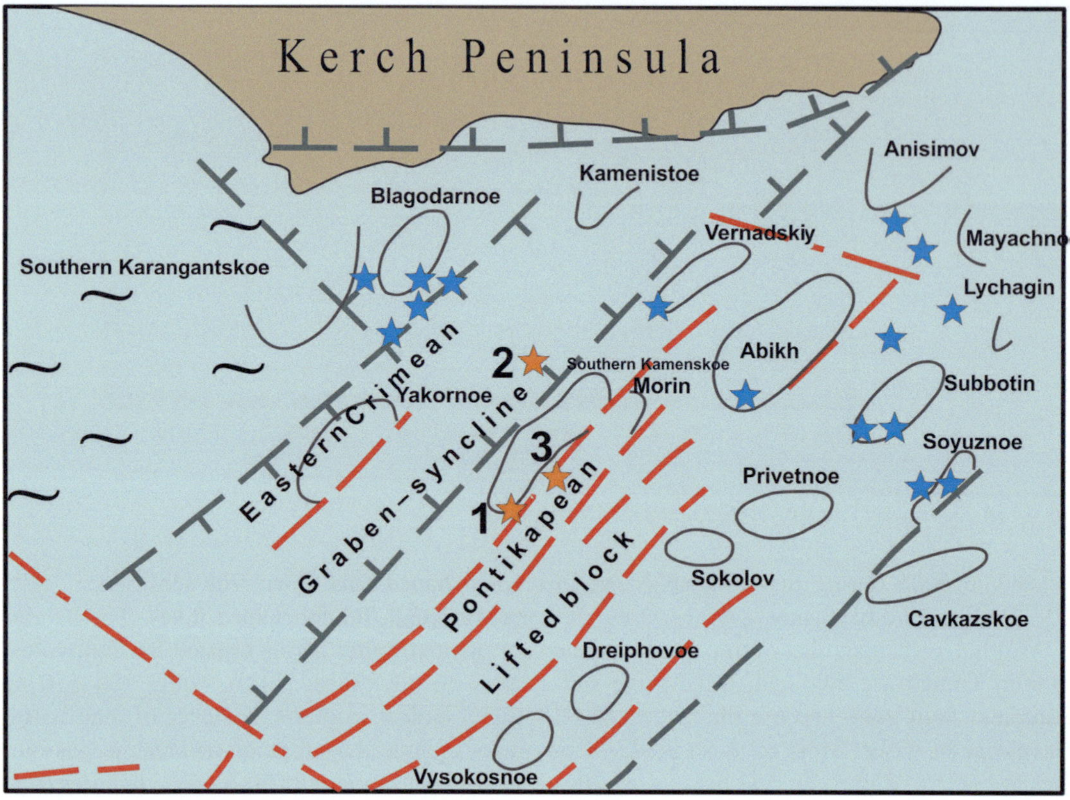

Fig. 6.161 Mud volcanoes plotted on a tectonic map of the Kerch-Taman shelf and continental slope (with use of materials from the industrial geological association "Ukrgeofizika," 2009). Mud volcanoes: (1) Mitin; (2) Naumenko; (3) OMGOR; blue stars – supposed mud volcanoes identified by geophysical methods

geological investigation has been hampered by thick shell banks that have formed over them.

The Mitin mud volcano is located on the Black Sea continental slope aligned with the meridian of Durmen mountain (Kerch Peninsula) at a depth 680 m bsl. It was discovered in 2004 during the 61st voyage of R/V "Professor Vodyanitskiy" and named in honor of the Head of the USSR Hydrography Branch of the Black Sea Fleet, Admiral L.I. Mitin (Shnyukov et al. 2005).

On the echo sound recordings, a gas flare and gas clouds were observed above this mud volcano (Fig. 6.163).

At the time of the study, the gas flare rose 560 m. Its localization was traced to the top of a flat hill 30 m high that resembled mud volcanic hills on land. Gas jets were carried by the water current, and the banding caused by the movement of bubbles is clearly visible. On the echograms, there were also separate gas clouds detached from the source. Perhaps in this case, we observed a combined cloud of several gas sources. Sometimes emission of gases is intermittent.

The diameter of the gas-producing vent, which has the shape of an oval, reached 90 m along its longer, west-east axis.

A lithological study of sediments recovered by gravity core at five stations revealed a peculiarity of the upper 3 m of the sedimentary cover. Typical mud volcanic breccia was not found, but gray, non-laminated, clayey mud typical of the vent facies of mud volcanoes was present. These facies made up almost the entire 3 m thickness in the area of the gas outlet and wedged out radially in all directions (Figs. 6.164, 6.165, and 6.166).

The mud volcanic facies include low-thickness lenses of coccolith, hydrotroilite, and silt with

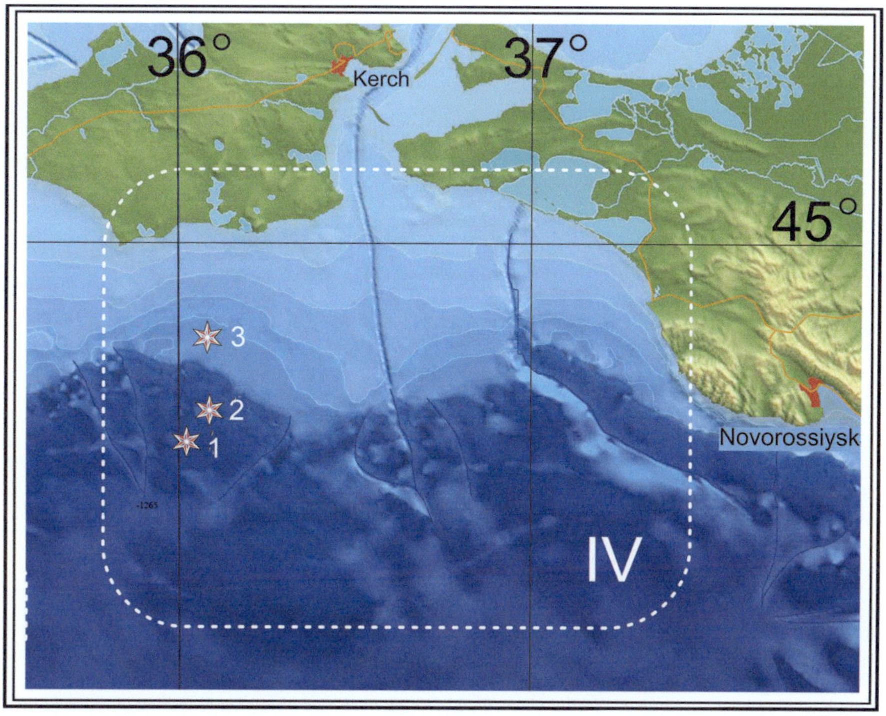

Fig. 6.162 Mud volcanoes of the Kerch-Taman shelf and continental slope. Key: (1) Mitin; (2) Naumenko; (3) OMGOR

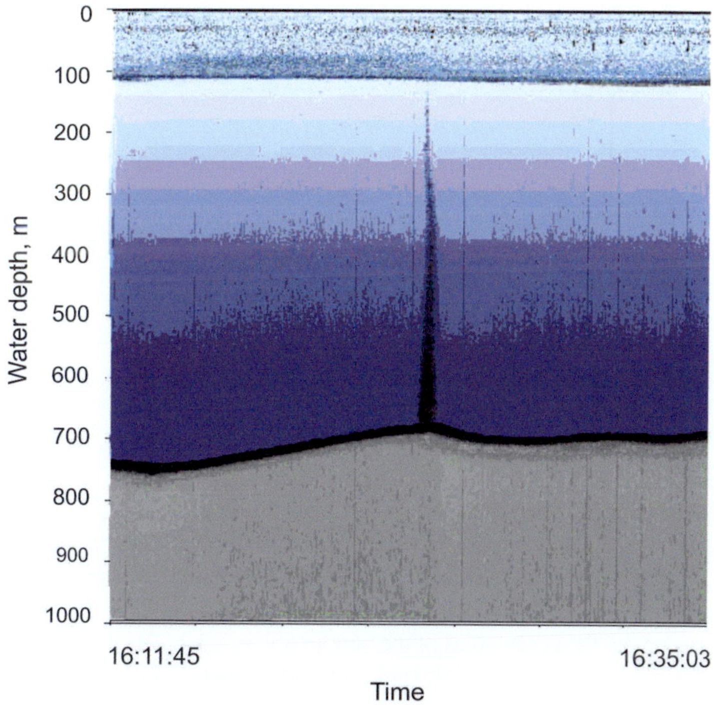

Fig. 6.163 Echogram of a gas flare above the Mitin mud volcano (recorded at a speed of 6 knots)

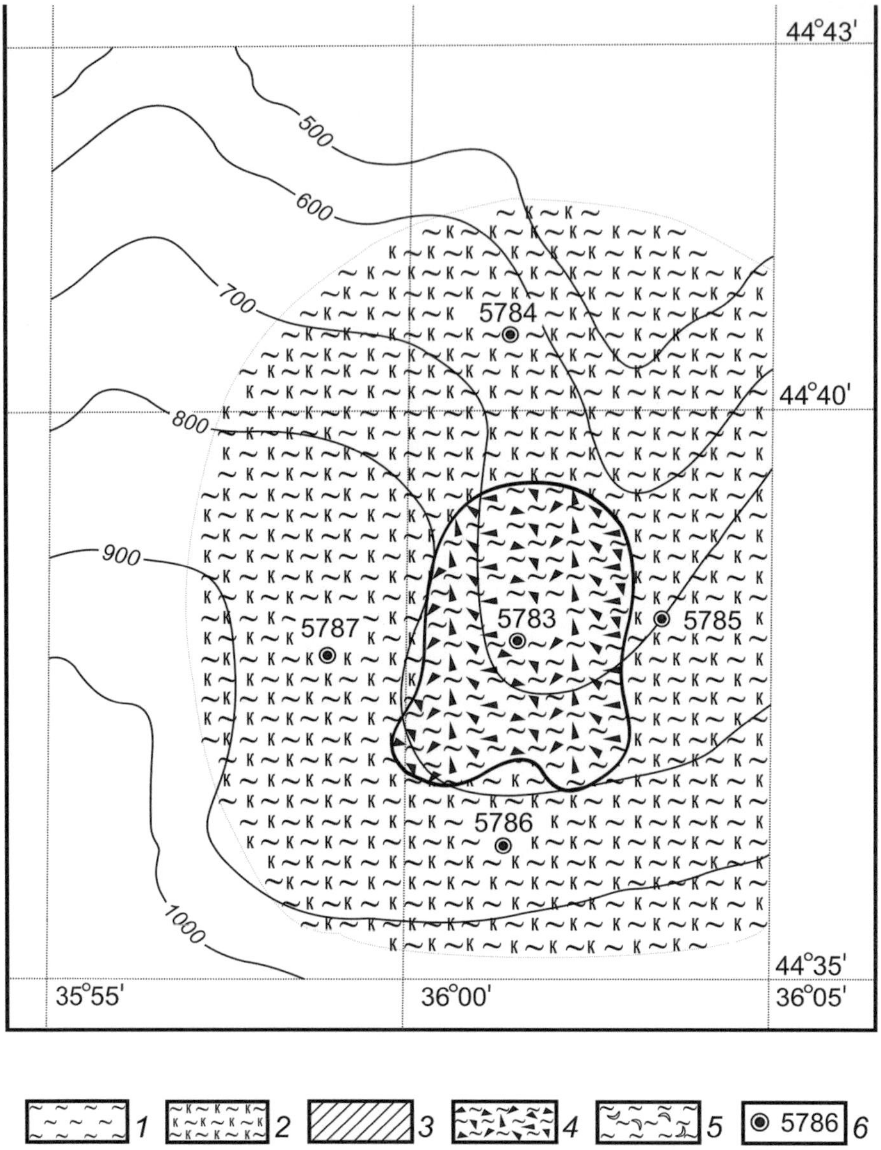

Fig. 6.164 Bottom relief, location of stations, and lithological diagram of the supposed mud volcanic field in the area of the Mitin mud volcano. Key: (1) liquid mud, (2) coccolithic silt, (3) hydrotroilite silt, (4) silt with selvages of hydrotroilite, (5) volcanic mud, (6) a sample locality

Micromelania lincta. On the sides of the volcano, they are wedged out and replaced by coccolithic silt. Judging by the section at St. 5785, the mud volcanic process was more powerful in the past. In accordance with the general bias of the bottom, the volcanic mud spread mainly southward. At St. 5784, there is volcanic mud at the foot of the core, suggesting the existence of another mud volcanic process during the Quaternary, but extinguished in the present. Sapropels are virtually absent in the section.

The temperature of the sediments varied from 9° to 11° C. Heat flow was within the range of 21–26 mW/m², and temperature anomalies were not detected.

Hydrochemical studies performed by Yu.N. Goryachkin yielded interesting results.

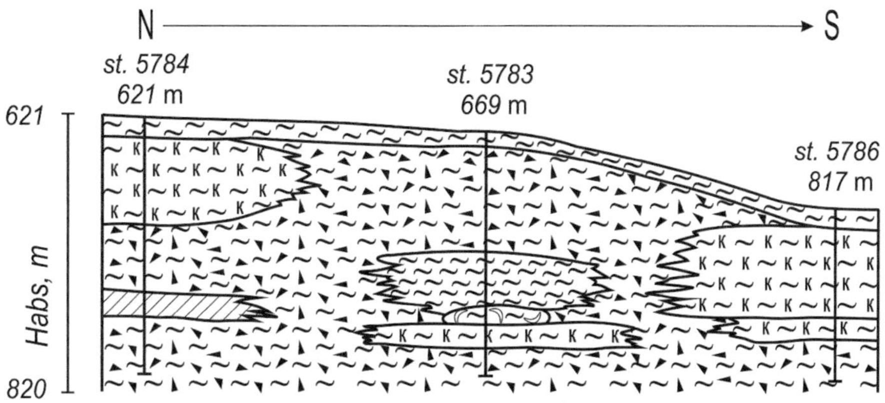

Fig. 6.165 Lithological section (north-south) of the supposed mud volcanic field in the area of the Mitin mud volcano. Legend in Fig. 6.164

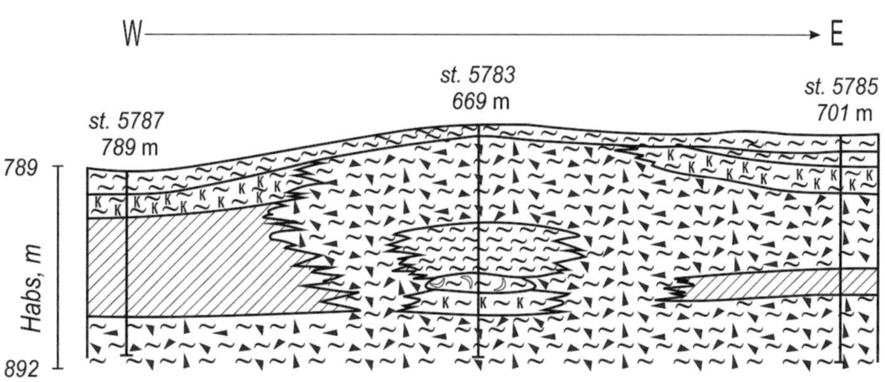

Fig. 6.166 Lithological section (west-east) of the supposed mud volcanic field in the area of the Mitin mud volcano. Legend in Fig. 6.164

At stations located in the center of the study area, at a distance of 5.5 km from the center of the volcano, a number of elevated vertical salinity gradients were detected. The difference in salinity ranged from 0.02 to 0.05 psu, reaching a maximum at a depth of 375 m (Fig. 6.167).

Anomalies in the concentration of hydrogen sulfide are more pronounced. Since the comparison of absolute values would not be appropriate due to the difference in depth, vertical gradients of hydrogen sulfide concentration in the bottom layer were analyzed. At central stations, they averaged about 1.0 μm per meter, which is more than an order of magnitude higher than the average gradient values for these depths. At background stations, it was 0.2 μm per meter to the southwest and 0.6 μm per meter to the northwest of the test site (Fig. 6.168).

Thus, the assumed participation of mud volcanic processes in the formation of hydrogen sulfide in the Black Sea is confirmed.

The Naumenko mud volcano was named in memory of the famous Soviet scientist of the Kerch mud volcanoes, P.I. Naumenko. It is located at a water depth of 500 m bsl at the edge of the protrusion of the continental slope (Fig. 6.169).

It is represented by a conical hill with a height of up to 30 m and a diameter (at the foot of the hill) of 80 m. On the northern slope of the hill, there are clearly visible two "satellite" cones and a plume of mud accumulations with smoothed hilly relief.

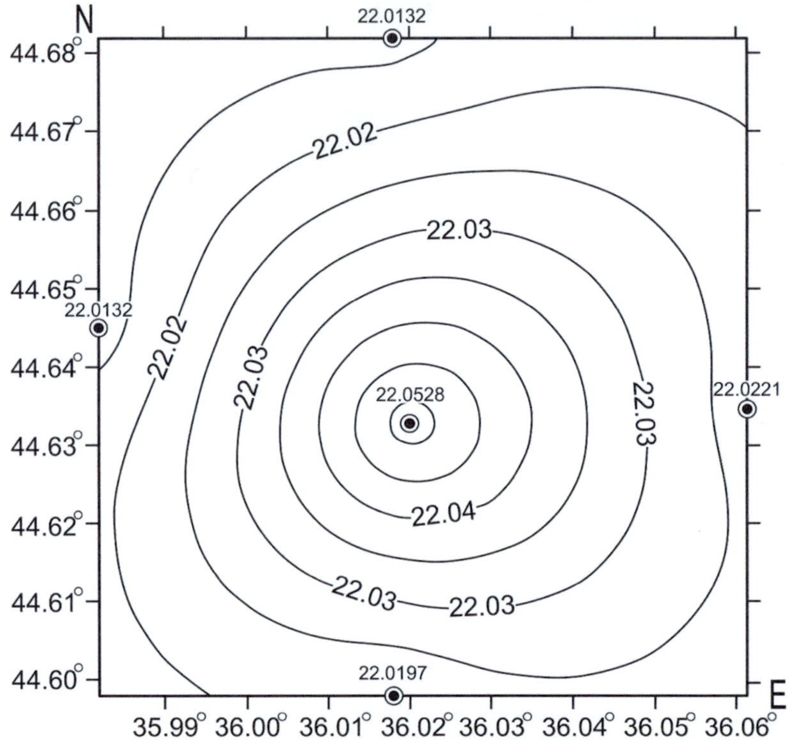

Fig. 6.167 Spatial distribution of salinity (psu) at horizon 500 in the area of the Mitin mud volcano. The numbers near the stations (black dots) are salinity values in psu

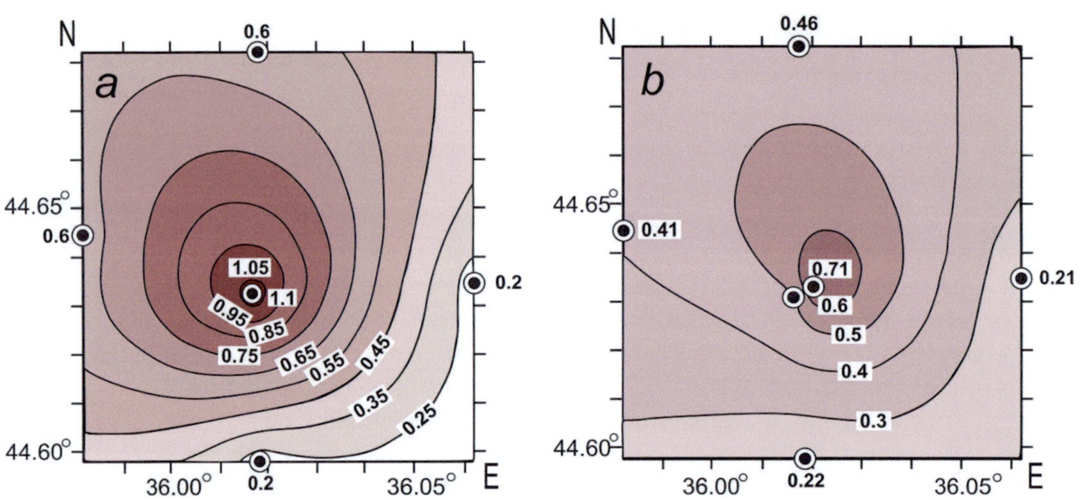

Fig. 6.168 Distribution of hydrogen sulfide concentration gradients in the water column at 5 m (**a**) and 35 m (**b**) from the bottom in the area of the Mitin mud volcano

The geological section of the core recovered in the area of the Naumenko mud volcano is shown in Fig. 6.170.

In 2010, geophysical observations and bottom sampling in the area of the Naumenko mud volcano were conducted during the 65th cruise of the R/V "Prof. Vodyanitsky," and two other manifestations of mud volcanism were discovered (Shnyukov et al. 2011). In particular, a presumed mud volcanic breccia was detected at St. 11 (water depth 203 m bsl,

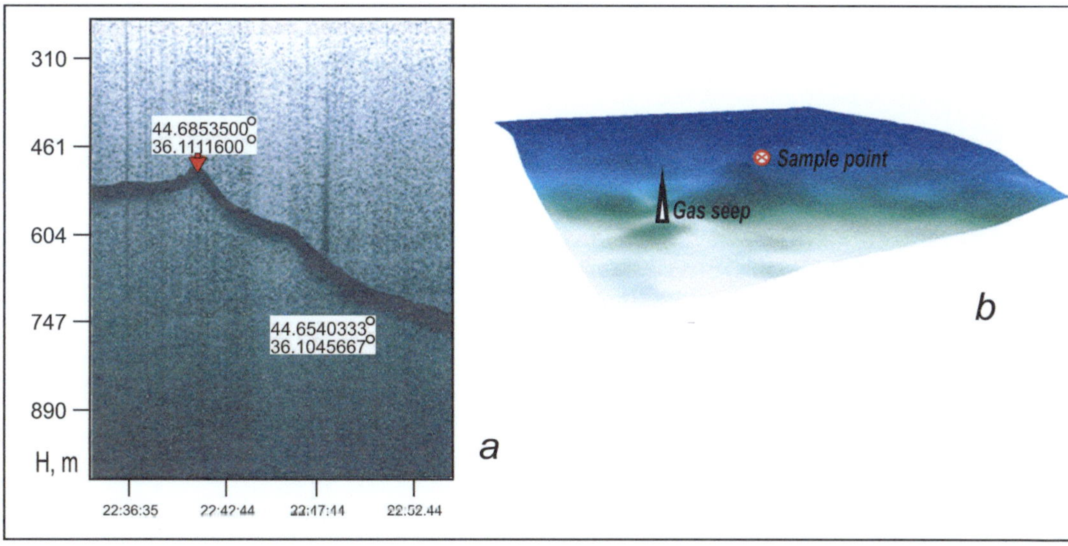

Fig. 6.169 The Naumenko mud volcano: (**a**) echogram; (**b**) 3D model

Interval, m	Lithology
0.0-0.15	Dark gray liquid aleurite-pelitic mud
0.015-0.60	Cocolithic mud
0.50-0.60	Green-grayish soft plastic mud with weakly expressed layering
0.60-0.90	Clayey mud with two interlayers (1 cm) of sapropel
0.90-1.0	Green-grayish mud with lens of greenish mud enriched with sapropel
1.0-1.04	Sapropel
1.04-1.52	Green-grayish unlayered clayey mud
1.52-1.66	Black hydrotroilite mudwith folded fine bedding in the upper part
1.66-1.85	Hydrotroilite mud with banded structure from the middle of interval upwards due to alternation of ligher microlayers on dark-colored background
1.85-2.35	Mud volcanic breccia soaked with hydrotroilite

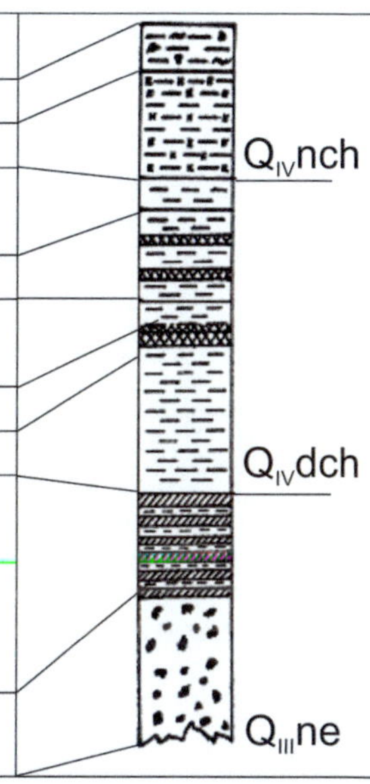

Fig. 6.170 Log of the core recovered at St. 105 (44° 39′ 21″ N; 36° 07′ 23″ E), at a water depth of 500 m bsl, in the area of the Naumenko mud volcano

44° 43′ 9″ N; 36° 04′ 00″ E) and St. 12 (depth 206 m bsl, 44° 43′ 9″ N; 36° 04′ 10″ E).

The OMGOR mud volcano was named after the organization *O*tdelenie *M*orskoy *G*eologii i *O*sadochnogo *R*udoobrazovaniya [Marine Geology and Mineral Resources] of the Ukrainian Academy of Sciences that has studied mud volcanism in the Black Sea for many years. The OMGOR mud volcano is located within the southern margin of the South Kerchian structure at a water depth of 83.4 m. It is expressed by a small depression (actually a pockmark of intensive gas release) with diameter and depth up to 50 and 4 m, respec-

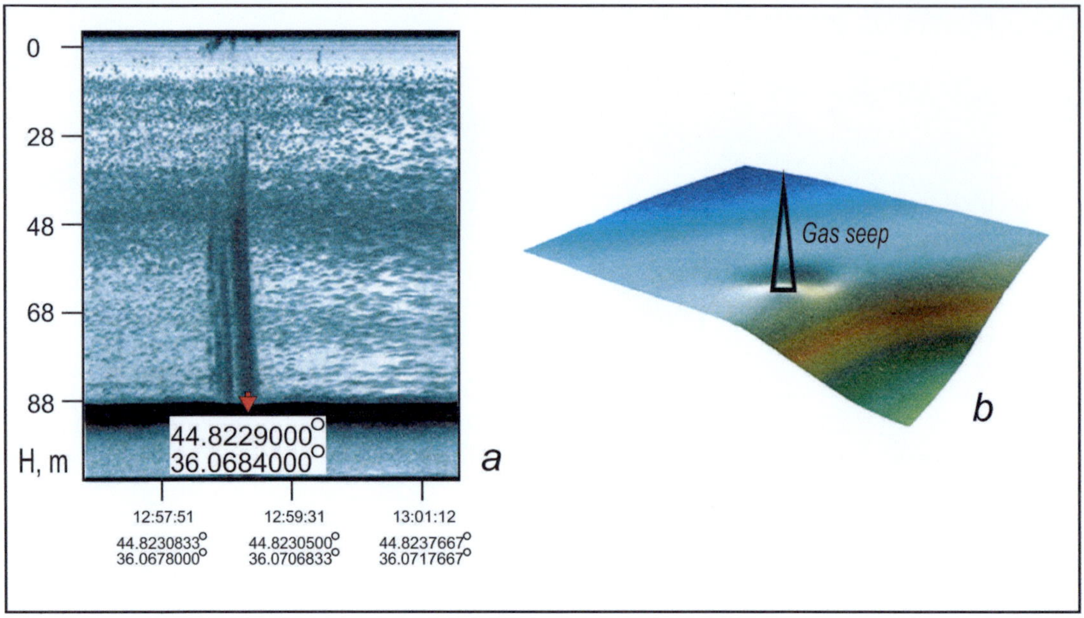

Fig. 6.171 The OMGOR mud volcano: (**a**) echogram; (**b**) 3D model

Interval, m	Lithology
0.0-0.20	Dark gray high-plastic mud with shells of *Modiola phaseolina*
0.20-0.45	Gray-greenish aleurite-clayey mud with shells of *Mytilus galloprovincialis*
0.45-0.70	Gray soft-plastic mud with admixture of aleurite and shells of *Dreissena polymorpha*
0.70-1.20	Gray condensed mud (clay) with weakly expressed spots of hydrotroilite and presumably inclusions of mud volcanic breccia

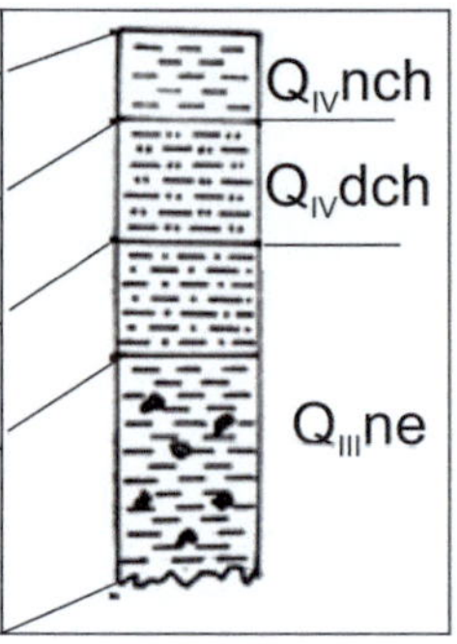

Fig. 6.172 Log of the core recovered at St. 5 (44° 49′ 5″ N; 36° 04′ E), at a water depth of 83.4 m bsl, in the area of the OMGOR mud volcano

tively, as well as by three powerful gas torches that rise up to 75 m in height (Fig. 6.171).

The geological section of a core recovered in the vicinity of the OMGOR mud volcano is shown in Fig. 6.172.

Thus, it can be seen that both the Naumenko and OMGOR mud volcanoes are hidden beneath Neoeuxinian sediments (Figs. 6.170 and 6.172).

According to Khortov and Neprochnov (2006), mud volcanoes of the Black Sea Russian sector do not have a direct connection with diapiric structures and are confined to faults. This picture can also be observed on seismograms of some mud volcanoes of the Kerch-Taman shelf and continental slope. The Mitin mud volcano is confined to a zone containing three major faults in the Pliocene-Quaternary subhorizontal sedimentary sequence. The thickness of this sequence in the region of the volcano is more than 2 km, within which are strongly folded and dislocated Maikopian and Mesozoic deposits. The faults can be traced, but diapir structures have not been observed (Fig. 6.173).

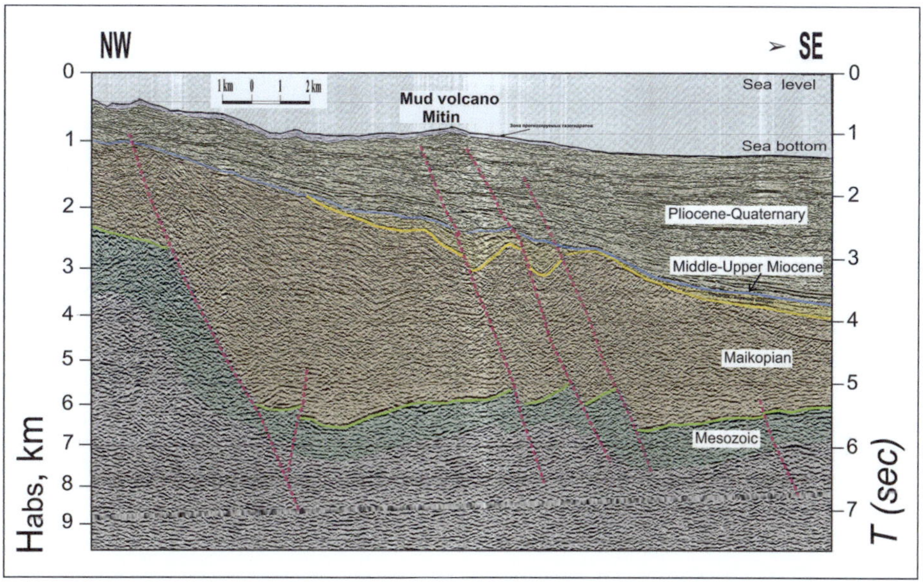

Fig. 6.173 The Mitin mud volcano. CDP time cross section showing the structure of the Cenozoic sequence and its boundary with the Mesozoic in the area of mud volcano activity. Red lines indicate faults (Shnyukov and Netrebskaya 2013)

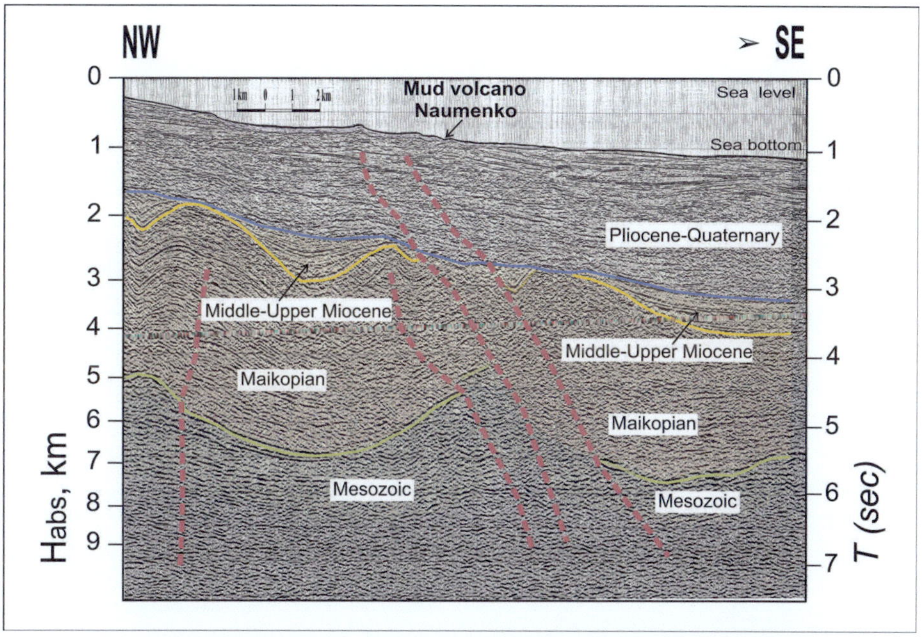

Fig. 6.174 The Naumenko mud volcano. CDP time cross section showing the structure of the Cenozoic sequence and its boundary with the Mesozoic in the area of mud volcano activity. In the near-bottom sediment layer, gas hydrates can be presumed according to the characteristic wave pattern (BSR zone). Red lines indicate faults (Shnyukov and Netrebskaya 2013)

A similar picture is observed with the Naumenko mud volcano, also associated with a deep fault in the absence of signs of diapirism (Fig. 6.174).

Perhaps, it is the tectonic origin of these mud volcanoes that explains the absence of coarse clastic rock material and typical mud volcanic breccias in their emissions.

Currently, only some of the potential but as yet unverified mud volcanoes in this area have been identified. Their exploration and identification, however, should be continued as they reveal signs of the possible presence of oil and gas.

6.10 Mud Volcanoes of the Kerch Strait and the Sea of Azov

Group VI includes eleven mud volcanoes localized within the Kerch Strait and the Sea of Azov (Fig. 6.175).

The Kerch Strait hides many secrets: sunken ships, submerged ancient settlements, and, of course, numerous geological puzzles, which include mud volcanoes. Andrusov (1926) denied the existence of mud volcanoes in the Kerch Strait, yet subsequent exploration has shown that, in fact, they do exist there. Much remains controversial, however. A number of rounded shallows

Fig. 6.175 Mud volcanoes of the Kerch Strait and the Sea of Azov. Key: (1) volcano in Lake Tobechik; (2) Tuzlinskiy; (3) mud volcano in the southern part of the Kerch Bay (without name); (4) Blevaka; (5) Peklo Azovskoe; (6) Temryukskiy (also Kazbek); (7) Golubitskiy; (8) Khakhalev

to the north of the Strait continue to be poorly understood and may be mud volcanoes. Zenkovich (1894) described a small island that appeared in Kerch Bay in 1880 but was washed away 2 weeks later. Its exact location is unknown, but in those years, its origin was associated with "volcanic forces."

According to recent understanding, the Kerch Strait lies within the boundaries of two structures: the Kerch-Taman transversal trough and the North-Taman zone of positive structures. By itself, the Kerch Strait is a large, deep sub-meridional fault intersected in several places by younger sub-latitudinal regional faults. The former is underlain by riverine sediments of the Paleo-Don river valley. Its Quaternary terraces are partially preserved along the shores of the strait, while the channel is filled by alluvial sands underlain by Neogene and Paleogene deposits that are crumpled within the folds oriented sub-meridionally and, in the north, latitudinally. Nodes of the fault intersections and latitudinal tectonic disturbances tend to favor the development of mud volcanic processes and the emergence of gas outlets or gas seeps (Fig. 6.176).

In the west are found the Bulganakskiy, Malo-Tarkhanskiy, Baksinskiy, Yenikalskiy, and Soldatsko-Slobodskoy mud volcanoes; in the east are the Peklo Azovskoe, Gora Gorelaya, Fontalovskiy, Karabetova Gora, and other mud volcanoes.

The mud volcano in Lake Tobechik (45° 09.21′ N; 36° 22.72′ E) is barely accessible for study. The liman is a fairly large (5.7 km^2) and shallow (1 m bsl) water body with a marshy bottom separated from the sea by a barrier beach. The liman is oriented diagonally to the southwest-northeast strike of the large Priozernaya anticline that occupies its eastern part, which is itself expanded up to 2 km in a north-south direction (Fig. 6.177). The western part of the liman lies above the Chorelekskaya syncline. The shores of the liman are composed of Sarmatian and Meotian sediments. On the southern shore is the Chongelekskaya recessed syncline composed of Pontian-Kimmerian (?) sediments (Lebedev 1968).

The well drilled by P.I. Naumenko into the barrier beach recovered Novochernomorian, Drevnechernomorian, Karangatian, and Post-Chaudian deposits underlain by Lower Sarmatian clay (Fig. 6.177).

Relying on the results of engineering and geological drilling conducted in the mid-twentieth century, Naumenko (1976) considered Lake Tobechik to be a multistage graben with a few mud volcanoes, most likely containing gryphons deriving from one mud volcano in its eastern part. In the summer of 2009, the liman was studied by the authors, who had the opportunity to verify the interpretations of P.I. Naumenko. The dry summer caused the precipitation of white and iron-colored salts on the extensive coastal shallows (Fig. 6.178).

In the eastern part of Lake Tobechik, black and gray-black spots stood out in sharp contrast to the white and iron-colored salts. The size of the largest of them reached 10 × 5 m (Fig. 6.179); others were smaller (Figs. 6.180 and 6.181).

Directly on the shore of the liman, near the village of Kostyrino, there is a small, regularly gurgling gryphon with a very gentle slope (Fig. 6.182); some other gryphons eject oil in addition to gas (Fig. 6.183).

Maps show several more mud volcanoes in the southern part of the Kerch Strait, but unfortunately, there are no new data on them. Several round shoals of unknown origin exist in the northern outlet from the Kerch Strait, and their nature must be explored. In general, the entire area of the Kerch Strait deserves further study.

The Tuzlinskiy mud volcano (45° 09.21′ N; 36° 22.72′ E) is located in the center of the Kerch Strait, to the west of Cape Tuzla (Fig. 6.179). It is the largest mud volcano in the area. Information about it was first obtained in 1914, when the steamship "Caesar" ran aground upon a 5 m cone-shaped sandbank that appeared at a water depth of 9 m bsl (Shepel' 1926). Samples recovered from the bank were identified as mud volcanic breccia. As Andrusov (1918) doubted the existence of mud volcanic manifestations in the Kerch Strait, he wrote that without denying the pos-

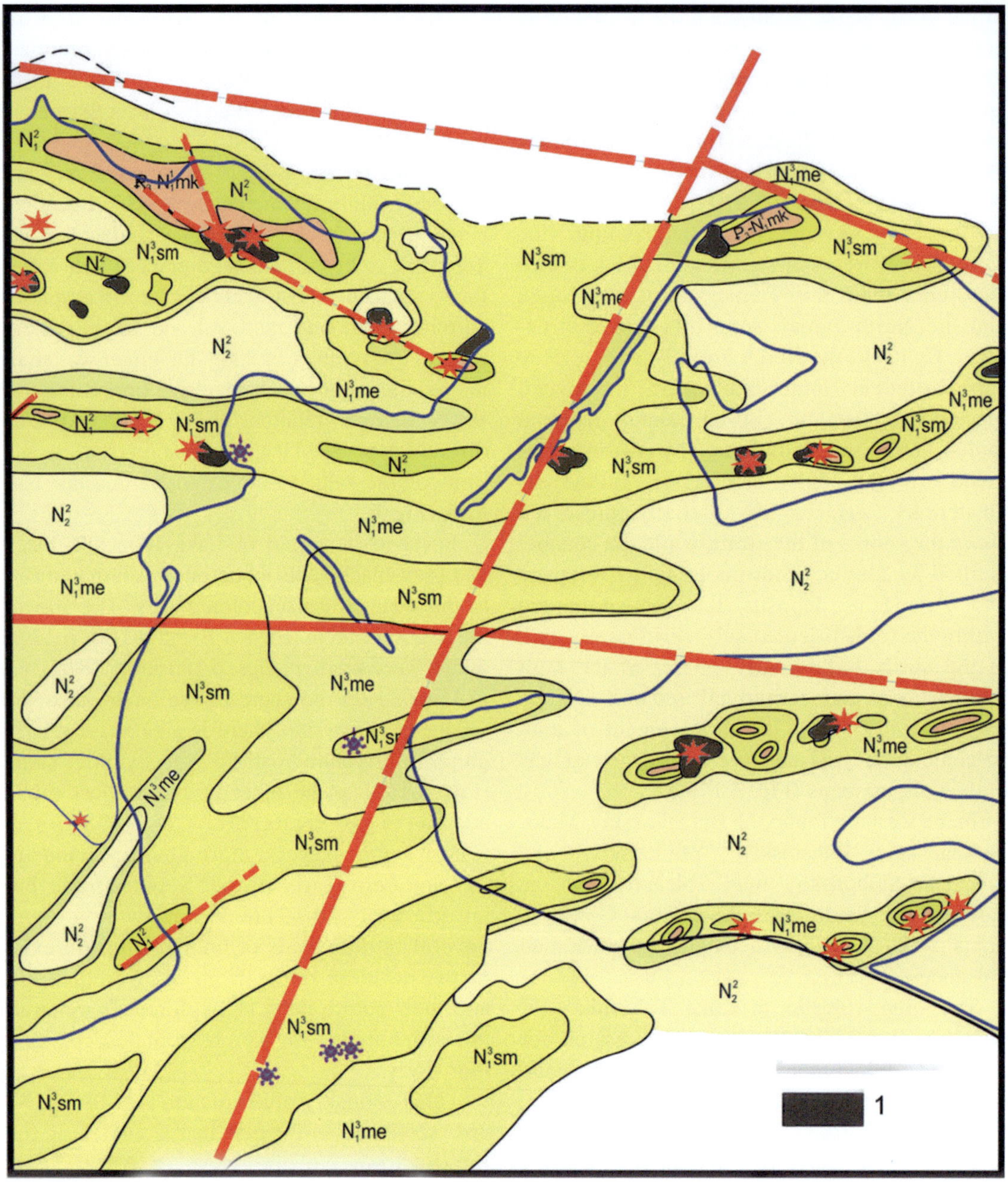

Fig. 6.176 Geological map of the Kerch Strait (simplified). Key: (1) mud volcanic breccia. For the rest of the legend see Figs. 6.3 and 6.4

sibility of CH_4 penetration into the sediments from the possible vents under the bottom of the strait, he nevertheless tended to explain the origin of gases in the wells of the Kerch Strait by their having been formed within the sediments themselves.

Our lithological studies of numerous cores extracted in the area of the Tuzlinskiy mud volcano demonstrated the presence of a few anomalous locations within the Neoeuxinian sediments that are characterized by the appearance of coarse sediments among fine ones (Fig. 6.184).

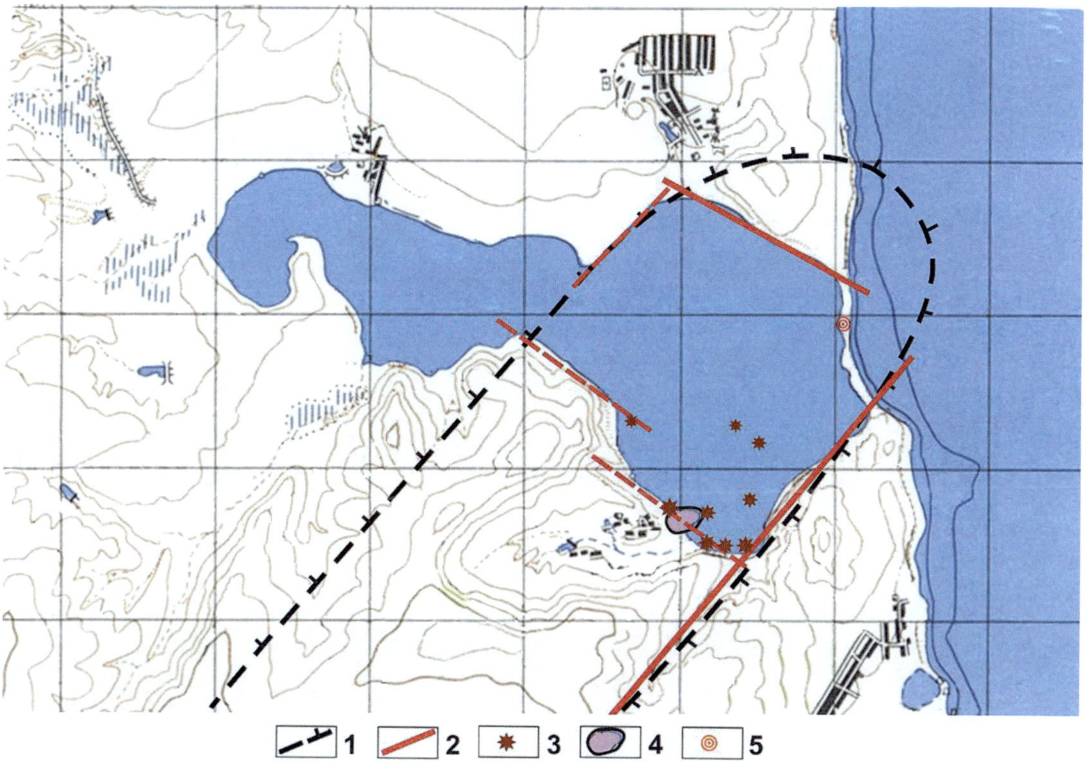

Fig. 6.177 Lake Tobechik and its position with respect to the Priozernaya (Chongelekskaya) anticline (for location see Fig. 6.7 and Table 6.2). Key: (1) contours of the Priozernaya anticline; (2) faults; (3) detected gryphons of mud volcanoes (known as of July 2009); (4) assumed recessed syncline; (5) borehole 1/71 (drilled by P.I. Naumenko)

Fig. 6.178 Precipitation of white and reddish (iron-colored) salts on the extensive coastal shallows of Lake Tobechik in summer of 2009

Fig. 6.179 Black and gray-black spots created by mud volcano outbursts on the white and reddish salts in Lake Tobechik in September 2009

Fig. 6.180 A black spot formed by a mud volcano outburst in Lake Tobechik in July 2009

These anomalies correspond to elevations (up to 20 m) in the bottom relief (Fig. 6.185).

This anomaly can be also seen in the spatial distribution of Drevnechernomorian (Fig. 6.186) and Novochernomorian (Fig. 6.187) sediments, i.e., it reveals an inherited character.

These lithologic data enable us to assume that the mud volcano has been active since Neoeuxinian time, when the Black Sea was a lake. Our attempts to study this mud volcano in greater detail from the R/V "Professor Vodyanitskiy" in 2005–2007 ended unsuccessfully. The entire volcano was covered by a mussel bank that did not allow coring, and maneuvering was quite dangerous due to heavy traffic in the Kerch Strait. Unfortunately, due to such navigational difficulties, geophysical study of

Fig. 6.181 A black spot formed by a mud volcanic outburst in Lake Tobechik in September 2009

Fig. 6.182 Small, regularly gurgling gryphon with very gentle slope on the shore of Lake Tobechik near the village of Kostyrino

this mud volcano was not possible. During geoecological research in 2007, yellowish-gray siliceous concretions, presumably of Chokrakian age, were recovered in the area of the mud volcano.

The Blevaka mud volcano (45° 21.54′ N; 36° 43.43′ E) was described by Felitsyn (1902) and by Belousov and Yarotskiy (1934) in the northern part of the Kerch Strait on the Chushka Spit, 7 km from its base. The hill is located on the seaward extension of the anticlinal zones lying on the Kerch and Taman peninsulas, roughly at the latitude of Gora Gorelaya (Table 6.2). The vol-

Fig. 6.183 Gas- and oil-ejecting gryphon located to the east of the gryphon shown in Fig. 6.182

cano appears as a cone of semiliquid mud with a height of 3 m asl. Its uniqueness consists of the absence of detrital material, particularly characteristic siderite fragments. The appearance or disappearance of those hills once, sometimes twice, in the early twentieth century suggests that mud volcanic processes are active. Given the fact that the Chushka Spit appeared relatively recently, it would not be unjustified to assume that formation of the hill occurred initially within the water area of the strait (Felitsin, 1902; Belousov and Yarotskiy 1936). Perhaps it is not a mud volcano, but a powerful gas seep, or a volcano sitting on a large tectonic disturbance. The question requires further study.

In the southern part of the strait, several other mud volcanoes are indicated on the maps. Unfortunately, we do not have new data about them. Several round shoals of unknown origin exist at the northern outlet from the Kerch Strait. The question of their nature remains to be studied. In general, the Kerch Strait is an area deserving of further exploration.

The first information about mud volcanoes in the Sea of Azov was provided by Pallas, who described the occurrence of mud volcanic islands in the southeastern part of the basin near the village of Golubitskaya in the vicinity of the town of Temryuk (Pallas 1795, 1883).

A chain of positive structures bearing mud volcanoes stretches along the southeastern shore of the Azov Sea. One of the structures contains two anticlines—Cape Kamenny and Cape Peklo—with two mud volcanoes, Peklo Azovskoe 1 (onshore) and its marine continuation Peklo Azovskoe 2, clearly seen in geophysical profile (Figs. 6.188 and 6.189).

The Peklo Azovskoe 1 ("Hell of Azov") mud volcano (45° 25.84′ N; 36° 55.30′ E) is significant in size and powerful in activity. Located within the anticline of the same name, it occurs at a place where landslips have created a semicircular scar in the cliffs. At its apex is a sharp, vertical cliff. A panoramic view of the coast shows masses of mud volcanic breccia forming powerful tongues that flow into the Azov Sea, standing out prominently against the light background of the surf (Fig. 6.190).

At the cliff, among the constantly ejected mud flows, there are detached loess-like rocks of

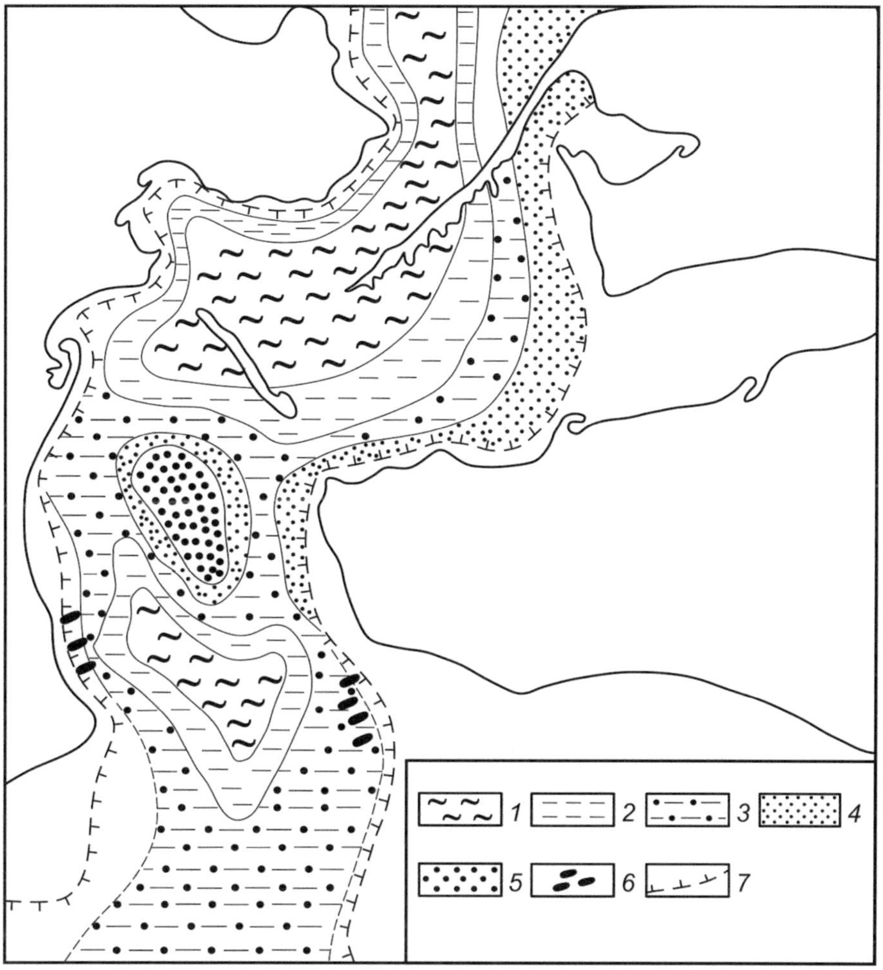

Fig. 6.184 Schematic lithological map of the Neoeuxinian deposits in the area of the Tuzlinskiy mud volcano. Key: (1) clayey silt, (2) silty mud, (3) fine silt, (4) silt, (5) sand, (6) shell detritus and pebbles, (7) boundary of maximal distribution of Neoeuxinian sediments

Quaternary age that lie higher upslope (Shnyukov et al. 2006). Numerous springs, some emitting hydrogen sulfide, are located mainly in the marginal parts of the mud tongues; they continuously pour out mud. In combination with the inclination of the relief, they create an extraordinary picture of mud volcanic breccia movement toward the sea (Fig. 6.191).

There are residues of Kimmerian iron ore at the end of the mud volcanic tongues on the beach (Shnyukov et al. 1986). This can be explained by the fact that the large and powerful Peklo Azovskoe 1 mud volcano breaks through 130 m of Kimmerian deposits that are known to contain iron ore (Fig. 6.192).

The underwater Temryukskiy mud volcano (also called Peresypskiy or Kazbek found at 45° 23′ 24″ N; 37° 10′ 48″) is located within the Temryukskaya anticline, which lies to the east of the Tizdarskaya positive structure (Table 6.2). The mud volcano is quite large and powerful (Fig. 6.193).

There is not much evidence of its activity, apparently due to its remoteness from human observation; it is located 5 km north of the coast and is designated by a buoy. The "Maritime manual of operations for commercial ships in the Sea

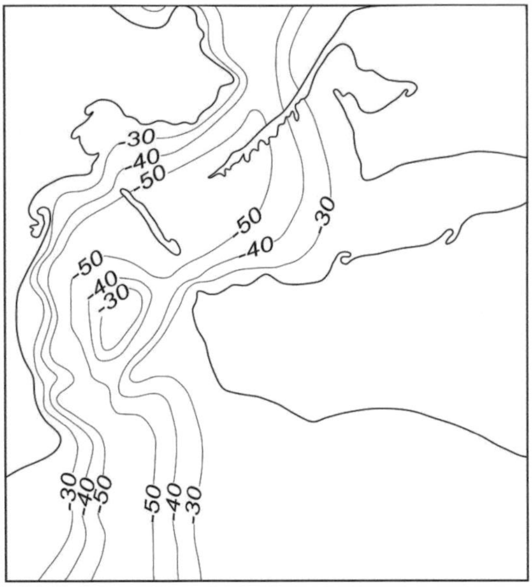

Fig. 6.185 Hypsometric elevation (up to 20 m) of the base of the Neoeuxinian sediments in the area of the Tuzlinskiy mud volcano

of Azov" draws the attention of seafarers to changes in depth and the potential emergence of temporary islands created by mud volcanic eruptions that occur quite often in that area. According to Naumenko (1976), a mud volcanic island appeared there in 1950. Shternov (1968) describes the eruption of a mud volcano on August 28, 1952, when an island composed of gray mud volcanic breccia and detrital material emerged and was soon washed out. In 1967 and 1968, Shnyukov et al. (1971) did not observe any island there, although the Temryukskaya bank was clearly expressed in the bottom relief. At a water depth of about 8 m, two small, cone-shaped ridges of elongated-round and sickle-shaped form were distinguished on the sea bottom. They were separated by a small hollow and together occupied about 1 km^2 of seafloor, reaching up to 7 m in height. In June 1968, a line of gryphons confined to the top of both ridges was observed. The group contained about 20 gryphons with different gas release intensities running in a west-east direction for about 100 m. In 1979, an explosion of the Temryukskiy mud volcano was accompanied by a subaerial column of water and smoke up to 100 m in height and the appearance of an island; these events recurred several times (verbal communication from I.D. Polovy, the inspector of the Temryuk lighthouse). In August 1986, a "clay island" was observed by sailors (verbal testimony of V.A. Nesterovskiy). On September 7, 1988, the island composed of mud volcanic breccia appeared again in the area of Temryuk bank. It was about 20 m in length, 10 m in width, and 0.5 m asl in height. The island was eroded and then reappeared several times above sea level until the end of October, when it was finally washed out. The authors observed a new explosion and the appearance of an island about 500 m seaward on August 21, 2001, and in March 2002. The island was subsequently washed out rapidly by waves (Shnyukov et al. 2006).

The underwater Golubitskiy mud volcano is located about 500 m seaward from the shore in front of the village of Golubitskaya (Shnyukov et al. 2006). In modern piloting charts of the Azov Sea, this area is marked by shallows. The mud volcanic activity varies over time (Fig. 6.194).

An eruption coincided with earthquakes that occurred in Krasnodar on September 5, 1799; the event was described by Pallas (1883), who wrote that about 300 m from shore, there was an underground buzz and then a deafening crash, followed by a column of fire and black smoke rising over the sea. The eruption continued for about 2 h and led to the formation of an island of mud with a diameter of 100 m and a height of 2 m.

Numerous eruptions of this mud volcano in 1804, 1814, 1862, 1880, 1906, and 1924 were mentioned by different authors. The eruptions were almost always accompanied by the emergence of an island that was quickly destroyed by the waves. The description of Shepel' (1926) is of particular interest. He wrote that in the first days of July 1924, near the village of Golubinka, an island appeared and constantly increased in size. Then, on July 15 at 4 pm, an explosion occurred, and a column of smoke, fire, stones, and small pieces of copper pyrite were ejected; this mud eruption lasted until 5 pm. Mud continued to erupt, increasing the area of the island. During the explosion, cracks up to 3 inches wide formed along the shore. By August 11, 1924, the

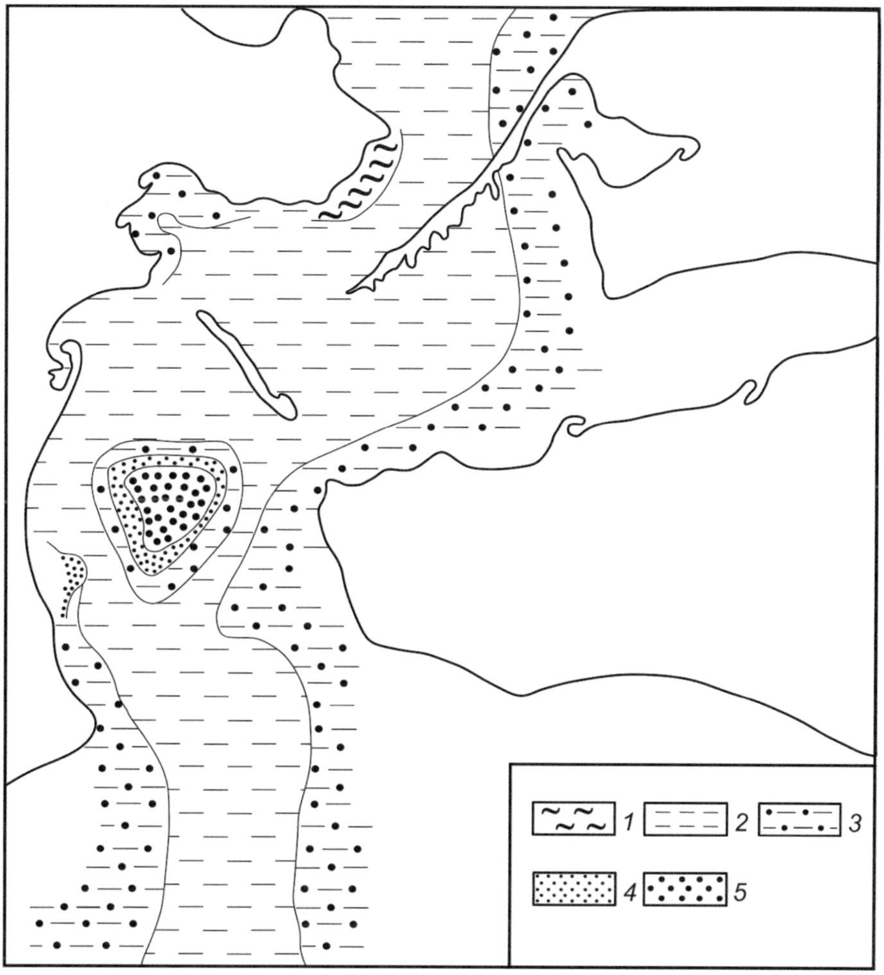

Fig. 6.186 Schematic lithological map of the Drevnechernomorian deposits in the area of the Tuzlinskiy mud volcano. Key: (1) clayey silt, (2) silty mud, (3) fine silt, (4) silt, (5) sand

length and width of the island had reached 38 and 28 fathoms, respectively. The island was quickly destroyed by waves and was completely washed out by October of the same year.

According to the custodian of the Temryuk lighthouse, the explosion of a mud volcano in the sea destroyed a balneological station on the shore in 1929.

Belousov and Yarotskiy (1936) surveyed the volcano during the period of its relative calm and found only a shoal of sand 20–30 m in diameter, which was elevated above the bottom by 4 m. No gassing from the volcano was detected.

This mud volcano was repeatedly described by different authors, both during its period of activity and at moments of relative dormancy in 1936 and 1967–1968, respectively.

In the following years, and especially after 1940, explosions began to occur quite often. A notable eruption was that of 1981, when an explosive discharge of mud, water, and detrital material was propelled up to 100 m in height creating a small islet as observed from the shore. During autumn and winter, storms attacked the island, decreasing its size to 7–8 m long and 3–4 m wide.

In August 1988, the volcano revived again with a very powerful eruption. On August 24, an island appeared in the sea, and on it was formed a clearly defined crater erupting gas, water, and

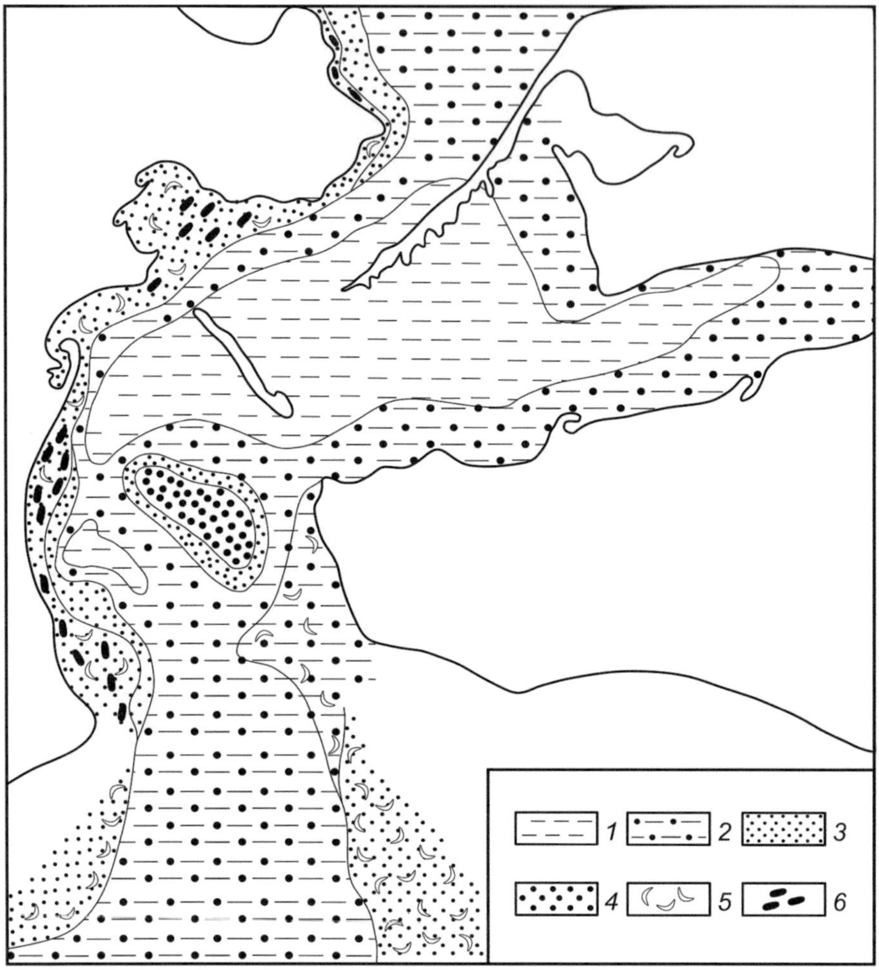

Fig. 6.187 Schematic lithologic map of the Novochernomorean sediments in the area of the Tuzlinskiy mud volcano. Key: (1) clayey silt, (2) silty mud, (3) fine silt, (4) silt, (5) sand, (6) shell detritus and pebbles

fragments of rock. On August 25, the size of the island was 15 m long, 7 m wide, and 0.5 m high. In early October, it was respectively 80 m, 40 m, and 2.5 m.

In June 1989, the mud volcano erupted thick mud, and the island continued to exist (Fig. 6.195) in October of that year, with mud still erupting from active gryphons.

Nesterovskiy (1990) described the eruption of 1988–1989 and admitted the coincidence of mud volcanic activation and seismic activity within the adjacent terrain, especially near the South Kuban fault. He claims that the Golubitskiy mud volcano became dynamic during the December 7, 1988, earthquake in Armenia.

Catastrophic eruptions of the Golubitskiy mud volcano occurred in 1994, 2000, 2002, 2008, and 2015; these events were repeatedly videotaped by different people (Figs. 6.196, 6.197, and 6.198).

The Khakhalev mud volcano (46° 12′ 50″ N; 38° 07′ 23″ E) (Fig. 6.175) was named after the famous geophysicist E.M. Khakhalev. It was discovered close to the mouth of the Beysugskiy liman during the course of a high-resolution seismic survey in 2008. It was described by Sheykov et al. (2013) and studied by lithological and petrographic investigation of three cores recovered at the structure

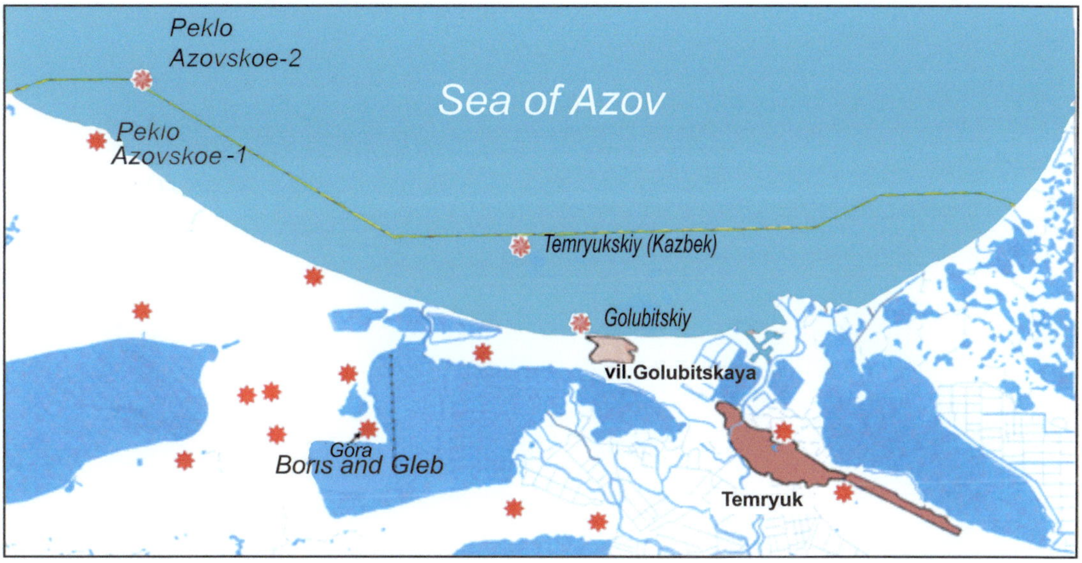

Fig. 6.188 Map of the southern coastline of the Azov Sea showing the track of a seismic profile (green line) from Cape Kamenny (left) to Temryuk Bay (right) and the locations of mud volcanoes (in stars). Note: the Peklo Azovskoe 1 mud volcano is located on the shore, whereas its continuation, Peklo Azovskoe 2, is underwater

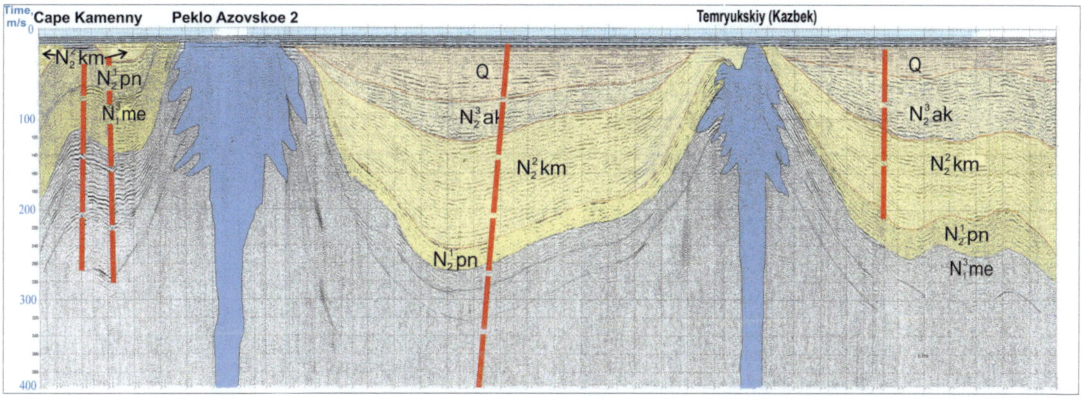

Fig. 6.189 Seismic profile from Cape Kamenny across Temryuk Bay; the Peklo Azovskoe 2 and Temryukskiy (Kazbek) mud volcanoes are clearly visible (Shnyukov et al. 2005)

(Shnyukov et al. 2015) that confirmed its mud volcanic origin.

Based on its round shape, Kovalevskiy (1960) assumed that the Zhelezinskiy bank located in the eastern part of the Azov Sea can be a mud volcano but did not provide any description.

New discoveries of mud volcanoes in the Azov Sea are quite possible. In general, when these mud volcanoes are in a calm state, they still produce gas; their location is revealed by the bubbles as well as fine mud entrained up to the surface. The eruptions of these volcanoes often lead to the formation of islands, which then are rapidly eroded by wave action. In place of the washed mud volcanic islands, clastic material remains and, over time, accumulates on the sea bottom. The debris becomes covered by seaweed and shells forming underwater banks of mixed mud volcanic and marine origin. The places at sea where mud volcanoes

Fig. 6.190 General view looking southward toward the terrestrial Peklo Azovskoe 1 mud volcano

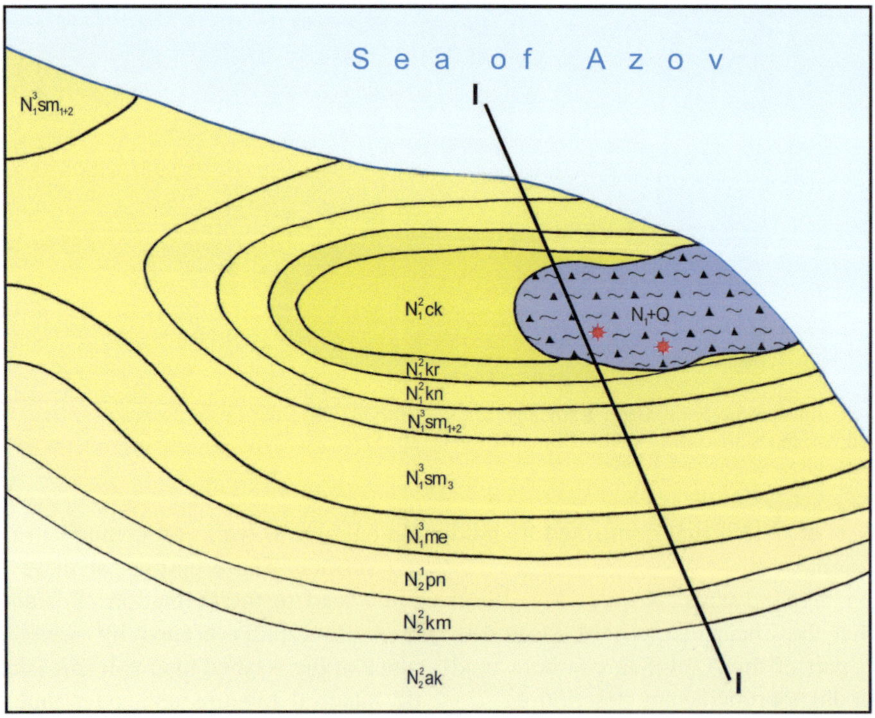

Fig. 6.191 Diagram of the geological structure of the Peklo Azovskoe 1 mud volcano. Two craters near the shore are marked with asterisks

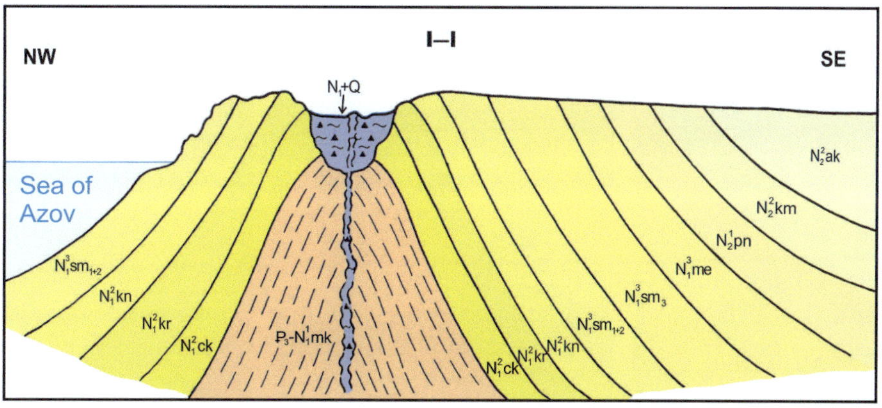

Fig. 6.192 Geological section of the Peklo Azovskoe 1 mud volcano

are developing may be dangerous for navigation, and for this reason, they are indicated in sailing directions, such as Temryukskaya and Golubitskaya banks.

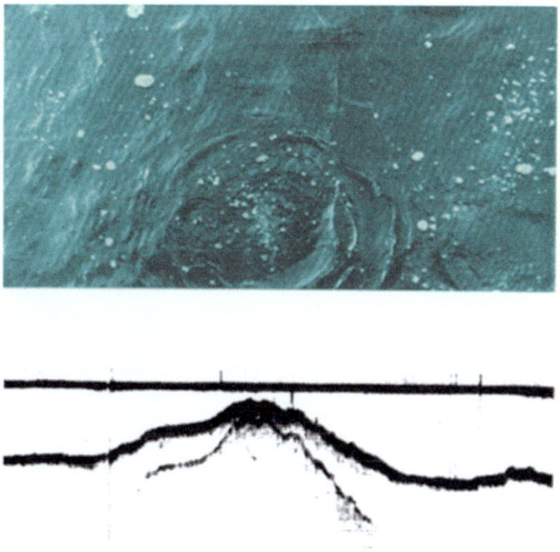

Fig. 6.193 The underwater Temryukskiy (Kazbek) mud volcano (Shnyukov et al. 2006)

6.11 Mud Volcanoes of the Tuapse Trough and Shatskiy Ridge

Group VI includes 16 mud volcanoes located in the Tuapse Trough and Shatskiy Ridge. The Tuapse Trough is located in the northeastern part of the Black Sea and south of the Kerch and Taman peninsulas (Fig. 6.126). It extends along the northwestern border of the Caucasian ridge from the city of Anapa to the city of Sochi.

Abundant oil and gas emissions have been recorded to the northeast of the coastal lands and shelf area, almost over the entire Sochi-Adler depression. They are especially frequent in the basins of the Khosta, Agura, and Matsesta rivers and within the lower reaches of the Sochi River. In all locations, they are encountered both in boreholes and in surface outcrops of various forms—manifestations ranging from the smell of bitumen to the seepage of liquid oil (Sokolov 1965). According to the results of a gas survey conducted by the "Yuzhmorgeologiya" within the context of ecological monitoring (1990–1993), pulsating natural anomalies of hydrocarbon gases were recorded in the rivers of Sochi district (Matsesta, Agura, Hosta, Herota, Kusdepta), Tuapse district (Loo, Yakornaya Schel', Uch-Dere), and Novorossiysk district (Dyurso) (Kruglyakova et al. 1998).

The Maikopian deposits of the Tuapse Trough form diapiric folds that are distinctly expressed in the roof and displayed in the overlying strata (Tugolesov et al. 1985; Limonov et al. 1994; Meisner et al. 1996). They are associated with upthrust-thrust deformations with a common surface of failure in the base of the sequence (Nadezhkin 2011) (Fig. 6.199).

The mechanism of mud volcanism within the Tuapse Trough is related to abnormal pressure buildup in clay sediments; this leads to the devel-

Fig. 6.194 Sequential images of an explosion of the Golubitskiy underwater mud volcano in 2008 (from video footage)

Fig. 6.195 Mud volcanic island in the Azov Sea near the village of Golubitskaya: (**a**) top view from October 1988; (**b**) top view from September 1989

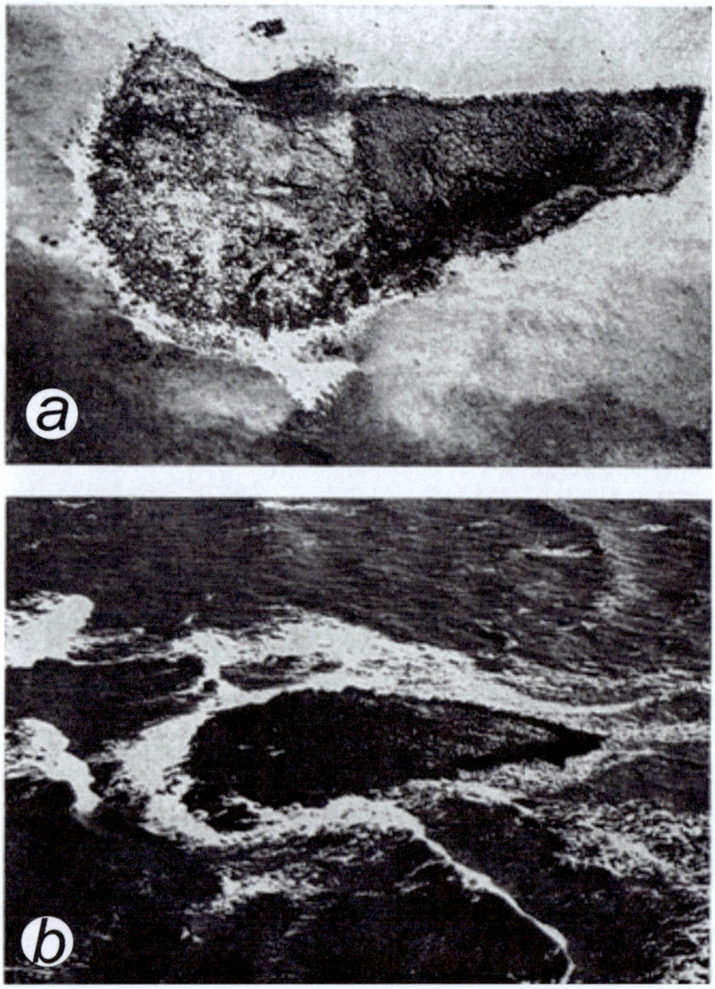

Fig. 6.196 Explosion of the Golubitskiy mud volcano in 2000. (Internet resource)

Fig. 6.197 Explosion of the Golubitskiy mud volcano in 2008. (Internet resource)

opment of mud volcanoes through which deep fluids can outflow to the surface (Almendiger 2011). Such processes are well studied in the majority of the deepest fold belts, and mud volcanoes of this type are known in the Tuapse Trough. The most complete information can be found in Andreev et al. (2006). Some details about them are summarized in Table 6.7, and their locations are plotted in Fig. 6.200.

In the western part of the Tuapse Trough, mud volcanoes are associated with anticlines, in particular with the large Manganari anticline (Fig. 6.201).

This anticline is located between the foot of the continental slope and the abyssal plain. In the bottom relief, it is expressed as a narrow, arcuate ridge, 200 m high and 25 km long at water depths of 900–1950 m; at the top, there are mud volcanoes. The bottom sediments sampled contained manifestations of oil and gases, such as methane (dominant), ethane (noticeable), ethylene, propane, propylene, n-butane, i-butane, butylene, n-pentane, i-pentane, and hexane (minor). Carbon dioxide is also present.

The mud volcanoes of the area can be exemplified by Manganari 1 and Manganari 2, which were discovered by Andreev et al. (2006) (Table 6.7, Fig. 6.202). The Manganari 2 mud volcano is smaller in size compared to Manganari 1.

Out of ten cores recovered from these mud volcanoes, eight revealed manifestations of oil coming from Novochernomorian and Neoeuxinian muds, which also smelled of hydrogen sulfide.

Fig. 6.198 Explosion of the Golubitskiy mud volcano in 2015. (Internet resource)

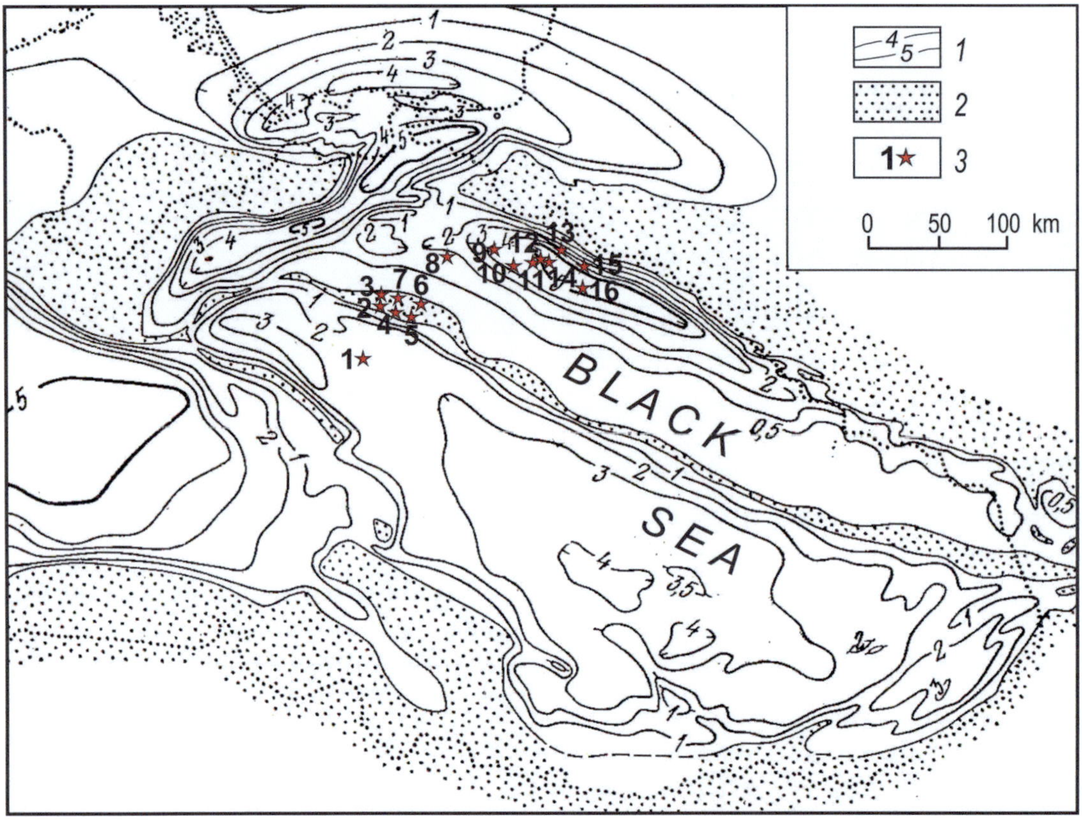

Fig. 6.199 Location of mud volcanic areas in the Tuapse Trough, Shatskiy Ridge, and Eastern Black Sea depressions (for location see Fig. 6.126). Key: (1) isopachytes of Maikopian sediments; (2) areas where Maikopian deposits are absent; (3) mud volcanoes identified in the key. Numbered mud volcanoes: (1) Gnom, (2) Udodov, (3) Irkutskiy, (4) Platformennyy, (5) Volokhin, (6) Panaev, (7) Dolgovskiy, (8) Arkhivnyy, (9) Neftyanoy, (10) Ekolog, (11) Manganari 1, (12) Manganari 2, (13) Periklinal, (14) Hydrograph, (15) Piter Gaz, (16) Bezymyannyy

Table 6.7 Coordinates and some characteristics of the mud volcanoes within the eastern part of the Black Sea

Mud volcano	Coordinates		Depth of the sea above the top of the cone, m	Cone height above the seabed, m	Area, m^2	Volume, m^3
	N	E				
Tuapse Trough						
Bezymyannyy	44° 07′ 22″	38° 03′ 22″	1750	10	250 × 250	No data
Ekolog	44° 14′ 38″	37° 27′ 20″	1860	30	600 × 500	No data
Manganari 1	44° 09′ 11″	37° 43′ 30″	1755	60	1000 × 600	No data
Manganari 2	44° 09′ 22″	37° 43′ 41″	1780	10	300 × 250	No data
Neftyanoy	44° 17′ 50″	37° 21′ 07″	1930	30	500 × 400	No data
Periklinal	44° 15′ 46″	37° 56′ 44″	1790	25	550 × 500	No data
Piter Gaz	44° 11′ 29″	38° 06′ 22″	1655	50	500 × 500	No data
Hydrograph (Simrad)	44° 12′ 43″	37° 52′ 51″	1610	20	350 × 350	No data
Shatskiy Ridge						
Arkhivnyy	44° 13′ 48″	35° 50′ 40″	1765	20	1000 × 1000	No data
Dolgovskiy	44° 01′ 08″	35° 41′ 22″	1940	45	700 × 600	No data
Gnom	42°41′16″	36°24′48″	2140–2150	80	200	No data
Irkutskiy	44°04′47′	36°33′67″	1880–1905	25	400ч400	
Irkutskiy	44° 04′ 28″	36° 33′ 40″	1880	25	400 × 400	No data
Panaev	44° 01′ 34″	36° 47′ 39″	1980	30	400 × 400	No data
Platformennyy	43° 59′ 43″	36° 40′ 28″	1975	25	350 × 300	No data
Udodov	44° 01′ 10″	36° 32′ 36″	1990	20	750 × 750	No data
Volokhin	43° 56′ 20″	36° 46′ 41″	2035	25	800 × 800	No data

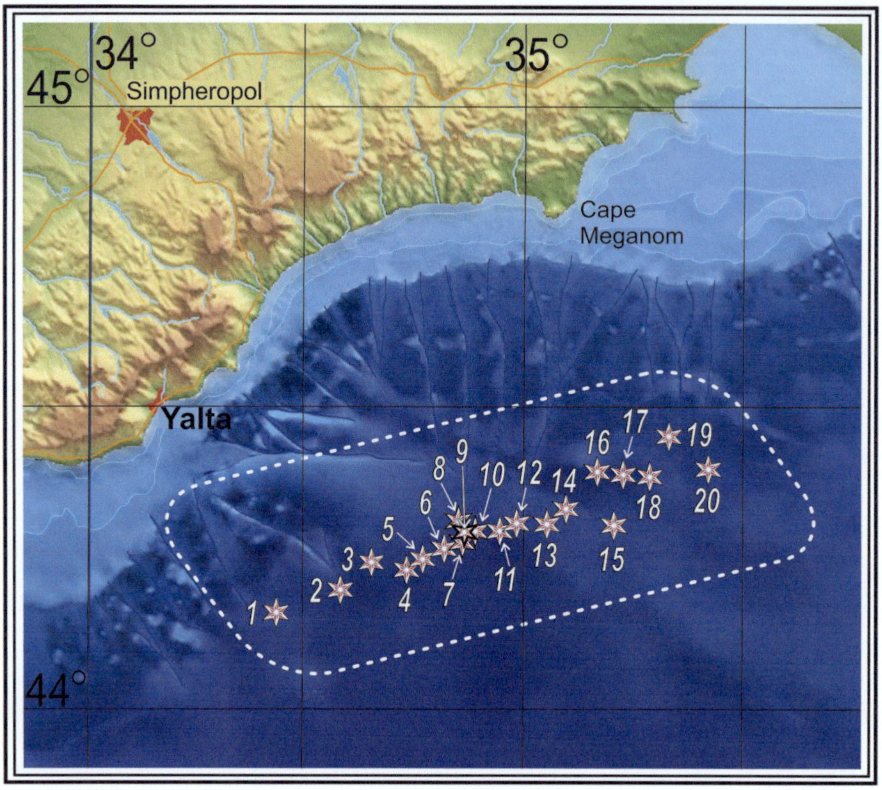

Fig. 6.200 Location of mud volcanoes in the Tuapse Trough and Shatskiy Ridge. Key: (1) Gnom, (2) Udodov, (3) Irkutskiy, (4) Platformennyy, (5) Volokhin, (6) Panaev, (7) Dolgovskiy, (8) Arkhivnyy; (9) Neftyanoy; (10) Ekolog; (11) Manganari 1; (12) Manganari 2; (13) Periklinal; (14) Hydrograph (Simrad); (15) Piter Gaz; (16) Bezymyannyy

This indicates that mud volcanoes were active in pre-Holocene times, and subsequently, the volcanic breccia was covered by marine sediments. Oil penetrated through nonstructured mud and breccia and was retained within the lower part of the sediment cover.

To the northwest of the Manganari anticline lies the Geoeko anticline, which is buried beneath the Kuban River deepwater cone. Within the Geoeko anticline, there are the "double-humped" mud volcanoes Ekolog and Neftyanoy (Table 6.7, Fig. 6.200).

The Neftyanoy mud volcano complicates the eastern part of the buried Geoeko anticline and was discovered in 2004 by S. Khrenov and D.A. Tugolesov (Andreev et al. 2006). The volcano is currently active; its hill is composed of a coarse-layered mud volcanic breccia represented by gray fluid-plastic mud with rock debris and gruss of strong gray, non-laminated clays.

Mud volcanism is also widely manifested in the western part of the Shatskiy Ridge on the southern slope of the North Black Sea uplift. Studies by "Yuzhmorgeologiya" (Andreev et al. 2006) established four groups of mud volcanoes that form gentle hills 15–30 m high and 300–800 m in diameter. They were named in memory of geophysicist E.Yu. Dolgovskiy, geologist V.A. Panaev, hydrographer V.A. Volokhin, and oceanologist, A.I. Udodova (Table 6.7; Fig. 6.203).

These groups are separated from each other by about 10–12 km, and each includes up to 5–7 mud volcanoes. The central part is occupied by the Platformennyy and Dolgovskiy mud volcanoes. The latter has a plateau-shaped surface that is

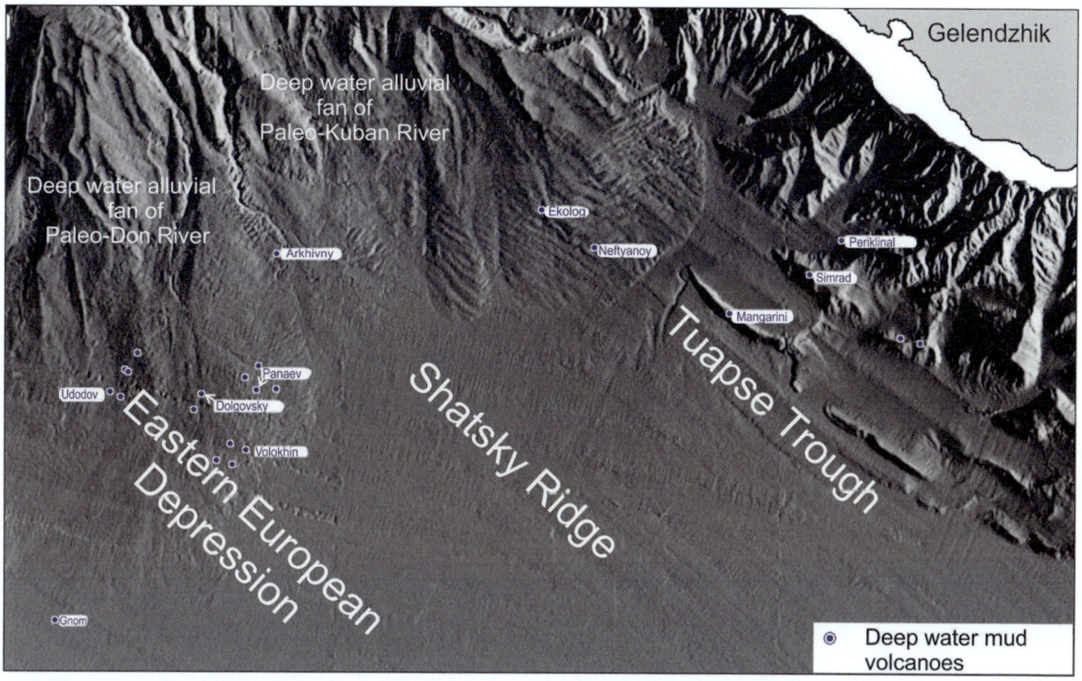

Fig. 6.201 Diagram of the main elements of relief and location of mud volcanoes in the Tuapse Trough (Andreev et al. 2006)

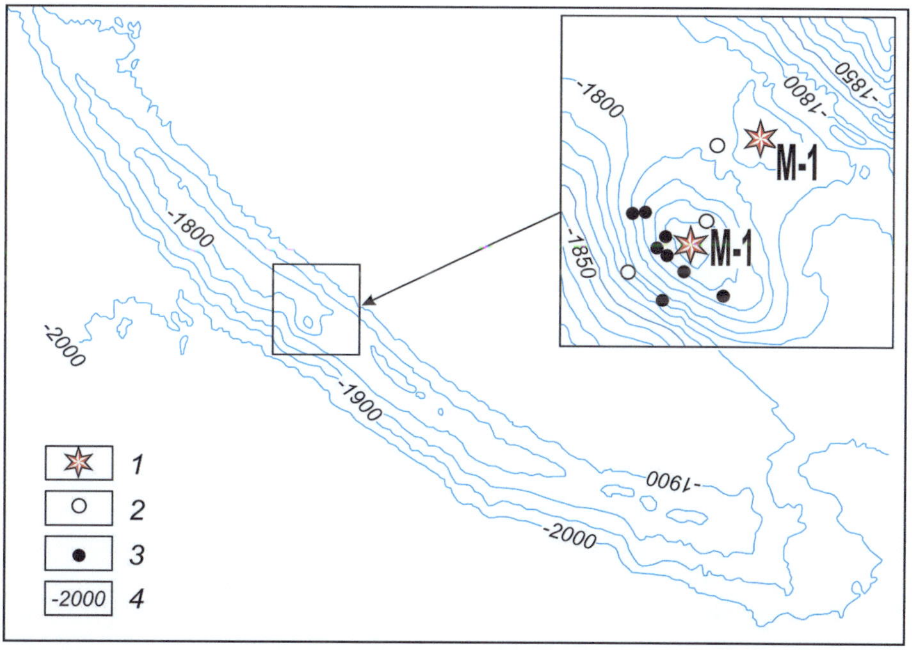

Fig. 6.202 The Manganari 1 and 2 mud volcanoes (Andreev et al. 2006). Key: (1) mud volcanoes, (2) core sampling stations, (3) core sampling stations with oil manifestations, (4) isobaths (m)

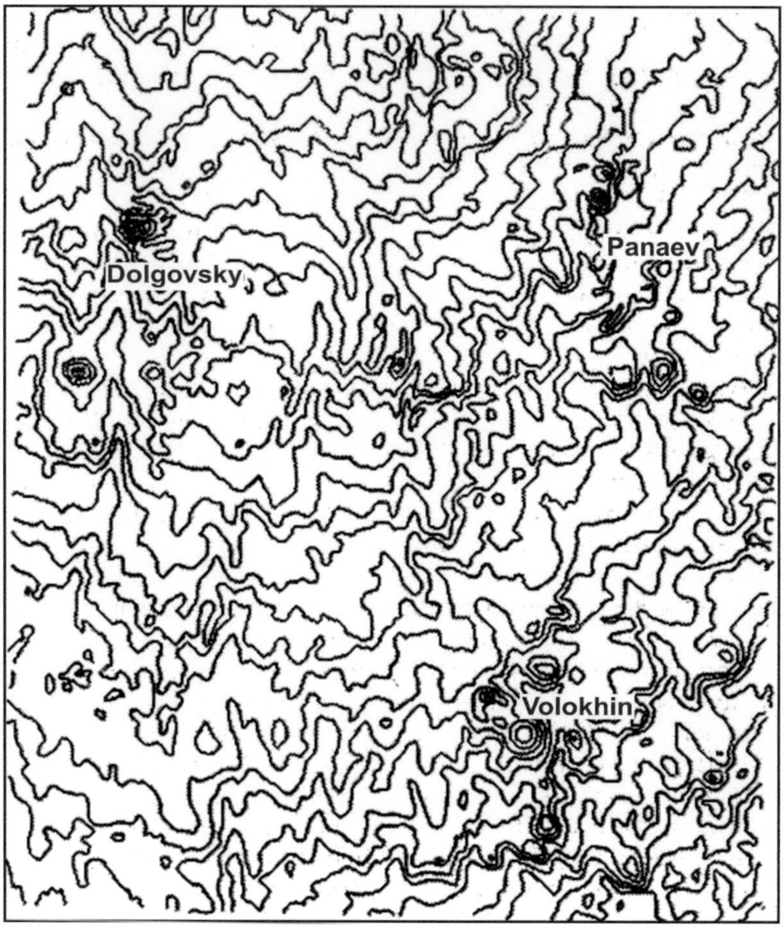

Fig. 6.203 Map of the location of mud volcanoes on the Shatskiy Ridge

15 m in diameter and rises 150 m above the seafloor (Figs. 6.204 and 6.205).

The hill of this mud volcano is composed of gray, terrigenous, aleuropelitic mud with a thickness exceeding 130 m and a conglomerate consisting of dark gray calcareous silt with 20% content of rounded inclusions consisting of gray calcareous true clay of 0.5–5.0 cm, and sometimes up to 10 cm in size. The mud contains coarse grains of quartz larger than 1 mm, concretions of authigenic pyrite, and Late Oligocene foraminiferal assemblages. The top of the mud volcano is surrounded by three peripheral hills, each 5–10 m high. The mud volcanic breccia and mud are sources of powerful methane emissions.

Farther to the east, on the meridian of the city of Gelendzhik, near the southern boundary of the Russian economic zone, there is a dome 4 km in diameter. About 15 km to the west, at a water depth of 400–500 m, a seismoacoustic profile revealed a Pleistocene mud volcano with height of about 100 m buried under the seabed (Andreev et al. 2006).

6.12 Mud Volcanoes of Other Sectors of the Black Sea

There is not much information available on mud volcanoes in other parts of the Black Sea. Two mud volcanoes, Iberia and Pechori, have been discovered in the Georgian sector of the Black Sea (Shnyukov et al. 2015, Fig. 2.231). The Iberia mud volcano (41° 52.340′ N; 41° 10.056′ E) is located at a water depth of 982–989 m; it was investigated from on board the R/V "Logachev" in 2005 (Akhmetzhanov et al. 2005), and the study was continued in the

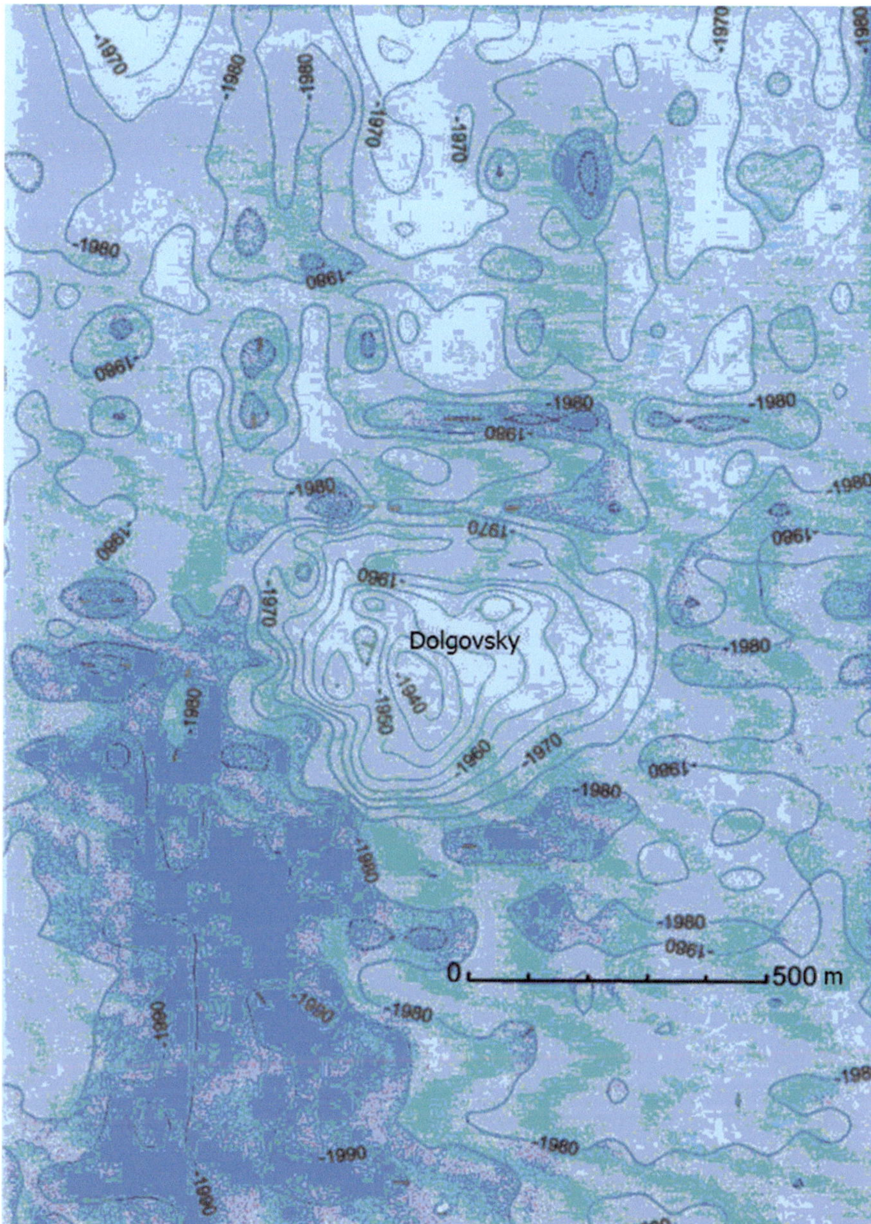

Fig. 6.204 Bottom relief map of the Dolgovskiy mud volcano (Shnyukov et al. 2015)

72nd cruise of the R/V "Meteor" in 2007. This mud volcano has the shape of a cone with a height of 30 m and a diameter of 1 km. Under a layer of bottom sediments, there is a layer of oil-saturated mud volcanic breccia samples of which were raised for the study on board the vessel. The Pechori mud volcano (41° 58.990′ N; 41° 07.595′ E) was discovered at a water depth of 1025 m during the 15th cruise of the R/V "Logachev" (Bohrmann et al. 2007). It forms a conical hill 75 m high and 2–2.5 km in diameter and ejects from 400 to 3000 tons of oil per year (Shnyukov et al. 2015).

In the Turkish sector of the Black Sea, in the Giresunkaya depression, three supposedly mud volcanic structures were explored during the

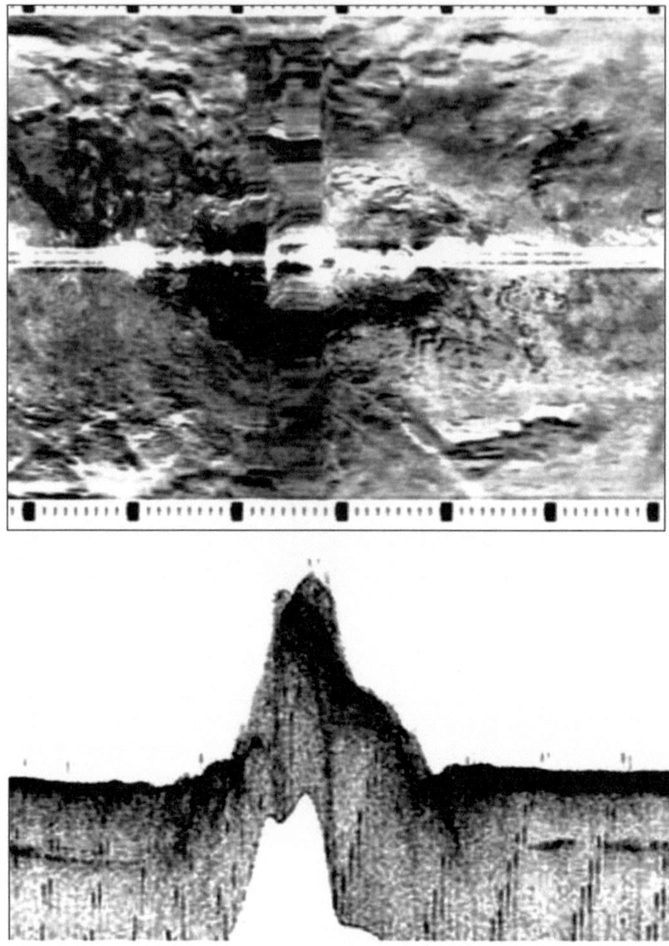

Fig. 6.205 Fragment of a sonogram and profilogram taken by MAK-1 deep-towed acoustic system through the Dolgovskiy mud volcano (Andreev et al. 2006)

voyage of the R/V "Logachev" (Akhmetzhanov et al. 2005), but mud breccia was not found. One of the mud volcanoes was described by Kruglyakova et al. (2009). In the same area, Sahling et al. (2009) described natural oil seeps that have been known for a long time. In course of the survey in the Trabzon-Giresun area, 1700 km high-resolution multichannel seismic and chirp data simultaneously were collected on board the R/V "K. Piri Reis." Two mud volcanoes were discovered and named Busan and Izmir. The observed fold belt is believed to be the main driving force for the growth of mud volcanoes (Okay et al. 2012).

During research cruises in 2010, 2012, and 2016, a total of approximately 4500 km of multichannel seismic (MCS) and chirp sub-bottom profiler datasets were acquired simultaneously along the Western Black Sea Turkish margin aboard the R/V "K. Piri Reis" of Dokuz Eylül University's Institute of Marine Sciences and Technology. The acoustic data indicate the existence of nine mud volcanoes, all located at the continental rise region of the shelf margin (Dondurur et al. 2008; Dondurur et al. 2018). Figure 6.206 shows the locations of the mud volcanoes deduced from seismic data, but only the westernmost one has a name: Sakarya mud volcano. No ground truthing has yet been obtained from the mud volcanoes, and thus no geochemical analysis of mud breccia has yet been performed.

The mud volcanoes tend to group in two areas: the five eastern ones (MV01 to MV05) offshore from Amasra, and the four western ones (MV06

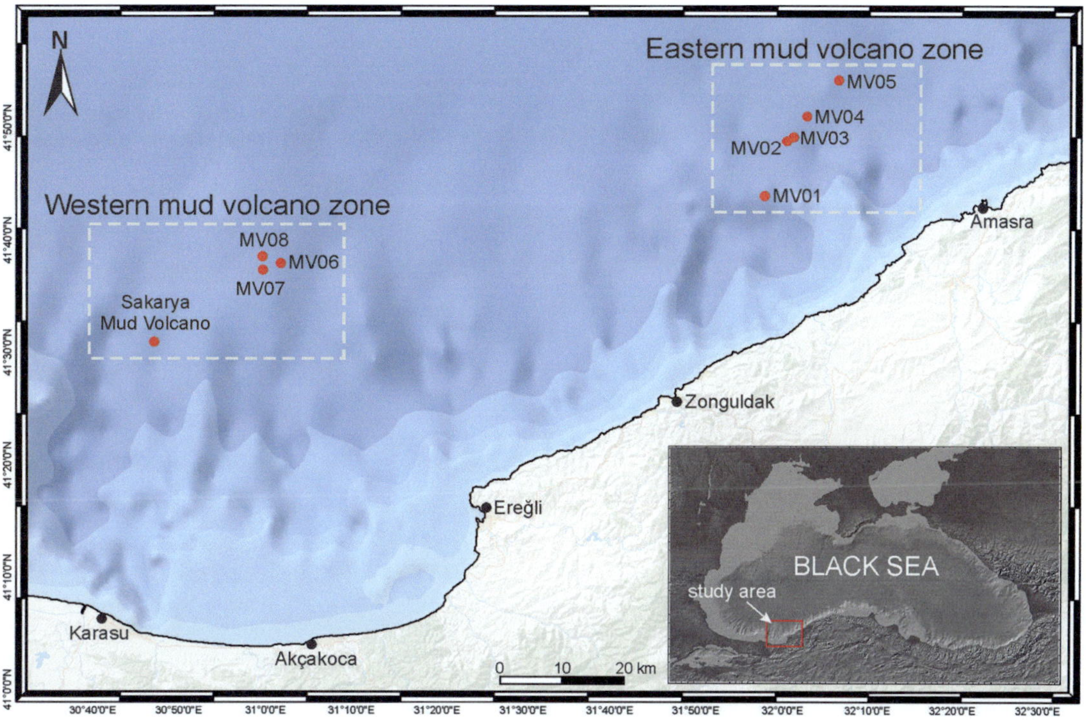

Fig. 6.206 Location of the mud volcanoes along the Western Black Sea Turkish margin obtained from multichannel seismic data (Dondurur, personal communication from October 27, 2019)

to MV08 and the Sakarya mud volcano) offshore from Akçakoca and Karasu (Fig. 6.206). Table 6.8 summarizes the general characteristics of the mud volcanoes in the areas.

Those in the eastern zone tend to follow a linear NE to SW trend, which strongly suggests that they occur along a fault trace within the deeper sediments. Of those in the western zone, however, all but one cluster within a small localized area situated over a relatively large sedimentary ridge structure; the Sakarya mud volcano by contrast lies in pelagic sediments of the flat sedimentary platform to the west. The feeder channel widths of the volcanoes vary from 150 to 1028 m, and their crater heights above the surrounding seafloor range between 8 and 63 m. Figure 6.2 shows seismic soundings of mud volcanoes from both eastern and western zones (Özel et al. 2015; Nasıf and Dondurur 2017). They reveal typical characteristics of the Black Sea mud volcanoes that have been encountered in the Sorokin Trough: (1) their transparent feeder channel with almost no internal reflections is quite distinct on the seismic data; (2) enhanced reflections are commonly observed on both sides of the feeder channels possibly indicating bright spots; (3) the interfaces at depth bend upward around the feeder channels for some of the mud volcanoes; and (4) there are almost always amplitude dimming zones within the deeper sediments below the mud volcanoes. In addition, the volcanoes, especially in the western zone, are commonly accompanied by gas chimneys (Fig. 6.207a).

Feeder channels crosscut the BSR reflections on the seismic data, which correspond to base of gas hydrate layers, whenever they exist. This situation suggests that the warmer fluids carried by the mud volcanoes from deeper sediments into the gas hydrate stability zone play an important role in the local dissociation of the gas hydrates.

Interestingly, MV03 and MV06 possess twin feeder channels which then coalesce to form a larger feeder channel at shallower depths. There are no "Christmas tree-type" feeder channels in

Table 6.8 Coordinates and some characteristics of the mud volcanoes within the Eastern Black Sea Turkish margin (Dondurur, personal communication from October 27, 2019)

Name	Width at the seafloor (m)	Crater height (m)	Characteristics
MV01	918	63	Located at the sedimentary ridge crest, relatively flat crater top, located along a possible fault
MV02	293	29	Located at the sedimentary ridge crest, conical crater shape, enhanced reflections at both sides of the feeder channel
MV03	1028	32	Located at the sedimentary ridge crest, relatively flat crater top, enhanced reflections at both sides of the twin feeder channels in deeper sediments
MV04	281	11	Located at the sedimentary ridge crest, relatively flat crater top, BSRs and enhanced reflections at both sides of the feeder channel
MV05	535	53	Located at the sedimentary ridge crest, conical crater shape, located along a possible fault
MV06	680	23	Twin conical craters and possible twin feeder channels, enhanced reflections at both sides of the feeder channel
MV07	150	8	Located within the pelagic sediments, relatively flat crater top, BSR and enhanced reflections at both sides of the feeder channel
MV08	730	18	Located within the pelagic sediments, conical crater shape, BSR and enhanced reflections at both sides of the feeder channel
Sakarya	800	15	Located within the pelagic sediments, conical crater shape, upward bending of deeper layers at both sides of the feeder channel

the area, which indicates that the eruptions are somewhat continuous as opposed to periodic large explosive events occurring after certain time intervals. We also suggest that the mud volcanoes with conical craters such as MV02, MV05, MV06, MV08, and Sakarya are possibly more active than those with flat-dome-shaped craters. The upward bending of the deeper layers around the feeder

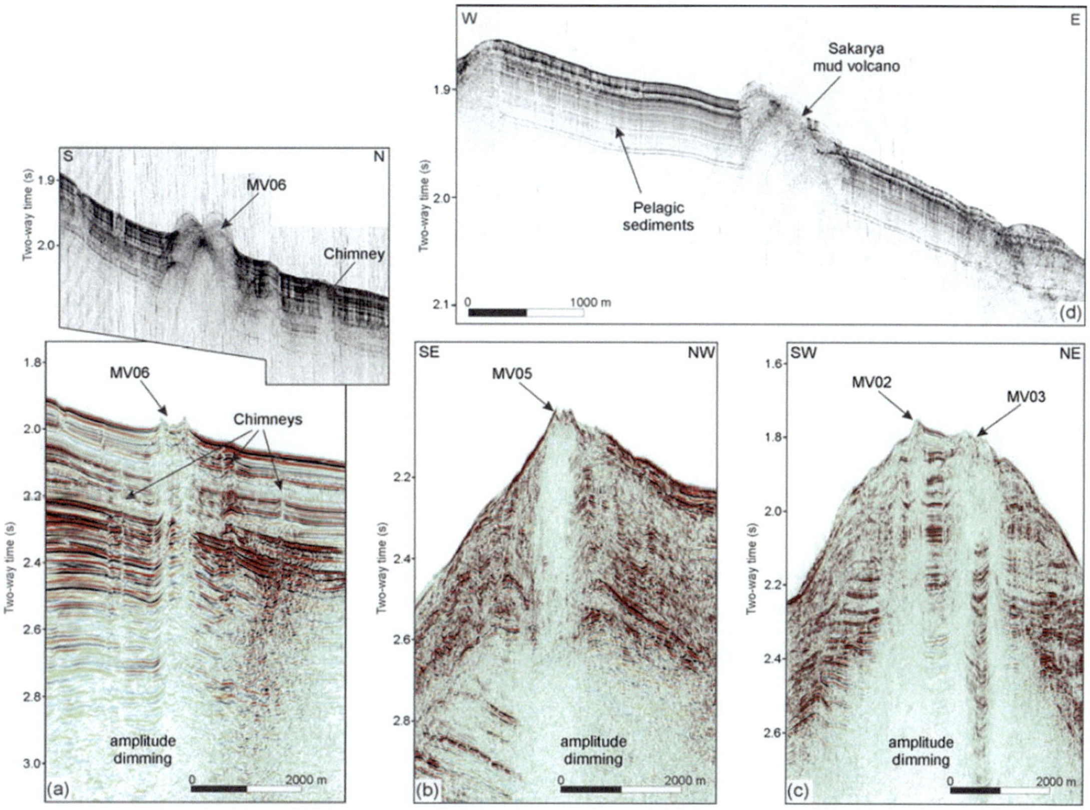

Fig. 6.207 Examples of seismic profiles for mud volcanoes discovered in the margin. (**a**) MV06 mud volcano in MCS (bottom) and chirp (top) seismic lines, (**b**) MCS line of MV05 mud volcano, (**c**) MCS line of MV02 and MV03 mud volcanoes, and (**d**) chirp seismic line of the Sakarya mud volcano (Dondurur et al. 2008, 2018; Özel et al. 2015; Nasıf and Dondurur 2017)

channels also supports this conclusion. Distribution of the mud volcanoes in the area shows that those in the eastern zone are fault-controlled since they appear along a linear trend, and we observe indications of faulting, especially below and around the feeder channels of MV01 and MV05. The volcanoes in the western zone, however, seem to occur within a local disruption zone, generally in pelagic or hemipelagic sediments.

On the Bulgarian shelf in the Trakya area, on a voyage of the R/V "Logachev," two cone-shaped mud volcanoes (41° 55.242′ N; 28° 55.495′ E and 41° 55.203′ N; 28° 55.479′ E) were observed at water depths of 880–883 m, and mud volcanic breccia together with fragments of methane gas hydrates were raised on board the vessel (Shnyukov et al. 2015).

References

Abich GV (1873) Geologicheskiy obzor poluostrovov Kerchi i Tamani (geological survey of the Kerch and Taman peninsulas). Zapiski Kavkazskogo otdeleniya russkogo geograficheskogo obshchestva 8:3–160 (in Russian)

Abryutskiy LP (1853) O deistvii gryazevogo vulkana na Tamanskom poluostrove v Avguste 1853 goda (on the activity of a mud volcano on the Taman Peninsula in August 1853). Gorny Zhurnal Book 10(Part 4):271–277 (in Russian)

Akhmetzhanov AM, Ivanov MK, Kenyon NH et al (2005) Deep-water cold seeps, sedimentary environments and

ecosystems of the Black and Tyrrhenian Seas and Gulf of Cadiz: preliminary results of investigations during the TTR-15 Cruise of RV Professor Logachev, June–August, 2005. IOC Technical Series 72. UNESCO
Alekseev P (1880) Zametki o gryazevykh vulkanakh (Notes about mud volcanoes). Zapiski Kievskogo Obschestva Estestvoispytateley 6(1):137–163 (in Russian)
Almendinger OA (2011) Trekhmernye modeli neogen-sovremennykh skladchatykh sedimentologicheskikh i erozionnykh protsessov (three-dimensional models of Neogene-modern folded sedimentological and erosion processes). Dissertation, Moscow (in Russian)
Alyaev SE (1947) Novye dannye o tektonike Kerchenskogo poluostrova (new data about tectonics of the Kerch Peninsula). Izvestya AN SSSR Seriya Geol 6:97–99 (in Russian)
Andreev VM, Tugolesov DD, Khrenov SN (2006) Gryazevye vulkany i nefteproyavleniya rossiyskogo sektora Chernogo morya (mud volcanoes and oil manifestations of the Russian sector of the Black Sea). Geologiya i poleznye iskopaemye mirovogo oceana 3:50–59 (in Russian)
Andrusov NI (1884) Geologicheskie issledovaniya na Kerchenskom poluostrove v 1862–1883 g. (Geological research on the Kerch Peninsula in 1862–1883). Zapiski Novorossiyskogo obshchestva estestvoispytateley X:1–76 (in Russian)
Andrusov NI (1893) O sostoyanii Chernomorskogo basseyna v pliotsenovuyu epokhu (On the state of the Black Sea basin in the Pliocene epoch). Mélanges géologiques et paléontologiques I:165–178 (in Russian)
Andrusov NI (1918) Geologicheskoe stroenie dna Kerchenskogo proliva (geological structure of the bottom of the Kerch Strait). AN SSSR Seriya 12(1):23–28 (in Russian)
Andrusov NI (1926) Geologicheskoe stroenie i istoriya Kerchenskogo proliva (geological structure and history of the Kerch Strait). Byull Mosk Obshch Ispyt Prir Otd Geol 34:294–332 (in Russian)
Arbuzov VA, Bobrushkin LG, Litvinenko AU (1967) Kerchenskiy zhelezorudnyy basseyn (Kerch iron ore basin). Nedra, Moscow (in Russian)
Artemov YG (2006) Software support for investigation of natural methane seeps by hydroacoustic method. Morskiy ekologichnuy zhurnal (Marine Ecol J) 5(1):57–71
Avdusin PP (1948) Gryazevye vulkany Krymsko-Kavkazskoy geologicheskoy provintsii (mud volcanoes of the Crimean-Caucasian geological province). AN SSSR, Moscow (in Russian)
Basov EI, Ivanov MK (1996) Pozdnechetvertichnyy gryazevoy vulkanizm v Chernom more (late quaternary mud volcanism in the Black Sea). Litol Pol Iskop 2:215–222 (in Russian)
Belousov AV, Yarotskiy LA (1934) Nekotorye obshchie voprosy tektoniki Kerchensko-Tamanskoy oblasti (some common questions on tectonics of the Kerch-Taman region). Problemy sovetskoy geologii 3:207–227 (in Russian)
Belousov AV, Yarotskiy LA (1936) Gryazevye sopki Kerchensko-Tamanskoy oblasti, usloviya ikh vozniknoveniya i deyatel'nosti (mud hills of the Kerch-Taman region, the conditions of their occurrence and activity). ONTI NKTP SSSR (in Russian)
Bogdanov YuA, Lisitsyn AP, Sagalevich AM et al (2006) Gidrotermalnyy rudogenez okeanskogo dna (hydrothermal rudogenesis of the ocean floor). Nauka, Moscow (in Russian)
Bohrmann G (2011) Report and preliminary results of RV MARIA S. MERIAN Cruise MSM 15/2, Istanbul (Turkey) – Piraeus (Greece), 10 May–2 June 2010. Origin and structure of methane, gas hydrates and fluid flows in the Black Sea. University of Bremen, Bremen
Bohrmann G, Ivanov M, Foucher JP et al (2003) Mud volcanoes and gas hydrates in the Black Sea: new data from Dvurechenskii and Odessa mud volcanoes. Geo-Mar Lett 23(3–4):239–249
Bohrmann G et al (2007) Report and preliminary results of R/V METEOR cruise M72/3, Istanbul–Trabzon–Istanbul, 17 March–23 April, 2007. Marine gas hydrates of the Eastern Black Sea. University of Bremen, Bremen
Borisyak AA (1907) Doklad po voprosu osmotra gryazevoy sopki bliz Vladislavovki (report on the inspection of the mud hill near Vladislavovka). Izvestiya Geolkoma 26(3):34–36 (in Russian)
Bouriak SV, Akhmetzhanov AM (1998) Origin of gas hydrate accumulations on the continental slope of the Crimea from geophysical studies. In: Henriet JP, Mienert J (eds) Gas hydrates: relevance to world margin stability and climate change, Geological Society, vol 137. Special Publications, London, pp 215–222
Brückner H, Kelterbaum D, Marunchak O et al (2010) The Holocene sea level story since 7500 BP – lessons from the Eastern Mediterranean, the Black and the Azov Seas. Quat Int 225(2):160–179
Chirvinskiy PN (1908) Zametka o gryazevykh sopkakh Kerchenskogo poluostrova (a note on the mud hills of the Kerch Peninsula). Zapiski Kievskogo obschestva estestvo ispytateley 20(3):791–797 (in Russian)
Dimitrov LI (2002) Mud volcanoes—the most important pathway for degassing deeply buried sediments. Earth Sci Rev 59:49–76
Dmitrievskiy AN, Valyaev BM (2010) Uglevodorodnaya degazatsiya Zemli i genezis neftegazovykh mestorozhdeniy: razvitie idey P.N. Kropotkina (Hydrocarbon degassing of the Earth and the genesis of oil and gas fields: the development of the ideas of P.N. Kropotkin). Materialy konferentsii "Degazatsiya Zemli…." GEOS, Moscow, pp 7–10 (in Russian)
Dubois de Montpéreux F (1837) Pis'mo o glavnykh geologicheskikh yavleniyakh na Kavkaze i Krymu Eli de Bomonu (letter on the main geological phenomena in the Caucasus and the Crimea to Eli de Beaumont). Gornyy zhurnal 3, 4:345–394 (in Russian)

Dvoychenko PA (1914) Izverzhenie gryazevoy sopki Dzhau-Tepe (eruption of the mud volcanic hill Dzhau-Tepe). Priroda 4:614–619 (in Russian)

Dvoychenko PA (1926) Estestvennye proizvoditelnye sily Kryma (natural productive forces of the Crimea). In: Ves' Krym. KrymTSIK, Simferopol, pp 1–18 (in Russian)

Eberzin AG (1933) Geologicheskie issledovaniya mestorozhdeniy Kerchenskogo poluostrova (geological studies of deposits of the Kerch Peninsula). Trudy VGRO, AN SSSR, Moscow (in Russian)

Eberzin AG (1940) Sredniy i verkhniy pliotsen Chernomorskoy oblasti. Kimmeriyskiy yarus (Middle and Upper Pliocene of the Black Sea. Kimmerian stage) Stratigrafiya SSSR. Neogen, pp 477–527 (in Russian)

Egorov VN, Artemov YuG, Gulin SB (2011) Metanovye sipy v Cherom more. Sredoobrasuyushchaya i ekologicheskaya rol'. Sevastopol' (Methane seeps in the Black Sea. Environmental and ecological role). EKOSI-gidrofisika, Sevastopol (in Russian)

Engibaryan AA (2006) Litologo-fatsialnye i tektonicheskie kriterii neftegazonosnosti mezokaynozoyskikh otlozheniy Tamanskogo poluostrova (Lithofacies and tectonic criteria for the oil and gas potential of the Meso-Cenozoic sediments of the Taman Peninsula). Dissertatsiya, Stavropol (in Russian)

Fedorov AN (1931) Geologicheskoe opisanie chasti Dzhiginsko-Varenikovskogo rayona (Geological description of part of the Dzhiginsko-Varenikovskiy district). Trudy neftyanogo geologo-razvedochnogo instituta, seriya B(4):1–33 (in Russian)

Felitsyn EV (1902) Nekotorye svedeniya o gryazevykh vulkanakh Tamanskogo poluostrova (some information about the mud volcanoes of the Taman Peninsula). Izvestiya obshchestva lyubiteley izucheniya Kubanskoy oblasti (in Russian)

Gembitskiy AA (1914) Izverzhenie Dzhau-Tepe (Eruption of Dzhau-Tepe). Yuzhnoe Inzhenernoe Obshchestvo 4:6 (in Russian)

Gerts K (1876) Istoricheskiy obzor arkheologicheskikh issledovaniy i otkrytiy na Tamanskom poluostrove (historical review of archaeological research and discoveries on the Taman Peninsula). Moscow (in Russian)

Ginsburg GD, Gramberg IS, Guliev IS (1988) Podvodno-gryazevulkanicheskiy tip skopleniy gazovykh gidratov (underwater mud volcanic type of gas hydrate accumulation). Doklady AN SSSR 300(2):416–418 (in Russian)

Ginsburg GD, Kremlev AN, Grigor'ev MN (1990) Fil'trogennye gazovye gidraty v Chernom more (Filtrogenic gas hydrates in the Black Sea). Geologiya i geofisika 3:10–20 (in Russian)

Golovkinskiy NA (1889) Otchet gidrogeologa Tavricheskoy zemskoy upravy za 1989 god (report of the hydrogeologist of the Taurian zemstvo government for 1889). Simferopol (in Russian)

Gonchar AI, Pisanko IN, Sobisevich LE et al (2004) Podvodnyy gryazevoy vulkanizm Azovo-Chernomorskogo basseyna (underwater mud volcanism of the Azov-Black Sea basin). Gidroakustichniy zhurnal 1:70–83 (in Russian)

Grechishkin LA (1931) Geologicheskoe opisanie Dzhanakbatskoy kotloviny na Kerchenskom poluostrove (Geological description of the Dzhanakbatskiy basin on the Kerch Peninsula). Trudy GRGU OSNKH "Geologo-razvedochnye issledovaniya v neftenosnykh raionakh Kerchenskogo poluostrova v 1926 godu" 38(1):85–89. Geolizdat GRGU, Moscow (in Russian)

Greinert J, Artemov Y, Egorov V et al (2006) 1300-m-high rising bubbles from mud volcanoes at 2080 m in the Black Sea: hydroacoustic characteristics and temporal variability. Earth Planet Sci Lett 244(1–2):1–15

Gubanov IG (1959) K paleogeografii severnoy gruppy zhelezorudnykh mul'd Kerchenskogo poluostrova (on the paleogeography of the northern group of iron ore deposits of Kerch). Izvesiya Krymskogo pedagogicheskogo instituta 34:23–37 (in Russian)

Gusakov IN (2004) Ritmika gryazevogo vulkanizma Tamanskogo poluostrova (The rhythm of mud volcanism on the Taman Peninsula). Azovskoe otdelenie "Kuban'heologiya," Temryuk, p 61–62 (in Russian)

Helmersen GP (1864) Issledovanie gryazevykh vulkanov i neftyanykh istochnikov v Krymu i ne Tamanskom poluostrove (study of mud volcanoes and oil sources in Crimea and on the Taman peninsula). Zapiski Imperatorskogo Mineralogicheskogo Obshchestva Part 1:294–295 (in Russian)

Herbin JP, Saint-Germès M, Maslakov NA et al (2008) Oil seeps from the "Boulganack" mud volcano in the Kerch Peninsula (Ukraine – Crimea), study of the mud and gas: inferences for the petroleum potential. Oil Gas Sci Technol – Rev IFP 63(5):609–628

Herz K (1876) Istoricheskiy obzor arkheologicheskikh issledovaniy i otkrytiy na Tamanskom poluostrove s kontsa XVIII stoletiya do 1859 goda. Preimushchestvenno po neizdannym istochnikam (Historical overview of archaeological research and discoveries on the Taman Peninsula from the end of the 18th century until 1859. Predominantly from unpublished sources). Sobranie sochineniy, vypusk 2. Tipografiya Imperatorskoy AN, Sankt-Peterburg (in Russian)

Ilyin SN (1931) Issledovanie neftyanykh mestorozhdeniy vostochnoy chasti Kerchenskogo poluostrova. Trudy Geologo-Razvedochnogo Upravleniya 39:37–48 (in Russian)

Ivanov MK (1999) Fokusirovannye uglevodorodnye potoki na glubokovodnykh okrainakh kontinentov (focused hydrocarbon flows in the deep-water margins of the continents). Avtoreferat doktorskoy dissertatsii, MGU, Moscow (in Russian)

Ivanov MK, Konyukhov AI, Kul'chitskiy LM et al (1989) V glubokovodnoy chasti Chernogo morya (mud volcanoes in the deep sea of the Black Sea). Vestnik Moskovskogo Univeriteta, Seriya Geologicheskaya 3:48–54 (in Russian)

Ivanov MK, Limonov AF, Cronin BT (eds) (1996) Mud volcanism and fluid venting in the eastern part of the Mediterranean Ridge. Initial results of geological, geophysical and geochemical investigations during the 5th "Training-Through-Research" Cruise of RV Professor Logachev (July–September, 1995). UNESCO Reports in Marine Science 68

Ivanov MK, Limonov AF, van Weering TCE (1997) Comparative characteristics of the Black Sea and Mediterranean Ridge mud volcanoes. Mar Geol 132(1–4):252–271

Izgaryshev NA, Sludskiy AF (1917) Gryazevye vulkany Kerchenskogo poluostrova i Temryuksko-Tamanskogo rayona (mud volcanoes of the Kerch Peninsula and Temryuk-Taman region). Rudnye vesti 2(3/4):97–112 (in Russian)

Kalinko MK (1960) O mekhanizme i usloviyakh obrazovaniya gryazevykh vulkanov (on the mechanism and conditions of formation of mud volcanoes). Problemy neftyanoy geologii. Gostekhizdat, Moscow, p 83–90 (in Russian)

Kenyon NH, Ivanov MK, Akhmetzhanov AM et al (2002) Geological processes in the Mediterranean and Black seas and North East Atlantic. Preliminary results of investigations during the TTR–11 of RV professor Logachev (July–September, 2001). UNESCO Reports in Marine Sciences 62

Khain Ye V, Zhabrev IP, Shapiro AR et al (1953) Tipy tektonicheskikh i gryazevulkanicheskikh brekchiy Kobystana (Types of tectonic and mud volcanic breccias of Kobystan). Azerbaijankoe neftyanoe khozyastvo 6:5–7 (in Russian)

Khortov AV, Neprochnov YuP (2006) Glubinnoe stroenie i nekotorye voprosy neftegazonosnosti yuzhnykh morey Rossii (deep structure and some issues of oil and gas potential of the southern seas of Russia). Okeanologiya 46(1):114–122 (in Russian)

Klepinin NN (1914) Gryazevye sopki Kerchenskogo poluostrova i izverzhenie sopki Dzhau-Tepe v 1914 godu (Mud hills of the Kerch Peninsula and eruption of the Dzhau-Tepe hill in 1914). Po Krymu No 2, Simferopol (in Russian)

Konstantov SV (1929) Otchet geologicheskogo komiteta za 1929–1927 gody (report of the geological committee for 1929–1927). Izvestiya Geolkoma (in Russian)

Konstantov SV, Kechek GA, Krasil'nikov LK et al (1933) Kerchenskie zhelezorudnye mestorozhdeniya (Kerch iron ore deposits). Trudy VGRO NKGP SSSR 325:1–128 (in Russian)

Konyukhov AM, Ivanov MK, Kulnitskiy AM (1990) O gryazevykh vulkanakh i gazogidratakh v glubokovodnykh rayonakh Chernogo morya. Litol Pol Iskop 3:12–23 (in Russian)

Kovalevskiy SA (1935) Gazovyy vulkanizm (vulkany i vulkanoidy) (Gas volcanism [volcanoes and volcanoids]). Azerbaydzhanskoe neftyanoe khozyaystvo 1 (in Russian)

Kovalevskiy SA (1960) Geologicheskie cherty lineamenta 38 meridiana v rayone Chernogo morya (geological features of the lineament 38 meridian in the Black Sea region). Doklady AN SSSR 130(6):1306–1309 (in Russian)

Krastel S, Spiess V, Ivanov M et al (2003) Acoustic investigation of mud volcanoes in the Sorokin trough, Black Sea. Geo-Mar Lett 23(3–4):230–238

Kruglyakova RP, Shevtsova NT, Zubova MV (1998) Ekologicheskoe sostoyanie rek Kavkaza v svyazi s neftyanym zagryazneniem pribrezhno-morskoy zony Chernogo morya (Ecological state of the rivers of the Caucasus due to oil pollution of the coastal-marine zone of the Black Sea). Geoinformatsionnyy sbornik "Geoekologicheskie issledovaniya i okhrana nedr", Geoinformark 3:11–17 (in Russian)

Kruglyakova RP, Byakov YA, Kruglyakova MV (2004) Natural oil and gas seeps on the Black Sea floor. Geo-Mar Lett 24(3):150–162

Kruglyakova RP, Kruglyakova MV, Shevtsova NT (2009) Geologo-geokhimicheskaya kharakteristika estestennykh proyavleniy uglevodorodov v Chernom more (geological and geochemical characteristics of the natural manifestations of hydrocarbons in the Black Sea). Geologiya i poleznye iskopaemye Mirovogo okeana 1:27–51 (in Russian)

Kudryashev EV (1939) Geologicheskiy ocherk gryazevogo vulkana Shugo (geological sketch of the Shugo mud volcano). Rezultaty issledovaniya gryazevykh vulkanov Krymsko-Kavkazskoy geologicheskoy provintsii. AN SSSR, Moscow-Leningrad, pp 45–56 (in Russian)

Kurishko VA, Mesyats IA, Tverdovidov AS (1968) Gidrogeologiya gryazevogo vulkanizma Kerchenskogo poluostrova (hydrogeology of mud volcanism of the Kerch peninsula). Geologicheskiy zhurnal 28(1):49–59 (in Russian)

Land F (1866) Neftyanye kolodtsy i gryazevye vulkany Tamanskogo poluostrova (Oil wells and mud volcanoes of Taman Peninsula) Meditsinskiy sbornik 1:14–24 (in Russian)

Lebedev YuS (1968) Novye proyavleniya zheleznykh rud v Chongelekskoy vdavlennoy sinklinali (new manifestations of iron ore in the Chonglek depressed syncline). Tezisy dokladov IV nauchno-teknicheskoy konferentsii po izucheniyu poleznykh iskopaemykh otlozheniy osadochnogo chekhla yuga Ukrainy. KGU, Kiev, pp 50–51 (in Russian)

Limonov AF, Woodside JM, Ivanov MK (eds) (1992) Geological and geophysical investigations in the Mediterranean and Black Seas. Initial results of the "training through research" cruise of R/V Gelendzhik in the Eastern Mediterranean and the Black Sea (June-July 1991). UNESCO Rep Mar Sci 56:173

Limonov AF, Woodside JM, Ivanov MK (eds) (1994) Mud volcanism in the Mediterranean and Black Seas and shallow structure of the Eratosthenes Seamount: initial results of the geological and geophysical investigations during the Third UNESCO-ESF "training-through-research" cruise of R/V Gelendzhik (June–July 1993). UNESCO Rep Mar Sci 64:173

Limonov AF, Kenyon NH, Ivanov MK et al (1995) Deep-sea depositional systems of the Western Mediterranean and mud volcanism on the Mediterranean Ridge: initial results of geological and geophysical investigations during the Fourth UNESCO–ESF "training-through-research" cruise of R/V Gelendzhik (June–July, 1994). UNESCO Rep Marine Sci 67:172

Maymin ZL (1951) Tretichnye otlozheniya Kryma (tertiary deposits of Crimea). Gostoptekhizdat, Moscow-Leningrad (in Russian)

Meisner LB, Tugolesov DA (2003) Opornye otrazhayushchie gerizonty v seysmicheskoy zapisi osadochnogo vypolneniya Chernomorskoy vadiny (reference reflecting horizons in seismic recording of sedimentary sequence of the Black Sea depression). Stratigrafiya, geologiya, korrelyatsiya 11(6):83–97 (in Russian)

Meisner LB, Tugolesov DA, Khakhalev EM (1996) Zapadno-Chernomorskaya gryazevulkanicheskaya provintsiya (Western Black Sea mud volcanic province). Okeanologiya 35(1):119–127 (in Russian)

Murzaev PM (1928) Izverzhenie sopki Dzhau-Tepe v 1927 godu (Eruption of the Dzhau-Tepe mud volcano in 1927). Trudy Krymskogo NII 1:87–91 (in Russian)

Nadezhkin DV (2011) Neftematerinskie svoystva maykopskikh otlozheniy i ikh rol v neftegazonostnosti vostochnoy chasti Chernogo morya (oil source properties of Maikop deposits and their role in the oil and gas presence of the eastern part of the Black Sea). Dissertation, Moscow (in Russian)

Naumenko PI (1976) Sovremennaya deyatelnost' gryazevogo vulkanizma Kerchenskogo poluostrova (modern activities of the mud volcanism of the Kerch Peninsula). Materialy po mineralogii, petrografii i geokhimii osadochnykh porod i rud 4:115–135. Naukova Dumka, Kiev (in Rusian)

Naumenko PI (1977) Nekotorye zakonomernosti razmeshcheniya rudnykh zalezhey Kerchensko-Tamanskoy oblasti v svyazi s osobennostyu ee tektonicheskogo stroeniya (some patterns in the placement of ore deposits of the Kerch-Taman region in connection with the peculiarity of its tectonic structure). Geologicheskiy zhurnal 37(6):28–37 (in Russian)

Nesterovskiy VA (1990) Aktivizatsiya gryazevykh vulkanov Kerchensko-Tamanskoy oblasti (activation of mud volcanoes of the Kerch-Taman region). Geologicheskiy zhurnal 1:138–143 (in Russian)

Nikitin PN (1955) Gryazevoy vulkan Shugo (the Shugo mud volcano). Priroda 10:109–110 (in Russian)

Okay S, Cifci G, Özel S (2012) New discovered Izmir and Busan mud volcanoes and application of seismic attributes and AVO analysis in the easternmost Black Sea. EGU general assembly 22–27 April, 2012, Vienna, Austria, p 13332

Pallas PS (1795) Kratkoe fizicheskoe i topograficheskoe opisanie Tavricheskoy oblasti (brief physical and topographical description of the Tauride region). St Petersburg (in Russian)

Pallas PS (1883) Poezdka vo vnutrennost Kryma vdol' Kerchenskogo poluostrova i na ostrov Taman (a trip to the interior of Crimea along the Kerch Peninsula and the island of Taman). Zapiski Imperatorskogo Odesskogo Obshestva Istorii i Drevnostey 13:35–108 (in Russian)

Pasynkov AA, Vakhrushev BA (2016) Gryazyevoy vulkanizm Kerchyenskogo poluostrova kak nyegativnyy faktor stroityelstva trassy "Kerch – Sevastopol'" (Kerch peninsula mud volcanism as a negative factor of the "Kerch – Sevastopol'" road construction). Uchenye zapiski Krymskogo federalnogo universiteta imeni V.I. Vernadskogo. Geografiya. Geologiya 2(68) 1:110–122 (in Russian)

Pokrovskiy NF (1933) Razvedka na bor i yod na gryazevykh sopkakh Tamanskogo poluostrova (exploration for boron and iodine in the mud hills of the Taman Peninsula). Geologiya na fronte industrializatsii 2:32–38 (in Russian)

Popov SP (1928) Gryazevye vulkany (Mud volcanoes). Priroda 6:541–554 (in Russian)

Prokopov KA (1931) Uslozhneniya v antiklinalyakh Kerchenskogo poluostrova (complications in the anticlines of the Kerch peninsula). Trudy GGRU 38:37–83 (in Russian)

Sahling H, Bohrmann G, Artemov YG et al (2009) Vodyanitskii mud volcano, Sorokin Trough, Black Sea: geological characterization and quantification of gas bubble streams. Mar Pet Geol 26(9):1799–1811

Samoylov Ya (1898) Yenikalskie gryazevye sopki (Yenikalskie mud volcanoes). Byulleten' impertorskogo obshchestva naturalistov 2/3:80–86 (in Russian)

Sedelshchikov VV, Kul'chavov GK (1914) Izverzhenie sopki Dzhau-Tepe (eruption of the Dzhau-Tepe mud volcano). Materialy po estestvenno-istoricheskomu obsledovaniyu rayona deyatelnosti Dono-Kubano-Terskogo obshchestva selskogo khozyaystva 1:53–59 (in Russian)

Shardanov AN, Malyshok VG, Peklo VP (1962) O kornyakh gryazevykh vulkanov Tamani. Trudy Krasnodarskogo filiala VNIIneft 5(1):53–66 (in Russian)

Shaulov MA (1967) Gazoneftenosnost' severo-zapadnoy periklinali Anastisievsko-Troitskoy skladki v svyazi s drevnim gryazevym vulkanizmom (gas-oil-bearing capacity of the northwest periclinal Anastisievsko-Troitskaya fold in connection with ancient mud volcanism). Izvestiya vuzov Neft' i gaz 7:5–8 (in Russian)

Shaulov MA (1969) Usloviya obrazovaniya diapirovykh skladok i gryazevykh vulkanov Zapadnoy Kubani i Tamanskogo poluostrova v svyazi s formirovaniem zalezhey nefti i gaza. Dissertation, Krasnodar (in Rusian)

Shepel' SA (1926) Kolebaniya dna Kerch-Yenikalskogo proliva (Fluctuations of the Kerch-Yenikal Strait bottom). Dekadnyy byulleten' pogody i sostoyaniya morya po Chernomorskomu i Azovskomy poberezhyu 30:11–18 (in Russian)

Sheykov AA, Glazyrina NV, Glazyrin Ye A (2013) K vydeleniyu gryazevogo vulkana Khakhaleva (Azovskoe more) (On the release of the Khakhalev mud volcano [Sea of Azov]). Geologiya morey i okeanov: Materialy XX Mezhdunarodnoy nauch-

noy konferentsii. GEOS, Moscow, pp 270–273 (in Russian)
Shnyukov EF (2016) Flyuidogennaya mineralizatsiya gryazevykh vulkanov Azovo-Chernomorskogo regiona (Fluidogenic mineralization of mud volcanoes of the Azov-Black Sea region). Logos, Kiev (in Russian)
Shnyukov EF, Fesyunov OE (1965) Geologiya i genezis Baksinskogo zhelezorudnogo mestorozhdeniya (Kerchenskiy basseyn) (geology and genesis of the Baksinskiy iron ore deposit [Kerch basin]). Litol Pol Iskop 5:28–37 (in Russian)
Shnyukov EF, Naumenko PI (1964) Kimmeriyskie zheleznye rudy vdavlennykh sinklinaley Kerchenskogo poluostrova (Kimmerian iron ores of recessed synclines on the Kerch peninsula). Krymizdat, Simferopol (in Russian)
Shnyukov EF, Naumenko PI (1966) Kimmeriyskie zheleznye rudy vo vdavlennykh sinklinalyakh Kerchenskogo poluostrova (Kimmerian iron ores of recessed synclines on the Kerch Peninsula). Soviet Geol 4:142–143 (in Russian)
Shnyukov EF, Naumenko PI (1982) Nakhodka Achinskoy rudonosnoy sinklinali na Kerchenskom poluostrove i ee paleogeograficheskoe znachenie (discovery of the Achinskaya ore-bearing syncline on the Kerch Peninsula and its paleogeographic significance). Geologicheskiy zhurnal 5:51–57 (in Russian)
Shnyukov EF, Netrebskaya EY (2013) Korni Chernomorskikh gryazevykh vulkanov (the roots of the Black Sea mud volcanoes). Geologiya i poleznye iskopaemye mirovogo okeana 2:66–80 (in Russian)
Shnyukov EF, Netrebskaya EY (2016) O glubinnom stroyenii eruptivnogo kanala gryazyevykh vulkanov (on the deep structure of the eruptive channel of mud volcanoes). Geologiya i poleznye iskopaemye mirovogo okeana 4:54–66 (in Russian)
Shnyukov EF, Schiptsov AA (1996) Geologicheskie issledovaniya NIS "Kiev" v Chernom more (6-y reys) (geological research of the R/V "Kiev" in the Black Sea [6th voyage]). OMGOR NAMIT, Kiev (in Russian)
Shnyukov EF, Naumenko PI, Lebedev YuS et al (1971) Gryazevoy vulkanizm i rudoobrazovanie (mud volcanism and ore formation). Naukova Dumka, Kiev (in Russian)
Shnyukov EF, Alenkin VM, Naumenko PI (1985) Rudonosnost' Armaelinskoy vdavlennoy sinklinali na Kerchenskom poluostrove (The ore-bearing capacity of the Armaelinskaya recessed syncline on the Kerch Peninsula). Doklady AN USSR, Seriya B:23–25 (in Russian)
Shnyukov EF, Sobolevskiy YuV, Gnatenko GI et al (1986) Atlas Gryazevye vulkany Kerchensko-Tamanskoy oblasti (mud volcanoes of the Kerch-Taman region). Naukova Dumka, Kiev (in Russian)
Shnyukov EF, Nesterovskiy VA, Alyonkin VM (1987) Geologichna budova i rudonosnist Bulganakskoï vdavlenoï sinklinali (geological structure and ore deposits of the Bulganak pushed syncline). Dopovidi AN USSR, seriya B (11):26–29 (in Ukranian)
Shnyukov EF, Gnatenko GI, Nesterovskiy VA et al (1992) Gryazevyy vulkanizm Kerchensko-Tamanskogo regiona (mud volcanism of the Kerch-Taman region). Naukova Dumka, Kiev (in Russian)
Shnyukov EF, Ivannikov AV, Pomozov AA et al (1994a) Geologicheskie issledovaniya NIS "Ikhtiandr" v Chernom more (geological research of R/V "Ichthyandr" in the Black Sea). OMGOR NAN Ukrainy (in Russian)
Shnyukov EF, Mitin LI, Tsemko VP (1994b) Katastrofy v Chernom more (disasters in the Black Sea). Manuskript, Kiev (in Russian)
Shnyukov EF, Vorob'ev AI, Grigorev AV et al (1995) Predpolagaemyy gryazevoy vulkan v glubokovodnoy chasti Chernogo morya. Doklady AN USSR 12:62–65 (in Russian)
Shnyukov EF, Inozemtsev YuI, Maslakov NA (1999) Geologiya, istoriya razvitiya rechnoy seti na severo-zapade Chernogo morya (geology, history of the development of the river network in the northwestern part of the Black Sea). OMGOR NAN Ukrainy, pp 238–244 (in Russian)
Shnyukov EF, Ivannikov AV, Inozemtsev YuI et al (2003) Litologo-stratigraficheskaya kharakteristika donnykh otlozheniy krymskogo shelfa i glubokovodnoy chasti Chernogo morya (lithologic-stratigraphic characterization of bottom sediments of the Crimean shelf and deep-sea part of the Black Sea). Geologicheskiy zhurnal 1:9–23 (in Russian)
Shnyukov EF, Starostenko VI, Ivannikov AV et al (2005) Gazovyy vulkanizm Chernogo morya (Black Sea gas volcanism). OMGOR NAN Ukrainy, Kiev (in Russian)
Shnyukov EF, Kutnyy VA, Maslakov NA et al (2006) K mineralogii karbonatnykh obrazovaniy gazovykh istochnikov Chernogo morya (on the mineralogy of carbonate formations of gas sources of the Black Sea). Geologiya i poleznye iskopaemye mirovogo oceana 2:69–81 (in Russian)
Shnyukov EF, Kutnyy VA, Naumenko SP et al (2007) Travertiny i drugie mineralnye obrazovaniya gazovodnykh istochnikov Kerchenskogo poluostrova (travertine and other mineral formations of gas sources of the Kerch Peninsula). Geologiya i poleznye iskopaemye mirovogo oceana 3:5–14 (in Russian)
Shnyukov EF, Emelyanov VA, Inozemtsev YuI et al (2011) Gazovo-gryazevyy vulkanizm Kerchensko-Tamanskogo sektora Chernogo morya. 65 reys NIS "Professor Vodyanitskiy," 2010 (Gas-mud volcanism of the Kerch-Taman sector of the Black Sea. 65 cruise of the R/V "Professor Vodyanitsky," 2010). Logos, Kiev (in Russian)
Shnyukov EF, Kobolev VP, Starostenko VI (2013) Gazovyy vulkanizm Chernogo morya (gas volcanism of the Black Sea). Logos, Kiev (in Russian)
Shnyukov EF, Stupina LV, Rybak EN et al (2015) Gryazevye vulkany Chernogo morya. Katalog (Black Sea Mud Volcanoes. Catalog). Logos, Kiev (in Russian)
Shternov AG (1968) Geologiya i genezis gryazevykh vulkanov Kerchensko-Tamanskogo rayona (geology and

genesis of mud volcanoes of the Kerch-Taman region). Dissertation, Kiev (in Russian)
Sivere L (1931) Izverzhenie Dzhardzhavskoy sopki (eruption of the Dzhardzha mud volcano). Priroda 1:98–99 (in Russian)
Slavyanov NN (1928) O nekotorykh maloizvestnykh mineralnykh istochnikakh Kubanskoy oblasti (on some little-known mineral springs of the Kuban region). Materialy po obshchey i prikladnoy geologii 82:15–25 (in Russian)
Sobisevich AL, Gorbatikov AV, Ovsyuchenko AN (2008) Glubinnoe stroenie gryazevogo vulkana gory Karabetovoy. DAN RAN Geogizika 422(4):542–546 (in Russian)
Sobisevich AL, Sobisevich LE, Tvertinova TYu (2014) O gryazevom vulkanizme v pozdnealpiyskom skladchatom sooruzhenii severobzapadnogo Kavkaza (na primere izucheniya glubinnogo stroeniya gryazevogo vulkana Shugo) (on mud volcanism in the Late Alpine folded structure of the Northwestern Caucasus [based on study of the deep structure of the Shugo mud volcano]). Geologiya i poleznye iskopaemye mirovogo okeana 2:80–93 (in Russian)
Sobisevich AL, Tvertinova TYu, Likhodeev DV et al (2015) Glubinnoe stroenie gryazevogo vulkana Dzhardzhava v predelakh Yuzhno-Kerchenskoy antiklinalnoy struktury (depth structure of the Dzhardzhava mud volcano within the South Kerch anticlinal structure). Voprosy inzhenernoy seysmologii 42(2):73–80 (in Russian)
Sokolov VA (1965) Protsessy obrazovaniya i migratsii nefti i gaza (the processes of formation and migration of oil and gas). Nedra, Moscow (in Russian)
Spadini G, Robinson A, Clothing S (1996) Western versus eastern Black Sea tectonic evolution: pre-rift lithospheric controls on basin formation. Tectonophysics 266(1–4):139–154
Starostenko VI, Makarenko IB, Rusakov OM et al (2010) Geofizicheskie neodnorodnosti Chernogo morya (geophysical heterogeneities of the Black Sea). Geofizicheskiy zhurnal 32(5):3–20 (in Russian)
Steber EA (1909/1910) Gryazevoy vulkan Karabetova gora bliz Tamani (Mud volcano Karabetova mountain near Taman). Izvestia Kavkazskogo otdeleniya Russkogo heograficheskogo obshchestva 20 (in Russian)
Steber EA (1912) Iezheuasa – gryazevoy vulkan severozapadnoy okonechnosti Kavkaza (Yezheuasa—the mud volcano of the northwestern tip of the Caucasus). Kavkazskogo Otdeleniya Russkogo Geograficheskogo Obshchestva 21(1):51–63 (in Russian)
Steber EA (1915) Proiskhozhdenie nefti (origin of oil). Priroda 4:1–30 (in Russian)
Subbota MI (1964) Novye dannye o gazakh i tipakh gryazevykh vulkanov rayona Zapadnoy Kubani (new data on gases and types of mud volcanoes in the area of the Western Kuban). Trudy Vsesoyuznogo nauchno-issledovatel'skogo geologo-razvedochnogo neftyaogo instituta nedr 41:156–178 (in Russian)
Tugolesov DA, Gorshkov AS, Meisner LB et al (1985) Tektonika mezo-kaynozoyskiy otlozheniy Chernomorskoy vpadiny (tectonics of the Meso-Cenozoic sediments of the Black Sea Basin). Nedra, Moscow (in Russian)
Turley GF (1930) Bulganakska grupa gryazovikh vulkaniv na Kerchenskomu pivostrovi ta produkti ïkhnogo vibukhu (Bulganaks'ka group of mud volcanoes on the Kerch Peninsula and products of their explosion). Trudy Kharkivs'kogo tovaryustva doslidnikiv pyrody, pp 14–22 (in Ukranian)
Tveritinova TYu, Sobisevich AL, Sobisevich LYe et al (2015) Strukturnaya pozitsiya i osobennosti stroeniya gryazevogo vulkana gory Karabetovoy (Structural position and features of the structure of the mud volcano of Karabetova Mountain). Geologiya i poleznye iskopaemye mirovogo okeana 2:106–122 (in Russian)
Vernadskiy VI, Popov SP (1899–1900) Yenikalskie gryazevye vulkany (Enikal mud volcanoes). Byulleten' moskovskogo obshchestva ispytateley prirody (Protokoly zasedaniy. Prilozheniya), pp 37–41 (in Russian)
Voskoboynikov NI, Gur'iev AV (1832) Geologicheskoe opisanie poluostrova Tamani, prinadlezhashchego k zemle voyska Chernomorskogo (geological description of the Taman Peninsula belonging to the land of the Black Sea army). Gornyy zhurnal 1 (in Russian)
Woodside JM, Ivanov MK, Limonov AF (1997) Neotectonics and fluid flow through seafloor sediments in the eastern Mediterranean and Black Seas. Intergovern Oceanogr Commission Tech Ser 48(I–II):226
Yanko-Hombach V (2007) Controversy over Noah's flood in the Black Sea: geological and foraminiferal evidence from the shelf. In: Yanko-Hombach V, Gilbert AS, Panin N, Dolukhanov PM (eds) The Black Sea flood question: changes in coastline, climate, and human settlement. Springer, Dordrecht, pp 149–203
Yanko-Hombach V, Gilbert AS, Buynevich I (eds) (2007) Field trip guide. Joint plenary meeting and field trip of IGCP-521 "Black Sea-Mediterranean corridor during the last 30 ky: sea level change and human adaptation," and IGCP 481 "dating Caspian Sea level change", September 8–17, 2007, Southern Branch of the Institute of Oceanology, Russian Academy of Sciences and Demetra Beneficent Foundation, Gelendzhik (Russia)-Kerch (Ukraine), Rosselkhozakademiya Printing House, Moscow
Zenin GN (1955) Gryazevye vulkany Tamanskogo poluostrova (mud volcanoes of the Taman Peninsula). Istoriko-arkheologicheskiy i geograficheskiy ocherk (historical-archaeological and geographical essay). Uchenye zapiski Krasnodarskogo pedagogicheskogo institute 14:54–59 (in Russian)
Zenkovich VKh (1894) Kerch v proshedshem i nastoyashchem: Ist.-arkheol. i geogr. ocherk.– Kerch (Kerch in past and present: historical-archaeological essay). Kerch (in Russian)

7 Other Types of Degassing in the Black Sea

Summary

This chapter provides an overview of other types of degassing in the Black Sea. In addition to mud volcanoes, there are numerous areas where gas bubbles escape from the Black Sea floor; these are called gas seeps, acoustic (i.e., obtained using echo sounders or sonars and visible on echograms) plumes, gas torches (due to their cone-like shape on eco sounder images), or gas flares; these are all referred to in the chapter as gas seeps. They are localized on the periphery of the Black Sea, particularly in its northwestern part, on the Bulgarian and Kerch-Taman shelf (including the Kerch Strait), and along the coast of the Caucasus. Quite often, they are confined to paleo-deltas of the largest Black Sea rivers, the Danube, Dnieper, Dniester, and Don, and within the canyons formed by them. The deepest seeps are associated with faults and mud volcanoes in the central Black Sea basin.

Gas seeps are generally found at water depths from 50 to 800 m, forming either single occurrences or grouped seeps; some seeps are quite large in area. Often, gas emissions cloud the water by dislocating bottom sediments that have been saturated with gas and water by mud volcanic activity and diapiric processes.

On average, the height of the rising bubbles above the surrounding seafloor is 100–200 m, and most of these bubbles do not reach the water surface. Sometimes it is difficult to distinguish gas seeps from mud volcanoes due to the location of seeps within small conical depressions on the sea bottom. As strong exhaust emissions of water-saturated gas pass upward through the sediment sequence, they can affect bottom morphology by forming rounded depressions or craters called pockmarks. Their locations indicate both the presence of gas seeps and methane biotopes on the seafloor.

Within the Black Sea, a kind of specific zoning emerges: the central deepwater part of the sea is occupied by gas hydrates, while its periphery contains abundant gas seeps. It is quite possible that gas or, more precisely, gas-water seeps are localized within the area of the land adjacent to the sea—for example, in the Kerch-Taman region—but for the most part, they are not fixed with certainty because of the difficulty of observation.

At least 4000 gas seeps are known in the Black Sea today. The maximum depth for most of the detected gas seeps is 725 m, which more or less corresponds to the stability zone for pure methane hydrate at a bottom temperature of 8.9 °C in this part of the Black Sea. This enables us to say that gas hydrates play a role as buffer for the ascending migration of gases and thereby prevent the seepage of huge amounts of methane into the water column.

Many areas where seeps have developed are confined to the places containing paleoalluvial strata, but not all. The authors are more inclined to link gas seeps with the Maikopian horizon of

E. Shnyukov, V. Yanko-Hombach, *Mud Volcanoes of the Black Sea Region and their Environmental Significance*, https://doi.org/10.1007/978-3-030-40316-4_7

the Black Sea geological section and with deep fluids. At the same time, they do not deny the participation of biochemical methane in the genesis of gas flares. Undoubtedly, this process accounts for gas seep generation in Quaternary bottom sediments, but its role, most likely, is subordinate.

Gas seeps were mentioned for the first time by Dimitrov et al. (1979) after they were observed in Balchik Bay, Bulgaria. Later, they were discovered at the northwestern shelf edge and upper slope in the anoxic zone of the Black Sea (Polikarpov and Egorov 1989). Since the end of the last century, there has been a sharp increase in multidisciplinary studies of gas seeps, and today, research is largely focused on their localization, spatial distribution, environmental role (Gevorkyan et al. 1991; Egorov et al. 2003; Shnyukov et al. 1999; Egorov et al. 1998; Yanko-Hombach et al. 2014, 2017), contribution to the formation of gas hydrates (Shnyukov et al. 1995, 2014; Naudts et al. 2006), quantitative estimate of methane emission into the seawater and atmosphere (Artemov et al. 2007; Egorov et al. 2003, 2011), bacterial oxidation of methane and formation of microbial buildups (diagenetic carbonates) in anoxic waters (Ivanov et al. 1991; Polikarpov et al. 1993; Michaelis et al. 2002), and determination of the age and genesis of methane (Lein et al. 2002; Gulin et al. 2003; Shnyukov et al. 2013, 2014). In recent years, gas flares have been found in shallow water (Sevastopol Bay, Bay of Laspi, Sudak Bay) and at depths of 2070 m (Egorov et al. 2003, 2011). The coordinates and characteristics of a large number of gas seeps together with a historical overview of their study were provided by Egorov et al. (2011). Sokolov (1965) noted that, in some cases, gas seeps can emerge on land as well, e.g., on the Kerch Peninsula, where their scale is insignificant, but as a rule, no gas seeps can be developed without water. Currently, several thousands of gas seeps are known (Fig. 7.1).

Gas seeps are localized on the periphery of the Black Sea, particularly in its northwestern part, on the Bulgarian and Kerch-Taman shelf (including the Kerch Strait), and along the coast of the Caucasus. Quite often, they are confined to paleodeltas of the largest Black Sea rivers, the Danube, Dnieper, Dniester, and Don, and within the canyons formed by them. The deepest seeps are associated with faults and mud volcanoes in the central Black Sea basin (Egorov et al. 2003; Krastel et al. 2003; Shnyukov et al. 1999, 2014). The southern part of the Black Sea remains poorly understood and will undoubtedly provide much new data. For example, geologists of "Yuzhmorgeologiya," who studied the route of the Russia-Turkey Blue Stream pipeline, discovered a number of gas flares, while Ergün et al. (2002) and Çifçi et al. (2003) described numerous pockmarks on the seabed and buried beneath it at water depths of 180–300 m along the Turkish coast.

An acoustic image of a gas flare can be seen on a computer monitor if it is constantly active, but if it has a pulsating character, the recording will be intermittent. The diameter of a single gas flare is calculated based on the vessel's speed above the sea bottom and varies between 10 m and 40 m on average, sometimes reaching as wide as 300 m.

Gas seeps are generally found at water depths from 50 to 800 m, forming either single occurrences (Fig. 7.2) or grouped seeps; some seeps are quite large in diameter and height (Fig. 7.3).

Quite often, gas emissions form a cloud (Fig. 7.4).

The clouds are created by the dislocation of bottom sediments that have been saturated with gas and water by mud volcanic activity and diapiric processes. Analysis of echograms has shown that the smallest gas bubbles have a low ascent rate, are extracted by bottom currents, and create gas clouds at the bottom (Ryazanov 1996; Artemov 2003).

A typical example of a gas cloud is the ring structure in the Western Black Sea basin (Fig. 6.145).

On average, the elevation of gas seeps above the surrounding seafloor is 100–200 m, and most of the rising bubbles do not reach the water sur-

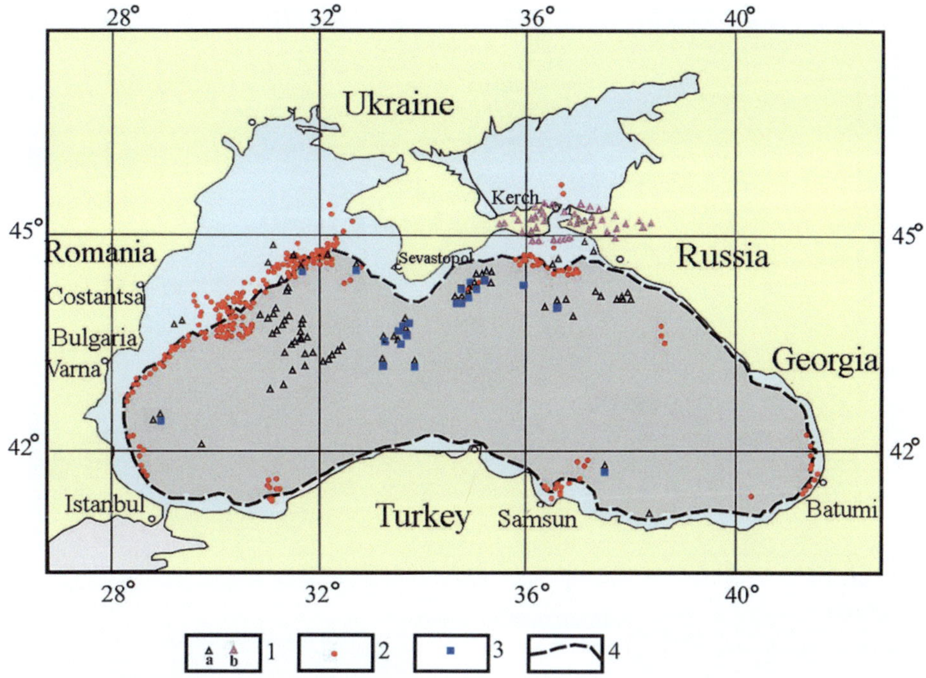

Fig. 7.1 Map showing the distribution of *1a*, offshore mud volcanoes; *1b*, onshore mud volcanoes; *2*, gas seeps; *3*, gas hydrates; *4*, shelf edge

Fig. 7.2 Echo sounder diagram of a gas flare on the submerged plain of the northwestern Black Sea shelf

face. Sometimes it is difficult to distinguish gas seeps from mud volcanoes due to their location within small conical depressions on the sea bottom. As they pass upward through the sediment sequence, strong exhaust emissions of water-saturated gas can affect bottom morphology by forming rounded depressions or craters called pockmarks (Ussler et al. 2003). Their locations indicate both the presence of gas seeps and methane biotopes on the seafloor (Hovland 2002) (Fig. 7.5).

Good descriptions of pockmarks can be found in Dimitrov and Doncheva (1994) and Çifçi et al. (2003), who discuss their occurrence on the Bulgarian and Turkish shelf, respectively.

According to Egorov et al. (2011), there is a clearly visible connection between the seeps and pockmarks on the shelf with depth of 1–2 m. Such pockmarks are observed in the depressions of many shelves around the world, and in these cases, they are confined to areas of fine sediment accumulation and are caused by scattered gas leaks.

In rare cases, tremendous, and even catastrophic, gas emissions from deep beneath the seabed occur during earthquakes. According to Kessler et al. (2006), the amount of released gas from the sea bottom ranges from 5.0×10^9 to 6.0×10^9 m^3 year^{-1}.

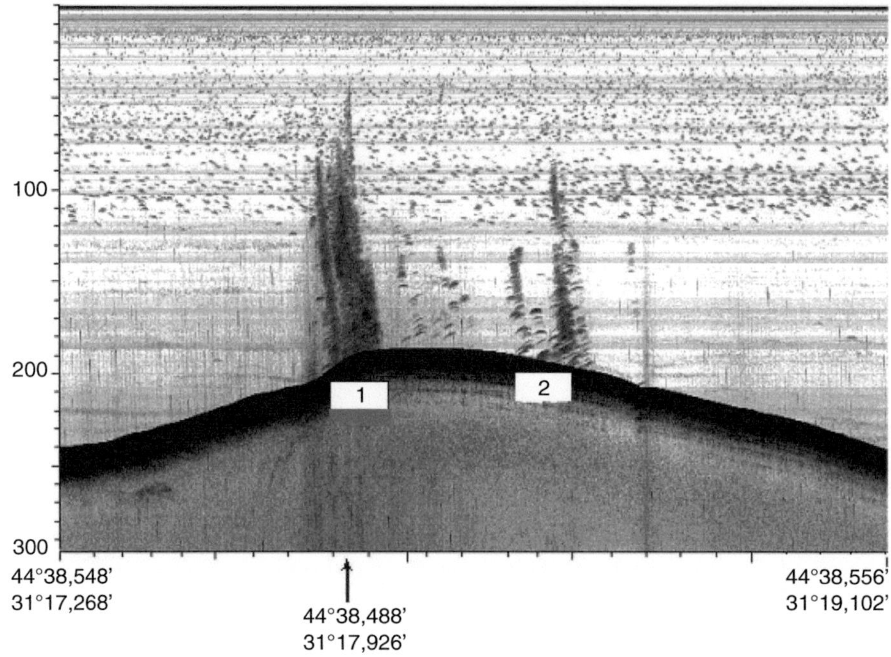

Fig. 7.3 Echo sounder diagram of two groups of gas flares on the northwestern Black Sea shelf

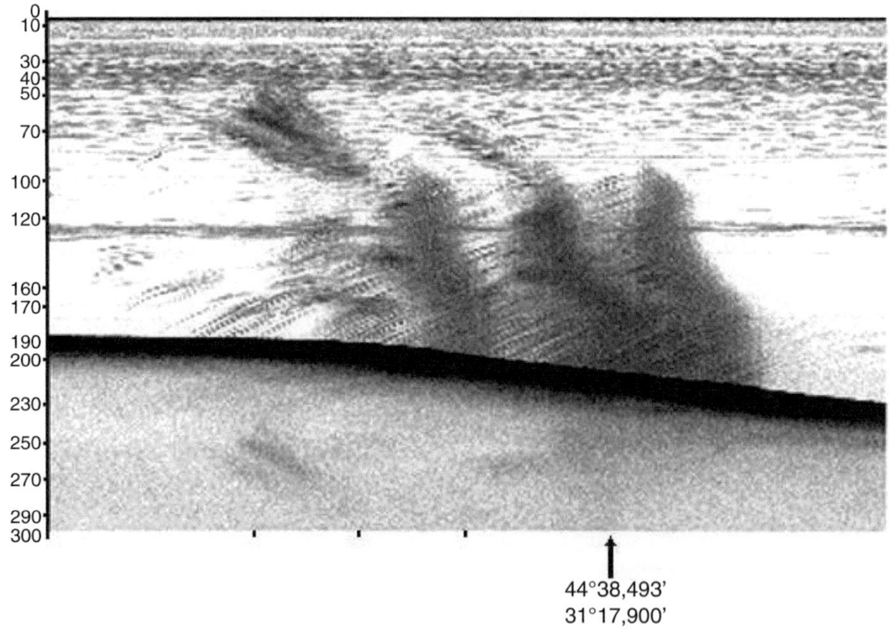

Fig. 7.4 Echo sounder diagram of a gas cloud. Imaged from the R/V "Professor Vodyanitskiy" in 2011

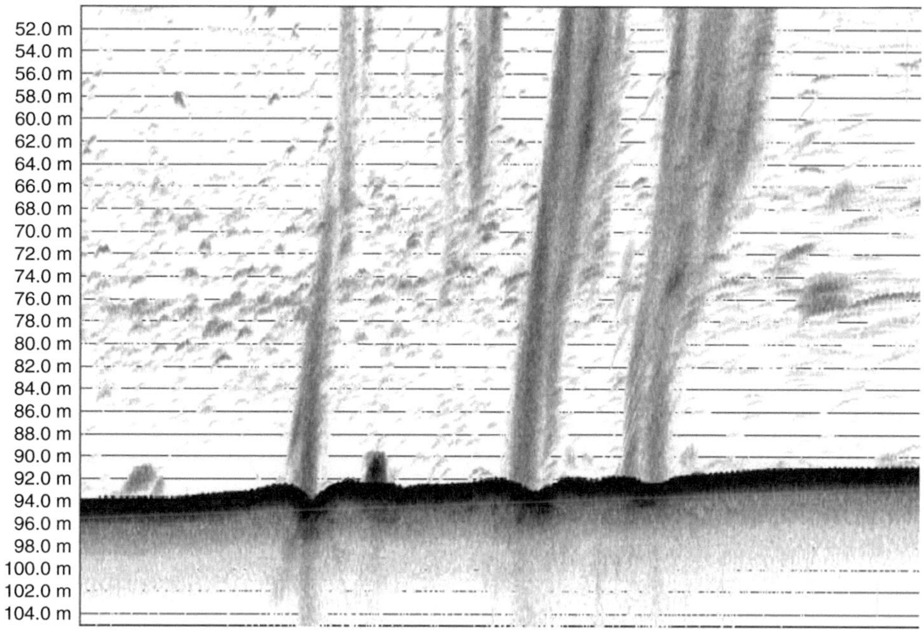

Fig. 7.5 Connection between gas seeps and pockmarks in the submerged paleodelta of the Dnieper River

7.1 Gas Seeps of the Northwestern Part of the Black Sea

Most of the gas flares identified to date are located within the northwestern part of the Black Sea, largely concentrated within the Western Black Sea morphostructural region of buried paleodeltas. The region encompasses a surface incorporating the "basin" and "watershed slopes" of the canyons formed by the Paleo-Danube, Paleo-Dniester, and Paleo-Kalanchak (Pra-Dnieper) rivers. Geomorphologically, the area is a continuation of the now flooded formerly subaerial plain of the northwestern shelf of the Black Sea. Formation of the main landscape types of the region is associated with the processes of marine sediment accumulation and their subsequent erosion and dismemberment by turbidity flows within zones of submarine unloading of groundwater over the buried valleys of paleorivers.

The fields of gas-releasing structures consist of successive elevations and depressions. The latter are represented by crumpled and fissured, gently sloping synclinal folds. Their central parts are complicated by ruptures, and the layered horizons of sediments are broken, creating vertical channels for ejecting gas.

The paleovalleys give rise to the underwater canyons that are etched into the continental slope. The shelf edge is expressed in the bottom relief by a sharp edge on the shelf surface, which is inclined from 1° to 10°, and even occasionally 20°, at depths between −100 and −200 m.

The region of the most active gas release is concentrated in the strip of sea bottom having coordinates between 44° 36′ to 44° 55′ N and 31° 00′ to 32° 30′ E (Shnyukov et al. 1999). It includes two areas: (1) the erosive-accumulative, gently sloping paleovalleys of the Danube-Dniester interfluve, and (2) those of the Paleo-Kalanchak (Pra-Dnieper) canyon (Pasynkova 2001).

The first area (1) encompasses the extreme western part of the Black Sea region from the canyon of the Paleo-Danube River to that of the Paleo-Kalanchak (Pra-Dnieper) River.

The bottom relief represents a relic of the subaerial erosive-denudative, pre-Holocene relief of the coastal plain. In its formation, the most important role was played by abrasion, underwater erosion, and sediment accumulation. The lower reaches of the large paleorivers were located within the plain, and therefore the flat watershed was dismembered by paleovalleys, creating separate elevations.

The canyons dissect the whole surface of the continental slope from its upper edge to its foot, extending onto the sea bottom, e.g., the Dniester Canyon with widths from 150 to 2500 m and depths from 300 to 500 m.

Most of the gas seeps are confined to the area of transition from the outer shelf to the continental slope (at its bend), and many appear in the Vityaz Canyon and the valley of the Paleo-Kalanchak River. Separate small gas seeps were also encountered in the abyssal part of the basin (Fig. 7.6).

Gas seeps are overwhelmingly located at water depths not exceeding 750 m. Almost half of them are found at depths from 100 to 200 m, and many of them occur even above 100 m. Below the compensation depth of gas hydrate formation (700–800 m), gas seeps are absent as a rule. From depths down to 250 m, methane can reach the sea surface and enter the atmosphere. When released from greater depths, the methane jets dissolve into the water column. Within the northwestern part of the basin, excluding the Bulgarian sector, a total of 303 gas seeps have been discovered at depths from 65 to 1030 m by Egorov et al. (2011) (Fig. 7.6).

No clear confinement of gas seeps to any form of bottom relief has been recognized. They are found on the erosive-denudative, pre-Holocene plain of the shelf (Figs. 7.7) and on slopes and

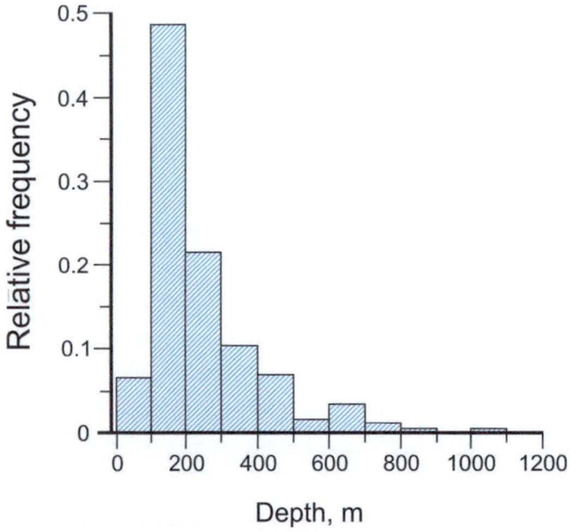

Fig. 7.6 Distribution of gas seeps along the water depth gradient within the northwestern part of the Black Sea along the Danube-Dniester interfluve (Egorov et al. 2011)

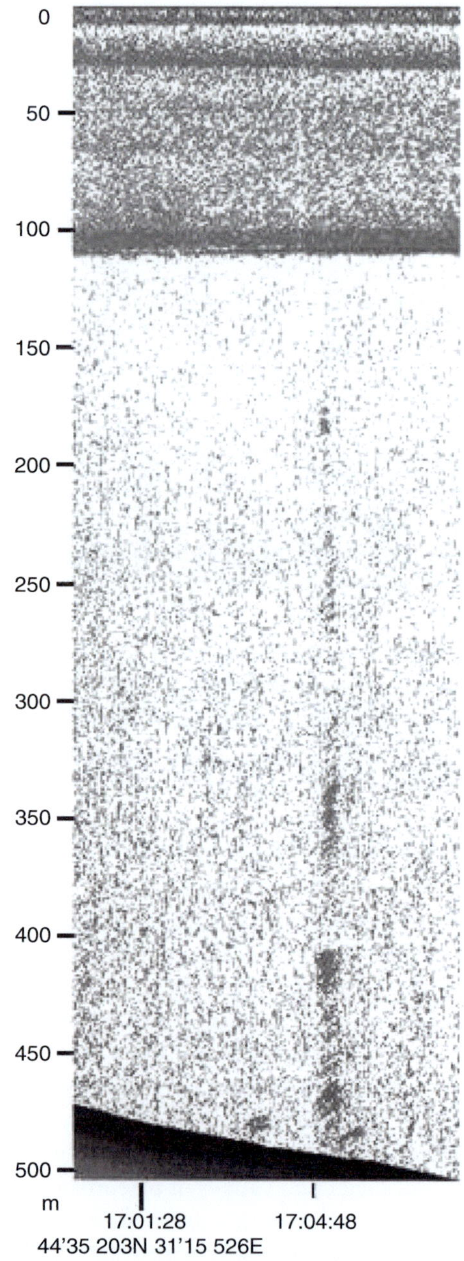

Fig. 7.7 Echo sounder diagram of a single gas seep with a height of 300 m on the plain of the northwestern shelf

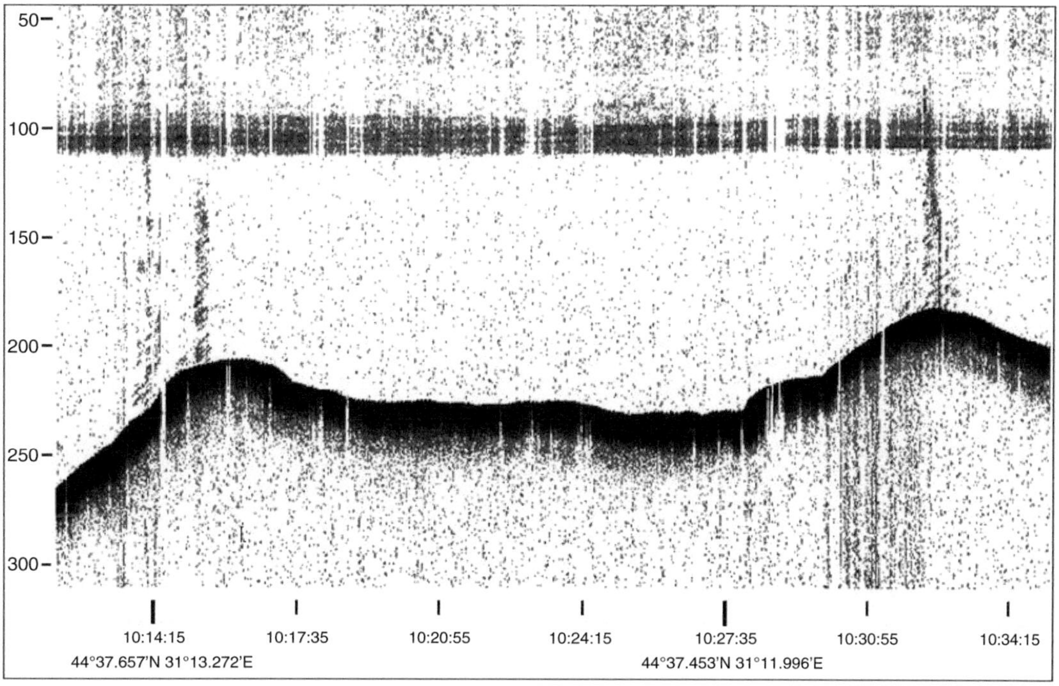

Fig. 7.8 Echo sounder diagram of a profile of the continental slope, showing gas seeps confined to the slopes and peaks of the hills within the valleys and canyons of the sea bottom

peaks of the hills within the valleys and canyons of the sea bottom (Fig. 7.8).

Most often, gas torches or flares reveal a conical appearance, gradually tapering toward the surface. Some of them form gas clouds with large gas cones. The height of gas torches varies between 10 and 300 m and sometimes reaches 400–500 m; they are often crowned by gas bubbles. As to their height and frequency, gas flares can be grouped as follows: flares with a height of 10–50 m (52%), up to 100 m (28%), up to 150 m (12%), up to 200 m (4%), up to 250 m (3%), and greater than 250 m (1%). Gas torches almost never reach the sea surface, even if they are located in shallow water, due to dissolution of the gas in sea water.

The second area (2) of erosive-accumulative canyons the Paleo-Kalanchak (Pra-Dnieper) River and covers the surface belonging to the "basin" and "watershed slopes" of the canyon.

Due to their predominantly eroded relief, the shelf and upper edge of the continental slope retreat far to the north in the form of a huge erosive circus. Its surface is dissected by branching and extended tributaries of the main channel of the canyon, and narrow ridges extend even to the abyssal part of the basin.

The Paleo-Kalanchak area is characterized by widespread development of methane seeps and manifestations of gas hydrate deposits.

Sometimes the gas seeps display a well-pronounced "dendritic" shape that looks like a vertical column about 100 m high in the center with branching to the sides (Fig. 7.9).

Gas emissions can have both a continuous and pulsating character (Fig. 7.10).

This area is also characterized by the development of gas seeps, gas-saturated silt, and gas hydrates (Shnyukov et al. 1999). To a large extent, gas seeps are located along the thalweg and watershed of the canyon network (Fig. 7.11).

The permanently operating deep-sea gas seeps have been investigated and photographed from a remote vehicle (Figs. 7.12, 7.13 and 7.14).

According to Egorov et al. (2011), 2650 gas seeps have been identified in the Paleo-Kalanchak canyon area, predominantly at water depths up to 725 m, yet some of them exist as deep as 850 m.

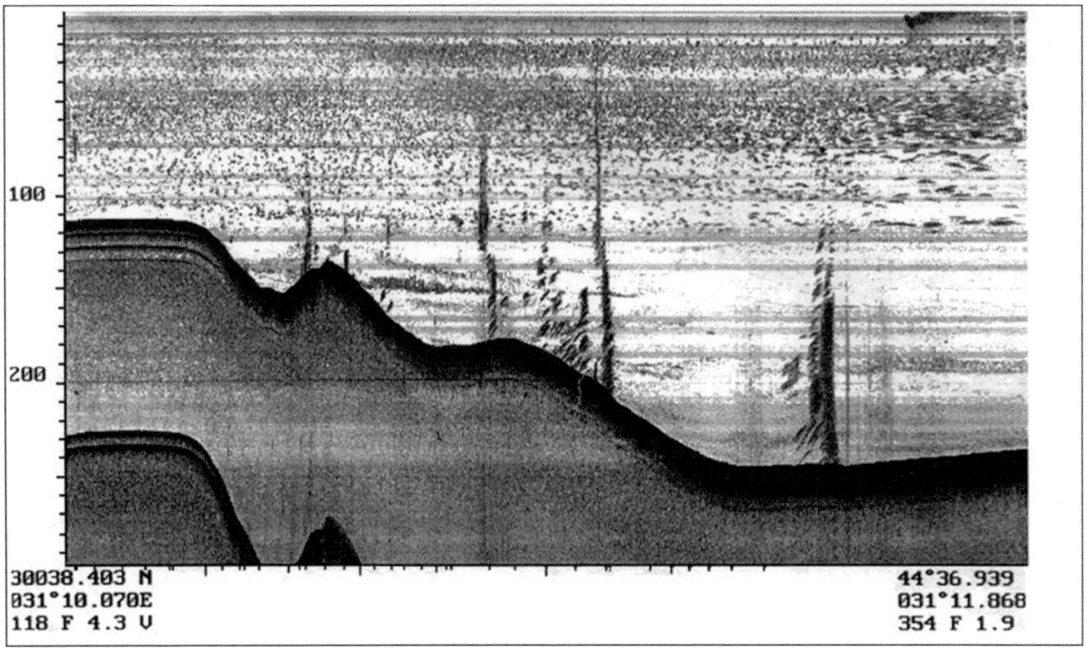

Fig. 7.9 Echo sounder diagram of "dendritic" gas flares in the Paleo-Kalanchak area

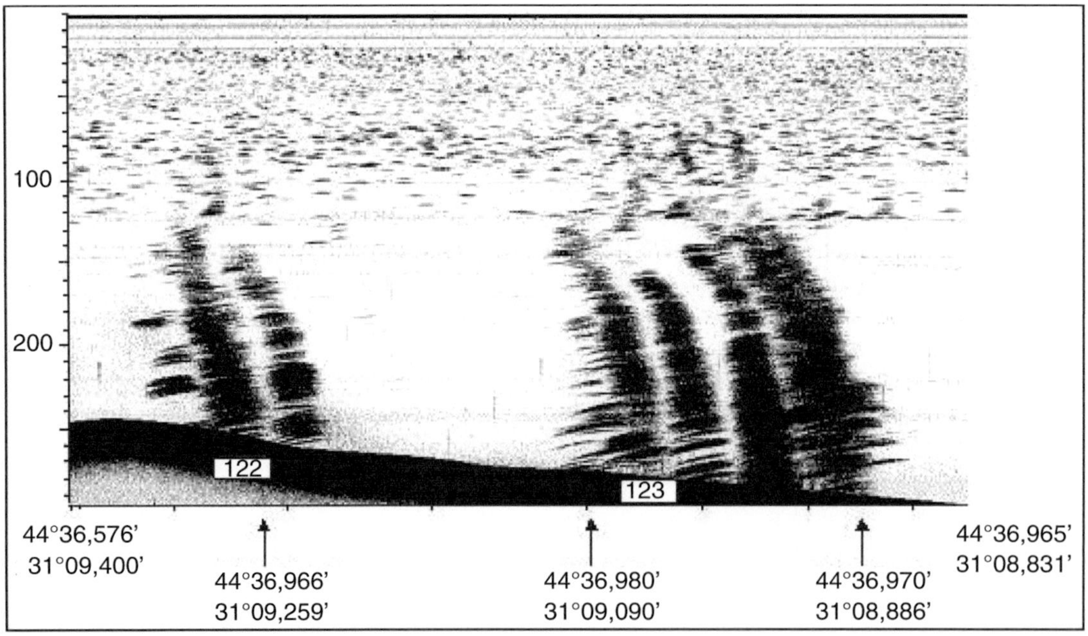

Fig. 7.10 Pulsating gas flares in the Paleo-Kalanchak area

The volume of methane flow (bubble transport) from the seeps in the Paleo-Kalanchak canyon is estimated at 212 liters of CH_4/m^2 per day. The proportion of different gases (in vol%) varies: CH_4 (between 95.0 and 99.6), H_2 (from 0.0 to 0.25), N_2 (between 0.0 and 5.1), and CO_2 (from 0.0 to 0.9%). The $\delta^{13}C$ ranges between 62.4‰

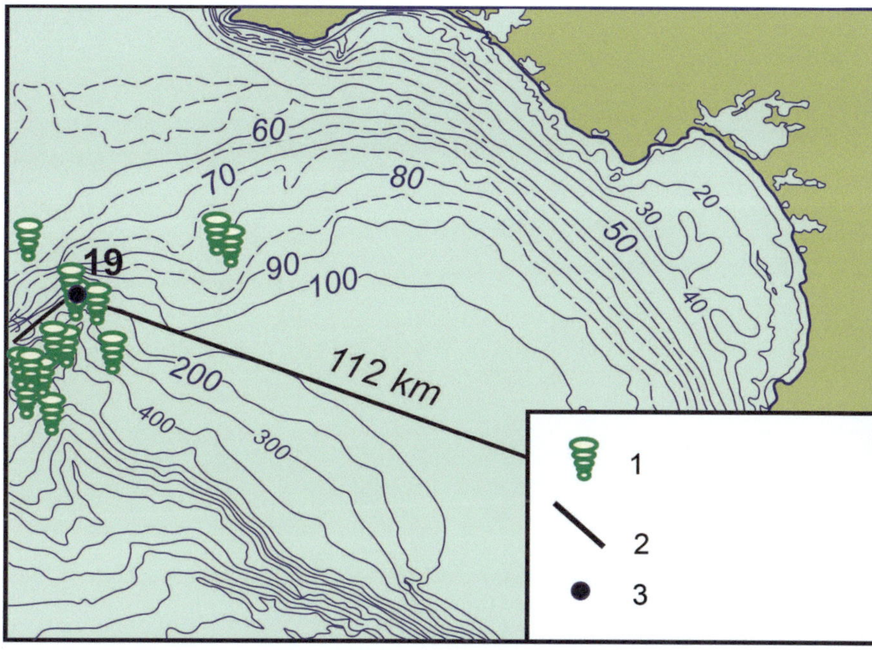

Fig. 7.11 Detailed bathymetric map of the Paleo-Kalanchak area with numerous gas flares on the sea bottom as observed by the underwater, self-propelled, television vehicle "Sophocles-1". Key: *1*, gas flare; *2*, route of vessel "Delta" movement; *3*, sampling point (Pasynkov et al. 2012)

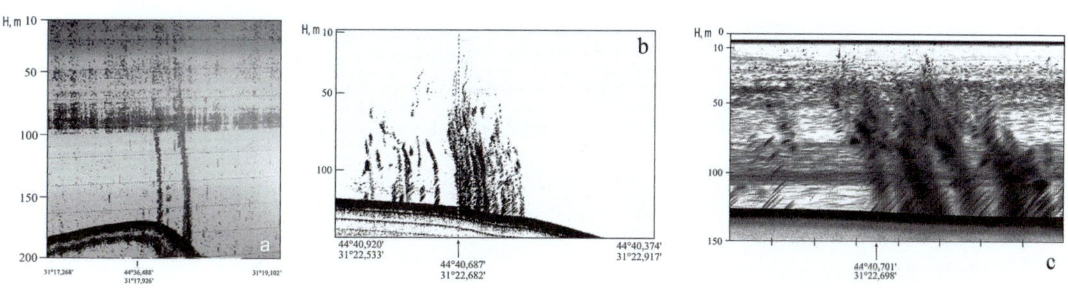

Fig. 7.12 Groups of gas flares in the Paleo-Kalanchak canyon. Imaged in 2001 (**a**, **c**) and 2011 (**b**)

and 70.5‰, and δD(SMOW) varies between 132 and 201‰ (Lein 2005).

Distinctive gas-releasing morphostructures were observed by A.A. Pasynkov during an underwater survey of the seabed in this region at depths of 95–101 m using the underwater, self-propelled, television vehicle "Sophocles-1" (Ryzhkov et al. 2011). During the survey, a video was recorded continuously with image output relayed to a service monitoring screen. The research site was previously observed by the R/V "Professor Vodyanitskiy" (Shnyukov et al. 1999) (Fig. 7.11). The new observations have confirmed that gas seeps are often localized within the river paleovalleys, and this indicates their common confinement to zones of major disturbance through which methane and other hydrocarbons rise from deep below the sea bottom, probably with the formation of intermediate traps in the sedimentary sequences. This particular site is located at the bend between the shelf and continental slope where flare heights do not exceed 100 m, and gas jets possess a pulsating character, eventually dissolving in the sea water. The echo sounder diagrams show both the vertical structure of the flares and the morphostructural pecu-

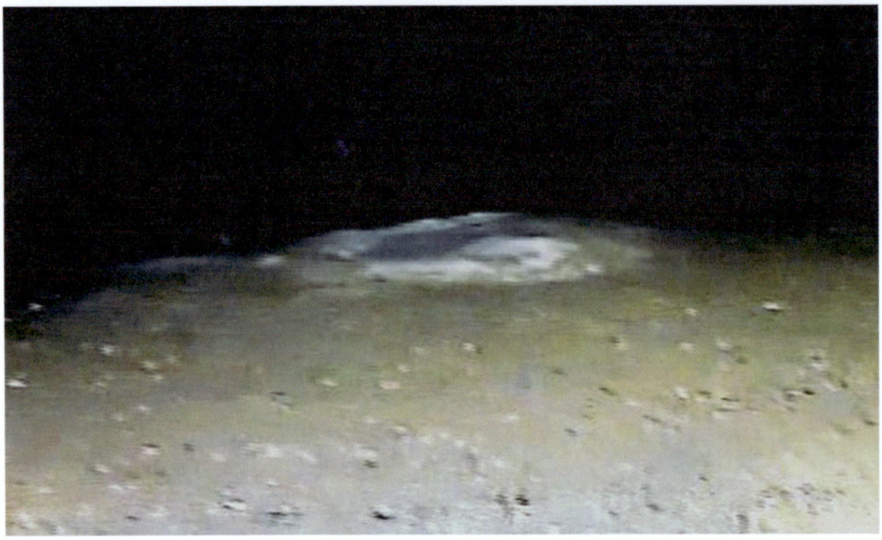

Fig. 7.13 Ring gas-releasing structure (video footage recorded by A.A. Pasynkov using "Sophocles-1," May 24, 2012). On the left ridge of the structure, a satellite salsa and two gryphons are visible

Fig. 7.14 Ring gas-releasing pulsating structure (video footage recorded by "Sophocles-1," May 24, 2012)

liarities of the sea bottom (e.g., concave, flat, comparatively hilly, stubby, ridged, etc.). The gas seeps vary significantly in their morphometrical characteristics (Figs. 7.13 and 7.14).

Monitoring of gas seeps in the region has been conducted over the years by the R/V "Kiev" (1995–1996) and R/V "Professor Vodyanitskiy" (2004–2011), but such comparisons are limited by the fact that the surveys of large numbers of gas seeps were carried out on different tacks using different echo sounders and at different speeds. These factors determine how the video footages of gas flares appear on the echo sounder monitor. Nevertheless, monitoring results have shown that all the groups of gas flares (gas fields) studied have been stable for the last 16 years and the foci of gas release are still located at the same points (within navigational error).

Gevorkyan et al. (1991) assumed that gas seeps relate to the deep Odessa fault. O.M. Rusakov drew attention to the confinement of gas seeps to tectonic ruptures in the northwestern part of the Black Sea (Shnyukov et al. 2005). In fact, it has been discovered that most often gas seeps are localized in the river valleys that follow the major tectonic disturbances, and this can be seen in the seismic profiles (Shnyukov et al. 2013) (Fig. 7.15).

On the seismograms, zones of general crushing and crumbling of sedimentary rocks that underlie gas seeps reach a visible depth of 500 m. The gas seeps are also confined to plicative and disjunctive tectonic zones. Such bruising of rocks is also characteristic of diapirism, which is often accompanied by mud volcanism; both require favorable tectonic conditions that vary somewhat for each phenomenon. The emergence of mud volcanoes requires (1) a thick (up to several km) buildup of sedimentary rocks that contain substantial clay overlying gas accumulations in the interior, (2) anomalously high (up to 300–400 atm) pressure affecting the gas deposits, and (3) tectonic disturbances in the capping layers that permit access to the surface (Shnyukov et al. 1971). There are no thick, clayey sediments under the gas seeps, and therefore, no mud volcanic breccia is present.

It is nevertheless possible in many cases to assume the nature of the deep processes leading to mud volcanism and gas release in the Black Sea. Findings of gas seeps at the foot of the continental slope, and even deeper, testify not only to their deep origin but also to their connection with the oil- and gas-bearing horizons exposed on the slope.

In general, the northwestern part of the basin is the largest region of development for gas seeps of different types and sizes in the Black Sea. As a rule, they are relatively stable over time and across space, and some of them have a pulsating character. The vast majority of gas seeps appear above the zone of methane hydrate formation, while a number of them are either gas streams from mud volcanoes or localized within zones of large tectonic disturbances.

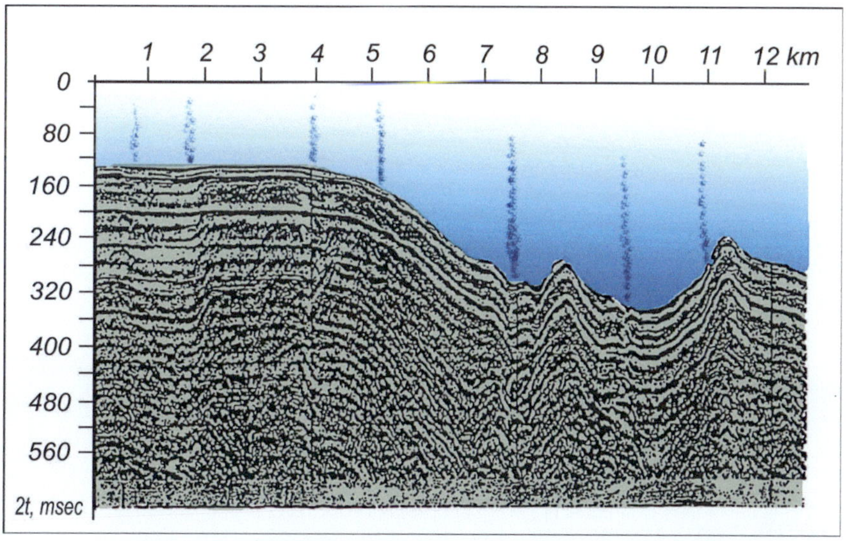

Fig. 7.15 Seismogram along the shelf-continental slope-subabyssal plain in the Paleo-Kalanchak area (Shnyukov et al. 2007)

7.2 Gas Seeps of the Crimean Shelf and Continental Slope

Gas seeps of the Crimean shelf and continental slope are extremely rare. Among those discovered, there are gas domes found at water depths of 1500–1700 m. They are confined to the zone of influence of the Lomonosov Transcrustal Fault: two gas torches are located on the traverse of the Forosskiy protrusion of the continental slope, and two others appear on the traverse of the Kara-Dag massif (Fig. 7.16).

The Lomonosov underwater massif (LUM) was discovered during geological research performed in 1989 (Shnyukov et al. 1997). It lies in the southwestern part of the underwater continuation of the Mountainous Crimea orogenic structures and runs in a sub-latitudinal direction. Current research has established that the LUM occupies a section of the continental slope restricted by the crystalline bedrock that is exposed on the sea bottom between the coordinates 44° 35′ N; 32° 35′ E and 44° 24′ N; 33° 04′ E.

The LUM is mainly composed of magmatic rocks of various depth facies. Sedimentary rocks are scarce and are represented by Carboniferous phyllite-shaped argillites, the three-component Tauride flysch, Upper Cretaceous carbonate clays and marls, and Lower Miocene limestones. The established length and width of the LUM along the 33° meridian is 44 km and 6 km, respectively. A widening of the slope surface occurs in the western direction; it is flattened by erosion and increased in thickness by its Quaternary sediments. The western and eastern parts of the slope are bounded by the isobaths at 1550–1600 m and 1700–1750 m on the south and east, respectively (Fig. 7.1).

The gas dome at the foot of the LUM was discovered during the 51st voyage of the R/V "Mikhail Lomonosov." It was originally called "an unidentified underwater object" that looked like a dense gas "cloud." The dome dimensions were 2 × 3 km in width and 270 m in height, and it took the shape of an uneven ellipse with a long axis oriented southwest to northeast (Shnyukov et al. 1990).

During subsequent research cruises, it was established that there was not just one but several gas domes possessing similar morphometric characteristics. Dredging of the sea bottom revealed carbonate plates and tubercles lying directly on the bedrock near the gas domes. Study of their carbon isotope composition revealed the presence of heavy hydrocarbon isotopes with $\delta^{13}C$ 8.2–8.5‰, indicating a deep origin for the carbonate buildup in anoxic water (Shnyukov et al. 1995).

On the 56th voyage of the R/V "Professor Vodyanitskiy," a "methane smoker" (fountain of methane) was raised from the sea bottom. It took

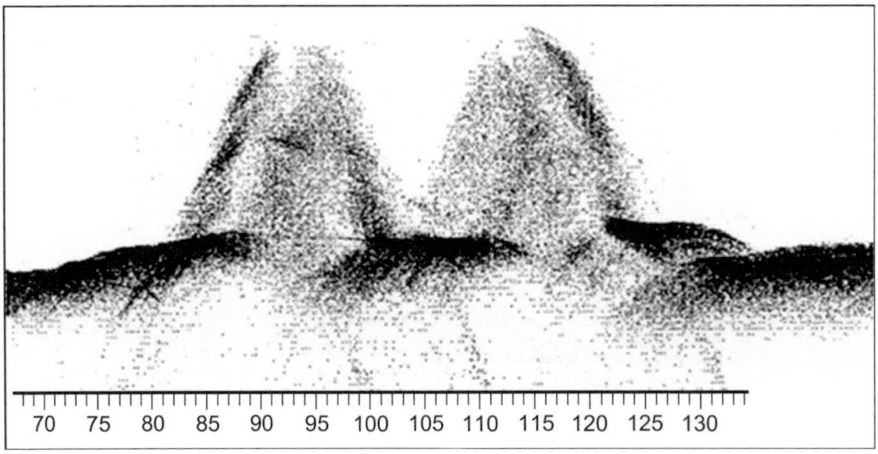

Fig. 7.16 Gas dome-shaped torches confined to the zone of the Lomonosov Transcrustal Fault

the form of a hollow carbonate structure with a diameter of 40 cm and a length of 1.5 m (Fig. 7.17).

In the Crimean part of the continental slope, the area of the Forosskiy protrusion is geomorphologically distinct. The Forosskiy protrusion covers the relatively steep slope located about 25 km from the southwestern coast of Crimea and close to the abyssal part of the basin (Pasynkova 1999). The outline of the protrusion mirrors the curvature of the southwestern coastline of Crimea, indicating the parallel relationship of this tectonic structure with Mountainous Crimea over the course of the geological development of the entire area. The structure of the protrusion is associated with the development of a system of sub-latitudinal and diagonal secant tectonic dislocations, in which contrasting immersion of separate blocks has occurred.

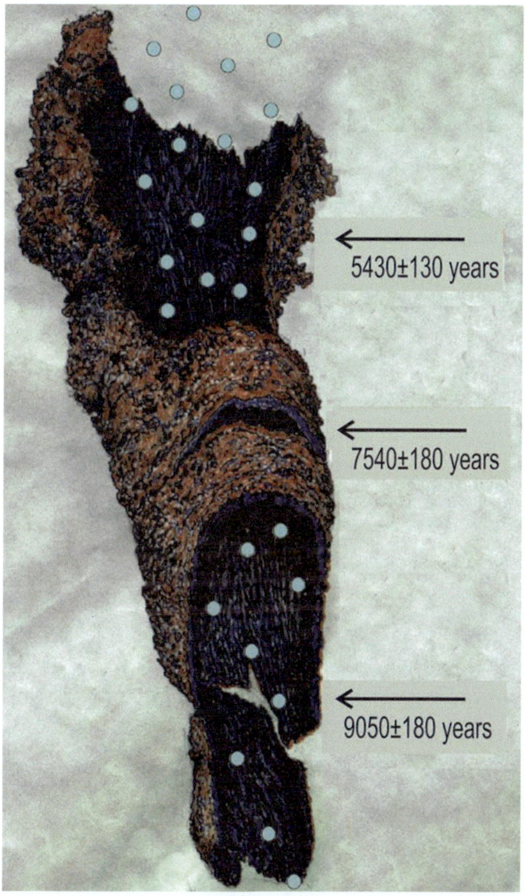

Fig. 7.17 Tubular carbonate structure ("methane smoker"). Reconstructed from fragments by V.A. Kutny. Absolute age was measured by ^{14}C method. Obtained during the 56th cruise of the R/V "Professor Vodyanitskiy" at coordinates 44° 27.4′ N; 32° 48.6′ E, at a water depth of 1600 m

The foot of the slope is bounded by the zone of the Circumchernomorian fault. During the 27th voyage of the R/V "Vladimir Parshin," two gas flares with heights up to 500 m were encountered at the foot of the Forosskiy protrusion and at a depth of 1600–1650 m. They were seeping from a shaft-like elevation on the seabed (Fig. 7.18).

During the same voyage, two gas flares of 400 m each were found to the south of the Kara-Dag massif at a depth of 687 m (coordinates 44° 39.57′ N; 35° 22.42′ E and 44° 39.24′ N; 35° 22.02′ E) (Fig. 7.19).

Gas outlets have also been recorded in coastal areas of the Crimean Peninsula. In Sevastopol Bay, one and a half dozen torches have been described (Eremeev et al. 2007), and they were also observed in Laspi and Sudak bays (oral communication by A. Pasynkov). In the coastal zone of Mountainous Crimea, they were described by Al'bov (1956), who summarized all archived and published materials over many years. According to these data, these gas manifestations are diverse

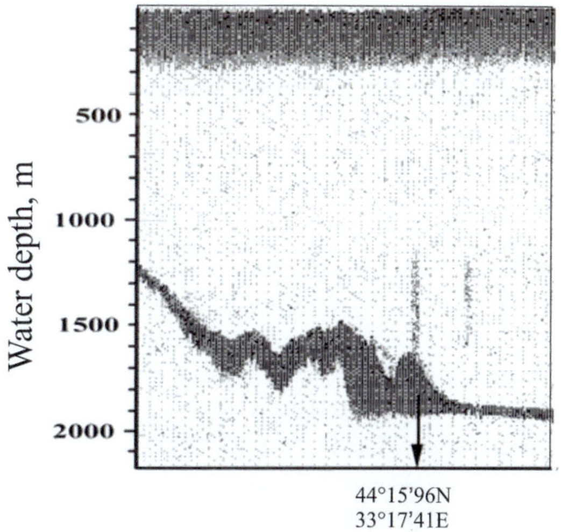

Fig. 7.18 Echogram of seabed relief obtained during the 27th voyage of the R/V "Vladimir Parshin" showing gas torches at the foot of the Forosskiy protrusion on the continental slope

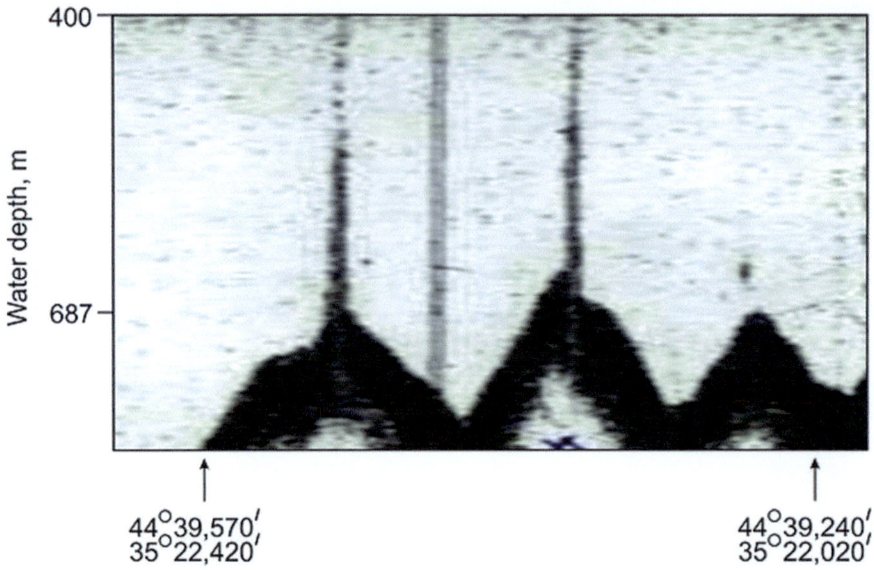

Fig. 7.19 Gas flares on the traverse of the Kara-Dag massif

in their gas composition, being largely represented by methane, carbon dioxide, nitrogen, and hydrogen sulfide, most often in their natural mixtures showing a predominance of one or another component. As a rule, there are no Quaternary deposits at the locations of these gas outlets.

For example, in the vicinity of Yalta—in Massandra, Magarach, Livadia, Oreand, between Alushta and Gurzuf in the Karabakh, and in the trenches made in the Tauride shales—carbon dioxide was detected. The deep well in Yalta (Industrial Report for Administrative Use) was faintly aerated. It penetrated downward more than 3 km and did not emerge from the Tauride sediments. Quite often, hydrocarbon gases are present. In 1950, in Alupka (above the upper highway) on the southern coast of Crimea, several wells drilled into the Tauride shales to a depth of 34 m released an outflow of gas that ignited from a cigarette and produced a column of flame that reached 6 m in height. This gas contained hydrocarbons (94.9–97.55%), nitrogen + rare gases (2.44–3.70%), and oxygen (0–1.39%). On land, Al'bov (1956) notes that carbon dioxide gas seeps in the vicinity of Yalta and Alushta are manifested in the air at low absolute elevations, while hydrocarbon gases are found at a height of several hundred meters above sea level.

At the same time, the well in Alupka, which was drilled in 1925, bubbled up with nitrogen including an admixture of combustible gases, hydrogen sulfide, and carbon dioxide. The spring at Ai-Iori in the western part of the Alushta amphitheater bubbled up with a natural mixture of nitrogen (92%) and some rare gases.

On the Kerch Peninsula, the waters within Maikopian deposits contain methane, hydrogen sulfide, carbon dioxide, etc.

Thus, according to Al'bov (1956), a definite stratigraphic "specialization" of sedimentary strata regarding gas seeps exists. Carbon dioxide—less often methane and even less often nitrogen—are confined to the Tauride deposits, while methane and other hydrocarbons, less often carbon dioxide and hydrogen sulfide, are confined to the Maikopian deposits.

All these strata are exposed on the outer shelf of Crimea as well. There are no grounds for assuming an abrupt change in the gas component of these strata over 20 miles outward from the shore. Therefore, it is necessary to check the composition of gases emerging from seeps within the sea; the thesis about their exclusively methane content needs additional verification. The total number of analyses for gas components coming from these seeps is still small, yet the literature refers a priori only to methane seeps. Mass analyses of these gases are necessary to confirm or reject the accepted wisdom. Especially important also is lithological study of the rocks where the seeps are located. Perhaps, some of them will turn out to be carbon dioxide or complex in the composition of their gas mixtures. Hence, there will likely be some corrections in the forecasts that estimate quantities of gas hydrate deposits.

7.3 Gas Seeps of the Kerch-Taman Shelf and Continental Slope

The Kerch-Taman region is located in a tectonic transition zone that separates areas of uplift from geological structures of Crimea and the Caucasus. The bottom relief is represented by huge, alluvial-proluvial cones of the Paleo-Kuban, Paleo-Don, and other ancient rivers as well as underwater canyons with numerous gas manifestations (Fig. 7.20).

The length of the shelf along the edge of the continental slope is about 75 km from Cape Chauda to the Kerch Strait. Its width, taking into account accumulative plumes, varies from 60 km to 90 km. The transverse profile of the continental slope that descends to the south is almost gentle, occasionally flat, or slightly convex. The upper part of the slope is inclined at angles of 3–50°, the slopes of the middle part at 1.1–1.5°, and the lower part at 0.1–0.6°. The upper ledge is clear, and in some areas, it inherits in part the position of sub-latitudinal tectonic disturbances. However, there are no sharp longitudinal folds of the slope, and the relief elements are linked together in a gradual transition.

Gas seeps within the Kerch-Taman shelf and the upper part of the continental slope were discovered by military hydrographers (Tretyakov 1993), although individual gas manifestations were known in the west and east of this region even earlier (Polikarpov and Egorov 1989; Polikarpov et al. 1989). Shimkus et al. (1998) noted that the sedimentary strata were saturated with gases, and Glebov et al. (2001) found areas with anomalously high gas saturation within the water column. However, the wide distribution of gas manifestations in this region was first highlighted by expeditions of OMGOR NASU (Shnyukov et al. 2003), when more than 200 gas seeps were observed within a narrow strip parallel to the Kerch-Taman coast (Shnyukov et al. 2003, 2010) (Fig. 7.21).

There are 657 known gas seeps in the region (Egorov et al. 2011), but this number is obviously not a final count. The majority of gas seepages occur along the shelf breaks in areas of fault scarps on the seafloor, shallow subsurface faults, small grabens, and conical depressions on domes of gently sloping anticlinal rises. Fault scarp lengths are up to several kilometers, and fault depths range from a few hundred meters to 1.5 km. In regions of methane seepages, specific biogeochemical processes are evident that relate to the activity of methane-oxidizing bacteria resulting in the formation of diagenetic minerals (e.g., carbonate buildups). The presence of gas hydrates is inferred from bottom-simulating reflections (BSRs). The distribution of gas facies shows major gas accumulations close to the seafloor in the coastal area and along the shelf break, as well as ubiquitous gas migration from the deeper subsurface on the shelf and gas hydrate occurrences on the lower slope (below 750 m water depth). Gas release at the seafloor is abundant at water depths shallower than 725 m, which corresponds to the minimum theoretical depth for methane hydrate stability, but it occurs only exceptionally at water depths where hydrates can form (Figs. 7.21 and 7.22).

According to the diagram, the greatest number of gas seeps is present within the edge of the outer shelf and at its bend into the continental slope at water depths of 100–150 m. Based on the evidence graphed in Fig. 7.22, it seems that seeps do not exist above isobath 50 m; however, this is not true. The distribution can be explained by the fact that Egorov et al. (2011) carried out their investigations onboard large ships that could not navigate in areas with a water depth shallower than 50 m. A.A. Pasynkov, working from smaller coastal boats, found a number of gas seeps within shallow water.

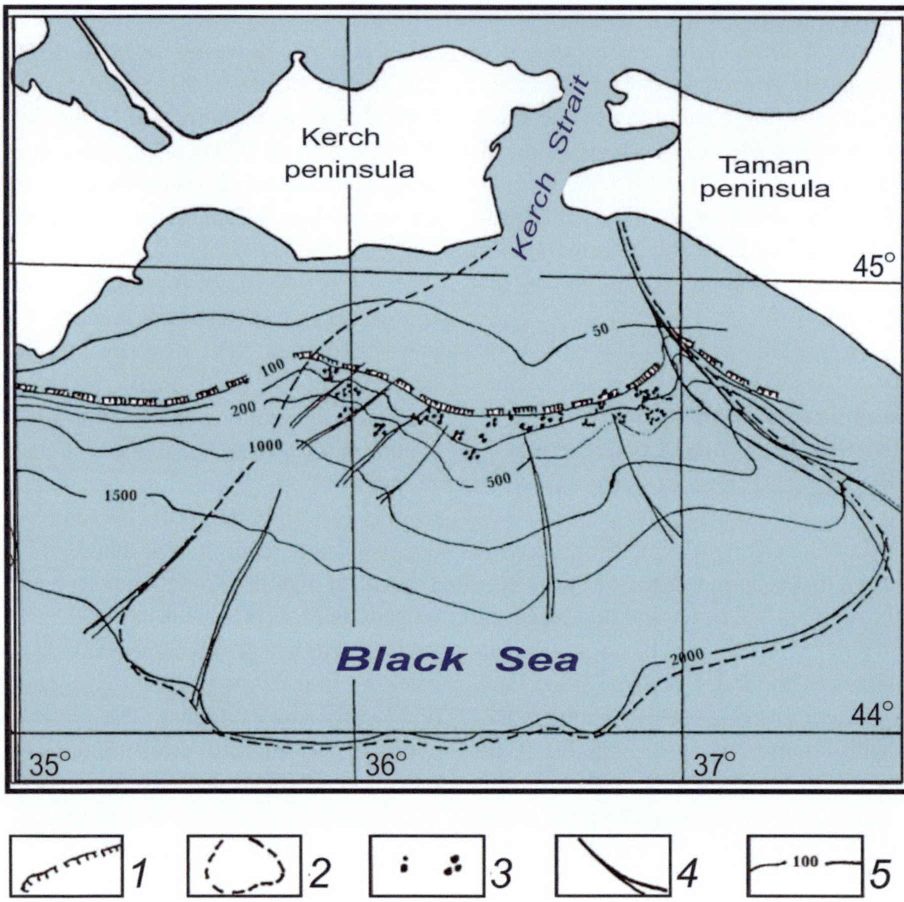

Fig. 7.20 Distribution of gas seeps on the Kerch-Taman shelf. Key: 1, roof of the continental slope; 2, contours of alluvial-proluvial cones; 3, areas of gas manifestations; 4, underwater canyons; 5, isobaths, m

Most of the discovered gas seeps within the continental slope at water depths of 200–500 m are confined to the positive oil-bearing Neogene structures (e.g., Anisimova and Yuzhnokercheskaya) and less often those confined to the Maikopian structures (e.g., Dreyfovaya, Drift, Visokosnaya, and Maykopskaya) that are located within the Kerch-Taman trough between Crimea and the Caucasus. The sediments in the trough transgressively overlap the Upper Cretaceous strata (Shnyukov et al. 2010) (Fig. 7.23).

The sedimentary cover within the trough contains numerous tectonic ruptures caused by compressive forces that apparently acted during the early stage (or stages) of mountain development. The gas seeps are quite abundant and exist either singly or in a group, as can be seen in the echo sounder diagrams obtained during the 65th voyage of the R/V "Professor Vodyanitskiy" in 2010 (Figs. 7.24, 7.25, and 7.26).

The most powerful gas flares are located on the continental slope at water depths of 400–500 m (Figs. 7.24, 7.25 and 7.26). At shallower depths, the gas seeps are weaker; their height hardly reaches 10 m, and their gas bubbles do not reach the surface (Fig. 7.27 and 7.28).

Within the Kerch Strait, which is part of the Kerch-Taman interperclinal trough (Sect. 6.2) that is in turn a part of the ancient Kerch-Mariupol fault, few gas seeps have been discovered so far because no proper search has been carried out

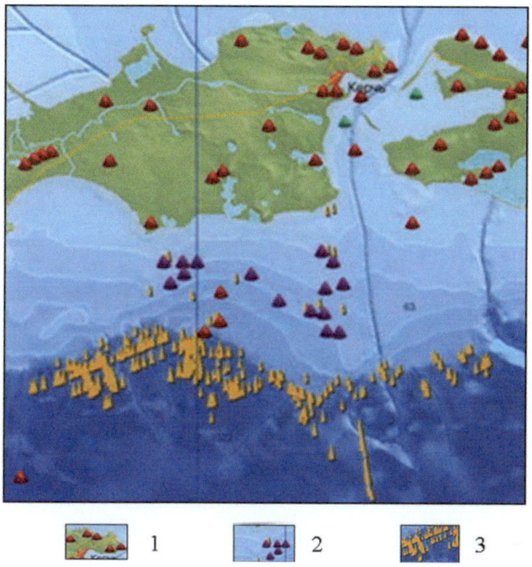

Fig. 7.21 Manifestations of gas volcanism on the Kerch-Taman outer shelf and the upper part of the continental slope. Key: 1 (onshore) and 2 (offshore) mud volcanoes identified on the base of geological and geophysical surveys, 3 gas seeps

due to technical reasons, e.g., heavy sea traffic, etc. It may be that the Blevaka mud volcano is, in fact, a strong gas torch.

7.4 Gas Seeps of the Caucasian Part of the Black Sea

The northern part of the Caucasus coast is about 400 km long. It has a southeastern strike, disturbed in the Gelendzhik Bay area, and includes the Idokonas, Pitsunda, Sukhumi, and Kodori peninsulas. The shelf is narrow and discontinuous with a width of only a few kilometers. Its changeable character is due to the geological structure of the coast, which is represented by young, folded mountains. The shelf itself belongs to the Colchis lowland and has a concave shape. Together with the adjacent continental slope, it is dissected by steep underwater canyons formed by the rivers Bzyb, Mzymta, Rioni, Chorokh, and

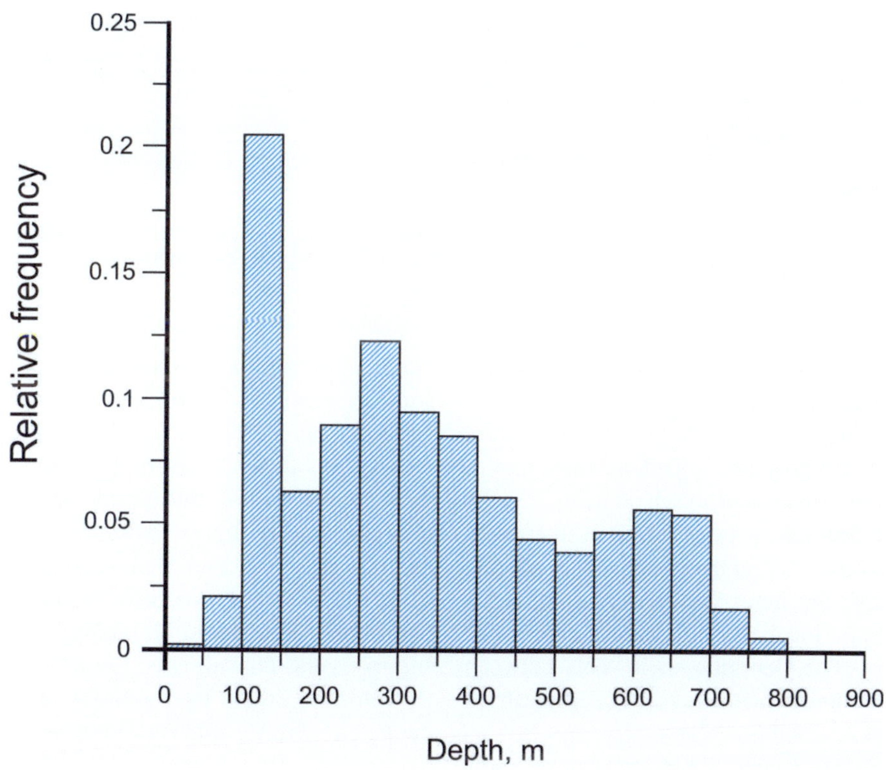

Fig. 7.22 Distribution of gas seeps along the water depth gradient within the Kerch-Taman shelf and continental slope (Egorov et al. 2011)

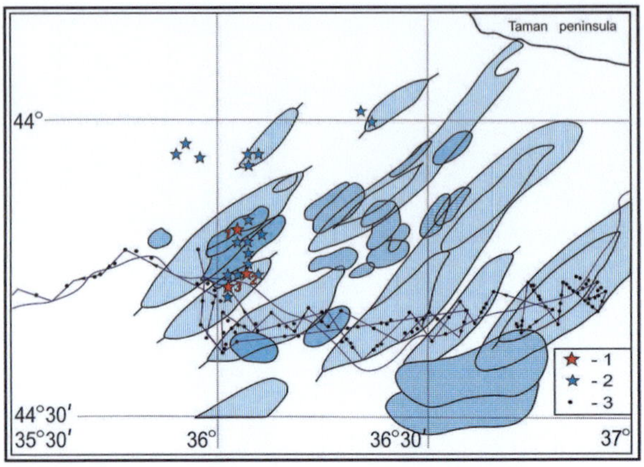

Fig. 7.23 Position of gas seeps relative to the positive structures of the Kerch-Taman trough. Key: 1, mud volcanoes (OMGOR, Naumenko, Mitin); 2, gas seeps; 3, sampling locations. Shades of blue indicate structures of different age

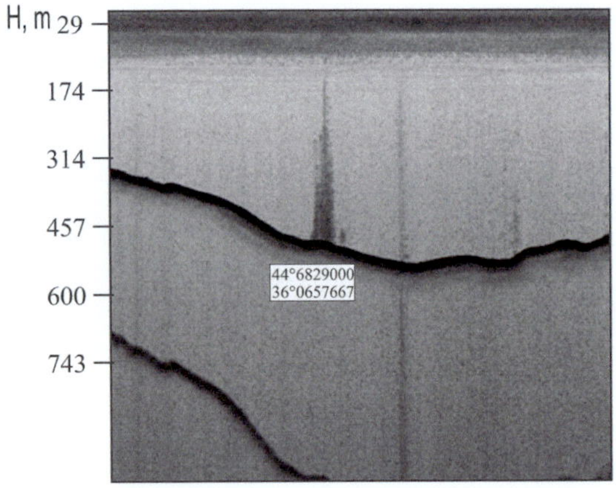

Fig. 7.24 Single gas seep with a height of about 300 m located at a water depth of 550 m

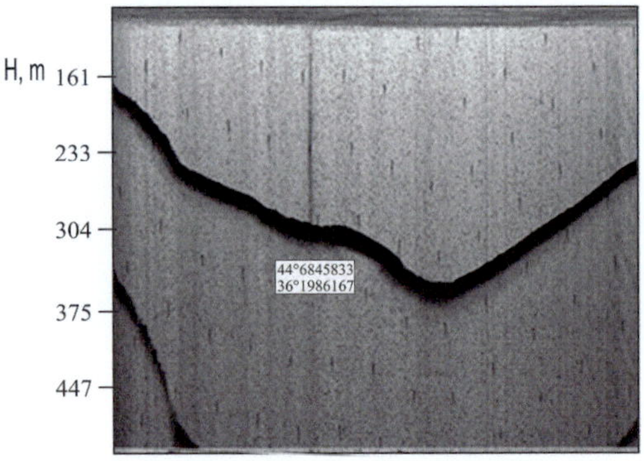

Fig. 7.25 Single gas seep with a height of about 90 m, located at a water depth of 300 m

some others. The shelf is composed of ancient marine terraces, e.g., the Karangatian (height 18–20 m), Chaudian, etc. There is a bench in the coastal part composed of folded, vertical, and in some sections, dislocated flysch deposits.

To date, 565 gas seeps of different intensities have been discovered in the region. They are distributed from the inner shelf to the continental slope, being localized at water depths of 14–1120 m. The Georgian shelf vents between 0.5 and 15 billion m^3 methane $yr.^{-1}$ (Tkeshelashvili et al. 1997).

Along with the seeps, the presence of oil spots has also been recorded. Most of the gas seeps are confined to depths not exceeding 300 m, while some of them are also present at depths of 600–1200 m (Fig. 7.29).

The gas seeps operating at depths below 725 m are within the limits of stability of methane gas hydrates. Attention should also be drawn to the fact that seep development occurs in the shelf areas near the mouths of the Chorokhi, Kintrishi, Natanebi, Supsa, and Inguri rivers. This allowed Egorov et al. (2011) to suppose that the genesis of methane occurred within alluvial sediments and to extrapolate this idea to the Dnieper and Danube rivers.

7.5 Gas Seeps of the Bulgarian Part of the Black Sea

The gas seeps of the Bulgarian part of the Black Sea have been described by numerous researchers. Palii et al. (1951) investigated gas seeps near the "Gold Sands" resort based on the stories of fishermen. Dimitrov et al. (1979) studied gas seeps in Balchik Bay, linking their origin to the migration of deep gases upward to the sea bottom. Khrischev (1981) and Chochov (1987) described gas-saturated Holocene silts and advanced the idea of their autochthonous genesis. Dimitrov (1990) observed gas seeps at depths of 100–120 m on the southern Bulgarian shelf and suggested their deep origin. He proposed that gas seeps were confined to tectonic disturbances and indicated the appearance of pockmarks on the seafloor in the places of gas emission at water depths of 220–325 m. The pockmarks had diameters of 50–210 m and an average depth of 4 m. They occurred in groups of 2–4 or more and possessed isometric shapes and a V-shaped profile. At shallower depths, pockmarks revealed an asymmetric, U-shaped profile. All pockmarks bore signs of a stream-like manifestation of ascending fluids, and the main gas component was methane.

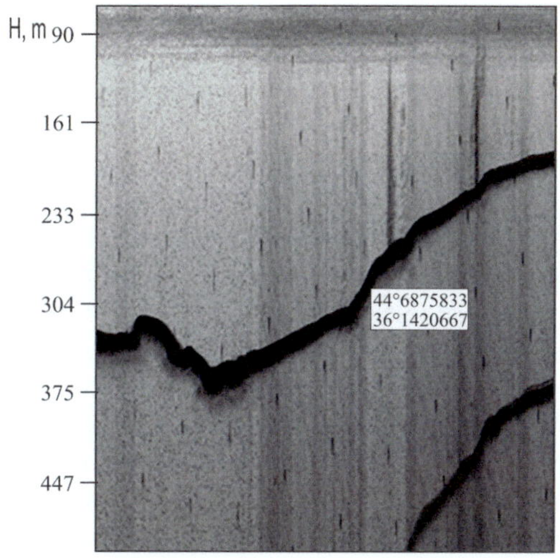

Fig. 7.26 A group of gas seeps with heights of about 150 m, located at a water depth of 647.8 m

Gas seeps in the Aladzha bank exist at depths of 10–15 m; they have been studied by Vassilev et al. (1996). They were also found on the Nessebar to Cape Emine traverse as well as within the southern shelf in the Burgas depression at water depths up to 325 m.

A comprehensive review of gas seep distribution within the Bulgarian part of the Black Sea was provided by Dimitrov (2002). He pointed out that, as of today, 482 targets have been identified as gas seepage plumes. Approximately 19,735 and 6020 individual seeps exist on the open shelf and coastal waters, respectively. This estimate is based on available public domain data, specific research, and results from a special questionnaire

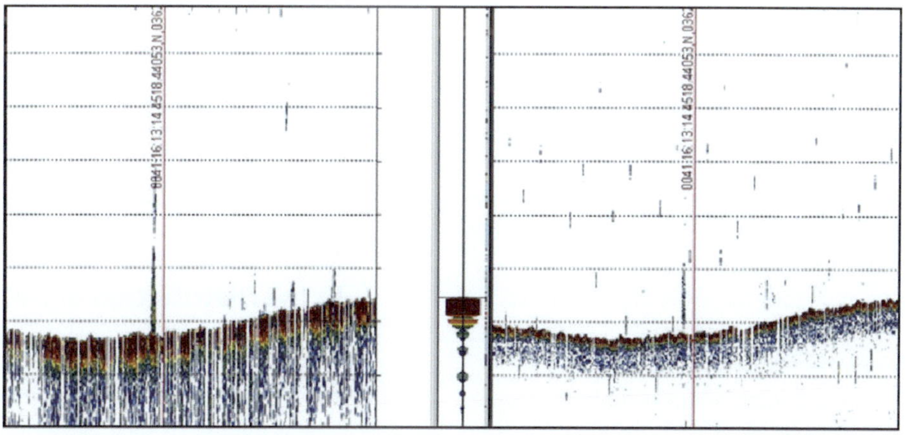

Fig. 7.27 Group of weak gas seeps with maximum heights of about 10 m, located at a water depth of 13 m (Pasynkov et al. 2009). For location, see Fig. 7.28

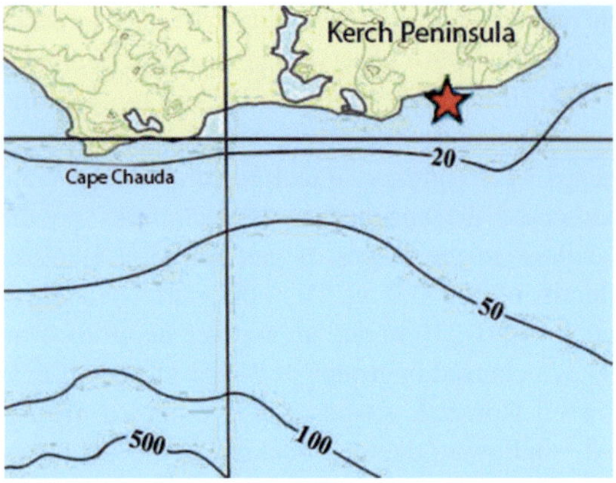

Fig. 7.28 Location of the low-power gas seeps (red star) off the southern coast of the Kerch Peninsula (Pasynkov et al. 2009)

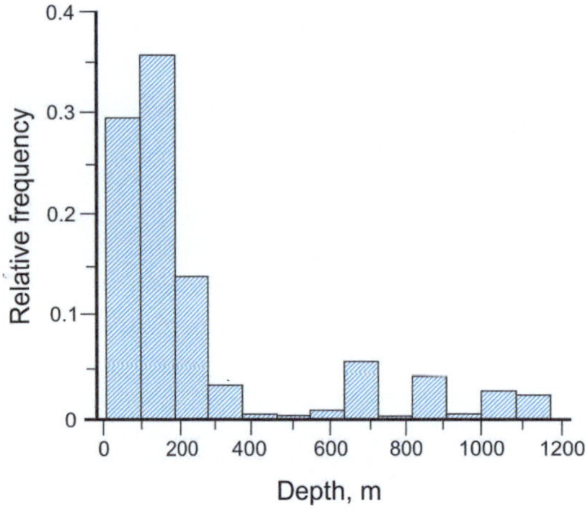

Fig. 7.29 Distribution of gas seeps along the water depth gradient within the Caucasian shelf and continental slope (Egorov et al. 2011)

that was distributed to a range of "seamen" (Fig. 7.30).

According to Dimitrov (2002, 2436) "There are three types of seepages which are the most representative of Bulgarian coastal waters. These have flux rates as follows: slow flux, 0.26–0.62 l/min; medium flux, 1.0–2.0 l/min; and fast flux, 2.5–3.8 l/min. (Fig. 7.31).

Quite often, gas seeps are confined to the areas of tectonic ruptures identified on echo sounder profiles (Fig. 7.32).

7.6 Gas Seeps of the Sea of Azov

During the 62nd voyage of the R/V "Professor Vodyanitskiy" (2006), an attempt was made to search for gas seeps in the Sea of Azov; however, the existing hydroacoustic equipment was not suitable for the shallow waters of the basin, and the echogram was "hammered" by overlapping noises, which "drowned out" the sonar recording.

In July 2008, on board the hydrographic vessel MGS-1845 "Vasily Zarudny" in the Sea of Azov, gas seeps were recorded for the first time using a mobile sonar Simrad EA-400R and StrataBox profiler (Pasynkov et al. 2009) in the central part of the sea at a water depth of 10.5 m (Figs. 7.33 and 7.34).

A central gas seep with a number of smaller seeps and individual bubbles is clearly visible in Fig. 7.33a. The height of the seep reaches 8 m, and judging by the echogram, the gas does not reach the water surface. Smaller low-power gas seeps form a gas cloud with heights up to 3 m from the seabed surface, and the cloud can be traced in the direction of the vessel's movement. In the lower "bottom" part of the echogram, a pattern typical for decompression of gas-saturated marine sediments is observed to a depth of 5 m. A profilogram (Fig. 7.33b) showing the upper part of the geological section to a depth of 10 m demonstrates that the gas release is confined to a clearly manifested break in the continuity of the Holocene and older Quaternary sediments.

Analysis of the tectonic position of the detected gas manifestation shows that the imaged gas flare is located at the junction of the zone of the South Azov and Kerch-Mariupol deep faults (Fig. 7.34).

7.7 Regularities in the Distribution of Gas Seepages

By studying the spatial distribution of gas seepages, a general picture of their development can be drawn. In the Georgian part of the basin, there are numerous gas seepages at water depths of

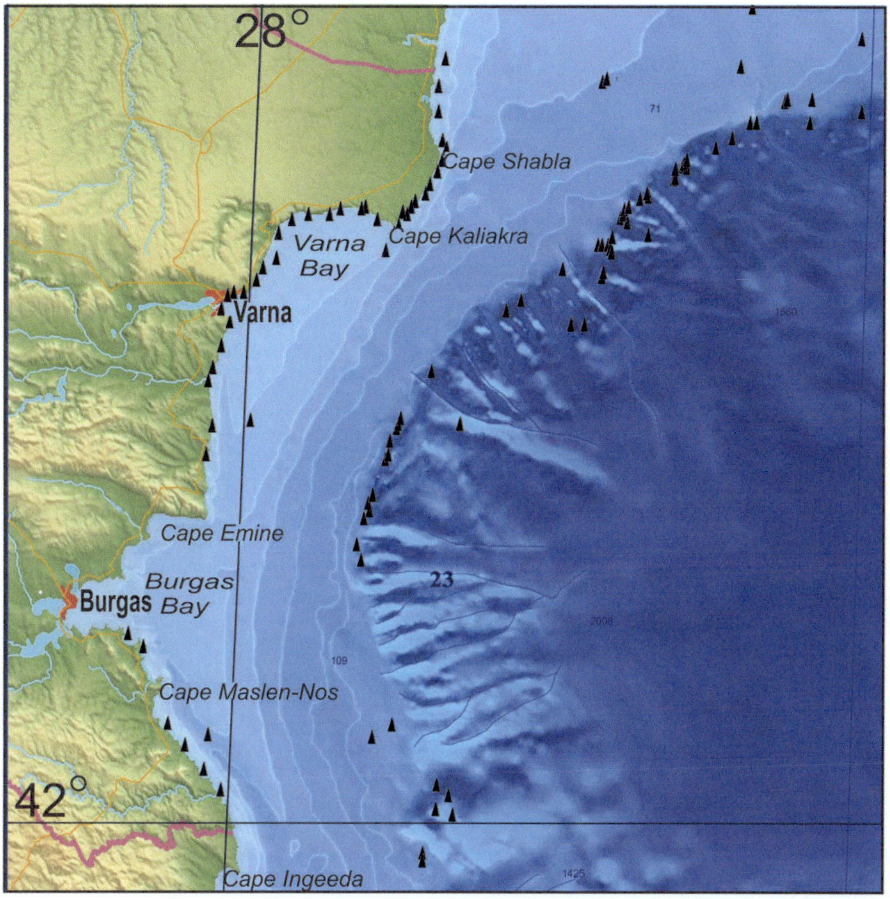

Fig. 7.30 Gas seeps (black triangles) within the Bulgarian part of the Black Sea. (Based on the data of Dimitrov 2002)

0–100 m, and a similar distribution is observed near the coast of Bulgaria, where about one third of the gas seepages are located near the shore (Dimitrov 2002). The majority of gas seepages are localized between water depths of 100 and 300 m, being largely confined to the shelf-continental slope boundary, although they are also present as deep as 700 m, and rarely 800 m. On occasion, they occur even at water depths of 1200 m and deeper, representing gas jets from mud volcanoes. Gas manifestations are also known on land, e.g., in Mountainous Crimea, where they appear in the sediment sequences of different ages as discovered by drilling (Table 7.1).

The question invariably arises about the reasons for the absence of gas seepages at the bottom of deep-sea depressions. Single finds of torches testify to the possibility that they do occur in the Black Sea depths.

In our opinion, the localization of gas seepages along the periphery of the sea and their actual absence in the deepwater zone have their own explanation.

Under a certain combination of temperature and pressure within the thermodynamic conditions of the seabed, there is a gas hydrate formation zone in the deepwater depression (Fig. 7.35).

When methane enters this zone, it is linked to gas hydrates and accumulates in sub-hydrate reservoirs. The colossal scales of gas emission at the periphery of the sea can be assumed to be indicative of a colossal degassing in the deep part. This in turn allows one to assume the scale dimensions of the gas hydrate deposits. It is evaluated at 288,100 km^2, representing about 68% of the total

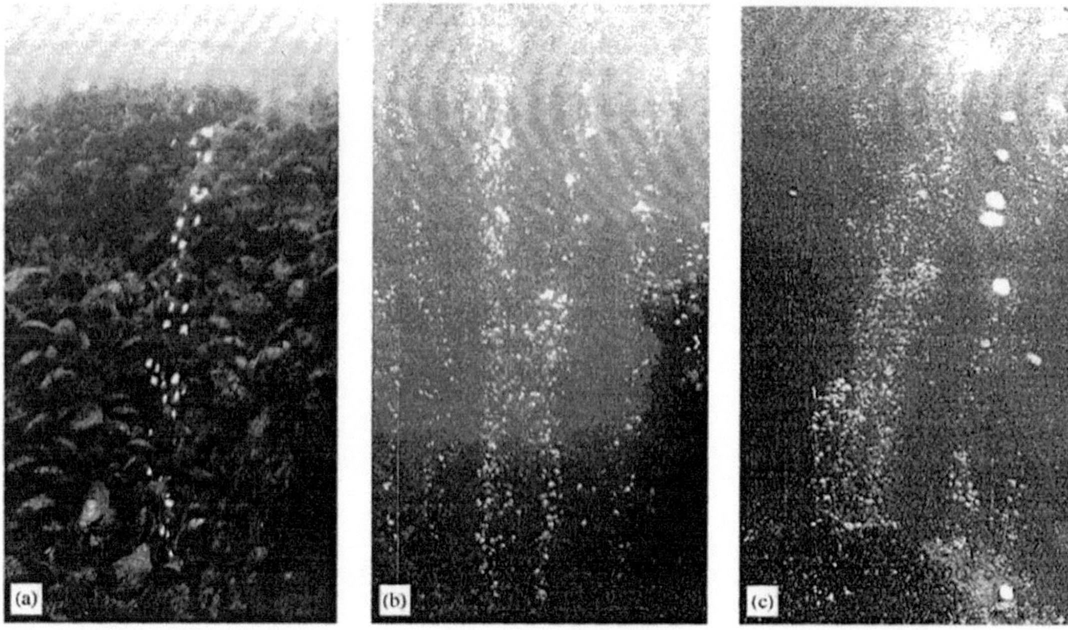

Fig. 7.31 Photographs of gas seepages in the "Golden Sands" area showing (**a**) a small seep producing up to 20 bubbles/s (bps) with a flux rate of 0.26–0.62 l/min; (**b**) two medium strength seeps, 40–60 bps, 1.0–2.0 l/min; and (**c**) a strong seep with more than 80 bps, 2.5–3.8 l/min (Dimitrov 2002)

Black Sea or almost 91% of the deepwater basin; the volume of gas hydrates has been estimated at 4.8 km^3, corresponding to $0.1–1 \times 10^{12}$ m^3 of free methane (Vassilev and Dimitrov 2002).

An uneven supply of gas from the depths within different regions is quite possible, and this would lead to uneven distribution of gas hydrate deposits within the gas hydrate stability zone (GHSZ), also called the methane hydrate stability zone (MHSZ). Single gas flares on the deep bottom indicate that not all gas was bound into gas hydrate. Part of this gas escaped, most likely, through tectonic ruptures or due to abnormal pressures on the seabed surface. According to Solovyev (2001), only 1–10% of the gases entering the GHSZ are stabilized.

Thus, within the Black Sea, a kind of specific zoning emerges: the central deepwater part of the sea is occupied by gas hydrates (Fig. 7.35), while its periphery contains abundant gas seeps. It is quite possible that gas or, more precisely, gas-water seeps are localized within the area of the land adjacent to the sea, for example, in the Kerch-Taman region, but for the most part, they are not fixed because of the difficulty of observation.

The question of the gas supply for the seeps is complicated because they are located within different geological structures, such as the Kalamitskiy, Lower Kamchiskiy, Kerch-Tamanskiy, and Gurian troughs as well as within the Western Black Sea basin, etc. (Fig. 7.36).

As a rule, gas seeps are confined to zones of tectonic ruptures. They do not form morphostructures, such as mud volcanoes, and are geologically simpler. It may be advisable to try to compare the development of the gas seepage fields with the development of these or other sedimentary sequences. Specialists who defend the biochemical origin of gases connect them with Holocene sediments and argue that the biochemical cycle of methane in the Black Sea is regulated by modern microbial processes. To them, it is only a question of the amount and nature of methane coming from mud volcanoes that remains to be discussed, since methane of other external sources (bottom sediments and cold seeps) is of microbial origin and young (no older

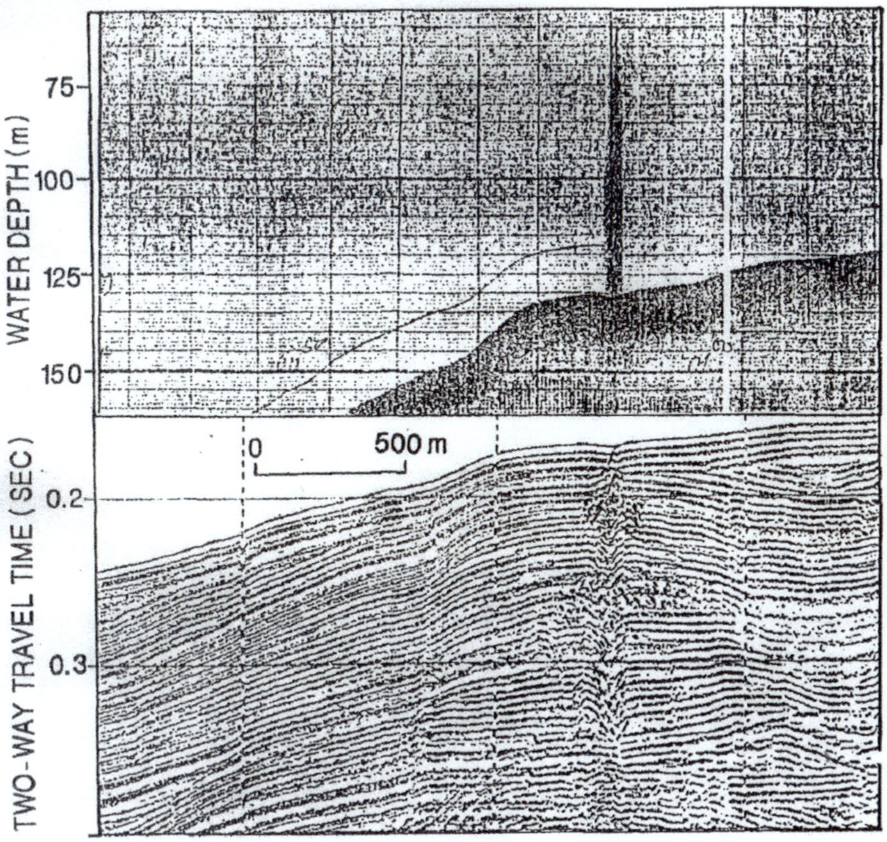

Fig. 7.32 Single gas torch on the outer Bulgarian shelf. Its confinement to a tectonic disturbance is visible on the echo sounder profile (Dimitrov and Doncheva 1994)

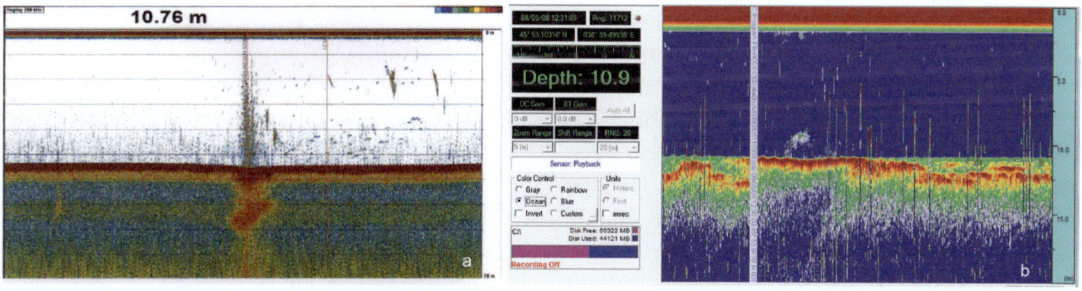

Fig. 7.33 Echo sounder diagram (**a**) and profilogram (**b**) of a gas flare and secondary gas jets in the Sea of Azov (Pasynkov et al. 2009)

than Holocene) (Ivanov et al. 2002; Lein and Ivanov 2005). This statement contradicts the fact that gas flares have been discovered emerging from crystalline rocks of the Lomonosov Massif, where there are no Quaternary deposits, and the same can be said about the gas outflows on land.

The sequences of Quaternary and, especially, Holocene deposits are not sufficiently thick and expansive as to provide so many gas seepages

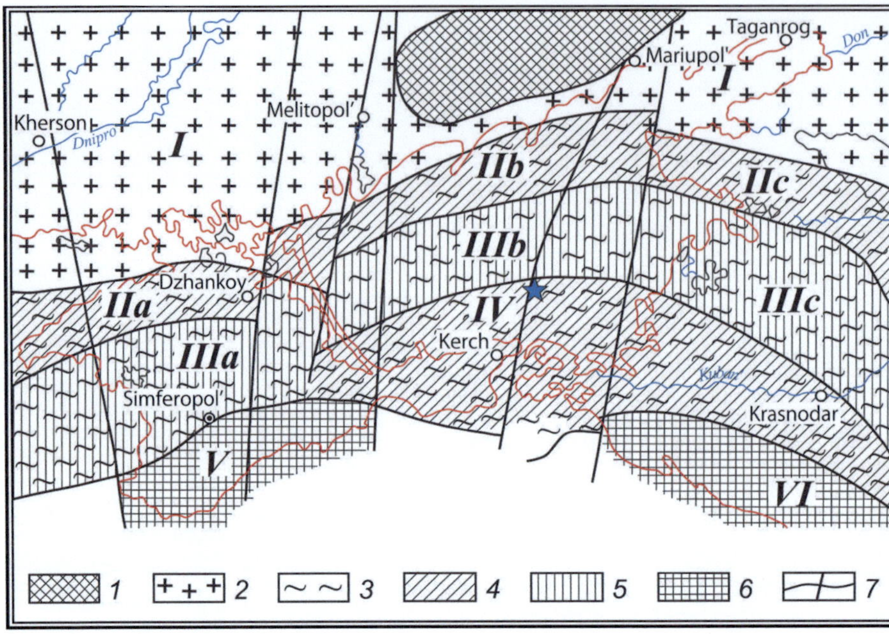

Fig. 7.34 Diagram of tectonic zoning of the Sea of Azov bottom. Key: *1*, Exposure of crystalline rocks on the bottom surface; *2*, submerged areas of Precambrian crystalline basement overlapped by Mesozoic-Cenozoic strata; *3*, area of Epihercinian foundation; *4*, trough; *5*, elevation; *6*, folded areas; *7*, boundaries of geostructures and tectonic ruptures. I, northern part of the Black Sea basin (northern side); II, zone of Black Sea troughs, (**a**) North Crimean, (**b**) North Azov, (**c**) Eyskiy; III, zone of Black Sea uplifts, (**a**) Central Crimean, (**b**) Middle Azovian shaft, (**c**) Kiev-Berezaskiy; IV, Indo-Kuban trough; V, Mountainous Crimea; VI, the Greater Caucasus. Blue star, location of the gas torch (Shnyukov et al. 1974)

with methane. Perhaps the more saturated paleoalluvial strata of rivers such as the Paleo-Dnieper, the Paleo-Dniester, and the Paleo-Danube might have played a certain role in the synthesis of methane. This has been proposed by Egorov et al. (2003, 2011), who note that the main areas for methane seepages are located in the paleovalleys and pre-estuarine zones of the Dnieper, Danube, Don, Chorokh, and other rivers. These authors stipulate, however, that the methane degassing is not confined exclusively to paleovalleys. Locations of gas flares show only a partial overlap with the areas of paleoalluvium (Fig. 7.37).

The authors of this book associate gas seeps with the Maikopian as well as Jurassic, Lower and Upper Cretaceous, Paleocene, and Neogene gas-bearing horizons because they manifest numerous (up to 10) small gas fields in the northwestern part of the Black Sea. Potential areas also include the northeastern part of the Black Sea and the Azov Sea.

A majority of the gas seeps in the northern part of the Black Sea are confined to the wedging of the Maikopian deposits (Fig. 7.38) that, in our opinion, serve as a stratum for unloading of methane from the underlying geological sequences.

Gas seeps may indicate the existence of petroleum reservoirs under the sea bottom. For example, according to Clark et al. (2003), more than half of the gas seepages in the Santa Barbara Strait in California are located over the South Elwood anticlinal, which possesses huge hydrocarbon reserves. Operations conducted at this richly productive oil field since the 1960s have led to a decrease by almost one half in the distribution of gas seepages, as was discovered by monitoring between 1973 and 1995. Gas seeps in

Table 7.1 Characteristics of oil and gas manifestations in Mountainous Crimea (Muratov et al. 1974)

Location	Age	Characteristics of oil and gas manifestations
Adhi-Su mineral water spring	Tauride series	Spring with flow of water 12 m^3/day; gas 1.5 m^3/day. The water and gas temperature is 25 °C. Chemical composition of released gases: H_2S (traces); CO_2 (0.3–0.6%); O_2 (0.1–1.4%); CH_4 (19.8–25.4%); N_2 + rare gases (73.8 74.5%); Ar, Cr, Xe (0.247–0.557%); He, Ne (elevated concentrations)
Magarach, Livadia, Oreanda Beshuyskoe coal deposit	Tauride series Jurassic	In the wells, the release of gas: CH_4 (94.2%); CO_2 (0.4%); O_2 (0.1%); N_2 + rare gases (5.3%); heavy rare gases (0.073%); light rare gases (0.004%)
Dvuyakornaya	Lower Jurassic	In the wells, sandstones are soaked in oil
Tunnel near the city of Yalta, Planer Square (Barakolskaya anticline)	Jurassic	Weak bubble up of sediments containing oil; gas release rate up to 10 m^3/day. Gas composition: CH_4 (81.2%); heavy hydrocarbons (1.6%)
Goncharovskaya area	Lower Cretaceous	Bubble up of sediments in deep wells
Near Sevastopol Beloglinskaya area	Jurassic	Bubble up of sediments and oil films in certain intervals recovered within wells
Ayan (Zatechnoe village)	Lower Cretaceous	Release of gas containing 35% hydrocarbons
Feodosia, hydrogeological well near the weaving factory, Belogorsk town	Lower Cretaceous	Mineral water spring with dissolved gas and rate up to 1 m^3/24 h from a depth of 25 m in the well. Weak bubble up of sediments in the wells from intervals 52–95 m, 190 m, and 345–451 m
Bakhchisaray town	Upper Cretaceous	With digging of the draw well, emission of flammable gas
Feodosiya city (near the water reservoir)	Paleocene	In the well from a depth of 330 m, a fountain of mineralized water with a flow rate of 100 m^3/day with combustible gas. The gas contains up to 91% hydrocarbons
Surroundings of Sudak city (mineral spring Shakir-Bumar)	Fossil alluvium	Spring with emission of non-flammable gas with a high nitrogen content
Simferopol, the valley of the river Salgir	Alluvium	Weak emission of marsh gas. Bubbles of combustible gas in the river bed

the area could appear under conditions of high pressure in the gas-bearing sediments.

Gas seepages, whether of Maikopian or other gas-bearing horizons, seem always to be restricted to tectonic ruptures. In general, they occur along the Circum-Black Sea fault as well as in the river paleovalleys that are caused by fault tectonics, and as such, they record deep ruptures on the ground. For example, there are numerous gas seepages within the mouth of the Chernaya River near Sevastopol (Eremeev et al. 2007).

Similarly, all gas and oil fields in the region are also attributed to tectonic disturbances (Chebanenko et al. 2003). This can be exemplified by Mountainous Crimea, which

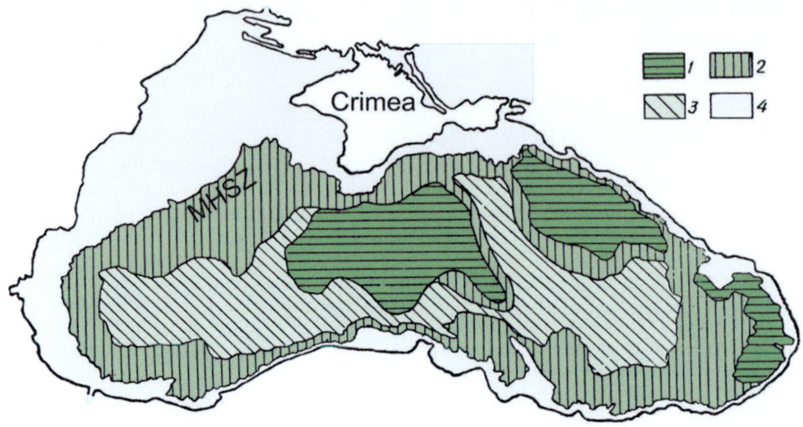

Fig. 7.35 Occurrence of gas hydrates within the Black Sea bottom (data from "Yuzhmorgeologia"). Key: *1*, high; *2*, moderate; *3*, small; *4*, none; MHSZ, methane hydrate stability zone (Korsakov et al. 1989)

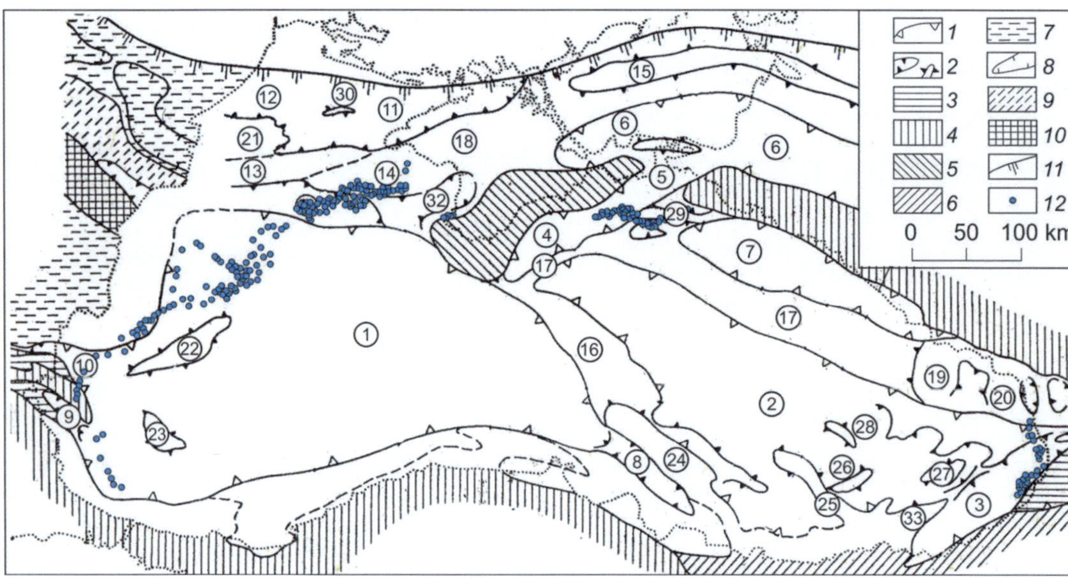

Fig. 7.36 Diagram of the distribution of gas seepages according to the geological structures of the Black Sea. Key: *1*, contours of the large Paleogene depressions and troughs; *2*, contours of the minor uplifts and depressions; 3, Paleocene-Eocene troughs of the Adzaro-Trialetskaya structure and the Balkans; *4*, folded structures of the Western Caucasus, Western Pont, and the Balkans; *5*, Crimean meganticlinorium; *6*, Eastern Pont; *7*, Paleozoic-Mesozoic sedimentary cover of the Mizia plate and Preddobrudzhinskiy trough; *8*, contours of Jurassic depressions within the Preddobrudzhinskiy trough; *9*, folded Paleozoic and Triassic of Northern Dobruja and the Priprutskiy protrusion; *10*, Riphean horst of Central Dobruja; *11*, the southern border of the ancient East European platform; *12*, gas torches. Depressions: *1*, Western Black Sea; *2*, Eastern Black Sea. Troughs: *3*, Gurian; *4*, Sorokin; *5*, Kerch-Tamanskiy; *6*, Indolo-Kubanskiy; 7, Tuapse; 8, Sinopskiy; 9, Bourgaskiy; 10, Lower Kamchian; *11*, Karkinitskiy; *12*, Krylovskiy. Shafts: *13*, Gubkin; *14*, Kalamitskiy; *15*, Azovskiy; *16*, Andrusov; *17*, Shatskiy. Vaults: *18*, Crimean; *19*, Gudautskiy; *20*, Ochamchirskiy. Uplifts: *21*, Kiliyskiy; *22*, Polshkovskiy; *23*, Druzhba; *24*, Archangelskiy; *25*, Muratov; *26*, Chikhachev; *27*, Dzhanelidze; *28*, Strakhov; *29*, Baryerny; *30*, Golitsyn; *31*, the boundary step; *32*, Alminskaya depression; *33*, Trabzon's ledge. The map and structural interpretations were borrowed from Tugolesov et al. (1985)

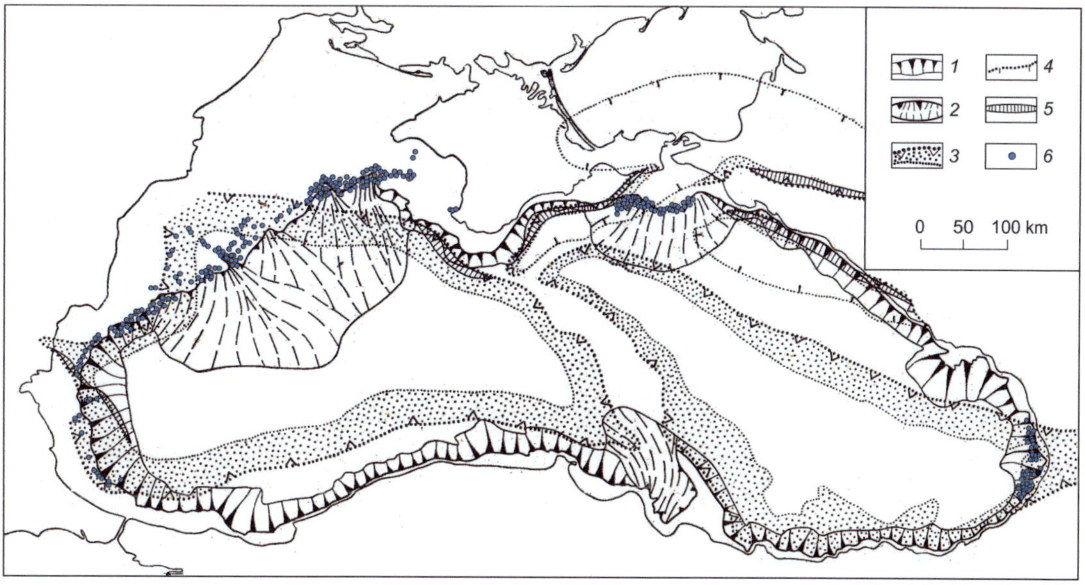

Fig. 7.37 Layout of the Black Sea continental slope today and slopes of the Paleogene depressions and gas flares. Key: *1*, denuded areas of the modern continental slope; *2*, accumulative areas of the modern continental slope (large river deltas and underwater alluvial fans); *3*, steep slopes of the Paleogene depressions; *4*, conditional position of the boundaries of the Paleogene troughs on gentle slopes; *5*, zones of possible development of faults on the slopes of troughs; *6*, gas torches. The geological map was borrowed from Tugolesov et al. (1985)

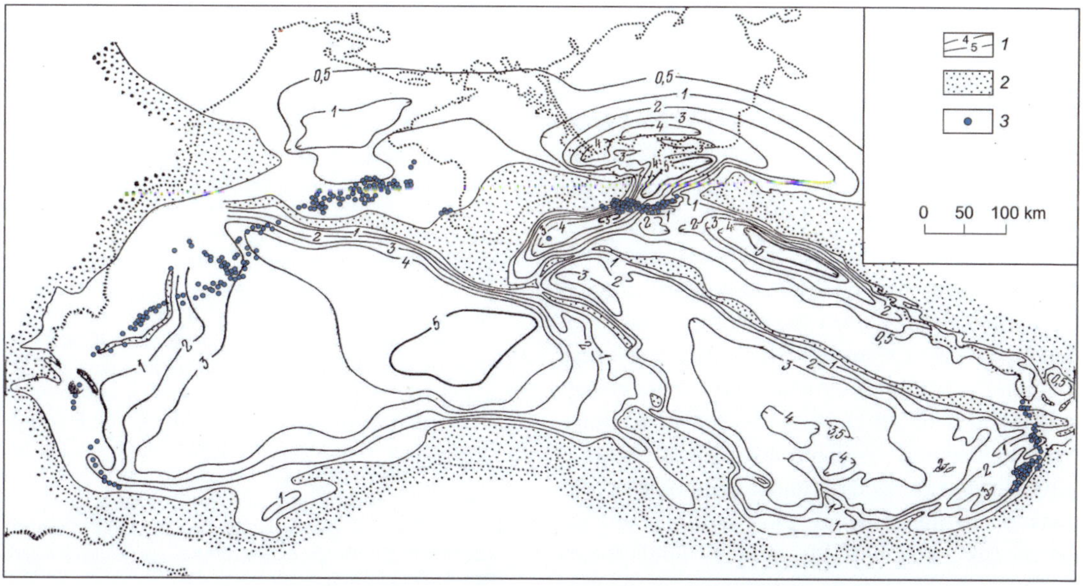

Fig. 7.38 Map of the thickness of the Maikopian deposits of the Black Sea with gas seepages. Key: *1*, isopachytes, km; *2*, areas showing an absence of Maikopian deposits; *3*, gas seepages. The geological map was borrowed from Tugolesov et al. (1985)

paleogeographically represents a Mesozoic island in a Maikopian Sea. Although there are no Maikopian deposits in Crimea, gas-bearing horizons are still present there. This has been pointed out by Chernyak et al. (1974), who demonstrated the existence of small, rare, and low-amplitude gas seepages from rocks of the Tauride series (Adgi-Su, Magarach) as well as Jurassic, Lower and Upper Cretaceous, Paleocene deposits, and alluvial strata (Table 7.1).

At present, gas flares are practically absent from the shelf and upper part of the continental slope of Mountainous Crimea. A Quaternary cover exists, but there are no gas torches because there is no recharge from the Earth's interior. It is very likely that small seepages will be found, but if so, they would be similar to those that are noted in Mountainous Crimea, with minor differences in their composition regarding the concentration of CH_4, N_2, CO_2, and H_2S. The different age and diverse origin of the gas seepages cause their complex and often hard-to-explain chemical and isotopic composition.

As a final conclusion, we reiterate that gas seepages are naturally located along the periphery of the sea, being confined to those areas of the shelf and slope where Maikopian deposits are developed.

Analysis of obtained data has shown that the lateral distribution of methane seeps is not accidental (Shnyukov et al. 2013). Their bulk within the Dnieper canyon is concentrated on the structural-accumulative, gentler slope to the west of the Nikolaevskiy Fault (Fig. 11.2a).

Gas outlets occur in places where concentrated fluid flows exceed the volume capacity of pore space, as a result of which methane seeps through the bottom sediments into the water column (Judd et al. 2002). Under normal conditions, most methane is oxidized by communities of sulfate-reducing bacteria under anaerobic conditions or in the zone of their aerobic activity (Michaelis et al. 2002). Migration of fluids is carried out through tectonic disturbances in the bottom sedimentary horizons, deep fault zones, and mud diapirs by diffusion of dissolved or free gas or in the form of geomorphologically focused flows. This gas can be partially a gas hydrate. With a change in pressure and/or temperature, the gas hydrates can dissociate and thereby feed the gas seepages on the sea bottom resulting in destabilization of the sediments.

The maximum depth for most of the detected gas seeps is 725 m, which more or less corresponds to the stability zone for pure methane hydrate at a bottom temperature of 8.9 °C in this part of the Black Sea (Naudts et al. 2006). This enables us to say that gas hydrates act as a buffer for the ascending migration of gases and thereby prevent the seepage of huge amounts of methane into the water column.

At the same time, it is not necessary to deny the participation of biochemical methane in the genesis of gas flares. Undoubtedly, this process accounts for methane generation in Quaternary bottom sediments; however, its role, most likely, is subordinate.

References

Al'bov SV (1956) Mineral'nye istochniki Kryma (Mineral springs of Crimea). Krymizdat, Simferopol (in Russian)

Artemov Yu G (2003) Acoustic observations of gas bubble streams in the NW Black Sea as a method for estimation of gas flux from vent sites. EGS-AGU-EUG Joint Assembly, Abstracts from the meeting held in Nice, France, 6–11 April 2003, abstract id. 9421

Artemov Yu G, Egorov VN, Polikarpov GG et al (2007) Methane emission to the hydro- and atmosphere by gas bubble streams in the Dnieper paleo-delta, the Black Sea. Marine Ecol J VI(3):5–26

Chebanenko II, Gozhik PF, Yevdoshchuk NI et al (2003) Skhema glubinnykh razlomov na uchastkakh Krymskogo i Kavkazskogo poberezhiy Chernogo morya (The scheme of deep faults in the areas of the Crimean and Caucasian coasts of the Black Sea). Geologicheskiy zhurnal 1:54–58 (in Russian)

Chernyak NI, Bogaets AT, Gordievich VA et al (1974) Neftegazonosnost. Geologiya SSSR. Krym. Poleznye iskopaemye (Oil and gas content. Geology of the USSR. Crimea. Minerals), v. 8. Nedra, Moscow (in Russian)

Chochov SD (1987) Methane-bearing clays in the western half of the Black Sea. Comptes rendus de l'Acad. Bulgare des. Science 40(3):81–84

Çifçi G, Dondurur D, Ergün M (2003) Deep and shallow structures of large pockmarks in the Turkish Shelf, eastern Black Sea. Geo-Mar Lett 23(3–4):311–322

Clark JF, Leifer I, Washburn L et al (2003) Compositional changes in natural gas bubble plumes: observations

from the coal oil point marine hydrocarbon seep field. Geo-Mar Lett 23(3–4):187–193
Dimitrov L (1990) Form tipa kraterov na dne Chernogo more. Bulgarian Acad Sci Bull 43(5):37–39 (in Bulgarian)
Dimitrov L (2002) Contribution to atmospheric methane by natural seepages on the Bulgarian continental shelf. Cont Shelf Res 22(16):2429–2442
Dimitrov L, Doncheva V (1994) Seabed pockmarks in the southern Bulgarian Black Sea zone. Bull Geol Soc Den 41(1):24–33
Dimitrov P, Dachev V, Nikolov H et al (1979) Estestvenni gazovi izvori v akvatoriyata na Balchshkiya zaliv (Natural gas seepages in the offshore area of Balchik Bay). Oceanologiya 4:43–49 (in Bulgarian)
Egorov VN, Luth U, Luth C et al (1998) Gas seeps in the submarine Dnieper Canyon, Black Sea: acoustic, video and trawl data. MEGASEEPS Gas Explorations in the Black Sea, Project Report. Zentrum für Meeres- und Klimaforschung der Universität Hamburg, pp 11–21
Egorov VN, Polikarpov GG, Gulin SB et al (2003) Sovremennye predstavleniya o sredoobrazuyushchey i ekologicheskoy roli struynykh metanovykh gazovydeleniy so dna Chernogo morya (Modern ideas about the environmental and ecological role of jet methane gas emissions from the bottom of the Black Sea). Morskoy ekologicheskiy zhurnal 2(3):5–26 (in Russian)
Egorov VN, Artemov YuG, Gulin SB (2011) Metanovye sipy v Chernom more. Sredoobrasuyushchaya i ekologicheskaya rol'. Sevastopol' (Methane seeps in the Black Sea. Environmental and ecological role). EKOSI-gidrofisika, Sevastopol (in Russian)
Eremeev VM, Egorov VN, Polikarpov GG et al (2007) Novi struminni gazovi vidilennya iz dna morya v akvatorii Sevastopolya (New stream gas flow from the sea bottom in the water area of Sevastopol). Visnuk NANU 4:47–50 (in Ukrainian)
Ergün M, Dondurur DC, Çifçi G (2002) Acoustic evidence for shallow gas accumulations in the sediments of the Eastern Black Sea. Terra Nova 14(5):313–320
Gevorkyan VH, Burakov VI, Isagulova UK et al (1991) Gazovmeshchayushchie postroyki na dne severo-zapadnoy chasti Chernogo morya (Gas-bearing buildings at the bottom of the northwestern part of the Black Sea). Doklady AN USSR 297(4):80–85 (in Russian)
Glebov AYu, Kruglyakova RP, Shelting SK (2001) Estestvennoe vydelenie gazov v Chernom more (Natural gas emissions in the Black Sea). Razvedka i okhrana nedr 8:19–23 (in Russian)
Gulin SB, Polikarpov GG, Egorov VN (2003) The age of microbial carbonate structures grown at methane seeps in the Black Sea with an implication of dating of the seeping methane. Mar Chem 84(1–2):67–72
Hovland M (2002) On the self-sealing nature of marine seeps. Cont Shelf Res 22(16):2387–2394
Ivanov MV, Polikarpov GG, Lein AY (1991) Biogeochemistry of the carbon cycle in the area of methane gas emission of the Black Sea. Doklady Acad Sci USSR 3(5):1235–1240
Ivanov MV, Pimenov NV, Rusanov II et al (2002) Microbial processes of the methane cycle at the north-western shelf of the Black Sea. Estuar Coast Shelf Sci 54(3):589–599
Judd AG, Hovland M, Dimitrov LI et al (2002) The geological methane budget at continental margins and its influence on climate change. Geofluids 2(2):109–126
Kessler JD, Reeburgh WS, Southon J (2006) Basin-wide estimates of the input of methane from seeps and clathrates to the Black Sea. Earth Planet Sci Lett 243(3–4):366–375
Khrischev Kh (1981) Geologicheskaya obstanovka metanoobrazovaniya v osadkakh Yuzhnobolgarskogo chernomorskogo shelfa (Geological conditions of methane formation in the sediments of the Southern Bulgarian Black Sea shelf). Geologica Balcanica 11(4):57–78 (in Russian)
Korsakov OD, Byakov UA, Stupak SN (1989) Gazovye gidraty Chernomorskoy vpadiny (Gas hydrates of the Black Sea basin). Soviet Geol 12:3–10 (in Russian)
Krastel S, Spiess V, Ivanov M et al (2003) Acoustic investigation of mud volcanoes in the Sorokin trough, Black Sea. Geo-Mar Lett 23(3–4):230–238
Lein AY (2005) Potoki metana iz kholodnykh metanovykh sipov Chernogo i Norvezhskogo morey: kolichestvennye otsenki (Methane flows from the cold methane seeps of the Black and Norwegian Seas: quantitative estimates). Geokhimiya 4:438–453 (in Russian)
Lein AY, Ivanov MV (2005) Krupneyshiy na Zemle metanovyy vodoem (The largest methane reservoir on Earth). Priroda 2:19–26 (in Russian)
Lein AYu, Ivanov MV, Pimenov NV (2002) Potoki metana iz kholodnykh metanovykh sipov Dneprovskogo kanyona v Chernom more (Currents of methane from cold methane seeps in the Dnieper canyon in the Black Sea). Doklady Rossiyskoy Akademii Nauk 387(2):242–244 (in Russian)
Michaelis W, Seifert R, Nauhaus K et al (2002) Microbial reefs in the Black Sea fueled by anaerobic oxidation of methane. Science 297(5583):1013–1015
Muratov MV, Sidorenko AV, Chernyak NI (eds) (1974) Geologiya SSSR. Mineraly (geology USSR. Minerals). Geologiya SSSR, v. 8. Nedra, Moscow (in Russian)
Naudts L, Greinert J, Artemov Y et al (2006) Geological and morphological setting of 2778 methane seeps in the Dnepr paleo-delta, northwestern Black Sea. Mar Geol 227(3–4):177–199
Palii A, Gelev S, Yankov K (1951) Geological framework and oil and gas perspectives of the Varna region. Open field work report, National Geofond, Sofia (in Bulgarian)
Pasynkov AA, Tikhonenkov EP, Smagin YuV (2009) Gazovye fakely na dne tsentralnoy chasti Azovskogo morya (Gas torches at the bottom of the central part of the Azov Sea). Geologiya i poleznye iskopaemye mirovogo okeana 1:77–79 (in Russian)
Pasynkov AA, Voronov SA, Blintsov VS et al (2012) Morfostrukturnye osobennosti morskogo dna na uchastkakh gazoproyavleniy v severo-zapadnoy chasti Chernogo morya (Morphostructural features of the

seabed in the gas manifestation areas in the northwestern part of the Black Sea). Geologiya i poleznye iskopaemye mirovogo okeana 4:82–87 (in Russian)
Pasynkova LA (1999) Obshchie kriterii geodinamicheskoy ustoychivosti landshaftnykh geosistem kontinentalnogo sklona Chernogo morya (General criteria for geodynamic stability of landscape geosystems of the continental slope of the Black Sea). Kultura narodov Prichernomorya Nauchnyy zhurnal 11:7–12 (in Russian)
Pasynkova LA (2001) Morfogenez i morfostrukturnoe rayonirovanie Ukrainskogo sektora kontinentalnogo sklona Chernogo morya (Morphogenesis and morphostructural zoning of the Ukrainian sector of the continental slope of the Black Sea). In: Ivanov VA (ed) Sbornik nauchnykh trudov "Ekologicheskaya bezopasnost pribrezhnoy i shelfovoy zon i kompleksnoe ispolzovaniya resursov shelfa," issue 2, MGI, Sevastopol, pp 202–211 (in Russian)
Polikarpov GG, Egorov VN (1989) Viyavleno aktivni gazovidilennya z dna Chornogo morya (Active gas releases from the Black Sea bottom have been discovered). Visnyk AN USSR 10:108–111 (in Ukrainian)
Polikarpov GG, Egorov VN, Nezhdanov AI et al (1989) Yavleniya aktivnogo gazovydeleniya iz podnyatiy na svale glubiny zapadnoy chasti Chernogo morya (Phenomena of active gas emission from uplifts at a sharp drop in depth of the western part of the Black Sea). Doklady AN USSR 12:13–16 (in Russian)
Polikarpov GG, Ivanov MV, Gulin SB et al (1993) Deponirovanie ugleroda metana v karbonatnykh bakterialnykh postroykakh na svale glubin serovodorodnoy zony Chernogo moray (Deposition of methane carbon in carbonate bacterial structures on the buildup of the hydrogen sulfide zone in the Black Sea). Doklady AN USSR 7:93–94 (in Russian)
Ryazanov AK (1996) Gaz i gazovye tumany na shelfe Chernogo morya (Gas and gas fogs on the Black Sea shelf). Doklady AN USSR 4:90–94 (in Russian)
Ryzhkov SS, Blintsov VS, Egorov YuD et al (2011) Stvorennya universalnikh transportnikh suden ta zasobiv okeanotekhniki (Creation of universal transport vessels and means of ocean engineering). NUK, Nikolaev (in Ukrainian)
Shimkus KM, Moskalenko VN, Ryan WB et al (1998) Gazonosnye chetvertichnye otlozheniya prikerchenskoy chasti Chernomorskogo shelfa i ikh svyaz s pribrezhnymi litofatsiyami (Gas-bearing quaternary deposits of the Prikerchenskaya part of the Black Sea shelf and their connection with coastal lithofacies). Byull Mosk Obshch Ispyt Prir Otd Geol 73(4):51–55 (in Russian)
Shnyukov EF, Naumenko PI, Lebedev YuS et al (1971) Gryazevoy vulkanizm i rudoobrazovanie (Mud volcanism and ore formation). Naukova Dumka, Kiev (in Russian)
Shnyukov EF, Orlovskiy GN, Usenko VP et al (1974) Geologiya Azovskogo morya (Geology of the Sea of Azov). Naukova Dumka, Kiev (in Russian)
Shnyukov EF, Ivannikov AV, Bezborodov AA (1990) Rezultaty geologicheskikh issledovaniy 51-go reysa v Chernoe more R/V "Mikhail Lomonosov" (The results of geological studies of the 51st voyage to the Black Sea of the R/V "Mikhail Lomonosov"). Preprint IGN No 90-8, Kiev (in Russian)
Shnyukov EF, Vorob'ev AI, Grigor'ev AV et al (1995) Predpolagaemyy gryazevoy vulkan v glubokovodnoy chasti Chernogo moray (Estimated mud volcano in the deep water of the Black Sea). Doklady AN USSR 12:62–65 (in Russian)
Shnyukov EF, Shcherbakov IB, Shnyukova EE (1997) Paleoostrovnaya duga severa Chernogo moray (Paleo-island arc of the north of the Black Sea). Chernobyl'inform, Kiev (in Russian)
Shnyukov EF, Inozemtsev YuI, Maslakov NA (1999) Geologiya, istoriya razvitiya rechnoy seti na severozapade Chernogo morya (Geology, history of the development of the river network in the northwest of the Black Sea). OMGOR NAN Ukrainy, pp 238–244 (in Russian)
Shnyukov EF, Ivannikov AV, Inozemtsev YuI et al (2003) Litologo-stratigraficheskaya kharakteristika donnykh otlozheniy krymskogo shelfa i glubokovodnoy chasti Chernogo morya (Lithological-stratigraphic characteristics of bottom sediments of the Crimean shelf and the deep part of the Black Sea). Geologicheskiy zhurnal 1:9–23 (in Russian)
Shnyukov EF, Starostenko VI, Rusakov OM et al (2005) Glubinnaya priroda gazovykh fakelov zapadnoy chasti Chernogo morya po rezultatam geofizicheskikh issledovaniy (The deep nature of the gas flares of the western part of the Black Sea according to the results of geophysical studies). Geologiya i poleznye iskopaemye mirovogo okeana 1:70–82 (in Russian)
Shnyukov EF, Ivannikov AV, Inozemtsev YuI et al (2007) Geologiya kontinentalnoy okrainy Chernogo morya (geology of the continental margin of the Black Sea). OMGOR, Kiev (in Russian)
Shnyukov EF, Pasynkov AA, Lyubitskiy AA et al (2010) Gryazevye vulkany na Prikerchenskom uchaste shelfa i materikovogo sklona Chernogo moray (Mud volcanoes on the Prikerchensky section of the shelf and continental slope of the Black Sea). Geologiya i poleznye iskopaemye mirovogo okeana 3:28–36 (in Russian)
Shnyukov EF, Kobolev VP, Pasynkov AA (2013) Gazovyy vulkanizm Chernogo morya (Gas volcanism of the Black Sea). Logos, Kiev (in Russian)
Shnyukov EF, Inozemtsev Yu I, Kukovskaya TS et al (2014) Geologo-okeanologicheskie issledovaniya v Chernom more (NIS "Professor Vodyanitskiy" 2013 g., 73 reys) (Geological and oceanological research in the Black Sea [R/V "Professor Vodyanitsky" 2013, 73rd voyage]). Logos, Kiev (in Russian)
Sokolov VA (1965) Protsessy obrazovaniya i migratsii nefti i gaza (the processes of formation and migration of oil and gas). Nedra, Moscow (in Russian)
Solovyev VA (2001) Gazogidratonosnost' nedr Mirovogo okeana (Gas hydrate potential of the World Ocean). Gazovaya promyshlennost' 12:19–23 (in Russian)

Tkeshelashvili GI, Egorov VN, Mestvirishvili ShA et al (1997) Metanovye vydeleniya so dna Chernogo morya v priustevoy zone r.Supsa u poberezhya Gruzii (Methane emissions from the Black Sea bottom in the mouth zone of the Supsa River at the coast of Georgia). Geokhimiya 3:331–335 (in Russian)

Tretyakov MB (1993) Novye nakhodki predpolagaemykh gryazevykh vulkanov v akvatorii Chernogo morya (New discoveries of alleged mud volcanoes in the Black Sea). Geologicheskiy zhurnal 3:141 (in Russian)

Tugolesov DA, Gorshkov AS, Meisner LB et al (1985) Tektonika mezo-kaynozoyskiy otlozheniy Chernomorskoy vpadiny (Tectonics of the Meso-Cenozoic sediments of the Black Sea Basin). Nedra, Moscow (in Russian)

Ussler W III, Paull CK, Boucher J et al (2003) Submarine pockmarks: a case study from Belfast Bay, Maine. Mar Geol 202(3–4):175–192

Vassilev A, Dimitrov L (2002) Spatial and quantity evaluation of the Black Sea gas hydrates. Russ Geol Geophys 43(7):637–649

Vassilev A, Dimitrov L, Avramov S et al (1996) Kompleksno geofizichno izsledvane na gazovite izvori v rayona na Aladzha banka (Complex geophysical study of the gas springs in the region of Aladzha bank). Izvestiya na Syuza na uchenite 24(1):30–34 (in Bulgarian)

Yanko V (also Yanko-Hombach V), Kulakova I, Kravchuk A (2014) Vliyaniye metana na donnyye ekosistyemy Chernogo morya (Influence of methane on bottom ecosystems of the Black Sea). In: Proceedings of the All-Russian Conference "Methane in marine ecosystems," 13–15 October 2014, Sevastopol, Crimea, pp 134–137 (in Russian)

Yanko V (also Yanko-Hombach V), Kravchuk AO, Kulakova II (2017) Meiobentos metanovykh vykhodov Chernogo morya (Meiobenthos of methane outlets of the Black Sea). Phenix, Odessa (in Russian)

8 Origin of Mud Volcanoes

Summary

*In this chapter, the origin of mud volcanoes in the Black Sea region is considered. This question has been debated for more than two centuries. Voskoboynikov and Gur'iev (*1832*) were the first researchers to study the Taman mud volcanoes; they proposed that the existence of mud volcanoes could be explained by the increasing pressure of gases due to their vertical migration into the sedimentary strata. Abikh (*1840*) put forward a fundamentally different idea by connecting mud volcanoes with lava volcanism, deep faults, and magmatic foci as a whole.*

Subsequent researchers either defended and detailed or refuted these hypotheses. Most specialists, however, tended to recognize the genesis of mud volcanoes exclusively in the sedimentary strata due to the accumulation of hydrocarbon gases and their migration along tectonic fractures. The connection of mud volcanoes with deep faults was debated, the differences in gas content between mud and magmatic volcanoes were recorded, and the connection of mud volcanoes with magmatic sources was denied.

*Gubkin (*1937*) proposed a connection between mud volcanism and diapirism.*

*Given these early ideas, the views of Kropotkin and Valyaev (*1980*), Dmitrievskiy and Valyaev (*2002, 2010*), Letnikov et al. (*2010*), Lukin (*2014*), and some others look quite new and fresh. These authors put forward a hypothesis about "cold" (nonmagmatic) hydrocarbon degassing of the Earth as the main reason for the development of mud volcanism. They propose that the scale of deep Earth degassing must be enormous and the planetary evolution of the Earth should be considered a monotonously fading and general process involving continuous depletion of fluid components upward through the sedimentary cover. The basis of the fluids is hydrogen, and two distinct fluid systems are distinguished, hydrogen carbon and hydrogen sulfide, which are nucleated at different depths within the liquid core. The release of gas flows, or plumes, can last millions of years. Essentially, hydrogen plumes interact with the oxygen matrix, give off heat, and reach the upper layers of the lithosphere. In their collective work, the above authors obtained evidence supporting their hypothesis.*

It turned out that the roots of the Black Sea mud volcanoes can be seismically traced into the Mesozoic sediments and even deeper. One of the volcanoes—Mantiynyy—has roots reaching the Moho surface. One can assume that as technical improvements proceed, the connection of mud volcanoes with the mantle will be clearly established.

Numerous mineralogical finds of accessory minerals with unusual shapes in mud volcanic breccia speak in favor of this hypothesis. Such minerals are represented by non-rolled individual particles, complex crystals, wires, spherules, native minerals (gold, silver, osmium, iridium, platinum, iron, lead, zinc, aluminum, etc.*),*

E. Shnyukov, V. Yanko-Hombach, *Mud Volcanoes of the Black Sea Region and their Environmental Significance*, https://doi.org/10.1007/978-3-030-40316-4_8

carbides, and sulfides (cinnabar, antimonite, realgar, orpiment, etc.*). Especially unusual are the spherules of iron-rare-earth combination alloys typical of industrial metallurgy. According to Lukin* (2006, 2009)*, native minerals and carbides trace the movement of fluids and anhydrous super-compressed multicomponent gas.*

These data suggest the crucial role of deep fluids in the formation of mud volcanoes. The thickness of the sedimentary cover above the Moho surface also plays a role in the origin of mud volcanoes. The greatest quantity and diversity of native minerals, sulfides, and oxides (except for cinnabar, arsenic sulfides, and antimonite) coincide with the shallowest sedimentary cover above the Moho surface (19 km) in the mud volcanoes of the Western Black Sea depression.

The Kerch-Taman mud volcanoes, separated by 40–45 km of sedimentary accumulation from the Moho surface, contain numerous and diverse native minerals (iron, nickel, gold, lead, iron alloys, cerium, lanthanum, neodymium, etc.*). Sulfides show less diversity among those present but contain cinnabar, realgar, and orpiment. There are also oxides, carbonates, sulfates, and chlorides. In the formulas of some minerals, the content of Cl, Br, and F is elevated.*

Mud volcanoes sitting on the Cretaceous basement are the poorest mineralogically. Native minerals in their erupted breccia are represented by iron (ferrite, kamacite, octibbehite, nickel, neodymium-ferrous alloys, copper, nickel-tantalum grains, lead, and zinc—the latter two being numerous). Moissanite is also characteristic of this type of mud volcano.

The deep origin of mud volcanoes is reflected in their appearance, location, and activity, the last of which reveals (1) a constantly intermittent return of hydrocarbons, (2) the formation of salsas and springs as well as the bodies of mud volcanoes, and (3) large, sometimes catastrophic emissions that determine their morphology. On this basis, different types of volcanoes were previously identified (Kalinko 1960*), but in fact, these types represent different episodes, or morphological stages, of their existence. The impact of exogenous factors completes their modern land appearance, in particular by creating numerous characteristic ravines, or barrancos, on their slopes.*

8.1 General Remarks

Questions about the origin of mud volcanoes in the Black Sea region have been asked and debated for more than two centuries. Voskoboynikov and Gur'iev (1832) were the first researchers to study the Taman mud volcanoes; they proposed that the existence of mud volcanoes could be explained by the increasing pressure of gases due to their vertical migration into the sedimentary strata. Abikh (1840) put forward a fundamentally different idea by connecting mud volcanoes with lava volcanism, deep faults, and magmatic foci as a whole.

Subsequent researchers either defended and detailed or refuted these hypotheses.

Thus, the views of Abikh were partially or fully supported by Kovalevskiy (1935), Steber (1914), Gorin (1952), Kudryavtsev (1963), Porfiriev et al. (1978), and others.

Most specialists, however, tended to recognize the genesis of mud volcanoes exclusively in the sedimentary strata due to the accumulation of hydrocarbon gases and their migration along tectonic fractures (e.g., Arkhangel'skiy 1925; Belousov and Yarotskiy 1936; Gubkin 1937; Fedorov 1939; Yakubov and Zeynalov 1964; Kalinko 1960; Aliyev and Bayramov 1999; Aliyev et al. 2004; Kholodov 2002a, b).

The connection of mud volcanoes with deep faults was debated, the differences in gas content between mud and magmatic volcanoes were recorded, and the connection of mud volcanoes with magmatic sources was denied.

Gubkin (1937) proposed a connection between mud volcanism and diapirism. This provision has received almost universal acceptance. The discussion, however, lasted for many years, and continuing research accumulated many new facts that were difficult to explain. Given these early ideas, the views of Kropotkin and Valyaev (1980), Dmitrievskiy and Valyaev (2002, 2010), Letnikov et al. (2010), Lukin (2014), and some others look quite new and fresh. These authors put forward a hypothesis about "cold" (nonmagmatic) hydrocarbon degassing of the Earth as the main reason for the development of mud volcanism. They propose that the scale of deep Earth degassing must be enormous, and the planetary evolution of the Earth should be considered a monotonously fading and general process involving continuous depletion of fluid components upward through the sedimentary cover. The basis of the fluids is hydrogen, and two distinct fluid systems are distinguished, hydrogen carbon and hydrogen sulfide, which are nucleated at different depths within the liquid core. The release of gas flows, or plumes, can last millions of years. Essentially, hydrogen plumes interact with the oxygen matrix, give off heat, and reach the upper layers of the lithosphere. In their collective work, the above authors obtained evidence supporting their hypothesis.

It turned out that the roots of the Black Sea mud volcanoes can be seismically traced into the Mesozoic sediments and even deeper. One of the volcanoes—Mantiynyy—has roots reaching the Moho surface. One can assume that as technical improvements proceed, the connection of mud volcanoes with the mantle will be clearly established.

8.2 Roots of Black Sea Mud Volcanoes

One of the most important problems in understanding the origin and nature of mud volcanoes is the question of the depth of their formation.

Over the course of their studies, scientists have recorded the roots of mud volcanoes at ever greater depths. In the late 1950s, the roots of mud volcanoes in Turkmenistan were traced to depths of 5–8 km (Sungurov 1958; Kalinko 1960).

In the Southern Caspian, which is the most active mud volcanic region in the world, the roots of mud volcanoes were recorded at depths of about 9 km (Kulakova and Lebedev 1990). Most recently, Azerbaijani geologists (Golubov and Ismagilov 2003; Mamedov and Guliyev 2003; Guliyev 2010) explored more than 100 subvertical geological bodies penetrating the entire sedimentary cover from the surface to the foundation (and probably deeper) using seismic profiling. Some of these bodies coincide with mud volcanoes that serve as "degassing pipes" (Kropotkin and Valyaev 1980) for the emergence at the surface of gas-enriched fluids and unconsolidated sedimentary material originating at a depth of several kilometers. The diameter of these bodies varies from 3–4 to 10 km, and their length varies from 8 to 20 km. The largest bodies are confined to deep faults (Khaustov 2011).

Data on the roots of the Black Sea and Sea of Azov mud volcanoes are scarce. If one were to hypothesize on the basis of the clastic material they erupt (see Sect. 4.5), one would assume that the roots of these mud volcanoes are located somewhere in the Maikopian and Lower Cretaceous sedimentary sequences on the Kerch and Taman peninsulas, respectively. Controversial fragments of Cretaceous rocks are sometimes found in the Kerch mud volcanoes.

In this regard, the detailed study of the Karabetova Gora mud volcano performed by Sobisevich et al. (2008) is extremely important. This mud volcano is directly adjacent to the Black Sea, and as was already mentioned, it can be compared structurally with submarine mud volcanoes. Under the mud volcano at depths of 4.5–9 km, there is a vertical low-velocity zone associated with a supplying channel and a feeding reservoir saturated by fluids. At a depth between 800 and 1500 m, there is an isometric, low-velocity zone interpreted as a near-surface reservoir. Below 9 km, the seismic pattern shows less contrast, but the anomaly associated with the feeding reservoir continues. In other words, the roots of mud volcanoes seem to reach at least 15 km in depth (Fig. 8.1).

Seismic studies and deep drilling enabled Sobisevich et al. (2008) to determine a Cretaceous age for sediments represented by terrigenous-carbonate oil- and gas-bearing strata within the geological sequence of the Taman Peninsula at depths between 4.5 and 9 km. Below the Cretaceous strata lie terrigenous Jurassic and transitional terrigenous-carbonate Permian-Triassic rocks. In the eastern part of Ciscaucasia, the Lower Triassic is also oil- and gas-bearing. In

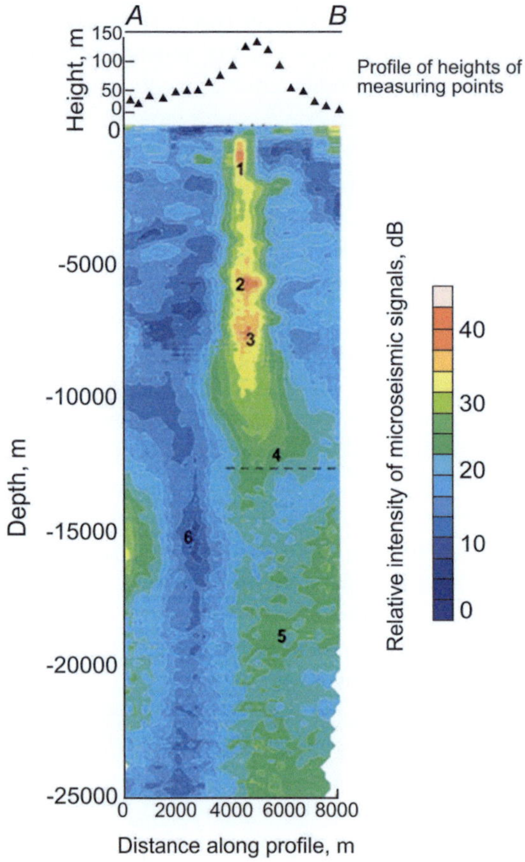

Fig. 8.1 Vertical microseismic section of the Karabetova Gora mud volcano. Key: *1*, near-surface reservoir; *2* and *3*, deep reservoirs; *4*, area of transition to the consolidated foundation; *5*, deep fluid-permeable structures; *6*, impermeable consolidated rocks. An increase in the intensity of Rayleigh microseismic waves corresponds to lower values of the velocities (lighter background); a decrease corresponds to higher values (darker background) (Sobisevich et al. 2008)

this connection, the generation of fluids at these depths is also likely. The crystalline basement within the Kuban is located at depths of 13–15 km, but the existence of oil and gas sources there is not considered likely.

Based on seismic data, Limonov et al. (1997) and Limonov (2004) concluded that Black Sea mud volcanoes can be associated with Maikopian clays, and as such, they can be a source of formation about diapiric structures. At the same time, the authors admitted the possibility that diapiric and mud volcanic structures could have developed independently.

Very interesting seismic data relating to mud volcanoes of the Black and Caspian seas were obtained by Ismagilov et al. (2006). Given that the mud volcanoes have a deep origin, they ask the reasonable question: how can Maikopian clay diapirs affect the development of mud volcanoes that have their roots in Mesozoic deposits. They emphasize that mud volcanoes are more complex geological phenomena than is commonly assumed and not necessarily related to the formation of clay diapirs. The analysis of regional seismic records carried out by the authors allows one to confirm in most cases the association of mud volcanism with Maikopian structures while fixing the roots of mud volcanoes not only in the Maikopian but in the Mesozoic strata as well. Such root depths have been determined by modern seismic exploration as, for example, in almost all mud volcanoes on the floor of the Western Black Sea to the south of Crimea (e.g., the Kazakov, Tbilisi, Istanbul, and some other mud volcanoes).

In this regard, the regional time seismic section of the underwater Mantiynyy mud volcano is of a special interest (Table 6.5, Fig. 6.138).

The amplification of the dynamics of the seismic waves accompanied by screening of energy in subsequent intervals, as well as clear changes in the polarity of the waves, all indicate a change in the acoustic stiffness of the reflecting boundaries in this section. In turn, this may be due to (1) fluid change (water behind the contour and gas within the contour) or (2) decompression of rocks, i.e., the destruction of their continuity and, accordingly, the weakening of the reflecting properties of the section boundaries. In our opinion, the first is more probable (Fig. 6.138).

For the Maikopian part of the section, the structural factor is clearly expressed. In seismic horizon I-IIa, there is a bend with 100 m amplitude along the Maikopian roof (Ia) and an 800 m amplitude along the Oligocene roof (Ib). Down the section, the structure is flattened: along horizon Ic, the bend is 600 m, and at the base of the Maikopian (IIa), it transforms into a negative structure. The "sagging" of the horizons corresponds to the arch and contrasts with the background of a larger positive form. Apparently, we are observing here a "false syncline," caused by the effect of a sharp decrease in velocity of the elastic waves. Usually, with a speed anomaly localized within a small vertical interval of the time seismic section, the "false syncline" ceases to be observed with increase of time (depth) due to wave refraction. In our case, the abnormal effect of low speed intensifies with sediment age down to the Cretaceous. This means that the entire section of Paleocene, Eocene, and Maikopian strata within the mud volcano zone, including its controlling structure, are saturated with gas. Regarding the second cause of the anomaly, such as the destruction of continuity, this is evident from the clear dynamics and coherence of wave Ib and other waves in the anomalous zone (Fig. 6.138).

This is in agreement with the conclusions of Meisner and Tugolesov (2003). They write that a high gas content in a sedimentary stratum leads to the slow passage of elastic waves through these uplifts. Therefore, the seismic horizons IIa and IIb above the uplifts give reverse records called "velocity synclines." This is the picture that we observe in Fig. 6.138.

In the Upper Maikopian, at time 5 s (depth 4.5 km) within the arch, we can see a clear local anomaly of low-frequency dynamic recording, which under detailed analysis enables us to predict the formation of a hydrocarbon reservoir due to the presence of good cap rocks (horizon in the interval Ia-IIb).

Deep seismic sounding (DSS) conducted in the area of mud volcanism within the Western Black Sea depression predicts a low-thickness suboceanic crust with reduced to complete wedging of the granite layer. The depth of the Moho boundary on the DSS profiles is located 19 km below the sea bottom, and it is also observed fragmentally on the regional profiles obtained by the common depth point (CDP) with sufficient length of signal recording (Shnyukov and Netrebskaya 2013).

The Mantiynyy mud volcano is extremely interesting, appearing as a kind of fluid plume piercing an almost 10-km thick sedimentary cover—Quaternary-Miocene (3 km), Maikopian (5 km), Eocene (2 km), Paleocene (1.2 km), and low-thickness Mesozoic (0.25–0.75 km) sediments—and then further deepening into the upper parts of the basement, presumably to the Moho boundary (Fig. 6.138) (Starostenko et al. 2010). It is quite possible that this mud volcano as well as other similar structures represents a channel for hydrocarbon intrusions associated with deep faults or injection zones (Dmitrievskiy and Valyaev 2010).

Presumably, all mud volcanoes will reveal a connection with the mantle, and this will become apparent with advancements in exploratory technology, which is only a matter of time. Specialists engaged in geochemistry sometimes express the idea that mud volcanoes are connected with the mantle and the deepest horizons of the Earth's crust, but unfortunately, these mentions are only assumptions or general provisions.

It is interesting that the concentration of mercury increases by a factor of 2.5 within water adjacent to methane seepages, and 64% of this chemical element is represented by its dissolved form. Near unamed mud volcano (43°37.5′ N and 31°42′ E) at a depth of 1868 m, the mercury concentration of the near-bottom layer has been recorded as 2676 ng/l, which is an order higher than the background. Similarly, on the Dvurechenskiy mud volcano, mercury content in the bottom water was seven times higher than the background concentration. A similar situation has been observed in close proximity to seepages but to a lesser extent. Experts believe that the flow of mercury into bottom water has a lithospheric genesis and is not directly related to the destruction of organic matter (Kostova et al. 2006; Egorov et al. 2011).

In 2013, Shnyukov and Netrebskaya wrote that it would be desirable to find other geochemical indicators of the deep origin of mud volcanoes. To this end, detailed mineralogical investigations of breccia and other ejected material were conducted in a search for exotic minerals and chemical elements that would better illuminate the connection between mud volcanoes and the deep crust of the earth.

8.3 General Features of Fluidogenic Mineralization in the Mud Volcanic Process

Mineralogical studies of mud volcanic breccia have unexpectedly yielded extensive information of significance for understanding the genesis of mud volcanoes.

Most of the previously studied accessory minerals in mud volcanic breccia are quite commonly found in the sedimentary strata of Maikopian and Lower Cretaceous age, on which the mud volcanoes of the Black Sea region are developed. These are typical terrigenous minerals existing in an aquatic environment. All of them are rounded to different degrees by abrasion. The new potential research capabilities provided by SEM and associated microprobe have allowed us to increase the number of minerals found in breccia and, most importantly, to clarify their chemistry as well as the forms of individual crystals and aggregates. The result of these efforts has been a sharp expansion in the range of minerals known from breccia. Whole classes of minerals have appeared, not just separate ones. First, all types of mud volcanoes show the presence of numerous native minerals: gold, copper gold, silver, osmium, iridium, platinum, iron, lead, zinc, copper, tin, aluminum, rare earth alloys, as well as graphite and sulfur.

Carbides of some elements have also been encountered, such as moissanite (SiC), cohenite $(Fe,Ni,Co)_3C$, and some others.

The character of these mineral segregations in breccia is noteworthy: non-rolled individuals and their aggregates, complex segregations, sproutings, wires and elongated aggregates (iron), and in some cases rounded or spherical melted aggregates or spherules. Sometimes they are hollow inside. There are no signs of their transport or rolling, except in individual cases, and these allow for different interpretations.

Many sulfides of lead, copper, and zinc have been identified as well as more complex new mineral formations. Pyrite is abundant, and other iron sulfides are often observed.

Cinnabar, antimonite, realgar, and orpiment are also present, and their appearance is often idiomorphic, sometimes forming veins. In general, sulfides are not transportable, as they are unstable during water transport and can hardly be transferred very far.

In the opinion of Rukhin (1969), all sulfides are very unstable and do not endure weathering, so there are no clastic sulfides in sedimentary rocks. Smirnov (1951) considers pyrite and galena to be relatively stable, but these minerals do not withstand redeposition.

There are quite often carbonates and sulfates. Most of the carbonates are represented by rounded debris, pieces of which are clearly reworked from destroyed Maikopian siderites. At the same time, in a number of cases, it is possible to observe druses of crystals as well as some large, well-formed, often rhombohedral crystals that clearly grew in situ. Occasionally, these minerals contain a noticeable Cl and Br content.

The mineralogy of accessory elements within the breccia is unusually diverse, even exotic. In many cases, breccia from the same mud volcano contained iron and neodymium (up to 50%), often with an admixture of nickel. These minerals clearly developed in the breccia itself, as there are no signs of their redeposition. Compounds of cerium and lanthanum, sometimes together with neodymium, are encountered.

The new mineralogical findings enable us to demonstrate the geochemical uniqueness of the mineral suite in mud volcano eruptions. These are gold, silver, osmium, iridium, platinum, lead, zinc, copper, tin, arsenic, antimony, cerium, neodymium, lanthanum, tungsten, and other elements. Geochemically, minerals formed from these elements are rarely products of sedimentary processes.

There is a dominance of specific geochemical associations in various types of mud volcano and in certain mud volcanic regions of the Black Sea. For example, in mud volcanoes developed over Lower Cretaceous deposits, native lead and zinc together with many minerals containing these elements are common. There are also rare earth elements. In mud volcanoes developed over Maikopian deposits, gold, platinoids, copper, zinc, mercury, antimony, arsenic, and rare earth elements are more common.

Mud volcanic breccia can be considered a "prigozhinit" because, unlike sedimentary rock, the breccia was produced by an atypical sedimentary process. The term "prigozhinit" was introduced by Lukin (2006, 2009) to describe a variety of morphologies and compositions of dispersed particles (e.g., native metals, carbides, silicides, and carbon-halogen compounds) in mud volcanic breccia, and he considers them to be indicators of the composition of the deep geosphere. In his opinion, these particles trace the movement of fluids: an anhydrous super-compressed polycomponent gas (H, C, and S with high iron, silica, nitrogen, phosphorus, and halogens) that can serve as a tracer for oil and gas traps in respective provinces. For rocks affected by fluids, he introduced the new term, "prigozhinits," that represents a special type of nonequilibrium mineral formation.

The study of the minerals mentioned above indicates that mud volcanic breccia serves not only as a transport medium but also as a medium of intense mineral formation, where many of these minerals have been created.

Processes of metallogeny in mud volcanic breccia may be indicated by the character and morphology of the mineral segregations (e.g., unrolled individuals and their intergrowths, complex growths that seem to germinate in the breccia, hollow spherules) together with the absence of any signs of reworking.

Lukin (2013) focuses special attention upon spherules. He has analyzed their development under various geological processes, such as deposition of cosmic dust, impact, explosive volcanism, diatremes, ores, and oil-gas deposits. He considered the presence of spherules in mud volcanic breccia in close connection with deep mineral phases (i.e., refractory substances) on the one hand and with the low-temperature environment of clay minerals on the other. He regards deep spherules formed by refractory minerals as indicators of super-deep fluids associated with different phases of the tectonic evolution of plumes.

Mineral aggregates of ball-like spherules are found in the ashes of Kamchatka volcanoes, such as the Koryaksky volcano, where they are represented by submicron (up to 40 μm) spherules of iron oxides, spherules of pyrite that are hollow inside and possess a dendritic surface pattern, and cavernous grains of sulfur. In ashes of the Karymsky volcano, recovered spherules are represented by a number of intermetallic compounds of Fe, Mn, Cr, Ni, Ti, Cu, and a single spherule of native silver. In ashes of the Icelandic volcanoes, spherules are represented by fused particles of acidic volcanic glass, native iron, and platinum (Karpov et al. 2012). These authors conclude that a large group of native minerals and intermetallic compounds from volcanic exhalations are derivatives of basalt melt ascending at high speed from the upper mantle.

Sandimirova (2008) studied spherules of the Kuril Islands and Kamchatka volcanoes. The results show that microspherules are complex polycomponent mineral aggregates consisting mainly of native iron, its oxides, and glass. They are in paragenetic association with native elements, metals, and intermetallic compounds and are formed in the fluid systems as a result of rapid gas transport reactions within a reducing medium. Perhaps, the spherules are drops of solidified melt.

Similar mineral aggregates are encountered in the industrial waste of metallurgical processes (Nesterenko et al. 2007). They are represented by spherules possessing sizes from 1–2 μm to 6–7 cm (more often up to 1 mm). Extremely fine-grained particles are typical for dust, sludge, and slags. The spherules are uniformly filled, zonal, radial-radiant, tangled-fibrous, hollow, and sometimes show inclusions of several gas bubbles. They are composed of many minerals, e.g., metallic iron, graphite, wüstite, magnetite, hematite, calcium ferrite, olivine, diopside, quartz, clay minerals, calcium silicates, spinel, silicate and iron-silicate glass, etc. They crystallize over a wide range of temperatures under conditions of silicate and metallic melting as well as in a gas stream and slime.

Published information about the formation of mineral spherules is abundant. We provide some examples here that reflect the diversity of their genesis. Formation mechanisms can be very diverse, but the essential process leading to their production is quite similar and based on the forces of surface tension. More often than other reasons, spherule formation occurs as a result of the fouling of gas bubbles, drops of metal, or silicate melt during the course of high-temperature processes. This explains their unusual shape and mineralogical composition, which contrast sharply with the surrounding sediments comprising mud volcanic breccia. The conditions for the formation of spherules occur within mud volcanoes only because of intrusion by deep fluids (Kropotkin and Valyaev 1984).

As our investigation shows, mud volcanoes of the Black Sea region contain abundant hollow spherules with different mineral compositions. Their sizes are approximately the same, about 30–50 μm and occasionally as large as 100 μm. Apparently, their dimensions are limited by the size of the gas bubbles on which the smallest particles of native elements settle. It is very likely that, in their development, they expand, giving rise to those minerals that we observe now.

In terms of mineralogical composition, the simplest are spherules of hematite, such as those discovered in the Western Tsymbaly and Shkolnyy mud volcanoes, which reveal a multilayered structure (Fig. 8.2a, b).

Their upper layer is characterized by the presence of numerous pits and a particular crystallographic pattern. The sizes of the spherules in Fig. 8.2a, b exceed 100 and 200 μm, respectively. Probably, particles of native iron originally settled onto the surface of the fluid ball; they were first transformed into magnetite and then martitized.

More complex in their chemical composition are hollow spherules composed of cerium, lanthanum, and iron compounds. They have been found in breccia from the Burashskiy mud volcano. In size, they reach 50 μm, and their surfaces are covered with multilayered "microstalagmites" adjacent to each other with apical diameters up to 1 μm. Usually, the tops of these "microstalagmites" are rounded (Fig. 8.3a), but sometimes they look like rounded triangles (Fig. 8.3b).

Spherules possessing an even more complex chemical composition have been found in breccias of the Burashskiy and Dzhau-Tepe mud volcanoes. These spherules have a cerium-lanthanum-neodymium-ferruginous content with a small admixture of alumina and silica (Figs. 8.4a, b and 8.5a, b).

All the spherules are hollow inside and have diameters of about 50 μm. Their surfaces are covered with "microstalagmite" sinters and appear very complicated. With magnification up to 6000x, it can be seen that the "microstalagmites" cover part of the spherule and have the appearance of outgrowths perpendicular to the surface that form shallow, lumpy fields. Such complicated morphology can be explained by a change in spatial position of the spherule during its aggregation. The mineralogical composition of the aggregates is unclear. Obviously, they are composed of oxides of rare earth elements. These minerals were formed under oxic conditions at the top of a plume, and at the beginning, the chemical elements were deposited on a fluid ball.

The same mechanism of formation can be assumed for particles of tin and copper alloy that make up spherules of bronze that were then oxidized by the fluid process (Fig. 8.6).

Microspherules of fused silicates are among the unusual findings within the light fraction of breccia from some mud volcanoes located in the Western Black Sea depression (Shnyukov et al. 2013a). Particular attention was paid to the constant presence of glasses. They have a fresh appearance, contain inclusions of gases, and form angular chips, balls, and fused particles. The glass colors are quite diverse and vary between gray, blue, black, and bright green, as well as gray with a blurry striped distribution resulting from fluid flow. Many particles of the breccia are covered with a thin film of bright red or dark brown color. According to the results of microprobe study, the particle composition does not differ from that of the bulk of the breccia. Brecciated textures are observed in some of the aggregates, including the presence of vitreous

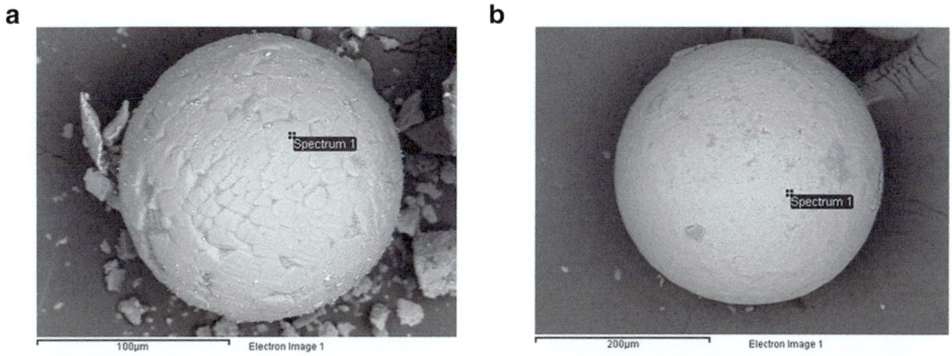

Element	Weight%	Atomic%
Fe	69.94	40.00
O	30.06	60.00
Totals	100.00	

Element	Weight%	Atomic%
Fe	69.86	39.91
O	30.14	60.09
Totals	100.00	

Fig. 8.2 Hollow spherules of hematite from breccia of the (**a**) Western Tsymbaly and (**b**) Shkolnyy mud volcanoes. Iron and oxygen assays determined at location of Spectrum 1 in each image

Spectrum	Concentration of chemical elements, wt.%						
	O	Al	Si	Fe	La	Ce	Total
Spectrum 1	43.32	0.41	0.57	5.95	12.13	37.63	100.00
Spectrum 2	42.79	0.46	0.66	9.57	9.72	36.80	100.00
Spectrum 3	41.62	0.63	1.08	6.22	12.54	37.92	100.00
Spectrum 4	47.88	0.47	0.77	11.69	11.42	27.77	100.00
Mean	43.90	0.49	0.77	8.36	11.45	35.03	100.00
Std. deviation	2.75	0.10	0.22	2.77	1.24	4.86	
Max.	47.88	0.63	1.08	11.69	12.54	37.92	
Min.	41.62	0.41	0.57	5.95	9.72	27.77	

Fig. 8.3 Hollow spherules of cerium, lanthanum, and iron compounds from breccias of the Burashskiy mud volcano: (**a**) general view, (**b**) magnified

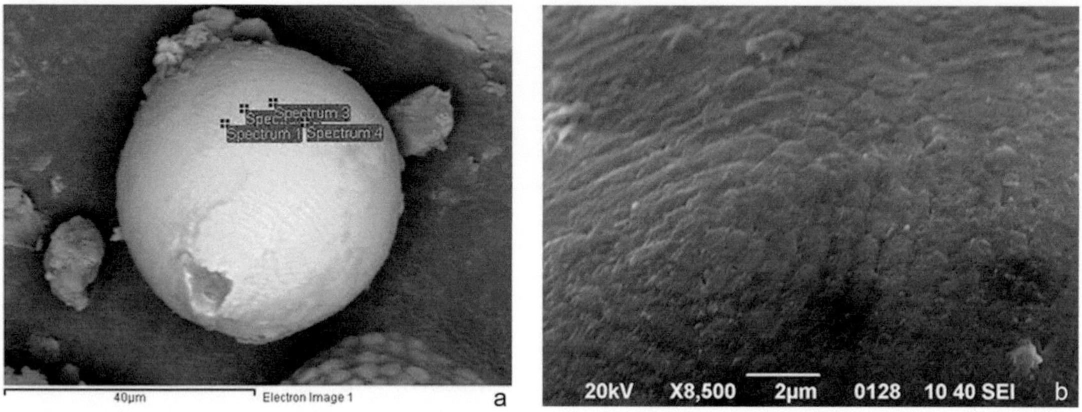

Spectrum	Concentration of chemical elements, wt.%							
	O	Al	Si	Fe	La	Ce	Nd	Total
Spectrum 1	35.04	0.66	0.80	1.80	15.72	39.25	6.73	100.00
Spectrum 2	36.52	0.47	0.51	1.47	15.39	39.87	5.77	100.00
Spectrum 3	38.12	0.55	0.95	1.69	16.10	36.26	6.33	100.00
Spectrum 4	34.99	0.00	0.60	1.81	15.46	39.31	7.84	100.00
Mean	36.17	0.42	0.71	1.69	15.67	38.67	6.67	100.00
Std. deviation	1.48	0.29	0.20	0.16	0.32	1.63	0.87	
Max.	38.12	0.66	0.95	1.81	16.10	39.87	7.84	
Min.	34.99	0.00	0.51	1.47	15.39	36.26	5.77	

Fig. 8.4 Hollow cerium, lanthanum, and iron spherules in breccia of the Burashskiy mud volcano: (**a**) – general view (**b**) – magnified

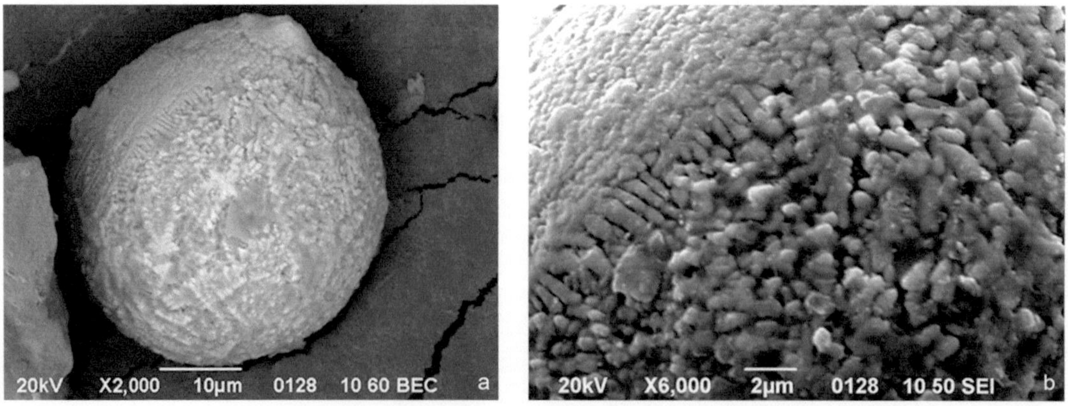

Fig. 8.5 Cerium-lanthanum-neodymium-ferruginous spherules with complex surfaces from breccias of the Burashskiy mud volcano: (**a**) general view, (**b**) magnified

masses, i.e., as if it was the initial stage of burning for the breccia.

It would seem that, with the upwelling of fluids from the deep crust through the entire sediment cover, very small debris of broken, burnt, and melted rocks gained entry, having been captured and entrained by the fluids in their upward movement. The study of such debris (shown below) confirms the existence of fluids and provides an estimate of the depth of their breakthrough.

Well-crystallized crystals of quartz, tens of μm in size, with one or two well-preserved pyramids, usually hexagonal heads, were discovered in the light fraction of the breccia (Fig. 8.7).

Fig. 8.6 Spherules of bronze from breccias of the Dzhau-Tepe mud volcano

Occasionally, the crystals or their clusters are partially eroded (Fig. 8.8) or dissolved (Fig. 8.9).

The process of quartz crystallization, and sometimes the dissolving of already formed crystals, is obvious (Fig. 8.9). Under artificial (experimental) conditions, these processes occur at temperatures of 300–400 °C. Similar processes can be assumed in the mud volcanic breccia when fluids are introduced.

It would seem that breccia is formed at the expense of the rocks that enclose mud volcanoes, and this suggests that it must contain only accessory terrigeneous minerals. But our studies reveal a different picture. Mud volcanic breccia is mineralogically much richer by perhaps a thousandfold. There are numerous minerals from different classes that are not characteristic of sedimentary rocks, such as native minerals, carbides, silicides, and certain sulfides—all formed under high temperatures and pressures. These minerals are evidently atypical for mud breccia geochemically; they are not rolled, and they often grow into the breccia, forming intergrowths with other minerals. They are clearly alien and appear as a result of some imposing processes. Thus, most native minerals that crystallize at high temperatures produce segregations in the form of wires, irregularly shaped dendrites, bizarre aggregates, etc. Despite the fact that they are high-temperature compounds, there are no traces of firing or other kinds of impact on the surrounding breccia,

Fig. 8.7 Quartz crystal from breccia of the MSU mud volcano

Fig. 8.8 Eroded crystal of quartz from breccia of the MGU mud volcano

Fig. 8.9 Complex cluster of quartz crystals from breccia of the MGU mud volcano

although these minerals often disperse within the silicate matrix.

This group of alien minerals described above is quantitatively insignificant compared to the terrigenous minerals of the sedimentary rocks within the mud volcanic breccia. Their percentage is not more than 1–2%, which is typical for accessories.

Grossgeim (1961) compared accessory minerals in mud volcanic breccia with terrigenous ones from Cretaceous and Jurassic sediments and showed that the former were enriched. Of course, such a finding might be explained by the technological limitations at the time that affected the level of detail possible in the examination of small objects in the breccia. But, in our opinion, the matter involves not only the better investigative opportunities enjoyed by geologists today. The object of our study is the mud volcano, and it is much richer mineralogically compared to ordinary sedimentary rocks. The association of resistant and semi-resistant terrigenous minerals in breccia is well preserved, and they are most often silicates (including high-temperature silicates). Most of the silicates (pyroxenes, amphiboles, micas, etc.) bear obvious traces of water transport. They are rounded, partly destroyed, and most likely, they were borrowed from sediments that eventually became the material for mud volcanic breccia. The native minerals, sulfides, oxides, and carbides are absolutely different. Many of them are absent from the list of typical terrigenous minerals prepared by Grossgeim (1961). Perhaps some of them may be grouped together under the term "black ores," but this is unlikely, especially since, in one case, sphalerite (a sulfide) was discovered at that time as well.

Descriptions of accessory minerals characteristic of the Lower Cretaceous and Maikopian sediments are given above, but the content of accessory minerals from mud volcanic breccia is quite different from those mentioned by Grossgeim (1961) and cannot be explained as belonging to terrigenous formations.

Thus, the mineralogical composition in mud volcanic breccias, especially compared with the Maikopian and Lower Cretaceous terrigenous-mineralogical complexes, is geochemically unusual. This includes the wide distribution of spherules of iron, tin, and rare earth elements, the presence of newly formed quartz crystals, and other indicators. All this enables us to consider the breccia from mud volcanoes of the Black Sea region as a medium for mineral formation occurring under the influence of fluid plume intrusions.

In addition to the most obvious signs of the effect of deep fluids on mud volcanic breccia (listed above), a systematic review of the mineralogical findings would be useful.

It is very difficult to distinguish a number of terrigenous minerals from those formed in breccia under the influence of fluids. Some subjectivity is invariably involved. To be on the safe side, we have decided to exclude silicates, some oxides, and carbonates from our consideration. We understand that such an approach is in a way conventional, but in the overwhelming majority of cases, these minerals bear signs of reworking and therefore represent most definitely terrigenous compounds.

It is true that some of the fluidogenic minerals can be missed using such an approach. But even while deliberately excluding controversial minerals, it is still possible to show the deep origin of mud volcanic processes and the influence of hot fluids on mineral formation in the breccia. Examples would be the presence of pyrite crystals and unique rare earth elements as well as iron spherules and fully formed or partially dissolved quartz crystals, etc.

Mineral formation occurs under specific conditions within different types of mud volcano. Each type is characterized by the age of the sediments on which it is formed, and the gas-transport reactions that occur between the fluids and sediments on their way to the surface. This makes genetic classification of mud volcanoes necessary.

8.4 Genetic Types of Mud Volcanoes

So far, about 100 onshore and 80 offshore mud volcanoes have been discovered in the Black Sea region. Based on the latest geophysical data on the deep roots of mud volcanoes, even reaching the Moho surface (Fig. 8.10), three main types of mud volcanoes can be distinguished (Fig. 8.11).

1. Mud volcanoes developed on Maikopian sediments in the Western Black Sea depression, where the Moho surface has been traced to a depth of about 19 km below the seafloor.
2. Mud volcanoes developed on Maikopian sediments within the Kerch and Taman peninsulas as well as the Sorokin Trough, where the Moho surface lies at a depth of about 44–45 km below the land surface.
3. Mud volcanoes developed on Lower Cretaceous sediments in the southeastern part of the northern Caucasus, where the Moho surface lies at a depth of about 44 km below the land surface.

Taking into account the ideas of P.N. Kropotkin, B.N. Valyaev, and some others on the role of fluid plumes in the development of mud volcanoes, it becomes evident how important the mud volcanic roots are in the formation of both the superimposed fluid component and the mineralization concentrated in mud volcanic breccia.

Though they have common features, mud volcanoes are quite diverse and, in general, reveal very individual characteristics that determine their geological succession, the lithological properties of their sediments and water content, the relief of their terrain, their tectonic activity, the power of the plumes that form within the mud volcano and associated apophyses, and the composition of their plume gases.

The individuality of a mud volcano is expressed in its external morphology, the volume of the cone formed by its breccia, the diameter of the mud volcanic focus, the presence/absence of other foci, etc. It is, however, natural to expect that, within the boundaries of large geological

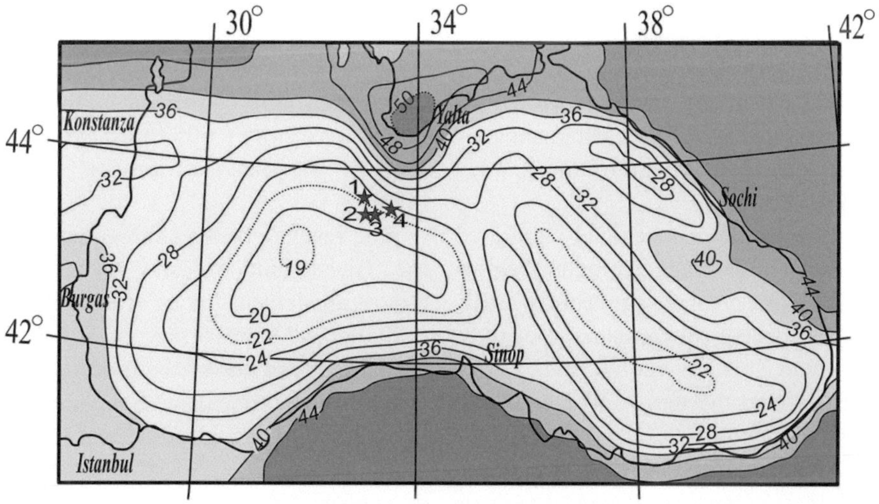

Fig. 8.10 Depth in km of the Moho surface in the Black Sea basin. Mud volcanoes: Tredmar (*1*), MSU (*2*), Yuzhmorgeo (*3*), Mantiynyy (*4*) (Starostenko et al. 2004)

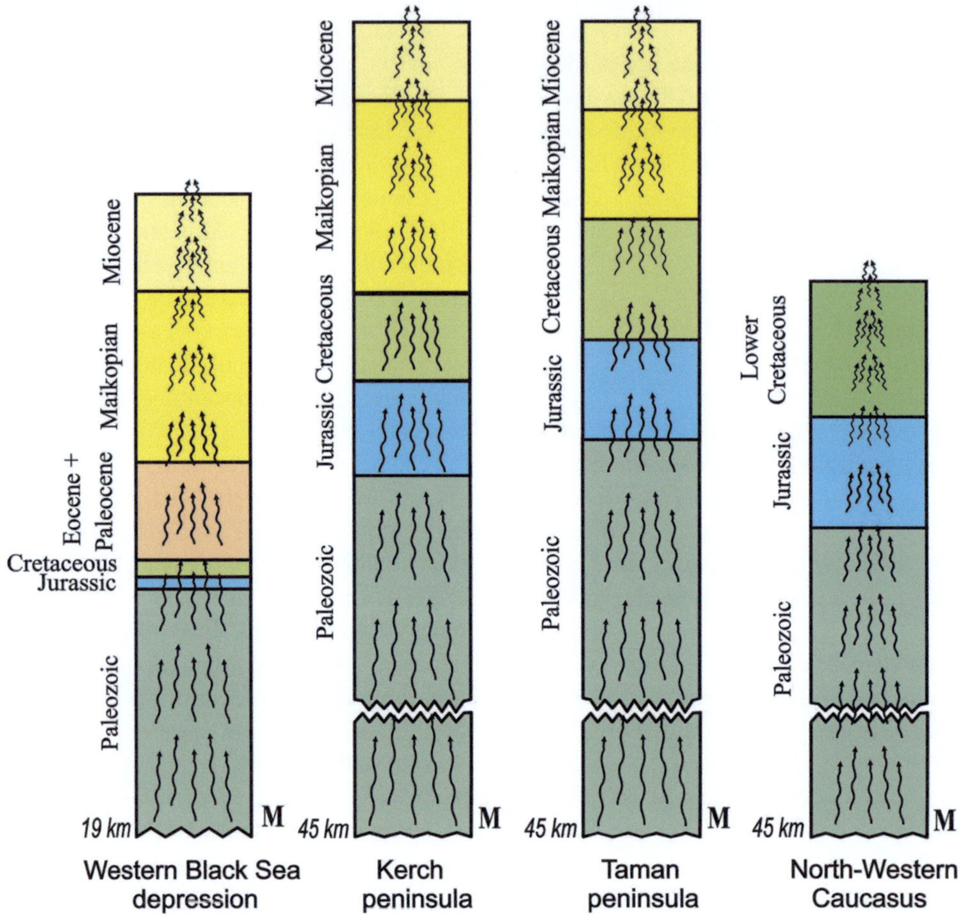

Fig. 8.11 Types of mud volcanoes

structures, mud volcanoes are similar not only in their appearance but in the enrichment of their breccia with minerals of certain chemical elements, the character of their manifestations, their originality, and diversity.

There is no complete geochemical picture of all mud volcanoes in the huge geological region under study. The data obtained by Fedorov et al. (2011) allow a geochemical comparison of only two types of mud volcano: the second (represented by Tizdar) and the third (represented by Semigorskiy and Shugo).

Fifty-nine chemical elements have been identified in mud volcanic breccia. Analysis has been conducted of a wide range of elements, many with low concentrations in the Earth's crust, using the Element 2 ICP-MS. Detection limits of uncommon, rare earth, and highly charged elements were from 0.005 to 0.1 g/t, and the accuracy of the analysis was on average 2–7% (Table 8.1).

The results obtained show that breccias from the Tizdar mud volcano reveal depletion of almost all elements by an order of 2–4 compared to Shugo and Semigorskiy; only the Al content at Tizdar exceeds that of the breccias at Shugo (by more than five times) and Semigorskiy (by about three times). The high content of Ba, Hg, As, and Li is considered by experts (Fedorov et al. 2011) as evidence of the significant depth of the mud volcano's roots and their connection with hydrothermal systems and magmatic foci.

Table 8.1 Elemental concentrations (g/t) in silt and water from the breccia of the Semigorskiy, Shugo, and Tizdar mud volcanoes (Fedorov et al. 2011)

	Mud volcano			Element	Mud volcano		
	Semigorskiy	Shugo	Tizdar		Semigorskiy	Shugo	Tizdar
	3rd type		2nd type		3rd type		2nd type
	Sample No.				Sample No.		
Element	vlk-1	vlk-3	vlk-4		vlk-1	vlk-3	vlk-4
Li	14.85	31.7	8.41	Te	0.27	0.23	0.15
Be	3.28	3.66	1.84	I	0.61	0.39	1.51
Na	19388.09	34683.08	23012.6	Cs	11.59	16.17	4.83
Mg	19263.06	17808.69	12512.15	Ba	1846.67	1346.04	479.62
Al	21605.66	14718.35	64248.66	La	67.69	62.2	24.46
P	1401.91	888.21	686.8	Ce	133.00	127.99	50.48
K	34320.01	34895.12	20275.05	Pr	16.89	15.28	6.07
Ca	99991.71	61076.26	35030.13	Nd	64.73	60.38	25.20
Sc	28.02	30.28	10.64	Sm	12.07	11.01	5.67
Ti	9233.01	9298.97	3649.86	Eu	3.08	2.57	1.15
V	207.13	286.82	91.20	Gd	10.77	9.17	4.43
Cr	202.08	257.93	83.15	Tb	1.56	1.33	0.63
Mn	2161.03	1073.49	698.00	Dy	8.53	7.56	3.8
Fe	85978.54	86754.54	36683.80	Ho	1.81	1.6	0.74
Co	38.97	31.92	13.96	Er	4.79	4.45	2.00
Ni	175.16	151.35	58.16	Tm	0.72	0.67	0.3
Cu	122.23	83.65	16.59	Yb	4.82	4.64	2.09
Zn	210.37	168.63	78.82	Lu	0.89	0.81	0.36
Ga	48.11	47.97	17.69	Hf	7.76	7.62	3.63
Ge	4.84	4.96	1.84	Ta	2.43	2.73	1.18
As	1.72	1.75	1.08	Re	0.02	0.03	0.02
Rb	195.63	270.6	74.49	Hg	0.17	0.15	0.11
Sr	928.89	719.51	314.42	Tl	1.22	19.94	0.41
Y	50.39	48.31	26.71	Pb	41.93	38.45	19.8
Zr	314.51	256.8	150.74	Bi	0.49	0.33	0.14
Nb	35.64	37.93	18.25	Th	26.73	29.1	8.75
Mo	2.38	2.2	0.88	U	6.47	5.24	1.79
Ag	0.08	0.07	0.02	Sum of rare earth elements (sum)	331.4	309.7	127.4
Cd	0.39	0.37	0.21	La_N/Yb_{N**}	9.48	9.05	7.92
In	0.11	0.11	0.06	Gd_N/Yb_{N**}	1.81	1.60	1.72
Sn	5.26	5.48	3.04	Eu/Eu*	0.83	0.78	0.70
Sb	0.51	0.52	0.4	Ce/Ce*	0.92	0.97	0.97

Eu/Eu* – measured content of Eu in the sample to theoretical Eu* calculated on the basis of a continuous spectrum with neighboring elements; analogical for Ce/Ce*; N** – ratios of elements, most likely according to compositions of chondrites or the mantle

Concentrations of the lanthanide rare earth elements Nd, Ce, La, and Dy, compounds of which are found in the breccia of the Semigorskiy and Shugo mud volcanoes developed on Cretaceous strata, are more than two times higher than those obtained at Tizdar.

8.5 Fluidogenic Mineralization of the Western Black Sea Mud Volcanoes

Among mud volcanoes of the Western Black Sea depression, breccia from three of them—MGU, Tredmar, and Yuzhmorgeo—has been studied;

within these samples, 23, 31, and 40 minerals, respectively, have been identified. A total of 60 minerals have been discovered to date in these mud volcanoes, enabling one to conclude that their fluidogenic mineralization is rich. Among them, there are native minerals, sulfides, oxides, carbides, carbonates, silicates, sulfates, phosphates, and arsenates. In these estimates, however, sampling error may be associated with the discovery process—e.g., random selection and vagaries in the fractionation enrichment (Table 8.2).

Among the **native minerals**, special attention is given to unusual grains of gold discovered in breccia from the Yuzhmorgeo mud volcano. Visual observation of the grains with the naked eye is not possible due to their very small sizes, which vary from tenths to hundredths and even thousandths of a millimeter. They can be best observed under binocular magnification. Chemical analysis performed by microprobe has shown that the amount of copper varies over a wide range (21–68%), and accordingly, the gold was classified either as copper gold (Fig. 8.12) or gold copper (Fig. 8.13).

Our observational data fit the experimental data of N. Kurnakov and S. Zhemchuzhnyy perfectly, as cited by Vernadskiy (1913), who specifically studied native minerals. These authors demonstrate that gold and copper produce alloys with different proportions. As such, Vernadskiy (1913) mentioned that copper-bearing gold should be considered an independent mineral that can contain not only copper but other elements, such as Ag, Pd, Zn, and Ni. This also corresponds with our data and shows that the amount of silver can be high, ranging between 9.6 and 14% (Fig. 8.14).

Native copper and copper minerals are prevalent in the Yuzhmorgeo mud volcano. In the Yuzhmorgeo breccia, it can be either a pure copper (Fig. 8.15) or copper with an admixture of gold, silver (Fig. 8.16), and zinc, also cuprite (Cu_2O) with admixtures of sulfides, tenorite (CuO), copper-bornite sulfides (Fig. 8.17), chalcopyrite, cubanite, and covellite (Fig. 8.18). Only chalcopyrite was occasionally observed in breccia from the Tredmar mud volcano.

Copper mineralization within the Yuzhmorgeo mud volcano is rich and unusual, as mineralogically diverse copper grains are often gathered together in isolated aggregates (Fig. 8.16). Bornite sometimes presents itself in colomorphic formations, and chalcopyrite appears in small irregular segregations. Both have clearly developed within the mud volcanic breccia.

Native lead, partly with an admixture of its oxide, massicot (PbO), forms small, angular segregations up to 10 μm (Fig. 8.19).

Massicot is a very rare mineral. Nowadays, it is produced during the industrial processing of lead and lead oxides, especially in the glass industry, which is the biggest user of PbO. In breccia, it is most likely a freshly developed mineral. By itself, lead is soft and malleable, and it could hardly be transported as a placer deposit. It is important to note that in placers, the lead is often associated with native gold, iron, and platinum, while in indigenous deposits, it is chemically almost pure (Vernadskiy 1913). In our case, the lead contains an admixture of Fe and Al. Perhaps this is simply contamination of the mineral with mechanical impurities.

According to Vernadskiy (1913), the emergence of native lead in indigenous deposits requires exceptional conditions that would bring about the disintegration of lead compounds. As a rule, native lead is accompanied by massicot, but native lead is not formed ordinarily when lead compounds disintegrate. In our opinion, such exceptional conditions have been created by powerful hydrocarbon fluids, which introduce a pronounced reducing environment.

Native iron is found in breccia from the Yuzhmorgeo and Tredmar mud volcanoes, and it is represented by ferrite (Figs. 8.20 and 8.21) and admixture of FeS (Fig. 8.21). According to Vernadskiy (1913), ferrite is a native iron containing up to 0.6% Ni, while kamacite is more enriched with Ni. In our case, the native iron contains 2.14% Ni, meaning it is kamacite (Fig. 8.22).

It is assumed that nickel forms a solid solution with iron. Grains of iron have a random shape, filling the voids between silicates in igneous rocks (Vernadskiy 1913). In our case, the microscopic (up to 5 μm) spherules have a spherical

Table 8.2 Minerals discovered largely in the heavy fraction (except for calcite, which was discovered in the light fraction) of breccia from mud volcanoes of the Western Black Sea[a]

Chemical elements and minerals in the breccia	mud volcanoes		
	MSU	Tredmar	Yuzhmorgeo
Native minerals			
Gold			Very rare, 3–5 grains (here and below in the field of view of the microscope)
Gold with Ag, Cu, Pd, Zn, Ni			"
Gold copper (up to 25% Cu)			"
Copper gold (up to 25% Au)			"
Copper	One grain		Rare, about 10 grains
Silver			Very rare, 3–5 grains
Sulfur			Minor
Iron		Very rare	Rare, about 10 grains
Nickel iron (taenite)		Very rare	
Chrome-plated nickel iron			Very rare
Carbides			
Cogenite (cementite) Fe_3C			Very rare
Sulfides			
Pyrite	Major	Major	Major
Marcasite	Major	Major	Major
Pyrrhotite	Rare	Rare	Rare
Bornite			Rare
Chalcopyrite		Rare	Rare
Cubanite			Very rare
Covellite, chalcocite			Very rare, 3–5 grains
Wurtzite	Rare	Rare	
Sartorite $PbAs_2S_4$		Very rare	
Oxides and hydroxides			
Corundum (leuco sapphire)	Very rare		
Wüstite			Very rare
Magnetite	Minor	Rare	Minor
Goethite	Minor	Minor	Minor
Chromite	Rare		Minor
Hematite		Rare	Minor
Ilmenite	Minor	Minor	Minor
Rutile	Secondary	Rare	
Chrome-magnesium spinel		Very rare	Rare
Leucoxene		Minor	
Franklinite			Rare, one skeletal crystal has been discovered as an admixture in magnetite
Perovskite			One skeletal crystal
Carbonates			
Calcite	Light fraction	Light fraction	Light fraction
Siderite			Minor
Smithsonite			Rare
Phosgenite $Pb_2(CO_3)C1$		Very rare	
Dolomite		Rare	
Sulfates			
Gypsum	Very rare	Very rare	

(continued)

Table 8.2 (continued)

Chemical elements and minerals in the breccia	mud volcanoes		
	MSU	Tredmar	Yuzhmorgeo
Barite	Very rare		Rare
Phosphates			
Apatite, including scheelite	Rare	Rare	
Pyromorphite $Pb_5(PO_4)_3C1$		Very rare	
Monazite		Very rare	
Fluorapatite	Rare		
Arsenates			
Carminite (?) $PbFe_2(AsO_4)_2(OH)_2$			Rare

[a]Based on their frequency of occurrence, the minerals are divided into major, minor, rare, and very rare (for the last, the number of grains is indicated for the Yuzhmorgeo mud volcano only; in other mud volcanoes, these minerals were not found); formulas are provided for the very rare minerals only

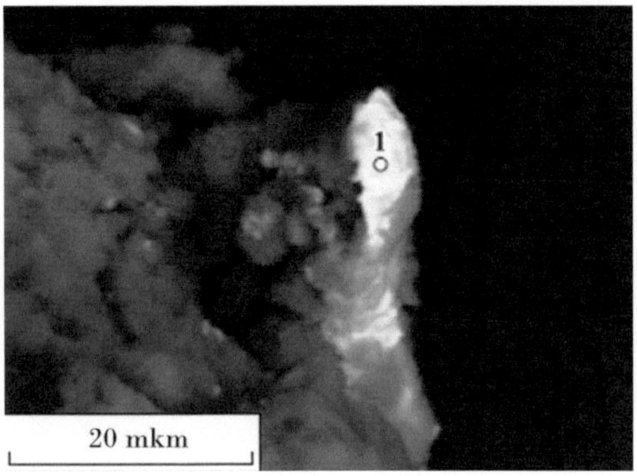

Spectrum 1		
Element	Weight%	Atomic%
O	12.08	50.45
Ni	3.92	4.46
Cu	19.26	20.25
Ag	10.29	6.37
Au	54.45	18.47
Totals	100.00	

Fig. 8.12 Copper gold with an admixture of silver and nickel from breccia of the Yuzhmorgeo mud volcano

Spectrum 1		
Element	Weight%	Atomic%
O	3.68	15.41
Al	1.08	2.68
Si	1.08	2.57
Cu	64.41	67.97
Ag	4.43	2.75
Au	25.33	8.62

Fig. 8.13 Gold copper from breccia of the Yuzhmorgeo mud volcano

shape and consist entirely of iron. Sometimes they create films on silica nodules (e.g., Fig. 8.20).

Vernadskiy emphasized that the formation of native iron from chemical compounds— oxides or sulfides—occurs under exceptional circumstances, for example, under the fires of coal deposits or the decay of organic matter. In his day, the possible influence of hydrocarbon gas plumes was not considered. The influence of such plumes on sulfides (pyrite), magnetite, iron hydroxides, siderite, etc. leading to the appearance of native iron cannot be excluded.

Sulfides are also numerous, both in the number of mineralogical findings and in their species composition. Among them, iron sulfides—pyrite and marcasite—predominate. Pyrite is less common in the breccias of submarine mud volcanoes compared to terrestrial mud volcanoes of the Kerch Peninsula. Most often, pyrite crystals have a cubic or octahedral appearance, and anti-

Spectrum 1			Spectrum 2		
Element	Weight%	Atomic%	Element	Weight%	Atomic%
O	3.78	23.28	O	7.92	39.55
Ni	2.66	4.46	Ni	4.37	5.94
Cl	16.97	26.35	Cu	16.86	21.20
Zn	1.75	2.64	Pd	3.83	2.88
Ag	13.96	12.77	Ag	9.63	7.13
Au	60.89	30.50	Au	57.40	23.29
Totals	100.00		Totals	100.00	

Fig. 8.14 Gold with an admixture of different elements from breccia of the Yuzhmorgeo mud volcano

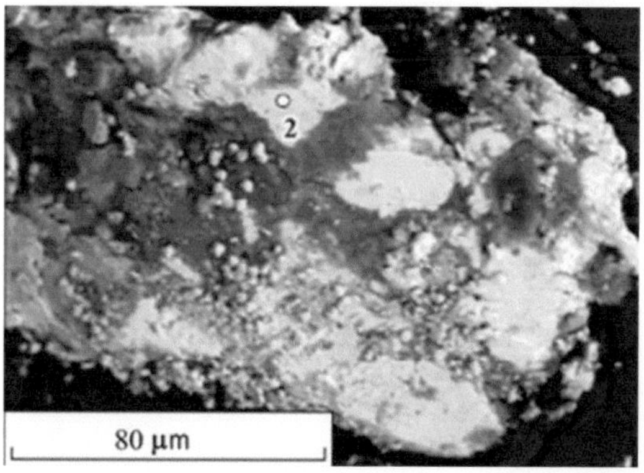

Spectrum 2		
Element	Weight%	Atomic%
Cu	100.0	100.0
Ag	0.00	0.00
Totals	100.00	

Fig. 8.15 Plate of native copper in a "shirt" of clay minerals from breccia of the Yuzhmorgeo mud volcano

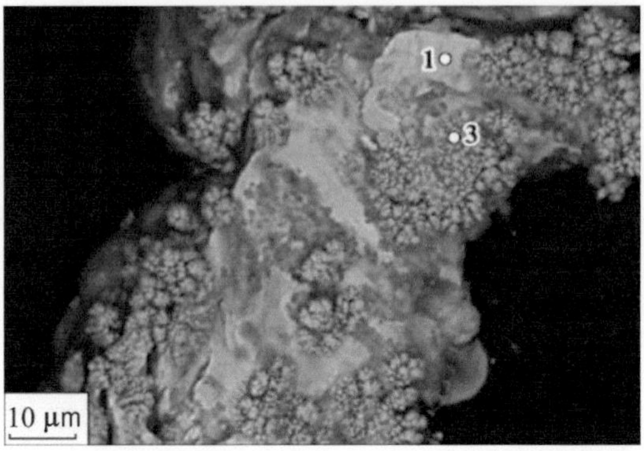

Spectrum 1			Spectrum 3		
Element	Weight%	Atomic%	Element	Weight%	Atomic%
Cu	92.29	94.12	Cu	89.67	92.18
O	7.46	5.71	O	9.90	7.64
S	0.25	0.10	Al	0.09	0
Totals	100.00		Si	0.13	0.06
			S	0.21	0.08
			Totals	100.00	

Fig. 8.16 Native copper with blooms of secondary cuprite from breccia of the Yuzhmorgeo mud volcano

Spectrum 2		
Element	Weight%	Atomic%
O	7.87	20.87
Al	2.10	3.30
Si	1.76	2.65
S	19.01	25.17
Cl	1.50	1.79
Fe	10.25	7.79
Cu	57.51	38.42
Totals	100.00	

Fig. 8.17 Segregations of bornite from breccia of the Yuzhmorgeo mud volcano

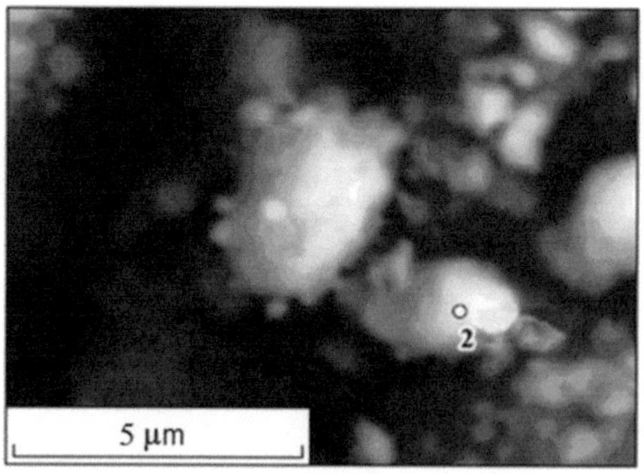

Spectrum 2		
Element	Weight%	Atomic%
O	5.64	15.49
Al	0.89	1.46
S	27.20	37.26
Cu	66.26	45.80
Totals	100.00	

Fig. 8.18 Rounded, melted grains of covellite with probable admixture of cuprite from breccia of the Yuzhmorgeo mud volcano

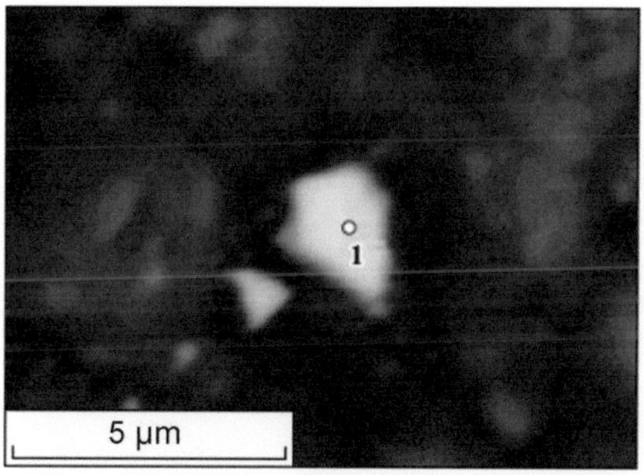

Spectrum 1		
Element	Weight%	Atomic%
O	20.39	73.41
Al	0.62	1.33
Fe	4.38	4.52
Pb	74.60	20.74
Totals	100.00	

Fig. 8.19 Native lead with an admixture of massicot from breccia of the Yuzhmorgeo mud volcano

Spectrum 1					
Element	Weight%	Atomic%	Element	Weight%	Atomic%
O	8.22	23.02	S	0.77	1.08
Al	1.28	2.12	Fe	87.47	70.17
Si	2.26	3.60	Totals	100.00	
Totals	100.00				

Fig. 8.20 Microscopic (up to 5 μm) spherules of iron, mostly ferrite, from breccia of the Yuzhmorgeo mud volcano. The presence of oxygen indicates a possible admixture of wüstite (FeO), while the sulfur indicates a possible admixture of iron sulfides; the silicate matrix includes Si, Al, and O

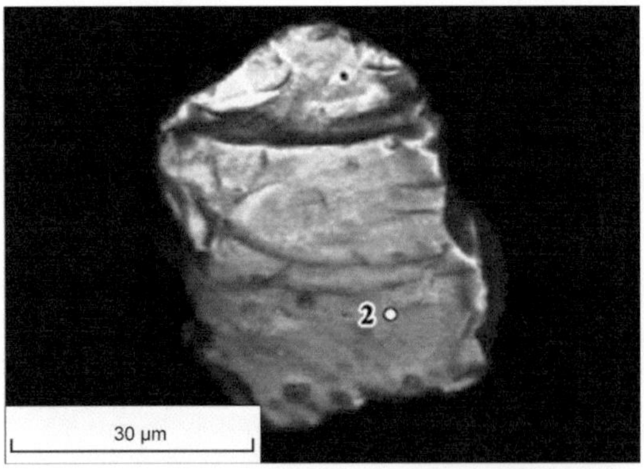

Spectrum 2		
Element	Weight%	Atomic%
Mg	0.82	1.68
Al	1.27	2.36
S	12.39	19.33
Fe	85.52	76.63
Totals	100.00	

Fig. 8.21 Native iron with an admixture of FeS (characteristic feature of ultrabasites) from breccia of the Yuzhmorgeo mud volcano

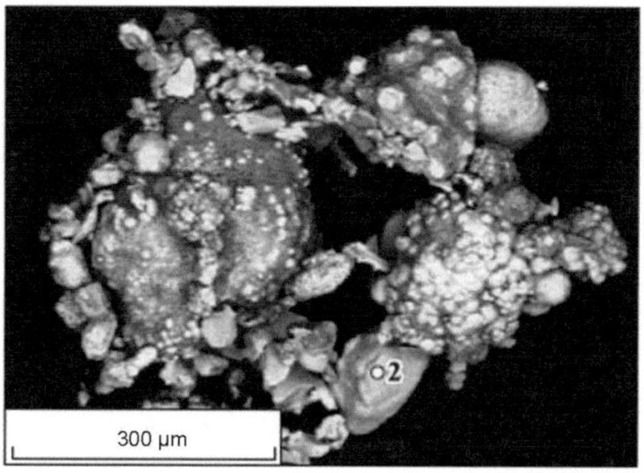

Spectrum 2		
Element	Weight%	Atomic%
O	2.81	8.21
Si	0.87	1.45
S	16.03	23.34
Fe	78.15	65.31
Ni	2.14	1.70
Totals	100.00	

Fig. 8.22 Kamacite with an admixture of iron sulfides from breccia of the Tredmar mud volcano

skeletal forms of growth are observed on well-defined crystals. On the faces, the central parts are, as it were, higher than the peripheral ones. Growths of pyrite with biological remains are visible. The sizes of the crystals are measurable in microns.

The fouling of ilmenite crystals by small octahedra of pyrite was observed. Sometimes, ilmenite is overgrown not by crystals but by rounded forms resembling framboids (Fig. 8.23).

Radially fibrous crystals of marcasite covered with pyrrhotite powder (Fig. 8.24) as well as various biomorphoses of iron sulfides have been encountered.

As was mentioned above, a variety of copper sulfide compounds was discovered as intergrowths with native copper.

There are also sulfides of zinc present, most often in the form of wurtzite. This mineral produces well-developed hexagonal pyramidal (Fig. 8.25) and skeletal (Fig. 8.26) crystals with dimensions of 0.15–0.4 mm. Sometimes sulfides of zinc form tiny (a few microns in size) xenomorphic, angular crystals of sphalerite.

Judging by the presence of frequent well-developed crystals of zinc sulfide, it is possible to assume that they crystallized directly within the mud volcanic flow.

Oxide minerals are quite often encountered in the heavy fraction of mud volcanic breccia. Of particular interest are the high-temperature oxides that retain their crystallographic morphologies, having been formed under conditions of high pressure. These include, for example, ferrigenous chromite crystals, which take the form of an octahedron with dissolved vertices (Fig. 8.27), as well as large angular fragments of chrome picotite, and fragmentary high-ferrigenous chromite. We also found a fragment of a glassy rock of ultrabasic composition in which numerous skeletal porphyry crystals of chromite were discovered (Fig. 8.28).

Among the other iron oxides, magnetite and perovskite ($CaTiO_3$) must be mentioned. A skeletal crystal of perovskite was observed. Its lower Ti content and light green color (Fig. 8.29) differentiate it from the original stoichiometric variety, which is reddish brown to yellow in color.

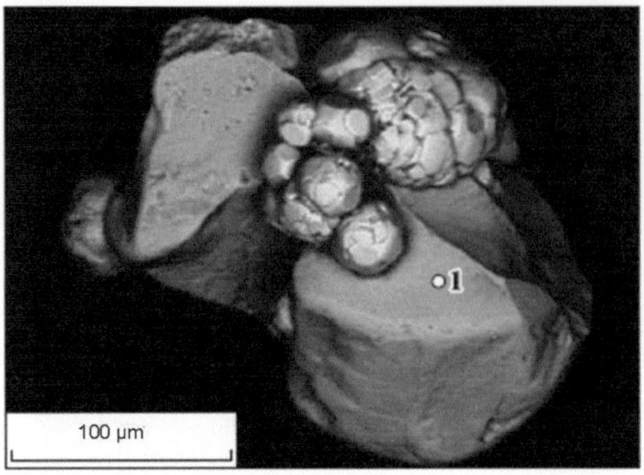

Spectrum 1		
Element	Weight%	Atomic%
O	41.71	69.46
Ti	34.39	19.13
Mn	1.12	0.54
Fe	22.77	10.86
Totals	100.00	

Fig. 8.23 Fouling of ilmenite with framboidal pyrite from breccia of the Tredmar mud volcano

Spectrum 1		
Element	Weight%	Atomic%
S	56.05	68.93
Mn	2.32	1.66
Fe	41.64	29.40
Totals	100.00	

Fig. 8.24 Radially fibrous aggregate of marcasite with a powder of fine-grained pyrrhotite (?) from breccia of the Tredmar mud volcano

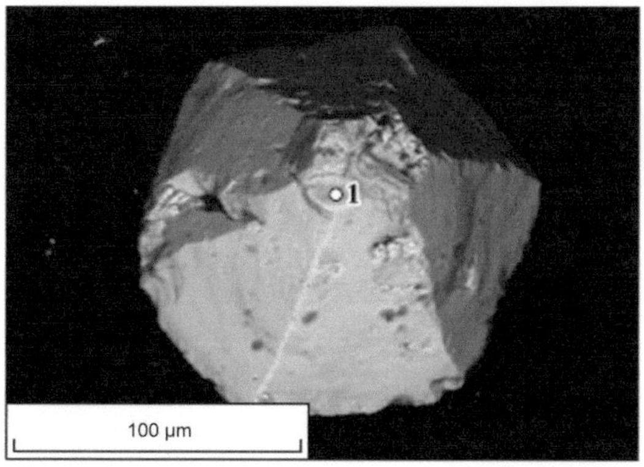

Spectrum 1		
Element	Weight%	Atomic%
S	39.20	56.22
Mn	2.44	2.03
Zn	58.36	41.35
Totals	100.00	

Fig. 8.25 Hexagonal-pyramidal crystal of wurtzite from breccia of the MSU mud volcano

Spectrum 1		
Element	Weight%	Atomic%
O	4.90	13.63
S	29.45	40.87
Fe	4.95	3.94
Cu	12.07	8.45
Zn	48.64	33.11
Totals	100.00	

Fig. 8.26 Skeletal crystal of copper wurtzite from breccia of the Tredmar mud volcano

Spectrum 1		
Element	Weight%	Atomic%
O	48.37	73.03
Mg	2.12	2.11
Al	5.38	4.81
Si	0.40	0.34
Cr	24.77	11.51
Fe	18.96	8.20
Totals	100.00	

Fig. 8.27 Octahedral crystal of chromite with dissolved vertices from breccia of the MSU mud volcano

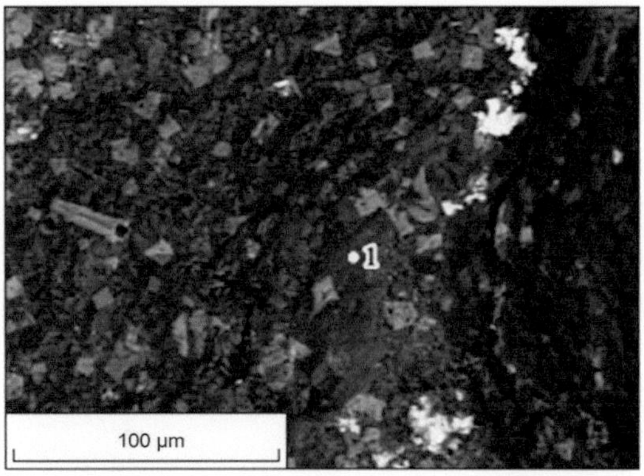

Spectrum 1		
Element	Weight%	Atomic%
O	43.23	60.76
Mg	6.58	6.08
Al	4.56	3.80
Si	23.89	19.12
Ca	8.40	4.71
Cr	3.13	1.35
Mn	7.41	3.03
Fe	2.80	1.13
Totals	100.00	

Fig. 8.28 Skeletal porphyry crystals of chromite in glassy rock of ultrabasic composition from breccia of the Yuzhmorgeo mud volcano

Spectrum 1		
Element	Weight%	Atomic%
O	44.50	67.71
Ca	41.12	24.98
Ti	14.38	7.31
Totals	100.00	

Fig. 8.29 Complex splicing of perovskite crystals from breccia of the Yuzhmorgeo mud volcano

Perovskite is a mineral typical of kimberlites and other basic rocks. Kimberlites develop within the mantle at depths between 150 and 450 km, potentially from anomalously enriched exotic mantle compositions.

Another mineral called sagenite has also been observed; it is represented by reticular mesh twins of rutile (Fig. 8.30). Well-crystallized corundum forms individual crystals in the habit of poorly expressed bipyramids (Fig. 8.31).

Corundum is a mineral typical of deposits occurring in nonsilicate rocks or those that are silicate poor and contain other ultrabasic, basic, alkaline, and nepheline rocks (Lazarenko 1970).

Silicates are often found among the accessory minerals of Black Sea mud volcanoes. Among them, there are feldspars, muscovite, biotite, diopside, enstatite, garnets (pyrope-almandine, andradite), hydromica, zircon, epidote, andalusite, rhodonite, and some others. In the overwhelming majority of cases, rolled individuals predominate. This indicates that we should regard them as redeposited terrigenous minerals deriving mainly from those sediments on which the mud volcanoes are sitting, mainly from the Maikopian clays.

Phosphates are quite rare among accessory minerals. Only two grains of apatite have been found. One of them, encountered in breccia of the MGU mud volcano, is represented by a glomercrystalline cluster of well-formed elongated crystals of fluorapatite with a size of 0.2 mm. It appears slightly melted, which is characteristic for metamorphic rocks (Betekhtin 1950). Small grains of scheelite have grown into the fluorapatite crystals (Fig. 4.41).

Another chamfered crystal of apatite (Fig. 4.42), discovered in breccia of the Tredmar mud volcano, is represented by a hexagonal prism of approximately the same dimensions as the previous example. This crystal is characterized by erosion and etching of the faces that occurred when the object entered a medium that partly attacked the crystal surfaces.

Both crystals contain fluorine and a small amount of chlorine. Presumably, their external appearance reflects conditions of their origin under the influence of mobile fluid flows.

Carbonates are scarce in the heavy fraction of breccia from marine mud volcanoes. Most often, they are siderites of detrital appearance, dolomite, very rare smithsonite ($ZnCO_3$ from the

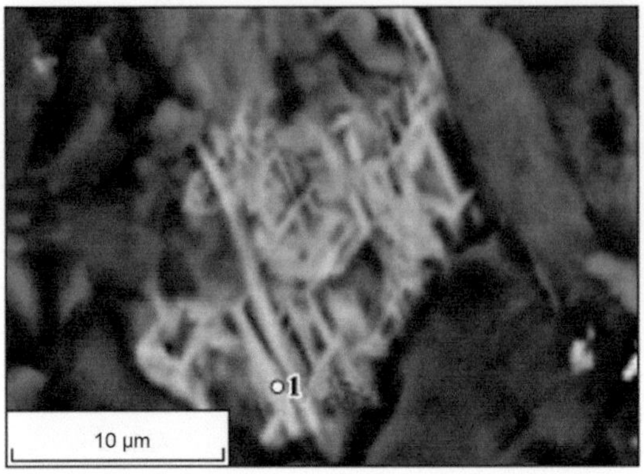

Spectrum 1		
Element	Weight%	Atomic%
O	44.20	60.76
Al	1.59	69.05
Si	2.99	1.47
K	0.80	2.66
Ti	50.43	0.51
Totals	100.00	

Fig. 8.30 Mesh twinnings of rutile from breccia of the of MSU mud volcano

Spectrum 2		
Element	Weight%	Atomic%
O	30.07	42.04
Al	69.93	57.96
Totals	100.00	

Fig. 8.31 Corundum (leucosapphire) from breccia of the MSU mud volcano; under binocular microscopy, it is colorless and transparent

Yuzhmorgeo mud volcano), and phosgenite ($Pb_2(CO_3)Cl$ from the Tredmar mud volcano). In the light fraction of all mud volcanoes, calcium carbonate is common. One can assume a mud volcanic origin for the dolomite, phosgenite, smithsonite, and some calcium carbonates.

Sulfates in marine mud volcanoes are also scarce. Only small grains of gypsum and barite in the form of minute debris have been observed. Barite could be considered a hypergenic mineral; however, in one of the samples, there was a twin of barite with a tungsten mineral (Fig. 8.32).

In summary, breccia from submarine mud volcanoes of the Western Black Sea depression is enriched with native minerals, most of which are absent within sedimentary rocks. These minerals are formed under the reduced oxygen conditions created by deep gaseous fluids. The latter destroy the rocks, elevate their remains through upwelling flows, and participate in the development of mud volcanoes. We therefore conclude that minerals found in breccia can serve as important indicators of the deep origin of submarine mud volcanoes in the Black Sea.

8.6 Fluidogenic Mineralization of the Kerch-Taman Mud Volcanoes

Mud volcanoes of the Kerch-Taman region have been studied in great detail (see Chap. 6), and mineralogical investigation of mud volcanic breccia has led to the discovery of many dozens of minerals that are difficult to classify as terrigenous. They are not rounded but instead have a specific habitus characteristic of neo-crystallizations.

Breccias from several mud volcanoes of Kerch (Dzhau-Tepe, Korolyovskiy, Bulganakskiy, Andreevskiy, Kayaly-Sartskiy, Soldatsko-Slobodskoy, Malo-Burashskiy) and Taman (Gladkovskiy, Gora Gnilaya, Akhtanizovskiy, Western Tsymbaly) have been examined. Their composition has already been described in a number of publications (Shnyukov et al. 1992, 2013a, b, 2016; Shnyukov 2016). Particular attention was paid to the composition of fragmentary calcite veins and new crystallizations; their visual appearance and composition show that they are made up predominantly of low magnesian calcite, inclusions of which are homogenized at temperatures ranging between 110–160 °C and 180–200 °C (Kalyuzhnyy et al. 1984).

Decrepitation of the inclusions occurs at temperatures of 110 °C and 150 °C (as observed in the Bulganakskiy and Gladkovskiy mud volcanoes, respectively). The inclusions contain mainly CO_2, and to a lesser degree CH_4, as well as some other gases. Their isotopic composition shows that $\delta^{13}C$ of CO_2 varies between +0.99‰ and + 1.17‰, while $\delta^{13}C$ of CH_4 varies between 3.33‰ and – 4.84‰. These values are quite close to those of volcanic gases from the Kuril Ridge. Fragments of calcite with inclusions having similar composition were also found in breccia of the Dzhau-Tepe mud volcano as well as some others.

As of today, 14 native minerals have been identified. They include sulfur, graphite, iron (mostly pure but sometimes with signs of partial oxidation), wüstite (not native but usually found alongside native Fe), nickel, gold, osmium, iridium, lead (partially oxidized into massicot), and titanium. Lukin (2006) discovered native tin, antimony, zinc, and aluminum in breccia of the Karabetova Gora mud volcano. We also found spherules of tin that were already oxidized to cassiterite as well as natural alloys of cerium, lanthanum, iron, and neodymium in the form of hollow spherules with dimensions of about 50–80 μm. Numerous minerals in mud volcanic breccias are partially oxidized, indicating the low oxygen conditions in the environments where they were formed as they approached the Earth's surface.

Judging by its chemical composition, the **native iron** is represented by ferrite and has been found in breccia of the Western Tsymbaly mud volcano. The mineral appears as deformed elon-

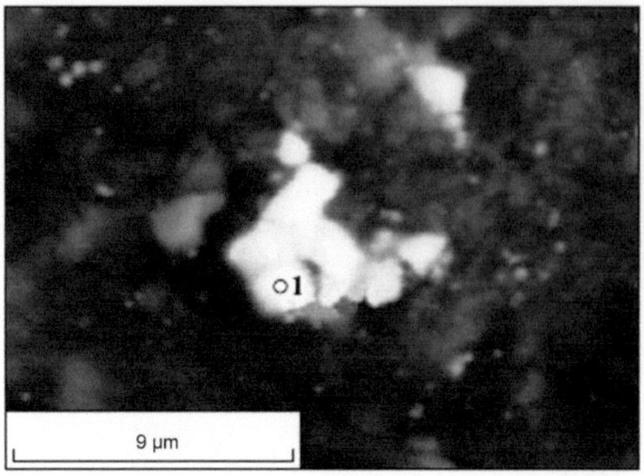

Spectrum 1		
Element	Weight%	Atomic%
O	31.26	67.96
Mg	0.65	0.93
Al	0.67	0.86
P	0.91	1.03
S	14.02	15.21
Ca	0.85	0.73
Fe	1.09	0.68
Ba	47.33	11.99
W	3.22	0.61
Totals	100.00	

Fig. 8.32 Barite with an inclusion of tungsten mineral from breccia of the Yuzhmorgeo mud volcano

gated individuals or aggregates, sometimes oxidized to wüstite, as well as in the form of wires, plates, and grains of bizarre shapes up to 0.5 mm in size (Figs. 8.33, 8.34, 8.35, 8.36, and 8.37).

Native nickel has been found so far only in breccia of the Gora Gnilaya and Western Tsymbaly mud volcanoes. In the former, the nickel is relatively pure and contains a small admixture of Fe (1.32%) as well as a mechanical impurity of Si (1.71%) (Fig. 8.38).

In breccia of the Western Tsymbaly mud volcano, nickel includes admixtures of Cu (5.49%) and Fe (3.33%), with Al and Si that represent most likely mechanical impurities from silicates (Fig. 8.39).

The nickel precipitates look like plates, i.e., flattened segregations with an uneven surface covered by a powder of fine copper sulfide crystals (Fig. 8.40a).

Unusual discoveries of **platinoids** have been made in the Andreevskiy and Soldatsko-Slobodskoy mud volcanoes. In the former, they are represented by a nevyanskite referred today as iridosmine (40% or more of iridium) as opposed to osmiridium, which has more osmium. Essentially, it is a natural alloy of osmium and iridium. Also, breccia from both volcanoes contains an admixture (low percentages) of the rare earth element dysprosium.

In the Andreevskiy mud volcano, the nevyanskite is represented either by elongated and slightly rounded grains up to 80 µm in length (Fig. 8.41) or by irregular spherules that are possible to interpret as hollow fluid bubbles over-

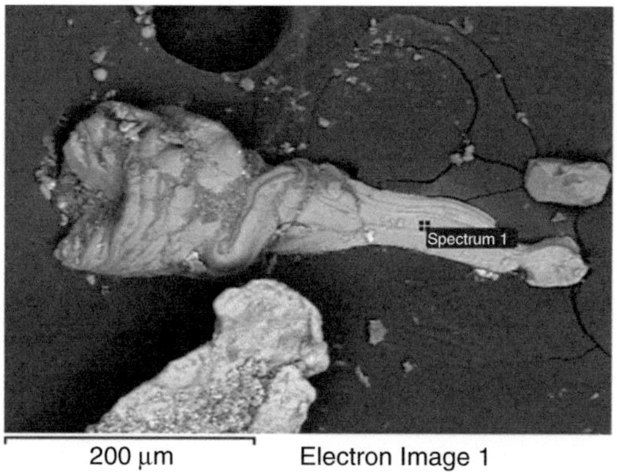

Spectrum 1		
Element	Weight%	Atomic%
Fc	100.00	100.00
Totals	100.00	

Fig. 8.33 Deformed aggregate of metallic iron from breccia of the Western Tsymbaly mud volcano

Spectrum 1		
Element	Weight%	Atomic%
O	22.10	49.75
Fe	77.90	50.25
Totals	100.00	

Fig. 8.34 Oxidized and deformed particle of metallic iron from breccia of the Western Tsymbaly mud volcano

grown with metal particles or as rounded grains (Fig. 8.42).

Minerals from breccia of the Soldatsko-Slobodskoy mud volcano are represented by several weakly crystallized grains and one well-pronounced crystal (Fig. 8.43). The figure clearly shows the tetragonal pyramidal peak of the crystal, while the mineral nevyanskite has a hexagonal syngony. This sample requires further study.

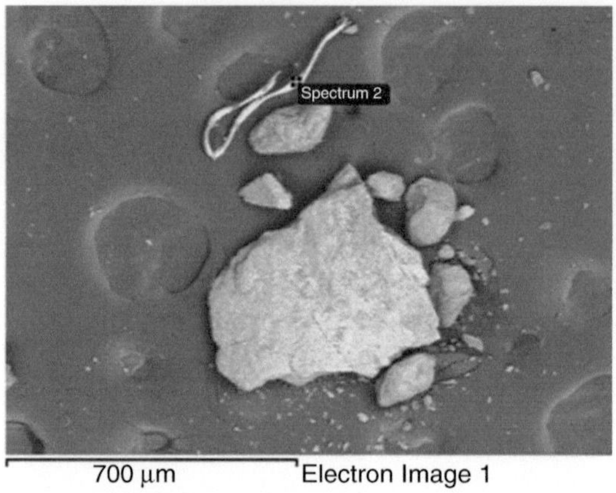

Spectrum 1		
Element	Weight%	Atomic%
Si	1.37	2.69
Fe	98.63	97.31
Totals	100.00	

Fig. 8.35 Filiform discharge of metallic iron from breccia of the Western Tsymbaly mud volcano

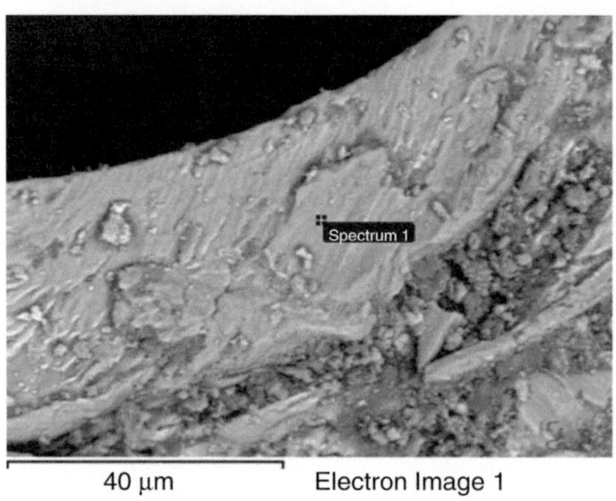

Spectrum 1		
Element	Weight%	Atomic%
Fe	100.00	100.00
Totals	100.00	

Fig. 8.36 Metallic iron from breccia of the Western Tsymbaly mud volcano

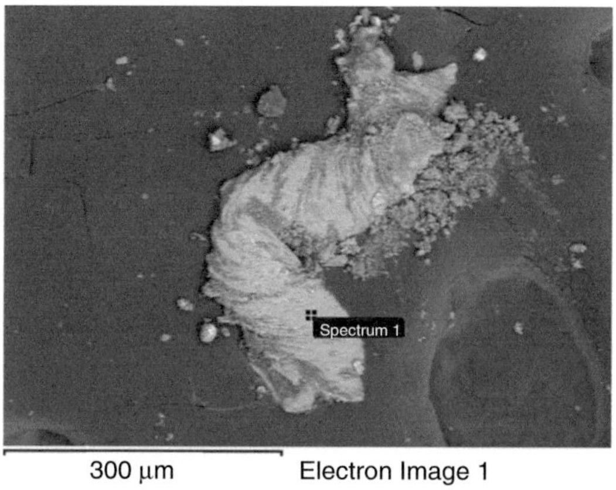

Spectrum 1		
Element	Weight%	Atomic%
Fe	100.00	100.00
Totals	100.00	

Fig. 8.37 Deformed aggregate of grains of native iron from breccia of the Western Tsymbaly mud volcano

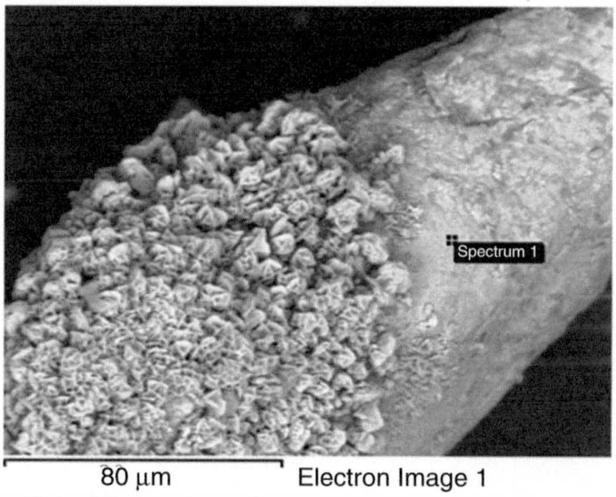

Spectrum 1		
Element	Weight%	Atomic%
Si	1.71	3.51
Fe	1.32	1.36
Ni	96.97	95.12
Totals	100.00	

Fig. 8.38 Native nickel with a crust of secondary nickel covellite from breccia of the Gora Gnilaya mud volcano

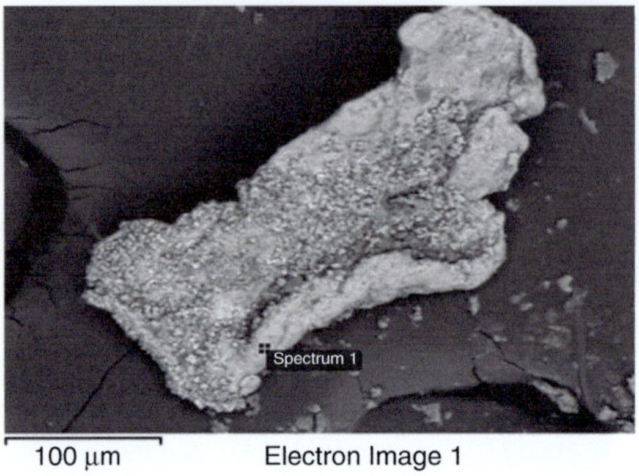

Spectrum 1		
Element	Weight%	Atomic%
Si	1.01	2.09
Fe	1.62	1.68
Ni	97.37	96.23
Totals	100.00	

Fig. 8.39 Plate of native nickel with a fine dusting of nickel sulfides from breccia of the Western Tsymbaly mud volcano

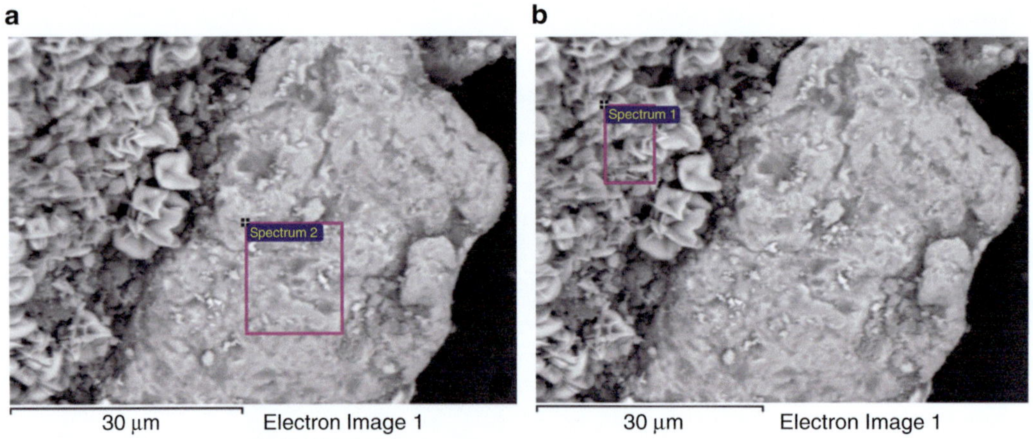

Spectrum 2			Spectrum 1		
Element	Weight%	Atomic%	Element	Weight %	Atomic %
Al	1.69	3.55	Si	1.02	2.18
Si	1.68	3.39	S	2.72	5.13
Fe	3.33	3.38	Fe	4.99	5.39
Ni	87.81	84.78	Ni	6.98	7.18
Cu	5.49	4.90	Cu	84.30	80.12
Totals	100.00		Totals	100.00	

Fig. 8.40 Native nickel (**a**) with an admixture of copper and (**b**) native nickeliferous copper with a probable admixture of pyrrhotite and other sulfides from breccia of the Western Tsymbaly mud volcano

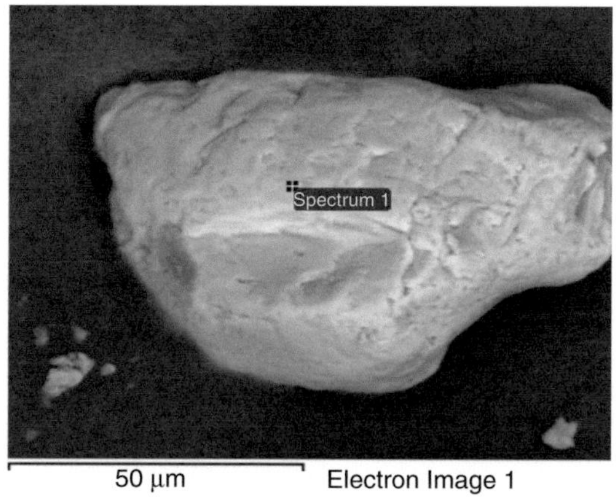

Spectrum 1				
Element	Weight%	Atomic%	Compound%	Formula
Ti	0.61	0.94	1.02	TiO_2
Dy	2.84	1.28	3.26	Dy_2O_3
Os	28.35	10.90	33.11	OsO_2
Ir	53.67	20.43	62.60	IrO_2
O	14.53	66.45		
Totals	100.00			

Fig. 8.41 Elongated and slightly rounded grain of nevyanskite from breccia of the Andreevskiy mud volcano

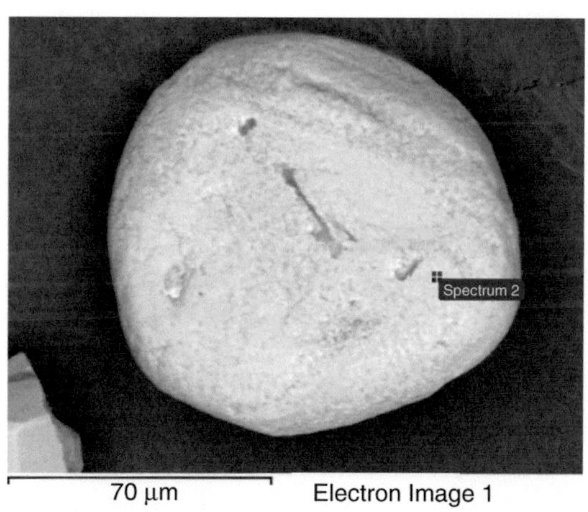

Spectrum 1		
Element	Weight%	Atomic%
O	50.34	91.66
Si	0.84	0.87
Ti	0.00	0.00
Dy	1.60	0.29
Os	16.56	2.54
Ir	30.66	4.65
Totals	100.00	

Fig. 8.42 Rounded grain of nevyanskite from breccia of the Andreevskiy mud volcano

Spectrum 1		
Element	Weight%	Atomic%
O	34.29	84.58
Si	1.17	1.64
Fe	0.97	0.68
Os	22.27	4.62
Ir	41.30	8.48
Totals	100.00	

Fig. 8.43 Nevyanskite from breccia of the Soldatsko-Slobodskoy (Nikitskiy) mud volcano

Other specimens of nevyanskite encountered within the breccia are usually elongated and do not exceed 0.1 mm in length (Figs. 8.44 and 8.45).

Based on the results of 12 analyses, the content of iridium ranges from 62.03 to 67.38% and osmium from 31.44 to 38.77%. This is in good agreement with results for the chemical composition of nevyanskite provided by Betekhtin (1950), in which the content of iridium and osmium varied between 46.8 and 77.2% and 21% to 49.3%, respectively. This author also mentioned Ru, Rh, Pt, Cu, and Fe as admixtures in breccia from the Soldatsko-Slobodskoy mud volcano; Dy, Gd, Ru, and Co were found in the sample from the Soldatsko-Slobodskoy (Nikitskiy) mud volcano; and Dy and Ti (just once) were present in the Andreevskiy mud volcano. Along with osmoid iridium, yttrium-iridium was also found (Fig. 8.46).

The origin of the platinoids in mud volcanic breccia is difficult to determine. The simplest hypothesis is to assume a breakthrough of some placers by upwelling fluid flow; however, the discovery of unrolled crystals and spherules excludes this assumption. In addition, there is the frequent presence of rare earth elements (e.g., Dy, Gd, etc.) in the composition of the platinoids. Most likely, somewhere deep in the Earth, there was a breakthrough by a powerful high-temperature, ultrabasite fluid flow. It could also involve the direct development of minerals by fluids followed by insignificant rolling during transportation of the breccia upward to the surface. The breakthrough of the ultrabasites would have had the effect of increasing the amount of ultrabasite minerals (e.g., Pb, S, Sb, Ag). Those are the components of ultrabasite, also called diaphorite, in the composition of the breccia fractions; this, in fact, has been observed.

The presence of **gold** in breccia of the Kerch-Taman mud volcanoes deserves special attention and further study. Despite the fact that breccia from numerous mud volcanoes was studied (e.g., Korolevskiy, Dzhau-Tepe, Nikitskiy, Kayaly-Sartskiy, Andrusov, Oldenburg, Obruchev, Tishchenko), gold particles were first discovered under a binocular microscope in breccia of the Korolyovskiy and then in the Tishchenko mud volcano. The main finds were made by

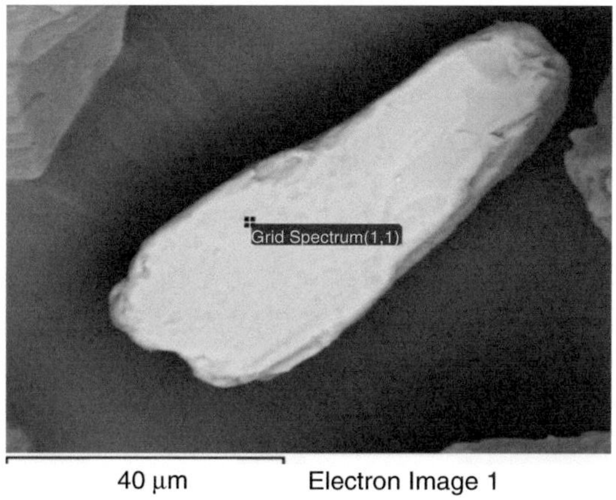

Spectrum 1		
Element	Weight%	Atomic%
Os	32.62	32.85
Ir	67.38	67.15
Totals	100.00	

Fig. 8.44 Nevyanskite from breccia of the Soldatsko-Slobodskoy (Nikitskiy) mud volcano

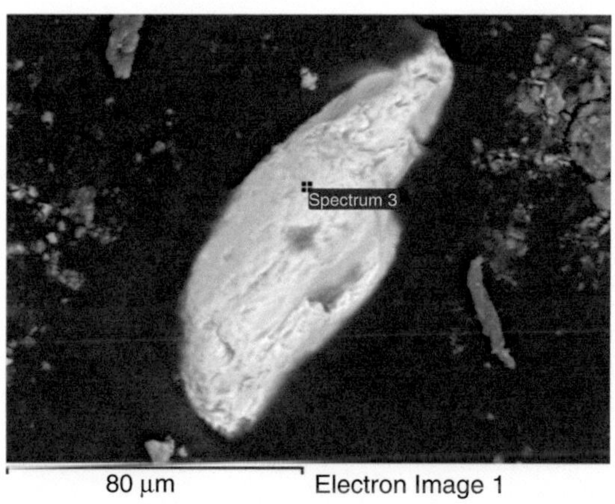

Spectrum 3		
Element	Weight%	Atomic%
Gd	1.40	1.69
Dy	3.59	4.20
Os	31.54	31.46
Ir	63.46	62.65
Totals	100.00	

Fig. 8.45 Nevyanskite from breccia of the Andreevskiy mud volcano

E.F. Shnyukov, E.V. Sokol, and E.N. Nigmatulina (Shnyukov et al. 2013b).

In addition to the Korolyovskiy mud volcano, gold particles were also discovered in breccia of the Tishchenko and Western Tsymbaly mud volcanoes. These particles are quite different in appearance and most probably in their origin as well.

In breccia from the Western Tsymbaly mud volcano, gold is represented by gold-bearing sand particles that are most likely of terrigenous origin (Fig. 8.47).

Spectrum 1		
Element	Weight%	Atomic%
Fe	1.51	3.55
Y	36.39	53.62
Dy	4.15	3.35
Ir	57.94	39.49
Totals	100.00	

Fig. 8.46 Yttrium-iridium from breccia of the Soldatsko-Slobodskoy mud volcano

Spectrum 1		
Element	Weight %	Atomic%
C	7.29	14.12
O	36.03	52.44
Al	1.05	0.90
Si	35.07	29.07
Fe	3.43	1.43
Au	17.14	2.03
Totals	100.00	

Fig. 8.47 Particle of gold-bearing quartz in a polymineral aggregate from breccia of the Western Tsymbaly mud volcano; the golden shade of the grains can be seen under a binocular microscope

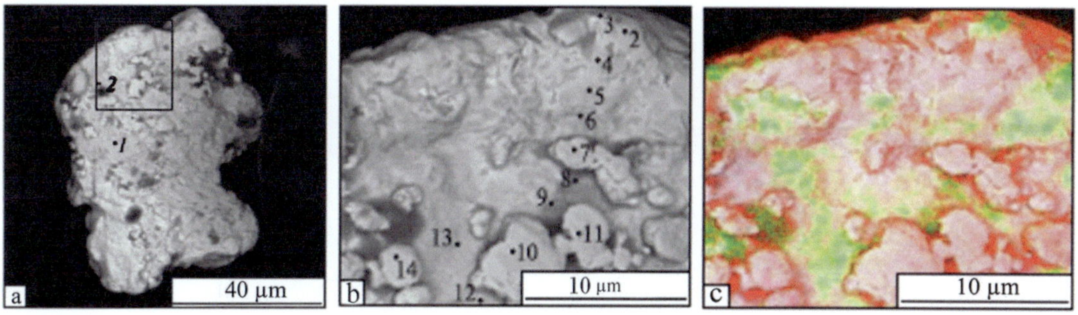

Fig. 8.48 Gold grain No. 1 from the Tishchenko mud volcano; (**a**) general view: 1 = Au-Ag (Ag 10 wt.%), 2 = Au-Ag (Ag 11 wt.%, Cu 1 wt.%); (**b**) magnified – numbers correspond to those in Table 8.3; (**c**) color map of the total fragment of the particle (Au = pink, Ag = green)

The most diverse gold particles have been discovered in breccia of the Tishchenko mud volcano, where they are concentrated in the size fraction 0.1–0.25 mm, and their sizes vary between 30 and 150 µm, mainly 60–80 µm. Massive and cloddy particles of irregular shape predominate. Crystals of pure gold are not found. Gold particles form thin-plated and other morphological types of aggregates indicating most probably a multistage formation process.

The gold particles studied to date do not bear visible traces of abrasion and transportation, nor do they reveal traces of leaching (e.g., pits and furrows from etching). Only one gold particle from breccia of the Korolyovskiy mud volcano is partially leached.

In the heavy fraction, gold is accompanied by newly formed crystals of siderite, rhodochrosite, baritocelestine, pyrite, ilmenite, rutile, monazite, opal, iron hydroxides, glauconite, gypsum, chlorite, as well as growths, single grains, and crystals of cinnabar.

The millesimal fineness of most gold particles is 800–950‰. The chemical composition of gold grains varies even within a single individual (Fig. 8.48, Table 8.3).

In gold grain No. 1, the outer rim contains 99.12‰ Au, the gold matrix contains 95.07 wt.% Au, and the overgrowth contains 99.36 wt.%. The content of Ag in the rim is 4.5 wt.%, usually 0.43–0.46 wt.%; in the matrix, it is 8.89 wt.%, more often 4–4.6 wt.%; in the growths, it is 2.83 wt.%, more often 0.00 wt.%. Cu is present only in the rim, measuring 0.83–0.90 wt.%; in the matrix and overgrowths, Cu is absent. In this gold particle, therefore, the rim and outgrowths are characterized by the highest levels of gold purity on the matrix surface. This can be interpreted as a result of the high dissolving capacity of the fluid leading to deposition on the surface of gold particles (0.5–5 µm) of the highest purity during the late generation and the removal of silver from the outer rim.

Additional photomicrographs show the measurements of gold grain No. 2, which represents a complex aggregate of small, fused gold particles with a higher (compared to gold grain No. 1) silver content (~20 wt.%) that remains constant within the grain (Fig. 8.49a–c, Table 8.3).

Gold grain No. 3 is also an aggregate of several small intergrown individuals containing up to 9–18% silver and about 1% copper (Fig. 8.50).

Gold grain No. 4 is a rounded particle containing up to 10% silver, 1% copper, and covered with outgrowths of pure gold (Fig. 8.51).

Gold grain No. 5 is a flattened plate and seems to have been broken into two parts. Along its outer border, it contains an average of 88.97% gold, up to 11.89% silver, and up to 0.98% copper. The center of the plate contains 79.83% gold and up to 19.70% silver (Fig. 8.52, Table 8.3).

Table 8.3 Spot analyzes of native gold grains extracted from the heavy fraction of breccia from the Tishchenko mud volcano (wt.%)

Points of measurements	Characteristics	Au	Ag	Cu	Total
Grain #1 (Fig. 8.48b)					
1–1	Total spectrum	94.90	4.50	0.60	100.00
1–2	Outer rim	98.66	0.43	0.90	100.00
1–3	Outer rim	98.71	0.46	0.83	100.00
1–4	Outer rim	100.00	0.00	0.00	100.00
Average millesimal fineness of gold in the rim, ‰		99.12			
1–5	Matrix	96.22	3.78	0.00	100.00
1–6	Matrix	92.64	7.36	0.00	100.00
1–8	Matrix	95.36	4.64	0.00	100.00
1–9	Matrix	100.00	0.00	0.00	100.00
1–13	Matrix	91.11	8.89	0.00	100.00
Average millesimal fineness of gold of early generation, ‰		95.07			
1–7	Outgrowth	97.17	2.83	0.00	100.00
1–10	Outgrowth	100.00	0.00	0.00	100.00
1–11	Outgrowth	100.00	0.00	0.00	100.00
1–12	Outgrowth	100.00	0.00	0.00	100.00
1–14	Outgrowth	99.64	0.36	0.00	100.00
Average millesimal fineness of gold of late generation, ‰		99.36			
Grain #2 (Fig. 8.49b)					
2–1	Total spectrum	80.04	19.31	0.65	100.00
2–2	Edge	73.61	25.70	0.70	100.00
2–6	Edge	83.62	16.38	0.00	100.00
2–7	Edge	80.08	19.92	0.00	100.00
2–8	Edge	80.24	19.76	0.00	100.00
Average millesimal fineness of gold in the edge, ‰		79.39			
2–3	Intermediate zone	76.99	23.01	0.00	100.00
2–4	Intermediate zone	80.40	19.60	0.00	100.00
2–5	Intermediate zone	83.32	16.68	0.00	100.00
Grain #5 (Fig. 8.52b)					
5–1	Total spectrum	87.25	11.89	0.86	100.00
5–2	Edge	87.98	11.17	0.85	100.00
5–3	Edge	88.67	11.33	0.00	100.00
5–4	Edge	88.58	11.42	0.00	100.00
5–5	Edge	91.53	8.47	0.00	100.00
5–6	Edge	89.82	9.20	0.98	100.00
Average millesimal fineness of gold in the edge, ‰		88.97			
5–7	Center	80.41	19.59	0.00	100.00
5–8	Center	79.24	19.70	1.06	100.00
Average millesimal fineness of gold of in the center, ‰		79.83			

As can be seen, the gold grains Nos 1–5 (all from the Tishchenko mud volcano) reveal approximately the same mineralogical processes, namely, infiltration of the outer rim by silver, and the appearance of rounded outgrowths of pure gold.

In gold grain No. 6 (Korolyovskiy mud volcano), half of the grain appears leached, and a more or less even distribution of silver and copper characterizes the entire grain (Fig. 8.53).

SEM investigations of the gold particles have enabled us to distinguish the sequencing of gold formation in mud volcanic breccia. The replacement of the earlier copper-silvery gold by later pure gold outgrowths on its surface is explained by E.V. Sokol (Shnyukov et al. 2013b) as a prod-

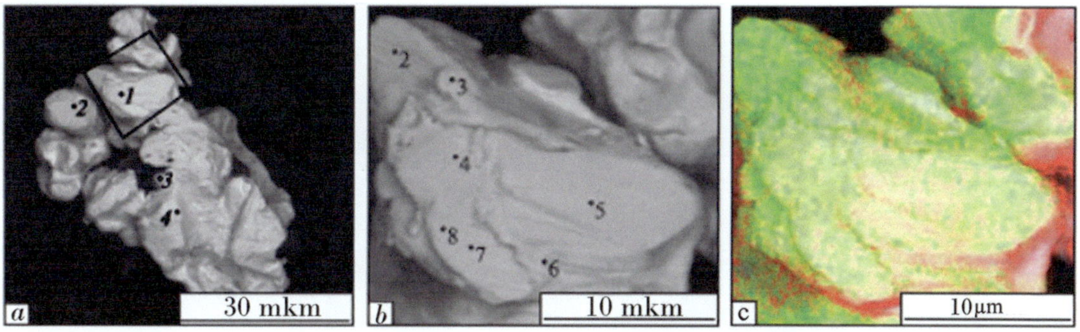

Fig. 8.49 Gold grain No. 2 from the Tishchenko mud volcano: (**a**) general view: 1, 2 = Au-Ag (Ag ~20 wt.%), 3 = quartz, 4 = Au-Ag (Ag ~10 wt.%, Cu ~1 wt.%); (**b**) measured points correspond to those in Table 8.3; (**c**) total color map of the area within the box in (a) (Au = pink, Ag = green)

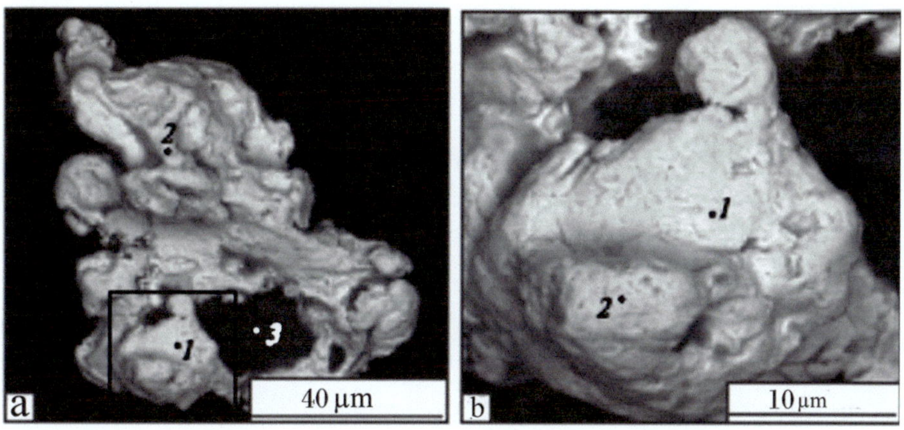

Fig. 8.50 Aggregate of intergrown individuals of gold grain No. 3 from the Tishchenko mud volcano: (**a**) general view: 1, 2 = Au, Ag (Ag ~9 wt.%, Cu ~1 wt.%), 3 = quartz; (**b**) fragment enriched with Ag: 1 = Ag ~10 wt.%, 2 = Ag ~18 wt.%

uct of gold purification by highly mineralized mud volcanic water followed by deposition of gold on the evaporation barrier due to destruction of colloids. How the original copper-silvery gold was formed is more complicated. Its aggregates are not found with typical mud volcanic neo-crystallizations (e.g., iron hydroxides, carbonates of Ca, Fe, Mn, Mg, barite, baritocelestine, and cinnabar). Instead, there are micro-inclusions of quartz and layered silicates. At the same time, there are no "pseudo facets" (braking surfaces) indicating the growth of crystals under the constraining conditions of the clay matrix. Given the low millesimal fineness of the gold grains, their chemical heterogeneity, and the presence of margins and rims, as well as the conclusions of Kalinin et al. (2009) and Hough et al. (2011), it is possible to link the generation of gold grains from mud volcanoes with endogenous processes.

One can assume either that gold particles from previous sources were reworked or that gold formation can be initiated in the depths of the mud volcanic system; in this particular case, the copper-silvery gold with impurities was removed on its way upward to the surface in the upper floors of a mud volcano (Shnyukov et al. 2013b).

The release of gold into mud volcanoes from an external source is unlikely. Ore deposits of alluvial gold located in the closest proximity to the studied mud volcanoes are within the Aktashskaya plain of the Kerch Peninsula. Gold mineralization is

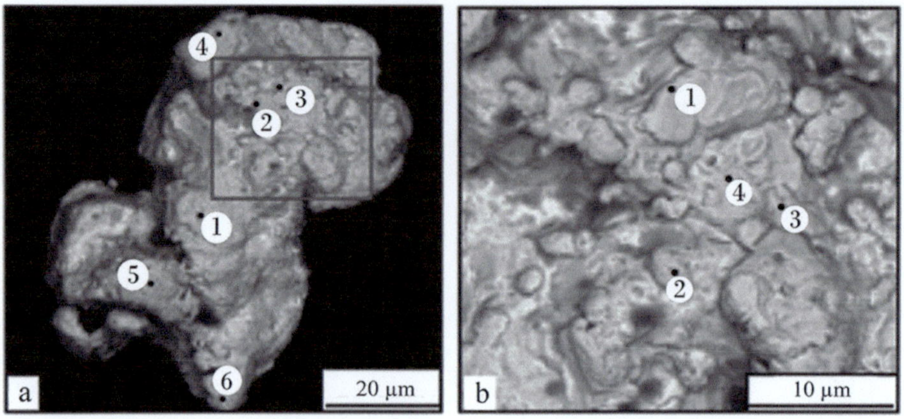

Fig. 8.51 Gold grain No. 4 from the Tishchenko mud volcano: (**a**) general view: 1–3, 5 = Au, 4 = Au-Ag (Ag ~10 wt.%, Cu ~1 wt.%), 6 = Au-Ag (Ag ~5 wt.%); (**b**) magnified fragment showing the rounded growths of pure Au (1, 2, 4) and Au with Ag ~1 wt.% (3)

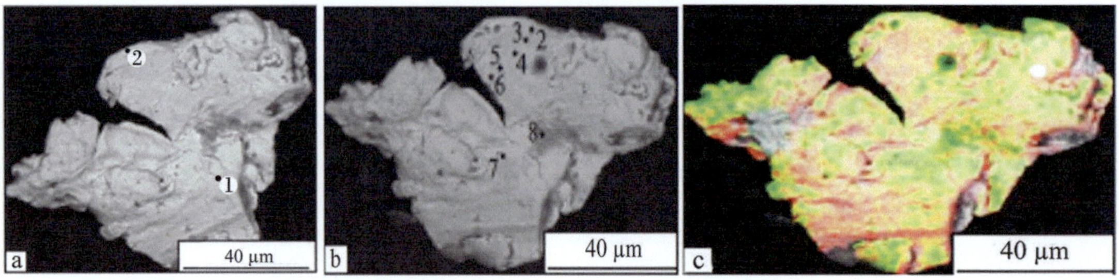

Fig. 8.52 Gold grain No. 5 from the Tishchenko mud volcano: (**a**) general view: 1 = Au-Ag (Ag ~1 wt.%), 2 = Au-Ag (Ag ~3 wt.%, Cu ~1 wt.%); (**b**) measurements corresponding to those in Table 8.3; (**c**) color map of the gold grain (Au = pink, Ag = green)

associated with redeposited Drevne- and Novoazovskiy (Holocene) alluvial deposits. The range of millesimal fineness of the Aktashskiy gold is much wider (680–990‰) compared to mud volcanic gold. Also, the composition of Aktashskiy gold contains up to 17% copper and numerous elements representing impurities, e.g., Hg, Bi, Fe, Pt, and Pd (Shnyukov et al. 2011). Clearly, the composition of mud volcanic gold, its morphology, and the micromorphology of gold minerals and mineral satellites all differ sharply from the Aktashskiy gold, which also bears numerous signs of mechanical impact during transport.

The mechanism of mobilization for dispersed gold with the formation of nano- and dust particles was considered by Hough et al. (2011). These authors noted that, in the active mud volcanic system formed by fluids, gas saturation and mineralization of water are high. Taking into account the abundance of iron hydroxides when gold occurs in situ, the restoration of Au^{3+} ions would be expected from thiosulfate, chloride, organic, and hydroxyl components, together with the formation of ferrite phases and hydroghetite. The association of gold with iron hydroxides has been established for numerous deposits in Mali, Ghana, and Brazil (Kalinin et al. 2009); however, a connection between iron hydroxides and neomineralizations with gold has not been identified so far in the Kerch mud volcanoes.

In recent years, a number of authors have linked the mobilization and accumulation of gold nanoparticles in near-surface contexts with their transportation and deposition within zones con-

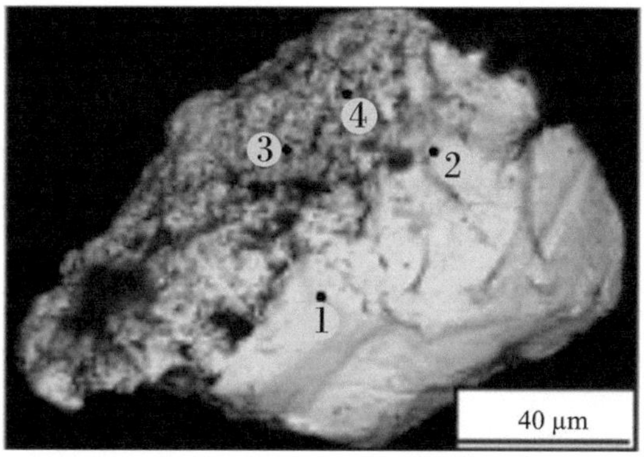

Fig. 8.53 Gold grain No. 6 from the Korolyovskiy mud volcano is characterized by a uniform distribution of Ag and Cu throughout the particle (1–4 = Au)

taining transit currents of gases (in particular, CO_2, CH_4, and sulfur gases) under conditions of oxygen deficiency (Aspandiar et al. 2006; Hough et al. 2011). This pathway to gold deposition along routes of reducing fluids rising to the Earth's surface is most likely what is happening in our case.

In the opinion of Kropotkin and Valyaev (1980), mud volcanoes are just locations of deep hydrocarbon degassing. Letnikov et al. (2010) believe that high-carbon fluids from deep lithospheric block systems are often associated with reduced hydrogen-sulfur fluids, on the basis of which clusters and segregations of noble (gold, silver, platinum) and other metals (e.g., Cu, Pb), carbides (e.g., moissanite, SiC), and sulfides are formed.

Moissanite was repeatedly encountered within the Korolyovskiy mud volcano in the form of small, isolated grains of blue color.

Copper minerals are rare in breccias of the Kerch-Taman mud volcanoes. More interesting are the findings of a supposedly natural alloy of copper and nickel with a probable admixture of pyrrhotite in breccia of the Western Tsymbaly mud volcano (Fig. 8.40b).

Brushes of small lamellar sulfide crystals are observed on the platey finds of native nickel (Fig. 8.40a), and these could be assumed to be covellite (Cu_2S), though their formula $(Cu,Ni,Fe)_{1,12}S$ indicates they are somewhat different.

Approximately the same composition of copper-nickel sulfides (Cu,Ni)S is also found in breccia of the Gora Gnilaya mud volcano; however, they are somewhat different in appearance and do not form the characteristic tabular crystals.

Native **lead** has been found on both the Taman and Kerch peninsulas. It forms irregular aggregations of small crystals dispersed throughout the clay. In the Western Tsymbaly mud volcano, the lead is chemically almost pure. The only admixture is Fe, which is present in the amount of 1.83% (Fig. 8.54).

Native lead found in the Burashskiy mud volcano is similar in composition but is more complex than that from the Western Tsymbaly mud volcano. In contrast, it contains impurities of Sb (3.43%) and Fe (2.22%) and is partially oxidized. Clay material also appears as an admixture (Fig. 8.55).

Native lead has been found in clay fragments of the Dzhau-Tepe mud volcano along with massicot, its oxidation product, as well as an admixture of Fe (2.07%) and some clay material (Fig. 8.56).

Vernadskiy (1913) noted the frequent transition of lead to massicot, which is also observed in our case.

Another lead mineral that is compositionally very complex has been encountered in breccia of the Western Tsymbaly mud volcano, but it is likely to be a mixture of several minerals. It is not described here because it requires further study.

Intermetallic compounds are also observed. Thus, in the Burashskiy mud volcano, an alloy with average content of Fe (8.36%), Ce (35.03%), and La (11.45%) has been discovered that forms hollow spherules with a diameter of 50 μm (Figs. 8.4 and 8.5). In the Dzhau-Tepe mud volcano, an even more complex alloy with average content of Ce (38.67%), La (15.67%), Nd (6.57%), and Fe (1.69%) was identified. This alloy also forms hollow spherules of 50 μm diameter (Fig. 8.3). The very small dimensions of these intermetallic compounds do not allow a precise mineralogical diagnosis.

A special place is occupied by the intermetallic compounds consisting, most likely, of initially native tin (56%), copper (13.5%), and rare earth elements that form spherules (Fig. 8.6). They may have originally been composed of a type of

Spectrum 3		
Element	Weight %	Atomic %
Fe	1.83	6.46
Pb	98.17	93.54
Totals	100.00	

Fig. 8.54 Metallic lead from breccia of the Western Tsymbaly mud volcano

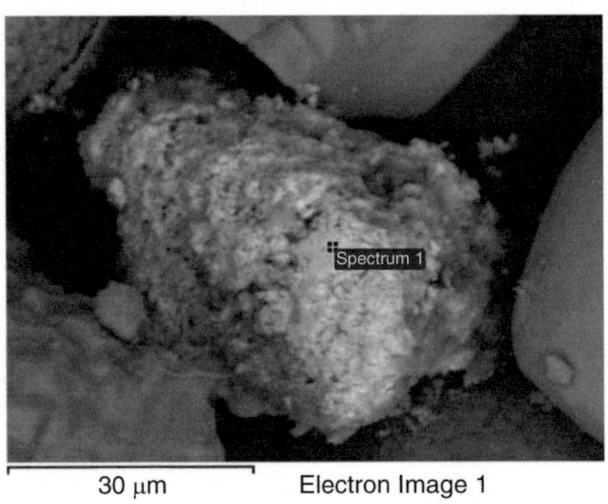

Spectrum 1				
Element	Weight%	Atomic%	Compound %	Formula
Fe	2.22	3.97	3.17	Fe_2O_3
Sb	3.43	2.82	4.11	Sb_2O_3
Pb	86.07	41.51	92.72	PbO
O	8.28	51.70		
Totals	100.00			

Fig. 8.55 Native lead with an admixture of Sb and Fe from breccia of the Burashskiy mud volcano

natural bronze alloy that was later oxidized. They have been found in breccia of the Dzhau-Tepe mud volcano.

Graphite was first identified in 1982 in breccia of the Korolyovskiy mud volcano. It was represented by small, unrolled, angular particles without any sign of transport in an aqueous environment (Sobolevskiy et al. 1982).

In 1969, Deryagin et al. (1969) reported on successful experiments involving the synthesis of diamond from methane under low pressure and high temperature. According to Portnov (1984,

Spectrum 1		
Element	Weight%	Atomic%
O	12.82	64.14
Fe	2.07	2.97
Pb	85.11	32.89
Totals	100.00	

Fig. 8.56 Native lead and its oxidation product, massicot, from breccia of the Dzhau-Tepe mud volcano

2015), these experiments confirmed the possibility that diamonds could crystallize from hydrogen-methane fluid. Kaminskiy et al. (1985) writes about the polygenetic origin of diamonds based on a number of patents in France and the USA for obtaining diamonds under certain atmospheric pressures.

Nikolskiy (1981) shows through calculations the possibility that diamonds can crystallize under partial oxidation of methane over a wide range of temperatures, pressures, and concentrations. Portnov (1984) drew attention to the emergence of two modifications of native carbon—diamond and graphite—during the process of mantle fluid diapirism with the accompaniment of burning hydride fluid. The constant presence of boron and nitrogen indicates that the diamonds crystallized from gas. Analysis of gas inclusions in diamonds also shows the presence of CO_2, H_2, CH_4, as well as the native minerals iron, nickel, chromium, silver, and iron, in addition to nickel sulfides.

The idea of an invasion of deep fluids causing the development of mud volcanoes enables one to suggest the possibility of diamond or nanodiamond formation during the course of the mud volcanic process, where many native minerals, boron, and rare earths are constantly present.

Sulfide mineralization within the Kerch-Taman mud volcanoes is relatively poor. It consists of ubiquitous pyrite and marcasite, rare pyrrhotite, rare galena, sphalerite, copper and nickel sulfide, cinnabar, realgar, orpiment, and antimonite.

Clearly pronounced pyrrhotite is found in breccia of the Gora Gnilaya mud volcano, where it forms peculiar excrescences up to 0.8 mm in size. The Fe/S ratio corresponds closely to the theoretical formula, while aluminum and silica represent an admixture within the clay matrix (Fig. 8.57).

A complex skeletal crystal of pyrrhotite was also discovered. The tops of the crystal are crowned by small pyrite subcrystals (Fig. 8.58).

Galena in the form of insignificant exudations is present together with native lead.

Sphalerite was first described in breccias of the Malotarkhanskiy, Novoselovskiy, Dzhau-Tepe, Illichivskiy, and Vladislavovskiy (Kerch Peninsula) and Gladkovskiy (Taman Peninsula) mud volcanoes. Sphalerite forms irregular grains, acute fragments, and less often tetrahedral crystals with sizes varying between 0.01 and 1 mm. The color of the sphalerite is yellow and brown with a strong adamantine luster. The mineral is

Spectrum 1		
Element	Weight%	Atomic%
Al	2.68	4.22
Si	5.06	7.64
S	31.89	42.23
Fe	60.38	45.91
Totals	100.00	

Fig. 8.57 Excrescence of pyrrhotite with illinition of clay minerals from breccia of the Gora Gnilaya mud volcano

fragile and has a hardness of 3–4. Under the microscope, it is isotropic. A Debye powder diagram coincides with the standard (Shnyukov et al. 1971).

Studies in recent years revealed sphalerite in two more mud volcanoes—Andrusov and Kayaly-Sartskiy. Microprobe analysis showed that this sphalerite is represented by small well-formed crystals sometimes corroded or twins of crystals with a size of 0.1 × 0.2 mm (Fig. 8.59).

In terms of their chemical composition, these crystals are represented by practically pure sulfides of zinc, and in one case, with a small admixture of iron. One of the samples is an aggregate with a crystal of baritocelestine. Most likely, this is a fragment of the vein secretion (Fig. 8.60).

The idiomorphic nature of the crystals and the fragility of their aggregates make for unambiguous evidence of the growth of sphalerite in mud volcanic breccia.

Cinnabar is found in most mud volcanoes of the Kerch-Taman region. It was first identified by Morozov in the Bulganakskiy, Tarkhanskiy, and Kayaly-Sartskiy mud volcanoes and was studied in detail during the twentieth century (Morozov 1965; Karasik and Morozov 1966). It was initially discovered in mud volcanoes of the northwestern Caucasus by Yakovleva (personal communication). These volcanoes are confined to, and thus controlled by, the Bezepskiy and other faults (Poyarkov 1967).

In breccia of the Boris and Gleb mud volcano (Taman), Shishkunov and Trufanov (1974) identified two genetic types of cinnabar. One of them is represented by rounded clastogenic cinnabar in sands of Kuyalnitskian sediments. The other is represented by bright scarlet, non-rounded crystals that, in the authors' opinion, have clear mud volcanic origins. In addition to these mud volcanoes, we observed cinnabar in the Korolyovskiy mud volcano, where its content reached 18 grains per 100 kg of breccia, as well as the Baksinskiy, Novoselovskiy, Dzhardzhava, and Dzhau-Tepe mud volcanoes, where the cinnabar content was also elevated in places (Figs. 8.61, 8.62, 8.63, and 8.64).

In the Caucasus, it was possible to confirm the presence of cinnabar in the Gladkovskiy and Gora Gnilaya mud volcanoes as well as elsewhere.

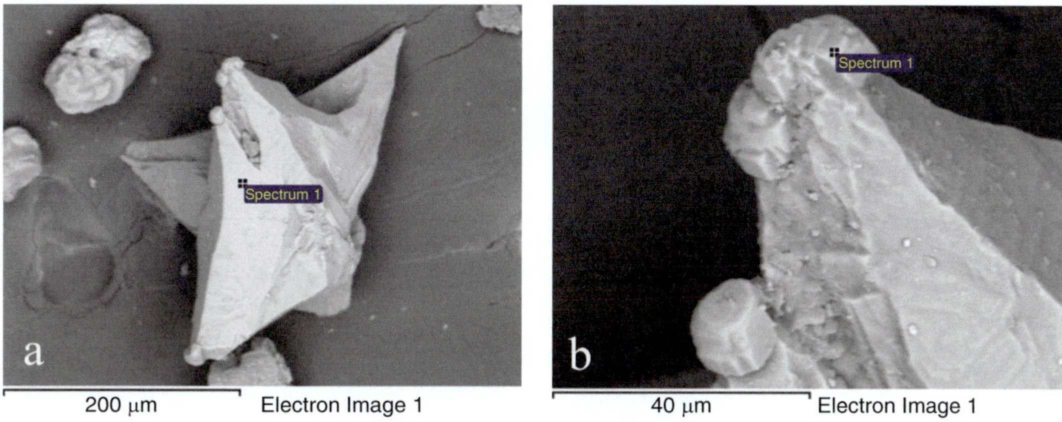

Spectrum 1			Spectrum 1		
Element	Weight%	Atomic%	Element	Weight%	Atomic%
S	34.76	48.14	S	58.86	71.36
Fe	65.24	51.86	Fe	41.14	28.64
Totals	100.00		Totals	100.00	

Fig. 8.58 Skeletal crystal of pyrrhotite from breccia of the Gora Gnilaya mud volcano: (**a**) general view; (**b**) the vertices of the skeleton are directed along the fourth-order axes of symmetry and end with growths of small spheroidal crystals of pyrite

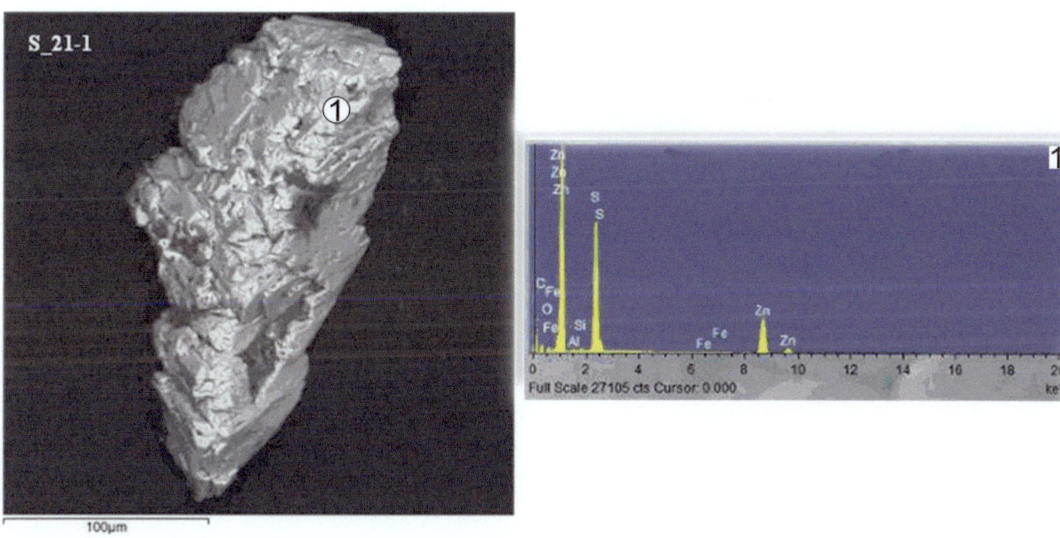

Fig. 8.59 Sphalerite from breccia of the Andrusov mud volcano plus powder diffractometry graph

Cinnabar grains in breccia from studied mud volcanoes display a visual appearance that is typical for this mineral: a red color, uneven or rough surface, glassy or diamond luster, and perfect cleavage. They appear as fragments of small, well-formed rhombohedral crystals (0.0–0.4 mm) or their growths. In one case, a crystal fragment of 0.5 mm was also observed. The smaller the individual, the more pronounced its crystallographic habit. In some mud volcanoes, e.g., Korolyovskiy, the crystals are heavily corroded (Figs. 8.65 and 8.66).

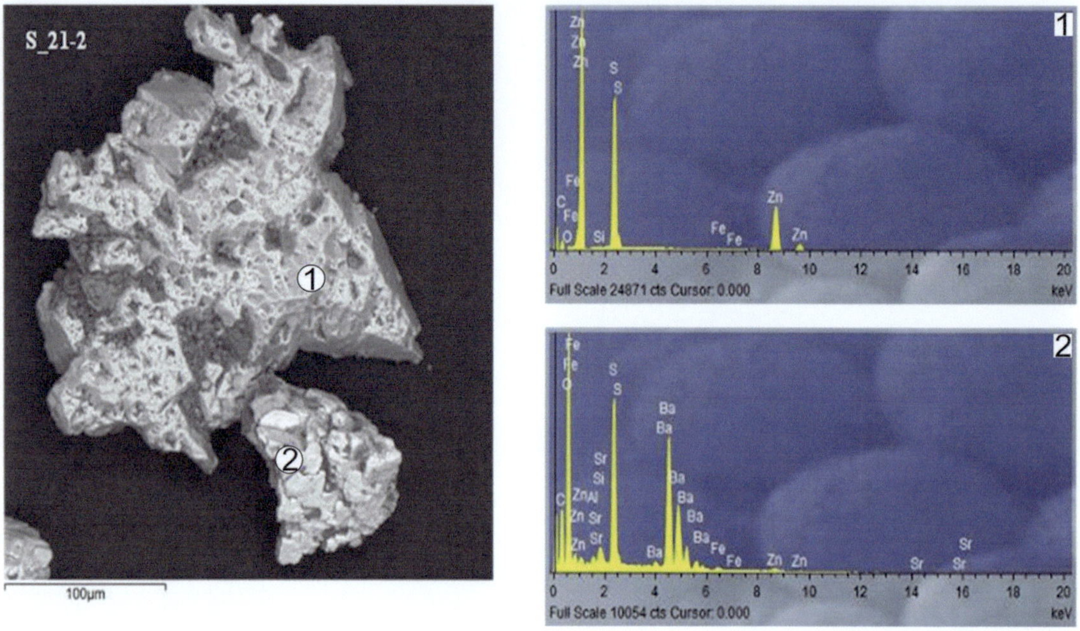

Fig. 8.60 Aggregate of sphalerite (1) with crystal of baritocelestine (2) from breccia of the Kayaly-Sartskiy mud volcano

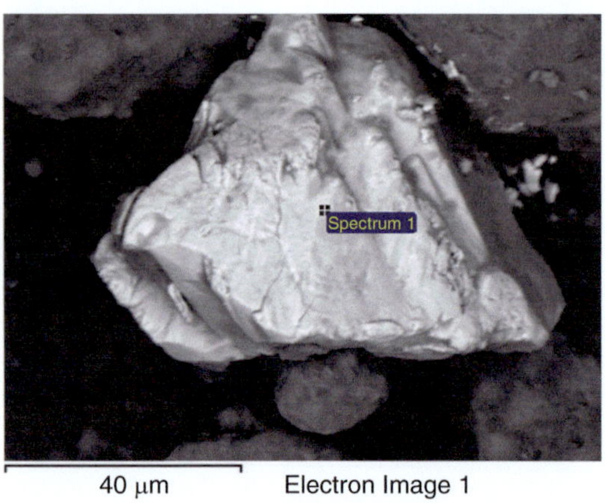

Spectrum 1		
Element	Weight %	Atomic %
S	15.20	50.13
Fe	2.71	5.13
Sb	4.32	3.75
Hg	77.77	40.99
Totals	100.00	

Fig. 8.61 Cinnabar from breccia of the Dzhau-Tepe mud volcano

Cinnabar from breccia of the Korolyovskiy mud volcano contains 77.77% mercury, 4.32% antimony, 2.71% iron, and 15.20% sulfur. Its calculated formula corresponds completely to that of cinnabar in breccia from the Dzhau-Tepe mud volcano (Fig. 8.61).

In other cases, an excess of sulfur over mercury and other admixtures is recorded more often. An admixture of iron up to 7.4%, but usually 1–2%, is always present.

Cinnabar was also found in mud volcanic breccias from the North Caucasus (e.g., Gladkovskiy) as well as some others.

Realgar in the iron ores of recessed synclines has been studied in detail (e.g., Shnyukov 1963; Shnyukov and Naumenko 1966; Shnyukov et al. 1971). It is often present together with iron ores in the layers of mud volcanic breccia. In the Rep'yevskaya recessed syncline, veins of realgar are represented by crystals elongated up to 7 mm, sometimes intergrown with each other. The crystals have the bright orange color that is typical for

Spectrum 1		
Element	Weight %	Atomic %
S	15.18	52.11
Fe	0.93	1.84
Hg	83.89	46.05
Totals	100.00	

Fig. 8.62 Growths of small rhombohedral cinnabar crystals from breccia of the Dzhau-Tepe mud volcano

Spectrum 3		
Element	Weight%	Atomic%
S	12.89	46.36
Fe	2.40	4.96
Hg	84.70	48.68
Totals	100.00	

Fig. 8.63 Cinnabar from breccia of the Dzhau-Tepe mud volcano

this mineral. More often, however, realgar is present together with orpiment-forming sooty excretions. There are almost no admixtures. The composition of realgar in mud volcanic breccia corresponds to the theoretical one.

A fouling of pyrite by realgar was observed in the Achinskaya recessed syncline. Realgar by itself was described in the Korolyovskaya and Gornostaevskaya recessed synclines (Chirvinskiy 1904–1905; Popov 1910), where the presence of iron ore is not proven. The fragility of realgar crystals excludes any possibility of their having been transported. This explains their absence in grain-size fractions of mud volcanic breccia.

The first discovery of **antimonite** (Sb_2S_3) in breccia from the Dzhau-Tepe mud volcano is of great interest (Fig. 8.67).

Antimonite is represented by a non-rounded fragment (150 × 50 μm) of a crystal. Its average chemical composition (based on three analyses) is Sb (74.74%) and S (25.26%) compared to the reference values of Sb (71.4%) and S (28.6%).

According to all mineralogical guidelines, antimonite is a low-temperature hydrothermal mineral, often paragenetic with cinnabar.

As a general feature of the sulfide mineralization within Kerch-Taman mud volcanoes, the prevailing development of complex low-temperature hydrothermal sulfides of mercury, arsenic, and antimony is notable.

Oxides are not rare in mud volcanic breccia. Many native elements yield minerals of this class during oxidation (e.g., massicot, Fig. 8.56). Findings of high-temperature oxides formed under conditions of considerable pressure are interesting. They often retain crystallographic forms, for example, a ferruginous chromite with no rounded octahedral crystals. Magnetite seems to display neo-crystallization, as its crystals appear to be immersed in a clay matrix. Chromite and magnetite crystals are not rounded; there is no evidence of their terrigenous origin; however, there is no evidence of their fluidogenic origin either.

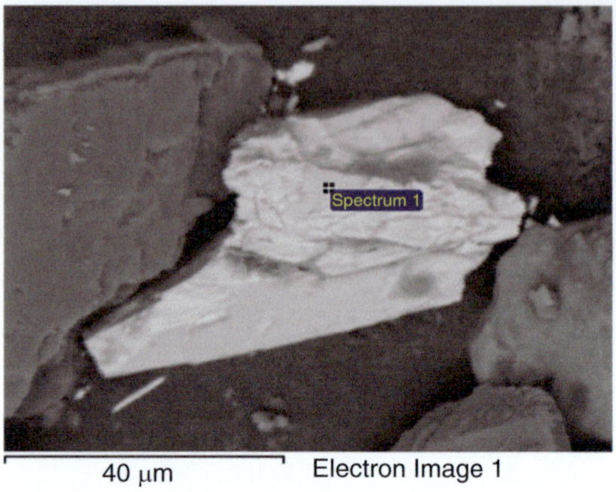

Fig. 8.64 Cinnabar from breccia of the Dzhau-Tepe mud volcano

Spectrum 1		
Element	Weight%	Atomic%
S	11.81	44.59
Fe	1.39	3.02
Hg	86.80	52.39
Totals	100.00	

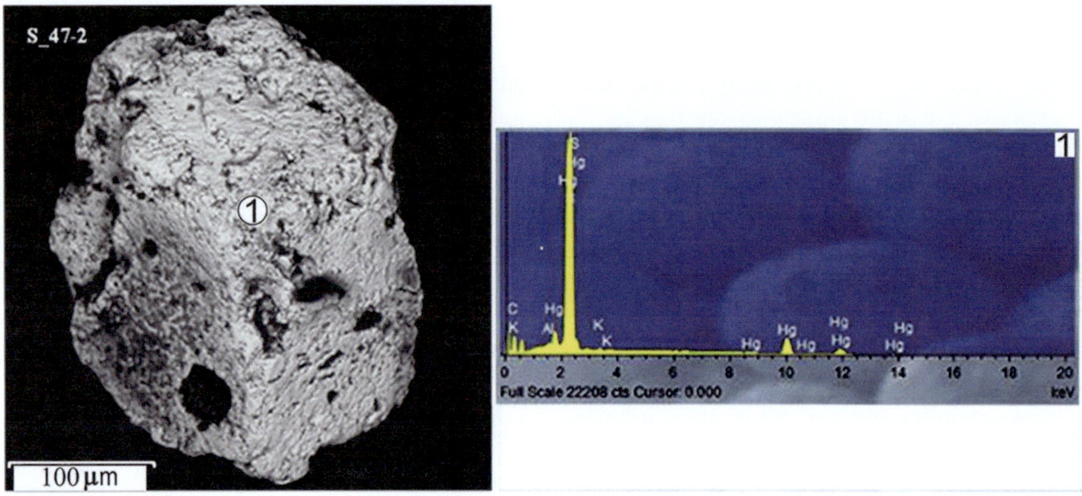

Fig. 8.65 Corroded grain of cinnabar from breccia of the Korolyovskiy mud volcano

An unusual mineral containing lead, iron, and arsenic was observed in breccia of the Dzhau-Tepe mud volcano (Fig. 8.68).

These concretions comprise fine-grained microscopic aggregates with sizes up to 50 µm each, and this form makes them difficult to study. By composition, this mineral resembles carminite $PbFe_2(AsO_4)_2(OH)_2$ and to a lesser degree beudantite $PbFe(AsO_4)(SO_4)(OH)_6$, but in contrast to beudantite, there is no (SO_4).

Particularly interesting among the oxides is the presence of minerals that are crystallized under conditions of high temperatures and pressures yet quite often retain their crystallographic appearance—e.g., magnetite, wüstite, and massicot (mentioned above).

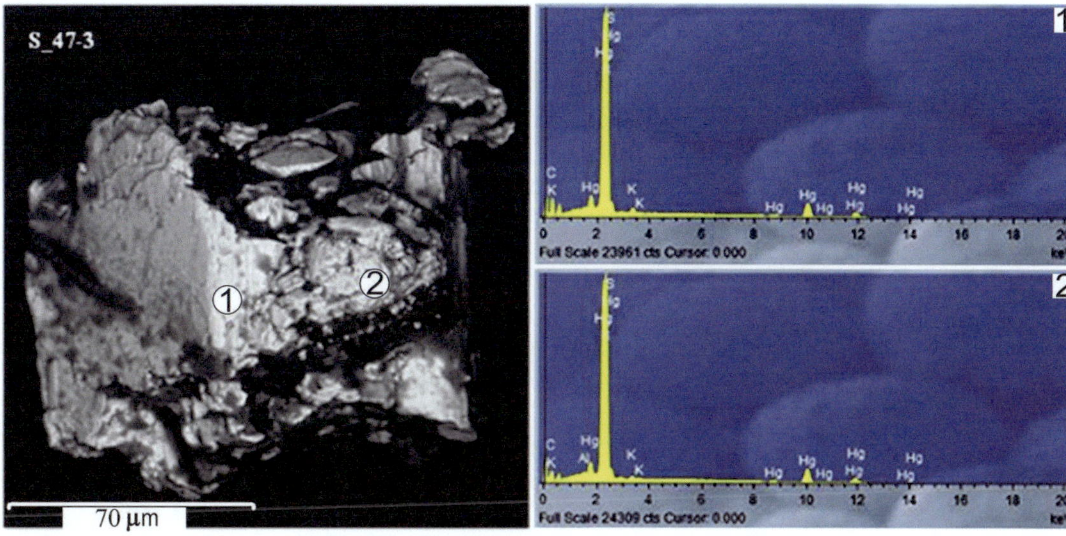

Fig. 8.66 Corroded grain of cinnabar from breccia of the Korolyovskiy mud volcano

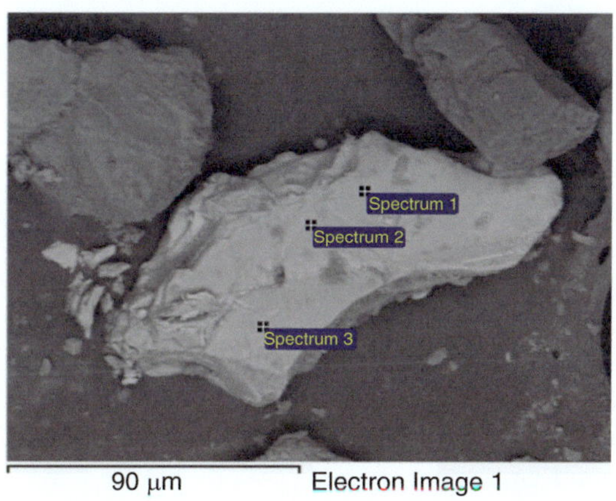

Spectrum	In stats.	S	Sb	Total
Spectrum 1	Yes	26.65	73.35	100.00
Spectrum 2	Yes	23.29	76.71	100.00
Spectrum 3	Yes	25.84	74.16	100.00
Mean		25.26	74.74	100.00
Std. deviation		1.76	1.76	
Max.		26.65	76.71	
Min.		23.29	73.35	

Fig. 8.67 Antimonite from breccia of the Dzhau-Tepe mud volcano

Along with iron oxides, manganese oxides are also encountered in the oxidation zone. In particular, psilomelane makes up small segregations in breccia of the Soldatsko-Slobodskoy (Nikitskiy) mud volcano.

Carbonates are widely distributed in breccias of the Kerch-Taman mud volcanoes. They are mostly represented by clastic material of sideritic Maikopian nodules as well as fragments of carbonate rocks, shells, and calcite precipitates.

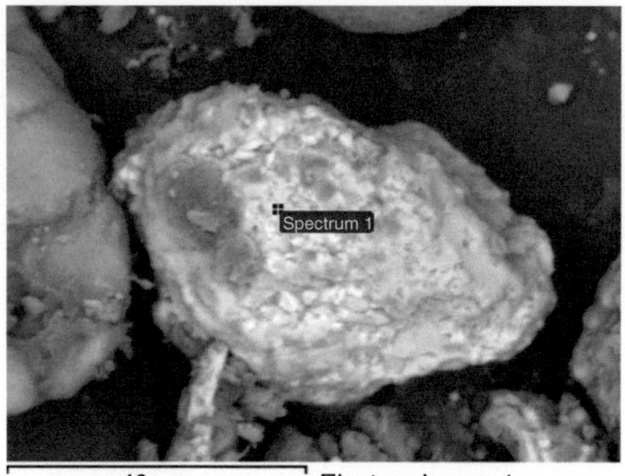

Element	Weight%	Atomic%	Compound%	Formula
Si	1.32	2.06	2.82	SiO_2
Ca	1.07	1.17	1.49	CaO
Fe	30.16	23.78	43.13	Fe_2O_3
As	11.89	6.98	15.69	As_2O_3
Pb	34.23	7.27	36.87	PbO
O	21.34	58.72		
Totals	100.00			

Fig. 8.68 Unusual mineral of lead, iron, and arsenic from breccia of the Dzhau-Tepe mud volcano

In mud volcanic breccias, however, carbonate neo-crystallizations are also found. In breccia of the Andrusov mud volcano, for example, peculiar sintered forms of siderite completely covering thc surface of the aggregate have been discovered (Fig. 8.69).

In some mud volcanoes (e.g., Trubetskoy), siderite forms druses composed of underdeveloped crystals. In other mud volcanoes (e.g., Korolyovskiy), siderite forms aggregates of mainly rhombohedral crystals (Fig. 8.70).

Sometimes siderite crystals seem to sit in clay matrix. In some cases, the carbonate corresponds to oligonite in terms of its composition. Crystals of dolomite and siderite differ somewhat, as can be seen in Fig. 8.71, where it was possible to observe their contact.

The dolomite contains a small admixture of iron. In some cases, the carbonate corresponds to ankerite (Fig. 8.72).

Dolomite recovered from breccia is as a rule slightly ferriferous (ankerite?), and it forms relatively large crystals up to 200 μm or their aggregates with well-defined cleavage. Such forms were observed in dozens of samples from different mud volcanoes (Figs. 8.73 and 8.74).

Rhodochrosite is a much rarer mineral in the Kerch-Taman mud volcanoes. It was found only once in breccia of the Korolyovskiy mud volcano. The mineral forms outgrowths with sizes up to 0.6 mm. These outgrowths are composed of very small crystals that most often cover clastic material from the breccia (Fig. 8.75).

There are also numerous and ubiquitous examples of the replacement of biological remains and terrigenous minerals by calcite.

Sulfates are minerals of the oxidation zone of mud volcanoes. Gypsum, which is rare in marine volcanoes, literally clogs the breccia of the Kerch-Taman mud volcanoes.

Barium sulfates are of much greater interest. Betekhtin (1950) noted that barite is a chemically resistant mineral, but like all minerals with good cleavage, it is easily susceptible to mechanical destruction and therefore rarely can be carried far.

This is an important clue that supports the formation of barite under conditions created by the

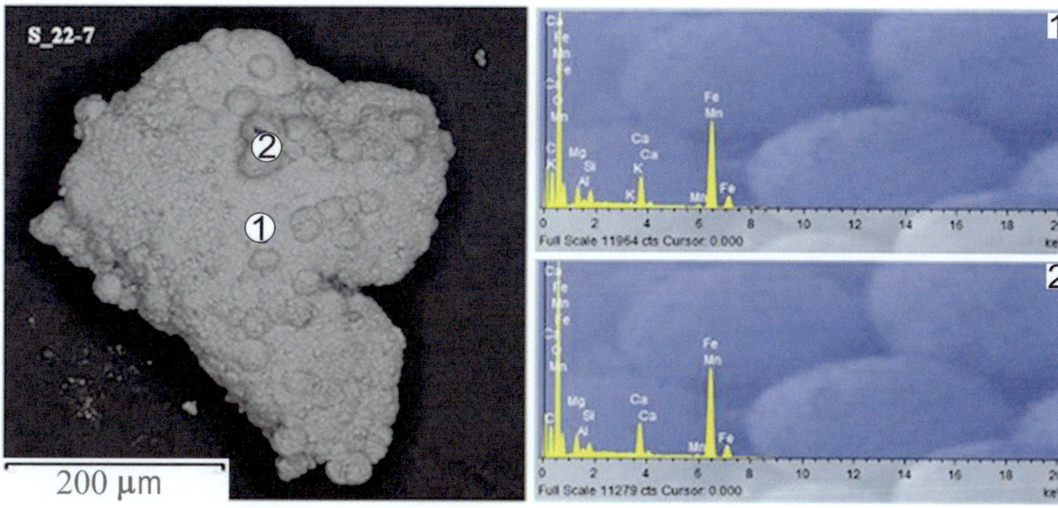

Fig. 8.69 Peculiar sintered forms of siderite from breccia of the Andrusov mud volcano

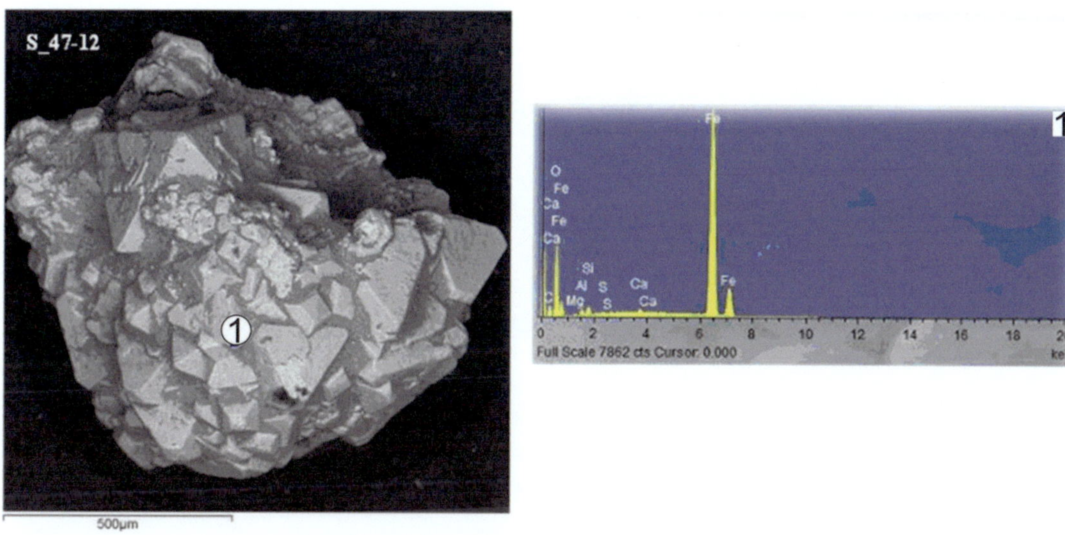

Fig. 8.70 Growths of rhombohedral crystals of siderite from breccia of the Korolyovskiy mud volcano

mud volcanic process. Undoubtedly, a kind of "barite flower," a small druse of crystals, which was discovered in breccia of the Trubetskoy mud volcano, could be formed only within a gas cavity in the breccia; even a loose breccia would not allow the development of fragile crystals (Fig. 8.76).

In most cases, however, barite forms homogeneous, often acute-angled aggregates (Fig. 8.77).

Baritocelestine occurs in intergrowths together with barite. In some places, baritocelestine is zonal, and sometimes there are inclusions of quartz. Perhaps, these are fragments of quartz-barite or quartz-baritocelestine veins (Fig. 8.78).

The presence of Cl or Br is often recorded in the formulas of many minerals. In the Korolyovskiy mud volcano, a growth of sylvite crystals on opal grains was observed. In general, many chlorides and other seasonal minerals (solid evaporite minerals that rapidly crystallize when aqueous solutions dry out on or near the

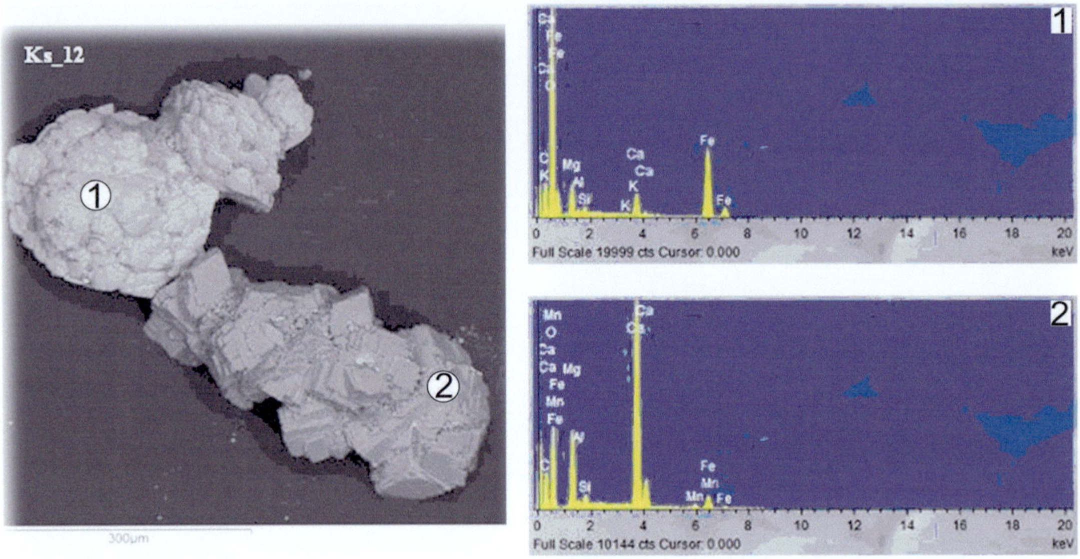

Fig. 8.71 Morphology of siderite (*1*) and dolomite (*2*) and their chemical differences (Korolyovskiy mud volcano)

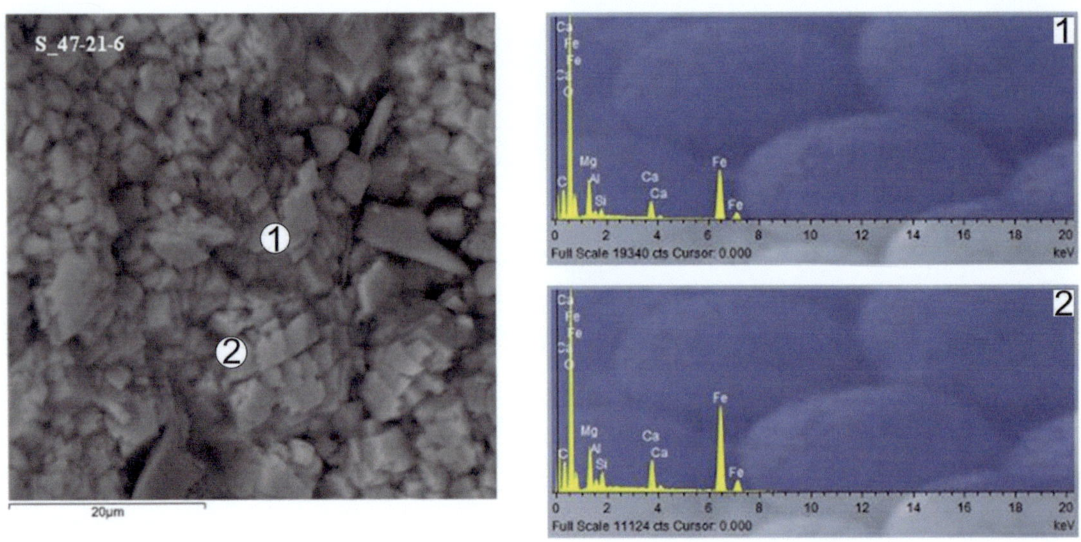

Fig. 8.72 Ankerite from breccia of the Trubetskoy (or Malotarkhanskiy) mud volcano

Earth's surface, then rapidly dissolve again when the surface is dampened) were studied in springs emanating from mud volcanoes by Deyak (2010).

Comparison of minerals discovered so far in breccias of the Kerch and Taman mud volcanoes shows that they are closely similar and reveal no fundamental differences.

8.7 Fluidogenic Mineralization of the North Caucasus Mud Volcanoes

A group of southeastern mud volcanoes located in the northern Caucasus (e.g., Semigorskiy, Shkolnyy, Gladkovskiy, and Shapsugskiy) is unique. Unlike all other mud volcanoes in the

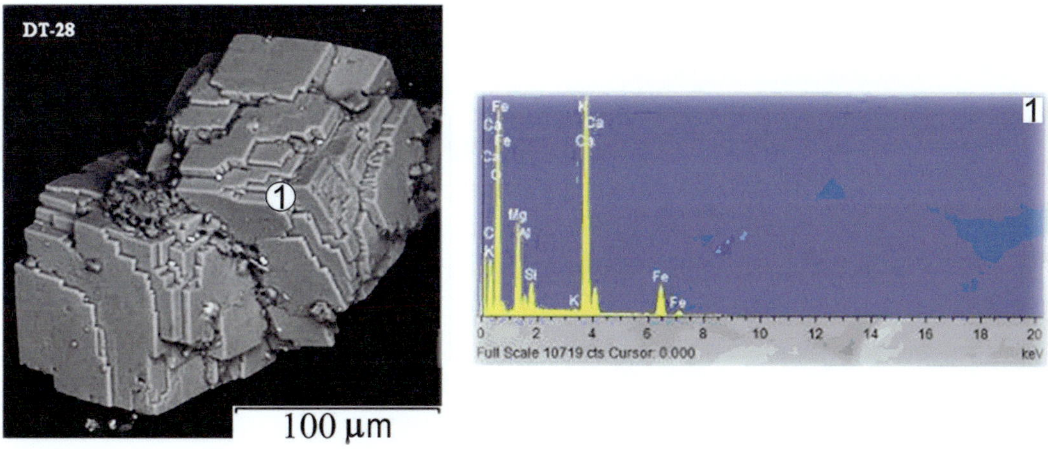

Fig. 8.73 Dolomite with an admixture of iron (ankerite?) from breccia of the Dzhau-Tepe mud volcano

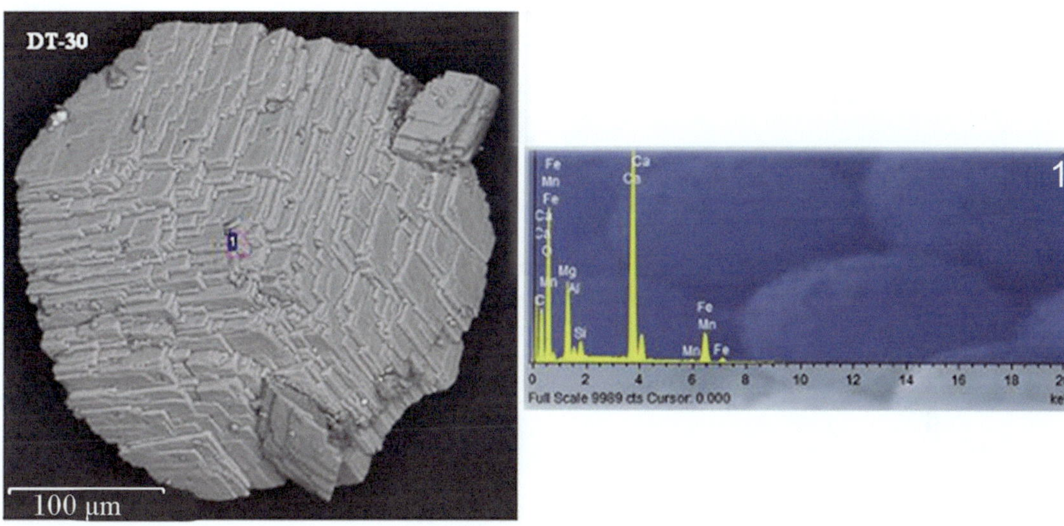

Fig. 8.74 Rhombohedral crystals of ferruginous dolomite from breccia of the Dzhau-Tepe mud volcano

region, they are developed on Lower Cretaceous deposits (Subbota 1964). Among them, the Semigorskiy mud volcano (Fig. 6.123) is the most typical for this kind of morphostructure. The Shkolnyy and Shapsugskiy mud volcanoes are not as large in size, and the Gladkovskiy mud volcano is fueled not only by Lower Cretaceous but perhaps by Maikopian breccia as well.

According to the data of Subbota (1964), the Semigorskiy mud volcano did not experience real eruptions. Its clastic material is composed of solid fragments of Lower Cretaceous gray marls and sandstones with numerous white efflorescences.

According to our observations, diversity in the composition of clastic material is much wider. Around its vent, the surface is covered by debris of dense, greenish-gray, fine-grained rocks, angular fragments of yellow and pink marls, and fragments of calcite veins (Fig. 8.79).

The fragments of greenish-gray rocks that were probably attributed by Subbota (1964) to sandstones are 1–3 cm in size and sometimes slightly rounded. They have a heterogenic structure, containing inclusions of white crys-

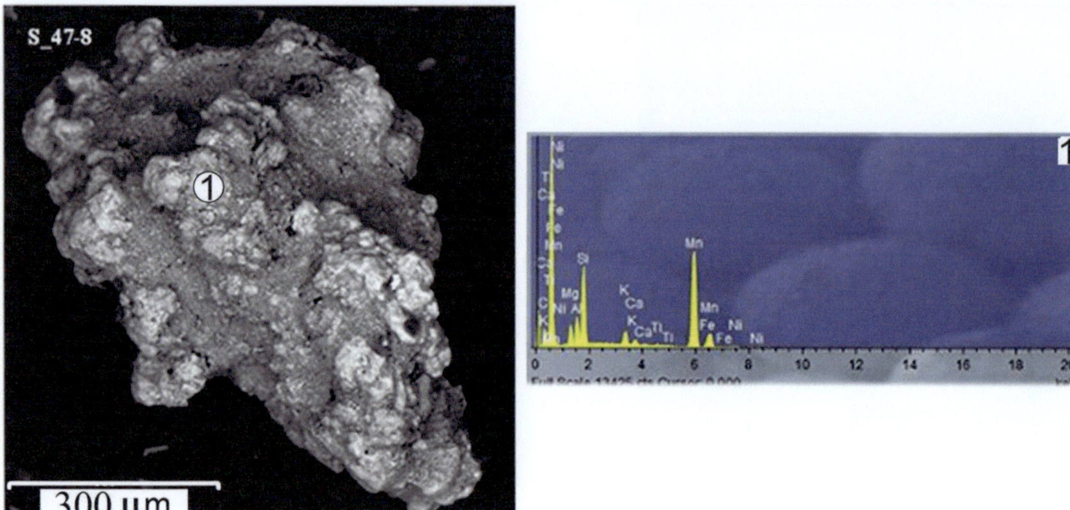

Fig. 8.75 Rhodochrosite from breccia of the Korolyovskiy mud volcano

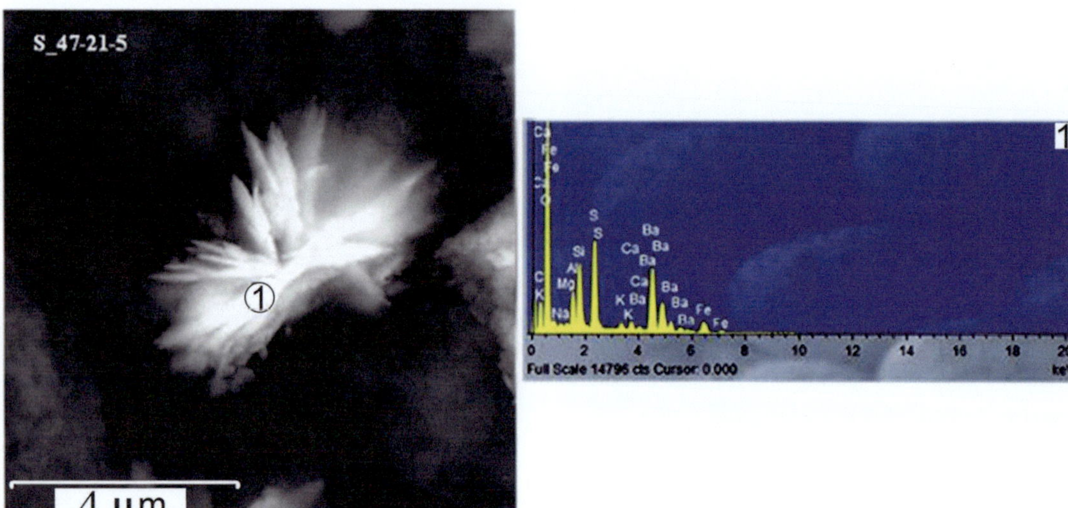

Fig. 8.76 Barite druse from breccia of the Trubetskoy mud volcano

talline aggregates in a fine-grained matrix. In most cases, the white inclusions are xenomorphic, possibly being altered crystallites of feldspar. In one of the samples, the white inclusions are represented by spherulites of a siliceous substance. The structure of the rock matrix bears obvious traces of metasomatic transformations, including also sulfide mineralization. As such, these fragments possibly result from the brecciation of rocks that underwent varying degrees of alteration (probably tuffs) in their fundamental composition. The determination of their geological origin requires more detailed studies.

The fragments of veined material are composed of calcite with sizes between 1 and 15 cm and show good preservation of their primary structure. Crystals of calcite enclose gaseous liquid inclusions that contain either a water solution (90–95%) or liquid hydrocarbons (Kalyuzhnyy

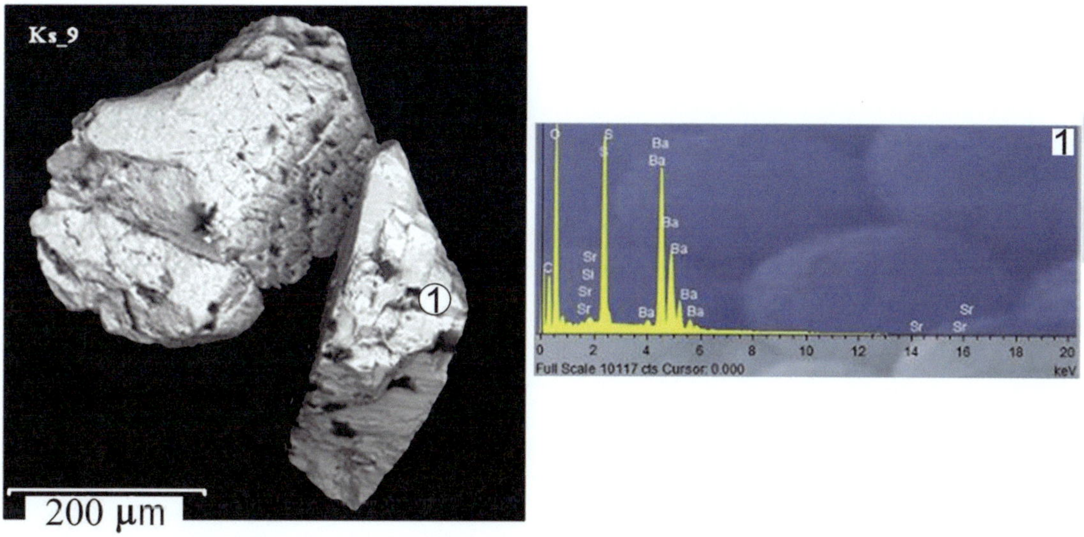

Fig. 8.77 Acute-angled aggregates of barite from breccia of the Kayaly-Sartskiy mud volcano

et al. 1984). Upon heating, a majority of the inclusions are depressurized at a temperature of 100–110 °C due to the presence within them of methane under high pressure. The rest of the inclusions are depressurized in the liquid phase (pH 7.0–7.2) at 100–130 °C, 150 °C, and rarely 200–210 °C.

The study of the inclusions shows that the calcite underwent two periods of formation (Naumko and Zankovich 2014): (1) crystallization in water-methane solutions at depths corresponding to the fluid regime of vein-imbedded mineralization in rocks enclosed within veins and (2) crystallization within the sedimentary matter during the course of its upward movement along the faults within which mud volcanoes are confined. It looks as if these inclusions were captured by calcite aggregates during intensification of mud volcanic processes when breccia saturated with a water-gas-pelite mixture was ejected onto the surface (Kalyuzhnyy et al. 1984; Naumko et al. 2008). The migration of methane together with water-pelitic volcanic breccia and calcite aggregates containing inclusions of methane speaks in favor of a genetic relationship between recent transitory canals and paleohydrocarbon systems (Naumko and Zankovich 2014).

According to Kalyuzhnyy et al. (1984), data obtained from inclusions suggest a hydrothermal origin for the calcite water-pelitic medium of mud volcanoes during conditions where separation (e.g., flows, bubbles) of liquid hydrocarbons and methane occurs. At the same time, the authors admit the possibility of further growth of calcite fragments trapped primarily by the breccia flow when crossing the sedimentary strata.

In our opinion, such calcite veins are formed during the late, relatively low-temperature phases of the mud volcanic process. At the same time, the possibility of rock fragment ejection, including veined material, from sediment horizons crossed by fluid clay flows on their way upward should be not excluded.

Earlier descriptions of the Semigorskiy mud volcano do not mention the observation of calcite veins. Therefore, our discoveries of large (up to 20 cm) fragments of pyrite-calcite veins are of particular interest. These fragments are represented by angular and irregularly shaped plates up to 3–5 cm thick, and they are composed of coarse-grained (1.5–2 cm), translucent to transparent, white calcite.

There are thin (0.3–1 mm) selvages on the host rock along the outer contacts of the veins, and only in single samples are there more massive (up to 8–10 mm) relics of such rocks (Fig. 8.80).

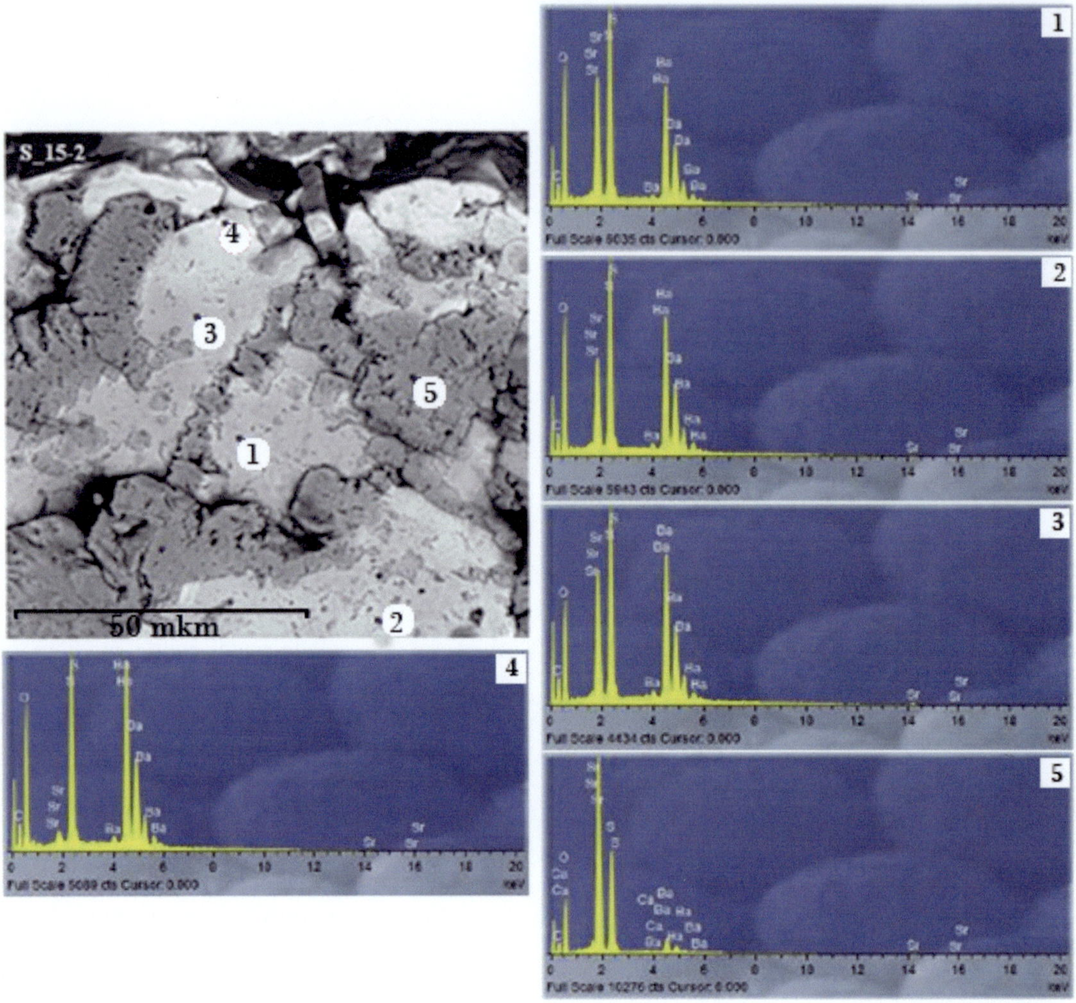

Fig. 8.78 Baritocelestine (*1*, *2*, *3*, *5*) and barite (*4*) from breccia of the Trubetskoy mud volcano with powder diffraction graphs of each of the five targets

Fig. 8.79 Breccia from the Semigorskiy mud volcano

In the crystalline mass of calcite that makes up veins, there are often signs of tectonization, which are expressed as:

1. Sharp changes in the thickness of veins caused most likely by the boudinage process, the possibility of which is indicated by the accumulation of voids in the carbonate matrix near the interboudinage pinches (Fig. 8.80a).
2. Development along the outer contacts of the veins of a series of thin subparallel cracks. These cracks can be traced using the dense aggregates of fine-grained gray calcite that form a so-called book structure along the contacts. Sometimes,

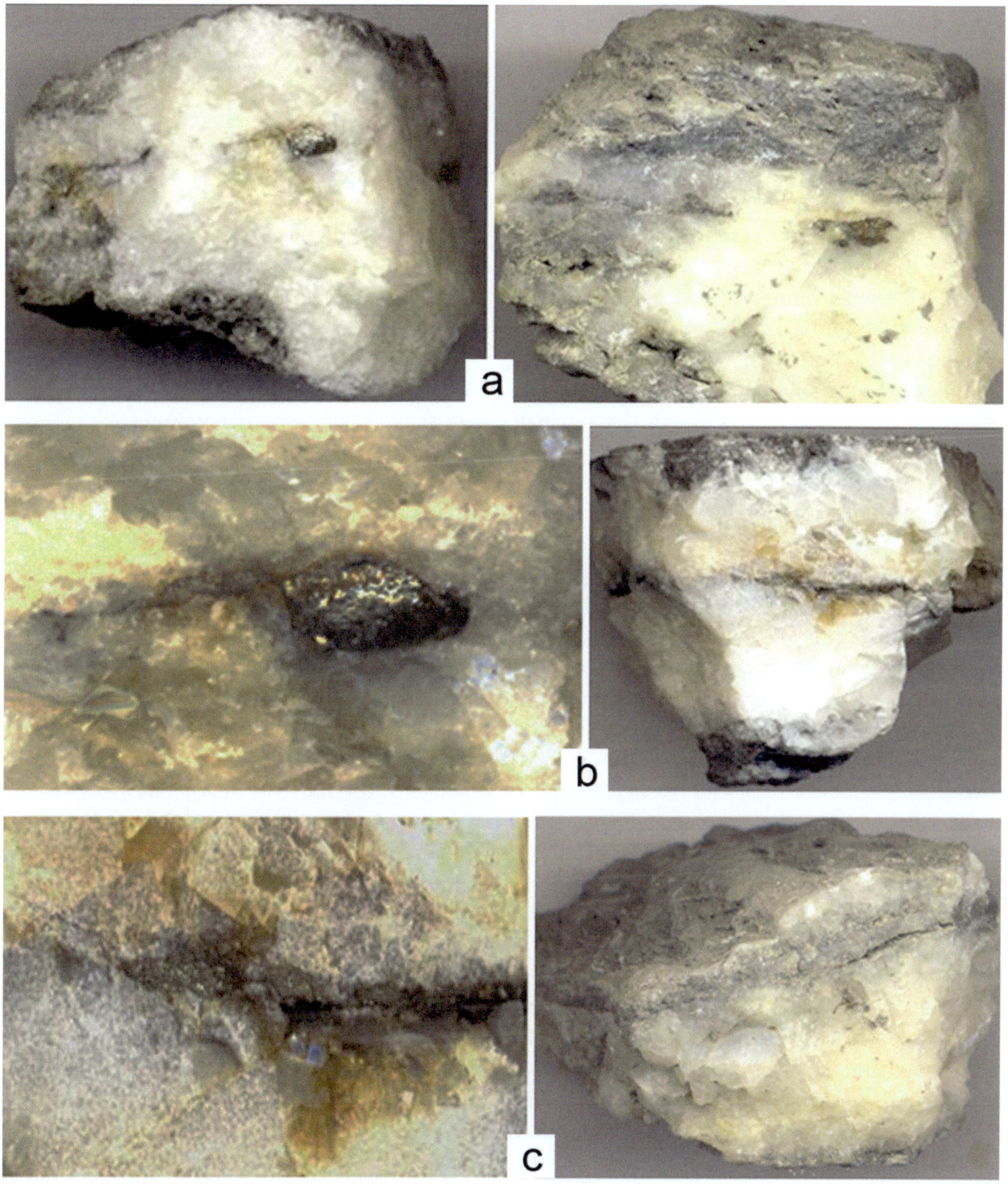

Fig. 8.80 Fragments of calcite veins from breccia of the Semigorskiy mud volcano: (**a**), an example showing enclosed rocks and fracture bands with gray, fine-grained carbonate as well as gaping cavities in the thickness of the carbonate veins (probably the result of boudinage); (**b**), (**c**), aggregates and individual crystals of pyrite developed along fissures in the central part of the veins

the cracks are not healed by fine-grained carbonate and remain gaping (Fig. 8.80b).

In many fragments along the central zone of the veins, there is a system of cracks running parallel to the contacts, and these cracks are filled with heterogeneous aggregates of pyrite (Fig. 8.80b, c). Sometimes these aggregates form small, individual nests near the contacts of the veins. In the main veinlets of sulfides, crystals of pyrite up to 3 mm in size predominate, and their morphology indicates they were formed under conditions of free growth. Electron microscopic study of small- and fine-grained (0.5–0.2 mm and less) sulfides isolated from the carbonate matrix of the veins also reveals the prevalence of idiomorphic and sub-idiomorphic crystals of pyrite exclusively. They form small crystals, and their intergrowths are most often of a cubic habit.

The chemical composition of the carbonate matrix of the veins is somewhat peculiar. The unusual nature of the composition of mineral associations within carbonate veins should be noted (Table 8.4).

In the monomineral fraction of coarse-grained carbonate, the main component is CaO; the mineral is calcite with a slightly elevated concentration of Mg. Insignificant impurities of SiO_2, Fe_2O_3, and Na_2O probably reflect the presence of fine-grained inclusions in the crystals, while MnO is an isomorphous impurity.

Despite the clearly hydrothermal appearance of calcite crystals, the search for gas-liquid inclusions has so far been unsuccessful. At the same time, the results of chemical analysis reveal an apparent excess of CaO in the crystal composition. This strongly suggests the presence of a mineral phase highly enriched with calcium (Table 8.5). Diagnostics of such a phase and elucidation of the formation conditions for such a mineral association require special studies.

The coarse-grained calcite is the "purest" in terms of element-impurity concentrations. The light gray carbonate crosscuttings along the contacts ("book texture") are somewhat enriched with zinc, while the gray carbonate of thin secant veins is characterized by slightly elevated concentrations of Ti, Cr, Cu, and Sc. Such a finding ties this type of carbonate together with the host rocks, for which elevated concentrations of Ti, Cr, Cu, Sc, Y, and P are typical (Tables 8.5, 8.6, 8.7, and 8.8).

The composition of elements and impurities in the pyrite monofraction is characterized by a distinctly high concentration of Cu-Ni and insignificant amounts of Co, Cr, and Ag. In general, such a composition of impurities is characteristic of many hydrothermal pyrites, including gold ore-bearing regions.

The high Ba content (<1%) indicates the presence of barite impurities in the sulfide concentrate, which also supports a hydrothermal genesis for sulfide mineralization. It is possible that calcite is represented by a manganous variety (rhodochrosite?). The presence of phosphates is also possible.

At the same time, the undoubtedly hydrothermal origin of the calcite of carbonate veins does not yet indicate a deep origin for the mineral-forming fluids. Some information on their origin can be obtained from the results of isotope-geochemical investigations. A preliminary study of the variations in isotopic composition of carbon and oxygen reveals that in calcite veins, the values of $\delta^{13}C$ and $\delta^{18}O$ for both coarse-grained and fine-grained gray calcite crosscuttings are practically identical (Table 8.8).

This suggests that the veins were formed by redeposition of carbonate from sedimentary rocks (Galimov 1967; Ustinov 1976). Enrichment of the ^{18}O isotope in fragments of fine-grained, altered, basic rocks indicates a different, though also not deep, oxygen isotope content in metasomatizing solutions.

Considering the role of deep gas-fluid flows in the geochemical paragenesis of mud volcanic breccias, we should again emphasize the above assumption that calcite veins can be either the product of the latest low-temperature phase of the gas-fluid flow or just fragments of rocks captured and raised to the surface during the upward migration of the clayey mass of sediments.

The mineralogy of this type of mud volcano is distinctive. Lead and zinc are characteristic of the Semigorskiy and Shkolnyy mud volcanoes, respectively. In addition, there are also native

Table 8.4 Elements and impurities in vein carbonates and host rocks of the Semigorskiy mud volcano (spectral analysis)

Samples	Concentration of chemical elements, g/t												
	Mn	Ni	Ti	V	Cr	Cu	Pb	Ag	Zn	Sc	Y	P	Ba
Host rocks	5000	6	1000	30	10	50	5	<1	–	20	50	>1%	–
Coarse-grained calcite of carbonate veins	>10,000	5	30	3	–	10	3	Trace	–	–	6	500	200
Calcite of "book structure"	>10,000	5	10	4	3	10	4	Trace	80	–	–	500	200
Veinlets of gray calcite	>10,000	5	100	5	10	30	4	Trace	–	10	10	600	300

Table 8.5 Chemical content (%) of carbonate within calcite veins from the Semigorskiy mud volcano (analyzed by V.V. Zagorodny)

Sample	SiO_2	TiO_2	Al_2O_3	Fe_2O_3	MnO	MgO	CaO	Na_2O	K_2O	P_2O_5	S	Cl	H_2O	Loss on ignition, %	Σ
Coarse-grained calcite	0.70	0.04	0.24	1.82	0.59	1.45	65.78	0.52	0.00	0.01	0.01	0.00	0.04	28.38	99.58

Table 8.6 Elements-impurities in the coarse-grained calcite of carbonate veins (X-ray fluorescence analysis performed by A.V. Andreev)[a]

	Concentration of chemical elements, g/t								
Sample	Fe	Mn	Ni	Mo	Cu	Pb	Zn	Sr	Ba
Calcite	6527	6821	37	<3	9	<2	7	1697	246

[a]Notes (also for Table 8.7): other elements discovered in the veins either reveal Clarke values (Zr, Nb, Bi, Sn, Ge, Ga Be, Yb, Li) or are below the detection limit (Ta, Hf, Sb, Cd, As, Th, Ce, La)

Table 8.7 Elements-impurities in sulfide concentrates from carbonate veins (emission spectral analysis)

	Concentration of chemical elements, g/t								
Sample	Mn	Ni	Co	Cr	Cu	Ag	Zn	P	Ba
Sulfide fraction from veined fragments	3000	50	5	6	300	1	–	1000	>10,000

Table 8.8 Isotopic composition of oxygen and carbon for various types of calcite in carbonate veins

Sample	$\delta^{13}C$, ‰ PDB	$\delta^{18}O$, ‰ SMOW
Transparent coarse-grained calcite	−2.37	24.14
Gray fine-grained calcite of thin crosscutting	−2.24	25.22
Carbonate from sample vlk-1 (see Table 8.1)	−1.48	28.24

iron and nickel, as well as intermetallic compounds of iron together with neodymium, tantalum, copper, and nickel. At the same time, there are no gold, platinoids, or copper minerals that are characteristic of the Western Black Sea and Kerch-Taman mud volcanoes, although a limited quantity of copper is occasionally present in the intermetallic compounds of both types of mud volcano.

As a rule, grains of native lead are comparatively rare in mud volcanic breccia but not in the volcanoes that develop on Lower Cretaceous sediments. Quite large grains, up to 0.4 × 0.2 mm but more often small and somewhat rounded individuals at sizes within several tens of microns, are abundant in the Semigorskiy mud volcano (Fig. 8.81).

In many cases, the large grains of lead are heterogeneous, as they also contain the products of lead oxidation, particularly massicot (PbO). In places within the grains, there are a number of lines. Since lead is a soft mineral, these lines are most likely traces of displacement that occurred in the medium of the mud volcanic breccia.

Another common feature of this type of mud volcano is the complete sterility of the chemical composition of lead. In all studied samples (more than 10), impurities are virtually absent, and in only one case has a supposedly slightly ferrous lead been observed.

Doubts concerning the natural origin of this native lead are likely unjustified. First, it is devoid of any impurities. Second, there is a variety of lead minerals—carbonates, phosphates, oxides, and arsenates—that strongly suggests the presence of native lead in the mud volcanic center. Third, fairly large plates of lead appear with a small admixture of massicot.

Native zinc is found in the form of relatively coarse grains in breccia from the Shkolnyy mud volcano. On the surface of the flat grains, essentially plates of zinc, some geometric figures are visible, as if this mineral had been crystallized inside the breccia. The mineral also contains 4% Al. This native zinc also reveals solid solutions of lead, either in its native form or as oxides (low-temperature litharge ά-PbO or high-temperature massotite β-PbO) that form decomposition structures with zincite about 1 μm in size (Fig. 8.82).

The presence of zinc minerals from other classes seems to confirm this special feature of the Shkolnyy mud volcano.

Native iron is characteristic of many mud volcanoes, including those that developed on Lower Cretaceous sediments. The iron forms curved

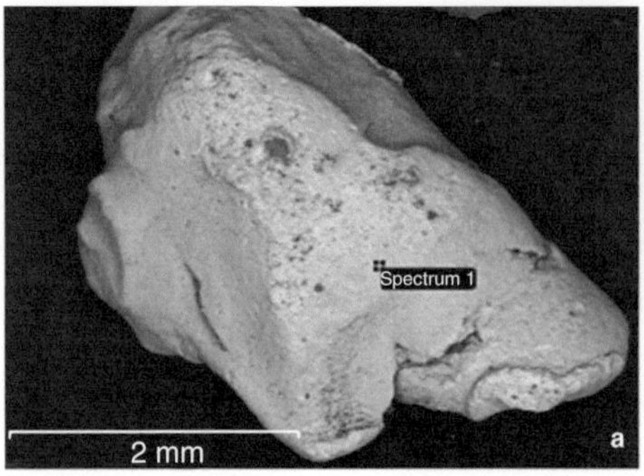

Spectrum 1			Spectrum 2		
Element	Weight%	Atomic%	Element	Weight%	Atomic%
Pb	100.00	100.00	Pb	100.00	100.00
Totals	100.00		Totals	100.00	

Fig. 8.81 Grains of native lead from breccia of the Semigorskiy mud volcano

wires (Fig. 8.83), flattened segregations (Fig. 8.84), and spherules (Fig. 8.85).

There are growths of iron with other minerals (e.g., silicates), but pure iron has also been observed (Fig. 8.86).

On the surface of metal particles, microsculptures of growth, traces of destruction expressed by vicinal faces, warts, striations, and crystal changes are visible. There are elongated filamentous grains with glide surfaces that reflect their growth under restrictive conditions, and there are also multidirectional furrows, scratches, and dents (Fig. 8.87b) that form due to the contact of this rather soft and malleable mineral with other harder minerals. All these traces indicate a prolonged rise to the surface of the metal particles from the considerable depths where they were formed.

Native nickel has been studied in breccia from the Semigorskiy mud volcano. It is represented by rounded plates/lumps often containing an admixture of iron (up to 48.47%). Their length and thickness reach 150 × 250 μm and 10–20 μm, respectively. The lumps are bent at the edges, heterogeneous, and contain inclusions of iron (up to 1.32–1.67 wt.%) and copper (up to 4.29 and even 12.4 wt.%) (Fig. 8.87).

During the course of magnetic separation, the iron-nickel particles were gathered together into a magnetic fraction in which magnetic flocs were concentrated with magnetite and other minerals (up to 0.2 mm in size). Their color varied between steel gray and black. In the interior parts of the grains, a semblance of crystals resulting from plastic deformations was preserved. They appear as isometric, elongated, prismatic, cubic, and tabular aggregates (Figs. 8.84 and 8.88).

The composition of grains that have been studied indicates perfect isomorphism of iron and nickel from 100.0% Fe to 99.28% Ni. Intermediate minerals in the Ni-Fe isomorphic series correspond to ferrite, kamacite, mean octibbehite, and native nickel. As elemental impurities, there are Cr, Ni, Nd, Ca, Al, and Si. One of the Debye powder diagrams obtained from a relatively large grain corresponded almost completely to awaruite (ferrous nickel with the formula Ni_2Fe).

Aggregates of goethite develop locally on the surface of native iron grains, but this phenomenon is not widespread.

Neodymium-ferruginous particles are found in breccia from the Semigorskiy mud volcano. In comparison with the segregations of other metals, they are smaller (up to 0.01 mm), reaching 0.16 mm only in their intergrowths with other metals. These particles have clearly pronounced magnetic properties and are fragile. They form complex microcrystals, prismatic secretions, and grains of irregular shape (Fig. 8.88), and they occur in the inner zones of complex polymetallic aggregates.

The aggregates are dominated by iron, and the content of neodymium ranges between 1.32 and

Element	Weight, %	Atomic, %
O K	32.02	64.96
Mg K	0.71	0.92
Al K	2.68	3.12
Si K	0.55	0.62
Fe K	0.33	0.19
Zn K	62.72	30.19
Totals	100.00	

a

b

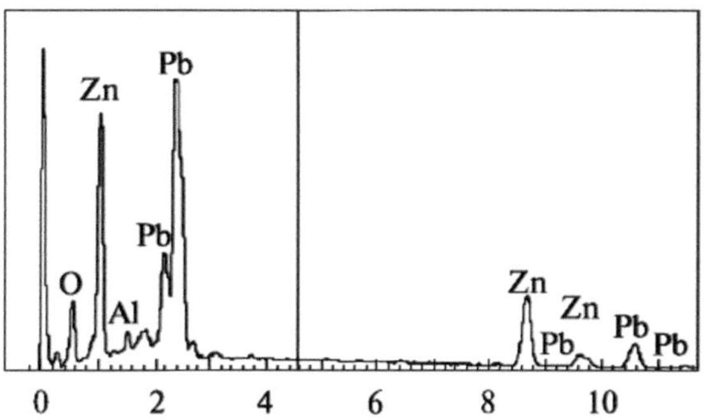

c

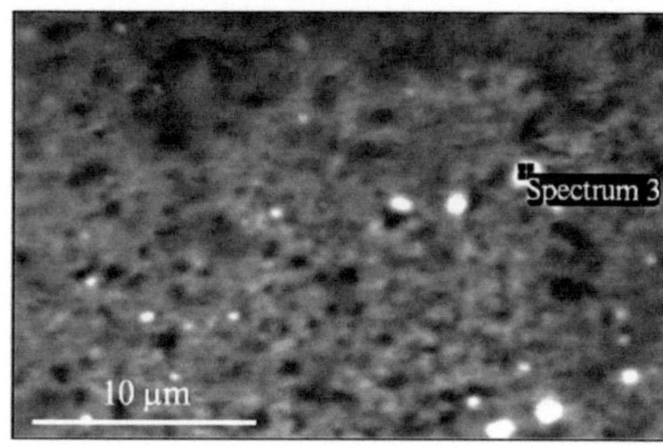

d

Fig. 8.82 Decomposition structures of solid solutions of zinc oxides (**a**, **b**) and lead (**c**, **d**) on the surface of a metal particle within breccia of the Shkolnyy mud volcano. Symbol K carries technical information (indicates the type of electrode)

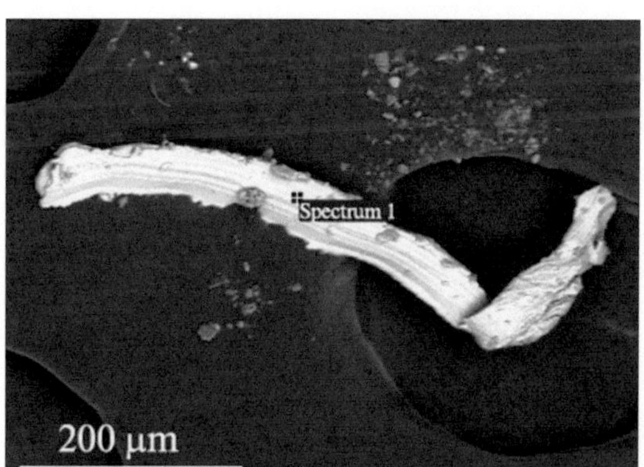

Spectrum 1		
Element	Weight%	Atomic%
Fe	100.00	100.00
Totals	100.00	

Fig. 8.83 Curved wire of native iron from breccia of the Gladkovskiy mud volcano

Spectrum 2		
Element	Weight%	Atomic%
Al	1.08	2.14
Si	2.82	5.39
Fe	96.11	92.47
Totals	100.00	

Fig. 8.84 Flattened segregation of native iron with micro-inclusions of aluminosilicates in a rock fragment. A magnetic Nd/Ni/Fe floc (coagulation) was formed around the sediment particle in breccia from the Semigorskiy mud volcano

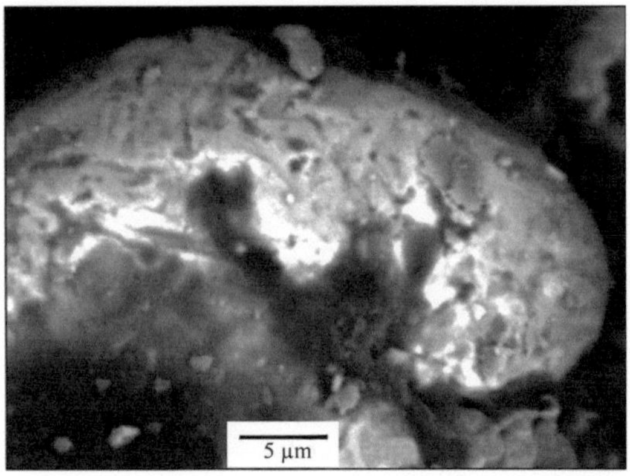

Fig. 8.85 Globular form of native iron in breccia from the Gladkovskiy mud volcano

36.98%; secondary changes (oxidation) are not observed (Table 8.9).

This is the first discovery of neodymium in appreciable quantities within mud volcanic breccia. The concentrations can reach 45.78–59.07%, but, as a rule, they are much lower and vary between 23.18 and 25.54 wt.%, more often 24–25 wt.% (Shnyukov and Ivanchenko 2015).

There are several grains with approximately the same content of neodymium and a Nd:Fe ratio of 1:3. The grains have specific forms, and one case gives the impression of a pronounced head on an elongated crystal (Fig. 8.89, Table 8.10).

Another case looks like a crystal fragment with small elongated particles adhering (Fig. 8.90).

In some places, the grains (200 × 150 μm) of this mineral are slightly rounded and framed on one side by smaller (20–100 μm) particles, one of which is well rounded, the rest are angular (Fig. 8.84). According to the scatter of particles, it seems that the mineral is magnetic. In large grains, the Nd content is 23.96%; in one of the small surrounding pieces, it is 23.5%. Correspondingly, Fe in the first case is 73.4% and in the second 75.13%.

There are also fine-grained aggregates composed of concretions of this mineral cemented by material containing 69.69 wt.% Fe, 23.65 wt.% Nd, and minor impurities of Si and Al. The grains seem to be overgrown by smaller needle-shaped

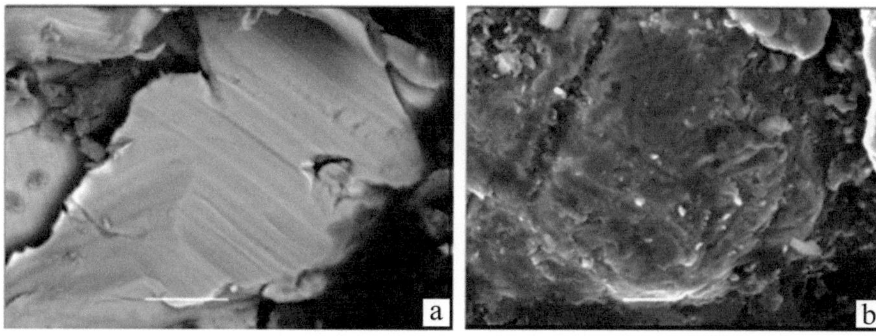

Fig. 8.86 Microsculptures of growth and destruction on the surface of native iron grains: (**a**) – growth striations, (**b**) – dissolution cracks and numerous scratches on the surface of a malleable particle of metallic iron in breccia from the Semigorskiy mud volcano

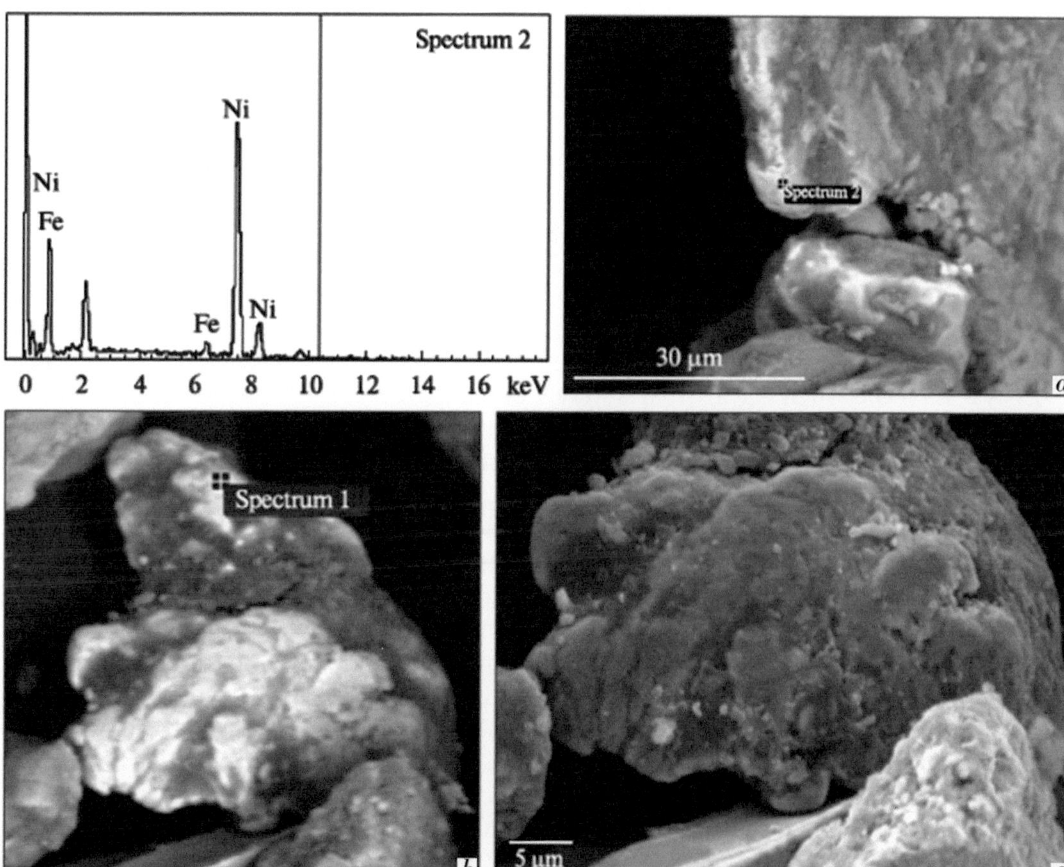

Fig. 8.87 Rounded plates/lumps of native nickel from breccia of the Semigorskiy mud volcano: (**a**) with isomorphic impurities of iron; (**b**, **c**) crystals and deformed grains of nickel-silver ferrite

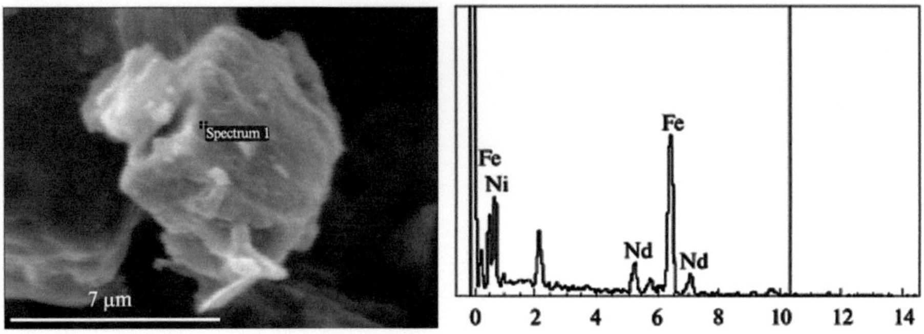

Fig. 8.88 Cubic crystal of Fe_9Nd from breccia of the Semigorskiy mud volcano

Table 8.9 Chemical composition of supposedly solid solutions of iron and neodymium from breccia of the Semigorskiy mud volcano

	Elemental content, wt.%			
Grain size, mm	Fe	Ni	Nd	O
0.012	79.62		20.38	
0.006	78.14		21.86	
0.315 × 0.12	98.69		1.32	
0.04 × 0.02	63.55		21.19	15.26
0.005 × 0.004	76.82		23.18	
0.16 × 0.1 × 0.0010	48.47	14.15	36.98	

crystals that give the impression of "hedgehogs"; these are visible everywhere but in places seem to be obscured by cement (Fig. 8.91).

The study of a Debye powder diagram (taken by analyst E.E. Grechanovskaya) from a very fine grain of the compound described above indicated an identification of iron carbide, or cogenite, $(Fe,Nd)_3C$, with a 25% admixture of Nd with no carbon. This determination is suggestive only because the grain in question is very small, and it crumbled partially during installation within the apparatus.

Minerals of nickel, copper, and tantalum are found in the outer zones of complex polymineral grains. They form a thin crust consisting of sheaf-like, paniculate aggregates on the surface of native nickel. In Fig. 8.92, maps of the spatial distribution of nickel, copper, and tantalum in a breccia sample from the Semigorskiy mud volcano are presented. It can be seen that copper and tantalum are localized in one area, and nickel forms separate concentration sites.

Data presented in Fig. 8.93 and its associated table (Spectra 1–6) indicate the saturation of the mineral-forming medium simultaneously for several metals. The regular replacement of metallic parageneses reflects the directed evolution of thermodynamic conditions in deep zones of the Earth's crust (Shnyukov and Ivanchenko 2015).

The internal structure of metallic aggregates is not homogeneous. A single grain can reveal several regularly changing segments composed of different native metals and their alloys. When the thickness of the plates is 15–20 µm, there are sometimes three segments of different chemical compositions proceeding from the center to the periphery.

Along with native lead and other native minerals, a number of other lead compounds including galena, cerussite, polyspharite, tsumebite, and mimetite were encountered.

Galena (PbS) is represented by a glandular variety called ferruginous galena. In these small (10–30 µm) and slightly rounded particles, the content of Fe varies from 2.56% to 6.72%, representing most likely an isomorphic impurity.

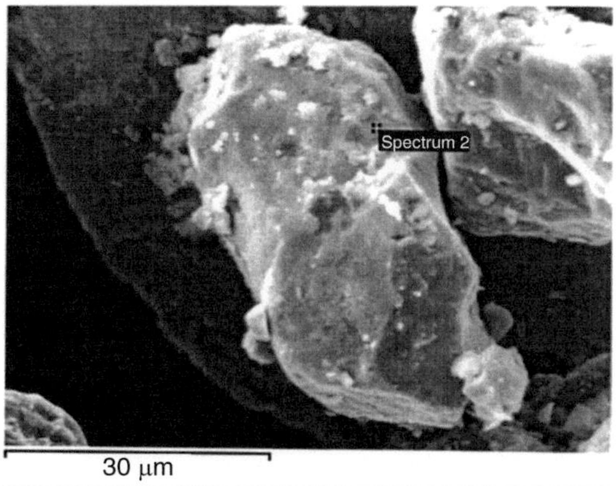

	Spectrum 2	
Element	Weight%	Atomic%
Fe	74.83	88.48
Nd	25.17	11.52
Totals	100.00	

Fig. 8.89 Semblance of a crystal with Nd-Fe composition

Table 8.10 Zonal distribution of native Fe, Ni, Nd, Cu, and Ta in the grain shown in Fig. 8.92

Chemical elements	Zonal content of chemical elements, %		
	Central	Intermediate	Outer
Fe	48.47	1.32–1.55	1.38–1.69
Ni	14.55	98.45–98.68	84.63–95.47
Nd	36.98		
Cu			4.29–12.40
Ta			3.49–5.12
Si			1.28–1.36
Al			0–1.49

Cerussite ($PbCO_3$) was discovered twice in the form of relatively large (up to 100 and even 300 μm) grains and aggregates.

Among the lead minerals are two phosphates. One of them is fairly close in composition to pyromorphite—more precisely to its calcium variety called polyspharite—that contains up to 8–9% CaO.

In three cases, the identification of minerals, even approximately, was not possible. Most likely, these cases represent mixtures of minerals, each containing from 13.82 to 32.26% Pb.

Because of the very small grain size, about 20 μm, the identification of another lead phosphate, tsumebite, is presumed. Its formula is $(Pb,Cu)_5(PO_4)_2 \times 6H_2O$, but in our case, the set of elements replacing Pb is different. They are Fe and Ca, so the first member in the mineral formula may need to be modified to (Pb, Fe, Ca).

Identification of mimetite, $Pb_5(AsO_4)_3C1$, is also presumed; it has been found among aggregates of the smallest concretions with dimensions of 70 × 40 μm and with iron content from 1.69 to 9.60%.

In some cases, the type of mud volcano developed on Lower Cretaceous sediments is characterized by an interesting feature: change in a chemical element and its mineralogy commensurate with changes in the geochemical and thermodynamic conditions characteristic of the fluid flow. For example, native lead is replaced by lead sulfide (galena) and then by massicot, cerussite, and mimetite.

Similarly, native zinc apparently gives rise to a number of zinc minerals: sphalerite, zincite (ZnO), and calamine $Zn(OH)_2Si_2O_7 \times H_2O$. The iron chain can be traced from native iron to pyrite and pyrrhotite, further to wüstite (FeO), magnetite, hematite, and iron hydroxides.

Spectrum 3		
Element	Weight%	Atomic%
Fe	75.07	88.61
Nd	24.93	11.39
Totals	100.00	

Fig. 8.90 Fragment of a crystal with Nd-Fe composition from breccia of the Semigorskiy mud volcano showing adhering small, elongated particles

Fig. 8.91 Aggregate of neodymium-iron composition from breccia of the Semigorskiy mud volcano

Sometimes, the geochemical features of minerals are inherited, as has been observed. Thus, native lead can be enriched with iron as well as galena within the same sample.

Native tungsten has not been found in Semigorskiy or any other mud volcanoes, yet its minerals and tungstates are present. In two small (10–20 μm) examples that have been studied, the elemental composition is similar to ferberite or iron tungstate. In both cases, this mineral is represented by ferberite with an admixture of cobalt. This is supposedly cobalt ferberite.

Moissanite (SiC) has been identified in breccia from the Gladkovskiy mud volcano. Under a binocular microscope, the mineral has a dark blue to black color, a strong adamantine luster, and translucency. It forms uniform grains of irregular shape and angular chips up to 0.2 mm in size. The identification has been confirmed using a Debye powder diagram (Fig. 8.94).

A strontium barite was found in breccia from the Semigorskiy mud volcano; it most likely comes from a vein (Fig. 8.95).

Weathering processes are reflected by such minerals as scorodite $Fe^{3+}(AsO_4) \times 2H_2O$, hydrobiotite, hematite, and iron hydroxides. There are also neo-crystallizations, such as pyrite and cerussite.

Most of the minerals described above are virtually devoid of any signs of horizontal water transport. Their segregations are angular, lamellate, curved, and tuberculate, but never rolled. It is probable that they did move but most likely traveled as a component of the mud volcanic breccia in its vertical rise due to the lifting by fluids.

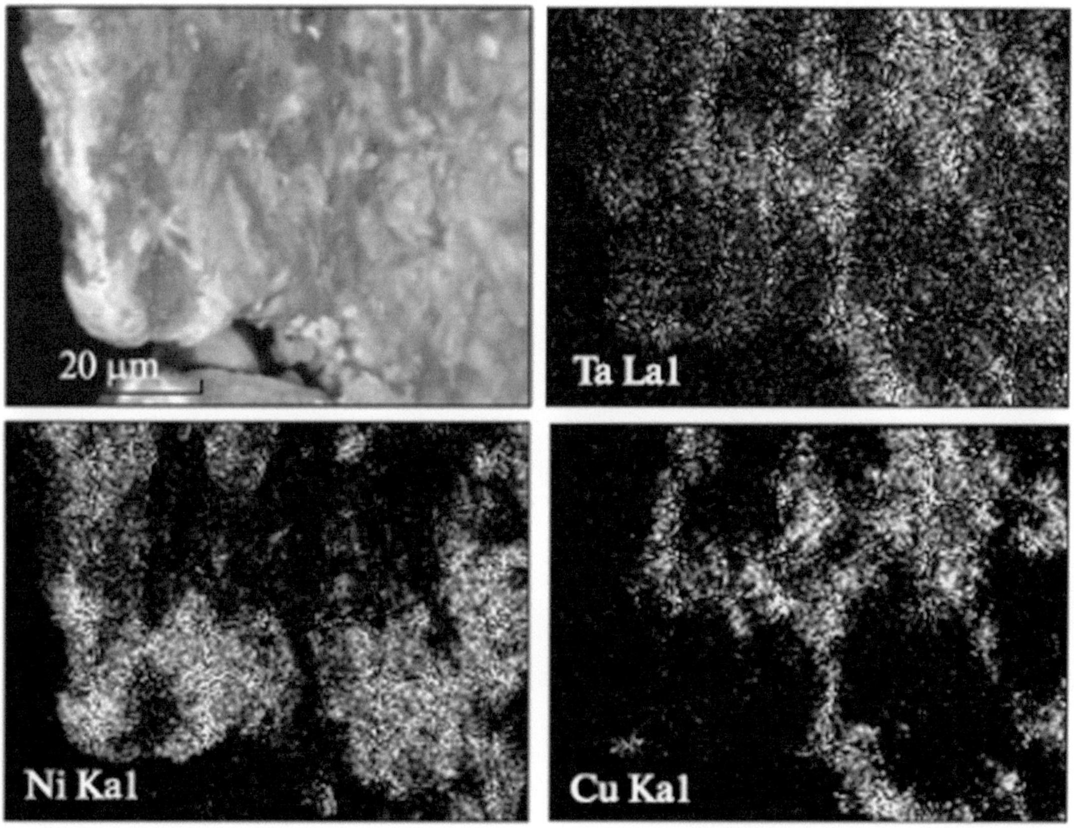

Fig. 8.92 Image (left upper corner) and maps of nickel, copper, and tantalum distribution in a metal grain from breccia of the Semigorskiy mud volcano

Attention should also be paid to the high-temperature character of many of the minerals described above. Their occurrence, as a rule, requires both high pressures and temperatures. Fluids impart highly reducing conditions that are necessary for creation of native minerals. The oxides formed in the rising breccia are later alterations caused by oxidizing conditions, etc. if not as fragments of the rocks that enclose the volcanic channel. In other words, they must be formed within breccias undergoing upwelling by deep fluids under compressive conditions caused by the rocks through which the fluids pass.

These minerals can be considered insignificant accessories, as their presence in mud volcanic breccia amounts to less than 1%. They appear as a result of the introduction into mud volcanic breccia of their constituent materials by deep fluids.

Mud volcanoes developed on Lower Cretaceous sediments typically show modern activity of very small scale, which is reflected in the smaller size of their morphostructures and mud volcanic features as well as the absence of catastrophic eruptions. Nevertheless, the participation of deep fluids in their formation is no doubt due to the mineralogical peculiarities of erupted material. These peculiarities include the presence of some native minerals, such as lead, zinc, and iron. In addition, other singular finds recorded in one of the volcanoes (Semigorskiy) are (1) an unusual mineral, presumably an alloy of iron and nickel with neodymium (the last having a content of about 25%), (2) an alloy of iron, nickel, neodymium, tantalum, and copper, and (3) a number of minerals of zinc, lead, and iron derivatives.

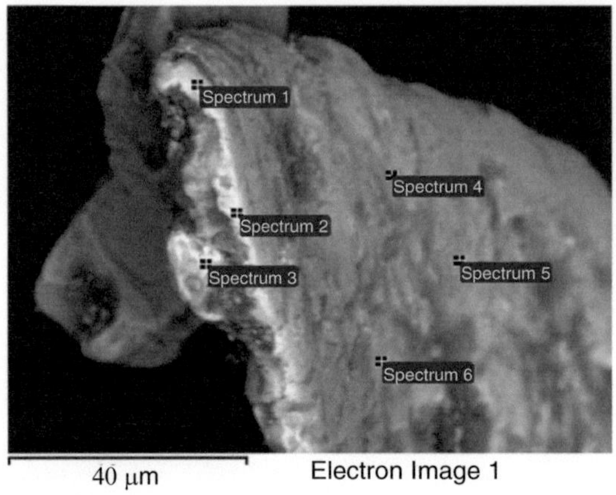

Weight%	Fe	Ni	Nd	Si	Cu	Al	Totals
Spectrum 1	1.55	98.45					100.00
Spectrum 2	1.32	98.68					100.00
Spectrum 3	48.47	14.55	36.98				100.00
Spectrum 4	1.69	84.63		1.28	12.40		100.00
Spectrum 5	1.67	95.47		1.36		1.49	100.00
Spectrum 6	1.38	94.32			4.29		100.00

Fig. 8.93 Metal plate consisting of several segments, each with a different composition of native metals; from breccia of the Semigorskiy mud volcano

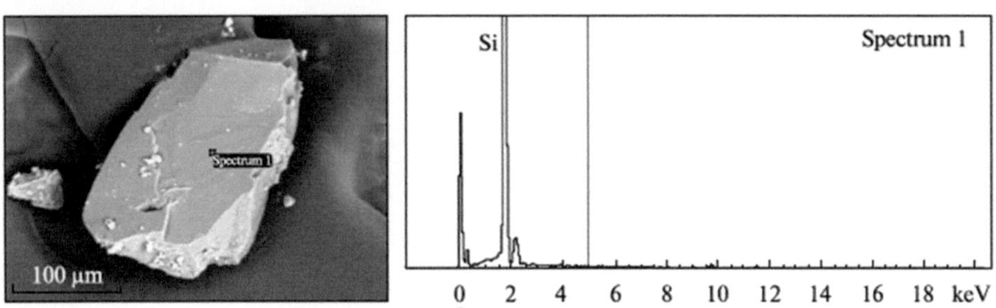

Fig. 8.94 Grain of moissanite in breccia of the Gladkovskiy mud volcano

More than 70% of the finds of native elements come from the Semigorskiy mud volcano, while the rest were discovered in the Gladkovskiy mud volcano that were most intensively studied. They are represented by four isomorphic rows (in which elements can completely replace each other) of native metals: Zn – Pb (20%), Fe – Ni (58%), Ni – Cu – Ta (8%), and Fe – Nd (12%) (Shnyukov and Ivanchenko 2015). Breccia from the Gladkovskiy mud volcano is dominated by iron, while that from the Semigorskiy mud volcano is represented by a much greater diversity of native minerals.

It should also be noted that mud volcanic breccia from volcanoes developed over Cretaceous deposits displays a relative poverty of allotigenic (clastogenic) minerals from the Cretaceous deposits. There are secondary minerals within the oxidation zone. Gold, platinum, and minerals of mercury, arsenic, and antimony are absent, which is quite unexpected given the localization of mercury deposits in Cretaceous sediments of the North Caucasus.

Spectrum 1				
Element	Weight%	Atomic%	Compound%	Formula
S	18.53	18.93	46.28	SO3
Sr	5.54	2.07	6.55	SrO
Ba	42.25	10.07	47.18	BaO
O	33.68	68.93		
Totals	100.00			

Fig. 8.95 Strontium barite from breccia of the Semigorskiy mud volcano

8.8 Development of the Mud Volcanic Process

The above review of the fluid mineralization observed in breccias from three different types of mud volcanoes shows that their composition reveals very different minerals that are distinctive in terms of the environmental conditions of their formation. They are represented by local minerals of different genesis, such as sulfides (magmatic—high, medium, and low temperature—or hydrothermal), oxides, carbides, silicides, silicates, etc. all developed under variable thermodynamic conditions. Most likely, these conditions incorporated moving and gradually changing waves of fluids interacting with different types of rocks under variable tectonic forces. Emerging local centers of mineral formation most probably retained some common features, for example, enrichment with hydrogen sulfide, chlorine, or fluorine that were reflected in the composition of some of the minerals previously described. Therefore, despite the fact that processes influencing the development of different types of mud volcanoes were fundamentally the same (they were caused by deep fluids), the fluid mineralization was markedly different from place to place, as were the different rocks crossed by these fluids. One cannot ignore the possible diversity of escaped fluids. Ultimately, the combined factors form complex and often inconsistent mineralogical compositions in the breccias within different types of mud volcano. As such, the mineralogical composition of breccias reflects many processes occurring at depth that affect mud volcanic development.

As shown by Kropotkin and Valyaev (1980), the functioning of mud volcanoes is possible only under conditions where deep fluids replenish mud volcanic gases.

Varied evidence suggests that mud volcanoes result from the activity of deep fluids: the composition of their breccias, the presence within breccias of new mineral formations that are unique in composition and structure (such as spherules of cerium, lanthanum, neodymium, iron minerals); the newly formed quartz crystals; the fused silica spheres; the inclusions within breccias showing unique mineralogical content (native minerals, sulfides, carbides, silicides, some oxides, etc.); the isotopic composition of methane (content of $\delta^{13}C$), etc.

The mechanism underlying the formation of mud volcanoes can be understood from the study of particular indicators of the mud volcanic process: fluidogenic minerals.

8.9 Mineralization in the Mud Volcanic Process

8.9.1 Statement of the Problem and the Origin of Authigenic Minerals in the Mud Volcanic Process

The problem of interpreting breccia mineralization that supposedly takes place within the eruptive channels of mud volcanoes is complicated. Background ideas significant for its understanding were laid out by Kropotkin and Valyaev (1980), Dmitrievskiy and Valyaev (2002, 2010), Pikovskiy (2002), and Lukin (2014), and these ideas were based on a hypothesis about nonmagmatic hydrocarbon degassing of the Earth. The development of the Earth is considered to be a monotonically fading process involving the intrusion of deep fluid components from the mantle plumes into the upper horizons of the lithosphere followed by periodic pulses of intense degassing (Letnikov et al. 2010).

The basis of deep fluids is hydrogen. Two fluid systems are distinguished: hydrogen-carbon and hydrogen-sulfur. They are located at different levels within the liquid core.

The gas flows of the plumes have initial temperatures of about 4000 °C and a pressure of about 1 million bars. They seem to burn the mantle and the lower crust. The mantle plumes spawn fluid flows, feed them with energy, and lead to degassing or defluidization of the lower part of the lithosphere that occurs via discrete subvertical zones, i.e., "degassing pipes." These zones are especially evident in the "multistory" oil and gas fields, deposits of which are located one above the other, throughout the stratigraphic section. The role of "degassing pipes" is most often played by mud volcanoes. Their activity is caused by the action of the hydrogen-carbon fluid system.

Previously, most geologists presumed that mud volcanic eruptions are forced by hydrocarbon gases due to tectonic-dynamic activity. They did not consider the possibility of a deep origin. Kropotkin and Valyaev (1980) showed that gases of mud volcanoes must originate from the mantle, and mud volcanic activity can be explained only by an outburst of deep fluids at a huge scale. In their opinion, mud volcanoes can function only when they are intensively fed from deep foci located at the base of the sedimentary section, in other words, fueled by deep degassing. They considered mud volcanism as the most important direct sign of the oil and gas potential of the Earth's interior. The fluid dynamic systems containing hydrocarbons are in fact gas-hydrothermal systems. Globally localized, fluid flows deliver hydrocarbons to natural traps. Ascending streams of superheated water-gas masses permeate the entire lithosphere, creating discrete manifestations in time and space.

A clear example of modern fluid plumes in the lithosphere can be seen in hydrothermal plumes in the oceans, especially in the "spreading zones." On the way up, deep plumes can extract part of the surrounding rocks from the breakthrough sequence, but this does not change their main deep nature.

As shown above, mud volcanoes of the Black/Azov Sea region are developed either on Maikopian or Lower Cretaceous sediments, and the distance to the Moho boundary is either 19 or 44–45 km, resulting in a sediment cover of varying thickness (Fig. 8.11). It may be that the effect of deep fluids breaking through either a thinner or thicker cover should be somehow different, and as such, there should be some difference in the mineralogical properties of breccias erupted by each type of mud volcano. In order to verify these assumptions, the most interesting mud volcanoes of each type were chosen for mineralogical investigation and further comparison.

8.9.2 Genesis and Chemical Environment of Mineralization in Mud Volcanoes

According to Pikovskiy (2002) and other researchers, plumes are one of the main factors in the evolution of the lithosphere during geological history. Deep fluids pulsationally move from the zone of generation to that of unloading. By fluid plumes, one should understand the ascending streams of superheated water-gas masses that are discrete in time and space and permeate all layers of the lithosphere. Lukin (2014) believes that, at their origin and initial motion, plumes correspond to a fluid (a gas with a liquid density expanding without limit). This is a multicomponent, super-compressed hydrocarbon-based gas-like substance similar to a real plasma in the modern sense; it is "loaded" with native metal particles and moves upward along deep faults and their intersections. In the sections of the crust that face these faults, there are numerous zones of decompaction-dilatancy, facilitating the invasion of plumes.

Pikovskiy (2002) considers the intrusion of deep hydrocarbon fluids into the upper floors of the lithosphere as a manifestation of the "cold" degassing of the Earth (according to P.N. Kropotkin). Summarizing the evidence, he considers the shape and structure of the fluid plumes of the lithosphere similar to the models of mantle and hydrothermal plumes.

In the lower parts of the lithosphere, plumes are relatively narrow zones filling the sections of deep faults and their intersections. The upper parts of the plumes "radiate" upward, spreading like a fan along the decompaction zones of the rocks to tens and even hundreds of kilometers across. He notes that fluid plumes are single phase, essentially aqueous solutions in the supercritical state (above 300–400 °C). They contain alkali metals, volatile acids, halides, CO_2, CO, CnHm, H_2, N_2, H_2S, He, as well as other chemical elements and compounds. The composition of the fluid and the level of its mineralization are determined by the mantle plume.

Fluids change as they move. Originally redox, they oxidize when interacting with the lithosphere. However, when rising under the influence of gravity and thermodynamic sorption, hydrocarbons and other light and volatile substances accumulate in the upper part of the plume, again creating a redox environment in the deep Earth. Reduction of pressure and temperature below critical levels leads to fluid decay into liquid and gas phases. This is facilitated by the density of sedimentary rocks through which the fluids are filtered.

With the introduction of fluids, minerals arise; we earlier indiscriminately attributed these minerals to accessories. These include native minerals, a series of sulfides and oxides, etc., and it is these minerals that are the consequence of the invasion of fluids.

Whether it is a native iron, or a native lead, or a native zinc, Vernadskiy (1913) writes about the extraordinary conditions of their occurrence. The same can be said about such minerals as wüstite, cogenite, and a number of others.

We concur with the opinion of Kropotkin and Valyaev (1984) that mantle hydrocarbon plumes are extraordinary. A remote view of one plume, obtained through artificial means, took place on April 20, 2010 in the Gulf of Mexico (Zharvin 2013). According to this author, the depth of a borehole drilled by British Petroleum on the Macondo Prospect reached deeper than 5.66 km as was declared, but, as was written a month before the disaster, 11.2 km. At this depth, the borer reached "asthenospheric oil," which caused a catastrophic release of hydrocarbon gases and oil. Two automatic valves, rated at 700 atm each, were broken instantly, indicating that the pressure of the escaping gases was much higher than 700 atm.

Almost all authors assume a high pressure in mud volcanoes: up to 300–400 atm. As one can see, the pressures of mantle emissions can reach colossal values, much higher than have been assumed for mud volcanoes. With regard to temperatures, the presence of fused particles in the breccias already indicates episodic values over a thousand degrees C. High temperatures are required to form many native minerals (iron, copper gold, etc.). This evidence once again confirms that mud volcanoes are structures that arise under the action of mantle hydrocarbon plumes.

It is precisely under the conditions of high pressures, temperatures, and the reducing environment of the mantle plume, that crystallization of many high-temperature and high-pressure minerals—in particular, native minerals, sulfides, some oxides, carbides, etc.—is likely to occur.

In the Black Sea, the kind of echo produced by a mantle plume was observed in 2002–2003 above the Dvurechenskiy mud volcano. The gas fountain here reached 850 m above the sea bottom with a jet diameter of up to 400 m; the temperature of the breccia was 16–20 °C (compared to 8 °C within the silt). Powerful fountains were also observed in neighboring mud volcanoes.

A similar event occurred in the Kerch Strait area in 63 BC when a gas plume was provoked by an earthquake. According to historians, the sea burned and leaped, as cities and villages collapsed under great waves (Shnyukov et al. 2013a). Obviously, these events are the outward manifestations of deep processes that take place in the Earth's interior.

The hundreds of catastrophic eruptions of mud volcanoes in the Black Sea region described in previous chapters are manifestations on the Earth's surface of deep fluid dynamic processes. The composition of gases is often the cause of ignition. The weakly manifested systematic activity of mud volcanoes is in fact the resulting echo of previously emitted gases accumulated in pockets and inflated chambers of the eruptive channels within mud volcanoes.

Pikovskiy (2002) believes that fluid dynamic systems in the Earth's lithosphere are nothing more than gas-hydrothermal systems.

In the modern mud volcanoes of Western Porsugel and Pink Porsugel in Turkmenistan, precipitation of Zn, Pb, Cd, Cu, As, and other chemical elements in significant amounts from chloride-bicarbonate-sodium waters has been established (Lazarenko 1970). This is one of the natural models of the fluid process, which deserves more detailed consideration.

Lebedev (1975) paid close attention to the hydrothermal systems of the Chelekenskaya anticline (Cheleken Peninsula, Turkmenistan) when considering modern hydrothermal systems. He noted the existence of numerous tectonic distortions in this structure, including the Main Chelekenskiy fault, and the presence of mud volcanoes. He remarked that, according to seismic data, the Pribalkhashskaya depression, of which the Chelekenskaya anticline is an integral part, is located on the Moho surface and along the roof of the crystalline basement. The dimensions of the Chelekenskiy fold are 35 km (length) and 15 km (width), and the mud volcanoes are from west to east, West Porsugel, Aligul, and Pink Porsugel.

On the Cheleken Peninsula, there are numerous mineral springs, including thermal brine, hydrogen sulfide, carbon dioxide (the area of the extinct Aligul volcano), and water-oil springs. Both the Western and Pink Porsugel mud volcanoes are characterized by powerful water eruptions from the Apsheronian horizon as well as metal-bearing brines recovered by numerous wells from the overlying red-colored sedimentary sequence (Lebedev 1975).

In our opinion, this case implicates a powerful influx of deep fluids penetrating from the depths along mud volcanic "degassing pipes." The fluids, mixing with aquifers, acquire a wide variety of chemical elements (Zn, Pb, Cu, Cd, As, B, Sr) that are present in metalliferous brines. It should be noted that Hg is not present.

In thermal brines, Pb and Zn are present in the form of the simplest complexes: $PbBr^+$, PbJ^+, $PbSO_4^{3+}$, $PbOH^+$, $ZnSO_4^0$, $ZnOH^+$, $ZnBr^+$, and ZnJ^+. In the Black Sea mud volcanoes, Br and Cl are recorded in minerals. Recalculations by Lebedev (1975) show that the overwhelming amount of Pb (94%) and Zn (84.7%) is present in the form of anionic complexes $PbCl_3^-$, $PbCl_4^{2-}$, and $ZnCl^{2-}$. The remaining lead is present in the form of $PbCl_2^0$ (4.2%) and $PbSO_4$ (2%) and the zinc as $ZnCl^+$ (0.84%), $ZnCl_2^0$ (1.9%), $ZnSO_4$ (7.8%), and $ZnOH^+$ (4.7% %).

In deep contexts (wells have reached depths of 1000–1250 m), it is likely that $PbCl_4^{2-}$ and $PbCl_4^-$, which are stable on the surface, are associated with other complexes or are secondary, arising from the decay of more complex ones. With temperatures of 87–97 °C and Eh +140 MV, the Pb content is 24–60 mg/l.

Under elevated concentrations of CO_2 and HCO_3^- in deep brines, the presence of the carbonate-chloride $Pb(CO_3)_2Cl^{3-}$ and the carbonate $Pb(CO_3)_2^{2-}$ is likely. It is assumed that a mixed carbonate-chloride complex of Pb may prove to be stable at elevated temperatures.

Iron, copper, and cadmium in brines were also studied.

In general, it can be assumed that the ore complexes in modern brines of the Cheleken Peninsula are present in the form of negatively charged chloride complexes $PbCl_4^{2-}$ (55.7%), $PbCl_3^-$ (37.6%), $ZnCl_3^-$ (84.7%), $CdCl_4^{2-}$ and $CdCl_3^-$ (95–100%), $CuCl^{2-}$ and $CuCl_2^-$ (90%), $FeCl_4^{2-}$, and $FeCl_3^-$ (86–96%).

Lead and zinc can form neutral sulfates, and zinc can form hydroxyl complexes (Lebedev 1975).

In deeper horizons, where sulfide-bicarbonate-sodium waters predominate, it is possible to find heavy metals. They will be present in the form of hydrosulfide or sulfide complexes in zones of contamination, to which conditions of mud volcanoes largely correspond.

Precipitation of mercury minerals, most often cinnabar, was considered by Saukov (1946). He showed that cinnabar precipitates from diluted alkaline solutions as well as (1) when alkaline ore-bearing solutions are neutralized, (2) when they change pressure, and (3) when they come into contact with carbonic acid and oxygen.

As one can see, the Chelekenskiy mud volcanoes provide an example of the development of newly formed minerals in connection with mud volcanic processes.

The entire Black Sea mud volcanic region is huge. There are regional variations in how the processes develop and function, and these differences are reflected in the way new mineralizations take place.

8.9.3 Parageneses of Autogenous Minerals of the Black Sea Mud Volcanoes

As was already noted, there are many dozens of mud volcanoes in the vast Black Sea region. The overwhelming majority of them are confined to Maikopian sediments. According to Pikovskiy (2002), a sharp drop in temperature and pressure occurs at depths of 1–3 km. For most mud volcanoes of the Black Sea region, this is the depth of the Maikopian sediments. Only one small group of mud volcanoes (Semigorskiy, Shkolnyy, Shapsugskiy, and possibly, Gladkovskiy) is located in the southeast of the region and emanate from Lower Cretaceous rocks.

Comparison of the native mineralization of mud volcanoes in the Kerch-Taman region ("Maikopian") with those of the northern Caucasus ("Lower Cretaceous") shows the more limited mineralization of the latter. "Lower Cretaceous" volcanoes produce native lead, zinc, iron, and rare earths, whereas volcanoes of the Kerch-Taman areas have yielded native iron, gold, osmium, iridium, silver, lead, copper, zinc, nickel, rare earths, sulfur, graphite, cinnabar, realgar, orpiment, antimonite, and their sulfides and oxides. Perhaps, the fact that more localities have been studied in the Kerch-Taman region (12) than in the northern Caucasus (2) plays some part in this difference, and perhaps there are also regional variations. Finally, the influence of the Maikopian layer may be the significant factor. Gold, platinoids, mercury, and arsenic are absent in volcanoes based upon chalks, or perhaps more correctly, they have not yet been found in such contexts.

Comparison of native mineralization across the three types of mud volcanoes shows that the North Caucasian mud volcanoes "sitting" on Lower Cretaceous sediments contain native minerals (lead, zinc, iron, and rare earths) and sulfides (rare ferruginous galena, pyrite, and presumed pyrrhotite). The Kerch-Taman mud volcanoes "sitting" on Maikopian sediments contain native minerals (gold, osmium, iridium, silver, copper, nickel, sulfur, and graphite), sulfides (pyrite, arsenic pyrite, marcasite, wurtzite, galena, cinnabar, realgar, and orpiment), and oxides (antimonite). The Western Black Sea mud volcanoes "sitting" on Maikopian sediments contain pyrite, marcasite, pyrrhotite, bornite, chalcopyrite, cubanite, covellite, chalcocite, wurtzite, and sartorite ($PbAs_2O_4$).

As noted, these differences may be related to the varying number of studied mud volcanoes (2 in the northern Caucasus versus 12 in the Kerch-Taman region), regional peculiarities, or the influence of the sediments from which the mud volcanoes emanate on the mineralization processes. For example, the relative impoverishment in sulfides and their limited variety in the northern Caucasian mud volcanic breccias may indicate a depletion of sulfur (hydrogen sulfide) in the deep fluids. The enrichment of the Kerch-Taman mud volcanic breccias with sulfides of hydrothermal origin can perhaps be related to the fact that the 3–5 km thick Upper Maikopian sediments from which they emerge represent a zone of cooling and pressure loss in the deep fluids that then attain the level of low-temperature hydrothermal fluids that produce conditions favorable for the crystallization of these minerals.

According to Lukin (2014), contamination with dispersed native metal particles indicates the participation of anhydrous fluid flows. The temperatures of deposition are quite high for many native elements as well as the formation temperatures of the carbides moissanite and cogenite.

The recovery of hollow spherules of rare earth elements, sometimes spherules of cassiterite and magnetite (hematite) discovered in the Kerch-Taman mud volcanoes, and well-expressed quartz crystals from marine mud volcanoes suggest something about the effect of hot fluids on the Maikopian sediments at their base and possibly the introduction of particles from even greater depths. Under artificial conditions, quartz crystallizes at temperatures up to 360 °C, and temperature fluctuations are caused by changes in pressure.

Attention should be paid to the absence of cinnabar in marine mud volcanoes, despite the fact that mercury, 64% of which is in a soluble form, is always present in their vicinity (Egorov et al. 2011). Apparently, the fluid flow in marine mud volcanoes is still too hot and therefore thermodynamically insufficiently cool and low in pressure for Hg mineralizations. The question of the absence of cinnabar in Semigorskiy and some other "Lower Cretaceous" mud volcanoes (except Gladkovskiy) deserves further study, especially since Karasik and Morozov (1966) recorded from 0.3 to 10.0 g/t (avg 3.2 g/t) of mercury as well as single grains of cinnabar in ten samples of breccia from the Gladkovskiy mud volcano. This range is much higher than the Kerch-Taman mud volcanoes, where mercury content is below 0.7–1.35 g/t.

Note, however, that judging by geological indications, the structure of the Gladkovskiy mud volcano does not exclude the participation of Maikopian deposits in the fluidogenic mineralization process within this volcano.

Comparison of the newly formed oxides, carbonates, and sulfates from the three groups of mud volcanoes is quite informative.

In the Kerch-Taman mud volcanoes, oxides are represented by rare magnetite, wüstite, massicot, and opal.

In the North Caucasian mud volcanoes, oxides are also not diverse, being represented by wüstite, hematite, massicot, zincite, and rutile. The latter gives the impression of a clastic mineral.

In the Western Black Sea mud volcanoes, oxides are most variable and represented by wüstite, magnetite, hematite, iron hydroxides, ilmenite, leucoxene, rutile, chromite, chromium-magnesium spinel, corundum, and franklinite.

In the North Caucasian mud volcanoes, carbonates are represented by cerussite (?) and calcite in indeterminate forms of palmate grains (Semigorskiy mud volcano). Particularly noteworthy are fragments of calcite veins with pyrite.

In the Kerch-Taman mud volcanoes, carbonates are represented by clusters of small (dozens of μm) crystals of siderite and ankerite, relatively large (200 μm) crystals of dolomite in the form of rhombohedra or elongated along cleavages, sintered molds of rhodochrosite, and sometimes fragments of calcite veins. It is noteworthy to mention that carbonate veins within mud volcanic breccias are formed under temperatures of

100–150 °C, which are characteristic of "cold discharges" according to Belenitskaya (2011).

In the Western Black Sea mud volcanoes, carbonates are represented by siderite, dolomite and calcite, smithsonite ($ZnCO_3$) (single findings), and phosgenite ($Pb_2CO_3Cl_2$).

In the North Caucasian mud volcanoes, sulfates are represented by only rarely occurring gypsum.

In the Kerch-Taman mud volcanoes, sulfates are represented by abundant gypsum, frequent barite, baritocelestine, and anhydrite. A kind of barite flower, obviously a product of mud volcanic formation, was observed in breccia of the Trubetskoy mud volcano.

In the Western Black Sea mud volcanoes, sulfates are represented by gypsum and barite. The former was observed in the MSU and Tredmar mud volcanoes, the latter in the MSU and Yuzhmorgeo mud volcanoes. Usually, small angular fragments are recovered. The presence of inclusions of tungsten minerals in barite indicates that one could consider it a vein mineral.

Phosphates are found in all types of mud volcanoes. However, in the Western Black Sea and Kerch-Taman mud volcanoes, the phosphate is a genetically controversial and often well-crystallized apatite, while in the North Caucasian Semigorskiy mud volcano, it is most likely mimetite of mud volcanic origin.

Newly formed minerals also include arsenates of lead (carmenite), as in the Yuzhmorgeo mud volcano.

Tungstates, compositionally corresponding to ferberite, were found in the Kerch and northern Caucasian (Semigorskiy only) mud volcanoes.

All mud volcanoes contain chalcedony, glauconite, and a number of other low-temperature minerals.

It is helpful for clarity to separate the minerals of the mud volcanic process from those of the oxidation zone of mud volcanoes. The latter is particularly pronounced in the Kerch-Taman mud volcanoes, where iron hydroxides, some silicates (e.g., chalcedony), carbonates, sulfates, etc. are encountered.

Numerous sources on mud volcanoes discuss the composition of fluids. By cation composition, they are represented by chloride-hydrocarbonate-sodium, hydrocarbonate-chloride-sodium, sulfate-chloride-sodium, chloride-hydrocarbonate-sulfate-sodium, and chlorine-sodium waters. The waters also contain boron (up to 928 mg/l), lithium, arsenic, mercury, bromine, iodine, and others with the ratio B > Br > J (Shnyukov and Naumenko 1982).

Avdusin (1948) wrote about precipitation of halite and lecontite, $(NH_4,K)NaSO_4 \times 2H_2O$, from mud volcanic waters.

A list of seasonal minerals is provided by Deyak (2010). It includes (1) borates: borax $Na_2B_4O_7$, tincalconite $Na_2B_4O_7 \times 5H_2O$, and probertite $NaCa(B_5O_7[OH]_4) \times 5H_2O$; (2) carbonates: calcite, dolomite, northupite $Na_3Mg(CO_3)_2Cl$, gaylussite $Na_2Ca(CO_3)_2 \times 5H_2O$, pirssonite $Na_2Ca(CO_3)_2 \times 2H_2O$, nesquehonite $Mg(CO_3)_2 \times 3H_2O$, and trona $Na_3H(CO_3)_2 \times 2H_2O$; (3) sulfates: gypsum, astrakhanite $Na_2Mg(SO_4) \times 4H_2O$, epsomite $MgSO_4 \times 7H_2O$, hexahydrite $MgSO_4 \times 6H_2O$, pentahydrate $MgSO_4 \times 5H_2O$, senderite $MgSO_4 \times 2H_2O$, tenardite Na_2SO_4, and mirabilite $Na_2SO_4 \times 10H_2O$; (4) nitrates: nitratite $NaNO_3$; and (5) chlorides: halite NaCl. All these are water-chemical minerals, usually polymineral and microcrystalline. They are called seasonal minerals because their crystallization depends on the temperature in which they either precipitate or dissolve.

8.10 Conclusions

As can been seen, mud volcanoes of the Western Black Sea basin and emanating from Maikopian sediments also possess the thinnest sedimentary cover above the Moho surface. This closeness to the Moho is reflected in the largest amount of finely fragmented material of unusual rocks broken by plumes, and the widest variety of native minerals, sulfides, and oxides; cinnabar, realgar, orpiment, and antimonite are, however, absent.

The Kerch-Taman mud volcanoes also emanate from Maikopian sediments, but they are separated from the Moho surface by a 40–45 km thick sedimentary cover. These mud volcanoes contain numerous and diverse native minerals (iron, nickel, gold, osmium, iridium, lead, tin, iron, cerium, lanthanum, neodymium iron, bronze in the form of hollow spherules, etc.). Sulfides are poorer in composition, but unlike the marine volcanoes, they include cinnabar, realgar, orpiment, and antimonite. Oxides are also present. There are numerous carbonates, sulfates, and chlorides. In the formulas of some minerals, such chemical elements as Cl and Br are prominent.

Mineralogically, the North Caucasian mud volcanoes emanating from Lower Cretaceous sediments are the poorest in mineralizations. Native minerals are iron (ferrite, kamacite, octibbehite), nickel, Ne-Fe, copper, Ni-Ta particles, lead, and zinc. The last two are numerous and typical for this type of volcano. The carbide moissanite is also found. The activity of the oxidation zone is poorly developed.

The deep development of mud volcanoes is reflected in their appearance, location, and activity. Their activity has the character of a constant, limited release of hydrocarbons that form salsas, springs, and volcanic accumulations resulting from large and sometimes catastrophic emissions that determine the morphology of the volcanoes. Types of mud volcanoes previously established by Kalinko (1960) represent in fact different episodes of their development. The impact of exogenous factors completes the image of their modern landforms, in particular, creating numerous characteristic ravines, such as barrancos, on their slopes.

A substantial number of mud volcanoes in the Black Sea region, primarily within the Black Sea itself, are still in need of closer study.

Many questions related to their origin remain controversial and continue to excite and puzzle geologists.

References

Abikh GV (1840) O poyavivshemsya na Kaspiyskom more ostrove (On the island that appeared in the Caspian Sea). Materialy k poznaniyu gryazevykh vulkanov Kaspiyskoy oblasti. Trudy Geologicheskogo instituta Azerbaydzhanskogo filiala Akademii Nauk SSSR 12/63:21–118 (in Russian)

Aliyev AdA, Bayramov AL (1999) Nekotorye aspekty tektoniki gryazevulkanicheskikh zon Gobustana (Some aspects of the tectonics of the Gobustan mud volcanic zones). Izvestiya NAN Azerbaijana, Nauki o Zemle 1:129–131 (in Russian)

Aliyev AdA, Bayramov AL, Mamedova AN (2004) Tektonika i perspektivy neftegazonosnosti gryazevulkanicheskikh oblastey Azerbaydzhana v svete novykh dannykh (Tectonics and oil and gas potential of mud volcanic regions of Azerbaijan in the light of new data). Izvestiya NAN Azerbaijana, Nauki o Zemle 1:32–43 (in Russian)

Arkhangel'skiy AD (1925) Neskol'ko slov o genezise gryazevykh vulkanov Apsheronskogo poluostrova i Kerchensko-Tamanskogo regiona (A few words about the genesis of the mud volcanoes of the Absheron Peninsula and the Kerch-Taman region). Bul. MOIP Geological Branch 33(3/4):269–285 (in Russian)

Aspandiar MF, Anand R, Gray D (2006) Mechanisms of elements dispersion through transported cover: a review. CRC LEME Report 230. Elsevier, Perth, pp 1–81

Avdusin PP (1948) Gryazevye vulkany Krymsko-Kavkazskoy geologicheskoy provintsii (Mud volcanoes of the Crimean-Caucasian geological province). AN SSSR, Moscow (in Russian)

Belenitskaya GA (2011) Rol glubinnykh flyuidov v osadochnom porodo- i rudoobrazovanii (The role of deep fluids in sedimentary rock and ore formation). Degazatsiya Zemli i genezis neftegazovykh mestorozhdeniy. GEOS, Moscow, pp 143–188 (in Russian)

Belousov VV, Yarotskiy LA (1936) Gryazevye sopki Kerchensko-Tamanskoy oblasti, usloviya ikh vozniknoveniya i deyatel'nosti (Mud hills of the Kerch-Taman region, the conditions of their occurrence and activity). ONTI, NKPT SSSR, Moscow (in Russian)

Betekhtin AG (1950) Mineralogiya (Mineralogy). Gosudarstvennoe izdatel'stvo geologicheskoy literatutury, Moscow (in Russian)

Chirvinskiy PN (1904–1905) Ob anapaite i drugikh mineralakh iz rudnykh plastov Kerchenskogo i Tamanskogo poluostrovov (On anapaite and other minerals from the ore layers of the Kerch and Taman peninsulas). Yezhegodnik po geologi i mineralogii Rossii XII:28–34 (in Russian)

Deryagin BV, Fedoseev DV, Buylov LA (1969) Sinteticheskie nitevidnye almazy (Synthetic thread diamonds). Fiziko-khimicheskie problemy kristallizatsii. Alma-Ata, pp 58–60 (in Russian)

Deyak MA (2010) Suchasni vodno-khemogenni utvorennya gryazovikh vulkaniv Kerchenskogo pivostrova (Modern water-chemogene formation of mud volcanoes of the Kerch Peninsula). Dissertation, Department of Marine Geology and Sedimentary Ore Formation of the National Academy of Sciences of Ukraine, Kiev (in Ukrainian)

Dmitrievskiy AN, Valyaev BM (2002) Uglevodorodnaya degazatsiya cherez dno okeana: lokalizovannye proyavleniya, masshtaby, znachimost' (Hydrocarbon degassing through the ocean floor: localized manifestations, extent, significance). Materialy vserossiyskoy konferentsii "Degazatsiya Zemli i genezis uglevodorodnykh flyuidov i mestorozhdeniy," GEOS, Moscow (in Russian)

Dmitrievskiy AN, Valyaev BM (2010) Uglevodorodnaya degazatsiya Zemli i genezis neftegazovykh mestorozhdeniy: razvitie idey P.N. Kropotkina (Hydrocarbon degassing of the Earth and the genesis of oil and gas fields: the development of the ideas of P.N. Kropotkin). Materialy vserossiyskoy konferentsii "Degazatsiya Zemli i genezis uglevodorodnykh flyuidov i mestorozhdeniy," p 710 (in Russian)

Egorov VN, Artemov YuG, Gulin SB (2011) Metanovye sipy v Cherom more. Sredoobrasuyushchaya i ekologicheskaya rol'. Sevastopol' (Methane seeps in the Black Sea. Environmental and ecological role). Ekosagidrofisika, Sevastopol (in Russian)

Fedorov SF (1939) Gryazevye vulkany Krymsko-Kavkazskoy geologicheskoy provintsii i diapirizm (Mud volcanoes of the Crimean-Caucasian geological province and diapirism). Rezul'naty issledovaniya gryazevykh vulkanov Krymsko-Kavkazskoy geologicheskoy provintsii. AN SSSR, Moscow-Leningrad, pp 5–44 (in Russian)

Fedorov YuN, Maslov AV, Ronkin YuL (2011) Sistematika elementov-primesey v produktakh izverzheniy ryada gryazevykh vulkanov Kerchensko-Tamanskoy oblasti (po dannym ICP–MS (Systematics of impurities in the products of eruptions of a number of mud volcanoes of the Kerch-Taman region [according to ICP–MS]). Lithosphera 5:117–123 (in Russian)

Galimov EM (1967) Ob evolyutsii ugleroda Zemli (On the evolution of the Earth's carbon). Geokhimiya 5:530–536 (in Russian)

Golubov BN, Ismagilov DF (2003) Truboobraznye tela pod dnom Severnogo Kaspiya i flyuidnyy rezhim ego nedr (Pipe-like bodies under of the northern Caspian Sea bottom and its fluid regime). Genezis nefti i gaza. GEOS, Moscow, pp 78–80 (in Russian)

Gorin VA (1952) Strukturnaya migratsiya uglevodorodov i ee rol v formirovanii neftyanykh i gazovykh mestorozhdeniy Azerbaydzhana (Structural migration of hydrocarbons and its role in the formation of oil and gas fields in Azerbaijan). Izvestiya AN Azrbajanskoy SSR, Seriya geologo-geograficheskih nauk 3 (in Russian)

Grossgeim MF (1961) Istoriya terrigennykh mineralov v mezozoe i kaynozoe Severnogo Kavkaza i Predkavkazya (History of terrigenous minerals in the Mesozoic and Cenozoic of the northern Caucasus and Ciscaucasia). Gostotekhizdat, Leningrad (in Russian)

Gubkin IM (1937) Neft' v SSSR (Oil in the USSR). Energeticheskie resursy SSSR, Akademiya Nauk, Moscow-Leningrad 1:169–194 (in Russian)

Guliyev IS (2010) Subvertikalnye geologicheskie tela – novye obekty poiskov mestorozhdeniy uglevodorodov (Subvertical geological bodies – new objects of hydrocarbon exploration). Vserossiyskaya konferentsiya "Degazatsiya Zemli: geodinamika, geoflyuidy, neft', gaz i ikh parogenezy," GEOS, Moscow, p 140–145 (in Russian)

Hough RM, Noble RP, Reich M (2011) Natural gold nanoparticles. Ore Geol Rev 42(1):55–61

Ismagilov DF, Krasikova TA, Terekhov AA et al (2006) Osobennosti formirovaniya osnovnykh strukturno-tektonicheskikh elementov rossiyskoy chasti Chernogo morya (Features of the formation of the main structural-tectonic elements of the Russian part of the Black Sea). Geologiya, geofizika i razrabotka neftyanykh mestorozhdeniy 2:4–11 (in Russian)

Kalinin YuA, Kovalev KR, Naumov YeA et al (2009) Zoloto kory vyvetrivaniya Suzdalskogo mestorozhdeniya (Kazakhstan) (Gold of the weathering crust of the Suzdalsiy deposit [Kazakhstan]). Geol Geofiz 50(3):241–253 (in Russian)

Kalinko MK (1960) O mekhanizme i usloviyakh obrazovaniya gryazevykh vulkanov (On the mechanism and conditions of formation of mud volcanoes). Problemy neftyanoy geologii. Gostekhizdat, Moscow, p 83–90 (in Russian)

Kalyuzhnyy VA, Telepko LF, Shchiritsa AS (1984) Sostav i R/T–parametry glubinnykh flyuidov po dannym izucheniya vklyucheniy v kaltsitakh iz vybrosov vulkanov (Composition and P/T – parameters of deep fluids according to the study of inclusions in calcites from volcanic emissions). Mineralogicheskiy sbornik L'vovskogo universiteta 2:32–38 (in Russian)

Kaminskiy FV, Blinova GK, Galimov EM et al (1985) Polikristallicheskie almazy s lonsdeylitom iz rossypey Yakutii (Polycrystalline diamonds with lonsdaleite from the Yakutian placers). Mineralogicheskiy zhurnal 7(1):27–36 (in Russian)

Karasik MA, Morozov VI (1966) Osobennosti rasprostraneniya rtuti v produktakh gryazevogo vulkanizma Kerchensko-Tamanskoy provintsii (Features of the distribution of mercury in the products of mud volcanism of the Kerch-Taman province). Geokhimiya 6:668–678 (in Russian)

Karpov GA, Anikin LP, Nikolaev AG (2012) Samorodnye metally i intermetallidy v peplakh deystvuyushchikh vulkanov Kamchatki i Islandii (Native metals and intermetallic compounds in the ashes of active volcanoes of Kamchatka and Iceland). In: Materialy konferetsii "Vulkanizm i svyazannye s nim protsessy" traditsionnaya regionalnaya nauchnaya konferentsiya, posvyashchennaya Dnyu Vulkanologa (k 50-letiyu IViS DVO RAN), Petropavlovsk-Kamchatskiy, 29–30 March 2012, p 18 (in Russian)

Khaustov VV (2011) O glubinnykh vodakh Yuzhno-Kaspiyskoy vpadiny (On the deep waters of the South Caspian depression). Uchenye zapiski: elektronnyy nauchnyy zhurnal Kurskogo gosudarstvennogo universiteta 2(18) (in Russian)

Kholodov VN (2002a) Gryazevovulkanicheskiye provintsii i morfologiya gryazevykh vulkanov (Mud volcanic provinces and the morphology of mud volcanoes). Geologiya i poleznye iskopaemye Mirovogo okeana 3:227–241 (in Russian)

Kholodov VN (2002b) Geologo-geokhimicheskie osobennosti i model' firmirovaniya gryazevykh vulkanov (Geological and geochemical features and model of mud volcanoes). Geologiya i poleznye iskopaemye Mirovogo 4:339–358 (in Russian)

Kostova SK, Popovichev VN, Egorov VN et al (2006) Raspredelenie rtuti v vode i donnykh otlozheniyakh v mestakh lokalizatsii struynykh metanovikh gazovydeleniy so dna Chernogo morya (Distribution of mercury in water and bottom sediments in places of localization of jet methane gas emissions from the bottom of the Black Sea). Morskoy ekologicheskiy zhurnal 5(2):47–55 (in Russian)

Kovalevskiy SA (1935) Gazovyy vulkanizm (vulkany i vulkanoidy) (Gas volcanism [volcanoes and volcanoids]). Azerbaydzhanskoe neftyanoe khozyaystvo 1 (in Russian)

Kropotkin PN, Valyaev BM (1980) Geodinamika gryazevulkanicheskoy deyatelnosti v svyazi s netegazonosnostyu (Geodynamics of mud volcanic activity due to oil and gas content). Geologicheskie i geokhimicheskie osnovy poiskov nefti i gaza. Naukova Dumka, Kiev, p 148–178 (in Russian)

Kropotkin PN, Valyaev BM (1984) Tektonicheskiy kontrol' protsessov degazatsii Zemli i genezis uglevodorodov (Tectonic control of Earth degassing processes and the genesis of hydrocarbons). In: Proceedings of the XXVII Geological Congress, vol 13, pp 13–25 (in Russian)

Kudryavtsev NA (1963) O kolichestvennykh sootnosheniyakh mezhdu uglevodorodami osadochnykh porod i nefti (On quantitative ratios between hydrocarbons of sedimentary rocks and oil). Geologiya nefti i gaza 9:45–48 (in Russian)

Kulakova LS, Lebedev LI (1990) Nekotorye cherty gryazevogo vulkanizma v Yuzhnom Kaspii (Some features of mud volcanism in the Southern Caspian.). Kaspiyskoe more. Voprosy geologii i geomorfologii. Nauka, Moskow, pp 96–103 (in Russian)

Lazarenko EK (1970) Kurs mineralogiï (Course of mineralogy). Vishcha shkola, Kiev (in Ukrainian)

Lebedev LM (1975) Sovremennye rudoobrazuyushchie gidrotermy (Modern ore-forming hydrotherms). Nedra, Moscow (in Russian)

Letnikov FA, Zaechkovskiy NA, Letnikov AF (2010) K voprosu o geokhimicheskoy spetsializatsii glubinnykh vysokouglerodnykh sistem (On the issue of geochemical specialization of deep-carbon systems). Doklady AN Rossii 433(3):374–377 (in Russian)

Limonov AF (2004) Gryazevye vulkany (Mud volcanoes). Soros Educational Journal 8(1):63–69 (in Russian)

Limonov AF, Ivanov MK, Meisner LB et al (1997) Novye dannye o stroenii osadochnogo chekhla v progibe Sorokina (Chernoe more) (New data on the structure of the sedimentary cover in the Sorokin trough [Black Sea]). Vestnik MGU 4(3):22–28 (in Russian)

Lukin AYe (2006) Samorodnye metally i karbidy glubinnykh geosfer (Native metals and carbides of deep geospheres). Geologicheskiy zhurnal 4:17–46 (in Russian)

Lukin AYe (2009) Samorodno-metallicheskie mikro- i nanovklyucheniya v formatsiyakh neftegazonosnykh basseynov – trassery sverkhglubinnykh flyuidov (Native-metal micro- and nano-inclusions in formations of oil and gas basins – tracers of ultradeep fluids). Geofizicheskiy zhurnal 31(2):61–92 (in Russian)

Lukin AYe (2013) Mineral'nye sferuly – indikatory flyuidnogo rezhima rudoobrazovaniya i naftidogeneza (Mineral spherules – indicators of the fluid regime of ore formation and naphthidogenesis) Geofizicheskiy zhurnal 35(6):10–53 (in Russian)

Lukin AYe (2014) Flyuidnyy litogenez – vazhneyshie napravleniya litologicheskikh issledovaniy v XXI stoletii (Fluid lithogenesis – the most important areas of lithological research in the 21st century). Geologicheskiy zhurnal 4:27–46 (in Russian)

Mamedov IS, Guliyev IS (2003) Subvertikalnye geologicheskie tela v osadochnom chekhle Yuzhno-Kaspiyskoy vpadiny (Sub-vertical geological bodies in the sedimentary cover of the South Caspian Basin). Izvestiya Azerbaydzhanskoy Akademii Nauk o Zemle 3:139–146 (in Russian)

Meisner LB, Tugolesov DA (2003) Opornye otrazhayushchie gerizonty v seysmicheskoy zapisi osadochnogo vypolneniya Chernomorskoy vadiny (Reference reflecting horizons in seismic recording of the sedimentary sequence of the Black Sea depression). Stratigrafiya, geologiya, korrelyatsiya 11(6):83–97 (in Russian)

Morozov VI (1965) Rtut' v kaynozoyskikh otlozheniyakh Kerchenskogo poluostrova (Mercury in the Cenozoic sediments of the Kerch Peninsula). Doklady AN SSSR 163(1):209–211 (in Russian)

Naumko IM, Zankovich G (2014) Vuglevodni flyuïdnikh vklyuchen u mineralakh naftogazonosnikh porodnikh kompleksiv Krosnenskoï zoni Ukraïnskikh Karpat (stan i prioriteti doslidzhen) (Hydrocarbons of fluid inclusions in minerals of the oil and gas bearing rock complexes of the Krosnenskoyi zone of the Ukrainian Carpathians [Status and research priorities]). Mineralogichnyy sbirnuk 641:134–154 (in Ukrainian)

Naumko IM, Kalyuzhnyy VA, Telepko LF et al (2008) Ob usloviyakh obrazovaniya kaltsita i gipsa iz vybrosov gryazevykh vulkanov Kerchenskogo poluostrova (po flyuidnym vklyucheniyam) (On the formation conditions of calcite and gypsum from outbursts of mud volcanoes of the Kerch Peninsula [fluid inclusions data]). In: XIII All-Russian conference on thermobarogeochemistry in conjunction with the IV APIFIS symposium, IGEM RAS, Moscow, 22–25 September 2008, pp 221–224 (in Russian)

Nesterenko T, Ivanchenko V, Tirshikina S (2007) Ontogeniya kulyastikh individiv agregativ tekhnogennogo pokhodzhennya (Ontogeny of spherical individuals of man-made units). Mineralogichnyy zbirnyk 57(1):76–80 (in Ukrainian)

Nikolskiy NS (1981) O litostabilnoy kristallizatsii prirodnykh almazov iz flyuidnoy fazy (On lithostable crystallization of natural diamonds from the fluid phase). DAN SSSR 250(4):1215–1220 (in Russian)

Pikovskiy YuI (2002) Flyuidnye plyumy litosfery kak model nefteobrazovaniya i neftegazonakopleniya (Fluid plumes of the lithosphere as a model of oil formation and oil and gas accumulation). Degazatsiya Zemli i genezis neftegazovykh mestorozhdeniy. GEOS, Moscow, pp 254–268 (in Russian)

Popov SP (1910) Mineraly rudnykh plastov Kerchenskogo i Tamanskogo poluostrovov (Mineral ore layers of the Kerch and Taman peninsulas). Trudy geologicheskogo muzeya imeni Petra Velikogo. Imperatorskaya Akademiya Nauk 4(7):99–198 (in Russian)

Porfiriev VB, Sollogub VB, Chekunov AV et al (1978) Stroenie i neftegazonosnost' severnoy chasti Chernogo morya i sopredelnykh territoriy (Structure and petroleum potential of the northern part of the Black Sea and adjacent territories). Naukova Dumka, Kiev (in Russian)

Portnov AM (1984) Flyuidnyy diapirizm i genezis almazov v kimberlitakh (Fluid diapirism and the genesis of diamonds in kimberlites). Byulleten' moskovskogo obshchestva ispytateley prirody, Otdelenie geologii 59:42–49 (in Russian)

Portnov AM (2015) Almazy – sazha iz trub preispodney (Diamonds – Soot from the pipes of the underworld). Nauka i zhizn' 7:1–7 (in Russian)

Poyarkov VE (1967) O poiskakh, razvedke, otsenke rtutnykh mestorozhdeniy (On prospecting, exploration, evaluation of mercury deposits). ONTI, Alma-Ata (in Russian)

Rukhin LB (1969) Osnovy litologii. Uchenie ob osadochnykh porodakh (Basics of lithology. Theory of sedimentary rocks). Nedra, Leningrad (in Russian)

Sandimirova EN (2008) Sfericheskie mineralnye obrazovaniya vulkanicheskikh porod Kurilskikh ostrovov i Kamchatki (Spherical mineral formations of the Kuril Islands and Kamchatka volcanic rocks). Dissertation, Far East Branch of the Russian Academy of Sciences, Petropavlovsk Kamchatsky (in Russian)

Saukov AV (1946) Geokhimiya rtuti (Mercury geochemistry). Trudy Instituta geologicheskih nauk AN SSSR, seriya mineralogo-geokhimicheskaya 78(17) (in Russian)

Shishkunov MG, Trufanov VN (1974) O nakhodke kinovari v neogenovykh otlozheniyakh Tamani (On the discovery of cinnabar in the Neogene deposits of Taman). Soviet Geol 6:141–143 (in Russian)

Shnyukov EF (1963) Geologiya ta rudonosnist Kezenskogo zalizorudnogo rodovishcha (Geology and ore content of the Kesen iron ore deposit). Pam'yati akad. V.I. Vernadskogo. AN URSR, Kiev, pp 10–14 (in Ukrainian)

Shnyukov EF (2016) Flyuidogennaya mineralizatsiya gryazevykh vulkanov Azovo-Chernomrskogo regiona (Fluidogenic mineralization of mud volcanoes of the Azov-Black Sea region). Logos, Kiev (in Russian)

Shnyukov EF, Ivanchenko VV (2015) Samorodnye mineraly v sopochnoy brekchii gryazevykh vulkanov Azovo-Chernomorskoy provintsii, razvitykh na nizhnemelovykh otlozheniyakh (Native minerals in the breccia of mud volcanoes developed on the Lower Cretaceous sediments in the Azov-Black Sea province). Geologiya i poleznye iskopaemye Mirovogo okeana 2:81–91 (in Russian)

Shnyukov EF, Naumenko PI (1966) Kimmeriyskie zheleznye rudy vo vdavlennykh sinklinalyakh Kerchenskogo poluostrova (Kimmerian iron ores in depressed synclines of the Kerch Peninsula). Soviet Geol 4:142–143 (in Russian)

Shnyukov EF, Naumenko PI (1982) Nakhodka Achinskoy rudonosnoy sinklinali (Discovery of the Achinskaya ore-bearing syncline). Geologicheskiy zhurnal 5:51–57 (in Russian)

Shnyukov EF, Netrebskaya EYa (2013) Korni Chernomorskikh gryazevykh vulkanov (The roots of the Black Sea mud volcanoes). Geologiya i poleznye iskopaemye Mirovogo okeana 14:87–92 (in Russian)

Shnyukov EF, Naumenko PI, Lebedev YuS et al (1971) Gryazevoy vulkanizm i rudoobrazovanie (Mud volcanism and ore formation). Naukova Dumka, Kiev (in Russian)

Shnyukov EF, Gnatenko GI, Nesterovskiy VA et al (1992) Gryazevjy vulkanizm Kerchensko-Tamanskogo regiona (Mud volcanism of the Kerch-Taman region). Naukova Dumka, Kiev (in Russian)

Shnyukov EF, Emelyanov VA, Inozemtsev YuI et al (2011) Gazovo-gryazevyy vulkanizm Kerchensko-Tamanskogo sektora Chernogo morya (65-reys NIS "Professor Vodyanitskiy," 2010 god) (Gaseous mud volcanism of the Kerch-Taman sector of the Black Sea [65th cruise of R/V "Professor Vodyanitsky"]). OMGOR NAN Ukrainy, Kiev (in Russian)

Shnyukov EF, Kobolev VP, Pasynkov AA (2013a) Gazovyy vulkanizm Chernogo morya (Gas volcanism of the Black Sea). Logos, Kiev (in Russian)

Shnyukov EF, Sokol EV, Nigmatulina YeN et al (2013b) Zoloto v gryazevykh vulkanakh Kerchenskogo poluostrova kak pokazate' glubinnosti gryazevulkanicheskikh flyuidov (Gold in the mud volcanoes of the Kerch Peninsula as an indicator of the depth of the mud volcanic fluids). Geologiya i poleznye iskopaemye Mirovogo okeana 4:79–89 (in Russian)

Shnyukov EF, Grishchenko NO, Deyak MA et al (2016). Prognoz novykh zhelezorudnykh proyavleniy v strukturakh Andreevskogo i Malotarkhanskogo gryazevykh vulkanov (Forecast of new iron ore manifestations in the structures of the Andreevskiy and Malotarkhanskiy mud volcanoes). Geologiya i poleznye iskopaemye Mirovogo okeana 1:57–60 (in Russian)

Smirnov SS (1951) Zona okisleniya sulfidnykh mestorozhdeniya (Sulfide deposit oxidation zone). Academy of Sciences of the USSR, Moscow (in Russian)

Sobisevich AL, Gorbatikov AV, Ovsyuchenko AN (2008) Glubinnoe stroenie gryazevogo vulkana gory Karabetovoy (The deep structure of the mud volcano Karabetova mountains). Doklady RAN 422(4):542–546 (in Russian)

Sobolevskiy YuV, Sobolevskiy YuV, Kutniy VA et al (1982) Graphite in the breccia of the Korolyovskiy mud volcano (Kerch Peninsula). Doklady AN USSR, seriya B 10:20–22 (in Russian)

Starostenko V, Buryanov V, Makarenko I et al (2004) Topography of the crust-mantle boundary beneath the Black Sea Basin. Tectonophysics 381(1–4):211–233

Starostenko VI, Makarenko IB, Rusakov OM et al (2010) Geofizicheskie neodnorodnosti megavpadinyi litosferyi Chernogo morya (Geophysical heterogeneities of the Black Sea). Geofizicheskiy zhurnal 32(5):3–20 (in Russian)

Steber EA (1914) Gryazevye vulkany i proiskhozhdenie nefti (Mud volcanoes and the origin of oil). Yuzhnyy inzhener 4:120–128; 6:144–153 (in Russian)

Subbota MI (1964) Novye dannye o gazakh i tipakh gryazevykh vulkanov rayona Zapadnoy Kubani (New data on gases and types of mud volcanoes in the area of the Western Kuban). Trudy Vsesoyuznogo nauchno-issledovatel'skogo geologo-razvedochnogo neftyaogo instituta nedr 41:156–178 (in Russian)

Sungurov AM (1958) Osobennosti stroeniya gryazevykh vulkanov Yugo-Zapadnoy Turkmenii (Features of the structure of mud volcanoes in Southwest Turkmenistan). Novosti neftyanoy tekhniki, seriya geologicheskaya 9:12–14 (in Russian)

Ustinov VI (1976) Fraktsionirovanie izotopov ugleroda v protsesse dekarbonatizatsii (Carbon isotope fractionation during decarbonization). Geokhimiya 1:120–124 (in Russian)

Vernadskiy VI (1913) Opyt opisatelnoy mineralogii (Experience of descriptive mineralogy). Vibrani naukovi pratsi V.I. Vernadskogo, Kyiv, 2012, pp 33–242 (in Russian)

Voskoboynikov NI, Gur'iev AV (1832) Geologicheskoe opisanie poluostrova Tamani, prinadlezhashchego k zemle voyska Chernomorskogo (Geological description of the Taman peninsula belonging to the land of the Black Sea army). Gornyy zhurnal 1 (in Russian)

Yakubov AA, Zeynalov MM (1964) Gryazevoy vulkanizm Azerbaydzhana (Mud volcanism of Azerbaijan). Ocherki po geologii Azerbaydzhana. AN Azerbaijanskoy SSR, Baku (in Russian)

Zharvin NA (2013) Uroki neftyanoy katastrofy v Meksikanskom zalive (Lessons from the Gulf of Mexico oil disaster). Energiya, ekonomika, tekhnika, ekologiya 4:63–67 (in Russian)

9 Mud Volcanism and Mineral Resources

Summary

In this chapter, the connection between mineral resources and mud volcanism is explored. Mud volcanoes can be widely used by humans to assist in the search for fossil fuels, iron ores, rare chemical elements, non-metallic raw materials, therapeutic mud, and much more.

In Azerbaijan, a number of discoveries of oil and gas fields in the areas of mud volcanoes show that they represent potential indicators in the search for oil and gas. This should be relevant for the Kerch Peninsula as well. A number of wells drilled to a depth of 500 m on the Bulganakskiy mud volcano were unsuccessful, however, probably because they did not reach the required depth. Given the geological history of the Kerch Peninsula, oil and gas fields should be located at much deeper horizons.

In the Black Sea, mud volcanoes located below isobath 700 m can be considered indicators for the presence of gas hydrates. Particularly valuable in this regard can be ring-shaped (isometric) structures that follow the outline of mud volcanoes and are complicated by subsidence phenomena.

On the Kerch Peninsula, deposits of oolitic sedimentary iron ores are often associated with recessed synclines developed near mud volcanoes. The largest ore deposit is Novoselovskoye with confirmed reserves of 125 million tons; less significant ore deposits are associated with the mud volcanoes of Baksinskiy, Rep'yevskiy, Achinskiy, and some others.

Significant concentrations of boron and iodine have been found in the mud volcanoes of the Kerch-Taman region. At the Bulganakskiy mud volcano, boron in the form of borax was often mined in the 1920s.

Manifestations of sulfur mineralization are frequently associated with mud volcanoes on the Kerch Peninsula.

Deposits of sand as well as limestones of Sarmatian and Chokrakian age from recessed synclines are also present in various mud volcanoes of the Kerch-Taman region. Mud volcanic breccia in many cases is suitable for use as a ceramic raw material as well as for the production of metallurgical pellets.

Mud volcanic breccia and mineral springs enriched with hydrogen can be used as balneological raw material.

9.1 Onshore and Offshore Mud Volcanoes as Possible Indicators of Petroleum and Gas Resources

Mud volcanoes appear in 25 oil and gas basins globally, while about 170 (87%) basins reveal no signs of them (Rakhmanov 1982). The question therefore arises whether mud volcanoes can be

E. Shnyukov, V. Yanko-Hombach, *Mud Volcanoes of the Black Sea Region and their Environmental Significance*, https://doi.org/10.1007/978-3-030-40316-4_9

used as indicators in the search for oil and gas resources.

By analyzing mud volcanic processes and oil and gas deposits in the Caspian region, Kholodov (2002a, b) suggested that mud volcanoes are genetically related to each other in those areas where they coincide geographically. Kalinko (1960, 1964) linked mud volcanism only to areas where the oil- and gas-bearing strata are characterized by anomalously high pressures. But this feature is not universal, as there are areas where oil and gas reservoirs have abnormally high pressure but mud volcanoes are absent. For mud volcanoes to occur, abnormally high pressures in oil- and gas-bearing formations are required as well as a number of other geological conditions that are found in many places within the Black Sea region.

The famous Russian petroleum geologist and president of the 1937 International Geological Congress in Moscow, I.M. Gubkin, wrote in the introduction to his book (1939) that in accordance with theoretical considerations, mud volcanoes can be used as primary indicators in the search for, and exploration of, oil reservoirs. His recommendation was reinforced by deep exploratory drilling conducted within the Lok-Batan mud volcano in Azerbaijan. Well no. 45, which had been drilled within the oilfield of the same name in 1933, was located 1.5 km east of the mud volcano crater and yielded a fountain of oil with an output of 20 thousand tons a day. Despite the fact that more than 20 eruptions have occurred at Lok-Batan since that time, this well continues to operate, while the drilling of new wells that remain in continuous use has not decreased the volume of hydrocarbons ejected by the mud volcano during its many eruptive events. Currently, over 30 million tons of oil, 800 million m^3 of free gas, and one billion m^3 of oil gas have been extracted, and correlations between the volumes of ejected gases and the frequency of mud volcano eruptions have been established. Mamedov and Guliyev (2003) associated the mud volcanoes with subvertical geological bodies descending to a depth of 20 km that ejected huge amounts of gas. They concluded that over 70% of all oil and gas fields, including giant ones in the Southern Caspian basin, can be associated with mud volcanoes.

Rakhmanov (1982), however, denied a direct connection between mud volcanoes and oil deposits and understood mud volcanic activities as limited to gas deposits only.

Within the Kerch-Taman region, seepages of oil from mud volcanoes are not uncommon. Usually they are small in size but sometimes quite effluent, forming oil lakes near volcanoes. Some mud volcanoes are even referred to as oil-bearing because of the associated seepages (Figs. 9.1, 9.2, and 9.3).

Within the Kerch Peninsula, seepages of oil from mud volcanoes are rarer, but they still occur (Figs. 9.4 and 9.5).

Occasionally, some rocks (called kirs, which are products of subaerial diagenesis of oil) formed by a mixture of thickened oil or asphalt with sandy or clayey material can be observed.

Attempts to transfer search criteria that work well in Azerbaijan to the Black Sea in general, and the Kerch-Taman region in particular, have

Fig. 9.1 Oil manifestations on the Northern Neftyanoy (oil-bearing) mud volcano, Taman Peninsula

Fig. 9.2 Oil manifestations on the Southern Neftyanoy mud volcano, Taman Peninsula

Fig. 9.4 Discharge of oil at the site of a bore hole to the west of the Enikalskiy mud volcano, Kerch Peninsula

Fig. 9.3 Oil manifestations in the vicinity of the Bugazskiy mud volcano, Taman Peninsula

Fig. 9.5 Oil derrick within the Kayala-Sartskaya recessed syncline, Kerch Peninsula

failed, however. About 20 half-kilometer-deep exploratory wells drilled in 1957 within the Bulganakiy mud volcano did not yield positive results, and other locations also failed to find evidence of underlying oil reservoirs (e.g., Gubkin 1939).

In the regional literature, conflicting opinions exist on whether mud volcanoes can serve as indicators for oil and gas deposits. In the Kerch-Taman area, mud volcanoes have not been considered to be a search indicator but instead as a sign of the impoverishment of gas deposits that were rich in the past. It has been thought that mud volcanic activity marks the final stages in the depletion of oil and gas deposits.

In order to extrapolate the above data on Azerbaijan to the Kerch-Taman mud volcanic region, two features of its geological development should be considered. (1) In Azerbaijan, maximum mud volcanic activity occurred in the Quaternary, while in the Kerch-Taman region, particularly within the Kerch Peninsula, maximum activity took place within the Neogene. (2) The Kerch-Taman region is characterized by a thick (tens of kilometers) sedimentary cover (see Chap. 6) represented by Triassic-Jurassic, Lower Cretaceous, Upper Cretaceous-Paleogene, and Oligocene-Neogene oil- and gas-bearing deposits (Dolenko et al. 1968). It may be that similar deposits containing oil and gas will be discovered in older sediments as well. Judging by the literature (e.g., Guliev 2010), the roots of mud volcanoes presumably originate as deep as 18–20 km and more. It must therefore be assumed that mud volcanoes represent a drain that contributes to the depletion of the upper parts of oil and gas deposits. At the same time, they also encourage the replenishment of gas and oil resources removed from deeper horizons. The role of mud volcanoes, therefore, is contradictory, being at the same time diminishing and restorative.

There are examples in which gases released from mud volcanoes have been directly used to satisfy human needs. In one case, gases emitted from the Central Lake of the Bulganakskiy mud volcanic hearth, possessing a flow rate of about 100 m^3 per day, were fed through a 150 m gas pipeline to serve as an auxiliary fuel for the kitchen of a boron plant during its operation in the late 1920s and early 1930s (Belousov and Yarotskiy 1936) (Fig. 9.6).

Flammable gases from the Gora Miska mud volcano on the Taman Peninsula were explored by approximately 40 wells with depths from 6 to 30 sazhens (1 sazhen = 2.16 m) in 1916–1917. A gas-bearing layer up to 1 sazhen thick was discovered at a depth of 6–8 sazhen (Zenin 1955). This work was continued by A.N. Semizashennyy, who drilled up to six more wells with depths of 60 sazhen and encountered two low-thickness, gas-bearing layers. Some of them were exploited. Their gas pressure was initially equal to 0.8 atmospheres, but it gradually fell afterward, and production was halted (Pyatnitskiy 1925).

In the Black Sea, several oil spots of non-technogenic origin are known that could have been ejected by mud volcanoes. The first report is from the nineteenth century and was described within the eastern part of the Black Sea approximately 20 km offshore from Rize city—the capital of Rize Province, Turkey. Although it was studied using various methods, not much information is available. The size of the oil spot varied between 3 and 21 km^2, and oil emission varied from 140 to 1000 tons/year. The highest oil emission was estimated at 2300 thousand tons/year and most likely was related to mud volcanic activity as was confirmed when the Blue Stream gas pipeline was under construction (Okay et al. 2012).

Fig. 9.6 Construction for the collection of gases from the Central Lake of the Bulganakskiy mud volcanic hearth. (Belousov and Yarotskiy 1936)

The second large occurrence of oil spotting consisted of 16 drifting spots on the sea surface; it was identified on satellite images in 2011 in the southeastern part of the Black Sea, close to the port city of Poti in Georgia. These spots were estimated as equivalent to the release of 7000 tons of oil per year. The main source of the oil was an elevation on the sea bottom at a water depth of 1050 m (Shnyukov et al. 2015).

In other areas of the Black Sea, the scale of oil emissions from underwater mud volcanoes is far more modest. These emissions were identified either by instrumental measuring of bottom water near mud volcanoes or by coring into bottom sediments and mud volcanic breccia. A significant amount of work carried out in the Russian part of the Black Sea has revealed some signs of oil released by the presently active Neftyanoy mud volcano as well as the Manganari 1 mud volcano that was active in pre-Holocene times. Of the ten samples of bottom sediments that covered the Manganari 1 mud volcano, oil manifestations were identified in eight of them. The pores and caverns of the sediment were mainly filled with oxidized oil, and only in one case was fresh unoxidized oil detected (Shnyukov et al. 2013). An oil spot was discovered at the water's surface in the vicinity of the Neftyanoy mud volcano (see Table 6.7, Fig. 6.200 for location), which is located at a water depth of 1930 m.

Generalizing from the data on oil manifestations in the Black Sea, N.A. Maslakov pointed out the problem of distinguishing natural oil spots from those of technogenic origin (Shnyukov et al. 2015); distinguishing them became possible only with the help of satellite radar.

Manifestations of oil and gas have frequently been recorded in Neoeuxinian mud volcanic breccia (Shnyukov et al. 2013).

In the sediments of the Periklinal mud volcano (Table 6.7, Fig. 6.200), an anomalously high concentration of heavy methane homologs was detected. One sample of the uppermost layer (0–5 cm) of bottom sediments was enriched with an extremely high content of petroleum products (7947.4 mg/l, with an average value of 41.8 mg/kg). However, even though the oil materials were not visually observed, their distinct smell was quite apparent. The deep origin of hydrocarbons from the Periklinal mud volcano as well as those of Udodov and Panayev was confirmed by geochemical data (Kruglyakova et al. 2012).

It is interesting that supporters of both the organic and inorganic origin of oil and gas consider mud volcanoes as confirmation of their views. Thus, Kalinko supported the organic origin of hydrocarbons when calculating the productivity of mud volcanic areas in Eastern Azerbaijan but concluded that the sedimentary cover of the South Caspian basin is capable of releasing the volumes of hydrocarbon gases necessary to generate mud volcanic activity (Kalinko 1960).

In contradiction, Kropotkin and Valyaev (1980) supported the inorganic origin of hydrocarbons and came to the directly opposite conclusion: no oil and gas deposits are sufficient for feeding the activity of mud volcanoes in Azerbaijan from the Neogene to the present. Consequently, the source of any mud volcano cannot be a previously formed gas deposit within the sedimentary cover. The source of hydrocarbons must be deep, lying below the foot of the sedimentary metamorphic section, and therefore, these authors rejected the idea of mud volcanoes as destroyers of deeply buried oil-gas deposits. The activity of mud volcanoes is possible only when there is a constant intensive feeding of their deep (intermediate) foci at the base of the sedimentary section due to deep degassing. Oil and gas accumulation and mud volcanic activity are paragenetically linked as feeding on products of deep degassing and having single supply channels: tectonic disruptive disturbances.

These views fall on favorable ears in the Black Sea region. Rusakov and Kutas (2014) and Rusakov (2016) reveal the crisis of the organic hypothesis of the origin of hydrocarbons: in the Black Sea, only 1 of the 12 recently drilled deep boreholes has provided evidence of substantial oil production. The basic paradigm (organic) of hydrocarbon genesis deserves a revision. In our opinion, it is necessary to return to the exploration of mud volcanoes, primarily those of the Western Black Sea basin and the Sorokin Trough. In the Western Black Sea basin, the seabed is only 17 km above the mantle—the Moho surface. The largest mud volcanoes in the Black Sea are developed here. In the Sorokin Trough, they are also large enough and numerous—26 are known. In 5–6 of them, gas fountains up to 300 m in diameter reach 800–1300 m in height. The gas emissions are enormous, and this is in a marine environment! Gas hydrates quite often are present in the eruption of mud volcanoes. In this way, drilling within mud volcanoes is unlikely to be disappointing and should hold some promise for oil and gas exploration.

9.2 Submarine Mud Volcanoes as Search Indicators for Methane Gas Hydrates in the Black Sea

Depletion of natural gas deposits, on the one hand, and the promise of significant nontraditional resources on the other make it of paramount importance to consider the industrial development of such new energy sources in the near future. One of the encouraging opportunities for increasing the resource base, along with coal methane, shale gas, and gas from tight reservoirs, is the search, exploration, and development of natural gas hydrate accumulations (Holder and Bishnoi 2000; Majorowicz et al. 2002; Bugay and Balakirov 2001; Beauchamp 2004; Glumov et al. 2005; Hancock et al. 2005; Osadetz et al. 2005; Mazurenko et al. 2006; Dmitrievskiy and Balanyuk 2006; Shchebetov 2006; Max et al. 2006; Yakushev 2008; Anfilatova 2008; Moridis and Sloan 2007; Boswell 2009; Walsh et al. 2009;

Basniev and Sukhonosenko 2010; Marcelle-De Silva and Dawe 2011; Grigoryev and Afanasyeva 2012; Bohrmann and Torres 2016; and many others).

"Gas hydrates are non-stoichiometric, solid compounds similar to ice crystals" (Sloan 1998). In these compounds, also called clathrates (Latin "clatratus" for cage), water molecules form cage-like structures in which low molecular weight gases are enclosed as guest molecules. The gas molecules interact with water molecules through van der Waals (nonpolar) forces. Since no bonding exists between the guest and host molecules, the guest molecules are free to rotate inside the cages, and this rotation can be measured by spectroscopic techniques (e.g., Gutt et al. 1999). "Gas hydrate can contain different types of gas molecules in separate cages, depending on the gas composition in the environment of formation. Methane is the main gas in naturally occurring gas hydrates; however, H_2S, CO_2, and, less frequently, other hydrocarbons, can also be found within the hydrate structure" (Schulz and Zabel 2006, 481).

Russian scientists were the first to recognize that methane in natural systems could form gas hydrate deposits wherever the pressure and temperature conditions were favorable. Gas hydrates were discovered first in the permafrost regions of Russia (Makogon 1974) and Canada's MacKenzie Delta (Bily and Dick 1974) and subsequently in sediments of the Caspian Sea and Black Sea (Efremova and Zhizhchenko 1974).

Gas hydrate stability depends primarily upon temperature and pressure; however, other variables, such as gas composition and ionic impurities in water, influence stability boundaries (Kvenvolden 1993). The existence and depth of a hydrate deposit are often indicated by the presence of a bottom-simulating reflector (BSR). A BSR is a seismic reflection indicating the lower limit of hydrate stability in sediments due to the different densities of hydrate-saturated sediments, normal sediments, and those containing free gas.

About 98% of natural gas hydrates are marine and concentrated on the continental slope and shelf of the World Ocean (Ginzburg et al. 1988; Ginzburg and Solovyev 1994; Solovyev 2001; Makogon and Holditch 2002; Beauchamp 2004; Klauda and Sandler 2005). Basically, they are confined to fault zones, alluvial or talus fans, and are often located near underwater mud volcanoes. Accumulations of gas hydrates are associated with the unloading of fluids at sea bottoms or they form at a considerable depth below the bottom, but they are also caused by flows of underground fluids (Valyaev and Dmitrievskiy 2009).

Among the southern seas of Eurasia, the Black Sea shows the most promise for harvesting this resource due to its significant potential reserves of hydrated gas and its proximity to commercial consumers of hydrocarbons. Gas hydrates are the least investigated energy source but the most problematic because of their involvement in commercial mining, geological conditions of occurrence, methods of production, etc.

According to various estimates, the gas hydrates of the Black Sea contain between 20 and 49 trillion m^3 of gas (Korsakov et al. 1989, 1991; Vassilev and Dimitrov 2002).

In the Black Sea, gas hydrates were found for the first time and described by Efremova and Zhizhchenko (1974); they occurred in the form of small ice-like, white crystals in voids within the mud. Subsequently, gas hydrates were discovered in various areas of the basin during the course of dozens of research cruises on the R/Vs of Ukraine, Russia, and some European countries (Korsakov et al. 1991; Ginsburg et al. 1990; Shnyukov et al. 1990; Byakov and Kruglyakova 2001). According to present knowledge, gas hydrates form small cakes and ice packs in the mud, irregular small discharges, veins, and thickenings. So far, no thick and uniform layers of gas hydrates have been observed, although small interlayers 1–2 cm thick have occasionally been encountered.

As of today, discoveries of gas hydrates in the Black Sea are confined to the Sorokin Trough, the Western Black Sea basin, and the Tuapse and Giresun depressions. They were also discovered offshore from Sevastopol and elsewhere. The total number of findings does not exceed 20, and most of these are associated with mud volcanoes.

Gas hydrates occur in paleodeltas but less frequently. There is obviously a general pattern in the location of gas hydrate deposits: all of them lie within zones of deep or regional faults and especially their intersections. Mud volcanoes seem to "sit" on faults; the river paleovalleys reflect deep faults on the surface of the sea bottom. From the standpoint of the deep origin of gases, this is an entirely explicable coexistence. In this regard, the study of mud volcanoes as search criteria for gas hydrates is important.

Geologically, gas hydrates are confined to mud volcanic breccia, freshwater outlets on the sea bottom, and submerged paleodeltas and paleovalleys, in particular of the Danube and the Dnieper rivers.

Geophysical studies in recent years have added additional information yielding a better understanding of gas hydrate development and its main characteristics. For example, Romanian geophysicists carrying out seismic studies offshore from the Danube delta traced the BSR boundary together with characteristic features of gas hydrates across an area of about 2900 km^2. They also investigated the possible influence of climate change on gas hydrate chemistry, development, and stability during glaciations (Popescu et al. 2007).

Russian scientists from Moscow State University and "Yuzhmorgeologiya" studied gas hydrates in the Tuapse depression at the request of the Rosneft Oil Company—the leader of Russia's petroleum industry and the world's largest publicly traded petroleum company. Several areas of gas hydrates have been discovered, and among them, several mud volcanoes emitting oil have been encountered. Sediment cores enriched with oil have been raised in several locations within the Tuapse depression. In the sediment sequence, the BSR boundary was repeatedly traced.

Geophysical investigations of gas hydrates have also been carried out in the economic zone of Ukraine by German specialists during the course of "biological" studies to the west of Sevastopol at the mouth of the Paleo-Dnieper (Paleo-Kalanchak) River and within the Sorokin Trough. The area within which gas hydrates are found has been estimated at 805 km^2, and the thickness of the gas hydrate-containing layer ranges up to 100 m at water depths of 700–1350 m (Lüdmann et al. 2004). In 2010–2011, a study focused on the same area was carried out by geophysicists of the Institute of Geophysics of the National Academy of Sciences of Ukraine (Kobolev and Verpakhovskaya 2014).

It was suggested that the existence of methane gas hydrates in the Black Sea is thermodynamically possible only at water depths below 600–700 m, meaning that only those mud volcanoes that are located in deeper places are suitable as indicators in the search for gas hydrates. These mud volcanoes are quite numerous, and their activities are often accompanied by large explosions, especially if one takes into account their lengthy period of operation over geological time. During one of the scientific cruises on the R/V "Professor Vodyanitskiy" in the area of the Dvurechenskiy mud volcano, we observed a gas fountain with a height and diameter up to 800 m and 300 m, respectively. Rakhmanov (1987) wrote that about 250 million m^3 of gas was emitted into the atmosphere during one mud volcanic eruption in Azerbaijan and an estimated 125×10^{12} m^3 during the entire Quaternary period.

As a result of huge sediment movement from a substantial depth upward to the seafloor, positive morphostructures usually emerge in close proximity to mud volcanoes. This causes a lowering of the geostatic pressure within the morphostructures, which then causes them to sag, leading to the development of depressions. As mud volcanic eruptions continue, the depressions gradually subside and fill in with marine sediments. Such depressions can give the impression of tectonically produced synclines, but in fact they are pseudotectonic geological morphostructures that form within mud volcanoes or nearby. Quite often, they are framed by rings forming graben-compensating synclines that are similar to the terrestrial compensatory recessed synclines of the Kerch Peninsula.

The depressions reveal an isometric shape and are not large in size; they range from hundreds of meters to 4 km in diameter but sometimes can cover an area as large as 27 km^2. The thickness of marine sediments (e.g., clays, coccolithic muds, and other marine facies—depending upon the depth of the basin) interlayered with mud volcanic breccia can reach a few hundred meters.

As the Black Sea basin is a remnant of the Paratethys, most mud volcanoes have been in operation for a long time, at least from the Neogene to the present. This means that depressions should be filled at least with Neogene and Quaternary sediments accumulated synchronously with mud volcanic eruptive material in a manner similar to recessed synclines of the Kerch and Taman peninsulas.

Repeated sea-level changes over such a long period of time could have affected the depth of the basin by a maximum of approximately 200 m. This would mean that most of the mud volcanoes now located below isobath 700 m on the Black Sea bottom would always have been covered by a thick water column, creating stable temperature and pressure conditions permitting the formation of gas hydrates.

Presumably, the development of similar compensatory depressions occurs in the Caspian Sea, where large mud volcanoes are numerous, as well as in other areas of mud volcanic activity in the World Ocean, for example, near Trinidad Island.

Pseudotectonic compensatory depressions most likely contain gas hydrates as well as ore deposits and other mineral resources. This appears likely because mud volcanoes constantly release natural gas mixtures in the form of torches within the hydrate formation zone. This usually occurs somewhere in the center of the top and/or in the weakened zones near volcanoes, i.e., in the depressions.

According to Solovyev (2001), (1) only 1 to 10% of the gases entering the hydrate formation zone are stabilized in gas hydrates; (2) gas hydrates do not form continuous layers; (3) gas hydrates are not exposed on the seafloor but have a discrete distribution within the sedimentary sequence; (4) gas hydrates occupy only 10% of the total area of the hydrate formation zone; and (5) deposits containing gas hydrates create a flow of outgoing gas-saturated water, rather than a flow of gases.

In this regard, the conclusion of Konyukhov et al. (1990) based on study of the MGU mud volcano is of interest. The authors write that a search for gas hydrates was carried out at the top of the MGU mud volcano in 1988 immediately after their discovery there by scientists from Moscow State University and "Yuzhmorgeologiya." The search was unsuccessful despite the dozens of sediment cores extracted and investigated using different methods. Apparently, the gas hydrates were confined to a narrow zone where the mud volcanic breccia comes to the surface. Since this process is highly localized, the zone of gas hydrate development is very limited in area and is probably connected with the activity of a peripheral, not central, crater in the western part of the volcano's peak.

Presumably, such a narrow zone of development for methane hydrates represents a compensatory deflection zone within which gas hydrate formation is confined. Sedimentary thickness in the Novoselovskaya recessed syncline on the Kerch Peninsula is 211 m. It is justifiable to assume that clay-gas hydrate deposits of at least that thickness exist within compensating deflections near the large mud volcanoes of the Black Sea. Near such mud volcanoes, the section of modern Black Sea sediments shows sharply increased thicknesses of hydrotroilitic silt, indicating an elevated intake of hydrogen sulfide from the Earth's interior. Hence, the development of hydrogen sulfide gas hydrates seems possible here.

The most likely to develop, however, are methanohydrates because they are slightly heavier; hydrates of carbon dioxide and hydrogen sulfide are lighter than water, and they therefore float upward. The adhering of clay particles to the gas bubbles and settling of these lighter CO_2 and H_2S hydrates to the bottom is nevertheless also possible.

A detailed study of some acoustic profiles of the Black Sea bottom allows us to demonstrate the existence of small depressions-grabens, described above in the sections of many mud volcanoes (Fig. 9.7).

Similar pseudotectonic compensatory depressions have been observed and are described in

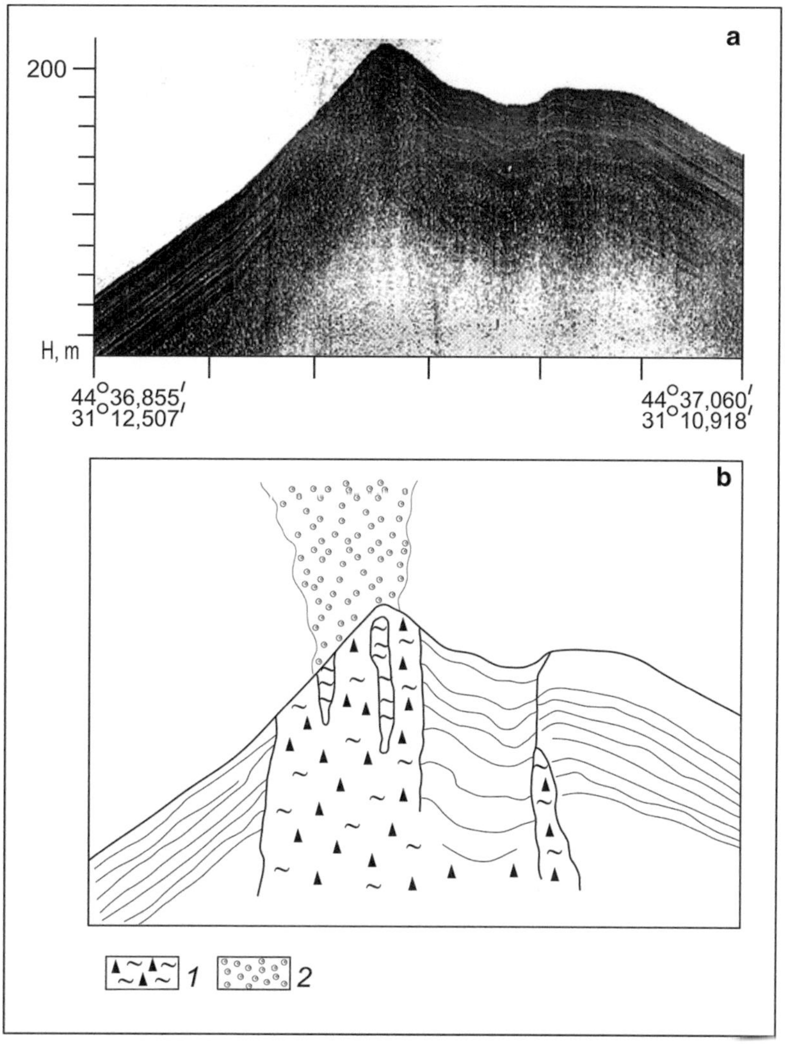

Fig. 9.7 Acoustic profile through one of the Black Sea mud volcanoes; (**a**) acoustic profile, (**b**) profile interpretation. Key: (1) mud volcanic breccia, (2) gas outlets

other publications (e.g., Gorchilin and Lebedev 1991). These can serve as traps for gas hydrate formation (Shnyukov et al. 2013).

Of particular interest is a large ring morphostructure (Fig. 6.145) in the central part of the Western Black Sea basin that was discovered during the course of research cruises by the R/Vs "Kiev" and "Professor Vodyanitskiy." Based upon a number of indirect signs, one can assume a mud volcanic origin. This is evidenced by (1) the relief, (2) the specific stratification and dislocation of sediments in the geological section as observed in hydroacoustic profiles, (3) some lithological features, and (4) the high gas saturation of the silt. Intensive degassing of the sediments indicates that the bottom of the basin where this morphostructure is located represents a so-called gas swamp, which is a characteristic feature of mud volcanic fields and gas-saturated silts in the basin.

Taking into account the already mentioned connection of gas hydrates with mud volcanic activities, the constant release of gases from the bottom of the morphostructure permits the assumption that gas hydrates are developing there.

Of greater interest to those studying the Black Sea is, most likely, not the modern activity of mud volcanoes but their past activities, specifically from the Neogene to the present. In this case, the pseudotectonic compensatory depressions filled with sediments of considerable thickness can reach considerable sizes and accumulate gas hydrates.

Gas emissions around mud volcanoes are not necessarily localized in connection with compensatory depressions. They can be developed over a larger area, and as such, their formation should not be limited to these morphostructures. Simply, in compensatory depressions, thickness of gas hydrates is much higher, and their age is older compared to other areas.

As gas hydrates are quite sensitive to changes in temperature and pressure, they very likely would not be preserved within the sequences of Neogene deposits. This issue is subject to further study.

Comparison of mud volcanic processes within the Kerch-Taman region to those in the Black Sea suggests that (1) mud volcanoes can be used as an indicator in the search for gas hydrates, as is correctly described in many publications, (2) submarine mud volcanoes, most likely, also have roots somewhere deep within the Mesozoic and even in the mantle, (3) pseudotectonic compensatory depressions related to mud volcanoes are similar to recessed synclines on the Kerch Peninsula, (4) the thickness of the hydrate-containing sediments is greater within pseudotectonic compensatory depressions than it is around mud volcanoes, (5) similar subsiding depressions have developed on many active mud volcanoes, and (6) the large ring morphostructure in the center of the Western Black Sea basin is likely to be a subsidence structure which quite likely contains gas hydrate deposits.

9.3 Mud Volcanism and Iron Ore Deposits

Some information on Kerch iron ore deposits is given in Sects. 6.2 and 6.3. Here, we provide more data on this subject.

The Kerch region of Neogene oolitic iron ores includes both the Kerch and Taman peninsulas. It was first described at the beginning of the nineteenth century and was continually studied throughout the nineteenth and twentieth centuries (Shnyukov 1999). The main iron ore deposits include the Kamysh-Burunskoye, Eltigen-Ortelskoye, Kyz-Aulskoye, and Katerlezskoye. In the 1930s, they were explored under the supervision of S.V. Konstantov from the USSR Geological Committee. In 1947–1954, investigations were conducted by the Ukrainian Geological Trust. Additional exploration of the northern group of Kerch ore deposits (Katerleskoye, Chegene-Salinskoe, etc.) was carried out in 1962–1963 by I.V. Dynovskiy, M. Rudman, and others. It was determined that the development of iron ore deposits took place within the large synclinal structures bearing the same names as the mud volcanoes. In general, the synclinal structures were 5–10 km in length and 3–6 km in width. The thickness of the iron ore deposits ranges between 5 and 10 m, with total reserves estimated at about 2 billion tons.

The iron ores are represented by oxidized (brown ores), by unoxidized (tobacco ores), and, to a lesser degree, by redeposited caviar ores. At present, the iron ores of the Kerch basin are not being developed.

Unlike typical Kerch iron ores localized within the synclines, those associated with mud volcanoes are confined to anticlines as well as compensated recessed synclines formed at their tops. They are similar to the Kerch ores, but their deposits retain original, ore-specific features. These ores form a new type of sedimentary iron ore deposit and are the most important mineral raw material in the sinking mud volcanic structures (recessed synclines). They have been discovered so far only on the Kerch Peninsula and have not been found in other areas of the Black Sea mud volcanic province. Separate outcrops of ores have been known for a long time and were identified as a new type of deposit and ore only in the 1960s (Shnyukov and Naumenko 1964; Shnyukov 1962). After that initial discovery, several new deposits and ore occurrences of this type were made on the Kerch Peninsula.

After many years of research, these ore deposits were located in the recessed synclines of Baksinskaya, Novoselovskaya, Uzunlarskaya, Rep'yevskaya, Achinskaya, Batalnenskaya,

Kamenskaya, Tarkhanskaya, Andreevskaya, Bulganakskaya, Kesenskaya, and Ossovinskaya (Table 6.2) (Fig. 9.8).

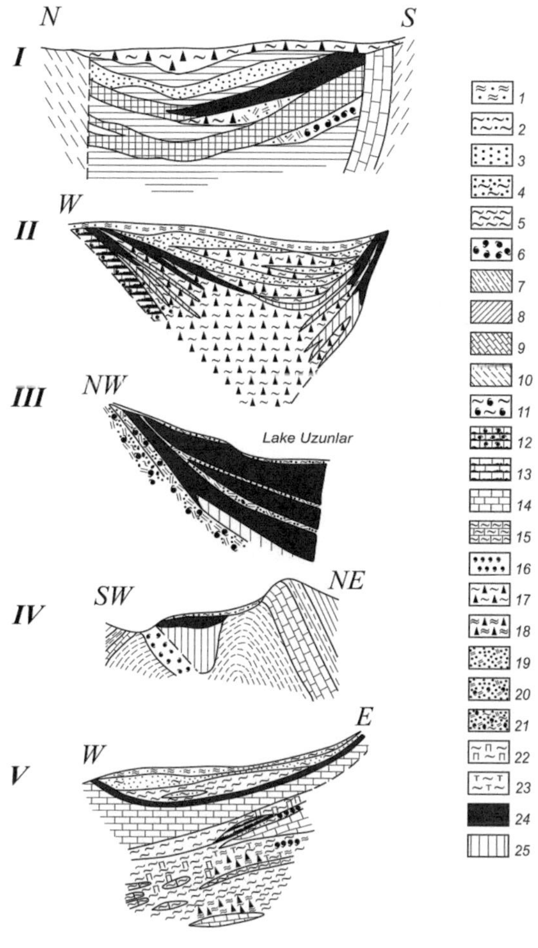

Fig. 9.8 The main types of ore deposit in recessed synclines with accentuated oxidation-reduction zones (scheme without scale). Mineral fields: (I) Novoselovskoye, (II) Baksinskoye, (III) Uzunlarskoye, (IV) Rep'yevskoye, (V) Kezenskoye. Key: (1) loess-like loams, (2) lake silt, (3) clayey sands, (4) quartz sand and aleurite with interbeds of sandy clay, (5) dense gray clays, (6) Pontian sandstones with fauna, (7) spaniodont clays with interbeds of marls and sands, (8) Lower Sarmatian clays, (9) Chokrakian limestones, (10) Maikopian clays, (11) clayey coquina, (12) organogenic limestones, (13) reef limestones, (14) calcareous limestones, (15) argillaceous limestones-shell rocks, (16) interlayers of limestone in clay, (17) redeposited mud volcanic breccia, (18) mud volcanic breccia, (19) clayey ferruginous conglomerate, (20) clayey ferruginous conglomerate with abundant fauna, (21) clayey ferruginous coquina, (22) washed clays, (23) fractured clay, (24) iron ore oxide (brown), (25) iron ore protoxide (tobacco)

According to our estimates, the total reserves of iron ore in the deposits of the recessed synclines on the Kerch Peninsula reach 400–450 million tons (MT), which include 125 MT in the Novoselovskiy deposit, 42.5 MT in the Baksinskiy deposit (Fesyunov 1968), 16.5 MT in the Uzunlarskiy deposit, 2.5 MT in the Rep'yevskiy deposit, 26 MT in the Kezenskiy deposit, and 6 MT in the Ossovinskiy deposit. The reserves contained within the Achinskiy, Batalnenskiy, Kamenskiy, Andreevskiy, Tarkhanskiy, and Bulganakskiy deposits were not counted, as they have been insufficiently explored to date.

The area of ore deposits within the recessed synclines is not large, being much smaller than that within the Kerch troughs. The largest ore deposit (5 km^2) is located in the Baksinskaya recessed syncline, and the smallest one (0.08 km^2) is in the Rep'yevskaya syncline. The thickness of typical ore deposits within recessed synclines is extremely significant, however. The iron ore layer with the greatest known thickness (92.7 m) was discovered within the Novoselovskoye deposit located in the Novoselovskaya recessed syncline.

The ore deposits of recessed synclines are heterogeneous in structure. As a rule, they incorporate lenses of clay or mud volcanic breccia in their central parts, and sometimes these lenses attain great thicknesses, resembling something of a complex "layered pic." Toward the periphery of the structures, the lenses of clay wedge out, the ore layer becomes uniform, and its thickness increases. As a result, the peripheries appear to be framed by a thick ring or semicircle of high-quality iron ore. The marginal zone of the ore deposits in recessed synclines is composed of ooid (oolitic-pseudo-oolitic) ores; this zone is characterized by the greatest bulges of the ore layer (Figs. 9.9 and 9.10).

Closer to the center of the deposits, the ooid ores are replaced by ferruginous clays and then by clays or mud volcanic breccias. In the Baksinskoye deposit, iron ores in the central part look as if they are being "eaten" by mud volcanic breccia: the ore layer, which is thick along the periphery, separates into thin intercalations that

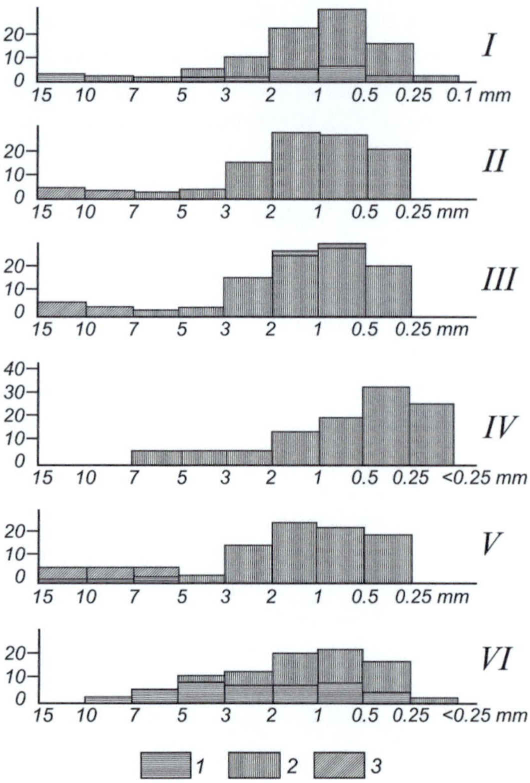

Fig. 9.9 Histogram of the content of oolites, pseudo-oolites, and pebbles in the ores of the most important deposits within recessed synclines of the Kerch Peninsula. Key: (1) oolites, (2) pseudo-oolites, (3) pebbles of clay. Deposits: (I) Baksinskoye, (II) Novoselovskoye, (III) Uzunlarskoye, (IV) Andreevskoye, (V) Rep'yevskoye, (VI) Kezenskoye

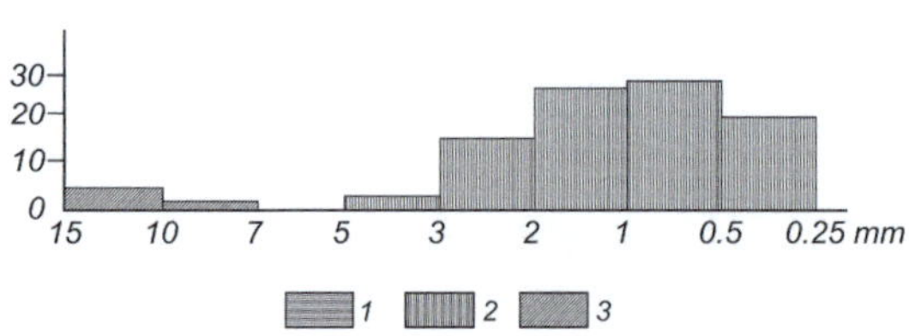

Fig. 9.10 Average grain-size composition of the Kimmerian iron ores of the typical recessed synclines on the Kerch Peninsula (based upon results of grain-size analysis of the Baksinskoye, Novoselovskoye, Uzunlarskoye, and Rep'yevskoye ore deposits). For key see Fig. 9.9

wedge out within mud volcanic breccia toward the center of the structure. In the Rep'yevskaya recessed syncline, due to subsequent erosion, only one part of the wing of ooid ore deposit is

Fig. 9.11 Iron ore sample from a recessed syncline with a weakly expressed layered texture

preserved. Therefore, the spatial distribution of iron ores is not as clear as in the Baksinskoye, Uzunlarskoye, and Novoselovskoye deposits. In general, the zonal structure of the deposits is expressed more sharply in recessed synclines than in other types of Kimmerian deposit. This may be related to the steep inclination of the layers, which accelerates the facies transitions in the ore layer.

The ooid ores constitute the main economic value of recessed synclines (Figs. 9.9 and 9.10). They are characterized by specific features that distinguish them as a special type of Kerch iron ore (Shnyukov 1962). These features are (1) a predominance of 0.5–3 mm, more often 1–3 mm, fractions, which, in their general appearance, resemble the Kerch caviar ores but differ from them in possessing a relatively high content of cementing matrix; (2) a prevalence of rounded ore debris, so-called pseudo-oolites; (3) a relatively low content of oolites; (4) the presence of large pebbles and gravel of clay, less often limestone, and other rocks; and (5) slightly lower iron content compared to the Kerch ores as well as a slightly increased content of manganese.

The ores of typical recessed synclines (Novoselovskoye, Rep'yevskoye, Uzunlarskoye, Baksinskoye, Bulganakskoye, and Andreevskoye) are also characterized by peculiar structural and textural patterns (Shnyukov and Naumenko 1964). As a rule, they reveal a weakly expressed layered texture (Fig. 9.11).

The textural pattern of the ores is determined by the dominance of clastic material, among which two dissimilar components are clearly expressed. The first component is represented by rounded ore fragments—pseudo-oolites up to 3–4 mm in diameter. The second component consists of gravel and pebbles, even clay pieces sometimes up to 2 cm and occasionally up to 5 cm in size. In general, the textures of ores in recessed synclines can be identified as oolitic-detrital varieties. Based on their structural and textural characteristics, the following types of ores can be distinguished: (1) pebble-oolitic-pseudo-oolitic, (2) pebble-pseudo-oolitic, (3) conglomerate of tobacco clay fragments, and (4) tobacco clays and other poor ores.

Sometimes clay pebbles are absent, and the structure bears a pseudo-oolitic character. The ratio of pseudo-oolites, oolites, and clay pebbles in the ores is quite changeable, but pseudo-oolites dominate in almost all fractions with sizes between 0.25 and 5 mm. The chemical composition of the Kimmerian iron ores from recessed synclines is defined by ratios of the abovementioned structural components.

In the Kimmerian iron ores, over 36 chemical elements have been identified (Shnyukov and Naumenko 1964). Chemical analyses have revealed the presence of Si, Fe, Al, Mg, Cu, Mn, C, P, K, Na, Ti, As, B, S, and V; spectral analyses further reveal Ni, Co, Cr, Zr, Zn, Ba, Sr, Se, Ga, Au, and some others. An insignificant admixture of iron ores includes Cu, Pb, and Zn. Zr and Sr are often present. These elements vary quantitatively: Fe, Si, and O are present in amounts exceeding 10%; Al, Mg, Ca, Mn, and C exceed 1%; P, K, and Na content ranges between 1 and 0.14%; and B, Ba, Sr, As, V, Ni, Co, and Cr are present in excess of 0.01%.

Pseudo-oolites and oolites are enriched with iron (35–49% and 39–45.5%, respectively). The cementing matrix is depleted in iron (23.2–26.2%). Clay pebbles dilute the ores because their iron content is only 8–10%.

Microprobe study of ore chemistry is particularly instructive. Oolites and pseudo-oolites of the Novoselovskoye, Uzunlarskoye, and some other ore deposits are characterized by sharp differences in iron content across phases of the material. Thus, a division can be seen between oolites and the transitions from cementing matrix to oolites (Fig. 9.12).

The distribution of the most important chemical elements in ore deposits from recessed synclines shows a certain structural zonality. Rich oolitic-pseudo-oolitic (sometimes pebble-oolitic-pseudo-oolitic) ores are localized on the periphery of the deposit, which causes elevated iron and manganese concentrations to form along the outer fringes. The central part of the deposits is conversely depleted of iron and other useful components. Such zonality is observed, for example, in the Novoselovskoye and Uzunlarskoye deposits.

The Kezenskoye and possibly the Ossovinskoye ore deposits were formed under conditions of nearby mud volcanic activity that had ceased in Kimmerian time. Therefore, the ores accumulated here did so under conditions of a rigid ore-bearing structure and do not differ qualitatively from the ores of typical Kerch deposits. Thickness of the ore layer in both deposits does not exceed 6 m (on average 2–3 m). The central part of the Ossovinskoye deposit is eroded, but judging from the Kezenskoye deposit, the ore structure in both deposits is also zoned. Along their periphery, there is a zone of oolitic-pseudo-oolitic ores that experienced replacement in their deepest central parts by tobacco clays. Oolitic-pseudo-oolitic ores are devoid of clay pebble admixtures that are characteristic of recessed synclines. Ores from the Kezenskoye deposit contain significant impurities consisting of terrigenous minerals, such as quartz, glauconite, microcline, plagioclase, and muscovite; the presence of these impurities accounts for a depletion of iron ores as well as other important chemical elements (Tables 9.1 and 9.2).

On their mineral composition, the ores of recessed synclines are reminiscent of the ores of Kerch, but they contain oxidation-reduction and oxidation zones. The former is represented by tobacco varieties, while the latter often consist of iron hydroxides. In tobacco ores, small inclusions of pyrite, marcasite, vivianite, kerchenite, neomesselite, calcite, rhodochrosite, as well as

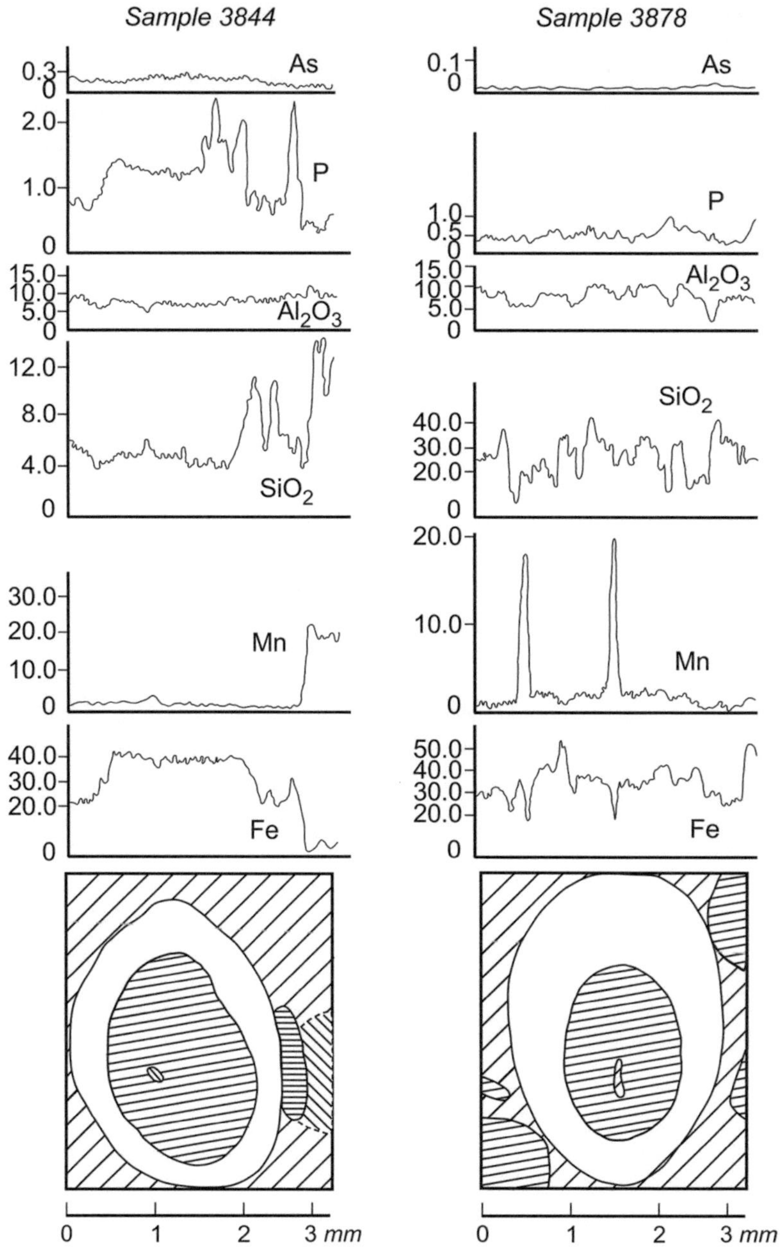

Fig. 9.12 Distribution of chemical elements within two oolites from the iron ores of the

realgar, orpiment, and some other minerals are present.

In the ores of recessed synclines, there are numerous admixtures of terrigenous minerals and pebbles of clay. The latter are composed of hydromica and montmorillonite. Of the terrigenous minerals, quartz, feldspars (microcline, occasionally plagioclase), and glauconite are widespread, with an insignificant amount of zircon, rutile, and other minerals. In the total balance of this type of mineral reserve, tobacco ores comprise up to 70%.

Judging from their structural and texture features, brown ores do not differ from the original tobacco ores because they develop due to oxidation of tobacco ores under the influence of exogenous factors. The oxidation zone within the deposits of recessed synclines develops in the

Table 9.1 Average content of the most important chemical elements of the Kimmerian iron ores found within recessed synclines of the Kerch Peninsula (weight %)

	Deposit							
Chemical element	Baksinskoye	Novoselovskoye	Uzunlarskoye	Rep'yevskoye	Bulganakskoye	Achinskoye	Kezenskoye	Kamenskoye
Fe	34.4	36.5	34.4	37.5	35.42	20.45	27.6	41.26
Mn	1.6	2.10	1.15	1.31	4.71	0.47	1.8	2.73
P	0.77	0.86	0.82	0.66	0.74	0.45	0.66	2.16
As	0.1	0.73	0.087	0.09	–	–	0.036	0.13
V_2O_5	0.09	0.10	0.084	0.112	–	0.024	–	–
SiO_2	21.75	22.16	24.5	15.9	–	–	32.22	–
$A1_2O_3$	6.51	7.26	7.64	4.9	–	–	–	–
CaO	1.71	2.25	5.25	2.24	–	–	–	–
MgO	0.66	1.56	1.09	1.41	–	–	–	–

Table 9.2 The content of iron, manganese, phosphorus, and arsenic in the most important ore types within recessed synclines of the Kerch Peninsula (weight %)

Deposit	Chemical element	Pebble-pseudo-oolitic			Pebble-oolitic-pseudo-oolitic		Tobacco clay (with pseudo-oolites and fauna)				Conglomerate		
		Max	Avg	Min	Max	Avg	Min	Max	Avg	Min	Max	avg	min
Baksinskoye	Fe	45.32	38.93	33.90	44.85	40.72	28.18	29.84	23.8	14.91	–	–	–
	Mп	4.62	2.07	0.28	2.27	0.72	0.18	4.54	1.12	0.10	–	–	–
	P	1.21	0.92	0.77	1.34	1.07	0.68	0.63	0.39	0.22	–	–	–
	As	–	–	–	0.74	0.13	0.12	–	–	–	–	–	–
Novoselovskoye	Fe	44.9	35.7	32.5	41.0	35.5	29.0	31.1	24.4	14.3	35.28	29.7	23.4
	Mn	11.01	3.13	1.11	3.85	3.42	1.25	2.06	1.26	0.54	1.72	0.89	0.48
	P	1.11	0.93	0.71	0.97	0.88	0.74	0.97	0.65	0.18	0.62	0.54	0.47
	As	0.132	0.104	0.085	0.137	0.110	0.070	0.078	0.038	0.01	–	0.023	–
Uzunlarskoye	Fe	39.5	35.0	30.9	–	–	–	25.4	21.2	15.6	23.0	19.9	16.7
	Mп	2.82	1.28	0.56	–	–	–	2.33	1.13	0.23	0.96	0.86	0.76
	P	0.90	0.80	0.60	–	–	–	0.44	0.40	0.37	0.52	0.43	0.34
	As	–	–	–	–	–	–	–	–	–	–	–	–
Rep'yevskoye	Fe	43.5	36.43	27.1	39.4	35.47	32.3	30.6	28.7	26.7	–	–	–
	Mn	2.66	0.94	0.30	2.76	1.92	0.82	2.19	1.27	0.34	–	–	–
	P	1.35	0.70	0.09	0.83	0.52	0.33	0.64	0.51	0.38	–	–	–
	As	–	–	–	–	–	–	–	–	–	–	–	–
Kezenskoye	Fe	–	–	–	37.42	27.6	17.42	–	–	–	–	–	–
	Mn			–	3.79	1.8	0.01				–	–	–
	P	–	–	–	1.38	0.66	0.37	–	–	–	–	–	–
	As	–	–	–	0.07	0.036	0.017	–	–	–	–	–	–

form of a blanket of different thicknesses in the upper parts of the ore deposits. The thickness of the blanket can reach 6–10 m. Only in the western part of the Novoselovskoye ore deposit do brown ores occur in places at a depth of 80–100 m. At the same time, the southeastern and central parts of this deposit are covered by a thick (sometimes up to 80 m) sequence of mud volcanic breccia and clay, and as a result, the ores are deprived of their oxide caps and are represented exclusively by tobacco varieties. The total reserves of brown ore in already known deposits, such as recessed synclines, make up about one third of the total amount of ore reserves. In the Novoselovskoye deposit, for example, this amounts to about 40–50 million tons. The highest proportion of brown ores is present in the Kezenskoye deposit, where almost all the ores are oxidized. At the same time, in some recessed synclines, the content of brown ores does not exceed 25%.

In 2015, ore oolites were found among the heavy fraction of the erupted material from the Andreevskiy mud volcano, and based upon this evidence, a forecast about mineralization within the Andreevskaya recessed syncline was made. In the same year, this structure was drilled by M.A. Vishnevetskiy, and the forecast was confirmed (Shnyukov et al. 2016).

On the basis of the finding of Kimmerian fauna in mud volcanic breccias, the literature suggested the presence of Kimmerian deposits in the recessed synclines of Burashskaya (Dvoychenko 1926), Soldatskaya, Mayakskaya, and Biyuk-Tarkhanskaya (Eberzin 1940). For the Burashskaya recessed syncline, P.A. Dvoychenko provided chemical analyses of the ferruginous rocks (Luchitskiy 1922).

Our studies of materials extracted from six boreholes did not confirm the presence of Kimmerian deposits in the Burashskaya recessed syncline, however. Only ferruginous clays of Pontian age, occasionally containing thin intercalations of chlorite composition up to 15 cm thick, were found. These intercalations contain up to 14% iron. Other signs of iron mineralization were not determined, and positive assessments of the deposit have not been made.

In the Mayakskaya recessed syncline, two boreholes up to 60 m deep were drilled along its axis. They recovered clayey coquina containing mollusk shells of the Lower Kimmerian (possibly Pontian) age and an iron content up to 25%. Both boreholes did not recover shells of Kimmerian mollusks in the mud volcanic breccia. Three more boreholes were drilled near the Soldatskaya recessed syncline. They also stopped within the mud volcanic breccia, at a depth of about 48 m. Thus, the question of ore content in these recessed synclines remains open. Tobacco clays were found in the Korolevskaya recessed syncline. In the Illichivskaya recessed syncline, Pliocene ferruginous sands seem to be present based on data from geological survey. More research regarding iron ore deposits in the Mayakskaya, Soldatskaya, Illichivskaya, and Korolevskaya (Naumenko 1967) recessed synclines is required.

In our opinion, iron ores are likely to be found within the Lake Kachik (Durmenskaya) anticline. Presumably, ore-bearing deposits can also be found within the Chongelekskaya and Korolevskaya (Naumenko 1967) recessed synclines (Lebedev 1968).

Under favorable paleogeographic conditions, the formation of recessed synclines is a general pattern of mud volcanic processes; however, accumulation of sediments in recessed synclines does not by itself create iron ores and other economically valuable deposits. Only if the area of mud volcanism is paleogeographically included within the area of development of certain mineral deposits can the latter be accumulated in morphostructures such as recessed synclines. So far, the only example of this is the Kerch Peninsula, which was a part of the Azov-Black Sea ore province during Kimmerian time when mud volcanism was especially prominent. That time was characterized by the coincidence of environmental conditions favorable for the development of oolitic sedimentary iron ores—a new and specific type of iron-bearing deposit.

In the literature, the mechanism by which the Kimmerian ore-bearing sediments were deposited was repeatedly considered to have been a complex geological process. The deposits often include the accumulation of chemogenous and clastic sediments together with mud volcanic breccias. The thick sediment sequence created pressures that, under conditions of continuous mud volcanic eruptions, led to subsidence of the rocks, which led to the creation of recessed synclines, their gradual growth, and increase in the thickness of their complex interbedding of sediments (Shnyukov and Naumenko 1964).

Deposition of iron ores and the development of recessed synclines in connection with past mud volcanic processes occurred under shallow-water conditions and was restricted to the narrow Kimmerian time interval in the morphostructures of Novoselovskaya, Baksinskaya, Rep'yevskaya, Uzunlarskaya, Andreevskaya, Achinskaya, Batalnenskaya, Kesenskaya, Ossovinskaya, and some others.

In typical and widely known Kerch iron ore deposits, the genesis of which has been well studied (Shnyukov 1962), ore formation took place under conditions similar to the shallow Kimmerian Sea. Nevertheless, the iron ores and structure of their deposits within typical recessed synclines differ significantly from those of Kerch by having a higher content of ore and non-ore clastic material, stunted development of oolites, somewhat reduced iron content, limited areas of ore deposit formation, and an enormous thickness of the ore layer with its peculiar structure. All these basic ore characteristics appear only in the recessed synclines. In our opinion, the main reason for the differences in lithological composition and structure between typical Kerch iron ores and those in recessed synclines lies not in the difference between the paleogeographical conditions but in the sharply manifested subsidence of the recessed synclines compared to the stable conditions within the Kerch troughs during Kimmerian time.

Being structures of mud volcanic origin, the Kesenskaya and Osovinskaya recessed synclines were characterized in Kimmerian time by the stability of their ore-bearing foundations. Therefore, ore deposits here resemble those of typical Kerch ore-bearing troughs.

Subsidence of recessed synclines is indisputable as evidenced by the monotonous lithological composition of iron ore deposits that reach many meters (up to 100 m) in depth within the synclines. Thickness of ore deposits within the Novoselovskaya recessed syncline, which is a part of the Sartskaya anticline, reaches 100 m. With the depths of the Kimmerian Sea at 4–10 m—optimal for ore formation according to published and archival data and the great thickness of the iron ores—one can assume that local subsidence had to have taken place in the recessed synclines. If this had been a result of sea-level changes, the ores would have been deposited and most definitely preserved everywhere within the Sartskaya anticline. But this is not the case.

At the same time, the lithological background of ore formation during Kimmerian times could vary significantly with the general shallow-water type of sedimentation. So, in the Novoselovskaya, Rep'yevskaya, and Uzunlarskaya recessed synclines, minimally ore-bearing gravels (clay conglomerates) at the foot of the ore deposit pass upward into ooid tobacco ores and then higher into the caviar-like ores with clay pebbles.

The origin of ore development in recessed synclines deserves some attention. In typical Kerch iron ore deposits, the lower contact of the Kimmerian ore deposits is always sharp. Even in the Yanysh-Takilskiy trough, where the Lower Kimmerian horizon is known, the ore layer is characterized by sharp contacts with the enclosing clays. A different picture is observed in recessed synclines. There is no sharp boundary between ore layers and underlying deposits; the iron content gradually increases from the Lower Kimmerian deposits to the Middle Kimmerian ore layer.

For Middle Kimmerian time, the rate of sedimentation within recessed synclines was high. As a result, chemogenic sedimentation (colloids of Fe_2O_3, Al_2O_3, SiO_2, MnO_2, etc.) appears partially suppressed, and typical oolites that are known to be formed in a mobile environment at the interface between sediment and water are stunted. The clastic material is partly represented by rounded fragments of Maikopian clays and mud volcanic breccia, but to a greater extent by eroded and coagulated fragments of iron ore particles—pseudo-oolites. The filling in of the iron ore sediment with pseudo-oolites is explained by the development of recessed synclines as well as uneven subsidence of their beds. As a result, the peripheral sections of the deposits remained within the zone of wave action in the shallow waters of the basin. The distribution of clastic material was subordinated to all the regularities of sediment accumulation in the coastal zones that form beaches, bars, berms, and accumulative terraces.

Under gradual sea-level fluctuations of the Kimmerian Sea, the movement of the shoreline left traces in the form of certain accumulative forms. A concentration of coarser-grained material composing ooid iron ores appeared near the shore, and dispersed particles forming tobacco clays occurred deeper. In other words, the usual facies zonation of coarse fractions gradually decreasing along the depth gradient was operating (Shnyukov and Naumenko 1964; Shnyukov 1962; Swift and Thorne 1991). Subsidence of the ore-bearing bed within the recessed synclines facilitated the development of fossil accumulations of enormous thickness. It is characteristic that in the Novoselovskaya, Baksinskaya, and Kesenskaya recessed synclines, the common feature typical of the entire Kerch basin is repeated: clastic ores concentrate mainly near the northern and northwestern slopes of the depressions forming a fringing deposit. The rate of subsidence apparently coincided with the rate of sedimentation, as evidenced by the findings of homogeneous strata of caviar-like ores with a thickness of 20–30 m in the Novoselovskoye deposit.

At a certain stage of iron ore formation, the continuity of the layers ruptured due to the continuous subsidence of the recessed synclines, and ring faults appeared around the dropping bottoms of the synclines. Sediment and iron ores remaining on the outside of the round synclinal depressions were washed away and brought down into depressions, sometimes forming intercalations of iron ores (2.7 m thick in the Novoselovskaya recessed syncline). Eruptions of mud volcanic breccia flooded the recessed synclines and repeatedly interrupted the ore-forming process. The weathering of the ore-bearing sediments occurred almost everywhere, creating the cap of oxidized brown ores up to 10–15 m thick. In the Novoselovskoye ore deposit, the cap of brown ores penetrated in places to a great depth, up to 100 m. Thus, weathering contributed to preservation of the iron ores, protecting them from further weathering and erosion.

The Novoselovskoye iron ore deposit is typical of recessed synclines and has been relatively well studied. This enables us to outline the main stages of its development because the evidence is much better expressed there in comparison to other structures of the same kind. They are as follows.

Within the whole Sartskaya anticline, mud volcanism was widely and fully manifested during Middle Sarmatian time. Then, the center of mud volcanic activity moved toward the location occupied by the modern village of Novoselovka. However, after the Middle Sarmatian, there was a long break in mud volcanic activity. It is difficult to say why this occurred. It is not clear whether it was a real break or an apparent one caused by some kind of subaerial mud volcanism in the Late Sarmatian, Meotian, and Pontian times. Among all the recessed synclines, this was most likely the longest break in the subsidence and accumulation of sediments. The old Sarmatian mud volcanic hearth actually remains unknown. The mud volcanic hill, which currently stands near Novoselovka, is already associated with post-ore outflow through a fault in the Middle Kimmerian ore deposit. This is evidenced by the presence of Kimmerian mollusk remains in the modern volcanic breccia. It is quite possible that in Kimmerian time, mud outflows were passing along this volcanic canal.

The scale of Kimmerian mud volcanism in this center was enormous. Further, during this time interval, part of the compensatory subsiding shifted toward the west, and the bowl of deflection expanded.

In Kimmerian time, subsidence within the Novoselovskaya recessed syncline became particularly significant, as huge masses of ore accumulated. The bowl of the structure itself was bent unevenly, and its central parts dropped faster than its periphery. In all likelihood, the Novoselovskiy mud volcanic hearth erupted in the shallow sea, but the recessed syncline was not exposed at the surface. Shallow deposits represented by clastic-chemogenic iron ores indicate the approximate depth of this sea as no more than 10–15 m. At the beginning of the Kimmerian, the transgression of the sea proceeded across the strait perhaps near the village of Adzi-Mende, which represented a kind of breakthrough in the Parpachskiy ridge. This led to a change in the paleogeographic situation in the region of the village of Novoselovka.

The complex of Kimmerian sediments records the apparent change in sedimentary conditions with the continued deepening of the basin: conglomerates of clayey pebbles, broken shells, and other detrital materials that make up the Lower Kimmerian deposits are gradually replaced by finer-grained sediments, such as clastic-chemogenic ores.

The unevenness of the subsidence within the recessed syncline affected the composition of the Kimmerian ore deposits. Interlayering of ore deposits, most often tobacco clays with fine-grained sediments in the central part of the depression and a clastic-ooid composition of multimeter homogeneous bundles within the deposits along the periphery, indicates the complexity and differences in the development of the recessed syncline, as well as shows its gradual deepening toward the center. Eventually, the basin became deeper everywhere as evidenced by an increase in more dispersed and fine-grained content spreading over the entire territory of the depression, which reflects the presence of Upper Kimmerian sediments. During Kimmerian time, the process of normal sedimentation within the deepened water body was temporarily interrupted by effusions of mud volcanic breccia. This led to the formation of a complex sedimentary sequence within the Novoselovskoye iron ore deposit, which was covered from above with clay and sand, and a predominance of mud volcanic breccia.

Up to 80 m thick in some places, the cover of sedimentary rocks and mud volcanic breccia indicate that the bending of the Novoselovskaya recessed syncline continued into subsequent time intervals, although it was uneven. This is evident from the presence of a lowered section of the morphostructure represented by a boggy meadow located to the east of the ore deposit contours, north of the Sarmatian limestone outcrop (Figs. 6.29, 6.30, 6.31, 6.32, 6.33, 6.34, 6.35, 6.36, and 6.37).

In our opinion, the Novoselovskaya recessed syncline in general is now emerging as a complex mosaic morphostructure consisting of several partial deflections of different ages, the first of which developed in the Sarmatian, the second in the Sarmatian-Kimmerian, and the third in the Quaternary. The last stage is characterized by its development in a subaerial environment. Outpourings of breccia continue to this day.

9.4 Mud Volcanism and Mining of Chemical Raw Materials

In the future, boron and iodine may become economically important extractive resources because they are present everywhere in mud volcanic breccia and other sediments as well as in the mineral springs of recessed synclines.

The high mobility of boron in solution contributes to its removal from deeply submerged zones of mud volcanoes.

Borax ($Na_2B_4O_7 \times 10H_2O$) is mentioned in the writings of Arab chemists of the eighth–ninth centuries due to its specific properties. It has been used since ancient times in blacksmithing when soldering precious metals, in the preparation of ceramic glazes, and as an antiseptic in medicine.

The uses of boron expand yearly as new applications of its compounds emerge. Industry demand for boron is therefore rapidly growing.

Boron and iodine were first discovered in volcanic mud on the Kerch Peninsula by V.I. Vernadskiy and S.P. Popov. Subsequently, these researchers confirmed this data, showing that the content of boron in mud volcanic waters often exceeds 1000 mg/l with a maximum of 6354 mg/l.

During the First World War, the boron contained within mud volcanic sediments attracted special attention. At that time, boron compounds were imported into Russia from America via Germany. Beginning in 1914, the war isolated

Fig. 9.13 Nodules of boronatrocalcite (ulexite) from mud volcanic breccia

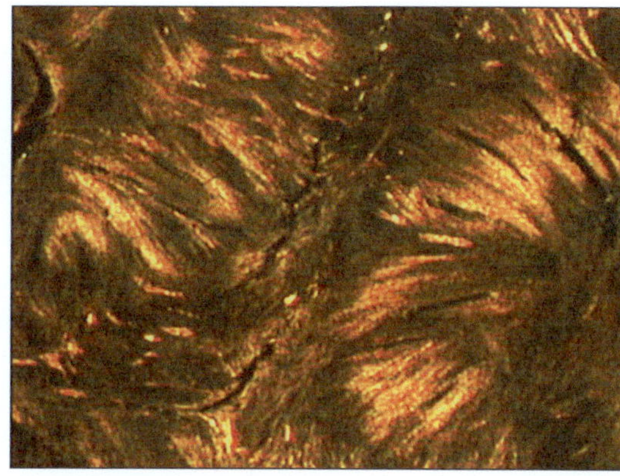

Fig. 9.14 Internal structure of boron-sodium-calcite. Thin section, plane polarized light, ×40

Russia from suppliers of boron and iodine at the very moment when demand for these elements sharply increased. Studies were conducted on the possibility of extracting boron and iodine from mud volcanic breccia. A small Bulganakskiy borate plant was built in 1914, but there is little information about its activities. The factory was not operating between 1917 and 1925, and according to Turley (1930), it produced about 5 tons of borax and several kilograms of iodine in 1925 and 3.15 tons of borax and 3.4 tons of soda in 1926. Originally, waters taken from the mud volcanic breccia were evaporated. In the early 1930s, only boronatrocalcite (ulexite), $CaNa(B_5O_9) \times 8H_2O$, from the mud volcanic sediments of the Bulganakskaya recessed syncline was developed (Figs. 9.13 and 9.14).

In 1932, 760 tons of boronatrocalcite were mined (Belousov and Yarotskiy 1936). By 1935, the richest areas were exhausted, and the discovery of borates in the Inder Lake (Kazakhstan) made further extraction in the Bulganakskiy basin uneconomical. Remains of the plant's buildings have survived to this day.

By the middle of the 1930s, significant geological exploration had been carried out to assess the reserves of boron and iodine in the mud volcanoes of the Kerch-Taman region.

According to Beketov (1916), one liter of water from mud volcanic breccia contains an average 30 mg of iodine, 2 g of anhydrous borax, 3.5 g of anhydrous soda, and 6 g of NaCl. The bromine content was estimated as insignificant. The author believed that with the use of all the waters of the Bulganakskiy mud volcano, up to 300 poods (4.8 tons) of borax, 2–3 poods (32–48 kg) of potassium iodide, and 400 poods (6.4 tons) of soda could be obtained annually.

Even greater opportunities were afforded by the use of dry mud volcanic breccia, reserves of which are colossal in the Kerch-Taman region. According to Izgaryshev and Sludskiy (1917), dry breccia almost always contains boron. They estimated the reserves of boron in the Shugo mud volcano as 390 tons of borax (132 tons of B_2O_3). Approximate estimates were also made for the Gnilaya Gora and Karabetova Gora mud volcanoes. Within the latter, between 1925 and 1930, and then in 1931–1932, Pokrovskiy (1933) carried out special prospecting for boron and iodine. We assess his results as the most reliable.

It is estimated that the reserves of iodine and borax in the mud volcanic sediments of Gnilaya Gora contain 140 tons of iodine and 13,200 tons of borax; in Karabetova Gora, there is an estimated 115,000 tons of borax (Fig. 9.15).

Reserves of the Dzhau-Tepe mud volcano were estimated to be 400 tons of borax and 4 tons of iodine (Steber 1913).

Pyatnitskiy (1925) estimated the reserves of iodine in the coarse sediments of Gnilaya Gora at 192 tons.

Taking into account all these data, the All-Union association "Himruda" designed and

Fig. 9.15 Efflorescences of boron minerals on the surface of mud volcanic breccia at the Karabetova Gora mud volcano

initiated the construction of a chemical processing plant for the extraction of borax, iodine, and soda based on the operations at the Bulganakskiy mud volcanic hearth in the 1930s, but the project was not implemented.

Minimally reliable data based on the memories of local residents indicate that German authorities extracted the boron-enriched deposits of mud volcanoes during their occupation of the Kerch Peninsula in 1942–1944. Local residents were driven to conduct this work and recalled mining "yellowish pieces of chalk" (Fig. 9.13), which were presumably boronatrocalcite.

It is likely that we will need to remember the past and again master the processes of extracting boron compounds from mud volcanic breccia and boron-containing waters. To do this, additional estimates of the ground flow of mud volcanic waters, their degree of mineralization, and the determination of other economically valuable components—such as iodine, bromine, lithium, etc., which are characteristic of mud volcanic waters—will be required, as well as a number of monitoring studies.

Nowadays, boron is found not only in mud volcanoes but also in the sediments of mud volcanic morphostructures: recessed synclines. This discovery was facilitated by a curious case. While conducting drilling operations at the Novoselovskaya recessed syncline to explore its iron ore deposits, some of the wells were placed within the village of Novoselovka. These operations were conducted during the hot days of July 1966 by the Ministry of Black Metallurgy of the USSR headed by P.I. Naumenko. One of the wells was drilled to a depth of 90 m on the western outskirts of the village. In the heat and aridity of its surroundings, the well began emitting a thin water stream and slime flowing up from the drilling rig. A number of domestic geese congregated around this stream to drink and shower their heads with the liquid. Immediately, the geese fell dead, and a few chickens that had also drunk from the stream died as well. After them, a group of sparrows also expired after taking a drink. The owner of the geese approached the drillers in anger, claiming: "You poisoned my birds!" The drillers did not understand what was happening, but becoming frightened, they picked up their equipment and departed rapidly.

We spent a long time analyzing this case and came to the conclusion that the poisoning was due to boron. Subsequently, the deposits of Novoselovskaya, and other recessed synclines, were studied to determine their boron content. It turned out that boron is consistently present in the sediments of all mud volcanic structures, and especially in the Kimmerian ore deposits.

Accessory impurities accompanying boron in the Kimmerian ore deposits of recessed synclines were first established by Shnyukov and Naumenko (1964) in the Novoselovskaya, Rep'yevskaya, and Uzunlarskaya recessed synclines. They also revealed widespread, regional contamination of these deposits with boron in almost all iron ore deposits of the Kerch Peninsula.

The content of boron in 537 analyses of the Kimmerian, Pontian, Meotian, Sarmatian, and Maikopian sediments from recessed synclines is shown in Table 9.3.

The table shows that, despite their differing ages, three maxima of accessory boron can be identified in the column of averages on the right. They are Sarmatian, the sharply distinguished Middle Kimmerian, and the somewhat weaker Kuyalnitskian. In our opinion, these peaks correspond to periods of intensive mud volcano activity and are connected to the lithological properties of the sediments accumulated within the recessed synclines. This provides a basis for establishing an order of lithologies from the most to least likely to contain boron: iron ores → mud volcanic breccias, clays → sands → limestones. Perhaps, it is the rock composition that accounts for the highest boron concentrations in the Middle Kimmerian deposits, where there are high values of iron hydroxides with high adsorption capacity. Undoubtedly, outbursts of mud volcanism occurred many times in the Neogene, but it is the lithological composition of the Kimmerian layers which, presumably, accounted for the greatest accumulation of accessory boron in the iron ore deposits. Boron-bearing properties of iron ores in various recessed synclines vary markedly from 0.034 B_2O_3 (Rep'yevskaya syncline) to 0.133% (Novoselovskaya syncline).

The typical Kerch iron ores also have boron as accessory impurities. Thus, the percentage of B_2O_3 for the Yanysh-Takylskoye deposit is 0.092 (0.028% B); for Eltigen-Ortelskoye, it is 0.062% (0.020% B); for Kamysh-Burunskiy, it is 0.095% (0.030% B); for Katerlezskoye, it is 0.136% (0.042% B); and for Chegen-Salynskoye, it is 0.041% (0.013% B). The Katerlezskoye deposit contains boron in approximately the same quantities as in that of Novoselovskoye.

The North Caucasian iron ores from Abinsk to Cape Zhelezny Rog contain elemental boron in the amounts of 0.001 to 0.005%, and only in a few samples does it decrease to 0.0001% (Orlovskiy 1971).

Typical Kerch iron ores reveal three sources of rocks or initiating processes. The first source type is represented by Precambrian magnetite-hematite quartzes; the second source consists of Paleozoic-Mesozoic and Cenozoic oolitic hematite-chamosite-siderite iron ores; and the third comprises Archaean massive or banded beds and lenses of magnetite ores (e.g., in Yakutia, USA, Sweden) that represent metamorphosed rocks of the shelf. Paragenesis of iron, boron, and rare earths is especially characteristic of the third type of ores (Serdyuchenko et al. 1963).

These data very clearly demonstrate the existence of geochemical paragenesis of iron-boron in the typical oolitic hydroferrichlorite iron ores of the Kerch Peninsula. As can be seen, the sedimentary iron ore layers of the second type are also boroniferous. Thus, there exist peculiar boron-iron ore formations in which the ore Clarke of concentrations for accessory boron reaches 70–250.

Judging from the ore Clarkes, the typical chemical suite for Kerch ores is arsenic-vanadium-manganese-iron. Thus, the Kerch iron ores should, in fact, be considered complex boron-arsenic-vanadium-manganese-iron ores.

Mineralogical studies of boron are hampered by the "invisible," scattered forms in which it exists within the Kimmerian deposits. Naturally, one assumes that this relates to the adsorptive character of boron and, in this connection, the adsorption of boron by iron-based minerals. This idea was expressed by Goldschmidt (1932), who studied ore samples from Lorraine. One could posit the existence of a proportional relationship between iron and boron; however, comparisons have not yielded evidence of a direct correlation between these elements, just as they have not between boron and manganese.

The phase analysis of iron ore samples from the Novoselovskoye deposit showed in many cases that 80–90% of the boron is in the form of compounds that are readily soluble in water.

According to Popov (1938), who examined the Bulganakskiy mud volcanic field, the significant difference between the forms of boron found in Kimmerian iron ore and mud volcanic breccia can be understood as a ratio between hydrochloric

Table 9.3 Content of B_2O_3 (%) ithe sediments of recessed synclines on the Kerch Peninsula[a]

Stage, horizon	Uzunlarskaya	Rep'yevskaya	Novoselovskaya	Kayaly-Sartskaya	Kara-Sidzheutskaya	Baksinskaya	Soldatskaya	Andreevskaya	Tarkhanskaya	Avg content	No. of analyses
Quaternary	–	–	0.03/6	–	–	–	0.02/1	–	–	0.03	7
Kuyalnitskiy	–	–	0.018/6	–	–	0.0132/8	–	–	0.064/46	0.052	45
Kimmerian (Upper)	–	–	0.041/94	–	–	–	0.011/11	–	0.03/3	0.028	108
Kimmerian (Middle)	0.067/12	0.034/12	0.133/213	–	–	0.045/29	–	–	–	0.12	254
Kimmerian (Lower)	–	–	0.035/33	–	–	–	–	–	–	0.035	33
Pontian	–	–	–	–	–	–	0.03/1	0.036/26	–	0.031	27
Meotian	–	–	–	–	–	–	–	0.03/2	–	0.03	2
Sarmatian	–	–	0.0025/2	0.047/20	–	–	–	–	–	0.045	22
Maikopian	–	–	Traces/18	0.052/2	0.023/19	–	–	–	–	0.01	39

[a]Cell contents: numerator = the content of B_2O_3; denominator = the number of analyses

acid and water. Hydrochloric acid extracts 1.5 times more boron than when it is extracted with water during mud washing. The author suggests that in volcanic mud, the insoluble boron compounds are represented by boronatrocalcite, while the water-soluble boron compounds are borax. The latter is extremely easily dissolved even by rainwater and is contained in clays mainly as adsorbed or mechanical impurities. Apparently, scattered admixtures of borax are also present in the Kimmerian ores, and borax is the basic mineral of boron in these ores. Luneburgite has recently been found in the iron ores. The distribution of boron minerals in ores is relatively uniform (Table 9.4).

In our opinion, boron mineralization in the Kimmerian iron ores of the Kerch Peninsula is associated with mud volcanic processes and migration of mud volcanic water. Given the widespread development of mud volcanism on the Taman Peninsula, one can assume the presence of boron in Taman iron ores as well. Unfortunately, the total volume of analyses of the Kimmerian iron ore is still insufficient to contour any zones of accessory boron enrichment in individual ore fields.

Until recently, the question of when boron mineralization developed has been unclear. Some arguments favor an epigenetic boron mineralization in the Kimmerian deposits of recessed synclines. The increased accessory concentrations of boron in the sediments of various ages, particularly in the Sarmatian, Kimmerian, Kuyalnitskiy, and Quaternary of the Kerch Peninsula, indicate most likely a later manifestation of boron mineralization. These data are not unambiguous and do not definitively assign a syngenetic or epigenetic character to boron mineralization within its host rocks. Perhaps, the outflow of mud volcanic waters into Kimmerian or Kuyalnitskiy water bodies and the adsorption of boron compounds to clay particles or chemogenic products took place. In such a case, the boron content of inflowing mud volcanic water would have been higher, and as a result, large amounts of boron could be adsorbed in adjacent zones.

One should recall that the Gladkovskiy, Shkolnyy, and Semigorskiy mud volcanoes developed on Lower Cretaceous clays and that they also contain boron. In other words, the boron was most likely introduced by mud volcanoes.

Boron mineralization seems initially to possess a syngenetic character. This is evident from the elevated boron concentrations in sediments of certain ages, e.g., Sarmatian and Kimmerian, which are characterized by intensive mud volcanic activity. Such activity brought boron with its erupted water and thereby enriched surrounding sediments. With time and continued eruptions, the boron concentration increased/changed and obtained its epigenetic character.

In summarizing the results of research on accessory boronization within the recessed synclines of the Kerch Peninsula, the first observation is the presence of several instances of boron mineralization in the Sarmatian, Middle Kimmerian, and overlapping strata.

From geological and mineralogical points of view, the most important Kimmerian iron ore deposits of recessed synclines on the Kerch Peninsula are those where the average boron content (based on 254 analyses) is 0.12% B_2O_3. These are the Baksinskoye and especially the Novoselovskoye deposits, where the average B_2O_3 content is 0.133%.

Among the typical Kerch iron ore deposits, the Katerlezskoye field contains the greatest amount of boron, although there is still too little data for a sufficiently reasoned assessment of these deposits.

Mineralogically, boron occurs in the Kerch iron ores in the form of luneburgite but more often as borax, as evidenced by the almost complete transition of boron from ores to water extract.

9.5 Mud Volcanism and Sulfur Mineralization

Another target of raw materials mining is native sulfur. In marine mud volcanoes, sulfur is not encountered, but on land, sulfur is associated with mud volcanoes in many areas. Kovalevskiy (1935) discovered sulfur in mud volcanic breccia of the Lok-Batan and Dysh-Gil mud volcanoes in

Table 9.4 Results of phase analysis of boron content from tobacco ores of the Novoselovskoye deposit. The table illustrates the extraction of borax (boron compound) from Kimmerian ores by the method of colorimetry with various solvents by fractions. The numbers show the proportion of borax being extracted

	Colorimetric determination of quinalizarin		Volume determination of boron mannitic acid titer	
Sample	B_2O_3	B	B_2O_3	B
I. General sample	0.80	0.24	0.87	0.27
Water extract	0.80	0.24	0.84	0.26
Acid extract	0.02	0.006	–	
Insoluble residue	0.04	0.01	–	–
Fraction, mm				
2–3	0.70	0.21	0.66	0.20
3–5	0.50	0.15	0.46	0.14
5–7	0.60	0.18	0.63	0.19
7–10	0.64	0.19	0.59	0.18
II. General sample	0.435	0.13	0.45	0.14
Water extract	0.38	0.11	0.38	0.11
Acid extract	0.04	0.013		
Insoluble residue	0.06	0.02	–	–
Fraction, mm				
0.25	0.60	0.18	0.56	0.17
0.25–0.5	0.44	0.13	0.38	0.11
0.5–1.0	0.60	0.18	0.52	0.16

Shnyukov et al. (1971)

Azerbaijan. Ladyzhenskiy et al. (1965) described sulfur from the Dzhau-Tepe mud volcano on the Kerch Peninsula, and there are many other examples. Unfortunately, the number of these findings is negligible. Sulfur is often found in numerous hydrogen sulfide springs near mud volcanoes, and these phenomena are noted most often on the Kerch and occasionally on the Taman peninsulas.

Quite interesting are the impregnations of sulfur into dolomitized limestones of Sarmatian (Cape Golubiny near Enikale in the vicinity of Sokolskiy village) and Chokrakian age (on the eastern shore of Lake Chokrak), as well as in the Burulkaiskiy cauldron, on the southern wing of the Seit-Eliyskaya anticline, in the village of Rep'evka, on Mount Konchek near Uzunlar Lake, and on the Parpachskiy Ridge to the north of the village of Marfovka, etc. (Popov 1938; Goryainov 1968; Sobolevskiy 1970). Thin sections of dolomitized limestone are characterized by quite complex interrelations between calcite, sulfur, chalcedony, and with gypsum established within the voids (Sobolevskiy 1970). In the voids, a calcite zone is confined directly to the walls. The inner part of the voids is filled with calcite that lines the voids, as well as gypsum, sometimes gypsum and sulfur, sometimes chalcedony (Mount Konchek). In some cases, the voids are filled with sulfur only, which is cryptocrystalline and crystalline, and has a brownish color due to a bituminous substance. Oil emissions in the voids are encountered quite often. Characteristically, all the sulfur-impregnated limestones are confined to recessed synclines, which indicates that this type of sulfur mineralization is somehow connected with the activity of mud volcanic fluids.

The most extensive sulfur mineralization is confined to gypsum-bearing Chokrak-Karagan strata located in particular on the Parpachskiy and Parabolic ridges of the Kerch Peninsula and developed in a number of isolated synclinal structures.

Within the Parabolic Ridge, Mokrinskiy (1916), Konstantov (1929), and Ishchenko and Kiselev (1967) discovered a site where the gypsum is partially replaced by sulfur: the Chekur-

Koyashskoye (Tarasovskoye) sulfur deposit. Within this deposit, sulfur is, as a rule, confined to gypsum strata and alternates with interlayers of gypsum. Sometimes, sulfur is contained in the gray clay underlying the gypsum layer. In the northern part of the deposit, sulfur is found only in calcareous clayey sandstones that are covered by clays.

Two more sites where similar manifestations have developed near Gornostaevka are known: (1) in the Chokrak gypsum deposits near the Seit-Eli carbon dioxide source on the southern wing of the same anticline and (2) on the southern wing of the Sultan anticline (Goryainov 1968; Sobolevskiy 1970).

The structure of the Chekur-Koyashskoye deposit is complex: the sulfur-bearing layer is broken by eight tectonic dislocations that give the deposit a blocky structure. These dislocations cause a change in the strike of the gypsum stratum from 10 to 15° and an inclination to the east along a strike of 40–50° with an inclination to the north in the southern part of the deposit. The dislocation zones are enriched in sulfur. Besides sulfur, gypsum, calc-silicate mass, and calcite are established within the sulfuric ore (Popov 1938). Sulfur reserves in explored areas have been estimated in different ways. Mokrinskiy (1916) arrived at a total of 16,000 tons of pure sulfur in the Chekur-Koyashskoye deposit, an additional 60,000 tons of probable reserves, and 130,000 tons of possible reserves (with a sulfur content of 20.49%).

Mokrinskiy (1916) interpreted the occurrence of sulfur in the composition of gypsum strata as a result of the influence of organic matter on the gypsum. In his opinion, the source of the organic matter was the vast endorheic plain to the north of Lake Koyash. Goryainov (1968) explained the deposition of sulfur differently. He believed that the gypsum-bearing rocks of the Chekur-Koyashskiy section of the Parabolic Ridge have undergone the regenerating influence of hydrocarbon gases from the mud volcano.

Geological exploration conducted by A.K. Markov in 1928–1929 yielded far more modest estimates of categories of reserves: A_2, 11278 tons; B, 6615 tons; and C, 4024 tons (Popov 1928). Each category is characterized by degree of exploration, quality of minerals, and mining and geological conditions of development. The quality of reserves declines sequentially from A to C2 (Biletsky 2004).

The deposit was worked out long ago. In our opinion, a more updated reappraisal of the ore is needed, especially since the prospectors indicated a new ore field east of the Chekur-Koyashskoye deposit.

Two more sites are known near Gornostaevka where similar manifestations of sulfur have developed. The first one is located in the Chokrak gypsum-bearing deposits near the Seit-Eliyskiy carbon dioxide spring on the southern wing of the anticline with the same name; the second one exists on the southern wing of the Sultanovskaya anticline (Goryainov 1968; Sobolevskiy 1970).

The noted sulfur manifestations at first glance are randomly scattered along the Kerch Peninsula. However, if we compare the locations of sulfur finds with the geological structure of the Kerch Peninsula, its connection with mud volcanoes then becomes apparent. Almost all hydrogen sulfide mineral springs are associated with mud volcanoes, and all inclusions of sulfur into dolomitized limestones occur in recessed synclines.

Even the Chekur-Koyashskoye sulfur deposit, which remained mysterious for a long time, is currently associated with the impact of mud volcanic fluids. The argument that this field belongs to gypsum-bearing Tortonian deposits only supports this assumption. The gypsum stratum extends along the entire Parabolic Ridge, but it is only sulfur-bearing in the region that is under the influence of mud volcanic fluids.

9.6 Mud Volcanism and Building Materials

Upper Kimmerian sands of the Bulganakskiy recessed syncline have been used as building materials in recent years. The scale of their extraction (Fig. 9.16) is still negligible, however.

At present, the area within which sand is currently being extracted extends to a few square

Fig. 9.16 Sand quarry in the Upper Kimmerian sediments of the Bulganakskiy recessed syncline

kilometers. Thickness of the sand layer reaches 200 m, as determined from one of the wells of the Kerch Geological Prospecting Party. In the lower part of the section, there is interbedding with mud volcanic breccia exposed on the hillside to the west of the Andrusov mud volcano. At the base of the bed, there is a layer of ferruginous sands approximately 4 m thick.

Lithologically, the sands are represented by fine-grained, predominantly quartz silty sands with spotty cementation. Sandy-silty and clayey particles constitute 61–73.8% and 11.9–27%, respectively (Shnyukov et al. 1971).

For many years, at various places on the Kerch Peninsula, Sarmatian and Chokrakian limestones from recessed synclines are being used as building materials. Many abandoned quarries are preserved (Fig. 9.17).

9.7 Mud Volcanic Breccia as Mineral Raw Material

By itself, mud volcanic breccia is not only an interesting subject of geological study for the minerals it contains, but in many cases, these minerals possess value as minable raw materials.

9.7.1 Mud Volcanic Breccia as Raw Material for Expanded Clay

One of the most progressive building materials—expanded clay—has many remarkable properties. It is light, strong, and low cost. It is used as a filler for concrete, thereby facilitating various kinds of structures, saving on cement consumption, and reducing the cost of construction. The use of expanded clay is increasing every year, and therefore the need for production of expanded clay is constantly growing (Yukhanov 1968).

The beginning of research on expanded clay dates back to 1956, when the "Ukrgeolnerud" trust carried out prospecting for this kind of material in the area of the Bulganakskiy mud volcanic hearth. Laboratory studies and subsequent semi-industrial tests showed that these breccia-bearing clays can be used as valuable raw materials for the production of expanded clay and ceramics. Evaluation of reserves led to an estimation of 3469.9 thousand m^3. Out of this number, 870.2 thousand m^3 were estimated as raw material suitable for tile production, and 962 thousand m^3 as claydite raw materials suitable for the production of expanded clay. Based on the Bulganakskoye deposit in Kerch city, a brick factory was built for the manufacture of expanded clays; it was commissioned in 1960.

Fig. 9.17 Fragment of a quarry for the extraction of limestone for construction purposes within the Burulkayskaya recessed syncline

Yukhanov (1968) conducted an extensive series of geological and technological studies of the largest fields of mud volcanic breccia on the Kerch Peninsula in order to evaluate their use for the production of expanded clay (Table 9.5).

The suitability of sediments such as claydite raw materials for the production of expanded clay was determined first by technological requirements: in particular, the degree of bulging during burning, the volume weight of the resulting bulged material, the lowest temperature at the initiation of bulging, and the heat interval during which the clay softens or bulges. The degree of bulging is described by the coefficient of bulging (K), which refers to the ratio of the volume of granules after burning (V_2) to the volume of granules por to burning (V_1): $K = \frac{V2}{V1}$.

According to the classification proposed by Onatskiy (1987), depending on the coefficient of bulging during burning, clays can be divided into three groups. Clays for which the bulging ratio is less than 2 may be grouped within the poorly bulging category, clays for which the coefficient of swelling is 2–4 are medium-bulging, and clays with a bulging ratio greater than 4 are well-bulging.

The bulk density of expanded clays is actually derived from the first property. For expanded clay from well-bulging clays, it is 200–600 kg/m^3, and for poorly bulging clays, it is 900–1200 kg/m^3.

One of the requirements for claydite raw materials is associated with the technical and economic side of production. The optimal temperature for the bulging of clays during burning should not exceed 1250 ° C, with a permissible interval according to the Intergovernmental Standard of at least 50°. Moreover, clayey rocks that bulge during burning should have a certain lithological-mineralogical and chemical composition; in particular, at least 20% of their constituent particles should be smaller than 0.005 mm. In other words, clays must have a certain plasticity so that granules can be obtained in the raw state.

Clays must also satisfy geological requirements. First, they should be sufficiently thick and homogeneous in order to avoid contamination of the future raw materials by impurities such as sand, marl, limestone, etc. Second, they must be easily accessible so that clays can be extracted without the expense of removing the overburden of substandard deposit. The ratio of overburden to the thickness of the mineral deposit should not exceed 1:2.

In the course of laboratory studies, samples represented by clay balls with a diameter of 15 mm after drying were fired in a sillite furnace with a maximum temperature of 1350–1400 ° C. The balls were fired at temperatures of 1020, 1050, 1080, 1110, 1140, 1170, 1200, and 1230 ° C. This temperature range was determined by the need to identify the temperature at which the clay begins to melt and that at which optimum bulging occurs, i.e., the heat at which maximum bulging takes place. The difference in these temperatures determines the interval of bulging. At temperatures between 1020 and 1170 ° C, the clays were kept in the furnace for 10 min, and between 1200 and 1230 ° C, the clays were exposed for 5 min.

Before firing, the samples were sometimes subjected to preliminary heat treatment, and various kinds of organic substances were added, etc. (our research).

It was discovered that the majority of mud volcanic fields of the Kerch Peninsula contain clays that conform to geological and technological requirements. Lithologically, mud volcanic breccia is fairly monotonous and completely dominated by fine-grained fractions. Comparison of breccia from different mud volcanic fields shows their homogeneity. The content of clayey particles ranges from 85.4 to 94.94%, most often 88–89%; the content of silt particles varies between 4.80 and 8.8%; the content of sand particles ranges between 0.18 and 6.8% with the bulk of the particles in this fraction measuring between 0.1 and 0.25 mm. The content of dispersed particles smaller than 0.005 mm varies between 58.8 and 74.2%, which is more than half of the deposit. The high content of fine particulate material determines the plasticity and good formation properties of mud volcanic clay.

Mud volcanic clays contain (%): 55–58 silica, 17–18 alumina, 6–10 iron oxide, 1–1.8 calcium oxide, up to 2.5 magnesium oxide, up to 2.4 alkalis, etc. At the same time, only fresh, dark gray and gray brecciated clays possess satisfactory qualities for claydite development. The loams, which arise due to redeposition of the breccia, bulge greatly during firing. Apparently, not only the chemical composition determines the claydite-forming properties of the clay, but their mineralogical composition, which is complex and diverse, is also influential. The clay minerals are dominated by hydromica, most likely illite, and occasionally by montmorillonite.

Fresh mud volcanic brecciated clays and pelites (Table 9.5) are characterized by good claydite-forming properties. The coefficient of bulging for breccias of different mud volcanic fields varies from 3.08 to 4.80, and volume weight varies from 320 to 560 kg/m^3. Fresh breccia from the Korolevskoye mud volcanic field perform poorly, as their oxidized mud volcanic breccia and redeposited loams are characterized by inferior characteristics: the bulging coefficient is up to 3.6, and the volume weight is over 660 kg/m^3. Brown loams from the Baksinskoye and Soldatsko-Slobodskoye mud volcanic fields do not bulge at all during firing.

The physicochemical and mineralogical nature of the process of claydite production from mud volcanic clays of the Kerch Peninsula was studied by Yukhanov (1968). It is generally accepted that the process of claydite formation consists of two stages. The first stage corresponds to the transition of clay at high temperature into a plastic state. The second stage occurs simultaneously with the transition of the clay to the plastic state and involves the release of gases that cause pore formation.

It is assumed that mainly CaO and MgO lower the temperature range for melting and viscosity reduction.

The transition of clays to a plastic state upon heating is due to their chemical composition. Any clay is a system of SiO_2-$A1_2O_3$ with impurities CaO, MgO, Fe_2O_3, Na_2O + K_2O, etc. The question of the influence of these impurities, called fluxing agents, on clay viscosity and fusibility when the clay is heated, is controversial. Most researchers suggest that CaO and MgO positively affect fusibility and viscosity and shorten the melting temperature range.

The question of bulging agents is still debatable. One of the first to attempt an explanation of the pore formation process was Jackson (1925). He suggested that the formation of pores occurred due to the dissociation of iron oxides:

Table 9.5 Technological properties of mud volcanic clays

Mud volcanic field	Sediments	No. of samples	Coefficient of swelling			Bulk weight, kg/m³		
			Max	Min	Avg	Max	Min	Avg
Bulganakskoye	Loams	12	3.59	3.17	3.35	570	470	480
	Breccia	30	3.94	2.63	3.23	720	360	510
Dzhardzhavskoye	Breccia	20	5.08	3.29	4.07	490	240	440
Soldatsko-Slobodskoye	Loams	10	Majority did not bulge					
	Brown clays	25	4.14	3.10	3.51	570	320	470
Baksinskoye	Brown clays	10	Under burning did not bulge					
	Breccia	20	3.66	2.65	3.08	620	480	550
Malobabchinskoye	Breccia	25	4.15	2.33	3.19	810	450	560
Novoselovskoye	Dark gray breccia	30	4.64	2.30	3.40	670	360	450
Andreevskoye	Breccia	10	4.00	2.56	3.10	800	420	550
Ilyichevskoye	Breccia	12	5.88	3.16	3.90	560	220	400
Kayaly-Sartskoye	Breccia	8	3.84	2.92	3.24	820	320	500
Korolevskoye	Breccia	6	Did not bulge					
Nasyrskoye	Breccia	10	6.52	3.80	4.80	450	210	320

$$6\,Fe_2O_3 \rightarrow 4Fe_3O_4 + O_2$$

$$2\,Fe_3O_4 \rightarrow 6\,FeO + O_2$$

The hypothesis that explains the process of claydite formation by the transition of iron oxides to the ferrous form was put forward by Kostyrko and Proshenitsyn (1931) and Shchemelev and Gervids (1934). The proportion of iron oxides sufficient to produce intensive bulging was determined to be within the limits of 6–10% and relates to the influence of iron and alkali oxides as well as explained pore formation by the volatility of alkalis. Earlier, Orton and Stely (1913) had shown that some samples containing oxides in apparently sufficient quantity do not bulge. Therefore, oxides play a decisive role in the bulging process together with impurities of organic substances. Wilson (1927) suggested that the bulging process is due to the combined effect of many gases: CO_2, CO, SO_2, H_2O, and O_2, which are products of clay decomposition, as well as air, which is in the pores of minerals.

Onatskiy (1987) described probable gas-forming substances at different stages: free and physically bound water; water that is bound with the crystal structure of the minerals; and gaseous products deriving from the dissociation of carbonates, sulfates and sulfides, carbon, and Fe_2O_3.

As a first approximation, visualizing the nature of mineralogical changes is often based on analysis of chemical changes in clays as they are heated. To obtain a detailed understanding of these changes, the chemistry of expanded and poorly bulging clays at different stages of heating was studied. Without exception, all such clays were characterized by the reduction of iron, dehydration, and an increase in the specific gravity of the remaining clay components. However, good and poorly bulging clays showed some differences in their chemical changes that manifested, in particular, in the speed and completeness of the release of water, which are somewhat slower in claydite clay. Even more revealing are the results of changes in the content of ferrous iron and other oxides, of which the most obvious are data showing sharp differences between good and poorly bulging clays in the changes in iron oxides at temperatures ranging between 1050 and 1200 ° C. The degree of reduction in iron oxides of claydite clays is usually much higher.

The following points should be emphasized: well-bulging clays are usually characterized by a predominantly hydromicaceous (illite) composition; changes in iron compounds proceed continuously toward the regeneration of this element and are completed by the formation not only of wüstite, but ά-iron as well; complete restoration of iron in claydite-forming clays occurs at the expense of a larger content of organic material than in non-bulging clays, in which the content of organic matter is less. Thus, it can be concluded that iron oxide influences the process of claydite formation.

Mud volcanic breccias of the Kerch Peninsula satisfy the necessary geological and technological requirements and can be considered as a very promising claydite raw material. Economically, the extraction of breccias is beneficial, especially for nearby processing enterprises. Of particular interest are the mud volcanic clays of the Novoselovskoye mud volcanic field. They overlap a large iron ore deposit, making it possible to engage in synchronous extraction of iron ore and claydites. Mud volcanic clays of the Kerch Peninsula can also be considered a very promising claydite raw material.

9.7.2 Mud Volcanic Clay as Raw Material for the Production of Metallurgical Pellets

In connection with the development of mining enterprises for ore concentration, metallurgy in Ukraine is in need of high-quality clays, necessary for pelletizing fine iron ore concentrates to produce pellets with a diameter of 10–15 mm. The raw pellets obtained from pelletizers must be strong enough to resist collapse during transportation and roasting. Such pellets should also resist compression in both wet and dried states. Finely dispersed mud volcanic clays of the Kerch Peninsula are promising for this purpose. Yukhanov (1968) obtained samples from various mud volcanic fields of the Kerch Peninsula in order to test mud volcanic clays for pelletizing of fine-grained ores. The preliminary data suggest that they can serve as a good raw material for the production of pellets.

A batch was prepared for testing the clays, which comprised a concentrate (99.5%) and clay (0.5%). By weight, this amounted to 15 kg of concentrate and 0.075 kg of clay. Dry concentrate (moisture is 2–4%) was passed through a 2 mm sieve, so that it would not contain lumps. The batch was thoroughly mixed in place with a shovel, and then for 1.0–1.5 min, it was blended in a spherical granulator. The diameter of the granulator was 1.0 m, the angle of inclination was 45°, the peripheral speed of the drum was 56.5 m/min, and the rotation of the granulator was 18 rotations per minute. During the 1.0–1.5 minutes in the granulator, a thin stream of water was added until an optimum moisture content of 7.5–9% was reached. The pellets were then formed for 3 min and then pelletized for another 2 min. Following this, a sample was taken to check for moisture content, and 10 pellets with a diameter of 15–18 mm were sampled to test dropping strength. Impact tests were carried out by dropping pellets onto an iron plate from a height of 300 mm. Rolling was then continued for another 5 min followed by a repeat of the same tests. Both cycles of tests were performed to determine the strength of the pellets at optimum and near-optimum moisture levels achieved at different times during the rolling.

The crushing strength of raw pellets was determined by applying a static load using a special apparatus. Since the strength of pellets changes during drying and before burning, studies were carried out with nearly dried pellets as well. To this end, they were placed in a drying cabinet and kept at a temperature of 110–120 ° C for 20–120 min. After drying, the pellets were tested for crushing. For comparison, pellets from concentrate without clay additives were also tested. Preliminary tests have shown that breccias of certain mud volcanic fields on the Kerch Peninsula can be used as additives to produce pellets instead of bentonite clays.

9.7.3 The Recreational Value of Mud Volcanoes

The recreational value of mud volcanoes, in particular of the Kerch-Taman region, has been known and exploited for a long time, mainly for balneological uses of the mud and hydrogen sulfide springs that accompany mud volcanic eruptions.

According to the Russian Scientific Center for Rehabilitation and Physiotherapy (Adylov et al.

1995), volcanic mud has analgesic, anti-inflammatory, bactericidal, spasmolytic, and hyposensitizing effects, normalizing the functional state of the central and autonomic nervous system and metabolic processes. Most balneologists consider volcanic mud gentler than other kinds of mud because it contains bromine, has an alkaline reaction, and is less mineralized.

Medical indications for the use of volcanic mud are for diseases of the nervous system, musculoskeletal tissues, digestive organs, the urinary tract, genital organs, ear and throat diseases, and skin. There is also information about favorable cosmetological effects of this mud.

One of the first appreciations of the balneological properties of volcanic mud from Taman volcanoes was by Land (1866). Also, resort facilities of Lake Chokrak, located in the north of the Kerch Peninsula, were repeatedly noted in the literature (Alenchikov 1899). In the nineteenth century, up to 20,000 tons of salt were extracted from this lake, which is supposedly the now flooded caldera of a mud volcano. The remarkable balneological properties of the lake's mud were subsequently described and exploited by many. Data on their quality are summarized by Ponizovskiy and Stashchuk (1974), who wrote that the therapeutic mud of Lake Chokrak has no equal in chemical, physical, and medicinal properties and that the nearby presence of hydrogen sulfide and carbon dioxide springs, the proximity of the Sea of Azov, and some other features have contributed to the creation of a resort of national significance.

In old reference books on Crimea published before 1917, the authors found advertising messages signed by Mr. Franz Tomazini, the owner of the balneological establishments near Lake Chokrak. They are quite curious and reveal the popularity of the Chokrak mud.

He wrote that the high season for its use lasted 3 months, from June 1 to September 1. The hottest months are June and July. "Persons suffering from rheumatism and related illnesses, whose treatment sometimes requires high-temperature baths (the sun can heat mud up to 40° C), are advised to visit the establishment mainly in the hottest months, in June and July; those suffering from scrofula and related illnesses, especially children, where hot baths are not required at all, less hot season (August) is recommended. The course of treatment for each patient is three weeks. In view of the short duration of the medical season and the large number of patients, people traveling to Chokrak from afar are advised to consult beforehand about a free hotel room in an institution … doctors provide free medical treatment."

Before the Second World War, a balneological resort was in operation here. During the war, the resort, which was near the front line, was completely destroyed and, unfortunately, has not been restored.

In the 1980s, Odessa Scientific Research Institute of Balneology (Ukraine) carried out balneological research on the volcanic muds of the Kerch Peninsula (Al'bov 1971). It constituted a continuation of balneological studies at the Bulganakskiy mud volcanic hearth that had begun in the early twentieth century. The mud was proven effective in medical treatments at a number of specialized institutions in Kerch and Feodosia. At the same time, V.A. Antonenko and K.V. Nesterov, employees of the Odessa Institute of Scientific and Applied Art, studied volcanic muds of the Soldatsko-Slobodskiy, Enikalskiy, Nasyrskiy, and Burashskiy mud volcanoes.

According to data from the Odessa Scientific Research Institute of Balneology, high plasticity values characterize the volcanic muds from the Bulganakskiy (Vernadskiy hillock) mud volcanic field, 22073 dynes/cm^2; Burashskiy, 92522 dynes/cm^2; and Nasyrskiy and Enikalskiy, 12263 dynes/cm^2. The average figure for plasticity is 41,500 dynes/cm^2. The moisture content of Bulganakskiy mud is 35.4%, for Nasyrskiy mud 48.02%, and for Soldatsko-Slobodskiy mud 34.6%. The stickiness of Bulganakskiy mud is 17,168 dynes/cm^2, for Soldatsko-Slobodskiy 15941 dynes/cm^2, and for Nasyrskiy mud 2453 dynes/cm^2.

Specific heat (the amount of heat released when cooling 1 gram of mud 10 ° C) varies from 0.584 cal/g/°C (Nasyrskiy volcano) to 0.477 cal/g/°C. The specific heat of Bulganakskiy mud is 0.483 cal/g/°C. The volume weight of all Kerch volcanic muds varies between 1.40 and 1.42 g/cm^3. Hydrogen sulfide in all volcanoes is contained in insignificant amounts (0.01–0.03%), and only the mud of Lake Chokrak reaches 0.20%.

Considering grain-size, volcanic mud contains no greater than 0.1–0.4% of coarse (vent facies) fraction; the rest is represented by clay and dust particles. This is a very important indicator in the medical application of volcanic mud, because the presence of particles larger than 0.25 mm can cause spot burning during the course of mud application.

It was suggested that mud from all mud volcanoes of the Kerch Peninsula can be used for balneological purposes, but the muds from the Bulganakskiy and Yenikalskiy volcanoes are particularly suitable (Antonenko and Nesterov 1968). For the Kerch Peninsula in general, the mud resources are huge, and for balneological purposes, they are virtually inexhaustible.

In the 1980s, mud treatment became very popular, and special mud treatment units were established in a number of hospitals in Kerch and Feodosia. Some of them have survived until now. There is some verbal evidence that, at present, the mud of the Chokrak Lake is even exported.

One of the first attempts at balneological use of volcanic mud on the Taman Peninsula was undertaken in 1858, when a pool for mud baths was dug on the southern side of the Gora Gnilaya (Taman) hill and a temporary hospital department was established (Belousov and Yarotskiy 1936).

Later, volcanic mud was studied by the official Russian Federation medical organizations "Geominvod" (1974) and "Egida" (1994). Particular attention was given to the Akhtanizovskiy and Tizdar mud volcanoes on the Kerch and Taman peninsulas, respectively. Based on these studies, the Russian Research Center for Rehabilitation and Physiotherapy of the Ministry of Health issued instructions for the medical use of volcanic mud accompanied by its balneological properties. It was noted that the emission of mud is accompanied by the release of methane. The mud of the Akhtanizovskiy volcano has a moisture content of 61%, a specific gravity of 1.35, a specific heat of 0.66 cal/g/°C, an alkaline reaction with a pH of 8.1, and only 0.5% content of particles larger than 0.25 mm. It is characterized by a poorly reducing environment and an organic matter content of 0.50%. Mineralization of the mud solution is 13.5–15.5 g/l. Bromine and iodine are constantly present in quantities up to 50 mg/l and 5 mg/l, respectively. Sanitary-bacteriological indicators for the mud corresponded in most cases to the norms, except for individual samples (Adylov et al. 1995).

Balneological properties of mud from the Tizdar mud volcano were studied in some Moscow laboratories based on materials from the "Egida" enterprise. The State Unitary Enterprise "Lechmedresursy" of the Ministry of Health of the Russian Federation concluded that the mud is especially useful in cosmetology, and its sanitary-bacteriological properties are close to those of Akhtanizovskiy mud (Kozlov et al. 1996). An exception is the elevated concentrations of bromine and iodine, which reach 323 mg/l and 85 mg l, respectively. Of the 19 common pesticides, 13 are not found; the rest as well as heavy metals and radionuclides are detected well below natural background concentrations. The mud is classified as iodine-bromine-weakly sulfide therapeutic mud. It is recommended for masks and applications that promote the regeneration of cells; it softens and moisturizes the skin and gives it freshness. All sorts of allergic dermatitis, acne, wrinkles, and aging of the skin can be treated. The mud is also exported to the mud baths of Anapa.

Clearly, volcanic mud is under-utilized within the entire Kerch-Taman region. We do not take full advantage of the natural wealth within our reach, most notably in developing further the spontaneous initiative people have taken in the use of mud volcanoes for medical purposes.

Buses and cars constantly arrive throughout the summer at the mud volcano known among the people as the "Azov Hill," otherwise, Tizdar. Its caldera is filled with water and located on top of a truncated cone. The diameter of this lake is 16 m, and its water, only 5–10 cm deep, covers a fresh, gray, semiliquid mud, which bubbles from time to time with methane emerging from a vent in the center. The depth of the semiliquid mud, which was medically studied in 1994, is not less than 6–14 m. The mud reserves are estimated at more than 2000 m^3, but these are estimated numbers.

The mud volcano is enclosed by a fence, and several guards sell tickets and regulate access to the lake. There are dozens of people in the water. Having taken a mud bath and smeared themselves with mud, people then run down to the Sea of Azov to wash off the mud and then change clothes in private booths specially equipped with freshwater (Fig. 9.18).

The bus then leaves, and the next group of vacationers from the Caucasian sanatoriums repeats the process. In summer, mud baths take in 700–800, and occasionally more than 1000 people daily, and up to 90,000 over the course of the summer!

The Tizdar mud volcano is also used by local authorities as a kind of balneological clinic, while the Shugo mud volcano is perhaps the most picturesque geological site of the North Caucasus. Lately, this mud volcano is also used for balneological purposes.

Bathing by thousands of people should be accompanied by careful monitoring of the medical and sanitary situation in the mud volcano caldera, yet this is not yet available. Mud within the crater does not flow, and although its adsorption capacity is high, the load on the crater produced by the people bathing is also great. Obviously, medical evaluation of the effectiveness of this activity should be provided by physicians. There is also a possibility (however, very limited) of sudden gas emissions, which can be dangerous for people taking mud baths.

As one can see, the balneological resources of the Kerch-Taman region are substantial, and since dozens of promising mud volcanoes have not yet been studied, the level of their balneological usefulness is unknown. This is a kind of scientific medical virgin territory.

Another direction of balneological development for mud volcanoes in the Kerch-Taman region is the utilization of mineral springs that often accompany mud volcanoes.

Perhaps the most interesting are hydrogen sulfide springs located near mud volcanoes (Figs. 9.19, 9.20, and 9.21).

Hydrogen sulfide springs are recorded near the majority of mud volcanoes on the Kerch Peninsula, while on the Taman Peninsula, they are less common (e.g., they are present near the Bugazskiy, Peklo Azovskoye, Yuzhno-Neftyanoy, Fontalovskiy, Shapsugskiy, and Karabetova Gora mud volcanoes).

Fig. 9.18 Mud bath in the caldera of the Tizdar mud volcano

Fig. 9.19 One of the hydrogen sulfide springs near the Syuyurtashskiy mud volcano

Fig. 9.20 Hydrogen sulfide spring at the foot of the Dzhau-Tepe mud volcano

Hydrogen sulfide springs near mud volcanoes have been studied by many well-known scientists, such as S.P. Popov, G.A. Lychagin, L.A. Yarotskiy, O.E. Fesyunov, and others, but only some authors have attempted to link them to mud volcanism, in particular G.A. Lychagin and O.E. Fesyunov. The role of mud volcanism has been underestimated by most researchers, and it has not been taken into account by others. In light of the established features of mud volcano structure, the association of many hydrogen sulfide and carbon dioxide springs is currently not considered surprising.

Lychagin (1952) believes that water in the springs of mud volcanoes rises from the depths of the Maikopian series and is enriched during its passage through the Chokrakian limestones. Fesyunov (1968) distinguishes between strong and weak hydrogen sulfide springs containing more and less than 150 mg/l, respectively. Strong

Fig. 9.21 Hydrogen sulfide spring on the Bugazskiy mud volcano 9 (in Russian)

hydrogen sulfide springs are developed in the northern part of the Kerch Peninsula, near the Azov coast. Weak hydrogen sulfide springs are found near mud volcanoes throughout the eastern part of the Kerch Peninsula.

In the water of hydrogen sulfide springs, the total hydrogen sulfide content reaches 250–730 mg/l. The springs are highly mineralized (up to 20–34 g/l) by sodium chloride or a mixture of sodium sulfate and chloride. They also contain carbon dioxide (100–800 mg/l), iodine (20–190 mg/l), bromine (10–130 mg/l), and boron.

The total reserves of hydrogen sulfide water within springs near the mud volcanoes of both the Kerch and Taman peninsulas have not been estimated. Only the hydrogen sulfide waters of Lake Chokrak's mud volcanic field are relatively well studied; they reach 400,000 l/day. The presence of carbon dioxide and boron in their composition, along with geological features of the area, enabled O.E. Fesyunov to associate them with fossil or presently active mud volcanoes.

Another kind of mineral water linked to mud volcanoes, both on the Kerch and Taman peninsulas, is represented by carbon dioxide springs. These were studied by Al'bov (1971, 1973) in the mineral springs of Bulganakskiy, Tarkhanskiy, Seit-Ely, and Sultanovskiy (near Gornostaevka), and Kayala-Sartskiy (discovered by N.I. Andrusov). All of them are confined to mud volcanic structures.

According to the ionic composition, they contain chloride-hydrocarbonate-sodium waters with a mineralization of 11 g/l reminiscent of the commercially distributed mineral water "Essentuki" No. 17. They contain iodine (10–25 mg/l), lithium (3 mg/l), bromine (10–30 mg/l), and boron. They also contain carbon dioxide (73–92%), methane (7%), nitrogen (1–4%), and small amounts of hydrogen sulfide (0.2%). The total content of carbon dioxide is 0.6–1.8 g/l.

The Kayala-Sartskiy spring contains up to 760 mg/l of carbon dioxide and the same content of methane.

Carbon dioxide springs near mud volcanoes of the Taman Peninsula have not been explored yet.

As one can see, the balneological prospects of the mud from volcanoes of the Kerch-Taman region have been insufficiently studied. Mud volcanoes are located near large resort areas, and their use as mud deposits would be a valuable addition to the touristic and medicinal resources of resorts in the Caucasus (primarily the Anapa region) and Crimea (Kerch, Feodosia). For this purpose, additional study of mud volcanoes should be conducted. Such study should be oriented toward identifying areas that are the most suitable for balneological use of therapeutic mud. Exploiting therapeutic muds should be dramatically expanded to provide greater economic benefits and health improvements for local inhabitants and visitors.

The mineral springs associated with mud volcanoes have received even less study. There is no estimate of the total flow rate as well as production rate of the springs, and there is no clear plan for their development and use. This includes both the hydrogen sulfide and carbon dioxide mineral springs. Assessment of the possibility of using them for therapeutic purposes should be carried out in the near future, so that mud volcanoes can become a powerful factor in balneological tourism for the Kerch-Taman resorts.

References

Adylov GT, Mansurova EP, Sigalov LM (1995) Issledovaniya i modifikatsiya bakorovykh ogneuporov, poluchennye v usloviyakh luchistogo nagreva (Studies and modification of bakor refractories obtained under conditions of radiant heating). AN UzSSR, Tashkent (in Russian)

Al'bov SV (1971) O Kerchensko-Tamanskoy gidrogeokhimicheskoy i gryazevulkanicheskoy oblasti (On the Kerch-Taman hydrogeochemical and mud volcanic region). Doklady AN SSSR 197(1):175–177 (in Russian)

Al'bov SV (1973) O geokhimicheskoy obstanovke v Kerchensko-Tamanskoy oblasti i blizlezhashchikh ploshchadyakh (On the geochemical situation in the Kerch-Taman region and nearby areas). DAN SSSR 208(1):184–187 (in Russian)

Alenchikov P (1899) Chokrakskoye solyanoye ozero (Chokrak salt lake). Trudy pervogo s'ezda deyateley klimatologii, gidrologii i balneologii (in Russian)

Anfilatova EA (2008) Analiticheskiy obzor sovremennykh zarubezhnykh dannykh po probleme rasprostraneniya gazogidratov v akvatoriyakh mira (Analytical review of current foreign data on the distribution of gas hydrates in the world's waters) Neftegazovaya geologiya. Teoriya i praktika 3:1–8 (in Russian)

Antonenko VA, Nesterov KV (1968) Sravnitelnaya kharakteristika gryazevykh vulkanov Kerchenskogo poluostrova kak obektov gryazelecheniya (Comparative characteristics of the mud volcanoes of the Kerch Peninsula as objects of mud treatment). Tezisy dokladov IV nauchnoy konferentsii po izucheniyu poleznykh iskopaemykh otlozheniy osadochnogo chekhla yuga Ukrainy. KGU, Kiev, pp 32–33 (in Russian)

Basniev KS, Sukhonosenko AL (2010) Perspektivy osvoeniya resursov gazogidratnykh mestorozhdeniy (Prospects for the development of gas hydrate deposits). Gazovaya Promyshlennost' 1:22–23 (in Russian)

Beauchamp B (2004) Natural gas hydrates: myths, facts and issues. Geoscience 336:751–765

Beketov VN (1916) Yod, brom, bornaya kislota v okrestnostykh Kerchi i Tamanskom poluostrove (Iodine, bromine, boric acid in the vicinity of Kerch and on the Taman peninsula). Izvestiya Akademii Nauk seriya 6 (11) (in Russian)

Belousov VV, Yarotskiy LA (1936) Gryazevye sopki Kerchensko-Tamanskoy oblasti, usloviya ikh vozniknoveniya i deyatel'nosti (Mud hills of the Kerch-Taman region, the conditions of their occurrence and activity). ONTI, NKPT SSSR, Moscow (in Russian)

Biletsky VS (ed) (2004) Malaya gornaya entsiclopedia (Small mining encyclopedia). 1, Donetsk, Donbass (in Russian)

Bily C, Dick JWL (1974) Naturally occurring gas hydrates in the Mackenzie delta, N.W.T. Bull Can Petrol Geol 22:340–352

Bohrmann G, Torres ME (2016) Marine gas hydrates. In: Harff J, Meschede M, Petersen S, Thiede J (eds) Encyclopedia of Marine Geosciences, Encyclopedia of earth sciences series. Springer, Dordrecht, pp 433–437

Boswell R (2009) Is gas hydrate energy within reach? Science 325(5943):957–958

Bugay YuN, Balakirov YuA (2001) Gazogidratnye mestorozhdeniya (Gas hydrate deposits). MNTU, Kiev (in Russian)

Byakov YuA, Kruglyakova RP (2001) Gazogydraty tolshchi Chernogo morya – uglevodorodnoe syr'e budushchego (Gas hydrates of the Black Sea stratum – the hydrocarbon material of the future). Razvedka i okhrana nedr 8:14–19 (in Russian)

Dmitrievskiy AN, Balanyuk IE (2006) Gazogidraty morey i okeanov – istochnik uglevodorodov budushchego (Gas hydrates of the seas and oceans – the source of hydrocarbons of the future). OOO "IRTs Gazprom," Moscow (in Russian)

Dolenko GN, Parilyak IP, Kopach IP (1968) Neftegazonosnost' Kryma (Crimea's oil and gas potential). Naukova Dumka, Kiev (in Russian)

Dvoychenko PA (1926) Estestvennye proizvoditelnye sily Kryma (Natural productive forces of Crimea). In: Ves' Krym. KrymTSIK, Simferopol, pp 1–18 (in Russian)

Ebersin AG (1940) Sredniy i verkhniy pliotsen Chernomorskoy oblasti. Kimmeriyskiy yarus (Middle and Upper Pliocene of the Black Sea. Kimmerian stage) Stratigrafiya SSSR. Neogen, pp 477–527 (in Russian)

Efremova AG, Zhizhchenko BP (1974) Obnaruzhenie kristallogidratov v osadkakh (Detection of crystalline hydrates in precipitation). Doklady AN SSSR 214(5):1179–1181 (in Russian)

Fesyunov OE (1968) O mineralnykh vodakh Kerchenskogo poluostrova (On the mineral waters of the Kerch Peninsula). Tezisy dokladov IV nauchnteknicheskoy konferentsii po izucheniyu poleznykh iskopaemykh otlozheniy osadochnogo chekhla yuga Ukrainy. KGU, Kiev, pp 117–119 (in Russian)

Ginzburg GD, Solovyev VA (1994) Submarinnye gazovye gidraty (Submarine gas hydrates). Vserossiyskiy nauchno-issledovatelskiy institut geologii i mineralnykh resursov Mirovogo okeana "VNIIOkeangeologiya," Sankt-Peterburg (in Russian)

Ginzburg GD, Gramberg IS, Ivanov VL et al (1988) Problemy gazogidratonosnosti nedr Mirovogo okeana (Problems of gas hydrate-bearing interior part of the World Ocean). Geologiya okeanov i morey. PGO Sevmorgeologiya, Leningrad, pp 27–29 (in Russian)

Ginsburg GD, Kremlev AN, Grigor'ev MN (1990) Fil'trogennye gazovye gidraty v Chernom more (Filtrogenic gas hydrates in the Black Sea). Geologiya i geofisika 3:10–20 (in Russian)

Glumov IF, Glumov AI, Kazmin YuB et al (2005) Gazovye gidraty mirovogo okeana (Gas hydrates of the World Ocean). Geologiya i poleznye iskopaemye Mirovogo okeana 2:30–40 (in Russian)

Goldschmidt VM (1932) Geochemische Leit-Elemente. Die Naturwissenschaften 20(51):947–948

Gorchilin VD, Lebedev LI (1991) O priznakakh gazogidratov v osadochnoy tolshche Chernogo morya i vozmozhnom tipe lovushek uglevodorodov (On the signs of gas hydrates in the sedimentary sequence of the Black Sea and the possible type of hydrocarbon traps). Geologicheskiy zhurnal 5:67–75 (in Russian)

Goryainov EP (1968) Nekotorye osobennosti geologicheskogo stroeniya Chekush-Koyashskogo sernogo mestorozhdeniya i novye proyavleniya sery na Kerchenskom poluostrove (Some features of the geological structure of the Chekur-Koyashsky sulfur deposit and new manifestations of sulfur on the Kerch Peninsula). Paper presented at the IVth Scientific conference on the study of mineral resources of fossil sediments of the aedimentary complex of the south of Ukraine. Izdatel'stvo Kievskogo Universiteta, pp 55–58 (in Russian)

Grigoryev GA, Afanasyova TA (2012) Pyerspyektivy promyshlyennogo osvoyeniya nyetraditsionnykh ryesursov gaza v Rossii (Prospects of industrial development of non-traditional gas resources in Russia). Neftegazovaya geologiya. Teoriya i praktika 7(2):1–21 (in Russian)

Gubkin IM (1939) Rezultaty issledovaniya gryazevykh vulkanov Krymsko-Kavkazskoy provintsii (The results of a study of mud volcanoes of the Crimean-Caucasian province). AN SSSR, Moscow (in Russian)

Guliev IS (2010) Subvertikalnye geologicheskie tela – novye obekty poiskov mestorozhdeniy uglevodorodov (Subvertical geological bodies – new objects of hydrocarbon exploration). Vserossiyskaya konferentsiya "Degazatsiya Zemli: geodinamika, geoflyuidy, neft, gaz i ikh parogenezy." GEOS, Moscow, pp 140–145 (in Russian)

Gutt C, Asmussen B, Press W et al (1999) Quantum rotations in natural methaneclathrates from the Pacific sea-floor. Europhys Lett 48:269–275

Hancock S, Okazawa T, Osadetz K (2005) A preliminary investigation of the economics of onshore gas hydrate production: status of unconventional gas in Canada and the U.S. 7th Annual Conference on Unconventional Gas, 8 November 2005, Calgary, Alberta, Canada

Holder CD, Bishnoi PR (eds) (2000) Gas hydrates: challenges for the future. Ann N Y Acad Sci 912

Ishchenko VD, Kiselev NV (1967) K voprosu o geologicheskom stroenii i obrazovanii gipsovykh mestorozhdeniy Kryma (On the question of the geological structure and the formation of the gypsum deposits of Crimea). Geologicheskiy zhurnal 27(1):68–76 (in Russian)

Izgaryshev NA, Sludskiy AF (1917) Gryazevye vulkany Kerchenskogo poluostrova i Temryuksko-Tamanskogo rayona (Mud volcanoes of the Kerch Peninsula and Temryuk-Taman district). Rudnye vesti 2(3/4):97–112 (in Russian)

Jackson FG (1925) The oxidation of ceramic wares during firing, VI. A laboratory study of the effect of varying the rate of gas flow and of heating on the decomposition of pyrite in clay. J Am Ceram Soc 8(8):534–540

Kalinko MK (1960) O mekhanizme i usloviyakh obrazovaniya gryazevykh vulkanov (On the mechanism and conditions of formation of mud volcanoes). Problemy neftyanoy geologii. Gostekhizdat, Moscow, pp 83–90 (in Russian)

Kalinko MK (1964) Osnovnye zakonomernosti raspredeleniya nefti i gazov v zemnoy kore (Basic regularities of distribution of oil and gas in the Earth's crust). Nedra, Moscow (in Russian)

Kholodov VN (2002a) Gryazevovulkanicheskiye provintsii i morfologiya gryazevykh vulkanov (Mud volcanic provinces and the morphology of mud volcanoes). Geologiya i poleznye iskopaemye Mirovogo okeana 3:227–241 (in Russian)

Kholodov VN (2002b) Geologo-geokhimicheskie osobennosti i model' firmirovaniya gryazevykh vulkanov. Geologiya i poleznye iskopaemye Mirovogo okeana 4:339–358 (in Russian)

Klauda JB, Sandler SI (2005) Global distribution of methane hydrate in ocean sediment. Energy Fuel 19(2):459–470

Kobolev VP, Verpakhovskaya AO (2014) Skopleniya gazovykh gidratov v paleodelte Dnepra kak obekt seysmicheskikh issledovaniy (Gas hydrate accumulations in the paleodelta of the Dnieper as an object of seismic studies). Geologiya i poleznye iskopaemye Mirovogo okeana 1:81–93 (in Russian)

Konstantov SV (1929) Otchet geologicheskogo komiteta za 1926–1927 (Report of the Geological Committee for 1926–1927). Izvestiya Geolkoma 15 (in Russian)

Konyukhov AM, Ivanov MK, Kulnitskiy AM (1990) O gryazevykh vulkanakh i gazogidratakh v glubokovodnykh rayonakh Chernogo morya. Litologiya i poleznye iskopaemye 3:12–23 (in Russian)

Korsakov OD, Byakov UA, Stupak SN (1989) Gazovye gidraty Chernomorskoy vpadiny (Gas hydrates of the Black Sea basin). Soviet Geol 12:3–10 (in Russian)

Korsakov OD, Stupak SN, Byakov YuA (1991) Chernomorskie gazogidraty – netraditsionnyy vid uglerodnogo syrya (Black Sea gas hydrates – an unconventional type of carbon raw material). Geologicheskiy zhurnal 5:67–75 (in Russian)

Kostyrko EP, Proshenitsyn PA (1931) Novye stroitelnye materialy. Keramzit (New building materials. Expanded clay). Stroitelnye materially 3:106–109 (in Russian)

Kovalevskiy SA (1935) Gazovyy vulkanizm (vulkany i vulkanoidy) (Gas volcanism [volcanoes and volcanoids]). Azerbaydzhanskoe neftyanoe khozyaystvo 1 (in Russian)

Kozlov AP, Trebukhov YaA, Bykova, NA (1996) Balneologicheskoe zaklyuchenie na ispolzovanie lechebnykh gryazey mestorozhdeniya "Sopka Azovskaya" Krasnodarskogo kraya v kosmetologicheskikh tselyakh (Balneological conclusion on the use of therapeutic mud from the Sopka Azov deposit of Krasnodar Territory for cosmetic purposes). GUP "Lechminresursy" MZ RF, Moscow (in Russian)

Kropotkin PN, Valyaev BM (1980) Geodinamika gryazevulkanicheskoy deyatelnosti v svyazi s netegazonosnostyu (Geodynamics of mud volcanic activity due to oil and gas content). In: Geologicheskie i geokhimicheskie osnovy poiskov nefti i gaza. Naukova Dumka, Kiev, pp 148–178 (in Russian)

Kruglyakova RP, Kruglyakova MV, Shevtsova NT (2012) Geoekologicheskie problemy uglevodorodnykh flyuidov i gazogidratov na dne Chernogo morya (Geoecological problems of hydrocarbon fluids and gas hydrates at the bottom of the Black Sea). Ratsionalnoe osvoenie nedr 6:48–57 (in Russian)

Kvenvolden KA (1993) Gas hydrates-geological perspective and global change. Rev Geophys 31(2):173–187

Ladyzhenskiy GM, Plotnikov AM, Gordievich VA (1965) Pro sirku na gryazovomu vulkani Dzhau-Tepe (Kerchenskiy pivostriv) (On sulfur in the Dzhau-Tepe mud volcano, Kerch Peninsula). Dopovidi AN URSR 5:640–643 (in Ukrainian)

Land F (1866) Neftyanye kolodtsy i gryazevye vulkany Tamanskogo poluostrova (Oil wells and mud volcanoes of the Taman Peninsula) Meditsinskiy sbornik 1:14–24 (in Russian)

Lebedev YuS (1968) Novye proyavleniya zheleznykh rud v Chongelekskoy vdavlennoy sinklinali (New manifestations of iron ore in the Chongelek depressed syncline). Tezisy dokladov IV nauchno-teknicheskoy konferentsii po izucheniyu poleznykh iskopaemykh otlozheniy osadochnogo chekhla yuga Ukrainy. KGU, Kiev, pp 50–51

Luchitskiy VI (1922) Kerchenskiy zhelezorudnyy basseyn (Kerch iron basin). Trudy Tsentral'nogo upravleniya promyshlennoy razvedki 1:1–39 (in Russian)

Lüdman T, Wang HK, Konerding P et al (2004) Heat flow and quantity of methane deduced from a gas field in the vicinity of the Dnepr Canyon, north-western Black Sea. Geo-Mar Lett 24:182–193

Lychagin GA (1952) Iskopaemye gryazevye vulkany Kerchenskogo poluostrova (Fossil mud volcanoes of the Kerch Peninsula). Byulleten' moskovskogo obshchestva ispytateley prirody, Otdelenie geologii 27(4):3–13 (in Russian)

Majorowicz AP, Hannigan PK, Osadetz KG (2002) Study of the natural gas hydrate "trap zone" and the methane hydrate potential in the Sverdrup Basin, Canada. Nat Resour Res 11(2):79–96

Makogon YuF (1974) Gidraty prirodnykh gazov (Natural gas hydrates). Nedra, Moscow (in Russian)

Makogon, YuF, Holditch, SA (2002) Gas hydrates as a resource and a mechanism for transmission. SPE Annual Technical Conference and Exhibition San Antonio, Texas, 29 September–2 October 2002

Mamedov IS, Guliyev IS (2003) Subvertikalnye geologicheskie tela v osadochnom chekhle Yuzhno-Kaspiyskoy vpadiny (Sub-vertical geological bodies in the sedimentary cover of the South Caspian basin). Izvestiya Azerbaydzhanskoy Akademii Nauk o Zemle 3:139–146 (in Russian)

Marcelle-De Silva M, Dawe R (2011) Towards commercial gas production from hydrate deposits. Energies 4:215–238

Max MD, Johnson AH, Dillon WP (2006) Economic geology of natural gas hydrate. Springer, Berlin/Dodrecht

Mazurenko LL, Solovyev VA, Matveeva TV (2006) Gazovye gidraty Mirovogo okeana (Gas hydrates of the World Ocean). Gazovaya promyshlennost', spetsvypusk "Gazovye gidraty," pp 2–6 (in Russian)

Mokrinskiy VV (1916) Chekur-Koyashskoe sernoe mestorozhdenie (Chekur-Koyashsky sulfur deposit). Otchety KEPS, Petrograd (in Russian)

Moridis GJ, Sloan ED (2007) Gas production of disperse low-saturation hydrate accumulations in oceanic sediments. Energy Convers Manag 48:1834–1849

Naumenko PI (1967) Geologiya vdavlennykh sinklinaley Kerchenskogo poluostrova (Geology of recessed synclines of the Kerch Peninsula). Dissertation, Odessa I.I. Mechnikov State University (in Russian)

Okay S, Çifçi G, Özel S et al (2012) New discovered Izmir and Busan mud volcanoes and application of seismic attributes and AVO analysis in the easternmost Black Sea. Geophysical Research Abstracts 14, EGU2012-13332-3

Onatskiy OP (1987) Proizvodstvo keramzita (3-e izdanie) (Manufacturing expanded clay, 3rd edn). Stroyizdat, Moscow (in Russian)

Orlovskiy GN (1971) Kimmeriyskie zheleznye rudy Tamanskogo poluostrova i severo-zapadnogo pogruzheniya Kavkaza (Kimmerian iron ores of the Taman Peninsula and the northwestern immersion of the Caucasus). Veshchestvennyy sostav osadochnykh rud i voprosy ikh obogashcheniya, Naukova Dumka, Kiev, pp 26–33 (in Russian)

Orton E, Stely X (1913) Brit Brick Manufactur Abs, pp 9–16

Osadetz K, Brent T, Chen Z et al (2005) Gas hydrates in Canadian Sverdrup Basin, Canadian Arctic Archipelago: a potential new focus for Canadian resource characterization. Adapted from oral presentation at AAPG Annual Convention and Exhibition, New Orleans, Louisiana, April 11–14, 2010

Pokrovskiy NF (1933) Razvedka na bor i yod na gryazevykh sopkakh Tamanskogo poluostrova (Exploration for boron and iodine on the mud hills of the Taman Peninsula). Geologiya na fronte industrializatsii 2:32–38 (in Russian)

Ponizovskiy AM, Stashchuk MF (1974) Solyanye ozera i Sivash (Salt lakes and Sivash). Geologiya SSSR, v. VIII Krym. Nedra, Moscow, pp 109–121 (in Russian)

Popescu I, Lericolais G, Panin N et al (2007) Seismic expression of gas and gas hydrates across the western Black Sea. Geo-Mar Lett 27(2–4):173–183

Popov SP (1928) Gryazevye vulkany (Mud volcanoes). Priroda 6:541–554 (in Russian)

Popov SP (1938) Mineralogiya Kryma (Mineralogy of Crimea). AN SSSR, Moscow (in Russian)

Pyatnitskiy PP (1925) Tamanskiy poluostrov v gornogeologicheskom otnoshenii (Taman Peninsula in geological terms). Izvestiya soveshchaniya po issledovaniyu i izucheniyu Kubanskogo kraya 3 (in Russian)

Rakhmanov RR (1982) Gryazevoy vulkanizm podvizhnykh poyasov i ego geotektonicheskie pozitsii (Mud volcanism of mobile belts and its geotectonic position). Dissertation, Baku (in Russian)

Rakhmanov RR (1987) Gryazevye vulkany i ikh znachenie v prognozirovanii gazoneftenosnosti (Mud volcanoes and their significance in forecasting of oil and gas). Nedra, Moscow (in Russian)

Rusakov OM (2016) V pogone za prizrakom biogennykh uglevodorodov v Chernom more (Chasing the phantom of biogenic hydrocarbons in the Black Sea). Geologiya i poleznye iskopaemye Mirovogo okeana 4:118–127 (in Russian)

Rusakov OM, Kutas RI (2014) Fatamorgana biogennoy doktriny uglevodorodov v Chernom more (Chasing the phantom of biogenic hydrocarbons in the Black Sea). Geofizicheskiy zhurnal 2:3–17 (in Russian)

Shchebetov AV (2006) Mestorozhdeniya gazogidratov: resursy i vozmozhnye metody razrabotki (Gas hydrate deposits: resources and possible development methods). Tekhnologii toplivno-energeticheskogo kompleksa 2:12–17 (in Russian)

Schulz HD, Zabel M (eds) (2006) Marine geochemistry. Springer, Berlin

Serdyuchenko DP, Glebov AV, Pavlov VA (1963) Lyudvigitovaya mineralizatsiya i (Fe-B-TR) paragenezis v drevnykh platformakh (Ludwigite mineralization and [Fe-B-TR] paragenesis in ancient platforms). AN SSSR, Moscow (in Russian)

Shchemelev PP, Gervids NA (1934) Keramzit i ego poluchenie (Expanded clay and its procurement). Stroitelnye materially X934(2):55–60 (in Russian)

Shnyukov EF (1962) Novyy tip kimmeriyskikh zheleznykh rud Kerchenskogo basseyna (New type of Kimmerian iron ore of the Kerch basin). Doklady AN SSSR 145(5):1127–1130 (in Russian)

Shnyukov EF (1999) Gryazevoy vulkanizm v Chernom more (Mud volcanism in the Black Sea). Geologicheskiy zhurnal 2:38–47 (in Russian)

Shnyukov EF, Naumenko PI (1964) Kimmeriyskie zheleznye rudy vdavlennykh sinklinaley Kerchenskogo poluostrova (Kimmerian iron ores of recessed synclines of the Kerch Peninsula). Krymizdat, Simferopol (in Russian)

Shnyukov EF, Naumenko PI, Lebedev YuS et al (1971) Gryazevoy vulkanizm i rudoobrazovanie (Mud volcanism and ore formation). Naukova Dumka, Kiev (in Russian)

Shnyukov EF, Ivannikov AV, Bezborodov AA et al (1990) Rezultaty geologicheskikh issledovaniy 51-go reysa v Chernoe more NIS "Mikhail Lomonosov" (The results of geological studies of the 51st cruise of the R/V "Mikhail Lomonosov" to the Black Sea). Preprint IGN 90–8, Kiev (in Russian)

Shnyukov EF, Kobolev VP, Starostenko VI (2013) Gazovyy vulkanizm Chernogo morya (Gas volcanism of the Black Sea). Logos, Kiev (in Russian)

Shnyukov EF, Deyak MA, Ivanchenko VV et al (2015) Neobychnaya mineralogiya gryazevykh vulkanov Kerchenskogo poluostrova (Unusual mineralogy of the mud volcanoes of the Kerch Peninsula). Geologiya i poleznye iskopaemye Mirovogo okeana 4:5–19 (in Russian)

Shnyukov EF, Grishchenko NO, Deyak MA (2016) Prognoz novykh zhelezorudnykh proyavleniy v strukturakh Andreevskogo i Malotarkhanskogo gryazevykh vulkanov (The forecast of new iron ore manifestations in the structures of the Andreevskiy and Malotarkhanskiy mud volcanoes). Geologiya i poleznye iskopaemye Mirovogo okeana 1:57–60 (in Russian)

Sloan ED Jr (1998) Clathrate hydrates of natural gases, 2nd edn, revised and expanded. Marcel Dekker, New York

Sobolevskiy YuV (1970) Mineralizatsiya samorodnoï sirki vkraplenogo tipu u neogenovikh vidkladakh Kerchenskogo pivostrova (Mineralization of native sulfur impregnated in the Neogene sediments of the Kerch Peninsula). Doklady AN URSR seriya B 7:585–587 (in Ukrainian)

Solovyev VA (2001) Gazogidratonosnost' nedr Mirovogo okeana (Gas hydrate potential of the World Ocean). Gazovaya promyshlennost' 12:19–23 (in Russian)

Steber EA (1913) K voprosu o proiskhozhdenii produktov izverzheniya gryazevykh vulkanov (On the question of the origin of the products of the eruption of mud volcanoes). Izvestiya Nauchno-Tekhnicheskogo obschestva pri Yekaterinoslavskom gornom institute (in Russian)

Swift DJP, Thorne JA (1991) Sedimentation on continental margins, I: a general model for continental shelf sedimentation. In: Swift DJP, Oertel GF, Tillman RW, Thorne JA (eds) Shelf sand and sandstone bodies: geometry, facies and sequence stratigraphy. Oxford, Special Publication 14, International Association of Sedimentologists, Blackwell Scientific Publications, pp 3–31

Turley GF (1930) Bulganakska grupa gryazovikh vulkaniv na Kerchenskomu pivostrovi ta produkty ïkhnyogo vybukhu (Bulganakska group of mud volcanoes on the Kerch Peninsula and products of their explosion). Trudy Kharkivs'kogo tovaryustva doslidnikiv pryrody, pp 14–22 (in Ukrainian)

Valyaev BN, Dmitrievskiy AN (2009) Masshtaby uchastiya glubinnykh uglevodorodov v genezise skopleniy gazogidratov (Scale of participation of deep hydrocarbons in the genesis of gas hydrate accumulations). Materialy mezhdunarodnoy konferentsii "Perspektivy osvoeniya resursov gazogidratnykh mestorozhdeniy" RGU Nefti i Gaza im. I.M. Gubkina, Moscow 17–18 November 2009 (in Russian)

Vassilev A, Dimitrov L (2002) Spatial and quantitative evaluation of the Black Sea gas hydrates. Russ Geol Geophys 43(7):637–649

Walsh MR, Hancock SH, Wilson SJ et al (2009) Preliminary report on the commercial viability of gas production from natural gas hydrates. Energy Econ 31:815–823

Wilson H (1927) Ceramics; Clay Technology. McGraw-Hill, New York

Yakushev V (2008) Prirodnye gazogidraty: gazovye resursy i geologicheskaya ugroza (Natural gas hydrates: gas resources and geological threat). Oil Gas J Russ 6:77–83 (in Russian)

Yukhanov IS (1968) Geologiya i veshchestvennyy sostav keramzitovogo syrya Kerchenskogo poluostrova (Geology and material composition of keramzite raw materials of the Kerch Peninsula). Avtoref dis kand geol-min nauk, Odessa I.I. Mechnikov State University (in Russian)

Zenin GN (1955) Gryazevye vulkany Tamanskogo poluostrova (Mud volcanoes of the Taman Peninsula). Istoriko-arkheologicheskiy i geograficheskiy ocherk (Historical-archaeological and geographical essay). Uchenye zapiski Krasnodarskogo pedagogicheskogo institute 14:54–59 (in Russian)

The Hazardous Nature of Mud Volcanoes

10

Summary
This chapter discusses the impact of mud volcanoes on the environment and, in particular, the dangers that may arise from mud volcanic activity.

There are historically known cases in which the destruction of villages and individual structures occurred as a result of volcanic eruptions. At present, many settlements sited near mud volcanoes have grown in size and importance (e.g., Kerch, Temryuk). With volcanoes now situated among their buildings, eruptions have become a real danger to the population. During eruptions, houses can be destroyed and gas pipelines ruptured (the ammonia pipe to Taman is a prime example of infrastructure at risk).

In the early 1990s, the public managed with great difficulty to stop the construction of the Crimean nucleur power plant near mud volcanoes.

The danger posed by eruptions extends not only to physically destructive processes but also to the systematic emissions of gases, which vary in composition and can include mercury, boron, and a number of other elements even during the daily activities of the volcano. For example, mercury content of the Kerch Peninsula volcanoes exceeds the background by 1–2 orders of magnitude.

In many locations, it seems prudent to create reserves around mud volcanoes to control development and thereby limit their threat.

Mud volcanoes can be dangerous in the sea as well. In shallow areas, shoaling occurs where vessels can run aground as a result of the activity of volcanoes (e.g., Kerch Strait, southeast of the Sea of Azov).

In deep-sea areas of the Black Sea, there are known cases of massive methane emissions with ignition of the gas, which also poses a navigational hazard.

Nowadays, mysterious maritime catastrophes have impacted vessels sailing in an area of developed underwater mud volcanoes (e.g., "Amira-1," "Memory of Mercury"). The possibility that large-scale gas emissions could cause the foundering and sinking of ships has been experimentally proven by Australian mathematicians D. May and J. Monaghan.

Mud volcanic activity can also damage gas pipelines laid in the sea as well as communication lines, etc.

Under all circumstances, the danger inherent in mud volcanic manifestations must be fully considered during the planning and implementation of infrastructure expansion and modification.

Although mud volcanoes do not spew out molten lava from the Earth's interior—usually (but not always), it is a cooler mixture of gas, water, and solids—they can still be quite dangerous to people, affecting their life and causing major impacts on the environment, both on land and at sea. The

E. Shnyukov, V. Yanko-Hombach, *Mud Volcanoes of the Black Sea Region and their Environmental Significance*, https://doi.org/10.1007/978-3-030-40316-4_10

influence of mud volcanoes on people and the environment can be visually observed on land, but it is difficult to record such events in offshore contexts, and information about underwater effects is scarce.

The largest terrestrial mud volcano in the world (informally called Lusi, which is an abbreviation of Lumpur Sidoarjo) is located in East Java near Madura Bay in Indonesia. Ejecta from this mud volcano has added to the adverse effects of human activities in the Porong River catchment, affecting the ecology and biogeochemistry of the estuary and Madura Strait coastal waters (Jennerjahn et al. 2013). Although the mud flow from Lusi has been contained by levees since November 2008, flooding regularly disrupts local highways and villages, and further breakouts of mud are still possible.

In Azerbaijan, there is a giant mud volcano called Goturdag; it is the only continuously erupting mud volcano in the world. There are mud volcanoes on the coast of the Caspian Sea: Bandovan, Hamamdag, and Bahar. The Bandovan mud volcano is particularly curious. At its foot, archaeologists have discovered the remains of a medieval city located about 100–200 m from the sea. This coastal city may have been destroyed not only by the rising level of the Caspian, but also by mud volcanic eruptions. During their excursion to the mud volcanoes of Azerbaijan, the authors of this book heard stories about the deaths of shepherds and their flocks during eruptions. In the late 1970s, the first author was able to observe a building for storage of weapons constructed on top of one of the mud volcanoes, which could be extremely dangerous in the event of an eruption.

In most places where they occur, mud volcanoes have been understood as potentially hazardous for humans and the environment since antiquity.

10.1 The Hazards of Terrestrial Mud Volcanoes

In the Black Sea region, there are cases in which villages inadvertently located near active mud volcanoes were completely destroyed during eruptions. Legends relate a story about the total destruction of a Circassian village situated near the Shugo mud volcano on the Taman Peninsula during the Middle Ages as well as the Tatar village of Dzhau-Tepe located near the Dzhau-Tepe mud volcano in Crimea during the seventeenth century.

A number of cases are known of people who died after carelessly approaching a mud volcano or settling at its top. One legend relates that the Boris and Gleb mud volcano on the Taman Peninsula was given its name in honor of two monks who imprudently built a cell on its top and died when it erupted. On the Tsentral Lake mud volcano (Bulganakskiy mud volcanic hearth), which was situated at the front line during a phase of World War II, a tank allegedly sank into the abyss while trying to cross a "puddle" (formed by mud volcanic eruptions?). More examples of the destructive capacities of mud volcanic eruptions can be found in Table 6.1.

Mud volcanic activity can also cause great damage to structures and communications facilities constructed in close proximity, and such errors in the location of buildings were already made by ancient Greeks. This can be seen from the fact that some of the Taman mud volcanoes (e.g., Gora Gorelaya, Gora Boris and Gleb) periodically eject architectural fragments of ancient temples, apparently once built on the tops of the hills.

Today, a number of mud volcanoes are located near the cities of Kerch, Temryuk, and some others. With time, these cities have grown, and mud volcanoes now appear inside the city limits, surrounded by urban expansion. Within the city precincts of Kerch, these mud volcanoes include Dzhardzhava, Enikalskiy, and Soldatsko-Slobodskiy. Perhaps the Dzhardzhava mud volcano, or more precisely one of its three hills, poses the greatest threat and thus garners the most infamy; during its 1930 eruption, the flow of breccia posed a direct hazard to the town of Voskhod and destroyed several houses (Fig. 10.1).

The flow of mud volcanic breccia during the 1982 eruption of Dzhardzhava threatened to block the Kerch-Kamish-Burun railway branch and destroy the radio and glass factories. In addition, the water conduit from the North Crimean

Fig. 10.1 Consequences of the 1930 eruption of the Dzhardzhava mud volcano on the town of Voskhod. (Belousov and Yarotskiy 1936)

Canal to Kerch was permanently destroyed over the course of 1 or 2 months in 1982 in association with subsidence caused by the mud volcanic eruptions.

Subsidence of sedimentary mass also occurred in the area surrounding the Enikalskiy mud volcano causing cracks in nearby building structures. Judging from years of observations, the peacefully gurgling Soldatsko-Slobodskiy mud volcano does not represent a danger of catastrophic eruption. However, it is located on the main traffic route from Kerch to the village of Arshintsevo, along which communications infrastructure has been laid and a wide asphalt road with trolleybus is planned to run almost right through the mud volcano. Near the Soldatsko-Slobodskaya recessed syncline, there is a block of residential buildings. Given such construction placed so close to a mud volcano, special engineering with geological sensitivity together with additional investigations for each house as well as geochemical monitoring of the habitat is absolutely necessary.

As reported in the literature (Shnyukov et al. 2005), V.M. Sheremetev, V.T. Levchenko, A.I. Chernytsyna, and I.V. Tselyutina carried out investigations in 1994–1995 to assess the risk of emergencies regarding mud volcanoes on the Taman Peninsula. Most attention was paid to the potential threat from the terrestrial Miska mud volcano to the city of Temryuk, in addition to the dangers posed by the shallow marine Golubitskiy mud volcano to the village of Golubitskaya.

Considering past catastrophic eruptions of the Miska mud volcano, the authors singled out three danger zones that could become hazardous to varying degrees in the event of further volcanic activity. The first danger zone covers about 200 private households with approximately 600–800 people and indicates a likelihood of 100% destruction of buildings and other constructions. In this zone are located also the Temryuk Television Center, the Military Hill Museum, the water supply canal, reservoirs with water reserves, and the Caucasus-Krasnodar motorway that runs along the inner part of the mud volcano's crater shaft. The probability of destruction of the abovementioned structures is high under the influence of mud flows, gas emissions, subsidence, etc. The second zone covers about 3500 one-story houses and most of the multistory buildings of the city with up to 25,000 people inhabiting them, and it suggests a 50% likelihood of building destruction as well as damage to other constructions by mud flows. The third zone includes 1500–2000 households with about 5000 people, the majority of the industrial enterprises, the railway station, etc., and it could sustain partial destruction of buildings and other constructions as well as the generation of landslides and flooding by the Kuban River. The picture, frankly speaking, is not entirely gloomy, as perhaps a few of the threats may be overstated.

In the village of Golubitskaya, eruptions around the underwater Golubitskiy mud volcano have created an ellipsoidal zone of underwater sediment accumulation 600 m long and 300 m wide. The contour of the zone corresponds to the first or second danger zone, and it includes part of the "Cinderella" recreational park, access roads, and balneological lake. The third zone includes the health resort "Chaika," and part of the "Blue Torch" radio-geodesic station, etc. Here, intensive landslides are quite possible.

The railway branch, which was built in 2002 for ongoing ammonia exports, to the bridge across the Kerch Strait has been laid at the foot of the Karabetova Gora mud volcano, and this location is also an area of potential destruction due to subsidence deriving from mud volcanic activity. This danger is amplified by the toxicity that would result from an ammonia spill capable of poisoning a territory up to 1 km in radius and inflicting partial damage over a zone reaching 4.2 km in diameter. The construction of this railway branch for ammonia transport in 2002 caused a wide wave of protests in the Temryuk district.

On the Taman Peninsula, cases of roads blocked by mud flows and destroyed communication lines are well known, as can be seen in an example of the eruption of Tsymbaly West mud volcano (Figs. 10.2 and 10.3).

Levchenko et al. (1996) rightly draw attention to projects involving large-scale construction of new ports and oil pipelines on the Taman Peninsula and the need for geological forecasting of mud volcanic eruptions to be included in the planning of these ventures.

There are examples in which foresight has been shown regarding mud volcanic hazards. In the early 1990s, public protests originating from the recommendations of geologists were successful in forcing the closure of the already completed Crimean nuclear power plant that was laid out close to the large latitudinal South Azov fault and near the associated mud volcanic and diapir structures. The Semenovskaya, Mysovaya, and Kamenskaya anticlines, each broken by tectonic dislocations, are located in the vicinity of the nuclear power plant. Within the Semenovskaya anticline, 800 m from the industrial site, mud volcanic breccia was recovered from a borehole at a depth of 250 m. In the 146–250 m interval along the borehole, gas discharges were detected with anomalous concentrations of helium and radon. In cores recovered at the Mysovaya anticline, two intervals of liquefied mud volcanic clay were recorded. The Kamenskaya anticline contains a recessed syncline and the active Nasyrskiy mud volcano in its eastern extremity. A special study by seismologists, geologists, archaeologists, and other specialists recorded very strong earthquakes up to magnitude 8.0–9.0 on the MMS scale that occurred in Panticapaeum (the capital of the Bosporan kingdom) in 63 BC and on the northern coast of the Kerch Peninsula in the third century BC. Given the impact of earthquakes on the activation of mud volcanic and diapiric processes, as well as the possibility of M_w 9.0 earthquakes, it was necessary to abandon the commissioning of the already completed nuclear plant due to fears of sudden catastrophic mud volcanic eruptions provoked by earthquakes. In

Fig. 10.2 A mud flow covering the road after an eruption of the Tsymbaly West mud volcano on February 13–18, 2002

Fig. 10.3 The same road 1 year after the eruption of the Tsymbaly West mud volcano on February 13–18, 2002

addition, the flooding and raising of groundwater level in the area decreased the building's resistance to seismic influence by 1–2 points.

There is another ecologically dangerous aspect to the influence of mud volcanoes on the environment: a geochemical one. Together with its gases, mud volcanoes emit mercury, bromine, iodine, boron, and possibly arsenic. As a rule, mud volcanoes release above them gas clouds of mercury, the concentration of which is 1–2 orders of magnitude higher than background (Karasik and Morozov 1966). Under certain conditions, emissions of hydrogen sulfide can be hazardous. Unfortunately, there are no medical statistics for those urban areas lying close to active mud volcanoes. If there were, the data would undoubtedly show the impact of geochemical anomalies caused by mud volcanoes and how they have affected humans. For example, it is necessary to check whether radon accumulates in the cellars of buildings adjacent to mud volcanoes, because some researchers have recorded the presence of radon among mud volcanic gases. These studies, including serious routine observations made in the vicinity of mud volcanoes, have yet to be performed by geologists and ecologists. Humans must know the dimensions of the danger and acquire the ability to predict occurrences in order to protect themselves from the potentially injurious effects of mud volcanoes.

Within the Kerch-Taman region, mud volcanoes are unusual in terms of terrain and appearance. Their "lunar" landscape, peculiar beauty, constant activity, and terrible catastrophic eruptions attract the attention of numerous visitors. In addition, mud volcanoes such as Shugo (Fig. 6.113, also on the cover) are located in picturesque areas and are quite scenic. It is not by chance that one of the names given to the Shugo mud volcano is Marvelous Mountain.

All this should serve as a warning to tourists and visitors about the dangers of mud volcanoes. Attempts to approach the volcano's center or multiple vents to view the semiliquid sedimentary deposits can result in great harm or worse, and therefore excessive curiosity can be quite dangerous.

Today, it is imperative to create protected zones in areas where mud volcanoes have developed. This requirement should be seriously considered in Azerbaijan, Turkmenistan, and especially for the Kerch-Taman region. In a few places, it has already been partly realized. For example, the efforts of Ad.A. Aliyev (Institute of Geology and Geophysics of the Azerbaijan Academy of Sciences) have led to the creation of a mud volcanic reserve in Azerbaijan. In western Caucasus, it has become painful to see how new roads are being laid through gradually thinning forests on Shugo mountain, how gryphons are being littered, and how cattle graveyards and dumps are growing within the mud volcanic calderas of the region. The beautiful Sinyaya Balka (Tizdarskiy) mud volcano was turned into a garbage dump in the center of the village, and only in recent years has it been cleared and used for balneological purposes (see Sect. 9.7.3). In the 1980s, the idea of building a sewage line from Kerch city into the Bulganakskiy mud volcanic hearth was seriously discussed and, fortunately, rejected.

Such actions not only destroy natural monuments, they are unsanitary and dangerous, since the activity of mud volcanoes is not incorporated into the planning. The process leading to dispersion of mud volcanic breccia includes not only the breccia but also all potential contaminants, and this fact is seldom considered.

Inattention to mud volcanoes will be much more expensive than understanding this natural phenomenon. When laying a road, it is easier to bypass the volcano than to try to fill and build over it. It is not just a matter of enacting geological statutes that exclude engineering work and construction, extraction of minerals, or burial of waste and garbage, but instead there must be a complex and comprehensive program of protection for the geological and living environment in the most interesting areas of mud volcanic development. Not all mud volcanoes can be preserved, but the largest and most interesting of them are definitely possible. This is necessary for the safety of curious visitors, conservation of the landscape for future generations, and progress in future scientific research. Such protections will not be particularly difficult to establish, since mud volcanic foci, as a rule, occupy marginal areas that are not suitable for agriculture. It would be advisable not to exploit the largest, most picturesque volcanoes as sources of building materials (expanded clay) and chemical raw materials. The existence of richer deposits of boron, iodine, and further elements in other areas has made the recovery of these elements within mud volcanoes unprofitable at the present time, a fact that helps preserve the volcanoes from exploitation. For small mud volcanoes, it is necessary to establish at least protective signage in order to urge the population to engage in appropriate stewardship of these natural monuments.

The best option for conserving mud volcanoes would be to create a series of reserve areas in the Kerch-Taman region in order to safeguard localities with high priority, of which there are many. One could name, for example:

1. Dzhau-Tepe, the majestic steppe giant mud volcano.
2. The "moon valley" of the Bulganakskiy mud volcanic hearth.
3. The "Kir" oil crust of the Nasyrskiy mud volcano. The term "Kir" has a Turkish or Iranian origin and represents a mixture of thickened oil or asphalt with sandy or clay material. The mixture develops where the oil-bearing formations reach the surface.
4. The Andreevskiy mud volcano.
5. The numerous gryphons of the Yenikalskiy mud volcano on the Kerch Peninsula.
6. The marvelous beauty of the Shugo and Gladkovskiy mud volcanoes.
7. The grandeur of the Boris and Gleb mud volcano.
8. The explosive activity of Karabetova Gora.
9. The Semigorskiy mineral source, important for its balneological attraction as well as the mud volcano itself.

All of these, and many others, could form the basis for a future reserve.

Some movement in this direction has already been made. The map titled "Natural protected sites of the Ukrainian SSR," published in 1977, reports that in 1963, the Dzhau-Tepe mud volcano and a surrounding area of 10 hectares were declared a geological monument of significance and taken under the protection of the republic. Unfortunately, during the last 30 years of nearly continuous visitation, we have not been able to see the promised protection. The map titled "Geological monuments of Ukraine," published in 1984, declares that the Andrusov, Obruchev, and Vernadskiy mud volcanoes, as well as other localities within the Bulganakskiy mud volcanic hearth, are geological monuments of local importance. In the 1980s, some measures to protect the

Bulganakskiy mud volcanoes were undertaken by the governmental agency Kamysh-Burunskiy Mine Geology and Exploratory Drilling under the supervision of P.I. Naumenko. No other real environmental steps taken to preserve mud volcanoes are known to us.

It is possible even at the present time to turn mud volcanoes into tourist destinations. Although they are as yet undeveloped and somewhat untidy, volcanic localities will attract the attention of those who value some of the most curious of natural landscapes, and the opportunity for organized tourism can also help to popularize a more protective and concerned attitude toward nature. Guidebooks, popular articles, maps, and diagrams explaining mud volcanic processes would need to be prepared and published in order to disseminate the specific details of particular sites as well as the larger picture of mud volcanoes in general.

Access to educational information on mud volcanoes would provide tourists and the Kerch-Taman population at large with knowledge about how to preserve this unique geological phenomenon and priceless gift of nature. In this direction, it would seem expedient to create a museum dedicated to mud volcanism within the city of Kerch using a small building, perhaps in the area of the Bulganakskiy mud volcanic hearth, staffed by several people. A reserve and museum in Temryuk could be located at Gora Gnilaya, where extremely important environmental research could be carried out, which would also help to halt the misuse and contamination of the most significant examples of mud volcanism. The museum would collect and display specimens representing all the interesting mineralogical and lithological finds, while illustrating the geological basis for mud volcanism within the Kerch-Taman region. Romania offers a positive example of a mud volcano preserve, which was created as early as 1924 at Piklele Miki.

Another important application of such reserves and the research conducted within them would be to draw the attention of government agencies to the region's mud volcanoes. Results of this research could facilitate the forecasting of dangerous volcanic manifestations, and it might convince authorities to stop underestimating the risks of construction in close proximity to mud volcanoes. Also important would be observations on hydrogen sulfide mineral resources that accompany mud volcanoes, some of which (e.g., the sources within the Lake Chokrak region) could be exploited immediately.

10.2 The Hazards of Marine Mud Volcanoes

Of the total number of subaqueous mud volcanoes (70) in the Black Sea and Sea of Azov (see Sects. 6.4, 6.5, 6.6, 6.7, 6.8, 6.9, 6.10, and 6.11), relatively few are active. This estimate may be incorrect due to the difficulties of direct observation of mud volcanic activity in marine environments; their signs are barely noticeable and cannot be easily recorded.

Mud volcanoes can pose a threat to ship navigation, both civil and military. The several large mud volcanoes of the southeastern Sea of Azov make this area particularly dangerous (Fig. 10.4).

Shepel (1926) first described a mud volcano located to the west of Cape Tuzla. According to him, the ship "Caesar" ran aground in the strait in 1914. It turned out that a cone-shaped 4 m bank appeared unexpectedly in this zone of 9 m depths, and soil samples taken from it were represented by mud breccia. Subsequent geological studies in this area show that lithologic and mineralogical anomalies and hypsometric elevation rises of Neoeuxinian, Old and New Black Sea sediments constantly occur in the vicinity of this eroded shoal. It gives the impression that a mud volcano has existed here from at least pre-Neoeuxinian times (e.g., about 30 ka) (Shnyukov et al. 1981).

Within the Temryuk anticline (also called the Kazbek bank or the Peresypskiy) is the underwater Temryukskiy mud volcano. It is located 5 km north of the coast and is marked by a buoy. In the "Navigation Map of the Sea of Azov," the attention of seafarers is drawn to the sudden change in depth and emergence of islands in the area.

Located to the south, the Golubitskiy mud volcano is situated within the anticlinal zone of Tsymbaly. It has been repeatedly described by

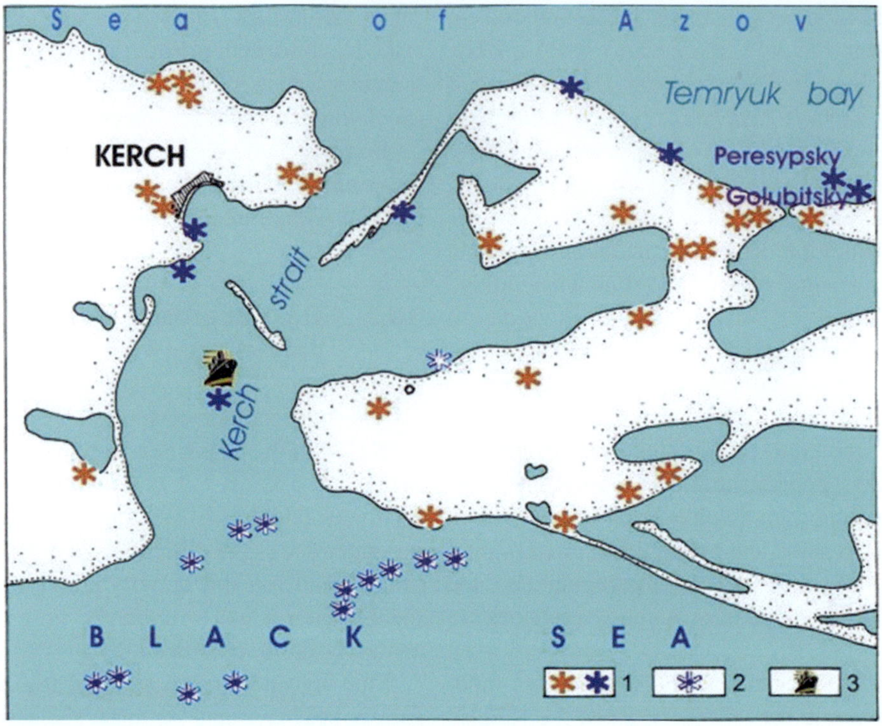

Fig. 10.4 Mud volcanoes of the Kerch Strait. Key: (1) onshore and offshore active mud volcanoes, (2) supposed mud volcanoes on the sea bottom, (3) the landing site of the ship "Caesar". (Shepel 1926)

different authors, both during periods of activity and at moments of relative rest.

Even more dangerous is the mud volcanism that exists within the Kerch Strait, which is navigated by up to 10,000 vessels annually. This is the second strait (after the Bosporus) of the Black Sea. The narrow water area of the strait still contains many mysteries, including mud volcanoes and sunken ships (Fig. 10.4).

The scale of mud volcanic activity is also enormous in the deepwater part of the Black Sea. Large-scale, one-time emissions of gases by mud volcanoes can pose a real danger to navigation. A huge bubble of released gas emerging at the surface can cause ships to founder. Even more dangerous are gas explosions and ignitions.

Egorov et al. (2011) calculated the conditions under which a vessel could lose its buoyancy due to reduction of the underlying water-gas density. According to their data, such a thing might occur if the critical value of the gas emission equaled some multiple of the vessel's area measured at the waterline and based on the critical specific gravity of the methane flow.

The same authors also calculated that flare ignition would be possible given a steady output of at least 150 liters of methane per second into the air layer immediately above the water's surface. Methane combustion can occur if its content in the atmosphere reaches between 5 and 15%. At a concentration of 15%, methane would burn as a calm and steady flame. Concentrations higher than 15% would likely lead to the flame transitioning into a flare.

During the Crimean earthquake of 1927, gas emissions were recorded to the southwest of Cape Khersones. In some places, the methane ignited, and the sea caught fire, posing an equal danger to large ships and small vessels bound for Crimean ports. Especially dangerous were explosions of gases to the west and southwest of Sevastopol, where the scale of gas emissions was simply enormous (Fig. 10.5).

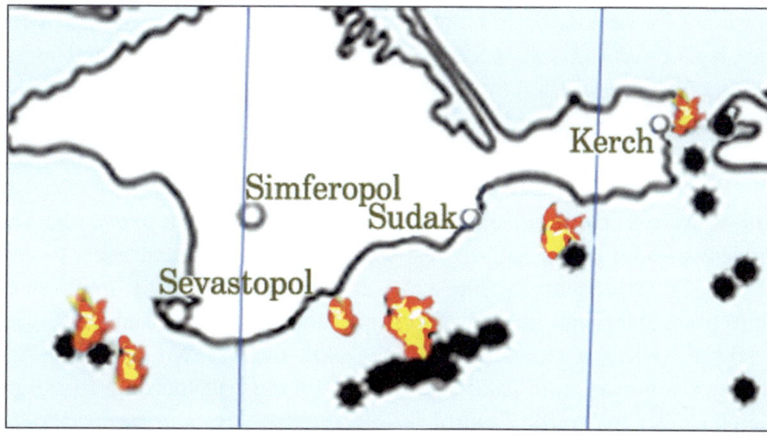

Fig. 10.5 Flashes of gases following the Crimean earthquake of 1927. Black dots, underwater mud volcanoes; yellow with red, gas flashes

On September 11, 1927, at 22:15 pm, immediately after the Mw 8 Crimean earthquake, the hydrographic station on Cape Lukull registered a column of flame that lasted 5 seconds. The lighthouse at Evpatoriya observed a flare of white fire at 2:48 am on September 12. The Sevastopol lighthouse (Konstantinovskiy Ravelin) recorded a flash of fire 500 m high and a mile and a half wide at 3:31 am. Gas explosions continued after the earthquake for almost 2 months. On October 14, 1927, a column of gases arose 7 km to the south of Alupka city. From the Evpatoriya lighthouse, burning clouds appeared over the sea starting at 3 am and continued for an hour as the records indicate.

On October 2, 3, and 4, 1927, burning fountains of gas were documented 30 km from Feodosiya to the east in the direction of Anapa. Over the same days, smoke and fire were visible in the sea as observed from Sudak toward the southwest in the direction of Alushta. On October 4, at 11:30 pm, a whitish strip flashed twice with a bright torch about 20–25 km to the west and southwest of Sudak (Nikonov 2002).

Ancient cases are also known. In the autumn of 63 BC, an earthquake comprising two shocks of magnitude 9 destroyed the capital of the Bosporan Kingdom, the city of Panticapaeum (present-day Kerch), and its surrounding territories. Numerous mud volcanoes within the Kerch Strait reacted to the seismic event with gas emissions (Petrov 2009).

In 1902, catastrophic events took place off the coast of Bulgaria near Balchik city.

Egorov et al. (2011) described an area of gas outlets off the Black Sea coast of Georgia (42° 42.2′ N; 41° 11.0′ E) at a depth of 530 m on the slope of a canyon that reached down to 800 m; the sea above was seen to burn during earthquakes and other geological cataclysms according to the local population, who referred to this manifestation as the anger of the "God of Ochamchire," after the Abkhazian coastal city.

Comparison of the location of underwater mud volcanoes to the recommended routes for seagoing vessels shows that many maritime passages navigate directly through areas of mud volcanic activity. Nowadays, up to ten ships per year sink in the Black Sea. As a rule, the causes of accidents are determined by special commissions, and most shipwreck incidents are thoroughly investigated to determine why they happened. Nevertheless, in recent years, there have been at least two catastrophes for which an explanation has been difficult, and the causes of the accident remain unclear.

The disappearance of the large cargo ship "Amira-1" deserves careful scrutiny, since it occurred in an area of mud volcanic activity. If we exclude the unlikely possibility of a criminal act associated with the ship's insurance, there is still much uncertainty. The "Amira-1" had a displacement of 10,182 tons and carried a cargo of coal weighing 9180.5 tons; it was manned by a crew of 24 people on board. It left the port of

Mariupol on January 5, 2003, at 12:10 pm, with the registration of all necessary documents, headed for the Turkish port of Gerze. The motor ship was built in 1975 and sailed under the flag of Tunisia; the shipowner was a Turkish company. Of the 24 team members, the captain and two sailors were Turkish citizens, three others were citizens of Azerbaijan, and the rest were citizens of Tunisia. On January 6, 2003, at 1:25 pm, the ship passed through the Kerch Strait, and at 2:10 pm, during its traverse of Cape Takil, the ship was removed from radar surveillance by the Kerch Center for Traffic Control.

On January 9, 2003, at 4:21 am, an American passenger airplane (Boeing 747) flying over the Black Sea picked up an emergency beacon signal at coordinates 43°41′ N; 33°02′ E and transferred it to the COSPAS-SARSAT satellite tracking station in Moscow. From there, the signal was transferred to Odessa where it was received by the State Rescue and Coordination Center of the Ministry of Transport of Ukraine. On the same day at 6:15 am, a navigational warning signal about the disaster was sent to all ships in the area, and a rescue operation was launched involving a Russian airplane and a Ukrainian helicopter. On January 14–15, 2003, a search was carried out by Turkish aircraft (Fig. 10.6).

During the search, an emergency buoy from the "Amira-1," a life preserver, and an oil stain 2 km long and 100 m wide (coordinates 43° 39′ 88″ N; 33° 45′ 11″ E) were found; not far away, a white board with the inscription "Amira-1," four life jackets, a canister containing solvent, a bag of pyrotechnics (coordinates 43° 40′ 8″ N; 33° 32′ 3″ E), two life rafts, and an oar were located. Experts came to the conclusion that the vessel either overturned or slowly sank. Apparently, emergency funds were in order. Why no one escaped is still a question, as bodies of people were not recovered.

On January 11, 2003, an investigatory delegation from Tunisia arrived in Kiev and then visited Mariupol and Sevastopol. It was noted that before the "Amira-1" left the port of Mariupol on January 5, the captain-inspector, A. Danilchenko, made a remark to the ship's captain that the ship was overloaded, creating a threat to the safety of navigation. As a result, 255 tons of coal were unloaded, and as a result, the ship's departure was delayed.

Judging by the fact that no radio signals were sent at the moment of the disaster or before, one can assume that the catastrophe was sudden. Most likely, the ship capsized. This would be unusual because coal is traditionally considered to be the optimum cargo in terms of maintaining vessel stability.

No data pinpoint the exact position of the ship at the time of the accident, but judging roughly by the sea surface finds (the buoy, etc.), the catastrophe occurred in an area of intense mud volcanism, 50 miles south of Cape Sarych. There are several very active mud volcanoes: MSU, Yuzhmorgeo, Kornev, etc. (see Fig. 10.6), and discovery of new mud volcanic morphostructures nearby is likely.

Discovery of a buoy 50 miles south of Cape Sarych is completely incomprehensible. The port of destination, Gerze, is located on the Turkish shore to the south of Sinope, and the route of a vessel with such a displacement should have passed farther to the east. Transfer of the buoy in a western direction is unlikely, since, according to the search results, it was generally carried to the east (the first signal was received from the point 43° 41′ N; 33° 02′ E; the buoy was raised at 43° 27′ N; 33° 43′ E).

If one assumes that the ship followed the Crimean coast and subsequently changed course to head due south, it is likely that it sailed over the Sorokin Trough, where almost two dozen mud volcanoes are known, and some of them, such as the Dvurechenskiy mud volcano located in the in the Sorokin Trough (see Sect. 6.7), are characterized by catastrophic explosions.

Two years earlier and not far from the area where a few buoys of the ship "Amira-1" were found, the steam ship "Memory of Mercury" sank. This catastrophe occurred on January 26, 2001, at the coordinates 43° 47′ 0″ N; 31° 43′ 0″ E. The vessel shuttled back and forth along the line between Istanbul and Evpatoriya. Before the accident, it had left Istanbul overloaded; instead of the maximum permissible load of 250 tons, it carried 327 tons, including an on deck cargo of

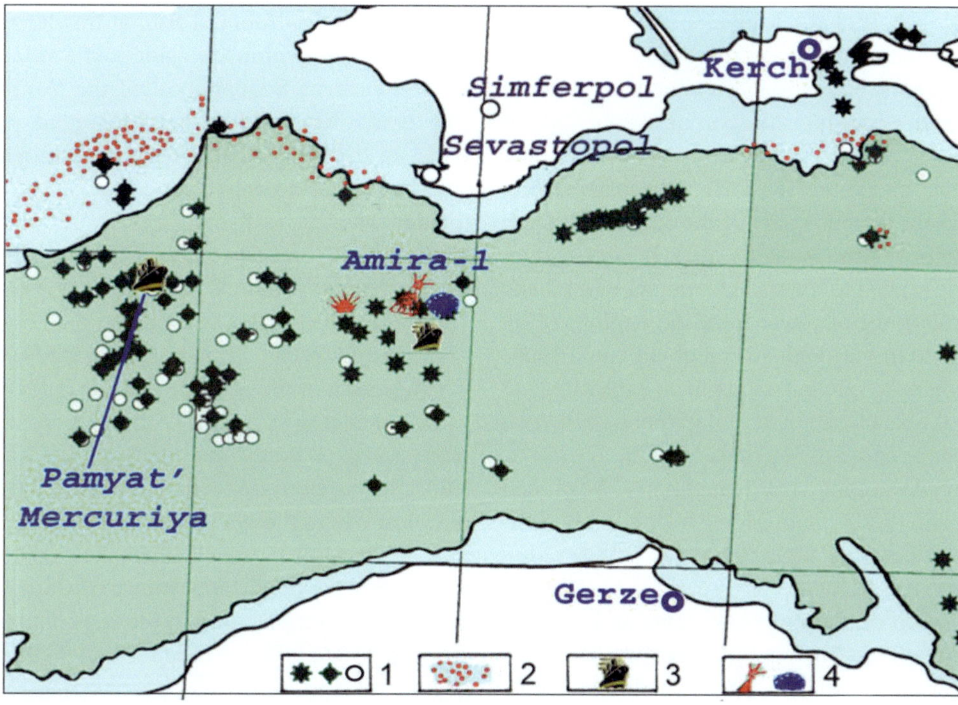

Fig. 10.6 Places of destruction of the vessels "Memory of Mercury" and "Amira-1." Key: (1) underwater mud volcanoes: apteryx – discovered, black dots with crosses – supposed; white circles, morphological elevations on the sea bottom that remind mud volcanoes; (2) gas seeps; (3) presumed locations of ship loss; (4) location where an emergency buoy and other items from the vessel "Amira-1" were found

237 tons (with an allowable maximum of 25 tons). The sea waves were moderate, as was the southeast wind. Because of the overload, the ship became unstable. Captain L. Ponomarenko contacted the shipowner to report on the situation. Aboard ship, there were 52 people (crew and passengers). The navigation area of the vessel was limited to 20 miles from the coast, yet the ship was at the time, and many times previously, crossing the Black Sea from Istanbul to Evpatoriya.

Disaster struck at 6:52 pm when the ship capsized over the course of a few minutes. In the opinion of the commission that later considered the case, the calamity occurred under bad weather conditions due to cargo displacement, but the "bad weather" was just a wave of 0.2 m or (0.66 ft) or 2 points on the Beaufort scale (https://simple.wikipedia.org/wiki/Beaufort_scale), which is not unusual for the Black Sea. Captain Ponomarenko was a very experienced navigator, who had sailed to Antarctica on ships of the same class, and he could hardly have allowed his cargo to be placed inappropriately. Some of the surviving members of the crew reported hearing a loud clap before the rollover began. Possibly, this was a gas blast?

The appearance of a large gas bubble can cause a vessel to tip and possibly founder. Ejected fragments of a mud volcano, including gas pockets within the sedimentary strata, can even stimulate weak earthquakes.

In pursuing this matter, the catalog of seismic signals (possible earthquakes) recorded by the Crimean Branch of the Institute of Geophysics of the National Academy of Sciences of Ukraine at the time of the sinking of "Amira-1" and "Memory of Mercury" was studied. A small earthquake seems to have occurred on the eve of the "Amira-1" disaster, that is, January 5, 2003. A giant gas bubble could have emerged from the seafloor as a secondary result of this earthquake.

Geologist Richard McIver (1982) suggested that methane hydrates on the ocean floor could break up and release giant methane gas bubbles that could endanger ships or airplanes. "Giant bubbles" would reduce the density of water and, when they reach the sea surface beneath a ship, could cause the buoyancy of the ship to decrease and thereby sink it. With aircraft, the danger could be related to the rise of a parcel of methane gas through the air; heat from the engines of an airplane flying through such a cloud could cause the methane to ignite, leading to an explosion and fire, or the methane could replace enough oxygen in the air to cause the engines to quit.

Two Australian mathematicians May and Monaghan (2003) from Monash University studied the effect of a single, large gas bubble rising underneath a floating body (a ship) based on a combination of simple experiments and numerical simulations. Their motivation was to assess the possible hazard arising from naturally occurring methane gas hydrates in the North Sea. The experiments enabled them to extend a benchmark of numerical simulations using smoothed particle hydrodynamics, and they obtained good agreement between the simulations and experimental results. In particular, it turned out that if the radius of a large bubble is comparable to or greater than the length of a ship's hull, the bubble can cause the ship to sink because a mound of water is raised directly above the bubble as it reaches the surface. "The flow from the mound creates a deep trough on each side of the mound, and the flow from the mound can carry a boat into the trough. Whether or not the ship will sink depends on its position relative to the bubble. If it is far enough from the bubble, it is safe. If it is exactly above the bubble, it also is safe, because at a stagnation point of the flow the boat is not carried into the trough. The danger position is between the bubble's stagnation point and the edge of the mound where the trough formed. If the ship is not at rest, then it would be possible for the ship to move into the region of the trough, or away from it, under its own power" (Fig. 10.7) (May and Monaghan 2003, 847).

Scientists have found some confirmation of these experiments. In 2000, one undersea exploration expedition found a fishing trawler resting on the bottom, within the caldera of a volcano in an area east of Scotland called the Wormhole. The crater arose, May and Monaghan (2003) write, in the process of a large release of methane, i.e., it was a mud volcano. The length of the trawler was 22 m; it had been in service from 1890 to 1930, and currently, it lies on its bottom in a horizontal position.

Similar phenomena can exist in many regions of the World Ocean, for example, in the Bermuda Triangle area of the Atlantic Ocean (Fig. 10.8).

A recent report suggests that extraordinarily large waves account for many of the missing ships in the Bermuda Triangle, but it begs the question where the waves come from. There are also suggestions about methane explosions responsible for the disappearance of ships.

It is believed that only in the last century, this sinister "triangle" has killed about a thousand people, leading to the disappearance of 50 ships and 20 aircraft. Among other reasons given for such losses (such as hurricanes, currents, whirlpools, magnetic anomalies, sudden obscurations of the consciousness of ship crews, aliens from outer space, the influence of sunken Atlantis, etc.), the possibility was also considered that giant methane bubbles may have been responsible for such disasters (e.g., Glasman et al. 2018).

Mud volcanoes can be hazardous in marine rescue operations. For example, a deep-sea sonar vehicle searching for Malaysia Airlines flight MH370 sank to the ocean floor after colliding with an underwater mud volcano in 2014. This impact against the 2200 m mud volcano broke the cable tethering the sonar vehicle, known as a towfish, to the Fugro Discovery search vessel. The Joint Agency Coordination Centre (JACC) has reported that the towfish and 4500 m of cable are now resting on the floor of the Indian Ocean. There were no injuries to the crew of the Fugro Discovery, and it is thought that the towfish can be recovered later (https://www.independent.co.uk/news/world/asia/mh370-search-vessel-sinks-after-crashing-into-volcano-a6832016.html).

Catastrophes associated with large emissions of gases from the sea bottom are possible in the

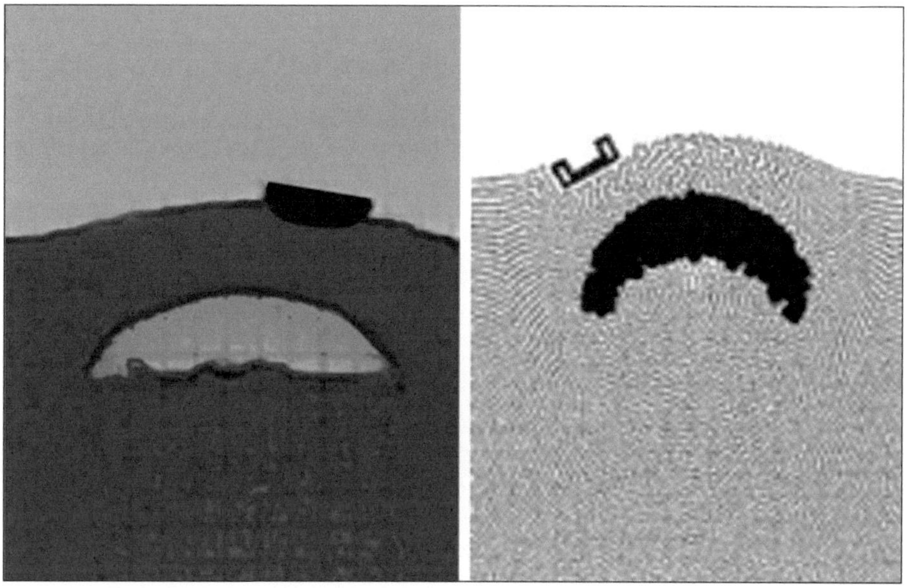

Fig. 10.7 Experiments by May and Monaghan (2003) with modeling of the loss of ships

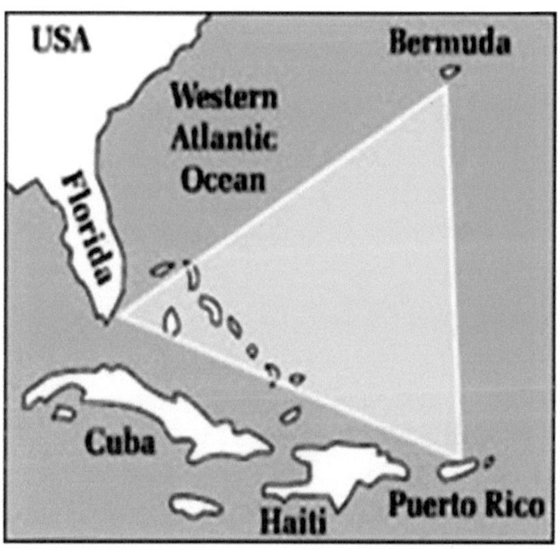

Fig. 10.8 Bermuda Triangle in the Atlantic Ocean

Black Sea. Beneath the basin's seafloor, there are remarkably large reservoirs of methane. Ruptures along active faults create mud volcanoes, which act as channels for the upward migration of the gas. The release of methane can change the density of the water column, and this can cause ships to "mysteriously" sink. Methane is also a powerful greenhouse gas and can produce adverse effects locally for ship crews and marine organisms. Gases released by mud volcanoes can also provoke the decomposition of gas hydrates, leading to increased methane emissions in the deep-sea area.

Mud volcanoes can be hazardous to offshore infrastructure, especially in deep water where the hydrate is stable (e.g., the gas pipeline "Blue Stream" that crosses the Black Sea basin at the traverse of Samsun). Theoretically, an increase in temperature and/or decrease in pressure can trigger the dissociation of gas hydrate and the release of free methanc, thereby weakening the sediment by removing hydrate as a cementing agent. In sediments of low permeability, hydrate removal would raise fluid pressure, making the sediment prone to submarine slumps that could be catastrophic in their consequences, such as the creation of a tsunami.

The latest research shows that methane in bottom sediments of the Black Sea released by mud volcanoes, gas seeps, etc. has a definite effect on biota. This subject is discussed in Chap. 11.

References

Belousov VV, Yarotskiy LA (1936) Gryazevye sopki Kerchensko-Tamanskoy oblasti, usloviya ikh vozniknoveniya i deyatel'nosti (Mud hills of the Kerch-Taman region, the conditions of their occurrence and activity). ONTI, NKPT SSSR, Moscow (in Russian)

Egorov VN, Artemov YuG, Gulin SB (2011) Metanovye sipy v Chernom more. Sredoobrasuyushchaya i ekologicheskaya rol'. Sevastopol' (Methane seeps in the Black Sea. Environmental and ecological role). Ekosagidrofisika, Sevastopol (in Russian)

Glasman G, Puigdevall F, López Trujillo F (2018) The secrets of the world's seas: Atlantis, the legend of the lost continent, and the Bermuda Triangle. Cavendish Square Publishing LLC, New York

Jennerjahn TC, Jänen I, Propp C et al (2013) Environmental impact of mud volcano inputs on the anthropogenically altered Porong River and Madura Strait coastal waters, Java, Indonesia. Estuar Coast Shelf Sci 130:152–160

Karasik MA, Morozov VI (1966) Osobennosti rasprostraneniya rtuti v produktakh gryazevogo vulkanizma Kerchensko-Tamanskoy provintsii (Peculiarities of mercury distribution in the mud volcanism products of the Kerch-Taman province). Geokhimiya 6:668–678 (in Russian)

Levchenko VT, Chernitsyna AI, Tselyutina IV (1996) Gryazevye vulkany Tamani – realnaya ugroza vozniknoveniya chrezvychaynykh situatsiy i katastrof (Taman mud volcanoes – the real threat of emergencies and disasters). Kubangeologiya (in Russian)

May DA, Monaghan JJ (2003) Can a single bubble sink a ship? Am J Phys 71(9):842–849

McIver RD (1982) Role of naturally occurring gas hydrates in sediment transport. AAPG Bull 66(6):789–792

Nikonov AA (2002) Krymskie zemletryaseniya 1927 goda. Neizvestnye yavleniya na more (Crimean earthquake in 1927. Unknown phenomena at sea). Priroda 9:13 (in Russian)

Petrov S (2009) Mitridatovo Tsarstvo (Mithridates Kingdom). Sekretnye materially (Secret Materials) 6 (in Russian)

Shepel SA (1926) Kolebaniya dna Kerch-Yenikalskogo proliva (Fluctuations of the Kerch- Yenikalsky strait bottom). Dekadnyy byulleten' pogody i sostoyaniya morya po Chernomorskomy i Azovskomu poberezhyu. Gidrometsentr Chernazmorey 30:11 (in Russian)

Shnyukov EF, Alenkin VM, Put AL et al (1981) Kerchenskiy proliv (Kerch' Strait). Naukova dumka, Kiev (in Russian)

Shnyukov EF, Starostenko VI, Ivannikov AV et al (2005) Gazovyy vulkanizm Chernogo morya (Black Sea gas volcanism). OMGOR NAN Ukrainy, Kiev (in Russian)

Black Sea Methane and Marine Biota (Case Study)

11

Summary

This chapter represents a case study that reviews research into the relationship between meiobenthos distribution and concentrations of hydrocarbon gases (HG), primarily methane, in the sediments of the northwestern part of the Black Sea, including gases released by mud volcanoes and gas seeps. Evidence forming the basis of this research comes from meiobenthos here represented by 29 species of benthic foraminifers, 7 species of ostracods, and 44 species of nematodes. The potential use of these meiobenthic organisms as indicators of gaseous hydrocarbons reservoirs existing under the seabed is evaluated according to two linked axes, namely, the dual analysis of abiotic factors (physical and chemical parameters of the water column, gasmetrical, geochemical, lithological, and mineralogical properties of the sediments) and biotic characteristics (quantitative and taxonomic composition of foraminifers, nematodes, and ostracods). Studies of this kind have been directed toward developing interdisciplinary methods to improve the search for HG accumulations, especially methane, under the seabed. Development of such methods might have substantial socioeconomic importance for the economy of Ukraine as well as that of other Black Sea countries, and such methods might also contribute to the sustainable development of Black Sea ecosystems.

11.1 Introduction

The Black Sea is the largest meromictic basin in the world (423,000 km^2). The structure of its water column is double-layered, in which the deep layer is salty and heavy, and the superficial one is desalinated, light, and enriched with oxygen. These two water layers do not mix with each other, and as a result, 90% of the deep water below the isobath at −220 m is oxygen-free. Aerobic life is concentrated mainly on the shelf, but it also occurs beyond (Sergeeva and Zaika 2013), forming fragile bottom ecosystems that are easily exposed and vulnerable to external influences.

The Black Sea basin is the largest methane reservoir on Earth (Lein and Ivanov 2005). Breakthrough tectonic disturbances form weakened zones in the sedimentary cover and act as "chimneys" through which filtration streams of fluids enriched mainly with methane rise to the surface of the seafloor. Powerful degassing of the deep Earth occurs due to jet emissions (cold seeps) and mud volcanoes, as well as microbiological processes throughout the depth range. According to different estimates, about 4.95–5.65 Tg (1 Tg = 10^{12} g) of methane are annually emitted from the entire Black Sea bottom into the water column (Kessler et al. 2006). The Black Sea basin therefore holds great promise for providing new non-traditional energy sources to surrounding countries, along with most of Europe,

E. Shnyukov, V. Yanko-Hombach, *Mud Volcanoes of the Black Sea Region and their Environmental Significance*, https://doi.org/10.1007/978-3-030-40316-4_11

due to the presence of huge methane resources stored in gas hydrates beneath the seafloor. These resources are estimated to be between 25 and 49 trillion m^3, which makes them several times greater than any other known gas reservoirs on Earth.

Two basic types of methane have been identified in bottom sediments of the Black Sea. The first type is characterized by simultaneous increase in the content of methane and adenosine triphosphate (ATP) in sediment cores, suggesting the existence of methanogenesis by bacteria with a chemolithoautotrophic type of metabolism that feed upon the streams of endogenous gases, in particular carbon dioxide. This type of methane is often widely scattered over the area and rarely forms large concentrations because it is essentially biochemical in nature and results from bacterial transformation of organic matter. This type of methane is most often confined to gas seeps.

The second type of methane is characterized by simultaneous increase of methane and decrease in ATP in the presence of carbon dioxide within sediment cores. This methane is conveyed by fluids moving upward from the deep Earth along tectonic ruptures that serve as constantly operating channels for its migration, and it is most often confined to the emissions of mud volcanoes, being more prevalent than biochemical methane. Under favorable thermobaric conditions and lithological sediment properties, this type of methane forms gas hydrates.

These two types of methane differ from each other. Biochemical methane almost never contains saturated homologues, while thermogenic methane does, in significant amounts and varieties (Lein and Ivanov 2005). One of these varieties, acetylene, speaks in favor of a deep origin for thermogenic methane. Acetylene can be formed under temperatures above 1400 ° C during dehydrogenation of two methane molecules. Such temperatures correspond to a theoretical depth of more than 45 km (Komornyy 1996), which exceeds significantly the depth of the basement in the Black Sea.

Currently, about 80 billion m^3 of methane is dissolved in Black Sea water (Sozanskiy 2013), and this condition persists despite a full cycle of water renewal due to the water exchange between the Black Sea and the Sea of Marmara through the Bosphorus Strait. According to various estimates, a full cycle of water renewal in the Black Sea varies between 410 and 2000 years; this means that waters entering the Black Sea from the Mediterranean as underflow as well as those brought in by the Dnieper, the Danube, and other rivers as overflow are constantly saturated with methane once they enter the Pontic basin. Keeping the methane content stable must require a powerful output of methane from the deep Earth (Shnyukov and Yanko 2014, 2017).

Bottom sediments saturated with methane and water should have an impact on biota, and especially on meiobenthos; however, the influence of methane on organisms is still poorly understood. In the case of fish, it has been shown that rapid penetration of methane into their bodies negatively affects the respiratory, nervous, and hematopoietic systems, as well as enzyme activity, and these effects result in a fatal outcome over a relatively short period of time (Patin 2004).

The Black Sea meiobenthos is characterized by substantial taxonomic diversity and by a huge density of colonizations (up to three million specimens/m^2). Meiobenthos biomass is equal to or greater than that of the macrobenthos. According to different authors, methane can have a positive as well as a negative effect, or it can have no effect at all on these organisms (Dando and Hovland 1992; Jensen et al. 1992; Polykarpov et al. 1998).

Luth and Luth (1998) show that methane has a positive effect on living organisms, especially if they live directly on methane seeps. Sergeeva and Gulin (2006) indicate the high density of meiobenthos (nematodes and polychaetes) at water depths between 182 and 308 m and attribute it to the presence of microbial mats formed around methane vents. Gulin et al. (2010) note that microbial mats are characterized by an alternation of unfavorable and favorable periods for the zoobenthic organisms to colonize and maintain themselves. They attribute this fact to the varying levels of oxygen and hydrogen sulfide. During periods of active HG emission from the seabed, the species diversity of meiobenthos is considerably impoverished in comparison with background areas.

In other basins, there is also no clear understanding of meiobenthos behavior in methane biotopes. Thus, in the Norwegian Sea, nematodes dominate quantitatively (<70%) among meiobenthos. The second largest group is represented by harpacticoids (<8%). The number of multicellular meiobenthos, in general, and nematodes in particular, on bacterial mats near methane vents is 1.3–1.5 times higher than in background sediments, and they are larger in size. Species diversity of nematodes is associated with the bacterial mats, but it decreases toward the methane outlets (Portnova et al. 2014).

In the methane outlets of the Adriatic Sea, a sharp decrease in the abundance and species diversity of foraminifera along with the presence of only cosmopolitan species has been noted (Panieri 2003). Kumar Bhaumik and Gupta (2005) pointed out the great stress on benthic foraminifera generated by bottom sediments with a high methane content in the gas hydrate fields of the Black Ridge, northwestern Atlantic Ocean. Epibenthic species adapted to life in oxic conditions could not survive there. At the same time, such an environment is suitable for some endogenous species adapted to life under oxygen-deficient conditions and even in anoxic or sulfidic conditions (Sen Gupta et al. 1997).

Pletnev et al. (2014) considered the possibility of using foraminifera as bioindicators of methane ecosystems on the island slope of eastern Sakhalin within the depth interval of 600–1500 m. They showed that foraminiferal density and diversity dramatically decrease in places of active methane release.

The main goal of this chapter is to examine the relationship between concentrations of hydrocarbon gases, primarily methane, and the distribution of meiobenthos with hard shells (foraminifers, ostracods) and those without them (nematodes) in sediments of the northwestern Black Sea.

To achieve this goal, we studied (1) tectonic features of the seabed, (2) physicochemical characteristics of the bottom water, (3) lithological and geochemical properties of the bottom sediments, (4) composition and concentrations of HG and CO_2 in the bottom sediments, and (5) the quantitative and qualitative (taxonomic) distribution of foraminifers, ostracods, and nematodes in bottom sediments. Correlation of these abiotic and biotic parameters (listed in 1–5 above) enabled us to integrate the results and compare them with other marine basins.

11.2 Study Area

The study area is located in the northwestern part of the Black Sea at water depths of 71–905 m, where 46 stations have been sampled (Fig. 11.1).

The Western Black Sea depression is underlain by suboceanic and oceanic crust; it is separated from the Eastern Black Sea depression (underlain by suboceanic and oceanic crust in the middle part and a refined continental crust on the sides) by the Andrusov Ridge. The latter is characterized by continental crust that is covered by sediments that are 5–6 km thick (Nikishin et al. 2001, 2011). The transitional zone from the Western Black Sea depression to the Andrusov Ridge is complicated by a system of tectonic fractures that incline at 30–45° on the west and east (Ermakov 2005).

11.3 Geomorphological and Tectonic Settings

Geomorphologically, the part of the basin that is included within the study area represents the underwater extension of a vast alluvial plain with a deep-lying bedrock covered by continental and marine sediments formed during the course of the Late Pleistocene-Holocene regression and transgressions, respectively (Yanko-Hombach 2007; Yanko-Hombach et al. 2014).

The width of the continental slope along with the sections of its submarine continuation toward the deepwater bed is 7–10 km, and the inclination gradient varies between 2° and 30°. On the continental slope lie the buried paleovalleys of the Danube, Dniester, Dnieper, and Kalanchak rivers. They give rise to the largest underwater canyons in the basin, which dissect the entire surface

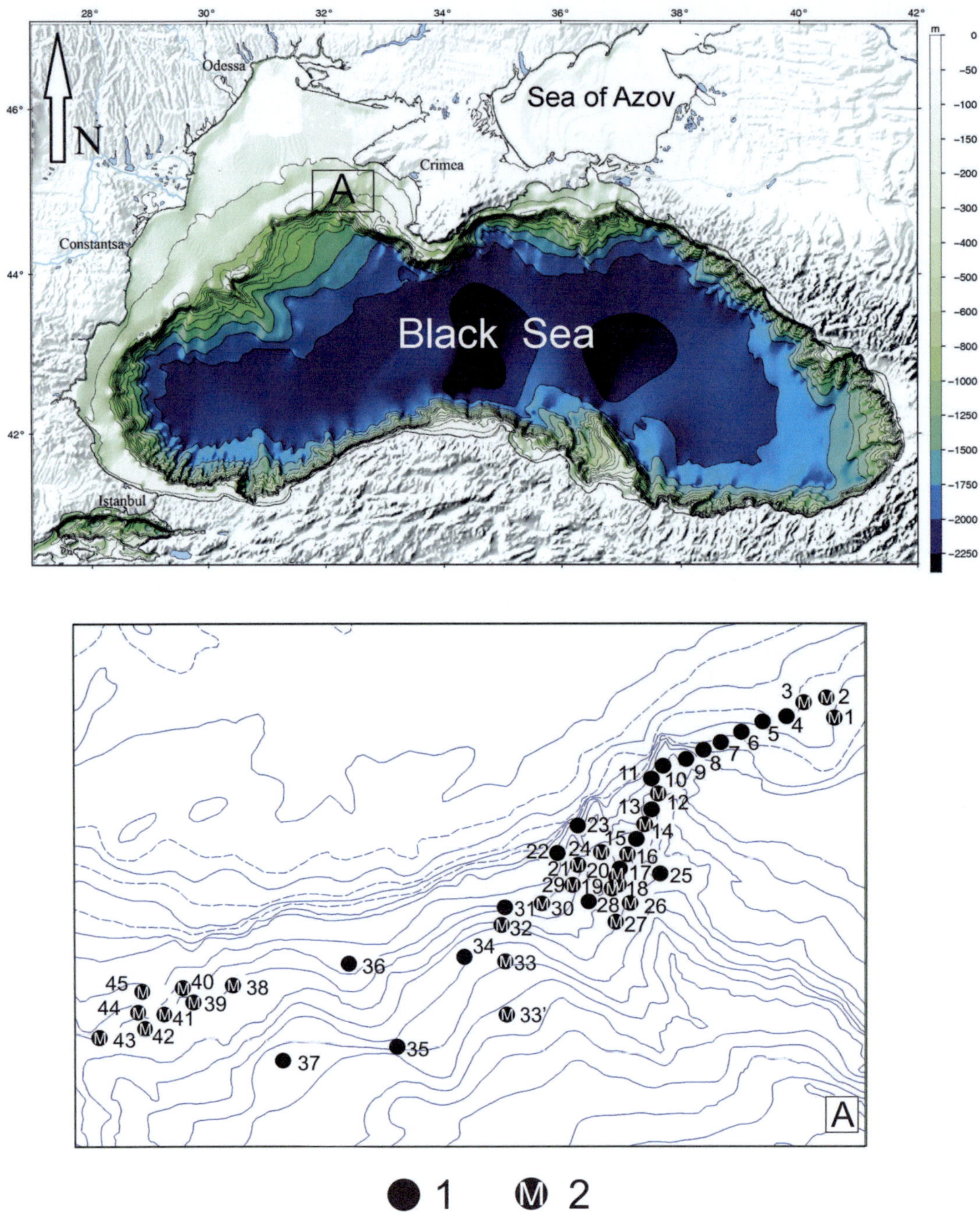

Fig. 11.1 Above: Map showing study area in rectangle A. Below: detail of A showing locations of sampled stations (*1*) and stations in which soft meiobenthos was also studied (*2*)

of the slope from its upper edge to its foot. The riverine paleovalleys inherit the strike of the faults and have mostly straight-line thalwegs and V-shaped bottom profiles, being filled with accumulative, accumulative-erosive, and structurally accumulative surfaces.

The study area is located largely within the Dnieper paleovalley and canyon, where a large number of gas seeps have been observed (Fig. 11.2b).

In terms of tectonics, the most significant feature of the region is the Odessa-Sinop Fault Zone (OSFZ). This zone consists of a series of subparallel tectonic disturbances with a minimum width of 30 km. Bounded on the east by the Western Black Sea depression, the OSFZ originates at the mantle and played a very significant role in the development of the basin (Fig. 11.2).

The OSFZ divides the study area into an eastern (with few gas seeps) and a western (enriched with gas seeps) part (Fig. 11.2a, b). The eastern part lies on the outer shelf area, while the western one, in addition to the outer shelf, includes also the continental slope. Both parts are here considered separately and compared with each other. In addition, the data of Sergeeva et al. (2014) on meiobenthos distribution are also considered (Fig. 11.2c, d).

11.4 Materials and Methods

Sampling was carried out by the faculty and staff together with undergraduate and graduate students of the Department of General and Marine Geology of Odessa I.I. Mechnikov National University over the course of the FR6 EU-funded HERMES-BS-ONU 01-2008 cruise performed September 19–27, 2008. The cruise was conducted together with the Ukrainian Scientific

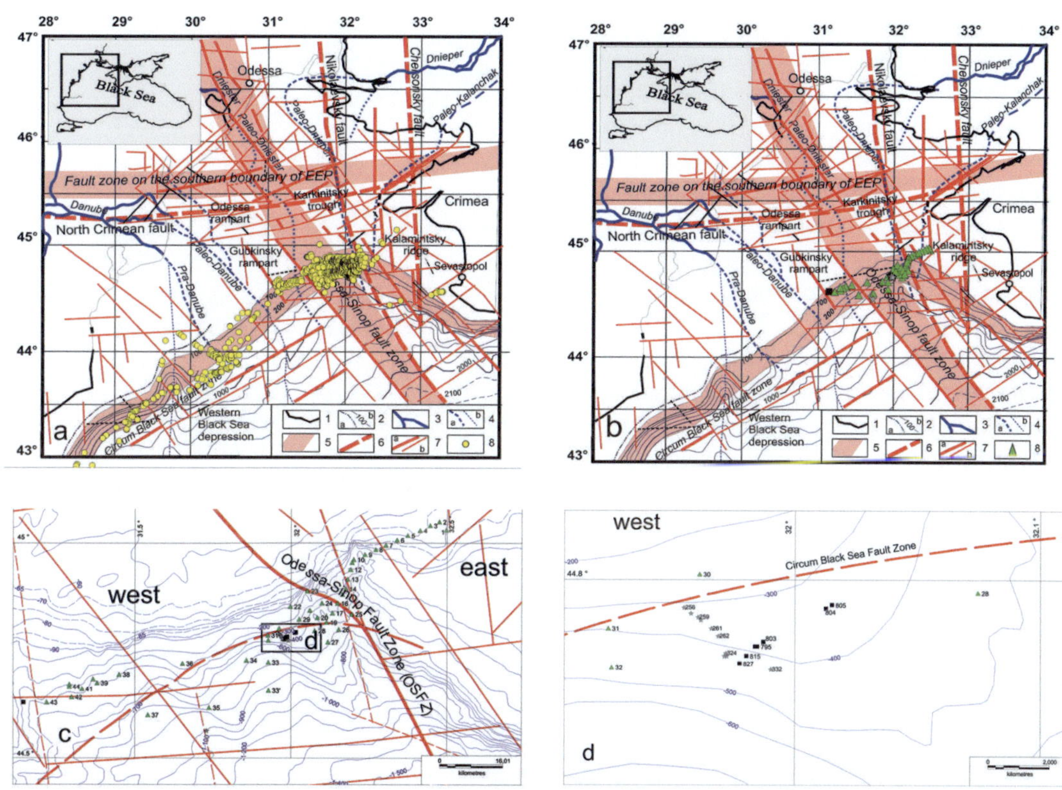

Fig. 11.2 (**a**) Structural-tectonic scheme of the northwestern part of the Black Sea against the background of the main geomorphological elements (the study area lies within the square on the Black Sea map). Key: (*1*) coastline; (*2*) shelf edge (*a*), isobaths (*b*); (*3*) river deltas; (*4*) paleoriver deltas: established (*a*), assumed (*b*); (*5*) fault zones going down to the mantle; (*6*) faults of the consolidated foundation; (*7*) tectonic dislocations of the first (*a*) and second (*b*) level; (*8*) gas outlets. The scheme was compiled by V.P. Kobolev, digitized by S.V. Kadurin. (**b**) The same map with sampled stations in green. (**c**) Numbers of sampled stations enlarged. (**d**) Part of the study area with stations investigated by Sergeeva et al. (2014) and Yanko et al. (2017)

Center of Ecology of the Sea in the northwestern part of the Western Black Sea basin using the Ukrainian Research Vessel "Vladimir Parshin" (Fig. 3.8).

The direction of sailing and position of the sampling stations were laid out in full accordance with the distributional patterns and coordinates of the cold seeps provided by Egorov et al. (2003) in order to capture the response of biota to HG emissions. An example of a gas seep in the vicinity of Station #39 (Table 11.1) is shown in Fig. 7.12c.

Echo-sounder mapping of bottom relief was carried out in conjunction with water and sediment observations.

The sailing route began on the shelf, 23 km to the east of the OSFZ, where 16 stations (Nos. 1–15, 23) were sampled at water depths between 71 and 170 m. The route continued to the west from the OSFZ, approximately along the traverse of the Dnieper canyon to the shelf edge and further to the continental slope. Here, 29 stations (Nos. 16–22, 24–45) were sampled at water depths between 86 and 905 m; almost half of the stations (Nos. 22, 24, 29, 30, 31, 36, 38, 39, 40, 41, 43, 44, and 45) were located on the outer shelf, i.e., within the same depth range as was sampled to the east of the OSFZ.

At each station, measurements of the water column were made to determine its physical and chemical characteristics. Use of the Neil Brown Instrument Systems (STD) complex with General Oceanic rosette and Niskin bottles provided water samples for further chemical analysis in the laboratory on board the vessel.

The following hydrological parameters were measured: water depth (D), salinity (S), temperature (T° C), pH, dissolved oxygen (DO), oxygen saturation index (SI), oxygen reduction potential (Eh), and H_2S (selectively).

Bottom sediments were recovered at each station by "Ocean-25" grab and gravity core. The upper 2 cm of undisturbed sediments were split for determination of the geochemical, grain-size, lithological, mineralogical, hydrocarbon, and meiobenthic properties of the bottom sediments.

Table 11.1 Coordinates and depth of sampled station (for their location, see Fig. 11.1)

No. station	Latitude, N	Longitude, E	Water depth, m
Stations located to the east from the OSFZ			
1	45.03083	32.49639	77
2	45.05028	32.47444	78
3	45.04148	32.44533	75
4	45.03017	32.41285	72
5	45.01823	32.37367	71
6	45.00691	32.33918	77
7	44.99546	32.30335	86
8	44.98507	32.2702	96
9	44.97363	32.23504	97
10	44.9618	32.19921	97
11	44.95515	32.19385	100
12	44.93855	32.18916	100
13	44.91642	32.18012	110
14	44.8939	32.17376	167
15	44.88109	32.16304	170
23	44.88889	32.05167	75
Stations located to the west from the OSFZ			
16	44.86035	32.1473	270
17	44.83777	32.1311	340
18	44.83123	32.11214	328
19	44.81583	32.11194	500
20	44.8272	32.08301	303
21	44.84418	32.05923	320
22	44.85278	31.99722	86
24	44.86056	32.0975	208
25	44.83541	32.19277	603
26	44.79944	32.15111	470
27	44.76992	32.11649	600
28	44.79634	32.0744	365
29	44.82323	32.02578	186
30	44.80167	31.96167	106
31	44.78583	31.92528	186
32	44.77444	31.92667	334
33	44.72274	31.92686	655
33*	44.6571	31.92753	905
34	44.72773	31.85644	620
35	44.6169	31.7379	643
36	44.7204	31.65374	109
37	44.60024	31.54174	746
38	44.69469	31.44987	192
39	44.675	31.38	143
40	44.68356	31.3674	112
41	44.66077	31.3319	190
42	44.6413	31.29909	234
43	44.62943	31.21805	195
44	44.66599	31.29105	113
45	44.67119	31.29172	107

Among the geochemical characteristics of bottom sediments, pH of pore waters, pH and Eh of sediments, C_{total}, C_{org}, and C_{carb} were measured using standard methods. In particular, C_{total} was determined according to Reference Methods for Marine Pollution Studies No. 63, UNEP, 1995. C_{carb} was measured by the titration method. These analyses were conducted by Yu.M. Dengya at the Laboratory of Chemical Analytical Research at the Ukrainian Scientific Center of Ecology of the Sea.

Grain-size analysis of the surface sediment layer (0–2 cm) was performed by sieving and elutriation methods described in Logvinenko and Sergeeva (1986). Based on the results, the Median Diameter (Md) and Coefficient of Sorting (So) were calculated for each sample.

Mineralogical analysis of the same sediment layer was carried out for sand (0.5–0.25 mm) and aleurite (0.25–0.1 mm) fractions. Each was weighed and viewed under a binocular microscope to determine the percentage of minerals found. The percentage of shell detritus was also calculated. Percentages obtained were converted into weight values while taking into account the weight of the individual fractions and that of the whole sample.

The hydrocarbon analysis included determination of saturated (ethane [C_2H_6], propane [C_2H_8], butane [nC_4H_{10}], pentane [iC_5H_{12}], neopentane [$neoC_5H_{12}$], i-pentane [iC_5H_{12}], and n-pentane [nC_5H_{12}]) and unsaturated (ethylene [C_2H_4], acetylene [C_2H_2], and propylene [C_3H_6]) HG in the bottom sediments. CO_2 was measured as well. The HG and CO_2 analyses were performed in a certified laboratory of the Geological Institute in Kiev for samples that were hermetically sealed on board using the patented technique of sample vibroacoustic degassing (Bagriy and Kizlat 2004; Yanko et al. 2017). Degassed samples were analyzed using the gaschromatographer "Crystal Lux-4000M" with certified calibration mixtures. In total, 46 samples were analyzed.

The quantitative and taxonomic compositions of hard-shelled meiobenthos (foraminifera and ostracoda) were determined according to the methods described in Yanko and Troitskaya (1987) and Yanko et al. (1998). The sediment samples were collected from undisturbed surface (0–2 cm) sediment recovered by grab and corer. The grab sampler used two rings, each 10 cm in diameter and 2 cm in height at each site. The sediment recovered by the corer was sampled using one ring due to the limited amount of sediment. In both cases, a wooden spatula was used. Samples for analysis were fixed with formalin buffered with seawater in the proportion 3:1 with the addition of 20 g of $Na_2B_4O_7$ per liter of solution. After the end of the cruise, they were sent to the micropaleontological laboratory of Odessa I.I. Mechnikov National University. In the laboratory, the samples were washed through a sieve with a mesh size of 63 μm and then put into distilled water buffered with rose bengal for 48 h. Wet samples were analyzed in Petri dishes under a binocular microscope.

For investigation of foraminifera and ostracoda at each station, the total assemblage, including live (stained) and dead (empty) tests of foraminifera and ostracoda, was determined and expressed as the number of tests (abundance) per 50 g of dry sediment. We report the total assemblage because (1) the scarcity of living specimens did not allow a relevant statistical study of their distribution, (2) the presence of living specimens among all identified species indicated that they likely live in the study area and therefore their empty tests are autochthonous, and (3) the total assemblage better characterizes the seasonal population dynamics (Debenay et al. 2001; Yanko-Hombach et al. 2017). Large samples were randomly split with a splitter into subsamples. At least 300 specimens, whenever possible, were picked by hand under a binocular microscope. Broken foraminiferal tests, fragments, and old tests (re-crystallized, worn down, filled with sediments) were considered to be reworked and excluded from the analysis.

All specimens were morphologically examined and taxonomically identified. Examples of all species were imaged using scanning electron microscopy (SEM) at the University of Manitoba, Winnipeg, Canada. In our taxonomic work, we followed the suprageneric classification in *Fundamentals of Paleontology* (Rauzer-Chernousova and Fursenko 1959), in combination with the generic classification of Loeblich and Tappan (1988, 1992) and some others (Krashenninikov 1960; Yanko 1989). Direct comparisons were made with the original collections of Black Sea foraminifera collected by Yanko (1989) and stored under the reference number 3906 in the Paleontological Museum of Odessa I.I. Mechnikov National University.

Foraminiferal and ostracodal analyses were performed in the Micropaleontological Laboratory of Odessa I.I. Mechnikov National University, Ukraine. For each species in each sample, the relative abundance was calculated and expressed as a percentage. The Shannon-Wiener coefficient of diversity (H′) was calculated for each sample.

For soft meiobenthos, recalculation based on the weight of dry sediment was not needed because it is not preserved in the fossil state. Therefore, its population density is characterized by the number of specimens per m^2. For identification, nematodes were extracted from the sample and placed in a liquid of Zainhorst (Seinhorst) solution—distilled water (70 parts), ethanol (29 parts), and glycerol (1 part)—to illuminate them. Then, they were laid on a slide with the addition of glycerin-gelatin and covered with a cover glass (Vincx 1996). The preparations were studied using a light microscope "Konus 5625 Biorex 3." Identification of nematodes was made by I.I. Kulakova (Institute of Marine Biology of the National Academy of Sciences) in accordance with the works of Filipyev (1918–1921), Platt and Warwick (1983, 1988), and the database of "NeMys" (Deprez et al. 2007).

To identify possible interrelations between meiobenthic assemblages and environmental parameters, correlation and multiple regression statistical methods were applied using the "Statistica 10" package.

11.5 Results

11.5.1 Physicochemical Characteristics of Bottom Water

Except for salinity (Fig. 11.3b), all other parameters decrease with depth. A sharp drop in temperature from 20 to 8 °C (Fig. 11.3a) occurs at depth −40 m (thermocline); at deeper levels, temperature remains constant. Water salinity increases gradually with depth from 18.9 to 21.8 psu (‰) at water depth −150 m and does not change further (Fig. 11.3b).

The content and saturation of water by oxygen decreases from 8.6 mg/l (103.4%) in the surface water layer to 0.46 mg/l (4.9%) at a water depth of −180 m (Fig. 11.3c). A sharp decrease in oxygen concentration (oxycline?) occurs at water depth −80 m. According to other data, the oxycline (with a steep gradient) is located in the Black Sea at −120 m (Sergeeva et al. 2014). This difference in values indicates that the position of the oxycline is unstable, which may be related to local conditions, seasonal variability, and other factors. With depth, the content of oxygen decreases while that of hydrogen sulfide increases (Fig. 11.4a, c).

Decrease in pH (Fig. 11.4b) together with an increase in H_2S (Fig. 11.4c) along the depth gradient is a result of oxidation-reduction reactions that occur at the oxycline boundary, leading to the release of hydrogen ions that lower pH (Konovalov and Ryabinin 1987).

11.5.2 Lithological and Mineralogical Characteristics of Bottom Sediments

Bottom sediments are represented by gravel, sand, silt, and clay fractions. Their distribution on both sides of the OSFZ is similar, although clay somewhat prevails to the west and sand to the east (Fig. 11.5a).

The average values of So are also similar, while Md is slightly higher to the west of the OSFZ (Fig. 11.5b). This difference is created by

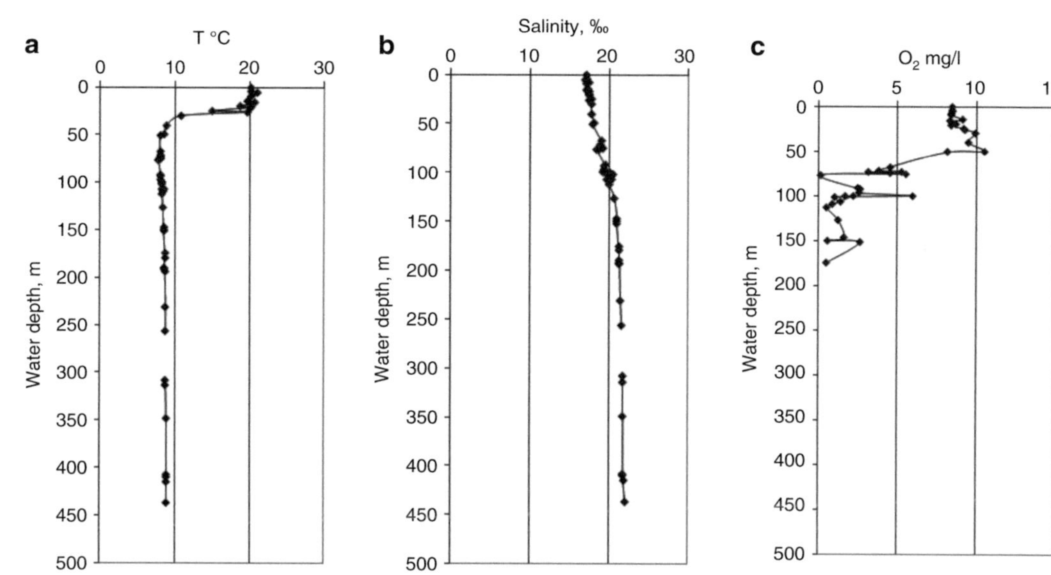

Fig. 11.3 Temperature (**a**), salinity (**b**), and dissolved oxygen (**c**) changes between water depths of 0 and 500 m

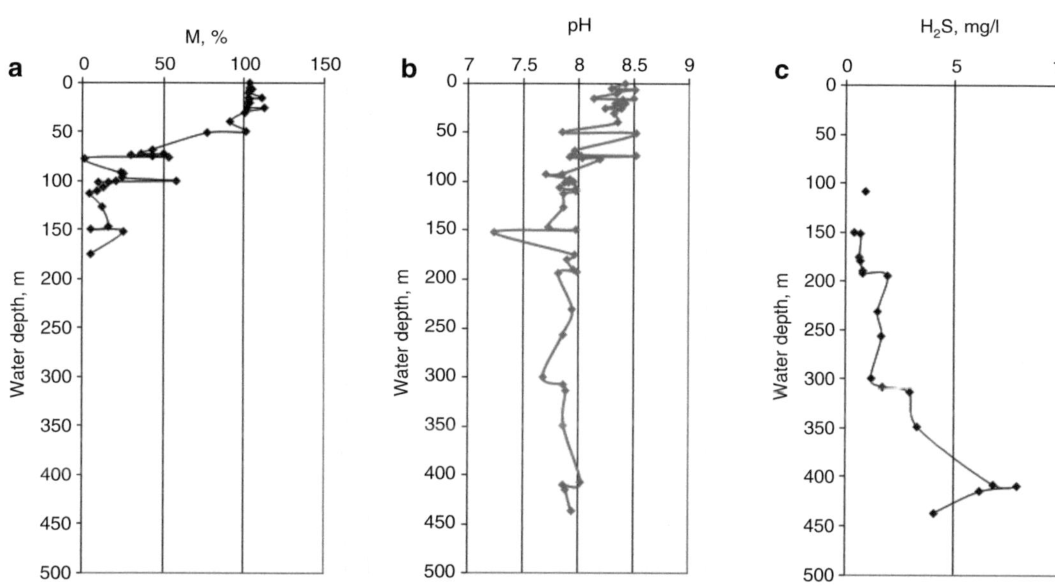

Fig. 11.4 Saturation with oxygen (**a**), pH (**b**), and hydrogen sulfide (**c**) of the water column between 0 and 500 m

Station #40 (water depth, −112 m), where Md = 9 due to the abundance of empty (dead) shells of mollusks. If this station is excluded from the calculations, the Md values to the east and west of the OSFZ become almost identical (Fig. 11.5c). Looking ahead, it should be noted that the bottom sediments at this station have one of the highest methane levels (22.28×10^{-2} vol%) for the study region, which may indicate an increased mortality for mollusks at this station.

Mineralogical analysis shows the presence at most stations of glauconite, hydrotroilite, pyrite, and quartz, the content of which varies by station, averaging 0.76%, 0.014%, 0.18%, and 0.15%, respectively. Their average contents (with the exception of quartz) are slightly higher to the

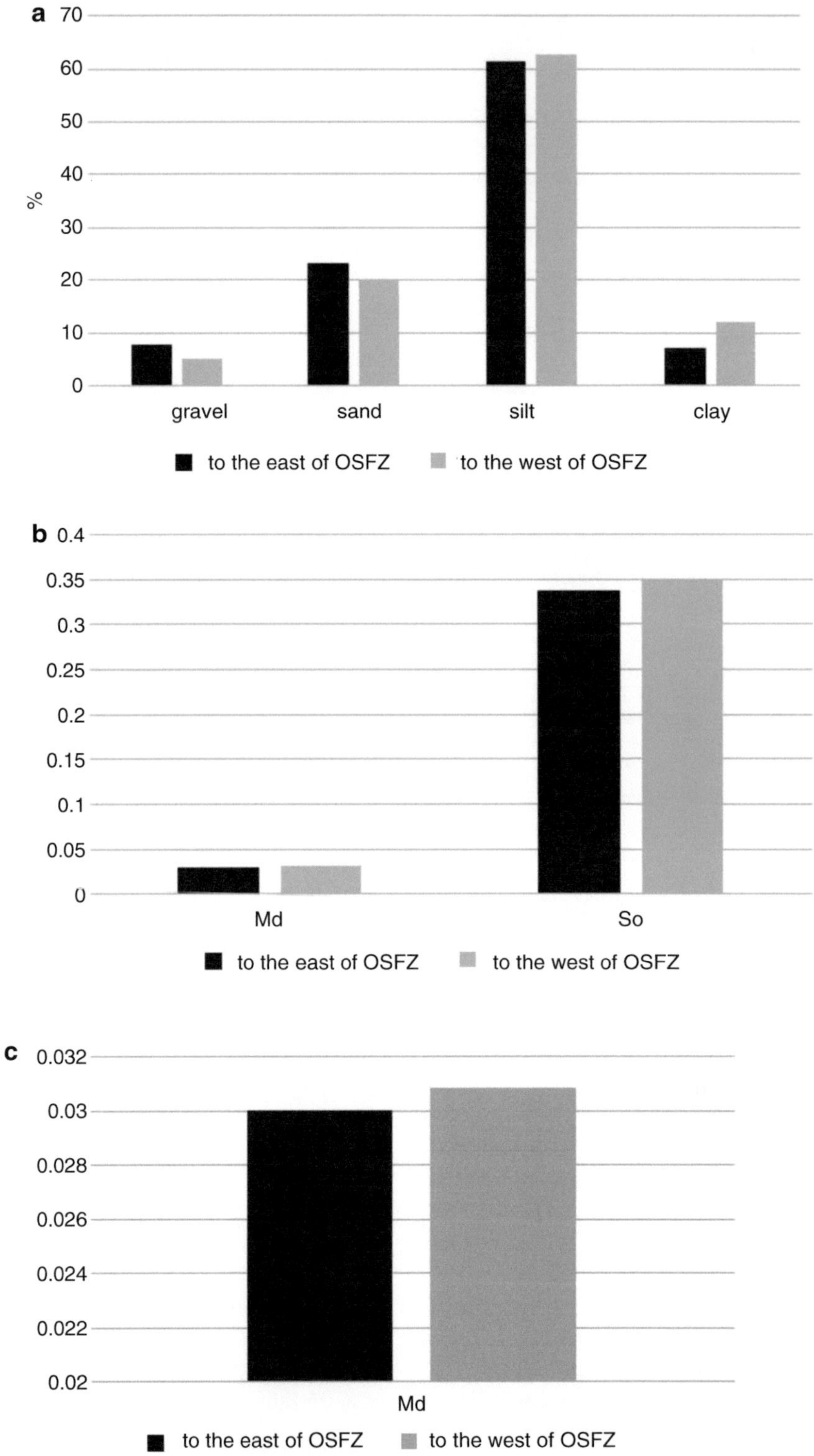

Fig. 11.5 Average values of grain-size parameters to the east and west of the OSFZ: (**a**) percentage of the grain-size fractions, (**b**) Md (the median diameter of fraction is the midpoint of the grain-size distribution: 50% (by weight) of the sediment is coarser and 50% is finer than the median grain size) and So (a coefficient used in describing the distribution of grain sizes in a sample of unconsolidated material), and (**c**) Md without station #40

west of the OSFZ. In one sample (Station #42, depth of the sea −234 m), greigite (0.35 g/t) was found. At this station, the content of methane (10.28×10^{-2} vol%) is almost ten times higher than its average content (1.67×10^{-2} vol%) to the east of the OSFZ. It is also above the mean value to the west of the OSFZ (8.4×10^{-2} vol%).

Glauconite can indicate neutral and slightly reduced conditions of sedimentation, which is associated with the formation of clay minerals in marine basins at considerable depth. Hydrotroilite can indicate an initial stage of sulfide formation, and pyrite development is possibly due to the removal of iron and sulfur by deep fluid flows. Greigite suggests an intensification of authigenic sulfide formation under local conditions. Quartz, whose crystals are largely rolled, indicates that terrigenous material plays an important role in the sedimentation process.

11.5.3 Geochemical Parameters of Bottom Sediments

The distribution of geochemical parameters in the samples of bottom sediments on both sides of the OSFZ is practically the same, despite the difference in depths at which they were taken (Fig. 11.6a).

Attention should be drawn to the negative values of Eh, which vary from −59.7 mV (Station #32, at water depth −334 m) to −125.5 mV (Station #42, at water depth −234 m) independently of station location, depth of the sea, and grain size of bottom sediments (Fig. 11.5b).

This suggests the presence of a reducing environment and is in good agreement with the findings of glauconite, hydrotroilite, and pyrite and also the sulfidization of organic linings of foraminifera.

All of the above allows us to state that environmental conditions of the sea bottom on both sides of the OSFZ are relatively similar.

The average content of carbon dioxide and methane in the bottom sediments, however, is much higher to the west of the OSFZ than to the east (Fig. 11.7).

A similar situation is observed for methane at the shelf stations located on both sides of the OSFZ (Fig. 11.8a). Methane content in bottom sediments to the west of the OSFZ is also much higher on the continental slope compared to the shelf (Fig. 11.8b).

Thus, the methane content in bottom sediments is much higher to the west than to the east of the OSFZ and does not depend on the depth of the sea.

The content of methane homologues does not exceed 4.5×10^{-4} vol%. The content of some of them (ethane, propane, acetylene, i-butane, propylene, and neopentane) is higher to the west than to the east of the OSFZ (Fig. 11.9a).

With regard to the comparison of stations located on the shelf and continental slope, the values of all measured methane homologues without exception are higher for the stations located on the continental slope (Fig. 11.9b).

The presence of methane homologues by itself suggests thermogenic methane (Sovga and Lyubartseva 2006), and the higher values of all HGs, without exception, to the west of the OSFZ, both on the shelf and on the continental slope, suggest a connection with the "chimneys."

Thus, the subdivision of all abiotic parameters according to the territorial principle, based on the tectonic layout of the area, enables us to form two data sets. In both data sets, the parameters of water and bottom sediments are very similar. The only exception is sea depth and content of HGs in the bottom sediments, which are much higher to the west than to the east of the OSFZ.

11.6 Meiobenthos of Methane Biotopes in the Black Sea

In the study area, ten groups of meiobenthos were recorded: Foraminifera, Nematoda, Harpacticoida, Polychaeta, Oligochaeta, Halacarida, Ostracoda, Turbellaria, Bivalvia, and Gastropoda. Within each group, rose bengal-stained specimens were present meaning that organisms most likely were alive at the time of sample collection.

Fig. 11.6 Average values of the geochemical parameters of bottom sediments to the east and west of the OSFZ: (**a**) salinity of pore water, pH of pore water (p.w.), pH of bottom sediments (b.s.), Eh of bottom sediments; (**b**), C_{total}, C_{org}, C_{carb}

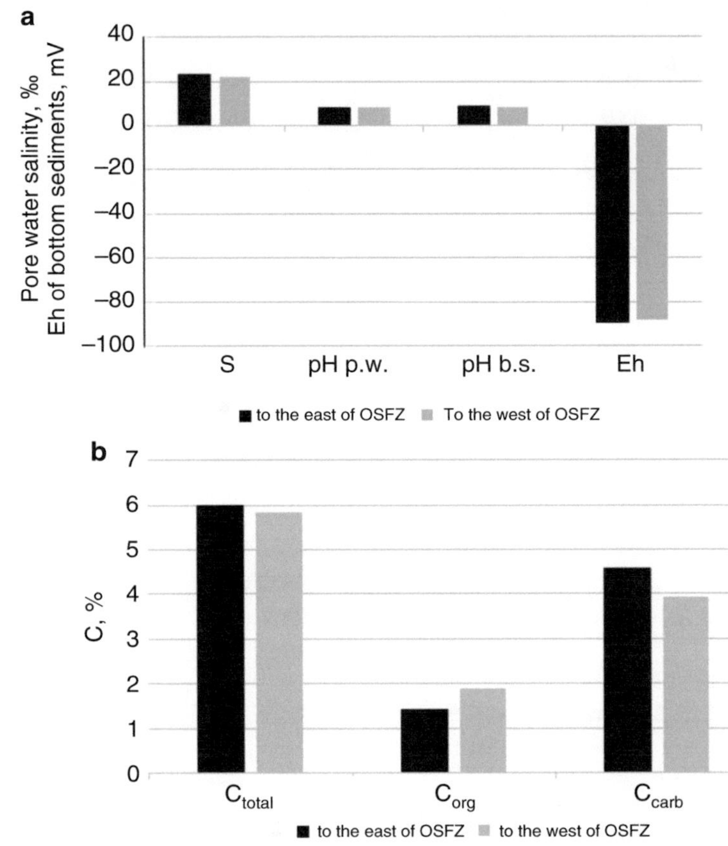

Foraminifera and Nematoda account for 33% of all studied meiobenthos. The next most common forms are present at only 6% each: Polychaeta, Oligochaeta, Ostracoda, and Harpacticoida. The share of Bivalvia is 4%. The content of Gastropoda, Turbellaria, and Halacarida is 2% of the total meiobenthos studied (Figs. 11.10 and 11.11).

11.6.1 Foraminifera

A total of 29 species of benthic foraminifera (one of them in the open nomenclature) from 20 genera, 10 families, and 5 orders (Lagenida, Rotaliida, Miliolida, Lituolida, and Textulariida) were discovered.

The most representative is the Order Lagenida, which includes 13 species. The second most common is the Order Rotaliida, which includes 12 species and subspecies. In third place is the Order Miliolida, represented by two species. The agglutinated orders Lituolida and Textulariida are represented by one species each (Fig. 11.12).

Foraminifera are represented by 20 polyhaline (18–26 psu), 7 stricteuryhaline (11–26 psu), and 2 holeuryhaline (1–26 psu) species. Oligohaline foraminiferal species are absent (Table 11.2).

Foraminiferal assemblages found to the east of the OSFZ resemble those of the Crimean outer shelf of the Black Sea, which are usually dominated by *Ammonia compacta* (up to 29%), *Canalifera parkerae* (up to 16%), and *Cribroelphidium poeyanum* (up to 13%), while *Nonion matagordanus* plays an accessory role (Yanko and Troitskaya 1987; Yanko-Hombach 2007).

The most and least abundant foraminifera were found at the stations located to the east (Fig. 11.13a) and west (Fig. 11.13b) of the OSFZ, respectively.

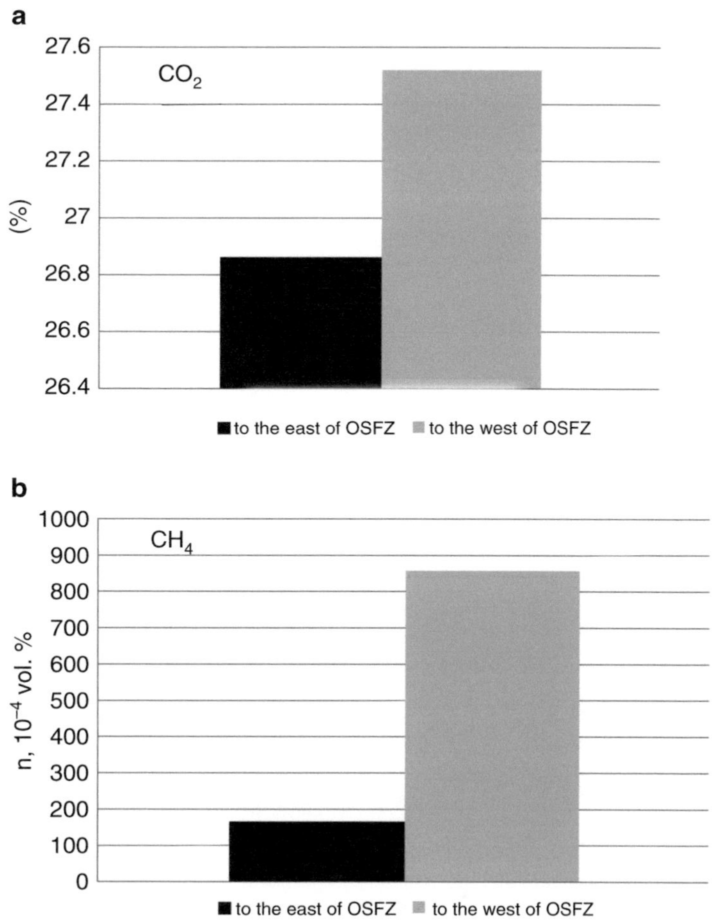

Fig. 11.7 Average content of CO_2 (**a**) and CH_4 (**b**) in bottom sediments to the east and west of the OSFZ

The highest number of foraminifera was encountered at St. 7 located at water depth −86 m, and the lowest was found at a number of stations, including St. 14 located at water depth −67 m that also possessed the highest content of methane in the bottom sediments of the study area.

The average abundance (Fig. 11.14a) and species diversity of foraminifera (Fig. 11.14b) are significantly higher at stations located to the east compared to stations located to the west of the OSFZ.

Even if we exclude the stations located on the continental slope and compare only those stations located on the shelf, the distribution of foraminiferal abundance would still be four times higher to the east than to the west of the OSFZ (Fig. 11.15a). This is so despite the fact that species diversity (H′) on the shelf and continental slope to the west of the OSFZ is practically the same, although on the continental slope it is slightly higher (Fig. 11.15b).

It is important to note that the average C_{org} content, which reflects the availability of food resources for foraminifera, is practically the same at all the stations, amounting to 1.43% and 1.51% to the east and west of the OSFZ, respectively. But at stations on the continental slope, where C_{org} is much higher (2.21%), the numbers of foraminifera are much lower than at stations located on the shelf. This is quite unusual because, in deep areas of the sea, C_{org} is the only source of food sustaining the life and reproduction of foraminifera (Loubere and Fariduddin 1999).

On both sides of the OSFZ, polyhaline species dominate among foraminiferal assemblages, while the number of holeuryhaline species within

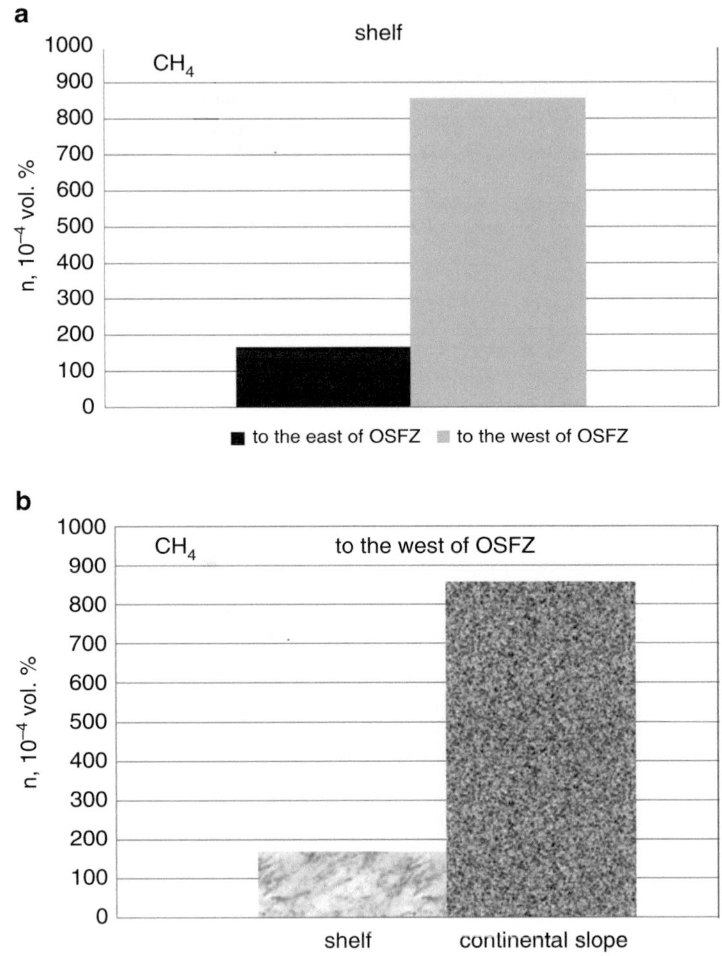

Fig. 11.8 Average content of methane in the bottom sediments (**a**) of the shelf stations located to the east and west of the OSFZ and (**b**) on the shelf and continental slope to the west of the OSFZ

the wider ecological spectrum is slightly higher to the west of the OSFZ (Fig. 11.16).

This is quite natural since holeuryhaline species adapt more easily to changes in conditions within the marine environment, and as our studies show, they adapt much better to stressful situations (Yanko and Troitskaya 1987; Yanko 1989; Yanko-Hombach 2007; Yanko-Hombach et al. 2017).

To the east of the OSFZ, the epibenthic polyhaline species *Ammonia compacta* prevails within foraminiferal assemblages, while to the west of the OSFZ, endobenthic polyhaline species such as *Lagena vulgaris*, *Fissurina fragilis*, and *Parafissurina* ex gr. *lateralis* are prevalent (Fig. 11.17).

In the organic linings of many foraminiferal tests in sediments at the stations to the west of the OSFZ, grains of sulfides (presumably pyrite) are present (Fig. 11.18).

11.6.2 Nematoda

Nematodes are represented by 44 free-living benthic species (9 of them in the open nomenclature) from 26 genera, 17 families, and 6 orders (Fig. 11.19).

Most species are marine, some of which can withstand brackish water conditions. Of these, 16 species are adapted to oxygen deficiency (Table 11.3).

In contrast to foraminifera, the average density of colonies (Fig. 11.20a) and species diversity (Fig. 11.20b) of Nematoda are significantly

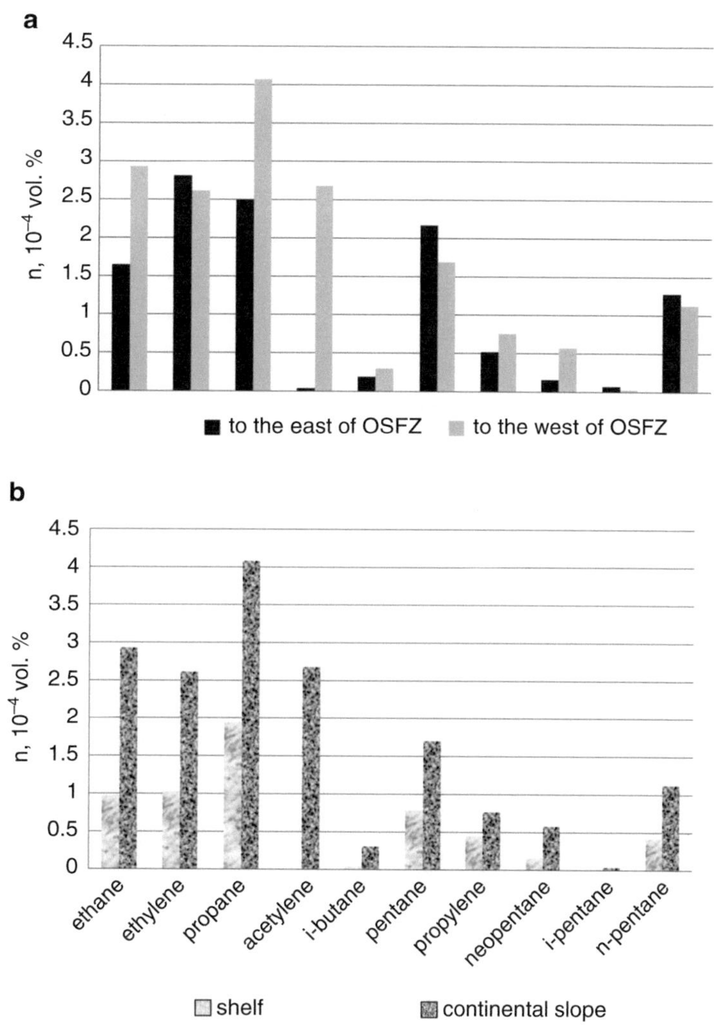

Fig. 11.9 Average content of methane homologues in the bottom sediments (**a**) at stations located on the shelf to the east and west of the OSFZ and (**b**) on the shelf and continental slope to the west of the OSFZ

higher to the west of the OSFZ than to the east of it.

To the east and west of the OSFZ, 6 and 39 species of nematodes were identified, respectively (Fig. 11.21).

To the east of the OSFZ, four species of nematodes are adapted to oxygen deficiency, while to the west, the number of such species is double (Table 11.3, Fig. 11.21).

The highest density of nematodes was noted at St. 40 (13,900 ind/m^2), where the methane content in the bottom sediments reaches 22.28 × 10^{-2} vol%. The Nematoda assemblage is dominated by the species *Sabatieria abyssalis*, which is well adapted to life under conditions of hypoxia-anoxia (Table 11.3, Fig. 11.22).

At the same time, the abundance of foraminifera at this station is very low (40 specimens per 50 g of sediment), as well as the number of species (5). A similar picture is observed with numbers (37 specimens/50 g of sediment) and the number of Ostracoda species (3).

11.6.3 Ostracoda

In the investigated material, seven species of benthic ostracods from seven genera were identified. Their average abundance (Fig. 11.23a) and species diversity (Fig. 11.23b) are higher to the east of the OSFZ than to the west.

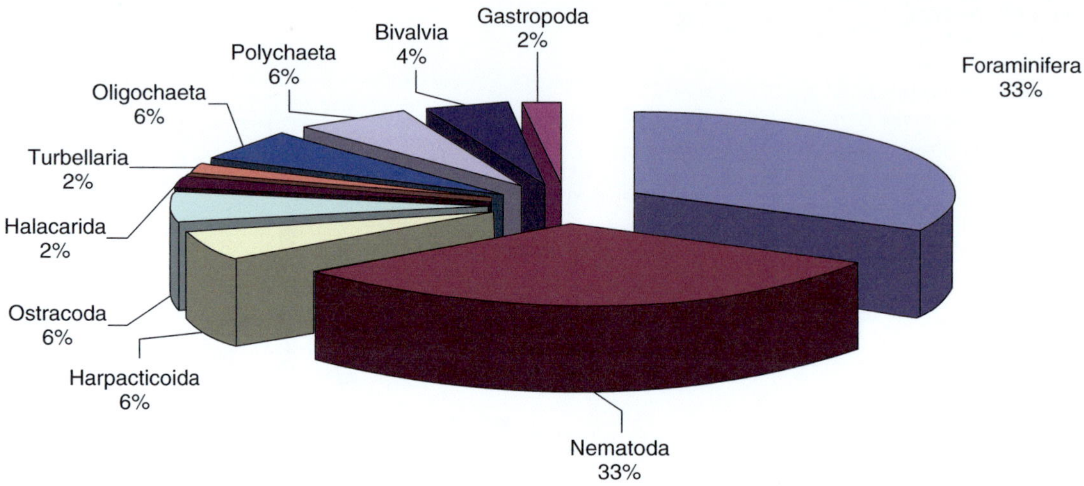

Fig. 11.10 Proportion of meiobenthos taxa in the total sample assemblage

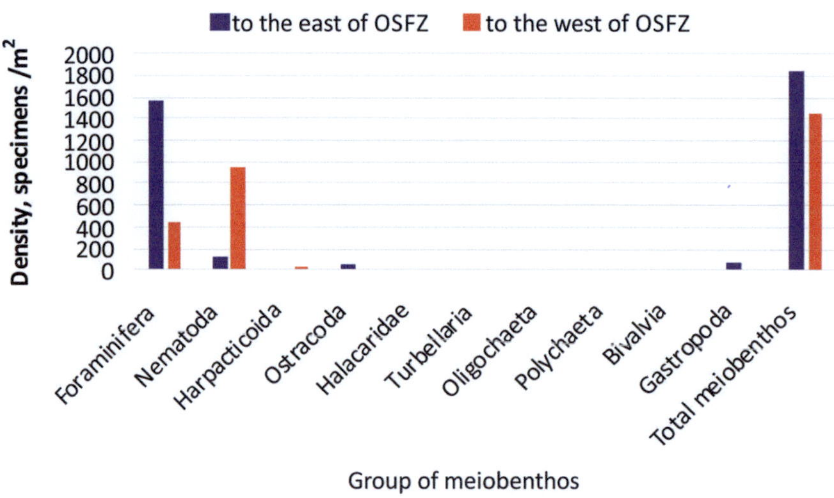

Fig. 11.11 Average density of colonies of individual groups of meiobenthos and all meiobenthos to the east and west of the OSFZ

Hiltermannicythere rubra and *Callistocythere diffusa* dominate in the ostracod complex to the east of the OSFZ, while there are no dominant species to the west. Instead, there are *Cytheroma* aff. *marinovi* and *Loxoconcha* (?) *sublepida*, which are present in only limited quantities (Fig. 11.24).

In general, the number of all identified ostracod species to the west is lower than it is to the east of the OSFZ. All species of ostracods (except for single, most likely redeposited freshwater-brackish *Graviacypris elongata* and *Loxoconcha* (?) *sublepida*) are stricteuryhaline (Yanko and Gramova 1990).

11.7 Correlation of Biotic and Abiotic Parameters

Statistical analysis of abiotic and biotic parameters showed that most distributions do not follow a normal (or Gaussian) distribution. As such, a

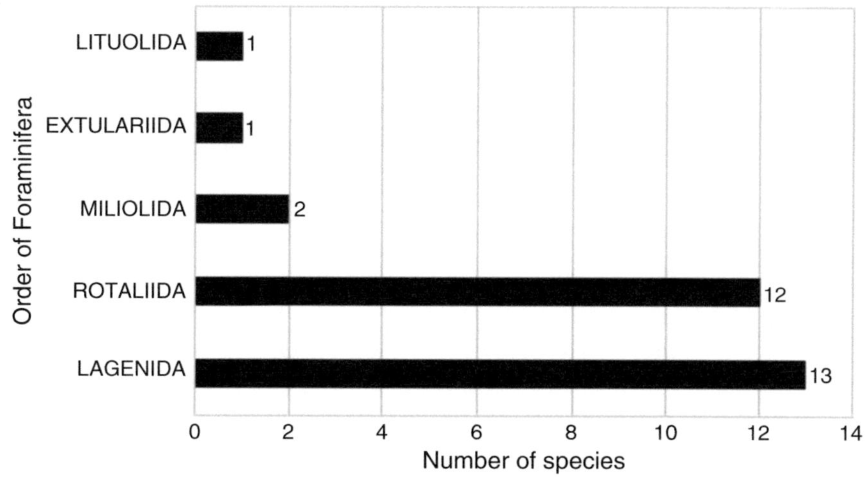

Fig. 11.12 Number of species in the represented orders of Foraminifera

correlation-regression analysis of such samples would be unacceptable. Therefore, all data were "reduced" to the same distribution law by "leveling." After that, they were transformed into a normal distribution by converting them into decimal logarithms with subsequent standardization. Calculation of the correlation coefficients was carried out for each group of data obtained from the eastern and western sides of the OSFZ. The correlation coefficients $r \leq 0.37$ were considered significant (under $P < 0.05000$ and confidence interval 95%). Such coefficients were taken into account only when comparing the groups of data. They are shown in bold in Tables 11.4 and 11.5.

Although the values of the abiotic parameters (with the exception of sea depth and the concentrations of methane and its homologues) are similar to the east and west of the OSFZ (Figs. 11.3, 11.4, 11.5, and 11.6), their correlation coefficients vary.

C_{org} and the content of the pelite fraction correlate negatively with the depth of the sea to the east of the OSFZ, and they correlate positively with the depth of the sea to the west. This suggests that in the east, the spatial distribution of sediments does not obey the law of normal distribution, while in the west, on the contrary, it appears to conform to this law. Such sediment distributions can be associated with different sources of sedimentary material and production processes. In the east, a significant component of the bottom sediments is represented by fragments and whole shells of mollusks.

Hence, the correlation trend shows grain-size fractions of gravel ($r = 0.37$) and sand ($r = 0.44$) increasing with depth. To the west of the OSFZ, shell material is much finer than it is to the east, being present mainly within the sand fraction, with which C_{carb} correlates positively ($r = 0.51$). In the silt fraction, it seems to be absent ($r = -0.44$).

The higher values of C_{carb} to the east of the OSFZ (Fig. 11.6b) indicate more favorable conditions for living organisms with calcium carbonate shells. This is confirmed by the much higher numbers of foraminifers (Figs. 11.13 and 11.14) and ostracods (Fig. 11.23), which may also be due to the difference in sedimentation conditions. To the east of the OSFZ lies the outer part of the shelf, where the living conditions for meiobenthos are better due to the presence of adequate amounts of oxygen, in any case up to a depth of 130 m.

To the west of the OSFZ, in addition to the outer part of the shelf, sedimentation settings also include the shelf edge and the continental slope, where (1) the processes of sediment transport and shell abrasion are more intense than on the shelf and (2) the environment changes from normoxic to suboxic and even hypoxic. The average content of C_{org} in the bottom sediments on both sides of the OSFZ does not exceed its average value

Table 11.2 Ecology of benthic foraminifera found in bottom sediments with information on their ecology and lifestyle

Order	Species	Ecology	Mode of life
LITUOLIDA Lankester, 1885	*Labrospira* sp.	Polyhaline	Endobenthos
TEXTULARIIDA Lankester, 1885	*Eggerelloides scaber* (Williamson, 1858)	Polyhaline	Epibenthos
MILIOLIDA Delage and Hérouard, 1896	*Miliolinella selene* (Karrer), 1868	Polyhaline	Epibenthos
	Pateoris dilatatus (d'Orbigny), 1839	Polyhaline	Epibenthos
ROTALIIDA Delage and Hérouard, 1896	*Ammonia tepida* (Cushman), 1928	Holeuryhaline	Epibenthos
	Haynesina anglica (Murray), 1965	Holeuryhaline	Epibenthos
	Ammonia ammoniformis (d'Orbigny), 1826	Stricteuryhaline	Epibenthos
	Canalifera parkerae Yanko, 1974	Stricteuryhaline	Epibenthos
	Cribroelphidium poeyanum (d'Orbigny), 1839	Stricteuryhaline	Epibenthos
	Cribroelphidium translucens (Natland), 1938	Stricteuryhaline	Epibenthos
	Elphidium ponticum Dolgopolskaja and Pauli, 1931	Stricteuryhaline	Epibenthos
	Nonion matagordanus Kornfeld, in Cushman, 1939	Stricteuryhaline	Epibenthos
	Porosononion martkobi ponticus Yanko, 1989	Stricteuryhaline	Epibenthos
	Ammonia compacta Hofker, 1969	Polyhaline	Epibenthos
	Bolivina variabilis (Williamson), 1858	Polyhaline	Endobenthos
	Bulimina marginata d'Orbigny, 1826	Polyhaline	Endobenthos
LAGENIDA Delage and Hérouard, 1896	*Entolingulina deplanata* Yanko, 1974	Polyhaline	Endobenthos
	Esosyrinx jatzkoi Yanko, 1974	Polyhaline	Endobenthos
	Esosyrinx undulosus (Terquem), 1878	Polyhaline	Endobenthos
	Fissurina fragilis Troitskaya, 1987	Polyhaline	Endobenthos
	Fissurina lucida (Williamson), 1858	Polyhaline	Endobenthos
	Fissurina solida Seguenza, 1862	Polyhaline	Endobenthos
	Lagena vulgaris Williamson, 1858	Polyhaline	Endobenthos
	Laryngosigma williamsoni (Terquem), 1878	Polyhaline	Endobenthos
	Ortomorphina calomorpha (Reuss), 1866	Polyhaline	Endobenthos
	Ortomorphina drammenensis Feyling-Hanssen, 1964	Polyhaline	Endobenthos
	Ortomorphina filiformis (d'Orbigny), 1826	Polyhaline	Endobenthos
	Parafissurina dzemetinica Yanko, 1974	Polyhaline	Endobenthos
	P. lateralis Cushman	Polyhaline	Endobenthos

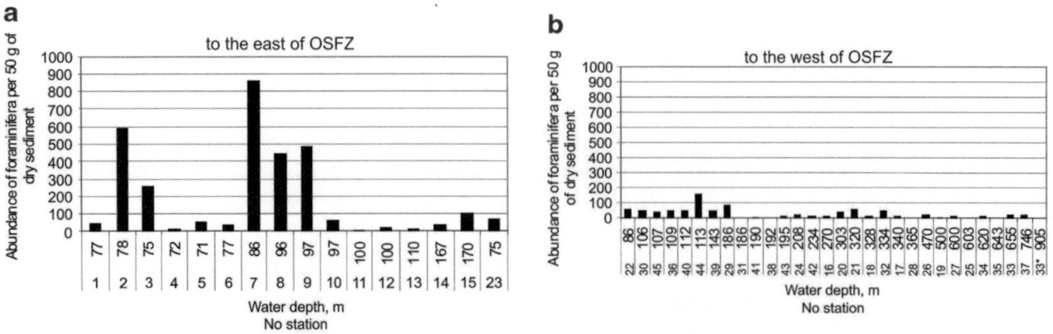

Fig. 11.13 Abundance of foraminifera at stations located to the east (**a**) and west (**b**) of the OSFZ

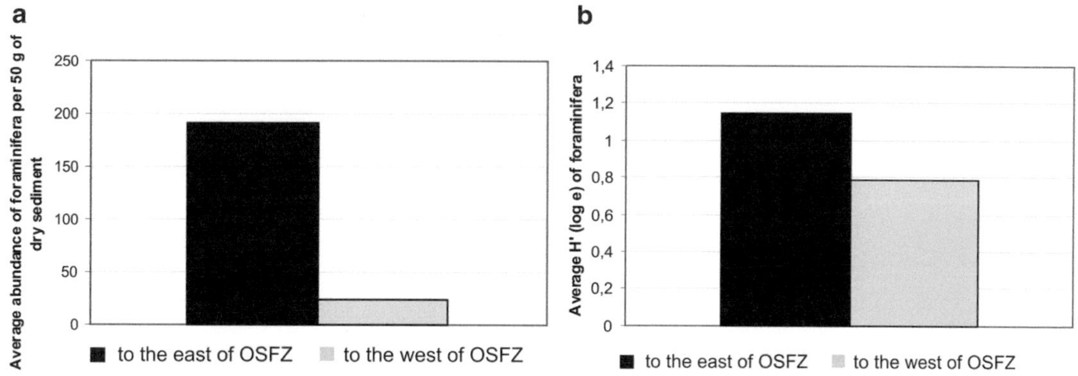

Fig. 11.14 Average abundance (**a**) and species diversity (**b**) of foraminifera to the east and west of the OSFZ

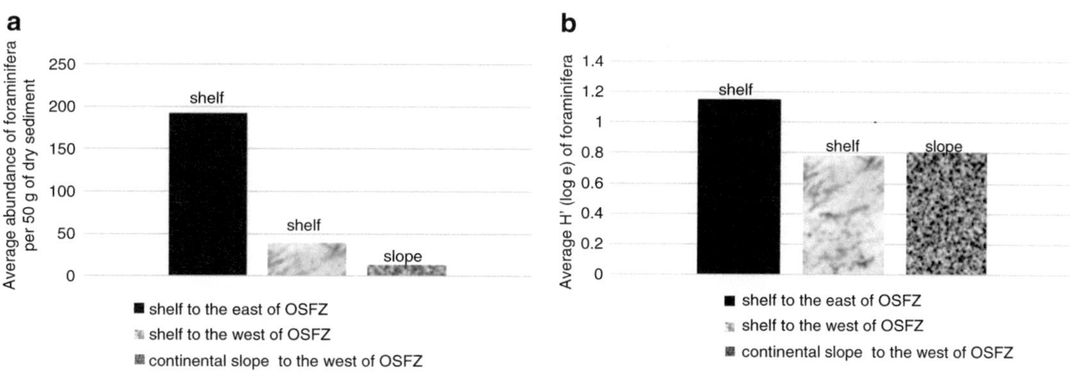

Fig. 11.15 Average abundance (**a**) and species diversity (**b**) of foraminifera on the shelf and continental slope to the east and west of the OSFZ

(2.15%) for the entire Black Sea (Gorshkova 1974). In Gorshkova's opinion, the main reason for the accumulation of organic matter is the reducing conditions of the water column below 200 m that contribute to the slow decay of dead plankton. With weak aeration in the bottom layer, the substantial quantities of organic matter enhance the general oxygen deficiency due to consumption for its own oxidation.

Negative correlation of C_{org} with depth ($r = -0.53$) to the east and positive correlation ($r = 0.69$) to the west of the OSFZ may indicate the activation of diatom decomposition. Diatoms are absent to the east of the OSFZ, while to the west, they are abundant and demonstrate a high correlation coefficient ($r = 0.99$) with depth (Yanko-Hombach et al. 2009).

The content of C_{org} also depends directly on the mechanical composition of the sediments (Gorshkova 1974), and, in our case, it is everywhere positively correlated with the silt and pelite fractions. This type of relationship is considered "classic" for all types of marine sediment and, in particular, for the entire Black Sea basin (Mitropolskiy et al. 1982). At the same time, the content of C_{org} and C_{carb} is subject to inverse dependence. As in the other areas of the Azov-Black Sea basin, the content of C_{carb} increases with an increase in the share of coarse fractions: gravel and sand (Ovsyanyy et al. 2015).

The correlation of abiotic and biotic parameters demonstrates the negative response of biota to an increase in depth and concentration of HG.

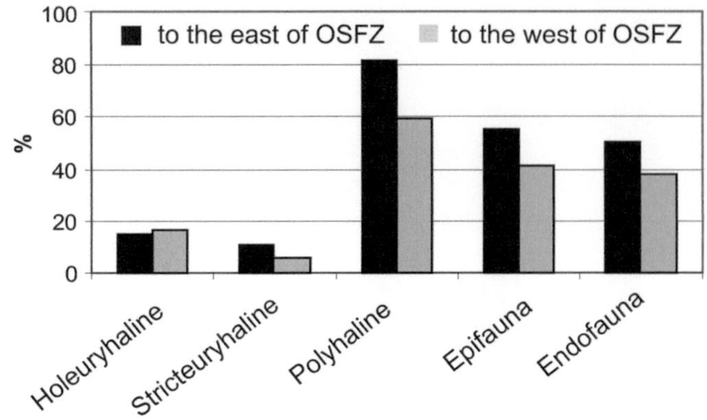

Fig. 11.16 Relative abundance of foraminiferal species showing the different ecological settings and lifestyles to the east and west of the OSFZ

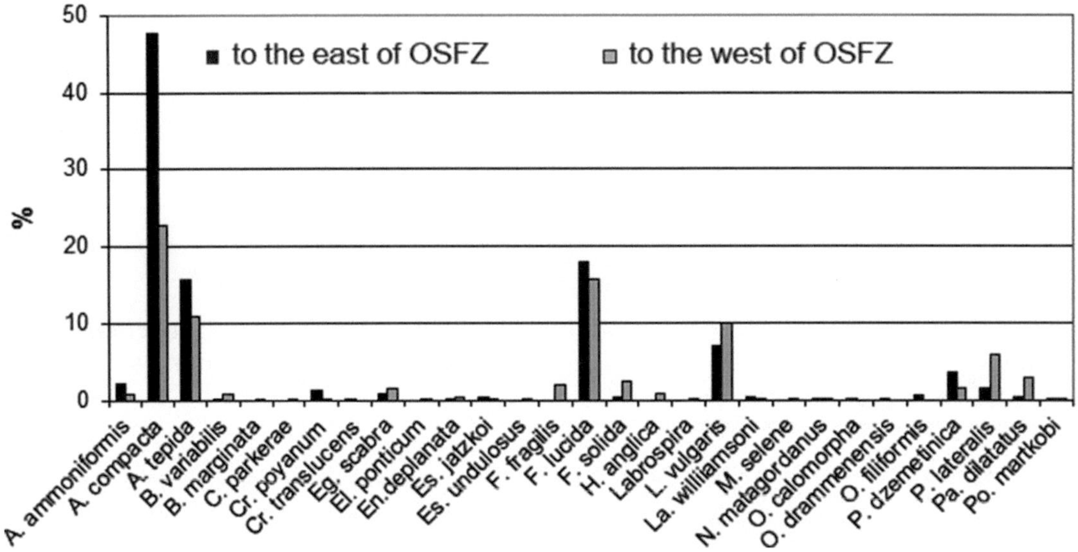

Fig. 11.17 Average relative abundance of foraminiferal species to the east and west of the OSFZ

11.8 Discussion

Three main problems must be discussed: (1) the origin of HG in the bottom sediments of the study area and their relationship to tectonics, (2) the impact of HG on the quantitative and qualitative (taxonomic) distribution of meiobenthic organisms on the Black Sea bottom and its comparison with other basins, and (3) prospects for using the distribution of meiobenthic organisms to determine HG distribution under the seabed.

Problem No. 1 is the origin of methane in the Black Sea. The presence of methane homologues in gas emissions (Fig. 11.9) suggests an endogenous (thermogenic) origin for methane, which can be formed only under conditions of high temperatures and pressures (Lein and Ivanov 2005). For example, acetylene can be formed by dehydrogenation of two methane molecules at a temperature above 1400 ° C, which corresponds to a theoretical depth of more than 45 km, i.e., much deeper than the basement of the Dnieper-Donets graben (Kononov et al. 1977), and of course the continental slope of the Black Sea. According to Sovga and Lyubartseva (2006), methane homologues cannot be formed when methane has a biochemical origin.

As was discussed in Chap. 7, a thermogenic methane source appears to be indicated by the localization of most gas flares on the

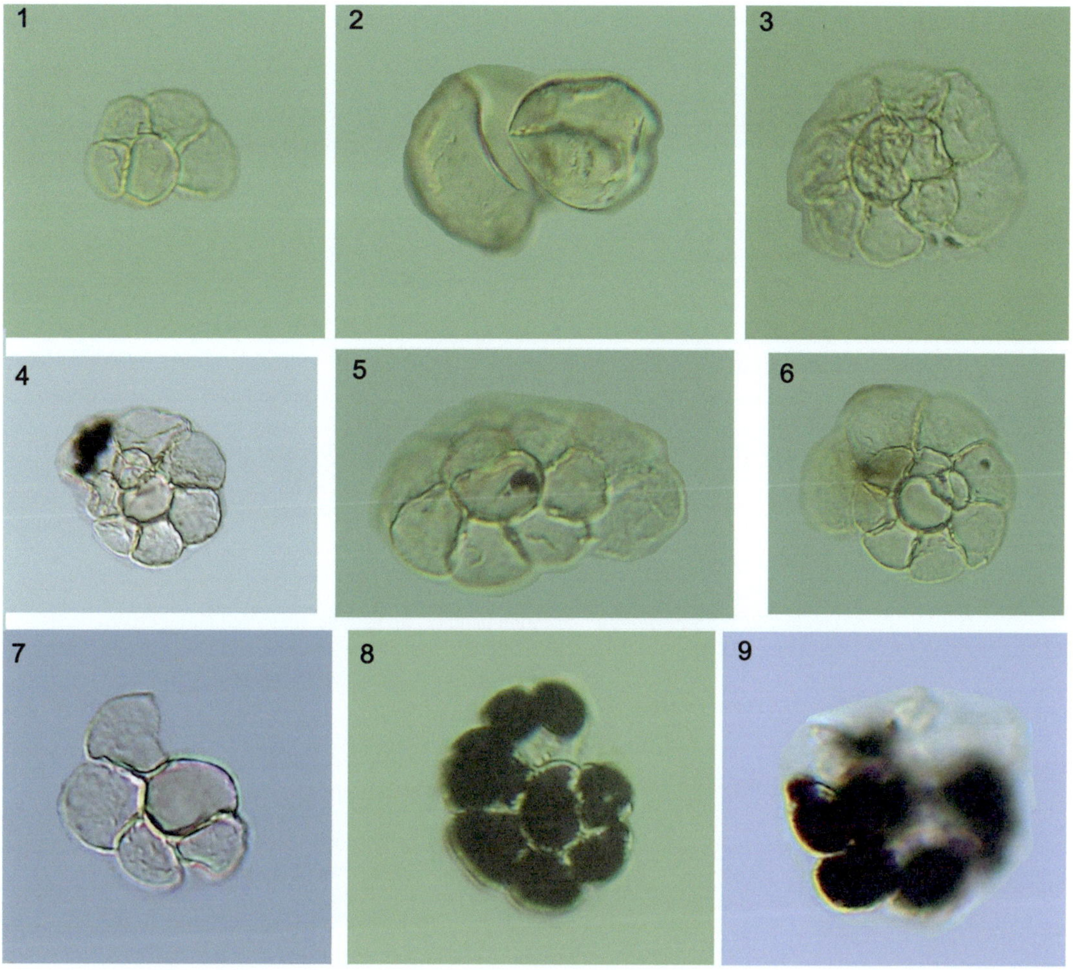

Fig. 11.18 Grains of sulfides, presumably pyrite (black), in organic linings of *Ammonia compacta* from stations 28, 33, and 45, located to the west of the OSFZ; photos taken in transmitted light. Courtesy of Prof. Petra Mudie, Canada

Fig. 11.19 Number of recovered species represented within the phylum Nematoda

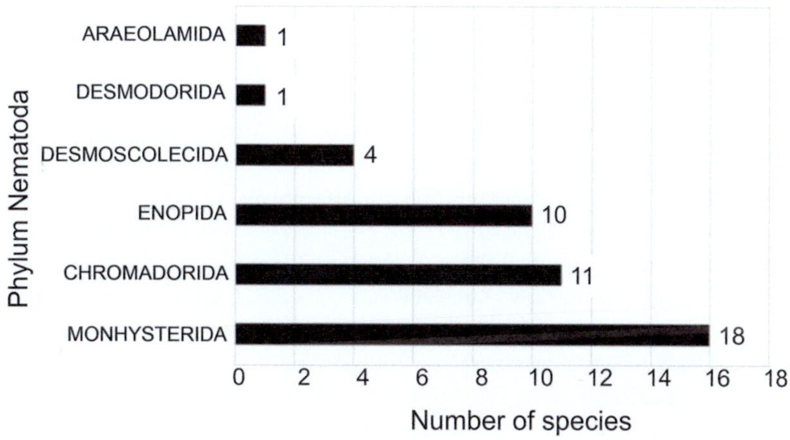

Table 11.3 Species of Nematoda known from the literature that are adapted to suboxic and dysoxic zones of the Black Sea

Group of Nematoda in relation to oxygen	Species	Mode of life Living depth, m	Source
Normoxic	*Chromadora nudicapitata* Bastian, 1865	Marine benthos, 80–140 m	Sergeeva (2003)
	Mohystera longicapitata Filipjev, 1922	Marine benthos, up to 150 m	Sergeeva (1973) and Sergeeva (2003)
	Prochromadorella mediterranea (Micoletzky, 1922)	Marine and brackish benthos, up to 140 m	Sergeeva (2000)
Normoxic-hypoxic	*Euchromadora striata* (Eberth, 1863)	Marine benthos	Sergeeva (2000)
	Metoncholaimus demani (Zuz Strassen, 1894)	Marine benthos, 300–365 m	Sergeeva and Kolesnikova (1996)
	Paracanthonchus caecus (Bastian, 1865)	Marine benthos, up to 150 m	Sergeeva (1992) and Sergeeva (2000)
	Paralinhomoeus filiformis (Filipjev, 1918)	Marine benthos, up to 150 m	Sergeeva (2000)
	Quadricoma loricata Filipjev, 1922	Marine benthos	
	Terschellingia longicaudata De Man, 1907	Marine benthos, up to 200–250 m	Sergeeva and Gulin (2006), Sergeeva and Kolesnikova (1996), and Mureşan (2014)
	Quadricoma steineri Filipjev, 1922	Brackish benthos, up to 150 m	
	Sabatieria abyssalis (Filipjev, 1918)	Marine and brackish benthos, up to 200 m	Sergeeva and Kolesnikova (1996), Sergeeva (2000), and Mureşan (2014)
	Sabatieria longicaudata Filipjev, 1922	Marine and brackish benthos, up to 200 m	Sergeeva and Kolesnikova (1996) and Sergeeva (2000)
	Sabatieria pulchra Filipjev, 1922	Marine and brackish benthos, up to 200 m	Sergeeva and Kolesnikova (1996), Sergeeva (2000), and Mureşan (2014)
	Sphaerolaimus dispar Filipjev, 1918	Marine and brackish benthos, up to 350–500 m	Sergeeva and Kolesnikova (1996) and Sergeeva (2000)
	Sphaerolaimus ostreae Filipjev, 1918	Marine and brackish benthos, up to 200–350 m	Sergeeva and Kolesnikova (1996)
	Terschellingia longicaudata De Man, 1907	Marine benthos, up to 200–250 m	Sergeeva and Gulin (2006), Sergeeva and Kolesnikova (1996), and Mureşan (2014)

shelf-continental slope boundary, where the OSFZ together with other faults can be traced. In our material, elevated concentrations of HG are recorded to the west of the OSFZ and are superimposed upon tectonic disturbances and methane flares described by Egorov et al. (2003).

HG content in bottom sediments does not correlate with any other parameter of bottom sediments, in particular with lithology, which is in full agreement with data known from the literature (Tkachenko et al. 1974). However, most HGs are positively correlated with each other, and to the west of the OSFZ, also with CO_2, suggesting that they possess a common origin. CO_2, in turn, positively correlates with fine grain-size fractions (silt and pelite) that prevail at stations located on the continental slope and outer shelf, again to the west of the OSFZ, which is broken by faults. All this implies an endogenous (thermogenic) origin for methane in the area under

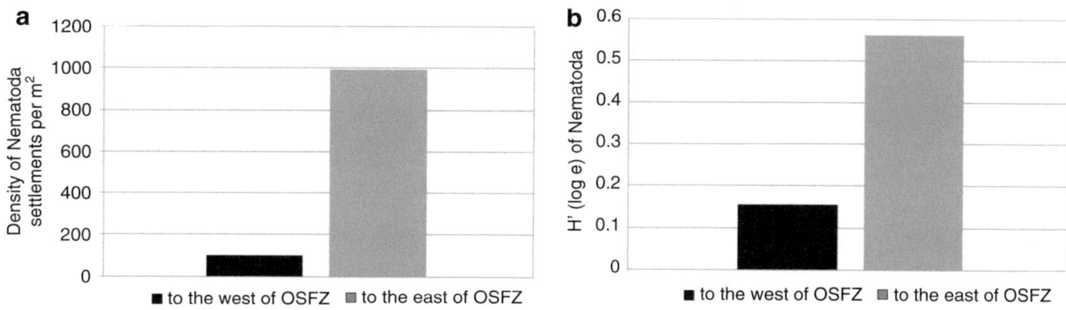

Fig. 11.20 Average density (**a**) and diversity (**b**) of Nematoda to the east and west of the OSFZ

study and corresponds to the mineralogical and other findings described in Chap. 8.

Problem No. 2 requires a discussion of the specific features developed by meiobenthos in areas of methane outlets. Compared with other regions of the Black Sea, the groups of meiobenthos studied here are characterized by low abundance and species diversity. In particular, out of 101 species and subspecies of benthic foraminifera known in the Black Sea (Yanko 1990; Yanko-Hombach 2007), only 29 were found in the study area. Their highest abundance (859 specimens per 50 g sediment) was found at only one station (No. 7) located to the east of the OSFZ; abundance figures are much lower at all the other stations. At the same time, other parts of the Black Sea that are free of HG-infused sediments reveal numbers of foraminifera reaching into the tens of thousands of specimens per 50 g of sediment (Yanko and Troitskaya 1987; Yanko 1989). Particularly impoverished are the foraminiferal assemblages located to the west of the OSFZ, including shelf stations but especially those of the continental slope, where the numbers are particularly low. This indicates the presence of highly stressful conditions for foraminifera, as well as ostracods, since their numerical and taxonomic characteristics are parallel.

One might initially suppose that this stress could be caused by the depth of the sea. However, depth by itself is not a controlling factor for the bathymetric sequence and numerical characteristics of benthic foraminifera, despite a high negative correlation coefficient ($r = -0.8$) between depth and foraminifera. Other factors related to the depth of the sea are responsible (Jorissen et al. 1992), and these factors include salinity (Sen Gupta 1999; Yanko-Hombach et al. 2017), oxygen (Kaiho 1999), food resources in the form of C_{org} (Fontanier et al. 2002; Wahyudi and Minagawa 1997; Yanko-Hombach et al. 2017), substrate (Basso and Spezzaferri 2000), pH of the bottom sediments (Saraswat et al. 2011), and others. However, as our studies show that these parameters are very similar on both sides of the OSFZ, another factor (or factors) may be responsible for the stress. This factor may well be methane, which is negatively correlated with the abundance of foraminifera ($r = -0.85$) and ostracods ($r = -0.96$), as well as the species diversity of the latter ($r = -0.65$). This means that, just as it impacts vertebrate organisms such as fish (Patin 2004), methane can negatively affect the metabolic and reproductive functions of invertebrates (e.g., foraminifera and ostracoda). Determining the mechanism of such influence requires experimental studies.

The inverse relationship between the numerical characteristics of foraminifera and the content of C_{org} in the bottom sediments is particularly unusual. In basins with a normal oceanic salinity and oxygen regime (e.g., in the Mediterranean Sea), the C_{org} current is the main supplier of nutrients for foraminifera at great depths (Loubere and Fariduddin 1999) because it ensures their high reproductivity and bathymetric sequence on the seabed (Bernhard 1992; Luth and Luth 1998). In our case, we find exactly the opposite. While

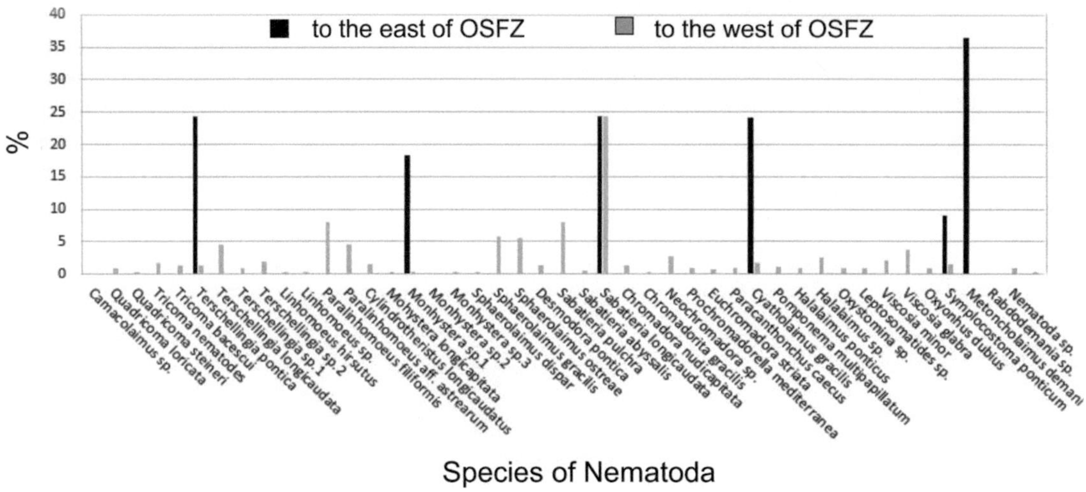

Fig. 11.21 Average relative abundance of Nematoda species to the east and west of the OSFZ

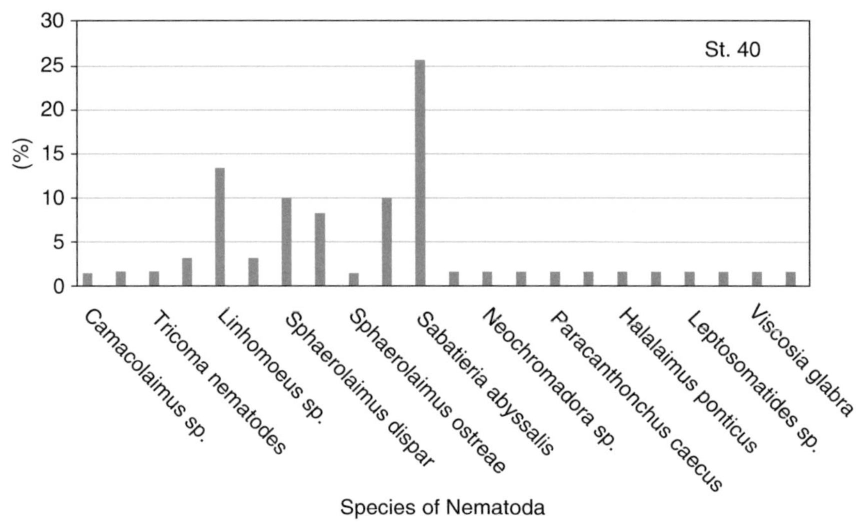

Fig. 11.22 Relative abundance of Nematoda species at St. 40, located to the west of the OSFZ at water depth –112 m

the content of C_{org} increases, the number of foraminifera decreases. Hence, either the source of nutrition is different, or the effect of endogenous methane and its homologues is so strong that it affects the foraminiferal need for food. Can this be because of oxygen deficiency? At first glance, the presence of sulfides (presumably pyrite) in organic foraminiferal linings may serve to support this assumption as is the case in other areas of the Black Sea (Kravchuk and Kravchuk 2007). However, pyritization of foraminiferal tests/organic linings is also common in bottom environments enriched in oxygen, for example, in Haifa Bay, Israel (Alve 1991; Yanko et al. 1994, 1998, 1999). Motnenko (1999) provides photographs of pyritized foraminiferal tests from Haifa Bay, whose waters are saturated with oxygen (Fig. 11.25).

Alve (1991) believes that pyritization of the foraminiferal tests is caused by stress, which then allows bacteria to attack the cytoplasm, creating a sulfide-favorable reducing microenvironment in

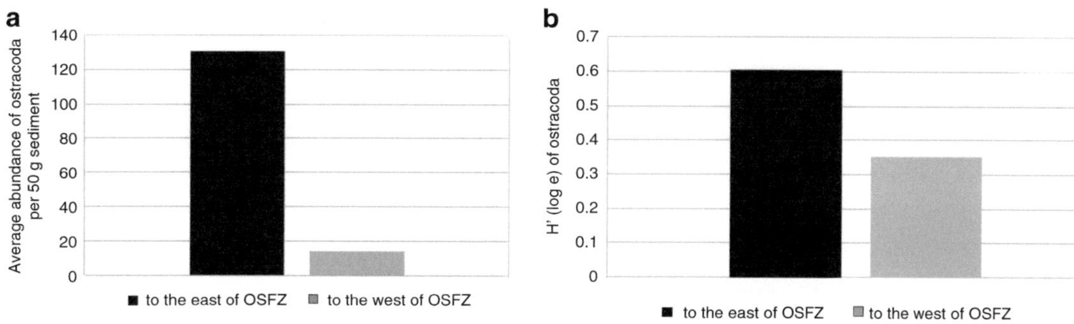

Fig. 11.23 Average abundance (**a**) and species diversity (**b**) of Ostracoda to the east and west of the OSFZ

Fig. 11.24 Abundance of Ostracoda species to the east and west of the OSFZ

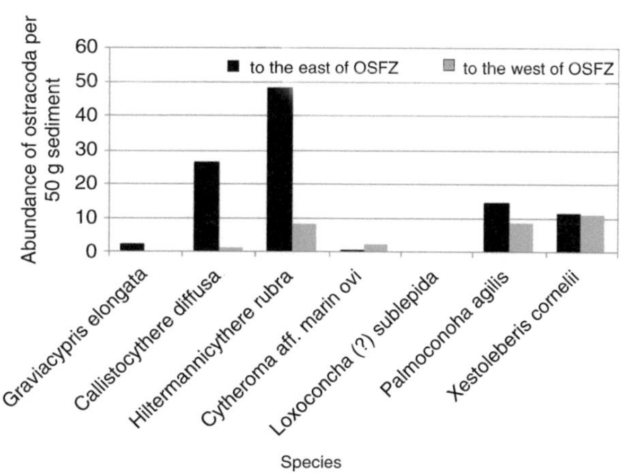

the "attacked" parts of organisms. Thus, pyritization of foraminiferal tests most likely indicates a violation of the organisms' defense mechanisms (Bresler and Yanko 1995a, b, 2000) due to poisoning—for example, by heavy metals, as is the case in Haifa Bay (Yanko et al. 1998), or HGs (Yanko et al. 2017), as is the case in our present study.

If it is not C_{org}, what other food source could be exploited by foraminifera to maintain their vital functions in areas with elevated HG concentrations? Some researchers, who have studied foraminifera near methane seeps in marine basins with normal oceanic salinity and oxygen concentration, believe that microbial mats formed around the seeps provide foraminifera with the necessary nutritional sources that would allow them to reproduce dramatically (Rathburn et al. 2000). Other scientists, however, point out that conditions within the microbial mats by no means remain constant and are associated with certain threshold levels of oxygen and hydrogen sulfide (Gulin et al. 2010). It should also be noted that all these authors refer to biogenic methane, while the effects of endogenous methane and its homologues may be different. This issue requires separate study, preferably under experimental conditions.

The dominance of endobenthic species of the Lagenida order that have also been found beyond the shelf within reduction biotopes contradicts our previous conclusion (Yanko 1989; Yanko-Hombach et al. 2007) about the absence of aerobic life in the Black Sea deeper than isobath −220 m and is consistent with the view of other researchers (Sergeeva and Gulin 2006; Sergeeva et al. 2014).

Like foraminifers and ostracods, nematodes are also characterized by an impoverished species diversity. In the investigated area, only 44 of

Table 11.4 Correlation coefficients between environmental parameters to the east of the OSFZ. Here and in Table 11.5, significant coefficients are shown in bold

Parameter	C_{org}	C_{carb}	CO_2	Methane	Ethylene	Acetylene	Propylene	Gravel	Sand	Silt	Clay	Md	So
Water depth	**−0.53**	−0.03	−0.11	−0.13	0.06	0.11	0.15	0.37	0.44	−0.4	**−0.86**	0.33	−0.03
C_{org}		**−0.6**	0.12	−0.14	−0.12	−0.21	−0.01	**−0.6**	**−0.71**	**0.67**	**0.63**	**−0.72**	0.38
C_{carb}			0.07	0.12	0.02	−0.11	−0.04	**0.63**	**0.64**	**−0.61**	−0.3	**0.68**	−0.5
CO_2				0.37	**0.66**	0.38	−0.27	0.21	0.27	−0.17	−0.01	−0.08	0.07
Methane					0.44	0.48	0.11	0.2	0.01	−0.11	−0.08	0.14	−0.08
Ethylene						**0.59**	−0.03	0.06	0.13	−0.19	−0.11	0.15	−0.08
Acetylene							0.35	−0.15	−0.12	0.21	−0.22	−0.15	0.27
Propylene								−0.0	−0.24	0.19	−0.23	0.03	0.09
Gravel									**0.79**	**−0.68**	−0.47	**0.62**	−0.4
Sand										**−0.87**	−0.46	**0.65**	**−0.56**
Silt											0.49	**−0.88**	**0.55**
Pelite												−0.52	−0.04
Md													−0.45

Table 11.5 Correlation coefficients between environmental parameters to the west of the OSFZ

Parameter	C_{org}	C_{carb}	CO_2	Methane	Ethylene	Acetylene	Propylene	Gravel	Sand	Silt	Clay	Md	So
Depth	**0.69**	−0.04	0.23	0.15	0.06	0.29	−0.21	−0.35	**−0.6**	**0.6**	**0.63**	−0.2	0.24
C_{org}		−0.15	0.03	0.06	−0.11	0.2	−0.17	−0.01	**−0.57**	**0.39**	**0.37**	0.11	0.22
C_{carb}			−0.02	0.15	0.12	0.03	0.24	0.11	0.51	−0.44	−0.26	0.03	−0.04
CO_2				**0.57**	**0.59**	**0.44**	**−0.48**	−0.28	−0.37	0.39	0.39	−0.16	0.29
Methane					0.19	0.2	−0.36	0.11	−0.12	−0.01	0.12	0.24	0.22
Ethylene						**0.37**	−0.1	**−0.39**	−0.19	0.36	0.23	−0.26	0.25
Acetylene							−0.18	−0.08	−0.34	0.23	**0.37**	0.03	0.28
Propylene								0.16	0.36	−0.35	−0.22	0.08	−0.22
Gravel									0.28	**−0.77**	**−0.58**	**0.9**	−0.24
Sand										**−0.81**	**−0.8**	0.08	−0.2
Silt											**0.72**	**−0.58**	0.33
Pelite												−0.32	0.03
Md													0.08

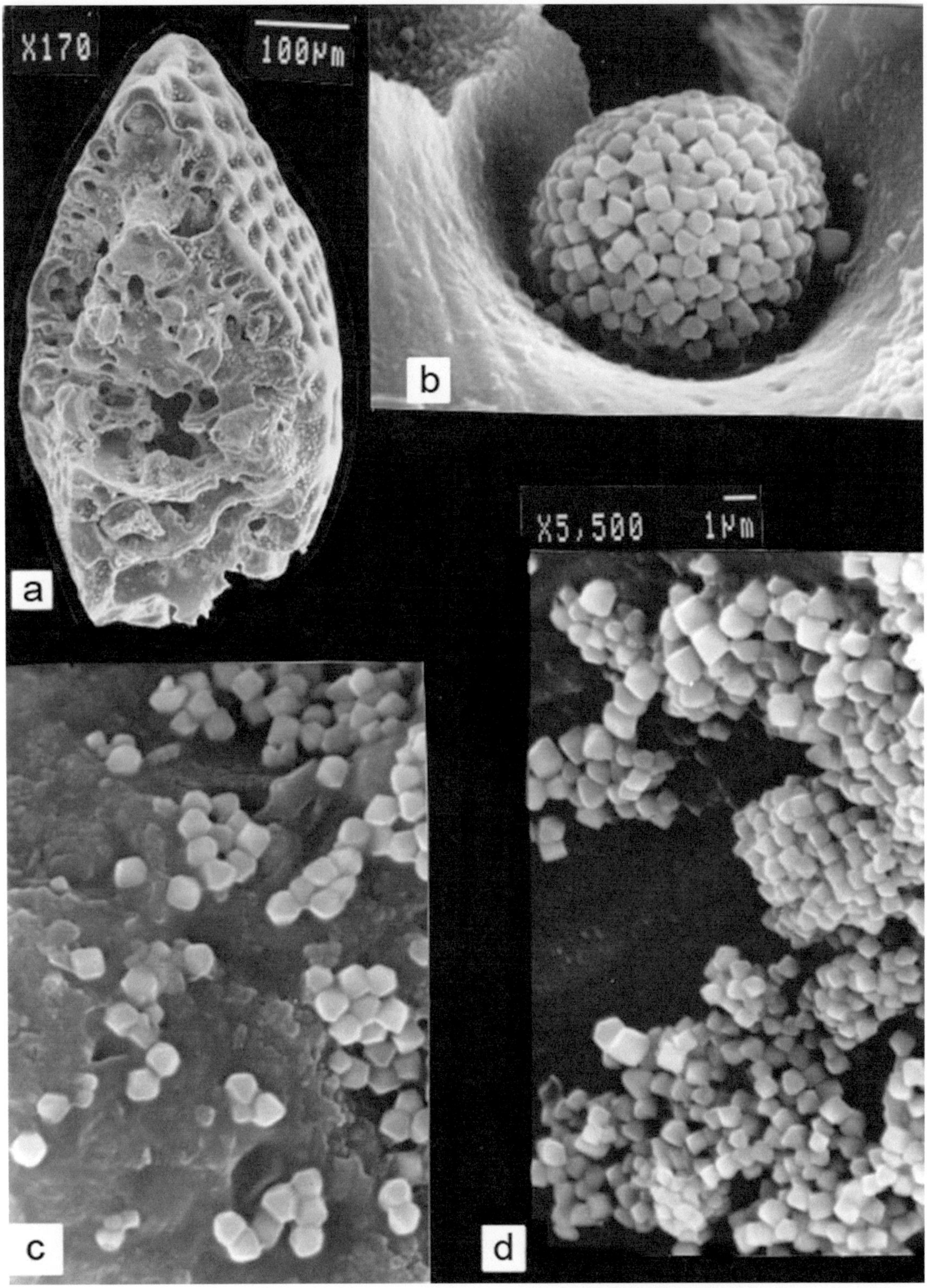

Fig. 11.25 Sulfide formation in the foraminiferal test of *Elphidium crispum* Linné from Haifa Bay, Israel (Motnenko 1999): (**a**) fragment of a test with clusters of pyrite; (**b**) microconcretion formed by tetrahedral and octahedral crystals of pyrite; (**c**) aggregation of crystals into three-dimensional groupings (to the right of the center, there are two cube-shaped units of crystals); (**d**) combination of tetrahedron and hexahedron aggregation into one-dimensional and two-dimensional groupings

143 species of Nematoda known from the Black Sea according to the literature (Sergeeva and Zaika 2013) were identified. A significant number of the identified species belong to the thiobiotic type (Jensen 1981, 1983, 1986) from the families Comesomatidae and Linhomoidae. They are characterized by a long, thin body and a small or completely undeveloped oral cavity, indicating their feeding on small particles of sediment or bacteria (Ott and Novak 1989).

An increase in population density, abundance, and species diversity of nematodes to the west of the OSFZ along with a decrease in these indices in foraminifers and ostracods suggests that elevated levels of HGs are not a hindrance, despite not being entirely favorable for the existence of nematodes. This is also indicated by the lack of correlation between the nematode parameters and environmental characteristics. Obviously, nematodes are able to occupy methane biotopes, which are neither suitable nor completely unsuitable for other meiobenthos.

It seems that a special nematode fauna is forming at the sea bottom to the west of the OSFZ, and some nematode species (e.g., *Desmodora pontica*, *Linhomoeus* sp., *Sabatieria abyssalis*, *Pomponema multipapillatum*, and *Terschellingia pontica*) can serve as indicators of methane biotopes. This observation agrees with the data of other researchers, who believe that nematodes may be the only multicellular animals occupying methane biotopes. Their assemblages, however, would be characterized as showing reduced species diversity and the dominance of specific taxa in comparison with those that inhabit biotopes lacking any methane discharge (Dando et al. 1994). Sergeeva and Gulin (2006) and Sergeeva et al. (2014) point out that some species of nematodes from methane biotopes are characterized by significant anomalies in their vital organs. Such anomalies are similar to those that are characteristic of nematodes occupying biotopes under strong anthropogenic impact. The reason or reasons for such anomalies are not clear in both cases, but the association of anomalies with toxicity of bottom sediments seems to be unquestionable (Sergeeva 1996). In our material, such anomalies were not observed, but the search continues.

By itself, the low number of ostracod species encountered on both sides of the OSFZ is indicative of unfavorable conditions for them; however, their higher abundance to the east compared to the west suggests more comfortable conditions for them there. The presence of two freshwater-brackish water species of ostracods in their assemblages suggests that they may have been redeposited, and this would not be surprising considering that the area under study is located near submerged river paleovalleys.

All data obtained on meiobenthos indicate that elevated concentrations of methane and its homologues in bottom sediments are unfavorable for living foraminifera and ostracods. They also provide encouraging evidence that at least among foraminifera, indicator species signaling the presence of HGs can be found among the endobenthic Lagenida. With regard to nematodes, such indicator species have already been found and can be used to identify places where HGs are being unloaded onto the seabed.

Problem No. 3 involves the possibility of using meiobenthic survey in the search for methane reservoirs concealed beneath the seabed. The proposed endogenous (thermogenic) origin of HG in the study area seems to be the most plausible one, and it may indicate territorial confinement of methane emissions to hydrocarbon deposits. Therefore, prospects look encouraging for using quantitative and qualitative (taxonomic) distribution characteristics of meiobenthos in order to delineate HG accumulations in bottom sediments. Meiobenthic surveys, being initially associated with geophysical and gas-geochemical surveys of bottom sediments, may in the future serve as a tool in the search for and mapping of HG resources under the seabed, thus reducing the cost and increasing the reliability of prospecting.

11.9 Conclusions

1. Twenty-nine species of benthic Foraminifera, 44 species of Nematoda, and 7 species of benthic Ostracoda (two of which, apparently, are redeposited) are described for localities near methane outlets on the Black Sea bottom.
2. Physicochemical parameters of the water column, with an emphasis on bottom water, grain-size, and geochemical parameters of bottom sediments, are similar throughout the study area and do not affect the distribution of meiobenthos.
3. The presence of methane and its homologues in elevated concentrations suggests that they share a thermogenic origin.
4. A decrease in the quantitative and taxonomic characteristics of foraminifers and ostracods along with an increase in these parameters among nematodes that are identified as species tolerant to elevated concentrations of HG in bottom sediments allows us to outline the presence of HG accumulations beneath the seabed. Among nematodes, species-indicators include *Terschellingia pontica*, *Linhomoeus* sp., *Sabatieria abyssalis*, *Desmodora pontica*, and *Pomponema multipapillatum*. To date, no species-indicators have been discovered among foraminifera that demonstrate HG emissions on the seabed however, an increase in the endobenthic representatives of Lagenida that are apparently tolerant of HG emissions can be considered a possible indicator of methane release on the bottom of the Black Sea.
5. Comparison of distributional patterns of meiobenthos in areas of methane outlets in the Black Sea shows a similarity to that found in other marine basins.
6. The influence of different concentrations of HGs on bottom ecosystems is significantly different and much more complicated than previously thought. It now seems that HG emissions into the marine environment are particularly unfavorable for Black Sea bottom ecosystems, which appear to be quite fragile and vulnerable.
7. The results obtained are absolutely new for science and require further investigation to be fully comprehensible. It is possible that the qualitative and quantitative characteristics of foraminifera and nematodes at the bottom of the Black Sea may prove to be useful search indicators for HG accumulations under the seabed. These results also enable us to assess the impact of HG on marine biota, a question that is currently open because the relevant information is extremely fragmentary and contradictory.
8. The study performed attempts to develop interdisciplinary methods in the search for accumulations of HG, including methane, under the seabed. Once developed, such methods might contain great socioeconomic importance for the economy of not only Ukraine but other Black Sea countries as well. It may also contribute to the sustainable development of Black Sea ecosystems.
9. This work should be considered a pilot. Its further development requires funding, including the involvement of experimental work in the laboratory.

Appendices

Appendix 1: Taxonomic List of Foraminifera

Regnum Chromista
- Subregnum Harosa
  - Interregnum Rhizaria
    - Type Foraminifera
      - Order Lituolida
        - Family Haplophragmoididae Maync, 1952
          - Genus *Labrospira* Höglund, 1947
            - *Labrospira* sp.
      - Order Ataxophragmiida Furssenko, 1958
        - Family Eggerellidae Cushman, 1937
          - Genus *Eggerella* Cushman, 1933

Eggerelloides scaber (Williamson), 1858

Order Miliolida Lankester, 1885

Family Miliolidae Ehrenberg, 1839

Genus *Pateoris* Loeblich et Tappan, 1953

Pateoris dilatatus (d'Orbigny), 1839

Order Miliolinella Wiesner, 1931

Family Hauerinidae Schwager, 1876

Genus Miliolinella Wiesner, 1931, emended Luczkowska, 1972

Miliolinella selene (Karrer), 1868

Order Lagenida Delage and Hérouard, 1896

Family Glandulonodosariidae Silvestri, 1901

Genus *Orthomorphina* Stainforth, 1952

Orthomorphina calomorpha (Reuss), 1866

Orthomorphina filiformis (d'Orbigny), 1826

Genus Lagena Walker et Jacob in Kanmacher, 1798

Lagena vulgaris Williamson, 1858

Family Glandulinidae Reuss, 1860

Genus *Esosyrinx* Loeblich et Tappan, 1953

Esosyrinx jatzkoi Yanko, 1974

Esosyrinx undulosus (Terquem), 1878

Genus *Laryngosigma* Loeblich et Tappan, 1953

Laryngosigma williamsoni (Terquem), 1878

Genus *Entolingulina* Loeblich et Tappan, 1961

Entolingulina deplanata Yanko, 1974

Genus *Fissurina* Reuss, 1850

Fissurina fragilis Troitskaja, 1987

Fissurina lucida (Williamson), 1858

Fissurina solida Seguenza, 1862

Genus *Parafissurina* Parr, 1947

Parafissurina dzemetinica Yanko, 1974

Parafissurina ex gr. *lateralis* (Cushman, 1913)

Order Rotaliida Delage et Herouard, 1896

Family Nonionidae Schultze, 1854

Genus *Nonion* Mortfort, 1808

Nonion matagordanus Kornfeld, 1931

Family Ammoniidae Saidova, 1981

Genus *Ammonia* Brunnich, 1772

Ammonia ammoniformis (d'Orbigny), 1826

Ammonia compacta Hofker, 1969

Ammonia tepida (Cushman), 1928

Family Canaliferidae Krasheninnikov, 1960

Genus *Canalifera*

*Canalifera parker*ae (Yanko), 1974

Family Elphidiidae Galloway, 1933

Genus *Porosononion* Putrja in Voloshinova, 1958

Porosononion martkobi ponticus Yanko, 1989

Genus *Haynesina* Banner et Culver, 1978

Haynesina anglica (Murray), 1965

Genus *Elphidium* Montfort 1808

Elphidium ponticum Dolgopolskaya et Pauli, 1931

Genus *Cribroelphidium* Cushman et Bronnimann, 1948

Cribroelphidium poeyanum (d'Orbigny)

Cribroelphidium translucens (Natland), 1938

Family Buliminidae Jones, in Griffith et al. (1875)

Genus *Bulimina* d'Orbigny, 1826

Bulimina elongata d'Orbigny, 1826
Family Bolivinidae Cushman, 1927
Genus *Bolivina* d'Orbigny, 1839
Bolivina variabilis (Williamson), 1858

Appendix 2: Taxonomic List of Ostracoda

Regnum Animalia
Type Arthropoda von Siebold, 1848
Subtype Crustacea Brünnich, 1772
Superclass Oligostraca Zrzavý, Hypša & Vlášková, 1997
Class Ostracoda Latreille, 1802
Order Podocopida G.O. Sars, 1866
Family Candonidae Kaufmann, 1900
Genus *Graviacypris* Baird, 1845
Graviacypris elongata (Schweyer, 1949)
Family Leptocytheridae Hanai, 1957
Genus *Callistocythere* Ruggieri, 1953
Callistocythere diffusa (Müller, 1894)
Family Trachyleberididae Sylvester-Bradley, 1948
Genus *Hiltermannicythere* Bassiouni, 1970
Hiltermannicythere rubra (Müller, 1894)
Family Cytheromatidae Elofson, 1939
Genus *Cytheroma* Müller, 1894
Cytheroma aff. *marinovi* Schornikov, 1969
Family Loxoconchidae Sars, 1925
Genus *Loxoconcha* Sars, 1866
Loxoconcha sublepida Stepanaitys, 1962
Genus *Palmoconcha* Swain and Gilby, 1974
Palmoconcha granulata (Sars, 1866)
Family Xestoleberididae Sars, 1928
Genus *Xestoleberis* Sars, 1866
Xestoleberis cornelii Caraion, 1963

Appendix 3: Taxonomic List of Nematoda

Regnum Animalia
Type Nematoda
Class Enoplea Inglis, 1983
Order Enoplida Filipjev, 1929
Family Oncholaimidae Filipjev, 1916
Genus *Viscosia* de Man, 1890
Viscosia minor Filipjev, 1918
*Viscosia glabr*a (Bastian, 1865)
Genus *Metoncholaimus* Filipjev, 1918
Metoncholaimus demani (zur Strassen, 1894)
Family Thoracostomopsidae Filipjev, 1927
Genus *Oxyonchus* Filipjev, 1927
Oxyonchus dubius (Filipjev, 1918)
Family Enchelidiidae Filipjev, 1918
Genus *Symplocostoma* Bastian, 1865
Symplocostoma ponticum Filipjev, 1918
Family Leptosomatidae Filipjev, 1916
Genus *Leptosomatides* Filipjev, 1918
Leptosomatides euxinus Filipjev, 1918
Family Oxystominidae Chitwood, 1935
Genus Oxystomina Filipjev, 1921
Oxystomina clavicauda (Filipjev, 1921)
Genus *Halalaimus* de Man, 1888
Halalaimus ponticus Filipjev, 1922
Halalaimus sp.

Family Rhabdodemaniidae Filipjev, 1934
Genus *Rhabdodemania* Baylis et Daubney, 1926
Rhabdodemania aff. pontica Platonova, 1965
Class Chromadorea Inglis, 1983
Order Chromadorida Chitwood, 1933
Family Chromadoridae Filipjev, 1917
Genus *Neochromadora* Micoletzky, 1924
Neochromadora sabulicola (Filipjev, 1918)
Neochromadora sp.
Genus *Chromadorita* Filipjev, 1922
Chromadorita gracilis Filipjev, 1922
Genus *Prochromadorella* Micoletzky, 1924
Prochromadorella mediterranea Micoletzky, 1922
Genus *Euchromadora* de Man, 1886
Euchromadora striata (Eberth, 1863)
Family Cyatholaimidae Filipjev, 1918
Genus *Cyatholaimus* Bastian, 1865
Cyatholaimus gracilis (Eberth 1863)
Genus *Paracanthonchus* Micoletzky, 1924
Paracanthonchus caecus (Bastian, 1865)
Genus *Pomponema* Filipjev, 1922
Pomponema aff. *multipapillatum* (Filipjev, 1922)
Order Desmodorida De Coninck, 1965
Family Desmodoridae Filipjev 1922
Genus *Desmodora* de Man, 1889
Desmodora pontica Filipjev, 1922
Order Monhysterida Filipjev 1929
Family Monhysteridae de Man, 1876
Genus *Monhystera* Bastian, 1865
Monhystera longicapitata Filipjev, 1922a
Monhystera sp.1
Monhystera sp.2
Monhystera sp.3
Family Xyalidae Chitwood, 1951
Genus *Daptonema* Cobb, 1920
Daptonema longicaudatum (Filipjev, 1922
Family Sphaerolaimidae Filipjev, 1918
Genus *Sphaerolaimus* Bastian, 1865
Sphaerolaimus gracilis de Man, 1884
Sphaerolaimus ostreae Filipjev, 1918
Genus *Parasphaerolaimus* Ditlevsen, 1918
Parasphaerolaimus dispar Filipjev, 1918
Family Linhomoeidae Filipjev, 1922
Genus *Terschellingia* de Man, 1888
Terschellingia pontica Filipjev, 1918
Terschellingia longicaudata de Man,1907
Terschellingia sp.1.
Terschellingia sp.2
Genus *Linhomoeus* Bastian, 1865
Linhomoeus hirsutus (Bastian, 1865)
Genus *Paralinhomoeus* de Man, 1907
Paralinhomoeus filiformis (Filipjev 1918)
Paralinhomoeus aff. *ostrearum* Filipjev, 1918
Paralinhomoeus sp.
Order Areolaimida De Coninck et Schuurmans-Stekhoven, 1933
Family Comesomatidae Filipjev, 1918
Genus *Sabatieria* Rouville, 1903
Sabatieria abyssalis (Filipjev, 1918)
Sabatieria pulchra (Schneider, 1906)
Sabatieria longicaudata Filipjev, 1922
Family Camacolaimidae Micoletzky, 1924
Genus *Camacolaimus* de Man, 1889
Camacolaimus aff. *zostericola* (Filipjev, 1918)
Order Desmoscolecida Filipjev 1929

Family Desmoscolecidae Shipleyr, 1896

Genus *Desmoscolex* Claparède, 1863

Desmoscolex bacescui (Paladin et Andriescu, 1963)

Genus *Tricoma* Cobb, 1894

Tricoma nematoides (Greeff, 1869)

Genus *Quadricoma* Filipjev, 1922

Quadricoma loricata Filipjev, 1922

Quadricoma steineri Filipjev, 1922

Nematoda sp.

References

Alve E (1991) Benthic foraminifera reflecting heavy pollution in Sorfjord, Western Norway. J Foramin Res 21(1):1–19

Bagriy ID, Kizlat AM (2004) Portativniy probovidbirnik-degazator (Portable sampler-degasser). Declarative patent for utility model issued on July 15, 2004. Bulletin no 7 (in Russian)

Basso D, Spezzaferri S (2000) Distribution of living (stained) benthic foraminifera in the Iskenderun Bay (Eastern Turkey): a statistical approach. Boll Soc Paleontol Ital 39(3):359–379

Bernhard JM (1992) Benthic foraminiferal distribution and biomass related to pore-water oxygen content. Deep-Sea Res 39:585–605

Bresler V, Yanko V (1995a) Chemical ecology: a new approach to study living benthic epiphytic foraminifera. J Foraminifer Res 25(3):267–279

Bresler V, Yanko V (1995b) Acute toxicity of heavy metals for benthic epiphytic foraminifera *Pararotalia spinigera* (Le Calvez) and influence of seaweed-derived DOC. Environ Toxicol Chem 14(10):1687–1695

Bresler V, Yanko-Hombach V (2000) Chemical ecology of foraminifera: parameters of health, environmental pathology and assessment of environmental quality. In: Martin R (ed), Environmental micropalaeontology. Kluwer Academic/Plenum Publishers, New York, pp 217–256

Dando PR, Hovland M (1992) Environmental effects of submarine seeping natural gas. Cont Shelf Res 12(10):1197–1207

Dando PR, Jensen P, O'Hara SCM et al (1994) The effects of methane seepage at an intertidal/shallow subtidal site on the shore of the Kattegat, Vendsyssel, Denmark. Bull Geol Soc Denmark 41:65–79

Debenay J-P, Tsakiridis E, Soulard R et al (2001) Factors determining the distribution of foraminiferal assemblages in Port Joinville Harbor (Ile d'Yeu, France): the influence of pollution. Mar Micropaleontol 43(1–2):75–118

Deprez T et al (2007) NeMys. http://www.nemys.ugent.be. Accessed 5 Dec 2018

Egorov VN, Polikarpov GG, Gulin SB et al (2003) Sovremennye predstavleniya o sredoobrazuyushchey i ekologicheskoy roli struynykh metanovykh gazovydeleniy so dna Chernogo morya (Contemporary ideas about the environmental and ecological role of methane jets from the bottom of the Black Sea). Morskoy ekologicheskiy zhurnal 2(3):5–26 (in Russian)

Ermakov AP (2005) Glubinnoe stroenie chernomorskoy vpadiny po rezultatam novoy interpretatsii seysmicheskikh dannykh (Deep structure of the Black Sea depression based on results of recent interpretation of seismic data). Dissertation, Moscow State University (in Russian)

Filipyev IN (1918–1921) Svobodnozhivushchie morskie nematody okrestnostey Sevastopolya (Free-living marine nematodes in the vicinity of Sevastopol). Trudy osob zool lab i Sevast biol st Ross AN, Serya 2, issue 4(I, II) (in Russian)

Fontanier C, Jorissen FJ, Licari L et al (2002) Live benthic foraminiferal faunas from the Bay of Biscay: faunal density, composition, and microhabitats. Deep-Sea Res 49(4):751–785

Gorshkova TI (1974) Biogeokhimiya sovremennykh morskikh osadkov i ikh biologicheskoe znachenie (Biogeochemistry of modern marine sediments and their biological significance). Trudy VNIRO 98, issue I, part 2:135–144 (in Russian)

Griffith JW, Henfrey A et al (1875) The micrographic dictionary, vol 1. London, pp 316–320

Gulin SB, Timofeev VA, Bondarenko LV (2010) Zoobentos v mikrobiotopakh metanovykh sipov shelfovoy zony Krymskogo poberezhya (Zoobenthos in microbiotopes of methane seeps of the shelf zone of the Crimean coast). Sistemy kontrolya okruzhayushchey sredy 14:225–229 (in Russian)

Jensen P (1981) Species, distribution and a microhabitat theory for marine mud dwelling Comesomatidae (Nematoda) in European waters. Cahiers de biologie marine 22:231–241

Jensen P (1983) Meiofaunal abundance and vertical zonation in a sublittoral soft bottom, with a test of the Haps corer. Mar Biol 74(3):319–326

Jensen P (1986) Nematode fauna in the sulphide-rich brine seep and adjacent bottoms of the East Flower Garden, NW Gulf of Mexico. Mar Biol 92:489–503

Jensen P, Aagaard L, Burke RAJ et al (1992) "Bubbling reefs" in the Kattegat: submarine landscapes of carbonate-cemented rocks support a diverse ecosystem at methane seeps. Mar Ecol Prog Ser 83(2/3):103–112

Jorissen FJ, Barmawidjaja DM, Puskaric S et al (1992) Vertical distribution of benthic Foraminifera in the northern Adriatic Sea: the relation with high organic flux. Mar Micropaleontol 19:131–146

Kaiho K (1999) Effect of organic carbon flux and dissolved oxygen on the benthic foraminiferal oxygen index (BFOI). Mar Micropalaeontol 37:67–76

Kessler JD, Reeburgh WS, Southon J et al (2006) Basin-wide estimates of the input of methane from seeps and clathrates to the Black Sea. Earth Planet Sci Lett 243:366–375

Komornyy AF (1996) Seysmicheskiy metod poiskov gazogidratov na akvatorii Chernogo morya (Seismic method for the search of gas hydrates in the Black Sea). Geofizicheskiy zhurnal 18(2):70–76 (in Russian)

Kononov SA, Ostrovskiy SA, Ustinyuk LF (1977) Novaya tekhnologiya nekotorykh sintezov na osnove atsetilena (A new technology for some acetylene-based syntheses). Nedra, Moscow (in Russian)

Konovalov SK, Ryabinin AI (1987) Vodorodnyy pokazatel (rN) vod Chernogo morya (Hydrogen indicator [pH] of the waters of the Black Sea). Meteorol Gidrol 10:75–81 (in Russian)

Krashenninikov VA (1960) Elfidiidy miotsenovykh otlozheniy Podolii (Elphidiids of the Miocene deposits of Podolia). GIN AN SSSR, issue 21, Moscow (in Russian)

Kravchuk AO, Kravchuk OP (2007) Autigennye sulfidy v rakovinakh foraminifer (Authigenic sulfides in foraminiferal tests). Vseukraïnska naukova konferentsiya "Vid geologiï do biosferologiï. Problemi sogodennya, maybutni perspektivi." KNU, IGN NANU, Kiïv, p 80–81 (in Ukranian)

Kumar Bhaumik A, Gupta A (2005) Deep-sea benthic foraminifera from gas hydrate-rich zone, Blake Ridge, Northwest Atlantic (ODP Hole 997A). Curr Sci 88(12):1969–1973

Lein AY, Ivanov MB (2005) Krupneyshiy na Zemle metanovyy vodoem (The largest methane reservoir on Earth). Priroda 2:19–26 (in Russian)

Loeblich AR, Tappan H (1988) Foraminiferal genera and their classification. Van Nostrand Reinhold Company, New York

Loeblich AR, Tappan H (1992) Present status of Foraminiferal classification. In: Takayanagi Y, Saito T (eds) Studies in benthic foraminifera. Tokai University Press, Tokyo, pp 93–102

Logvinenko NV, Sergeeva EI (1986) *Metody opredeleniya osadochnykh porod* (Methods for determination of sedimentary rocks). Nedra, Leningrad (in Russian)

Loubere P, Fariduddin M (1999) Benthic Foraminifera and the flux of organic carbon to the seabed. In: Sen Gupta B (ed) Modern foraminifera. Kluwer Academic Publishers, Dordrecht, pp 181–199

Luth U, Luth C (1998) Benthic meiofauna and macrofauna of a methane seep area south-west of the Crimean Peninsula, Black Sea. In: Luth U, Luth C, Thiel H (eds), MEGASEEBS-Methane gas seeps exploration in the Black Sea. Berichte aus dem Zentrum für Meeres- und Klimatoforschung 14:113–126

Mitropolskiy AY, Bezborodov AA, Ovsyanyy YI (1982) Geokhimiya Chernogo morya (Geochemistry of the Black Sea). Naukova dumka, Kiev (in Russian)

Motnenko IV (1999) Litologiya sovremennykh osadkov shelfa Izrailya i ikh antropogennoe zagryaznenie (Lithology of modern Israeli shelf sediments and their anthropogenic contamination). Moscow State University, Dissertation (in Russian)

Mureşan M (2014) Diversity and distribution of free-living nematodes within periazoic level on the Romanian shelf of the Black Sea. Geo-Eco-Marina 20:19–28

Nikishin AM, Korotaev MV, Bolotov SN et al (2001) Tektonicheskaya istoriya chernomorskogo basseyna (Tectonic history of the Black Sea basin). Byul. mosk. o-va ispytateley prirody., otd. Geoliya 76(3):3–18 (in Russian)

Nikishin AM, Ziegler P, Bolotov SN et al (2011) Late Palaeozoic to Cenozoic evolution of the Black Sea-Southern Eastern Europe region: a view from the Russian Platform. Turk J Earth Sci 21(5):571–634

Ott JA, Novak R (1989) Living at an interface; meiofauna at the oxygen/sulphide boundary of marine sediments. In: Ryland JS, Tyier PA (eds) Reproduction, genetics and distribution of marine organisms. 23rd European marine biological symposium, Olsen, Fredensborg, Denmark, pp 415–422

Ovsyanyy EI, Konovalov SK, Mitropolskiy AY et al (2015) Organicheskiy uglerod i karbonatnost sovremennykh donnykh otlozheniy Kerchenskogo proliva (Organic carbon and carbonate of modern bottom sediments of the Kerch Strait). Geokhimiya 12:1120–1131 (in Russian)

Panieri G (2003) Benthic foraminifera response to methane release in an Adriatic Sea pockmark. Rivista Italiana di Paleontologia e Stratigrafia 109(3):549–562

Patin SA (2004) Assessment of anthropogenic impact on marine ecosystems and biological resources in the process of oil and gas field development in the shelf area. Water Resour 31(4):413–422

Platt HM, Warwick RM (1983) Free-living marine nematodes. Part I. British Enoplids. Synopsis of the British Fauna No 28, Cambridge: The Linnaean Society of London and the Estuarine and Brackish-Water Sciences Association, Leiden

Platt HM, Warwick RM (1988) Free-living marine nematodes. Part II. British Chromadorids. Synopsis of the British Fauna (New Series) No 38. The Linnaean Society of London and the Estuarine and Brackish-Water Sciences Association, Leiden

Pletnev SP, Annin VK, Yu V et al (2014) Foraminifery i izotopiya (O^{16}/O^{18} i S^{12}/S^{13}) ikh rakovin v mestakh vykhoda metana na vostochnom sklone o. Sakhalin (Okhotskoe more) (Foraminifera and isotopes O^{16}/O^{18} and C^{12}/C^{13} of their tests in methane outlets on the eastern slope of Sakhalin peninsula, Sea of Okhotsk. Izvestiya TINRO 178:180–190 (in Russian)

Polikarpov GG, Tereschenko NN, Gulin MB (1998) Chemoecological study of the bivalve *Modiolus phaseolinus* in habitats near the oxic/anoxic interface near methane gas seeps in the Black Sea. In: Luth U, Luth C, Thiel H (eds) MEGASEEBS-methane gas seeps exploration in the Black Sea. Berichte aus

dem Zentrum für Meeres- und Klimatoforschung 14:92–100
Portnova DA, Mokievskiy VO, Khaflidason Kh et al (2014) Mnogokletochnyy meyobentos i taksotsen nematod v rayone metanovykh vykhodov Negga (Norvezhskoe more) (Multicellular meiobenthos and taxocene of nematodes in the area of methane outlets of Niega (Norwegian Sea). Biologiya morya 40(4):268–278 (in Russian)
Rathburn AE, Levin LA, Held Z et al (2000) Benthic foraminifera associated with cold methane seeps on the northern California margin: ecology and stable isotopic composition. Mar Micropaleontol 38:247–266
Rauzer-Chernousova DM, Fursenko AV (eds) (1959) Osnovy paleontologii. Obshchaya chast. Prosteyshie (Fundamentals of Paleontology. General Part. Protozoans). AN SSSR, Moscow (in Russian)
Saraswat R, Kouthanker M, Kurtarkar S et al (2011) Effect of salinity induced pH changes on benthic foraminifera: a laboratory culture experiment. Biogeosci Discuss 8:8423–8450
Sen Gupta BK (1999) Foraminifera in marginal marine environments. In: Sen Gupta BK (ed) Modern foraminifera. Kluwer Academic Publishers, Dordrecht, pp 141–160
Sen Gupta BK, Platon E, Bernhard JM et al (1997) Foraminiferal colonization of hydrocarbon-seep bacterial mats and underlying sediment, Gulf of Mexico slope. J Foraminifer Res 27:292–300
Sergeeva NG (1973) Novye vidy svobodnozhivushchikh nematod otryada Enoplida. Soobshchenie I (New species of free-living nematodes of the order Enoplida. Post I). Zool Zhurnal 52(11):1710–1714 (in Russian)
Sergeeva NG (1992) Kharakteristika fauny svobodnozhivushchikh nematod Yaltinskogo zaliva v periody funktsionirovaniya melkovodnogo i glubokovodnogo vypuskov khozbytovykh vod (Characteristics of the fauna of free-living nematodes of Yalta Bay during the periods of shallow and deep-water releases of domestic water). In: Zaika YeV (ed) Mnogoletnie izmeneniya zoobentosa Chernogo morya. Naukova Dumka, Chapter 5, pp 170–184 (in Russian)
Sergeeva NG (1996) Zonalnoe raspredelenie meyobentosa i ego vazhneyshego komponenta – svobodnozhivushchikh nematod Chernogo morya (The zonal distribution of meiobenthos and its most important component – the free-living nematodes of the Black Sea). Dissertation, Institute of Biology of the Southern Seas, Sevastopol (in Russian)
Sergeeva NG (2000) Biologicheskoe raznoobrazie v donnykh osadkakh serovodorodnoy zony Chernogo morya: raspredelenie po glubinam, stratifikatsiya v tolshche grunta (Biological diversity in the bottom sediments of the Black Sea hydrogen sulfide zone: distribution by depth, stratification in the sediment thickness). In: Shnyukov YeF (ed) Geologiya Chernogo i Azovskogo morey. Geoekologicheskie aspekty izucheniya Chernogo i Azovskogo morey. Nadra, Kiev, pp 314–331 (in Russian)
Sergeeva NG (2003) Meyobentos rayona s metanovymi gazovymi sipami (Meiobenthos of the area with methane gas seeps). In: Yeremeev VN, Gaevskaya AV (eds) Sovremennoe sostoyanie bioraznoobraziya pribrezhnykh vod Kryma (Chernomorskiy sektor). EKOSI-Gidrofizika, Sevastopol, pp 258–267 (in Russian)
Sergeeva NG, Gulin MB (2006) Meiobenthos of methane seepage areas in the upper anoxic zone of the Black Sea (underwater Dnieper canyon). In: Yanko-Hombach V, Gilbert A, Martin R (eds) Extended Abstracts of the Second Plenary Meeting and Field Trip of IGCP-521 Project "Black Sea-Mediterranean corridor during the last 30 ky: sea level change and human adaptation," Odessa I.I. Mechnikov National University, Odessa, Ukraine. 20–28 August 2006. Astroprint, Odessa, pp 152–153
Sergeeva NG, Kolesnikova YA (1996) Rezultaty izucheniya meyobentosa Chernogo morya (Results of studying the meiobenthos of the Black Sea). Ekologiya morya 45:54–62 (in Russian)
Sergeeva NG, Zaika V (2013) The Black Sea Meiobenthos in permanently hypoxic habitat. Acta Zoologica Bulgarica 5(2):139–150
Sergeeva NG, Mazlumyan DA, Lichtschlag A et al (2014) Benthic Protozoa and Metazoa living in deep anoxic and hydrogen sulfide conditions of the Black Sea: direct observations of actively moving Ciliophora and Nematoda. Int J Mar Sci 4(42):1–11
Shnyukov EF, Yanko VV (2014) Gazootdacha dna Chernogo morya: geologo-poiskovoe, ekologicheskoe i navigatsionnoe znachenie (Degassing of the Black Sea bottom: significance for geological exploration, ecology and navigation). Vestnik Odesskogo universiteta, Seriya geograficheskikh i geologicheskikh nauk 19(3, No 23):225–241 (in Russian)
Shnyukov EF, Yanko VV (2017) Problemy uglevodorodnogo potentsiala Chernogo morya i puti ego osvoeniya (Problems in the hydrocarbon potential of the Black Sea and its development). Geologiya i poleznye iskopaemye Mirovogo okean 4:41–53 (in Russian)
Sovga EE, Lyubartseva SP (2006) Istochniki, stoki i perenos metana v Chernom more (Sources, drains and methane transport in the Black Sea). Ekologichna bezpeka priberezhnykh ta shel'fovykh zon ta kompleksne vikoristannya resursiv shelfu 14:530–545 (in Ukranian)
Sozanskiy VI (2013) Vidnovlennya svitovikh zapasiv nafti i gazu yak strategichna problema suchasnosti (Recovery of world oil and gas reserves as a strategic problem of the present). Geologichniy zhurnal 2:68–74 (in Ukranian)
Tkachenko GG, Derkach YI, Yeyfa MM et al (1974) Gazopodibni vuglevodni donnikh vidkladiv zakhidnoï chastini pidnyattya Golitsina ta oznaki ïkh epigenetichnosti (Chorne more) (Gaseous hydrocarbon sediments of the western part of Golitsyn's rise and signs of their epigeneticity [Black Sea]). Geologiya uzberezhzhya i dna Chornogo ta Azovskogo moriv u mezhakh URSR 7:106–108 (in Ukranian)

Vincx M (1996) Meiofauna in marine and freshwater sediments. In: Hall GS (ed) Methods for the examination of organismal diversity in soils and sediments. Cambridge University Press, Cambridge, pp 187–195

Wahyudi MM (1997) Response of benthic foraminifera to organic carbon accumulation rates in the Okinawa Trough. J Oceanogr 53:411–420

Yanko VV (1989) Chetvertichnye foraminifery Ponto-Kaspiya: klassifikatsiya, ekologiya, biostratigrafiya, istoriya razvitiya (Quaternary foraminifera of the Ponto-Caspian: classification, ecology, biostratigraphy, history of development). Dissertation, Moscow State University (in Russian)

Yanko V (1990) Stratigraphy and paleogeography of marine Pleistocene and Holocene deposits of the southern seas of the USSR. Mem Soc Geol Ital 44:167–187

Yanko VV, Gramova LV (1990) Stratigrafiya chetvertichnykh otlozheniy Kavkazskogo shelfa i kontinentalnogo sklona Chernogo morya po mikrofaune (Stratigraphy of Quaternary deposits of the Caucasian shelf and the continental slope of the Black Sea based on the microfauna). Sovetskaya geologiya 2:60–72 (in Russian)

Yanko V, Troitskaya T (1987) Pozdnechetvertichnye foraminifery Chernogo morya (Late Quaternary Foraminifera of the Black Sea). Nauka, Moscow (in Russian)

Yanko V, Kronfeld A, Flexer A (1994) The response of benthic foraminifera to various pollution sources: implications for pollution monitoring. J Foraminifer Res 24(1):1–17

Yanko V, Ahmad M, Kaminski M (1998) Morphological deformities of benthic foraminiferal tests in response to pollution by heavy metals: implications for pollution monitoring. J Foraminifer Res 28(3):177–200

Yanko V, Arnold A, Parker W (1999) The effect of marine pollution on benthic Foraminifera. In: Sen Gupta BK (ed) Modern foraminifera. Kluwer Academic Publishers, Dordrecht, pp 217–238

Yanko V, Kravchuk A, Kulakova I (2017) Meyobyentos myetanovykh vykhodov Chyernogo morya (Meiobenthos of methane outlets of the Black Sea). Phenix, Odessa (in Russian)

Yanko-Hombach V, (also Yanko V) (2007) Controversy over Noah's Flood in the Black Sea: geological and foraminiferal evidence from the shelf. In: Yanko-Hombach V, Gilbert AS, Panin N et al (eds) The Black Sea flood question: changes in coastline, climate and human settlement. Springer, Dordrecht, pp 149–203

Yanko-Hombach (also Yanko V), Shnyukov E, Konikov E et al (2009) Response of biota to methane emissions in the Black Sea: preliminary results from complex geological, geochemical, palaeontological, and biological study. In: Gilbert A, Yanko-Hombach V (eds) Extended Abstracts of the Fifth Plenary Meeting and Field Trip of IGCP-521-INQUA 0501, Izmir-Çanakkale (Turkey), August 22–31, 2009. DEU Publishing House, Izmir, pp 181–184

Yanko-Hombach V (also Yanko V), Kondariuk T (also Kondaryuk T), Motnenko I (2017) Benthic foraminifera indicate environmental stress from river discharge to marine ecosystems: example from the Black Sea. J Foramin Res 47(1):70–92

Yanko-Hombach V (also Yanko V), Gilbert A, Dolukhanov P (2007) Critical overview of the Flood Hypotheses in the Black Sea in light of geological, paleontological, and archaeological evidence. Quat Int 167–168:91–113

Yanko-Hombach V (also Yanko V), Mudie PJ, Kadurin S et al (2014) Holocene marine transgression in the Black Sea: new evidence from the northwestern Black Sea shelf. Quat Int 345:100–118

Correction to: Mud Volcanoes of the Black Sea Region and their Environmental Significance

Correction to:
E. Shnyukov, V. Yanko-Hombach, *Mud Volcanoes of the Black Sea Region and their Environmental Significance,* https://doi.org/10.1007/978-3-030-40316-4

The original version of the book was inadvertently published with incorrect figures (Fig. 3.6 and Fig. 6.27) and caption (Fig. 4.4). These have been corrected now.

The updated versions of the chapters can be found at
https://doi.org/10.1007/978-3-030-40316-4_3
https://doi.org/10.1007/978-3-030-40316-4_4
https://doi.org/10.1007/978-3-030-40316-4_6

E. Shnyukov, V. Yanko-Hombach, *Mud Volcanoes of the Black Sea Region and their Environmental Significance*, https://doi.org/10.1007/978-3-030-40316-4_12

Main Conclusions

1. Mud volcanoes in the Black Sea region have been known for a very long time. People of the early Paleolithic would have been aware of them and could have used breccia, hot water, and gas flares from mud volcanoes in their daily life. From ancient historic times, there exist singular and haphazard observations of anomalous phenomena associated with mud volcanoes. Serious scientific study began at the end of the eighteenth century, and by the nineteenth century, mud volcanoes had already been described by many researchers. These early descriptions were scattered and sporadic. Only in the twentieth century were terrestrial mud volcanoes of the Kerch-Taman region systematically studied by V.V. Belousov and L.A. Yarotskiy. In a number of subsequent works, the presence of recessed synclines near mud volcanoes was noted by K.A. Prokopov, G.A. Lychagin, Z.L. Maimin, and others. In the second half of the twentieth century, iron ore deposits were discovered in recessed synclines (e.g., Novoselovskaya, Kamenskaya, Uzunlarskaya, Baksinskaya, and others) by E.F. Shnyukov and P.I. Naumenko. In the last quarter of the twentieth century, as a result of numerous marine expeditions conducted by different organizations, up to 70 mud volcanoes located in different underwater areas of the Black Sea were discovered and most of them described. A number of marine mud volcanoes located below the ~700 m isobath in the Black Sea erupted fragments of gas hydrates, thereby identifying the existence of HG gas beneath the sea bottom.
2. From the modern point of view, mud volcanism is considered as a form of local gas flow involving deep hydrocarbons moving upward to the Earth's surface accompanied by emissions of mud breccia, rock fragments, gases, and water. The flow travels through eruptive channels that have numerous voids in their upper parts and complex forms in their depths, and the ejecta often produce peculiar forms of relief such as cone-shaped hills that reveal craters or round cavities at their tops filled with liquid mud. Exogenous factors then complete the final appearance of these modern landforms, in particular, by creating numerous characteristic ravines, such as barrancos, on their slopes. The roots of the Black Sea mud volcanoes have been seismically traced into sediments of the Mesozoic and even deeper – down to the Moho surface (e.g., the Mantiynyy mud volcano). It is most likely that with improved technologies, the connection of mud volcanoes with the mantle will be clearly established in the future. A deep origin for mud volcanoes is also supported by the presence of unusual minerals, such as native metals (gold, silver, osmium, iridium, platinum, iron, lead, zinc, aluminum, etc.), as well as carbides, sulfides (cinnabar, antimonite, realgar, orpiment), and peculiarly shaped formations, such as non-rolled individual particles, complex crystals, wires, spherules, etc. Especially unusual are the spherules of iron, iron oxides, alloys of iron and rare-earth elements, and iron-silicate spherules—all typical by-products of industrial metallurgy and

E. Shnyukov, V. Yanko-Hombach, *Mud Volcanoes of the Black Sea Region and their Environmental Significance*, https://doi.org/10.1007/978-3-030-40316-4

indicating that they were formed under high temperature. According to A.Ye. Lukin, native minerals and carbides trace the movement of fluids and anhydrous super-compressed multi-component gas from great depths upward to the Earth's surface.

3. Mud volcanic breccia displays a primarily tectonic origin, having been formed through layer-by-layer rubbing, disharmonic crushing caused by diapiric bulging, and finally through reworking by water and gases that eject this breccia at the surface. The volume and size of fragments of erupted breccia vary greatly although very large pieces show up much less frequently. Lithologically, mud volcanic breccia is a clastic rock consisting predominantly of argillaceous sediments cemented by clay; it is often porous. Quite often, breccia is flamed and burned, occasionally kaolinized and carbonated. Marine mud volcanoes frequently erupt very small fragments of Jurassic coals, and possibly ferruginous quartzites, rounded fragments of porphyrite, andesite, and effusives. Accessory minerals in the composition of breccia are similar to those of the original rocks of the breccia, namely, Maikopian and, to a lesser degree, lower Cretaceous clay. Mud volcanic gases are unstable in composition. They contain predominantly methane, carbon dioxide, less often nitrogen, and heavy hydrocarbons, while helium, hydrogen, hydrogen sulfide, and carbon monoxide are present in small quantities. The waters of terrestrial mud volcanoes are characterized usually as chloride-hydrocarbonate-sodium, sulfate-sodium chloride, and sodium chloride with impurities of boron, bromine, and iodine.
4. As a result of the eruption from mud volcanoes of huge volumes of gases, mud breccia, detrital rock material, and water, weakened zones are formed nearby with dimensions proportional to the volume of ejected masses. On the surface, such zones are expressed by oval depressions called recessed synclines, and they can be divided into two main structural types. The first type includes those recessed synclines in which erupted material accumulates in the center of the structure, i.e., where the mud volcano itself is located. In this case, the calderas represent, in fact, the compensatory troughs (i.e., recessed synclines) that are filled with Neogene sediments that delineate its boundaries. The rocks composing the recessed synclines themselves form the frame of the caldera. The second type includes those recessed synclines in which erupted material accumulates in the vicinity of the mud volcano, not in the center of the recessed syncline. When this occurs, the development of compensatory troughs occurs to the side, and accumulated sediments do not outline the calderas, which apparently are eventually destroyed by weathering processes. As of today, 39 recessed synclines are known on the Kerch and Taman peninsulas. All of them are located in the arched parts of anticlines composed of Maikopian clay rocks. Many of them contain layers of Kimmerian iron ore deposits in their section. Undoubtedly, the mud volcanoes of the Black Sea must also be accompanied by recessed synclines but with a different sedimentary thickness compared to terrestrial recessed synclines. This question requires future investigation.
5. At least 4000 gas seeps are known in the Black Sea today. They are generally present at water depths from 50 to 800 m, forming either single occurrences or grouped seeps. The maximum depth for most of the detected gas seeps more or less corresponds to the stability zone for pure methane hydrate: a bottom temperature of 8.9 ° C in this part of the Black Sea. This enables us to say that gas hydrates play a buffering role for the ascending migration of gases and thereby prevent the seepage of huge amounts of methane into the water column. Some seeps are quite large in area, but most are localized on the periphery of the Black Sea, particularly in its northwestern part, on the Bulgarian and Kerch-Taman shelf (including the Kerch Strait), and along the coast of the Caucasus. Quite often, they are confined to paleo-deltas of the largest Black Sea rivers, the Danube, Dnieper, Dniester, and Don, and within the canyons formed by them. The deepest seeps are associated with faults and mud

volcanoes in the central Black Sea basin. On average, the height of the gas seeps above the surrounding seafloor is 100–200 m, and most of the emitted gases do not reach the water surface. Sometimes, it is difficult to distinguish gas seeps from mud volcanoes due to the location of seeps within small conical depressions on the sea bottom. As strong exhaust emissions of water-saturated gas pass upward through the sediment sequence, they can affect bottom morphology by forming rounded depressions or craters called pockmarks. Their locations indicate both the presence of gas seeps and methane biotopes on the seafloor. Within the Black Sea, a kind of specific zoning emerges: the central deepwater part of the sea is occupied by gas hydrates, while its periphery contains abundant gas seeps. It is quite possible that gas or, more precisely, gas-water seeps are localized within the area of the land adjacent to the sea—for example, in the Kerch-Taman region—but for the most part, they are not fixed with certainty because of the difficulty of observation. Many areas where seeps have developed are confined to the places containing paleoalluvial strata, but not all. The authors are more inclined to link gas seeps with the Maikopian horizon of the Black Sea geological section and with deep fluids. At the same time, they do not deny the participation of biochemical methane in the genesis of gas flares. Undoubtedly, this process accounts for gas seep generation in Quaternary bottom sediments, but its role, most likely, is subordinate.

6. For many years, the idea of a connection between mud volcanoes and associated recessed synclines with mineral ore content was explored during geological studies. This hypothesis was proposed in the early 1960s by Shnyukov and others and has been tested for many years, even up to the present day. It turned out to be accurate and led to the discovery and exploration of seven new iron ore deposits and manifestations and also revealed the elevated boron content of some of them. The largest ore deposit, Novoselovskoye, contains 125 million tons of ore reserves deposited within its recessed syncline. Less significant ore deposits are associated with the mud volcanoes Baksinskiy, Rep'yevskiy, Achinskiy, and some others. This idea that underwater Black Sea mud volcanoes could contain useful resources was shown to be true as well, as one can expect the accumulation of methane gas hydrates in recessed synclines lying at depths of more than 700 m—a thermodynamically favorable environment for their development. Particularly valuable are ring-shaped (isometric) structures that follow the outline of mud volcanoes and are complicated by subsidence phenomena. In this way, many mud volcanoes and their associated recessed synclines, terrestrial and offshore, can be used as indicators in the search for various mineral resources (gas, oil, gas hydrates, iron ore, sulfur, chemical raw materials, building materials, raw mineral and clay materials, etc.) as well as for production of metallurgical pellets and recreation. In Azerbaijan, a number of discoveries of oil and gas fields in the areas of mud volcanoes indicate their great potential for such prospecting, and this potential should be relevant for the Kerch Peninsula as well. A number of wells drilled to a depth of 500 m on the Bulganakskiy mud volcano were unsuccessful, however, probably because they did not reach the required depth. Given the geological history of the Kerch Peninsula, oil and gas fields should be located at much deeper horizons.

7. The activity of mud volcanoes on land can be dangerous for humans, as eruptions can cause the death of people and animals as well as great damage to homes and other infrastructure constructed in close proximity to mud volcanoes. There is another ecologically dangerous aspect to the influence of mud volcanoes on the environment: a geochemical one. Together with its gases, mud volcanoes emit mercury, bromine, iodine, boron, and possibly arsenic. As a rule, mud volcanoes release gas clouds of mercury, the concentration of which is 1–2 orders of magnitude higher than safe background levels. Under certain conditions,

emissions of hydrogen sulfide can be hazardous. The clear risks should serve as a warning to tourists and visitors that attempts to approach the volcano's center or multiple vents to view the semi-liquid sedimentary deposits can result in great harm or worse, and therefore excessive curiosity can be quite hazardous. Mud volcanic activities in the shallow sea are often catastrophic. Breccia and mud emissions can rapidly produce islands and shoals along the routes of seagoing vessel movement. One case of a vessel running aground in the Kerch Strait has been recorded. In the deeper areas of the Black Sea, there are known cases of emission and ignition of large masses of gases, events that pose great danger for navigation and can lead to the foundering of ships.

8. Mud volcanism and the related degassing of methane from the seabed affects living organisms (exemplified by meiobenthos), their metabolism, and activities. A case study conducted in areas of methane outlets within the northwestern part of the Black Sea examined the relationship between the distribution of meiobenthos and concentrations of hydrocarbon gases, primarily methane, in the sediments. The dual analysis of abiotic characteristics (physical and chemical parameters of the water column, gasmetrical, geochemical, lithological, mineralogical properties of the sediment) and biotic ones (quantitative and taxonomic composition of foraminifers, nematodes, and ostracods) shows the noxious effects of methane upon organisms but also afforded a means by which the taxonomic and quantitative distribution of foraminifera, ostracoda, and nematodes could be used as indicators of potential methane reservoirs lying beneath the seabed.

In this paper, we have transliterated Cyrillic letters into the Latin alphabet according to the BGN/PCGN Romanization system for Russian used by Oxford University Press. Exceptions are the names of authors, which we have left in their own preferred transliterations, as well as geographical names as presented most commonly in the majority of English papers.

Acknowledgments

The authors are grateful to all of the colleagues that have supported their work. Particular gratitude must be given to N. Maslakov and A. Paryshev for their assistance in drawing the illustrations. We are grateful to Prof. Allan Gilbert from Fordham University, USA, for editing the English text and for valuable comments. We are also grateful to Alexander Motnenko, University of Manitoba, for initial editing of Chaps. 1, 4, and 5. Prof. Avdil Aliyev and Prof. Vladimir Kobolev are thanked for their scientific review of this manuscript. Dr. Revinder Sidhu, Microscopy and Materials Characterization Facility Manager (Manitoba Institute for Materials, University of Manitoba, Canada), is sincerely thanked for her help in imaging of microfauna by SEM.

Index

E. Shnyukov, V. Yanko-Hombach, *Mud Volcanoes of the Black Sea Region and their Environmental Significance*, https://doi.org/10.1007/978-3-030-40316-4

MIX
Papier aus verantwortungsvollen Quellen
Paper from responsible sources
FSC® C105338

If you have any concerns about our products,
you can contact us on
ProductSafety@springernature.com

In case Publisher is established outside the EU,
the EU authorized representative is:
Springer Nature Customer Service Center GmbH
Europaplatz 3, 69115 Heidelberg, Germany

Printed by Libri Plureos GmbH
in Hamburg, Germany